Computer-Aided Verification '90

DIMACS

Series in Discrete Mathematics and Theoretical Computer Science

Volume 3

Computer-Aided Verification '90

Proceedings of a DIMACS Workshop
June 18–21, 1990

E. M. Clarke
R. P. Kurshan
Editors

NSF Science and Technology Center
in Discrete Mathematics and Theoretical Computer Science
A consortium of Rutgers University, Princeton University,
AT&T Bell Labs, Bellcore

American Mathematical Society
Association for Computing Machinery

The DIMACS Workshop on Computer-Aided Verification was held at the Center for Discrete Mathematics and Theoretical Computer Science on June 18–21, 1990.

1980 *Mathematics Subject Classification* (1985 *Revision*). Primary 68Q60, 68M15, 68Q05.

Library of Congress Cataloging-in-Publication Data

DIMACS Workshop on Computer-Aided Verification (1990: Center for Discrete Mathematics and Theoretical Computer Science)

Computer-aided verification '90: proceedings of a DIMACS workshop, June 18–21, 1990/E. M. Clarke, R. P. Kurshan, editors.

p. cm.—(DIMACS series in discrete mathematics and theoretical computer science, ISSN 1052-1798; v. 3)

"The DIMACS Workshop on Computer-Aided Verification was held at the Center for Discrete Mathematics and Theoretical Computer Science on June 18–21, 1990"—CIP t.p. verso.

Includes bibliographical references.

AMS: ISBN 0-8218-6594-3 (AMS: alk. paper).—ISBN 0-89791-386-8 (ACM: alk. paper)

1. Digital integrated circuits—Design and construction—Data processing—Congresses. 2. Computer-aided design—Congresses. I. Clarke, E. M. II. Kurshan, R. P. III. Title. IV. Series.

TK7874.D553 1990 91–8815

621.381′5—dc20 CIP

To order through AMS contact the AMS Customer Services Department, P.O. Box 6248, Providence, Rhode Island 02940-6248 USA. For VISA or MASTERCARD orders call 1-800-321-4AMS. Order code DIMACS/3.

To order through ACM contact ACM Order Department, P.O. Box 64145, Baltimore, Maryland 21264. Phone 301-528-4261. Order number 222913.

Printed in the United States of America.
The paper used in this book is acid-free and falls within the guidelines
established to ensure permanence and durability. ♾
This volume was printed directly from author-prepared copy.

Portions of this publication were typeset using $\mathcal{A}_{\mathcal{M}}\mathcal{S}$-TEX,
the American Mathematical Society's TEX macro system.

10 9 8 7 6 5 4 3 2 1 95 94 93 92 91

Contents

Preface

This volume is devoted to the proceedings of the second workshop on "Computer-Aided Verification" (the first to use this specific title). The motivation for a specific workshop in computer-aided verification (CAV) is to bring together researchers working on effective algorithms or methodologies for formal verification (as distinguished, say, from attributes of logics or formal languages). The considerable interest generated by the first workshop (held in Grenoble, June 1989) motivated this meeting. As this interest continued to grow, it was decided to continue CAV on an annual basis, and for that purpose the CAV Steering Committee was formed (see below). In view of this, we take the opportunity here to state the focus of CAV and briefly sketch the history of research in formal verification, as it relates to CAV.

It is the intention of the CAV Steering Committee that future CAV meetings will continue the current focus on the problem of making formal verification feasible for various models of computation. Present emphasis is on models associated with distributed programs, protocols, and digital circuits. A good test of algorithm feasibility is to embed it into a verification tool and exercise that tool on realistic examples. Thus, we have promoted special sessions for the demonstration of new verification tools. For the technical sessions, we seek theoretical results that lead to new or more powerful verification methods. We expect that there will be less emphasis at CAV meetings on purely theoretical results in program logics—not because fundamental results of this type are unimportant but because this research is adequately covered in other conferences. Since we expect that a number of the results presented at CAV actually will be used by hardware and software designers, we seek to maintain a balance between presentations by researchers and practitioners.

Proofs of correctness of algorithms such as the Euclidean algorithm go far back into history. The importance of such proofs in computing apparently was realized by Turing. However, it was not until the 1960s and early

1970s that provably correct computation began to attract much attention as a self-contained area of research. Fundamental contributions in this period established the vehicles through which formal proofs of program correctness could be constructed from axioms and rules of inference in the same way that proofs in mathematics are constructed.

The first proofs of programs were hand constructed and, therefore, quite short and easy to follow. As a general approach, however, manually constructed proofs of correctness were beset by two fundamental problems. First was the problem of scalability: real programs tend to be long and intricate (compared with the statement of a mathematical theorem), so a proof of correctness could be expected to be correspondingly long. Under these circumstances, it was unclear to what extent a methodology based upon manually constructed proofs could be expected to be successful. The second problem was credibility: unlike published mathematics which may be expected to undergo extensive peer review, proofs of programs are more likely to be read only by the author.

Much interesting work has continued in this direction, however, and through the middle 1980s, most of the weight of research on formal verification (as this area of research became known) remained focused upon manual proofs of correctness. Applications of the work mainly did not overcome these two fundamental problems.

The purpose of CAV is to feature research specifically directed at overcoming these two problems of scalability and credibility. Presently, this thrust has become synonymous with computer-aided verification.

Initially, researchers thought that computer programs for theorem-proving could be used in automatic program verification. Logics emerged as a mechanism to formalize the manipulation of the properties to be proved. Out of fundamental work in logic and mathematics, automatic theorem-proving advanced rapidly. Automated theorem-proving had its own problems, however, stemming largely from its nonalgorithmic nature and basic problems of tractability and decidability. These difficulties seemed to provide obstacles which were much too difficult for early theorem-provers to overcome. Many of the pioneers in program verification became disillusioned and moved into other research areas where progress was more rapid.

Recent advances and the maturation of theorem-provers (and more specifically, proof-checkers) has renewed interest in their application to practical tasks such as hardware verification. Currently, this direction has demonstrated applicability for proving properties of data paths in hardware designs. Theorem-proving has been less successful in verifying properties related to *control*, particularly when concurrency and process synchronization are involved.

Together with the early disillusionment in theorem-provers emerged an intense interest in restricted logics for which formula satisfiability, used to test $f \Rightarrow g$, is decidable. Preeminent among these logics was temporal logic.

However, this work had two significant deficiencies of its own. First, these logics invariably constituted a substantial abstraction of a restricted class of programs. In fact, abstraction was so great that formulas in the logic lost their connection to the programs they were meant to abstract. Second, as a purely computational matter, decision procedures still were largely intractable, being exponential in the size of the formulas.

This second problem was undercut in 1980 through the introduction of model-checking as an alternative to checking formula satisfiability. Not only was linear-time model-checking demonstrated (for branching-time temporal logic), but perhaps the first computer implementations of practical formal verification algorithms were produced, as well. Computational complexity nonetheless remained an issue: while model-checking could be done in time linear in the size of the model, the model itself grows exponentially in the number of model components; for "real" models, complexity still was the gating issue.

This problem and the problem of bridging the gap from model to implementation were addressed soon after through the introduction of homomorphic reduction. This permitted checking complex models indirectly through checks on reductions which are relative to the respective properties under test. Homomorphism also served as a mechanism for stepwise refinement, relating implementations to design models. This led to compositional and hierarchical verification, as well as specialized reduction methods involving certain types of induction. Complexity still remained an issue, however, as homomorphic reductions may be difficult (or impossible) to produce, especially in the case of large data path models; even small data path models with many inputs are not made readily tractable by homomorphic reduction. The same difficulties applied in some degree to induction.

Significant in-roads into these difficulties have been made through the work on symbolic model checking using binary decision diagrams (BDD's), as introduced at the first session of this workshop in Grenoble (1989). While not scalable (effective use of BDD's currently appears to be limited to around 150 binary variables, while applications often require thousands), use of BDD's in conjunction with homomorphic reduction and induction appears extremely promising and has generated considerable excitement. Introduced at this same meeting was work on real-time verification, which continues to generate much interest, as well.

The current workshop (CAV '90) presents some very substantial progress using BDD's. As well, this workshop introduces reductions based upon partial order representations; these reductions offer a new potential for dealing with the tractability problem inherent in state-based models. Other papers presented here explore theorem-proving and proof-checking in controller verification, and the remaining papers present much vital work which continues and extends current verification methods.

We have witnessed a migration from theorem-proving to model-checking, and a recent renewed interest in proof-checking. This may represent the start of a swing back toward theorem-proving, especially through the combination of model-checking and symbolic techniques. If so, we may expect more general theorem-proving to become integrated into existing verification tools, providing a basis for static and dynamic reasoning from the same platform.

For example, verification of a (dynamic) property of a model through expansion of the model state space or BDD evaluation may be simplified by exploiting a symmetry or inductive property in the model; the symmetry or inductive property upon which the simplification is based, may be verified through a (static) syntactic check on the model specification, using theorem-proving techniques.

Whatever the future may hold, our perception is that computer-aided verification has emerged from a polymorphous adolescence into a very exciting and promising adulthood.

Three more CAV workshops have been planned; the next will be held in Aalberg, Denmark in July 1991 under the direction of Kim Larsen.

We would like to thank the other members of the Steering Committee and the members of the CAV '90 Program Committee for their invaluable help in making this workshop a success. We are appreciative of DIMACS for their sponsorship of the workshop and the contribution of their facilities at Rutgers University for the duration of the workshop (June 18–21). In particular, we thank Pat Toci of DIMACS for her extensive work organizing and managing the entire on-site logistics of the workshop, and we thank Thelma Pickell of AT&T Bell Laboratories for administrative management of the meeting.

Steering Committee: E. Clarke, R. Kurshan, A. Pnueli, and J. Sifakis.

Program Committee: H. Barringer, G. Bochmann, R. Bryant, C. Courcoubetis, S. Dasgupta, D. Dill, A. Emerson, R. Gerth, B. Gopinath, Z. Har'El, G. Holtzmann, G. Milne, R. Platek, P. Sistla, M. Stickel, C. Stirling, P. Wolper, and M. Yoeli.

E. M. Clarke
R. P. Kurshan
Program Co-Chair
CAV '90

DIMACS Series in Discrete Mathematics
and Theoretical Computer Science
Volume 3, 1991

Temporal Logic Model Checking: Two Techniques for Avoiding the State Explosion Problem

EDMUND M. CLARKE, JR.

Abstract

Verifying the correctness of finite-state concurrent programs has been an important problem for a long time. However, lack of any formal and efficient method of verification has prevented the creation of practical design aids for this purpose. Since all known techniques of simulation and prototype testing are time-consuming and not very reliable, there is an acute need for such tools. We describe an automatic verification system for such systems in which specifications are expressed in a propositional temporal logic. We use a procedure called temporal logic model checking to determine automatically if a specification is true of a large state-transition graph. In contrast to most other mechanical verification systems, our system does not require user assistance and will produce a counterexample trace to show how an error occurs if this is possible.

The main difficulty with our approach is that the state space we have to search may grow exponentially with the number of circuit components. Recently, however, we have made significant progress on this problem. In this talk we outline how our verification method works and describe two of the techniques that we have developed for dealing with the state explosion problem.

The first technique is an automatic procedure for reducing the state-space of a modular system by abstracting away details of a module's behaviour that are not relevant to the specification being checked. This method relies on a description of the system as a hierarchy of concurrently executing modules. To provide a vehicle for such descriptions, we have extended the facilities of the language we use for describing synchronous hardware controllers, to allow the design of modular systems. We illustrate the use of this approach to verify the controller of a CPU with decoupled access and execution units.

We have also modified the model checking algorithm to represent a state graph using binary decision diagrams (BDD's). Because this representation captures some of the regularity in the state space of sequential circuits with data path logic, we are able to verify circuits with an extremely large number of states. We demonstrate this new technique on a synchronous pipelined design with approximately 5 times 10^{20} reachable states. We give empirical results on the performance of the algorithm applied to other examples of both synchronous and asynchronous cirucits with data path logic.

SCHOOL OF COMPUTER SCIENCE, CARNEGIE MELLON, PITTSBURGH, PENNSYLVANIA 15213

DIMACS Series in Discrete Mathematics
and Theoretical Computer Science
Volume 3, 1991

Automatic Verification of Extensions of Hardware Descriptions

Hans Eveking
Institut für Datentechnik
Technische Hochschule Darmstadt
D-6100 Darmstadt, Fed. Rep. of Germany

The extension of a hardware description is a description where all properties of the original one are maintained. The concept applies to a variety of design and verification problems including logic-verification and the verification of behavioral vs. structural descriptions. For a systematic discussion, several classes of temporal behavior and HDL-constructs for their representation are introduced. The verification tool LOVERT is surveyed which allows for the automatic verification of several types of extensions.

1 Extensions of Descriptions

In the following, the correctness of finite state systems is discussed in terms of an HDL-based hardware specification technique [3]. One hardware description, the *specification*, defines the meaning of correctness for another one, the *implementation* (Fig. 1).

$T(d1) \longrightarrow$ (**Description d1**) $\longleftarrow$ Specification

$T(d2) \longrightarrow$ (**Description d2**) $\longleftarrow$ Implementation

Fig. 1: Basic situation of a hardware specification technique

Hardware specification techniques are based on the concept of the *axiomatiziation* of HDL descriptions [4]. The axioms associated with a hardware description d have two sources (i) the *model-specific axioms* which are due to the hardware-model involved, e.g., the axioms of boolean algebra, (ii) the *description-specific axioms* reflecting the properties of the specific description d. The model- and description-specific axioms associated with a hardware description characterize a *theory*, i.e., a formal system of the predicate calculus.

Definition 1: A formula A is a correct statement about a hardware description d iff it is a theorem of the associated theory $T(d)$, i.e.,

$$\vdash_{T(d)} A. \tag{1-1}$$

The relationship between the specification and the implementation can be discussed in terms of the relationship between the associated theories. We study classes of a particularly simple relationship between two descriptions d1 and d2:

Definition 2: A description d2 is an ***extension*** of d1 iff for all axioms of $T(d1)$, i.e., for $A(T(d1))$ holds

$$\vdash_{T(d2)} A(T(d1)). \tag{1-2}$$

Note that on the basis of 1-2 *all* correct statements about d1 are correct about d2, too. The limitations of the concept of extension are due to the fact that the underlying modelling concepts of d1 and d2 have to be the same since the model-specific axioms of $T(d1)$ have also to be theorems of $T(d2)$ as required by 1-2. Problems involving temporal abstraction or value homomorphisms for which an interpretation of the theory $T(d1)$ is necessary [3,5] are not covered by the concept of extension.

In Section 2, the semantics of finite state systems is defined in terms of some concepts of mathematical systems theory [9]. HDL-representations of several types of temporal behavior are proposed. The HDL-constructs are taken from the CONLAN family of HDL's [8]. An axiomatiziaton of all HDL-constructs will be given.

In Sections 3-5, several types of extensions will be introduced, and proof-procedures will be discussed.

2 Classes of Temporal Behavior

We study systems that can be characterized by means of *time-functions*. A time-function represents the values that can be observed at a *carrier*, i.e., a point of observation. We consider time-functions to be functions from the set of natural numbers representing the time into some range, e.g., the set of boolean values $\{L, H\}$. Let P be a set of n time-functions $f_1, \ldots, f_n$. $P(t)$ denotes the

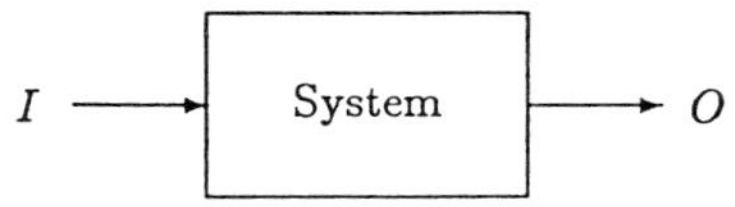

Fig. 2: A system with inputs and outputs

n-tupel $(f_1(t), \ldots, f_n(t))$, i.e., the tupel of all n values at point t of time. $P(< t_1, t_2 >)$ denotes the $(t_2 - t_1 + 1)$-tupel of all $P(t)$ in the interval $t_1 \leq t \leq t_2$.

A set P of time-functions associated with a system can be partitioned into a set I of input-functions and a set O of output-functions (Fig. 2). The classification of the temporal behavior of a system is based on the question:

Which information about $I(< 0, t >)$ *and* $O(< 0, t-1 >)$ *determines* $O(t)$ *uniquely ?*

In the rest of this Section, HDL-constructs for the representation of three classes of behavior will be presented.

2.1 Static behavior

A system has ***static*** behavior iff $O(t)$ is determined uniquely by $I(t)$ for all t. A typical example is the behavior of an AND-gate (Fig. 3).

L (low) and H (high) are the boolean constants. Static behavior is described by means of connections to carriers of type `btml` (boolean terminal)

```
g .= e
```

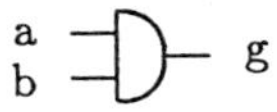

t	0	1	2	3	4	...
$a(t)$	L	L	H	L	L	...
$b(t)$	H	L	H	H	L	...
$g(t)$	L	L	H	L	L	...

Fig. 3: Static behavior of an AND-gate

with the meaning $\forall t: \; g(t) = e(t)$.
For the boolean functions & (and), | (or), || (exor) and ~ (not), corresponding boolean functions are defined in the predicate calculus, e.g.

$$(and(a,b) = r) \iff (a = H) \wedge (r = b) \vee (a = L) \wedge (r = L).$$

The meaning of an HDL-expression is defined by means of a time-function, too. For instance, the time-function $\lambda(t)(and(a(t),\; b(t)))$ is associated with the boolean expression `a & b` of Fig. 3. As a result, the meaning of the statement `g .= a & b` of Fig. 3 is

$$\forall t: \; g(t) \;=\; (\lambda(t)(and(a(t),\; b(t))))\,(t) \;=\; and(a(t),\; b(t)).$$

2.2 Transitional behavior

A system has ***transitional*** behavior iff $O(t)$ is uniquely determined by $I(t-1)$ and $O(t-1)$ for $0 < t$. The behavior of a Moore-machine [6] is an example of transitional behavior. This type of behavior is described by conditional transfers (Fig. 4) into carriers of type `budvl` (boolean unit delay variable). If the transfer condition a is H at point $t-1$ then the value of x at point t becomes

`IF a THEN x <- y ENDIF`

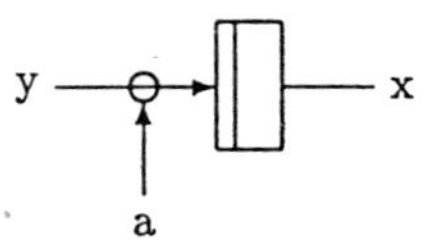

t	0	1	2	3	4	5	...
$a(t)$	L	L	H	L	L	H	...
$y(t)$	H	H	H	L	L	L	...
$x(t)$	L	L	L	H	H	H	...

Fig. 4: Transitional behavior of a transfer

the value of $y(t-1)$; otherwise, the old value of x is maintained. A default value L is assumed for point 0 of time.
The semantics of a conditional transfer is thus

$$\begin{aligned} &x(0) = L, \\ &\forall t: \; (0 < t) \wedge (a(t-1) = H) \;\Rightarrow\; (x(t) = y(t-1)), \\ &\forall t: \; (0 < t) \wedge (a(t-1) = L) \;\Rightarrow\; (x(t) = x(t-1)). \end{aligned} \qquad (2\text{-}1)$$

There may be several conditional transfers into one carrier. Assume n conditional transfers into the carrier x:

```
IF a1 THEN x <- y1 ENDIF,
...,
IF an THEN x <- yn ENDIF
```

If transfer collisions are excluded then the meaning is:

$$\begin{aligned}
&x(0) = L,\\
&\forall\, t:\ (0 < t) \wedge (a1(t-1) = H) \ \Rightarrow\ (x(t) = y1(t-1)),\\
&\ldots\\
&\forall\, t:\ (0 < t) \wedge (an(t-1) = H) \ \Rightarrow\ (x(t) = yn(t-1)),\\
&\forall\, t:\ (0 < t) \wedge (a1(t-1) = L) \wedge \ldots \wedge (an(t-1) = L) \ \Rightarrow\ (x(t) = x(t-1)).
\end{aligned} \qquad (2\text{-}2)$$

If a transfer condition or source expression is a boolean expression, e.g.,

```
IF a & b THEN x <- y ENDIF
```

then an anonymous time-function is associated with the boolean expression (Section 2.1). The time-functions of all carriers are thus bound to point $t-1$ of time. In the example, we obtain

$$\begin{aligned}
&x(0) = L,\\
&\forall\, t:\ (0 < t) \wedge (and(a(t-1), b(t-1)) = H) \ \Rightarrow\ (x(t) = y(t-1)),\\
&\forall\, t:\ (0 < t) \wedge (and(a(t-1), b(t-1)) = L) \ \Rightarrow\ (x(t) = x(t-1)).
\end{aligned}$$

2.3 Quasi-transitional behavior

A system has *quasi-transitional* behavior iff $O(t)$ is uniquely determined by $I(t)$ und $O(t-1)$ for $0 < t$. An example is the behavior of a latch (Fig. 5) described by a conditional assignment to a

```
IF a THEN r := y ENDIF
```

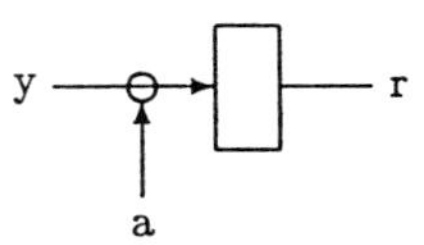

t	0	1	2	3	4	5	...
$a(t)$	L	L	H	L	L	H	...
$y(t)$	H	H	H	L	L	L	...
$r(t)$	L	L	H	H	H	L	...

Fig. 5: Quasi-transitional behavior of a latch

carrier of type `bvarl` (boolean variable). Note that in the example of Fig. 5, the output r follows the input y directly at points 2 and 5 of time.
The meaning of the conditional assignment of Fig. 5 is defined by

$$\begin{aligned}
&x(0) = L,\\
&\forall\, t:\ (0 < t) \wedge (a(t) = H) \ \Rightarrow\ (x(t) = y(t)),\\
&\forall\, t:\ (0 < t) \wedge (a(t) = L) \ \Rightarrow\ (x(t) = x(t-1)).
\end{aligned} \qquad (2\text{-}3)$$

The input/output behavior of an automaton of Mealy-Type [6] is also quasi-transitional.

3 Static Descriptions vs. Static Descriptions

The first type of extensions applies to situations where the specification as well as the implementation are given by static descriptions.

Definition 3: A description is called *static* iff all carriers have static behavior.

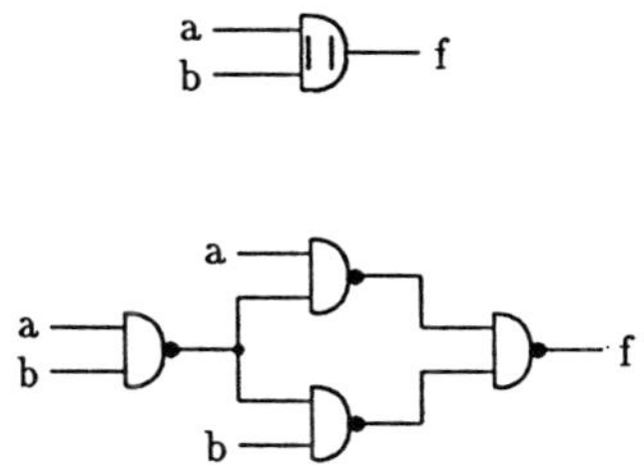

Fig. 6: An example of logic-verification

EXOR-function is implemented by means of a network of NAND-gates. To prove the correct implementation, the boolean terms of the output f are derived for the specification as well for the implementation; then the equivalence $\simeq$ of both terms is shown:

$$\texttt{a || b} \quad \simeq \quad \texttt{\~{}(\~{}(a \& \~{}(a\&b)) \& \~{}(b \& \~{}(a \& b)))}$$

Efficient procedures for the equivalence-proof of complex boolean expressions are available due to the work of Bryant [2] and Madre/Billon [7]. Since most HDL's provide bit-vectors as a basic type for the convenient description of complex circuits, the problem of logic-verification involves also the equivalence-proof of *vector-expressions*.

Definition 4: Two vector-expressions are equivalent, a $\cong$ b, iff for all elements i holds a[i] $\simeq$ b[i].

The semantics of vectors and vector-operations can be defined by means of lists and list-operations [4]. Three examples of increasing complexity are shown in Fig. 7. The first problem is easily solved if the commutativity of the boolean and-function is extended for vector-functions. In the

```
a[1:8] & b[1:8]            ≅      b[1:8] & a[1:8]

adc( a[1:8], a[1:8], 0)    ≅      a[1:8] # 0

gt(a, b)                   ≅      ~(adc( 0#a, 1#~b, 0)[2])
```

Fig. 7: Three examples of equivalent vector-expressions

second problem, the function adc is used which adds two n-bit vectors and a carry-input returning a normalized $n+1$-bit vector $1:n+1$. The left-most bit is the most-significant bit. For instance,

```
adc( a[1:8], b[1:8], c)
```

is a 9-bit vector; the carry-output is adc(a[1:8], b[1:8], c)[1]. The second equivalence of Fig. 7 is based on the fact that a plus a is equivalent to a multiplication of a by 2, i.e., a left-shift of

a or the concatenation # of a and 0.[1] The third problem is even more difficult: in order to compare two vectors a and b, b is subtracted from a adding the complement of b to a; the most-significant bit of the sum has to be inverted.
To address such a variety of proof-complexity, the LOVERT approach [1] follows a two-step procedure:

- Step 1: two expressions are rewritten using a number of rewrite-rules. If the rewriting results in textually identical expressions then the equivalence is proven (Fig. 7). An example of a

Fig. 8: Transformation of an equivalence-proof into an identity-proof

  rewrite-rule is the concatenation of the adc-function: two concatenated adders

```
adc( x, y, adc( v[1:n], w[1:n], cin)[1]) #
adc( v[1:n], w[1:n], cin)[2:n]
```

  are equivalent to one adder with catenated inputs:

```
adc( x # v[1:n], y # w[1:n], cin).
```

- Step 2: if the rewrite-technique fails, the vector-expressions are compiled into the basic boolean functions and, or and not. Vector expressions are sliced into single bits. The Madre/Billon tautology checking technique [7] is then applied (Fig. 9). Since the rewrite-rules used in Step 1 are not confluent, the second step ensures the completeness of the approach.

  The following table shows the cpu-time of a SPARC-station needed to solve the third problem of Fig. 7 depending on wordlength:

Wordlength	8	16	32	64	128
CPU-time	0.8 sec.	0.9 sec.	1.0 sec.	1.8 sec.	3.4 sec.

The two-step procedure has a significant advantage in a situation where a design error is detected. The behavior of a verification tool in an error situation is an important aspect for its acceptance by a designer. In the example of Fig. 10, a 16-bit adder is implemented by means of four 4-bit adders; however, the carry chain is broken at the carry input of the instance a2 since the carry input is erroneously set to 0 rather than to the carry output a1.co of the first adder.
If the implementation of Fig. 10 is compared with the specification of an 16-bit adder

```
adc( a[1:16], b[1:16], 0)
```

[1]For convenience, the HDL constants 0 and 1 are overloaded and represent the boolean constants L and H as well as the integers 0 and 1

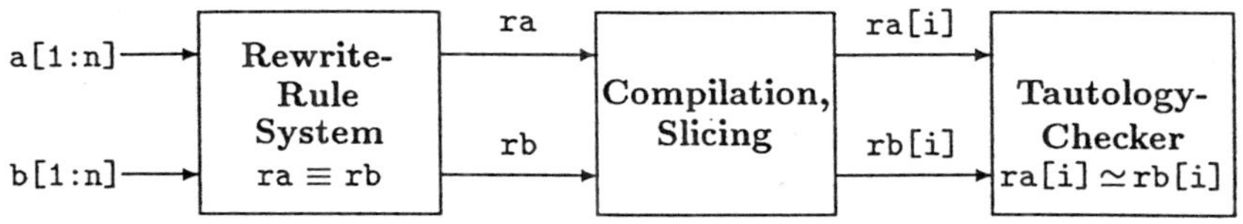

Fig. 9: Equivalence-proof of vector expressions

```
DESCRIPTION rt7483(IN  x, y: btml[1:4]; ci: btml;
                   OUT s: btml[1:4]; co: btml) BODY
            s  .=  adc( x, y, ci)[2:5],
            co .=  adc( x, y, ci)[1]
ENDrt7483
USE  a1( a[13:16], b[13:16], 0,     sum[13:16]),
     a2( a[9:12],  b[9:12],  0,     sum[9:12]),
     a3( a[5:8],   b[5:8],   a2.co, sum[5:8]),
     a4( a[1:4],   b[1:4],   a3.co, sum[1:4]): rt7483 ENDUSE
```

Fig. 10: Incorrect implementation of a 16-bit adder

then the response of LOVERT is the simplified expression of the implementation:

```
adc( a[1:12], b[1:12], 0) # adc( a[13:16], b[13:16], 0)[2:5]
```

The rewrite-rule system is able to simplify the expression for the three correctly chained 4-bit adders applying the simplification-rule for adders shown above; this results in an 12-bit adder which is concatenated with the last (and erroneously uncoupled) 4-bit adder. The expression gives thus a hint to the place where the problem is located.
This example shows also that LOVERT is able to cope with the problems involved in the aggregation of bit-sliced circuits.

4 Transitional vs. Structure-Oriented Static/ Transitional Descriptions

The second type of extensions applies to systems with sequential behavior.

Definition 5: A description is called *transitional* iff all carriers have transitional behavior.

Clearly, a description consisting of a collection of conditional transfers is transitional.
The main purpose of a transitional description is to display which transfers take place under which mutually exclusive conditions. The class of transitional descriptions comprises representations of simple state-diagrams as well as specifications of complex microprograms.
A further class of descriptions is the classical implementation of finite state machines as a composition of a transitional and a static subsystem (Fig. 11).

Definition 6: A *static/transitional description* is a combination of a static and of a transitional description.

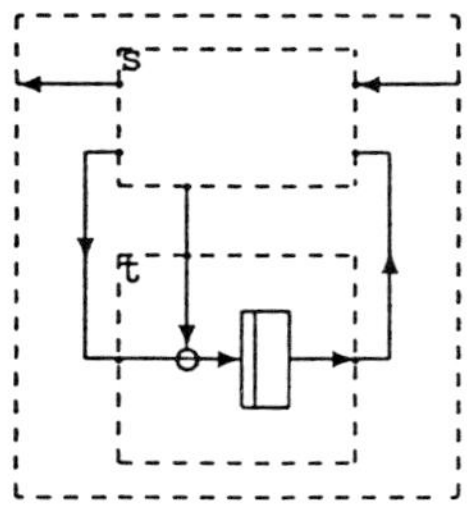

Fig. 11: Mixed transitional/static description

Definition 7: A *structure-oriented description* is a static/transitional description where each carrier of the transitional part occurs exactly once as destination.

Structure-oriented descriptions are amenable to an implementation by hardware since each transfer and each connection refers to one substructure.
We will now discuss a situation where the specification is a transitional description and where the implementation is a structure-oriented static/transitional description.
Assume several transfers into one carrier x in the transitional description d1:

```
DESCRIPTION d1 ...
   IF a1 THEN x <-  y1 ENDIF,
   ...,
   IF an THEN x <-  yn ENDIF
ENDd1
```

According to Definition 7, there is only one transfer into x in the description d2:

```
DESCRIPTION d2 ...
   IF ta THEN x <-  ty ENDIF,
   ta .=  ...,
   ty .=  ...,
   ...
   ASSERT c1,  ...,  cm END
ENDd2
```

The assertions c1, ... cm are used to specify don't care conditions, e.g., to exclude unreachable states or input combinations which must not occur.
To demonstrate 1-2, it must be proven for each transfer

```
IF ai THEN x <-  yi ENDIF
```

of d1 that

1. the condition ai of d1 implies the condition ta of d2:

$$\vdash_{T(d2)} \forall t: (c1(t) = H) \wedge \ldots \wedge (cm(t) = H) \wedge (ai(t) = H) \Rightarrow (ta(t) = H) \tag{4-1}$$

2. the condition ai implies the equality of the sources yi and ty, respectively:

$$\vdash_{T(d2)} \forall t: (c1(t) = H) \wedge \ldots \wedge (cm(t) = H) \wedge (ai(t) = H) \Rightarrow (ty(t) = yi(t)). \quad (4\text{-}2)$$

3. in the ***storage situation*** (see 2-2) the content of x remains unchanged if there is no ai active:

$$\vdash_{T(d2)} \forall t: (c1(t) = H) \wedge \ldots \wedge (cm(t) = H) \wedge (a1(t) = L) \wedge \ldots \wedge (an(t) = L) \Rightarrow (ta(t) = L) \vee (ty(t) = x(t)). \quad (4\text{-}3)$$

Note that 4-1, 4-2 and 4-3 refer to the point t of time only and amounts thus to an equivalence proof of vector-expressions. The LOVERT system is also able to cope with this type of extensions. Typical examples of application are the proof of the correct implementation of microprograms by means of a register-bus structure and a microprogram-sequencer including a ROM. An average verification time of 1 - 2 sec. per transfer on a SPARC-station was observed. The correct implementation of a complete microprogram is thus proven in a few minutes.

5 Transitional vs. Quasi-Transitional Descriptions

In the third type of extensions, transitional ("buffered") behavior of transfers is implemented by means of quasi-transitional ("unbuffered") assignments.
An example-specification of transitional-behavior is given in Fig. 12.

```
DESCRIPTION d1 ...
   DECLARE r1, r2: budvl END
   IF p2 & b1(r2) THEN r1 <- v1(r2) ENDIF,
   IF p1 & b2(r1) THEN r2 <- v2(r1) ENDIF
ENDd1
```

Fig. 12: Specification of transitional behavior

The implementation by means of a quasi-transitional description using conditional assignments is shown in Fig. 13.

Definition 8: A description is called ***quasi-transitional*** iff all carriers have quasi-transitional behavior.

As an example of the proof of 1-2, we study one of the axioms associated with r1 of the specification (see 2-1):

$$\forall t: (0 < t) \wedge (p2(t-1) = H) \wedge (b1(r2(t-1)) = H) \Rightarrow (r1(t) = v1(r2(t-1))). \quad (5\text{-}1)$$

The corresponding axiom of the implementation of r1 is according to 2-3

$$\forall t: (0 < t) \wedge (p1(t) = H) \wedge (b1(r2(t)) = H) \Rightarrow (r1(t) = v1(r2(t))). \quad (5\text{-}2)$$

In order to accommodate 5-1 and 5-2, we have to require:

$$\forall t: (0 < t) \wedge (p1(t) = H) \Rightarrow (p2(t-1) = H) \quad (5\text{-}3)$$

and

$$\forall t: (0 < t) \wedge (p1(t) = H) \Rightarrow (r2(t) = r2(t-1)). \quad (5\text{-}4)$$

```
DESCRIPTION d2 ...
   DECLARE r1, r2: bvarl END
   IF p1 & b1(r2) THEN r1 := v1(r2) ENDIF,
   IF p2 & b2(r1) THEN r2 := v2(r1) ENDIF
ENDd2
```

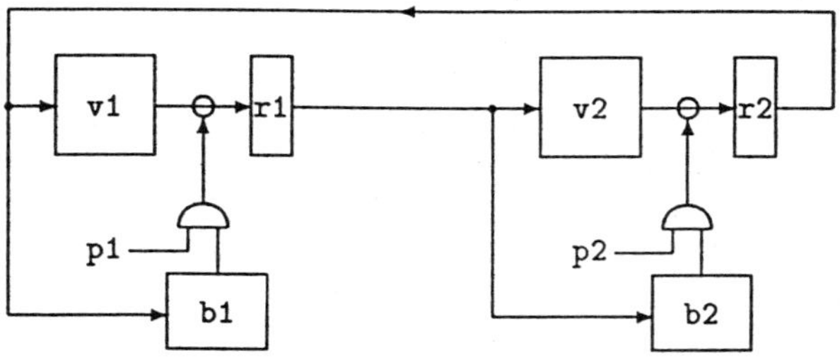

Fig. 13: Quasi-transitional implementation

The axioms associated with $r2$ are (see 2-3)

$$r2(0) = L,$$
$$\forall\, t:\ (0 < t) \wedge (p2(t) = H) \wedge (b2(r1(t)) = H) \Rightarrow (r2(t) = v2(r1(t))),$$
$$\forall\, t:\ (0 < t) \wedge ((p2(t) = L) \vee (b2(r1(t)) = L)) \Rightarrow (r2(t) = r2(t-1)).$$

Hence 5-4 is satisfied by

$$\forall\, t:\ (0 < t) \wedge (p1(t) = H) \Rightarrow (p2(t) = L). \tag{5-5}$$

Considering 5-3 and 5-5 we see that the correct implementation is ensured by a two-phase clock with the two phases p1 and p2. An example behavior is given in Fig. 14.

t	0	1	2	3	4	...
$p1(t)$	H	L	H	L	H	...
$p2(t)$	L	H	L	H	L	...

Fig. 14: Two-phase clock

As a result, it was shown that the quasi-transitional description of Fig. 13 has also transitional behavior. Note that the specification d2 of Fig. 12 is an abstraction from real hardware: there is no circuit that corresponds, e.g., to the expression `p2 & b1(r2)`.

6 Summary

The relatively simple concept of extension covers a number of relevant design and verification problems including the problem of logic-verification, the problem of the correct implementation of state-diagrams by means of structural resources, and the problem of the implementation of state-diagrams by means of two-phase clocked systems.

The introduction of several classes of temporal behavior represented by appropriate HDL-constructs allows for a systematic discussion of several types of extensions.
HDL-based hardware specification techniques offer a number of advantages: (i) hardware is represented in a convenient and user-friendly way, (ii) relevant classes of verification problems can be discussed in terms of the relationship between different forms of HDL-descriptions, (iii) the designer does not need proof-expertise since specialized verification tools support fully automatic verification.
The proof problems are complicated by the vector-functions provided by most HDL's for the compact representation of hardware. The LOVERT-approach proposes a combination of rewrite and tautology-checking techniques which makes the automatic verification of complex designs feasible.

References

[1] A. Bartsch, H. Eveking, H.-J. Faerber, M. Kelelatchew, J. Pinder, and U. Schellin. LOVERT - a logic verifier of register-transfer level descriptions. In L. Claesen, editor, *Proc. IMEC-IFIP Workshop on Applied Formal Methods for Correct VLSI Design*, pages 522–531, 1989.

[2] R.E. Bryant. Graph-based algorithms for boolean function manipulation. *IEEE C-35*, 677–691, 1986.

[3] H. Eveking. The application of CHDL's to the abstract specification of hardware. In Koomen/Moto-Oka, editor, *Proc. CHDL '85 (Tokio)*, pages 167–178, North-Holland, 1985.

[4] H. Eveking. Axiomatizing hardware description languages. *International Journal of VLSI Design*, 1990.

[5] H. Eveking. Formal verification of synchronous systems. In G. Milne and P.A. Subrahmanyam, editors, *Formal Aspects of VLSI Design*, pages 137–152, North-Holland, 1985.

[6] J. Hartmanis and R.E. Stearns. *Algebraic Structure Theory of Sequential Machines*. Prentice Hall, 1966.

[7] J.C. Madre and J.P. Billon. Proving circuit correctness by formally comparing their expected and extracted behavior. *Proc. 25th Design Autom. Conf.*, 205–210, 1988.

[8] R. Piloty, M. Barbacci, D. Borrione, D. Dietmeyer, F. Hill, and P. Skelly. *CONLAN Report*. Springer-Verlag, Berlin Heidelberg New-York Tokio, 1983.

[9] T. G. Windeknecht. *General Dynamical Processes*. Academic Press, 1971.

DIMACS Series in Discrete Mathematics
and Theoretical Computer Science
Volume 3, 1991

USING PARTIAL-ORDER SEMANTICS TO AVOID THE STATE EXPLOSION PROBLEM IN ASYNCHRONOUS SYSTEMS

David K. Probst and Hon F. Li
Department of Computer Science
Concordia University
1455 de Maisonneuve Blvd. West
Montreal, Quebec Canada H3G 1M8

ABSTRACT

We avoid state explosion in model checking of delay-insensitive VLSI systems by not using states. Systems are networks of communicating finite-state nonsequential processes with well-behaved nondeterministic choice. A specification strategy based on partial orders allows precise description of the branching and recurrence structure of processes. Process behaviors are modelled by pomsets, but (discrete) sets of pomsets with implicit branching structure are replaced by pomtrees, which have finite presentations by (automaton-like) behavior machines. The latter distinguish both concurrency and branching points, and define a finite recurrence structure. Safety and liveness checking are integrated. In contrast to state methods, our methods do not require enumeration or recording of states. We avoid separate consideration of execution sequences that do not differ in their partial order, and ensure termination by recording only a small number of system loop cutpoints -- in the form of system behavior states. In spite of the name, behavior states are not states.

Keywords delay-insensitive system, model checking, state explosion, partial-order semantics, branching point, recurrence structure, behavior machine, behavior state.

1. Introduction

There is considerable interest in asynchronous hardware systems, fueled by concerns about clock distribution and component composition in clocked systems [1,3,6,8]. Clearly, there is considerable conceptual overlap between asynchronous hardware systems and asynchronous distributed systems. Here, we focus on hardware systems, although we make our verification assumptions clear, as a first step towards extending the work to other application areas. Most theoretical asynchronous systems research is based on a formal representation strategy that underlies efforts in both verification and synthesis, and can have a major impact on efficiency (for example, on the time needed for verification, or on the size of synthesized objects). In delay-insensitive VLSI systems, system correctness is independent of delays in both asynchronous circuit components and transmission media. Both are specified as asynchronous processes. Delay-insensitive systems are modelled as networks of processes that communicate by direct contact. Since all communication is asynchronous and unbuffered, only local protocols are available to eliminate undesirable inputs. We avoid state explosion in model checking of delay-insensitive VLSI systems by not using states. Our specification strategy allows precise description of the branching and recurrence structure of processes. Processes are

This research was supported by the Natural Sciences and Engineering Research Council of Canada under grants A3363, A0921 and MEF0040121.

modelled by pomtrees, which differ from (discrete) sets of pomsets in making implicit branching structure explicit. Pomtrees have finite presentations by (automaton-like) behavior machines that distinguish both concurrency and branching points, and define a finite recurrence structure. Our slogan is, combine true concurrency with true nondeterminism -- and a finite recurrence structure -- to achieve efficient algorithmic processability. Our model checking strategy is, evaluate graph predicates on a system pomtree rather than state predicates on a system state graph. A small set of loop cutpoints in a finite presentation of the system pomtree is discovered during model checking. The core novelty of our approach is that (i) causality is checked directly, and (ii) behavior states (which are not states) are used only for termination.

2. Abstract specification of asynchronous processes

Abstract specifications define sets of externally-visible infinite computational behaviors, as well as branching points and a finite recurrence structure. Our behaviors are called complete to emphasize that they correspond to some maximal safe use of the process by an environment. In event structure terms, a complete behavior is a maximal conflict-free set of events. A process P has a set of input ports I and a set of output ports O. A process action is a (port, token) pair -- in VLSI, tokens correspond to signal (voltage) transitions. tI (tO) is the set of input (output) actions, and $\Sigma = tI \cup tO$ is the set of process actions. An event at an input or output port performs a process action defined by the value of the token received or sent. Since the focus in this paper is on systems with cleanly defined control states that are verified separately, tokens are essentially colorless. P's input actions are under the exclusive control of P's environment. P's output actions are under the exclusive control of P. This asymmetry of control is central to the model.

Safety properties (invariance properties) specify what the process is allowed to do; they also specify what the environment is allowed to do. A safety violation is an occurrence of a proscribed input or output event (in automata-theoretic terms, an undefined transition). After an input safety violation, a process becomes undefined. Liveness properties (inevitability properties) specify what the process is required to do; in our model, they do not specify what the environment is required to do. A liveness violation is a nonoccurrence of a prescribed output event. We group liveness properties into two camps, viz., (1) those related to progress (that is, response), and (2) those related to fairness of conflict resolution. Symmetric (asymmetric) specification of safety (liveness) properties has an interesting structuring effect on model checking (the examination of a special closed network of processes): although safety violations can show up in any process, liveness violations can only show up in the specification.

2.1. Primitive notions

At the level of (infinite) pomtrees, there are two primitive notions: (1) complete behavior (maximal conflict-free set of events), that is, any maximal infinite path through the pomtree, and (2) nondeterministic choice between mutually exclusive sets of process actions, by either process or environment, that is, selection of a particular pomtree branch. Complete behaviors are abstractions of infinite executions of the process that (i) correspond to some maximal safe use, and (ii) contain only necessary temporal precedences between external (interface) events. Concurrent events have no specified temporal relationship. A behavior "contains" any execution state that arises in any execution abstracted by the behavior. Execution state is affected by the performance of process actions in the usual way. P is input nondeterminate when P's environment can choose; P is output nondeterminate when P can choose.

At the level of (finite) behavior machines, there is a third primitive notion: recurrence (looping) in behavior space. Actually, the major precondition for the application of our techniques is the existence of a clean, finite recurrence structure. Behavior machines describe how commands (socket-extended finite pomsets) define transitions between selected pairs of behavior states. The essential use of behavior states is to identify cutpoints in loops of computational behaviors. A behavior state contains all the information in an execution state plus some additional information about how events in the past are temporally related to events in the future. A behavior state

is not a state in the sense that recording system execution states is not sufficient to discover a finite presentation of the system pomtree.

2.2. Pomsets and pomtrees

Pomsets and pomset operations have been studied extensively [4]. A labelled partial order (lpo) is a 4-tuple (V, Σ, Γ, μ) consisting of (i) a countable set V of events in a computational behavior, (ii) a finite set Σ of process actions, (iii) a partial order Γ on V that expresses the necessary temporal precedences among the events in V, and (iv) a labelling function $\mu : V \rightarrow \Sigma$ mapping each event $v \in V$ to the process action $\sigma \in \Sigma$ it performs. In our model, events at the same port in a given behavior are linearly ordered (concurrent events must be at distinct ports). Since every process has an initial state in which no events have occurred, and since the partial order Γ expresses action enabling (causal dependency among events), Γ must be well-founded (axiom of finite causes). Formally, a pomset (partially ordered multiset) is the isomorphism class of an lpo, denoted $[V, \Sigma, \Gamma, \mu]$.

Let p and q be pomsets, and let all partial orders Γ be written as "<". We say that q is a ρ-prefix of p when q is obtainable from p by deleting a subset of the events of p, provided that if event u is deleted and $u < v$, then v is also deleted. We say that α is a π-prefix of p when α is a finite ρ-prefix of p. $\pi(p)$ is the set of π-prefixes of p. We say that q is an augment of p when q differs from p only in its partial order, which must be a superset of that of p. If P is a set of pomsets, then $\pi(P)$ is $\cup$ $\pi(p)$, $p \in P$. Let $p = [V, \Gamma, \Sigma, \mu]$ be a pomset and let $V' \subseteq V$. The projection onto V' is $p' = [V', \Gamma', \Sigma, \mu']$, where Γ' and μ' are the restrictions of Γ and μ to V'.

We introduce the notion of prefix envelope. Let p be a pomset. ${}^{\circ}p$ is the set of action labels of initial events of p, that is, the set of $\mu(v)$ of v in p such that $\nexists u$ in p with $u < v$. If $\alpha \in \pi(p)$, then $p - \alpha$ is the projection obtained by deleting the events in α from p. The envelope of α in p, denoted $E_p(\alpha)$, is ${}^{\circ}(p - \alpha)$. Define $in_p(\alpha) = E_p(\alpha) \cap tI$ and $out_p(\alpha) = E_p(\alpha) \cap tO$. $E_p(\alpha)$ is the set of process actions that are concurrently enabled in behavior p after the events of partial execution α. We sketch the notion of branch point. Consider the following skeleton presentation of a simple finite pomtree. Let α, β_1 and β_2 be finite pomsets, with ${}^{\circ}\beta_1 \cap {}^{\circ}\beta_2 = \{ \}$. Lay down pomset α. Now, both concatenate β_1 with set Γ_1 of (α, β_1) precedences, and concatenate β_2 with set Γ_2 of (α, β_2) precedences. The pomtree contains one copy each of α, β_1 and β_2. β_1 and β_2 are the two branches. Let p and q be the two (maximal) paths through the pomtree. By construction, $\alpha \in \pi(p)$, $\pi(q)$ is a maximal (under prefix ordering) common prefix of p and q. Also by construction, $E_p(\alpha) \cap E_q(\alpha) = \{ \}$. We say that α is a branch point. If $E_p(\alpha), E_q(\alpha) \subseteq tI$ (tO), then α is an input (output) branch point. Pomtrees are like computation trees except that arcs are maximal determinate behavior segments, and vertices are input or output branch points.

2.3. Determinate processes

Fig. 1 shows a representation of a complete computational behavior p. Process P = {p} could be a C-element together with an unconnected wire. The use of pluses and minuses (from rising and falling signal transitions) is partly redundant; once the initial voltage level of a port has been fixed by a reset transition, each new transition at that port is the opposite of the preceding transition. Output port names have been underlined. The complete behavior is obtained by concatenating infinitely many copies of this figure, superimposing adjacent (a^+, b^+, d^+) triples. The figure is intended to illustrate both maximal safe use and necessary temporal precedence. For example, process P is a multithreaded process (it has two threads); the environment may elect to use one thread and not the other. A process with a single thread has essentially only one use. Necessary temporal precedence is explained below.

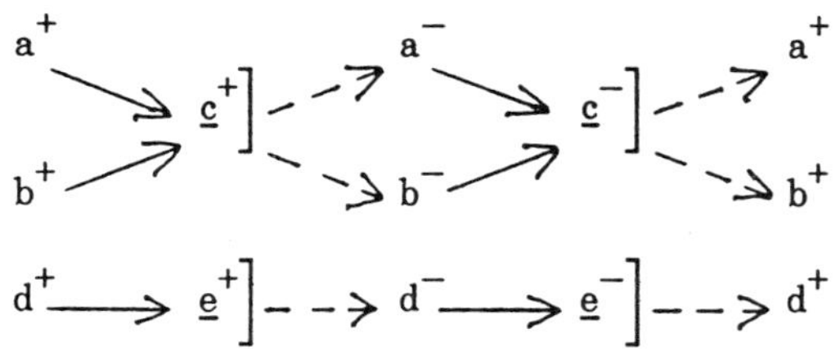

Fig. 1 Representation of a complete computational behavior.

The complete behavior precisely describes when the process (environment) is allowed to perform output (input) actions. In all processes, as illustrated in Fig. 1, the behavior partial order Γ in $p = [V, \Sigma, \Gamma, \mu]$ is the transitive closure Ω^+ of a nontransitive successor relation $\Omega = N \cup \Xi$, where the output protocol N is a relation from input events to output events, and the input protocol Ξ is a relation from output events to input events. In model checking, we use the fact that N (Ξ) is the causal (noncausal) part of Ω, by asymmetry of control. In short, the process (environment) enforces N (Ξ). We also say that the actions in $out_p(\alpha)$ [$in_p(\alpha)$] are causally [noncausally] enabled at α. The semantics of successor arrows is as follows. The process may perform an output action when all of its solid-arrow predecessor events have occurred. The environment may perform an input action when all of its dashed-arrow predecessor events have occurred. Violations of the input (output) protocol are called failures (errors) [6]. After a failure, a process may behave arbitrarily. This means that system pomtrees of (closed) networks of processes are undefined if there is a failure in any component process. Brackets are explained momentarily.

2.4. Delay insensitivity

In general, well-behavedness conditions are motivated by restrictions that allow asynchronous processes to be used as components of delay-insensitive systems [6]. Here, we present a minimal set of assumptions that are used in the model checking approach of this paper. Assumptions that guarantee delay insensitivity are distinguished from more general assumptions -- not shown in this paper -- that guarantee the finite-state character of asynchronous processes, and the bounded-size encoding of behavior states. These general assumptions are what guarantee the existence of a clean, finite recurrence structure in an asynchronous hardware or software system.

Well-behavedness conditions

Rule 1 There is no autoconcurrency. Formally, any two events at the same port in $p \in P$ are separated in Ω by at least one event at some other port.

Rule 2 There is no specified successor relationship either between two input events or between two output events. Formally, each line in $p \in P$ consists of an infinite sequence of strictly alternating input and output events.

2.5. Progress requirements

We specify progress requirements in a behavior p by bracketing output events. After partial execution $\alpha \in \pi(p)$, a determinate process $P = \{p\}$ is required to perform all bracketed output actions in $out_p(\alpha)$; this subset is denoted $req_p(\alpha)$. These actions must be performed eventually, after some arbitrary but finite delay. To simplify model checking, from now on we only consider systems in which $req_p(\alpha) = out_p(\alpha)$; this defines restricted process theory. We capture this notion in a rule.

Rule 3 In the absence of branching, there is never a choice between performing and not performing an output action. Formally, if $p \in P$, then all output events in p are bracketed.

2.6. Nondeterminate processes

Branching in an asynchronous process arises in either of two ways. There may be free choice (in the Petri net sense) in the environment to apply one of several mutually exclusive access operators, and there may be arbiter choice in the process to respond to one of several concurrent requests to perform a mutually exclusive operation on a shared object. Pomtrees adequately model the branching structure arising from these two sources. The semantics at input (output) branch points is straightforward. An input branch point α defines $\mathbf{in}(\alpha) = \{in_j(\alpha) : j \in J\}$, where J indexes the finite input branching at α. This is the family of disjoint sets of input actions that are concurrently enabled at α, respectively, in each of the futures that branch from α. An output branch point α defines $\mathbf{out}(\alpha) = \{out_k(\alpha) : k \in K\}$, where K indexes the finite output branching at α. This is the family of disjoint sets of output actions that are concurrently enabled at α, respectively, in each of the futures that branch from α. The semantics at nonbranch points is equally straightforward. If α is within a determinate behavior segment, then there is precisely one enabled set. If α straddles a branch point, then the enabled sets are not disjoint.

Consider an output branch point α in restricted process theory. If P advances to α, then P is required to perform all actions in some out_k within finite time. Explanation is required for arbiter choice, concerning progress requirements at prefixes of α. Suppose that output action u is required at β and that, before P performs u, P advances to α in which u and v are disjunctively required; P may then perform v and not u. This is an acceptable confusion situation (in the Petri net sense): in an arbiter, a required output action may become disjunctively required after the receipt of new input.

2.7. Behavior machines

Behavior machines are finite presentations of pomtrees. A behavior machine consists of a set of selected behavior states, and a set of commands. A command can be applied in certain behavior states, and produces a new behavior state. Commands are essentially finite pomsets with additional machinery to define the nonsequential serial concatenation -- modulo branching -- of the command to the finite pomtree generated so far. One fact of model checking is that components of systems are subject to nonmaximal safe use, requiring the encoding of behavior states that were not originally selected; for this reason, we encode all behavior states at specification time. Behavior machines specify both safety and progress properties; fairness properties must be specified by a supplementary condition constraining output choice.

Formally, a behavior machine is a set S of selected behavior states (with a distinguished start state), and a set C of commands. The machine describes the possible state transitions (c takes s to t) produced by each command. For each $c \in C$, there is a transition relation $\Delta(c)$ on S, where $(s, t) \in \Delta(c)$ if conceptual execution of command c in state s produces state t.

Fig. 2 shows a behavior machine for a C-element. Each command has the form: behavior state label, socket-extended finite pomset, behavior state label. In constructing behaviors, a new command may be applied provided the command prelabel matches the postlabel of the previous command. Sockets (denoted $\circ$) help describe how future behavior follows past behavior. If there is a successor arrow $\circ \rightarrow$ u, then $\circ$ is filled by some predecessor of u according to a well-defined rule. More precisely, if command c can follow partial execution α, then there is an injection (produced by simple labelling) from the set of sockets in c to the set of events in α that can still participate in (new) successor arrows. By convention, there is an imaginary initialization event $\bullet$ with postlabel 0. As a presentational device, we make the vertical placement of symbols in commands significant: a socket is always filled by an action that appears on the same horizontal line. For example, in Fig. 2, the unique socket $\circ$ can be filled by $\bullet$ (first application of the command) or by $\underline{c}^-$ (all subsequent applications of the command). Assume that the default vertical placement of $\bullet$ is "middle".

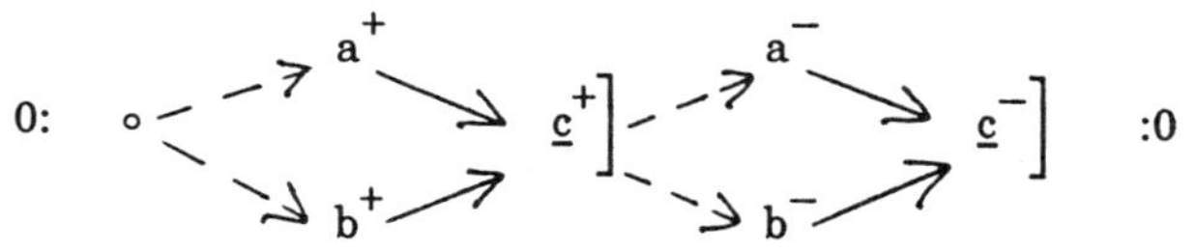

Fig. 2 Behavior machine for a C-element.

Fig. 3 shows a reduced behavior machine for a delay-insensitive arbiter. For conciseness of presentation, we have included several distinct behavior states under label 1. The abuse of notation is intentional. Label 1 in Fig. 3 is an equivalence class of behavior states. The two "commands" on the right in Fig. 3 are equivalence classes of commands. This is explained below. Each of two clients follows a four-cycle protocol, where $\langle A \rangle = \underline{c}^+]$ - -> a^- and $\langle B \rangle = \underline{d}^+]$ - -> b^- are the critical sections. A two-arrow socket that is always filled by • has been suppressed from command 1 (extreme left). As shown, sockets exist on three horizontal levels. The top ∘ (in command 2) is always filled by a^+ and the bottom ∘ (in command 3) is always filled by b^+. Either middle ∘ can be filled by •, a^- or b^-, perhaps redundantly.

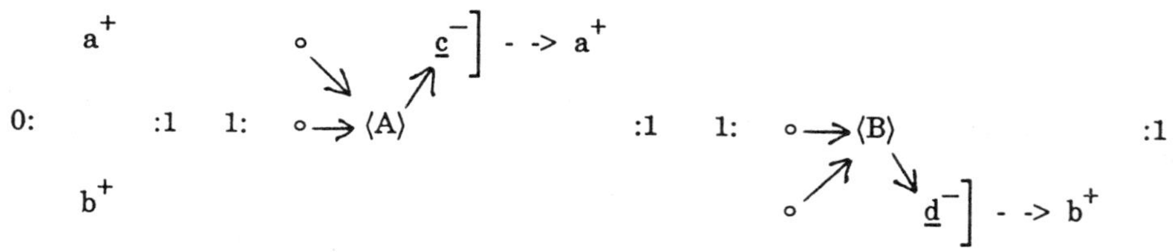

Fig. 3 Reduced behavior machine for a delay-insensitive arbiter.

The reduced behavior machine in Fig. 3 groups three distinct behavior states under label 1, all corresponding to the same execution state. The three may be distinguished as: (1.1) no critical-section entry has occurred (both middle sockets become redundant), (1.2) A's critical-section exit was the most recent (A's middle socket becomes redundant), and (1.3) B's critical-section exit was the most recent (B's middle socket becomes redundant). Behavior states 1.1, 1.2 and 1.3 are output branch points. Behavior states 1.2 and 1.3 form a complete set of loop cutpoints in the full behavior machine; such a set is called a dominator set [2,7]. The full behavior machine is easily obtained. It has four selected behavior states (0, 1.1, 1.2 and 1.3) and seven transitions -- but only five commands (distinct socket-extended pomsets). For example, behavior state 1.2 has a self-loop produced by command 2′ (command 2 minus its middle socket), and a transition to behavior state 1.3 produced by command 3.

The simplest way to encode behavior states is to use names of successor arrows. This is done by recognizing distinct arrows in a behavior machine, and assigning labels. For a finite-state determinate process P = {p}, the behavior state $s(\alpha)$ that corresponds to partial execution $\alpha \in \pi(p)$ is encoded as the set $\Omega(\alpha, p - \alpha)$ of successor arrows with source in α and target in $p - \alpha$. This encoding strategy works equally well for finite-state nondeterminate processes. For example, any behavior state of the arbiter can be encoded by three (small) integers representing the (virtual) successor arrows currently offered by the past to the future. Thus, behavior state 1.2 is encoded as: a^+ (b^+) is currently available to enable the next $\underline{c}^+$ ($\underline{d}^+$) that appears in any downward path through the pomtree, and a^- is currently available to enable the $\underline{d}^+$ that appears in the immediately adjacent B branch.

We say that two partial executions $\alpha, \beta \in \pi(P)$ are execution equivalent, written α

$\equiv_e$ β, when their sets of possible futures are equal, that is, $f_P(\alpha) = f_P(\beta)$ [5]. This is standard pomset (or pomtree) equality, based on lpo isomorphism. Each $\equiv_e$ equivalence class of partial executions is an execution state of process P. In the same way, each $\equiv_b$ equivalence class ($\alpha \equiv_b \beta$ when their sets of possible socket-extended futures are equal) is a behavior state of process P. Formally, if α and β are prefixes of infinite paths through pomtree P, then α and β result in the same behavior state of P precisely when the two pomtrees P/α and P/β descending from α and β are equal, where P/α and P/β have been extended to include the $(\alpha, P/\alpha)$ and $(\beta, P/\beta)$ precedences, and isomorphism now requires matching both event labels and arrow labels. In spite of the name, behavior states are not states.

3. Correct implementation

An implementation may exceed the minimum requirements of the specification. For example, it may be more liberal in accepting input and more conservative in producing output, but only if all progress requirements are satisfied. In model checking, a straightforward way to define correctness is to use the mirror of the specification as a conceptual implementation tester [1]. That is, one forms an imaginary closed system by linking mirror mP of specification P to the implementation -- in the general case, a network of processes *Net* --, and then examining certain properties of the resulting system. In the partial order world, the mirror is formed merely by inverting the causal/noncausal interpretation of P's successor arrows. Many things may now be examined. Is there a failure somewhere, causing the system to become undefined? Does the system just stop, violating fundamental liveness? Is some progress requirement violated? Is some conflict resolved unfairly? To study the closed system $S = mP|P'$ -- in the general case, $S = mP|Net$ -- by partial order methods, we first regroup all successor arrows (as shown below) to obtain the system causal and noncausal successor relations. Graph predicates are then evaluated on a system pomtree that has two distinct successor relations.

In closed system S, each action is attributed to precisely two processes as a joint action of those two processes. System behaviors are projected onto component alphabets to yield component behaviors. Consider first the trivial case that the implementation is a process. Conceptual execution of $S = mP|P'$ produces corresponding partial executions of specification P and implementation P′. These partial executions correspond by containing the same events, but may not agree on which temporal precedences are necessary. For $\alpha \in \pi(P)$ and $\alpha' \in \pi(P')$, $\alpha \Longleftrightarrow \alpha'$ denotes correspondence. Each α,α' pair is a "doubly-ordered" finite pomset $[V, \Sigma, \Gamma, \Gamma', \mu]$. Corresponding pairs are produced by conceptual execution of S only as long as no safety or liveness violation is detected.

Safety checking occurs on two levels. (i) Any action u that is causally but not noncausally enabled at α,α' is an immediate safety violation. Sets of concurrently causally enabled actions must be concurrently noncausally enabled. (ii) Moreover, while visiting event u, each noncausal arrow (t, u) is checked by a (breadth-first) search for a supporting chain [t, u] of causal arrows. The nonexistence of such a chain is a safety violation. Liveness checking also occurs on two levels. (i) Any external output action v that is noncausally but not causally enabled at α,α' is an immediate liveness violation -- unless v is only disjunctively required (see above). Each set of concurrently causally enabled external output actions must contain some set of concurrently noncausally enabled external output actions. (ii) Moreover, while visiting external output event v, each causal chain [t, v] from an external input event t is checked by a "search" for a matching noncausal arrow (t, v). The nonexistence of such an arrow is a liveness violation.

We give a state-based definition of correctness for $S = mP|P'$. First, consider determinate P and P′. Correctness of safety properties means: $\forall \alpha,\alpha' : \alpha \Longleftrightarrow \alpha' : in_P(\alpha) \subseteq in_{P'}(\alpha')$ and $out_P(\alpha) \supseteq out_{P'}(\alpha')$. Correctness of progress properties means: $\forall \alpha,\alpha' : \alpha \Longleftrightarrow \alpha' : req_P(\alpha) \subseteq req_{P'}(\alpha')$. In restricted process theory, this reduces to: $out_P(\alpha) =$

$out_{p'}(\alpha') = req_p(\alpha) = req_{p'}(\alpha')$. Next, extend consideration to nondeterminate P and P′. In restricted process theory, correctness now means:

$$\forall \alpha,\alpha' \ : \ \alpha \Longleftrightarrow \alpha':$$
$$\forall j \ \exists j' \ : \ in_j(\alpha) \subseteq in_{j'}(\alpha') \ \wedge \qquad (*)$$
$$\forall k' \ \exists k \ : \ out_k(\alpha) = out_{k'}(\alpha') \qquad (**)$$

These subscripts index sets of actions that are concurrently enabled at α, respectively, in each of the futures that branch from α.

We are now ready for a genuine partial order view that defines correctness without reference to states; in particular, the $\forall \alpha,\alpha'$ quantifier will be dispensed with. Consider a pair α,α', $\alpha \Longleftrightarrow \alpha'$. We regroup the causal and noncausal parts of $\Omega(\alpha)$ and $\Omega'(\alpha')$ to obtain the system causal and noncausal successor relations. We define the <u>system causal</u> successor relation $\tilde{\Omega} = (N' \cup \Xi)^{+-}$, where the superscript denotes transitive closure followed by transitive reduction. Similarly, we define the <u>system noncausal</u> successor relation $\hat{\Omega} = (N \cup \Xi')^{+-}$. When the implementation is a network *Net* of components P_i, $i \in I$, $\tilde{\Omega} = [(\cup_i N_i) \cup \Xi]^{+-}$, and $\hat{\Omega} = [N \cup (\cup_i \Xi_i)]^{+-}$.

Consider the general case. System pomtree S is undefined if a component fails. Suppose there is no failure (safety violation) in S. Each event v in S is classified as an input event in precisely one (say P_i) of the two processes to which the action performed by event v is attributed. From P_i's specification, there is a well-defined noncausal preset in P_i (say $\Pi_i(v) \subseteq \Sigma_i$) of the P_i action performed by event v. By definition, each event occurs in S because it is causally enabled. In order that P_i not "explode", v must also be noncausally enabled in P_i. This is equivalent to: (i) $\forall u \in \Pi_i(v) : u \in S$, and (ii) for each u, there is a chain of causal arrows [u, v]. This is safety correctness. Some events v that are noncausally enabled in S are, from specification P's point of view, classified as (external) output events. From P's specification, there is a well-defined noncausal preset in mP (say $m\Pi(v) \subseteq m\Sigma$) of the mP action performed by event v. In order that (the nonoccurrence of) event v not be a progress violation, we must have: (i) $v \in S$, and (ii) there exists a chain of causal arrows [u, v] iff $u \in m\Pi(v)$. Condition (ii) means that, after projection of S on mP, precisely $m\Pi(v)$ enables the mP action performed by v. Again, v may be only disjunctively required (see above). This is progress correctness.

4. <u>Model checking</u>

We sketch direct model checking of networks of processes. Given are specification process P and some number of implementation component processes P_i, $i \in I$. Network links map each internal output action of each component process P_i to precisely one internal input action of some other component process P_j [6]. Each internal event is attributed to two component processes. Conceptual links map each action of specification process P, whether input or output, to precisely one action (of the same type) of some component process P_i. Each external event is attributed to the specification process and one component process. External input events are caused by specification process P, while all other events are caused by some component process P_i.

The <u>verification procedure</u> shares superficial structure with the algorithm in [1]. Let D be a complete set of loop cutpoints (dominator set) for P. We enumerate system pomtree S recursively. We maintain (1) a <u>stack</u> to postpone the examination of system pomtree branches (maximal determinate behavior segments), and (2) a <u>table</u> to detect cycles in the enumeration of S. Recording the system behavior state each time P advances to some d ∈ D (at P command completion) leads to discovery of a loop cutpoint for each loop in a finite presentation of system pomtree S.

Fig. 4 shows a typical result of applying a P command to a network, here, in its initial state.

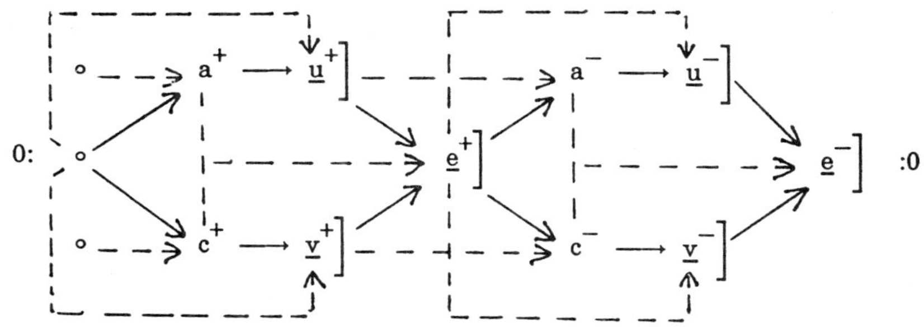

Fig. 4 Direct verification of a network.

Conceptual execution of system S produces n-tuples $<\alpha, \alpha_1, \alpha_2, \ldots>$ obtained by projection of system behavior prefixes onto processes P and P_i, $i \in I$. In the general case, corresponding partial executions no longer contain the same events. Each $\beta \in \pi(S)$ maps onto a system behavior state, which is the standard vector of behavior states of P and all P_i. Distinguish a "red" problem and a "green" problem. The red problem is, identify each maximal determinate behavior segment of system pomtree S. Branch points of S are scheduled for expansion in the usual way. The green problem is, identify each maximal behavior prefix $\beta \in \pi(S)$ that corresponds to a selected behavior state of specification P. System behavior states of such β's are candidates for entry into the table. The solution to both problems is, conceptually mark the following events in the indicated manner: (1) each initial event of each P command emanating from an input (output) branch point is marked with both a red and a green dot (with a green dot only), (2) each initial event of each P_i command emanating from an output branch point is marked with a red dot only, and (3) each initial event of each P command emanating from a determinate loop cutpoint is marked with a green dot only.

We move forward cleanly to a branch point of S by deferring visiting any red event as long as there are nonred events still enabled. Similarly, we move forward cleanly to a system behavior prefix $\beta \in \pi(S)$ that corresponds to a P rule completion by deferring visiting any green event as long as there are nongreen events still enabled. If the (selected) P behavior state of β has been identified as a loop cutpoint in P, then the system behavior state of β is entered into the table. If this state is already in the table, then β is not extended further.

Let $r : \pi(S) \to \pi(P)$ be the standard projection from system behavior prefixes to specification behavior prefixes. Refinement mapping is nontrivial because r has too many inverse images. The function of green dots is to select only a few "extremal" inverse images of selected behavior states of P -- to create (potential) table entries. For example, some P commands terminate in output-branch loop cutpoints. This is the case for command 1 in the full arbiter behavior machine, with a typical implementation being a ring of DME elements. Command 1 is applied to this implementation by moving downward along all paths of system pomtree S to include any event -- that is causally enabled by a^+ and b^+ -- up to but not including any green event (here, $\underline{c}^+$ and $\underline{d}^+$). With this implementation, command 1 of P corresponds to a finite initial pomtree prefix of S rather than a determinate system behavior segment. In our table, we record the two system behavior states that correspond to the <u>endpoints</u> of the two maximal paths through this finite system subpomtree.

Here is the algorithm summary. Recursively generate the system pomtree S by extending all system behavior prefixes $\beta \in \pi(S)$ as long as no safety or liveness violation is detected. Starting from the initial global state, apply P commands to the implementation by generating all causally enabled events in all possible futures up to but not including any green event. As events are generated, evaluate the graph predicates discussed in section 3. At P command completion, if the postlabel t in s: c :t belongs to the dominator set D of P, then record all pairs (t, N), where N is any network behavior state that corresponds to t by the construction in the previous paragraph. Do not apply a command with prelabel t to the implementation in network behavior state N if (t, N) is already in the table. Terminate the algorithm when it is no longer possible to apply any P command.

5. Conclusion

Behavior machines -- finite presentations of pomtrees -- precisely describe the branching and recurrence structure of processes, with benefits for both analysis and synthesis. Partial-order model checking allows us -- whenever there is a moderate amount of true concurrency but not an excessive amount of true nondeterminism -- to avoid the state explosion that frequently haunts verifiers based on state graphs, and limit any combinatorial explosion to that created by the branching structure of processes. Loop cutpoint detection in behaviors appears to require behavior states; such states contain information about how future behavior follows past behavior. Based on an understanding of the state space, the specifier/designer can create a useful separation of concerns for the verifier by structuring processes into (i) determinate behavior segments that are either completed or eliminated during conceptual execution (a pretending atomicity property), and (ii) input or output branch points. We obtain a dramatic performance improvement in model checking, and this even though only the specification process P gets to pretend atomicity.

References

[1] D.L. Dill, "Trace theory for automatic hierarchical verification of speed-independent circuits", Ph. D. Thesis, Department of Computer Science, Carnegie Mellon University, Report CMU-CS-88-119, February 1988. Also MIT Press, 1989.

[2] Z. Manna and A. Pnueli, "Specification and verification of concurrent programs by ∀-automata", Proc. of 14th ACM Symposium on Principles of Programming Languages, January 1987, pp. 1-12.

[3] A.J. Martin, "Compiling communicating processes into delay-insensitive VLSI circuits", Distributed Computing, Vol. 1, No. 4, October 1986, pp. 226-234.

[4] V.R. Pratt, "Modelling concurrency with partial orders", Int. J. of Parallel Prog., Vol. 15, No. 1, February 1986, pp. 33-71.

[5] D.K. Probst and H.F. Li, "Abstract specification of synchronous data types for VLSI and proving the correctness of systolic network implementations", IEEE Trans. on Computers, Vol. C-37, No. 6, June 1988, pp. 710-720.

[6] D.K. Probst and H.F. Li, "Abstract specification, composition and proof of correctness of delay-insensitive circuits and systems", Technical Report, Department of Computer Science, Concordia University, CS-VLSI-88-2, April 1988 (Revised March 1989).

[7] D.K. Probst and H.F. Li, "Partial-order model checking of delay-insensitive systems". In R. Hobson et al. (Eds.), Canadian Conference on VLSI 1989, Proceedings, Vancouver, BC, October 1989, pp. 73-80.

[8] J.v.d. Snepscheut, "Trace theory and VLSI design", Lect. Notes in Comput. Sci. 200, Springer Verlag, 1985.

DIMACS Series in Discrete Mathematics
and Theoretical Computer Science
Volume 3, 1991

A STUBBORN ATTACK ON STATE EXPLOSION

Antti Valmari
Technical Research Centre of Finland
Computer Technology Laboratory
PO Box 201
SF-90571 Oulu
FINLAND

Tel. +358 81 509 111

ABSTRACT

The paper presents the ***LTL preserving stubborn set method*** *for reducing the amount of work needed in the automatic verification of concurrent systems with respect to linear time temporal logic specifications. The method facilitates the generation of* ***reduced state spaces*** *such that the truth values of a collection of linear temporal logic formulas are the same in the ordinary and reduced state spaces. The only restrictions posed by the method are that the collection of formulas must be known before the reduced state space generation is commenced, the use of the temporal operator "next" is prohibited, and the (reduced) state space of the system must be finite. The method cuts down the number of states by utilising the fact that in concurrent systems the nett result of the occurrence of two events is often independent of the order of occurrence.*

1. INTRODUCTION

The automatic verification of temporal properties of finite-state systems has been a topic of intensive research during the recent decade. A typical approach is to generate the state space of the system and then apply a model checking algorithm on it to decide whether the system satisfies given temporal logic formulas [Clarke & 86] [Lichtenstein & 85]. A well known problem of the approach is that the state spaces of systems tend to be very large, rendering the verification of medium-size and large non-trivial systems impossible with a realistic computer. This problem is known as the *state explosion* problem.

Concurrency is a major contributor to state explosion. It introduces a large number of execution sequences which lead from a common start state to a common end state by the same transitions, but the transitions occur in different order causing the sequences to go through different states. This phenomenon has been recognized long ago and the choice of a coarser level of atomicity has been suggested as a partial solution (see [Pnueli 86]). Unfortunately, the power of coarsening the level of atomicity is limited. Consider a system consisting of n processes which execute k steps without interacting with each other and then stop. The system has $(k+1)^n$ states. Each of the processes of the system can be coarsened to a single atomic action. Coarsening reduces the number of states to 2^n which is still exponential in the number of the processes [Valmari 88c]. However, it seems intuitively that to check various properties of the system it would be sufficient to simulate the processes in one arbitrarily chosen order, thus generating only $nk+1$ states.

To our knowledge the first person to suggest a concurrency-based state space reduction method potentially capable of changing a state space from exponential to polynomial in the number of processes was W. Overman [Overman 81]. Overman's work is little known, perhaps because he considered a very restricted case (the terminal states of systems consisting of processes which do not branch or loop), and the algorithm he gave as part of his method for finding certain sets was not efficient enough from the practical point of view. He suggested also a modified method with a faster algorithm, but the modification destroyed the ability of changing exponential state spaces to polynomial.

The problems in Overman's approach were effectively solved by Valmari when he presented the so-called *stubborn set* method [Valmari 88a, 88b]. [Valmari 88a] presents the method in the framework of ordinary Petri nets. [Valmari 88b] sacrifices some reduction power and presents the theory in a more general *variable/transition* framework which is directly applicable to concurrent programs etc. Overman's method can be thought of as a special case of the more powerful and more widely applicable stubborn set method as presented in [Valmari 88b]. The computational problem which caused Overman to modify his method now reappeared as the problem of finding "good" *stubborn sets.* [Valmari 88a] gives a stubborn set search algorithm which (if required) finds the sets in Overman's method and is linear in the size of the system under analysis. However, the sets found by the algorithm (including Overman's sets) are not necessarily the "best" stubborn sets in the sense of leading to the best state space reduction results. [Valmari 88b] develops a quadratic (under reasonable assumptions) stubborn set search algorithm which is capable of finding in certain sense "best" stubborn sets. These results are summarised, compared to Overman's work and some details are refined in [Valmari 88c].

The above mentioned papers treat properly only the detection of deadlocks and terminal states using the reduced state spaces, although they anticipate the analysis of more properties. Their key theorem is "every reachable state without enabled transitions is present in the reduced state space". It is shown in [Valmari 89a] that the basic stubborn set method also preserves the existence of infinite execution sequences. It is thus suitable for detecting nontermination. Furthermore, the paper essentially gains back the part of the reduction power lost in the shift from [Valmari 88a] to [Valmari 88b].

The difference between the theories in [Valmari 88a, 89a] and [Valmari 88b] is clarified by the distinction of the *weak* and *strong* theories of stubborn sets in [Valmari 89b]. The weak theory is more complicated and more difficult to implement, but it leads to better reduction results. The above mentioned linear stubborn set search algorithm cannot exploit the extra freedom offered by the weak theory, but the quadratic algorithm can. [Valmari 89b] also opened the way to the use of stubborn sets in the verification of other than termination oriented properties by solving the so-called *ignoring problem.* As a concrete result, the paper shows how the reachability of a state satisfying a given state predicate can be decided using stubborn sets. This renders possible the verification of system invariants.

The present paper extends the stubborn set method to almost full linear temporal logic — almost, because the operator "next state" is forbidden. For simplicity, we have chosen to use the strong stubborn set framework, although with minor refinements the results are valid in the weak theory as well. The theory is developed in Chapter 2. Chapter 3 discusses how the theory may be implemented and Chapter 4 contains an example.

2. DEFINITIONS AND BASIC THEOREMS

2.1 Variable/Transition Systems

To develop the stubborn set method we look at concurrent systems as systems consisting of a finite set **V** of *variables* and a finite set **T** of *transitions*. Each variable v has an associated set called *type* and denoted by *type*(v), and at every instant of time v has a unique *value* belonging to its type. Assuming an ordering of **V**, the Cartesian product of the types of the variables is the set of *syntactic states* and is denoted by **S**. The value of variable v at syntactic state s is denoted by $s(v)$. There is a partial *next state* function *next* from $\mathbf{S} \times \mathbf{T}$ to **S** which defines when a transition is *enabled* and what is the result of the *occurrence* of an enabled transition. Transition t is enabled in state s, denoted by $en(s,t)$, iff $next(s,t)$ is defined. If $next(s,t) = s'$ we say that t may *occur* at s producing s' and often write $s -t\rightarrow s'$. We often merge the states between successive transition occurrences and write $s_0 -t_1\rightarrow s_1 -t_2\rightarrow \ldots -t_n\rightarrow s_n$ instead of $s_0 -t_1\rightarrow s_1 \wedge s_1 -t_2\rightarrow s_2 \wedge \ldots \wedge s_{n-1} -t_n\rightarrow s_n$. A sequence like this is called a *finite execution sequence* and its *length* is n. The *concatenation* of the execution sequences $\sigma = s_0 -t_1\rightarrow \ldots -t_n\rightarrow s_n$ and $\rho = r_0 -d_1\rightarrow \ldots -d_m\rightarrow r_m$ where $s_n = r_0$ is defined by $\sigma^\circ\rho = s_0 -t_1\rightarrow \ldots -t_n\rightarrow s_n -d_1\rightarrow \ldots -d_m\rightarrow r_m$. "$\rightarrow$" is defined by $s \rightarrow s' \Leftrightarrow \exists\, t \in \mathbf{T}: s -t\rightarrow s'$. "$\rightarrow^*$" is the reflexive transitive closure of "$\rightarrow$". There is a distinguished state $\mathbf{s}_0$ called the *initial state* of the system. A *variable/transition system* or *v/t-system* is the 5-tuple $(\mathbf{V},\mathbf{T},type,next,\mathbf{s}_0)$, where the components of the 5-tuple are as just explained.

Section 4.1 contains a non-trivial example of a v/t-system.

The stubborn set method relies on the analysis of certain relationships between transitions. Let us define and explain some necessary concepts.

Definition 2.1 Let $t, t' \in \mathbf{T}$, $s \in \mathbf{S}$, $V \subseteq \mathbf{V}$ and $T \subseteq \mathbf{T}$.

- t is *enabled with respect to* V at s, denoted by $en(s,t,V)$, iff
  $$\exists\, s' \in \mathbf{S}: en(s',t) \wedge \forall\, v \in V: s'(v) = s(v).$$

- T is a *write up* set of t with respect to V, iff for every $t' \in \mathbf{T}$ and $s' \in \mathbf{S}$
  $$\neg\, en(s,t,V) \wedge s -t'\rightarrow s' \wedge en(s',t,V) \;\Rightarrow\; t' \in T.$$

- t *accords with* t', denoted by $t \leftrightarrow t'$, iff for every $s \in \mathbf{S}$
  $$en(s,t) \wedge en(s,t') \;\Rightarrow\; \exists\, s', s_1, s'_1 \in \mathbf{S}: s -t\rightarrow s_1 -t'\rightarrow s'_1 \wedge s -t'\rightarrow s' -t\rightarrow s'_1.\ \Box$$

The intuition behind the definition of $en(s,t,V)$ is perhaps best understood by noticing that if t is *not* enabled w.r.t. V, then it is necessary to modify the value of at least one variable in V to enable t. The definition has the following rather obvious consequence: $en(s,t) \Leftrightarrow en(s,t,\mathbf{V}) \Leftrightarrow \forall\, V \subseteq \mathbf{V}: en(s,t,V)$.

A write up set of the transition t with respect to the set V of variables is any set of transitions containing at least the transitions which have the potential of modifying the status of t from disabled w.r.t. V to enabled w.r.t. V. We do not require the write up set to be the smallest such set, because minimality is not needed in the theory of stubborn sets and the smallest set may be difficult to find in practice. For instance, if the specification of transition t' contains a writing reference to a variable in V but has a complicated enabling condition evaluating to `false`, t' does not belong to the smallest write up set of t w.r.t. V because it is never enabled. However, it may be difficult to see that t' can be ruled out. Our definition allows the use of a write up set which can be easily computed and is an upwards approximation of the minimal set. Every transition t and subset of variables V has at least one write up set, namely the set of all transitions $\mathbf{T}$. From now on we assume that a unique write up set is defined for every (t,V)-pair. We denote it by $wrup(t,V)$.

The definition of according with can be illustrated graphically:

$$\begin{array}{ccc} s & -t'\rightarrow & s' \\ | & & \\ t & & \\ \downarrow & & \\ s_1 & & \end{array} \quad \Rightarrow \quad \begin{array}{ccc} s & -t'\rightarrow & s' \\ | & & | \\ t & & t \\ \downarrow & & \downarrow \\ s_1 & -t'\rightarrow & s'_1 \end{array}$$

According with is a static, symmetric commutativity property of transition pairs. Concurrent systems typically contain several pairs of transitions according with each other. Let us call the set of variables tested, read or written by a transition its *reference set*. Two transitions accord with each other if their reference sets are disjoint. The same is true even if there are common variables in the reference sets, as long as the transitions never write to them. Transitions writing to a fifo queue accord with transitions reading from it, unless there are other variables in common. Transitions corresponding to different locations in the code of a sequential process accord with each other independent of to what variables they refer, because they are never simultaneously enabled.

2.2 Stubborn Sets

We are ready to define *semistubborn* sets of transitions:

Definition 2.2 Let $T \subseteq \mathbf{T}$ and $s \in \mathbf{S}$. T is *semistubborn* at s, iff $\forall\, t \in T$:

(1) $\neg\, en(s,t) \Rightarrow \exists\, V \subseteq \mathbf{V}: \neg\, en(s,t,V) \wedge wrup(t,V) \subseteq T$

(2) $en(s,t) \Rightarrow \forall\, t' \notin T: t \leftrightarrow t'$ □

Part (1) of the definition guarantees that a disabled transition belonging to a semistubborn set can be enabled only by the transitions in the set. Part (2) says that transitions in a semistubborn set accord with outside transitions. It is not difficult to prove that at least the empty set and the set of all transitions $\mathbf{T}$ are semistubborn at every state, thus semistubborn sets always exist. Furthermore, if a transition outside a semistubborn set occurs, the set remains semistubborn. This is a justification for the peculiar term semi*stubborn*. The significance of semistubborn sets is in the following theorem:

Theorem 2.3 Let $T \subseteq \mathbf{T}$ and $s_0 \in \mathbf{S}$ such that T is semistubborn at s_0. Let $n \geq 1$ and

$$\sigma = s_0 -t_1\rightarrow s_1 -t_2\rightarrow \ldots -t_{n-2}\rightarrow s_{n-2} -t_{n-1}\rightarrow s_{n-1} -t_n\rightarrow s_n$$

be a finite execution sequence such that $t_1, \ldots, t_{n-1} \notin T$ and $t_n \in T$. There is the finite execution sequence

$$\sigma' = s_0 -t_n\rightarrow s'_0 -t_1\rightarrow s'_1 -t_2\rightarrow \ldots -t_{n-2}\rightarrow s'_{n-2} -t_{n-1}\rightarrow s'_{n-1},$$

where $s'_{n-1} = s_n$. □

The theorem can be illustrated graphically. σ corresponds to the top and right edges of the figure, while σ' corresponds to the left and bottom sides.

$$\begin{array}{ccccccccc}
s_0 & -t_1\rightarrow & s_1 & -t_2\rightarrow & \ldots & -t_{n-2}\rightarrow & s_{n-2} & -t_{n-1}\rightarrow & s_{n-1} \\
| & & & & & & & & | \\
t_n & & & & & & & & t_n \\
\downarrow & & & & & & & & \downarrow \\
s'_0 & -t_1\rightarrow & s'_1 & -t_2\rightarrow & \ldots & -t_{n-2}\rightarrow & s'_{n-2} & -t_{n-1}\rightarrow & s'_{n-1} \\
 & & & & & & & & = s_n
\end{array}$$

The proof is a straightforward application of the definitions, and can be found in [Valmari 88b, 89a, 89b].

Theorem 2.3 allows us to move the occurrence of a transition belonging to a semistubborn set from future to the current state. However, this is of course of no use if no transition in the semistubborn set is going to occur in the future. This is certainly the case if the set is empty. Therefore we augment the definition by the requirement that there must be an enabled transition in the set:

Definition 2.4 Let $T \subseteq \mathbf{T}$ and $s \in \mathbf{S}$. T is *stubborn* at s, iff T is semistubborn at s, and $\exists\, t \in T$: $en(s,t)$. □

A stubborn set exists exactly when there is an enabled transition, because then the set of all transitions **T** is stubborn. However, as will soon become obvious, it is advantageous (but not mandatory) to use stubborn sets containing as few enabled transitions as possible.

2.3 Reduced State Spaces

A variable/transition system defines a labelled directed graph called its *state space* in a natural way:

Definition 2.5 The *state space* of the *v/t-system* $(\mathbf{V},\mathbf{T},type,next,\mathbf{s}_0)$ is the triple $(\mathbf{W},\mathbf{E},\mathbf{T})$, where

- $\mathbf{W} = \{s \in \mathbf{S} \mid \mathbf{s}_0 \rightarrow^* s\}$
- $\mathbf{E} = \{(s,t,s') \in \mathbf{W} \times \mathbf{T} \times \mathbf{W} \mid s -t\rightarrow s'\}$ □

Let TS be a function from **S** to the set of the subsets of **T** such that $TS(s)$ is stubborn if $\exists\, t \in \mathbf{T}$: $en(s,t)$, and $TS(s) = \varnothing$ otherwise. We call the function TS a *stubborn set generator*. The stubborn set method uses a stubborn set generator to generate a *reduced state space* as follows. The generation starts at $\mathbf{s}_0$. Assume s has been generated. In ordinary state space generation, every transition enabled at s is used to generate the immediate successor states of s. In the stubborn set method, only the enabled transitions in $TS(s)$ are used. If $TS(s)$ contains less enabled transitions than **T**, the number of immediate successors is reduced. This often leads to a reduction in the total number of states. To distinguish between reduced and ordinary state space concepts we use underlining notation as follows:

- $\underline{en}(s,t) \Leftrightarrow en(s,t) \wedge t \in TS(s)$
- $s \;\underline{-t\rightarrow}\; s' \Leftrightarrow s -t\rightarrow s' \wedge t \in TS(s)$
- $s \;\underline{\rightarrow}\; s' \Leftrightarrow \exists\, t \in TS(s)\colon s -t\rightarrow s'$

- "$\underline{\rightarrow}^*$" is the reflexive and transitive closure of "$\underline{\rightarrow}$".

Definition 2.6 Assume a stubborn set generator TS is given. The *reduced state space* of the v/t-system $(\mathbf{V},\mathbf{T},type,next,\mathbf{s}_0)$ is the triple $(\underline{\mathbf{W}},\underline{\mathbf{E}},\mathbf{T})$, where

- $\underline{\mathbf{W}} = \{s \in \mathbf{S} \mid \mathbf{s}_0 \underline{\rightarrow}^* s\}$
- $\underline{\mathbf{E}} = \{(s,t,s') \in \underline{\mathbf{W}} \times \mathbf{T} \times \underline{\mathbf{W}} \mid s \underline{-t\rightarrow} s'\}$ $\square$

The definition implies that $\underline{\mathbf{W}} \subseteq \mathbf{W}$ and $\underline{\mathbf{E}} \subseteq \mathbf{E}$. The ordinary state space is a special case of a reduced state space, because we may choose $TS(s) = \mathbf{T}$ for states with enabled transitions. However, our goal is to keep the reduced state space small, to save effort in its generation and the model checking afterwards. Perhaps surprisingly, always choosing the stubborn set with the smallest number of enabled transitions does not necessarily lead to the smallest reduced state space [Valmari 88c]. However, it is obvious that if T_1 and T_2 are stubborn and the set of enabled transitions in T_1 is a proper subset of the corresponding set of T_2, then T_1 is preferable. We say that a stubborn set is *optimal* if it is the best possible in this respect. [Valmari 88a, 88b] give a linear and an (under certain reasonable assumptions) quadratic algorithm for finding almost optimal and optimal stubborn sets, respectively. The linear algorithm is particularly attractive because its best case complexity is better than linear; if it finds a stubborn set close to its starting point, it optimises it as much as it can and stops without investigating the rest of the v/t-system. The linear algorithm is briefly described in Chapter 3.

As a side issue we mention that the use of the stubborn set method in the analysis of deadlocks and terminal states arises from the following theorems, which say that the terminal states of the ordinary state space are terminal states of the reduced state space and vice versa.

(1) Let $s \in \mathbf{W}$. If $\forall\, t \in \mathbf{T}$: $\neg\, en(s,t)$, then $s \in \underline{\mathbf{W}}$ and $\forall\, t \in \mathbf{T}$: $\neg\, \underline{en}(s,t)$.

(2) Let $\underline{s} \in \underline{\mathbf{W}}$. If $\forall\, t \in \mathbf{T}$: $\neg\, \underline{en}(\underline{s},t)$, then $\underline{s} \in \mathbf{W}$ and $\forall\, t \in \mathbf{T}$: $\neg\, en(\underline{s},t)$.

The analysis of nontermination is based on a theorem saying that there is an infinite execution sequence in the reduced state space if and only if there is an infinite execution sequence in the ordinary state space.

2.4 Execution Sequences in Reduced State Spaces

This section is devoted to a construction which, given a finite execution sequence of the system under analysis, finds an execution sequence which is present in the reduced state space and is, roughly speaking, a permutation of an extension of the former. The construction is in the heart of most proofs in the stubborn set theory. It proceeds in steps. Each step appends a transition occurrence to the end of the constructed sequence. The transition occurrence is either picked and removed from the original sequence, in which case we say that an original transition occurrence is *consumed*, or a fresh new transition occurrence is found for the purpose. The construction may be continued until the original sequence is exhausted. It may happen that only fresh transitions are used from some step onwards, in which case the construction may be continued endlessly.

The construction is presented formally below. σ is the original finite execution sequence. The constructed sequence after step i is denoted by $\underline{\sigma}_i$ and its last state by $\underline{s}_i$. The execution sequences ρ_i and σ_i correspond to the unconsumed part of the original sequence and the original sequence appended by the occurrences of the fresh transitions used by the construction. $k(i)$ is the length of ρ_i, that is, the number of still unconsumed transition occurrences. It may be helpful to notice that $\underline{\sigma}_i \circ \rho_i$ exists and its first and last state are the same as the first and last state of σ_i. Furthermore, the transition occurrences of $\underline{\sigma}_i \circ \rho_i$ are the same as the transition occurrences of σ_i, but not necessarily in the same order.

Construction 2.7 Let TS be a stubborn set generator and $\sigma = s_0 -t_1\rightarrow \ldots -t_n\rightarrow s_n$ be a finite execution sequence. The states $\underline{s}_i$ and execution sequences $\underline{\sigma}_i$, σ_i and $\rho_i = r_{0,i} -d_{1,i}\rightarrow \ldots -d_{k(i),i}\rightarrow r_{k(i),i}$ are defined recursively as follows. $k(i)$ is defined as the length of ρ_i.

- $\sigma_0 = \rho_0 = \sigma$ and $\underline{\sigma}_0 = s_0 = \underline{s}_0$.

- If $k(i) = 0$, that is, ρ_i consists of one state and no transition occurrences, the construction cannot be continued.

- If $k(i) > 0$, $d_{1,i}$ is enabled at $r_{0,i}$, thus $TS(r_{0,i})$ is stubborn. There are two cases.

  (1) ρ_i contains at least one occurrence of a transition belonging to $TS(r_{0,i})$. We define $k'(i) = k(i)$, $\sigma'_i = \sigma_i$ and $\rho'_i = \rho_i$.

  (2) ρ_i contains no occurrences of transitions belonging to $TS(r_{0,i})$. By Definition 2.4 $TS(r_{0,i})$ contains an enabled transition t'_i. By (2) of Definition 2.2 t'_i is enabled at $r_{1,i}, \ldots, r_{k(i),i}$. We define $k'(i) = k(i)+1$, $d_{k(i)+1,i} = t'_i$, $\sigma'_i = \sigma_i -t'_i\rightarrow r_{k'(i),i}$ and $\rho'_i = \rho_i -t'_i\rightarrow r_{k'(i),i}$.

  By construction, ρ'_i contains at least one occurrence of a transition belonging to $TS(r_{0,i})$. Let $1 \leq j(i) \leq k'(i)$ be chosen such that $d_{1,i} \ldots d_{j(i)-1,i} \notin TS(r_{0,i})$ and $d_{j(i),i} \in TS(r_{0,i})$. By Theorem 2.3 there is the sequence $\rho''_i = r_{0,i} -d_{j(i),i}\rightarrow r'_{0,i} -d_{1,i}\rightarrow \ldots -d_{j(i)-1,i}\rightarrow r_{j(i),i} -d_{j(i)+1,i}\rightarrow \ldots -d_{k'(i),i}\rightarrow r_{k'(i),i}$. By the $(i-1)$th construction step $r_{0,i} = \underline{s}_i$. Hence $d_{j(i),i} \in TS(\underline{s}_i)$ and we may define $\underline{s}_{i+1} = r'_{0,i}$, $\underline{\sigma}_{i+1} = \underline{\sigma}_i \underline{-d_{j(i),i}\rightarrow} \underline{s}_{i+1}$ and $\sigma_{i+1} = \sigma'_i$. ρ_{i+1} is defined by $\rho''_i = r_{0,i} -d_{j(i),i}\rightarrow \rho_{i+1}$. After this i:th construction step $r_{0,i+1} = r'_{0,i} = \underline{s}_{i+1}$. □

In the future we will need the fact that in the above construction $d_{x,i} = d_{x,i+1}$ for $x = 1, \ldots, j(i)-1$. Also the order of transition occurrences in $\underline{\sigma}_{i+1} \circ \rho_{i+1}$ is otherwise the same as the order of transition occurrences in $\underline{\sigma}_i \circ \rho_i$, but either the occurrence of $d_{j(i),i}$ has moved to a different location, or a new transition occurrence has been introduced.

2.5 Linear Temporal Logic Preservation Theorem

In this section we state and prove the theorem underlying the linear temporal logic (LTL) preserving stubborn set method. LTL formulas state properties of infinite sequences of states. Infinite execution sequences of a v/t-system give naturally rise to infinite sequences of states. We also consider *stopping* execution sequences which are finite and end at a state without enabled transitions. As usual, we extract infinite sequences of states from stopping execution sequences by letting the last state repeat forever. We say that the infinite or stopping execution sequence σ *satisfies* the LTL formula φ iff φ is a true statement about the infinite sequence of states extracted from σ. We say that φ is *valid* in state s in a given state space, iff there is no infinite or stopping execution sequence σ in the state space starting at s such that σ satisfies $\neg\varphi$.

Let $\Phi = \{\varphi_1, \ldots, \varphi_f\}$ be a collection of LTL formulas. We associate with Φ the set $vis(\Phi)$ of *visible transitions*. Transition t is *properly visible*, iff there are syntactic states s and s' such that $s -t\rightarrow s'$ and the truth value of at least one state predicate appearing in at least one formula in Φ is different at s and s'. $vis(\Phi)$ is a set containing at least the properly visible transitions. The reason why we allow in $vis(\Phi)$ the presence of transitions which are not properly visible is the same as the reason of allowing the write up set to be an upwards approximation of the smallest set with the required property. Transition t is *visible* if $t \in vis(\Phi)$, otherwise t is *invisible*. The property we will use in the future is that when an invisible transition occurs the truth values of the state predicates of the formulas in Φ do not change.

The LTL preserving stubborn set method works only with *stuttering-invariant* formulas. Informally, stuttering-invariance means that the truth value of a formula on an infinite sequence of states does not change if one or more or even all of the states of the sequence are multiplicated, where multiplication means the replacement of the state by a positive finite number of copies of the state. Among the common LTL operators, "next state" (O) and "previous state" may introduce formulas which are not stuttering-invariant. Formulas containing no other temporal operators than "henceforth" ($\Box$), "eventually" ($\Diamond$), "until" ($\mathcal{U}$), and their corresponding past operators and the operators derived from them are stuttering-invariant (see [Lamport 83]).

We call the sequence of states $s_0s_1\ldots s_n$ an *elementary cycle*, if $n > 0$, $s_0 = s_n$, $s_i \neq s_j$ where $0 \leq i < j < n$ and $s_i \rightarrow s_{i+1}$ where $0 \leq i < n$. We can now formulate the theorem underlying the LTL preserving stubborn set method:

Theorem 2.8 Let $(\mathbf{V},\mathbf{T},type,next,\mathbf{s}_0)$ be a v/t-system and $\mathbf{S}$ and $(\mathbf{W},\mathbf{E},\mathbf{T})$ its set of syntactic states and its state space, respectively. Let Φ be a collection of stuttering-invariant LTL formulas such that the domain of the state predicates in the formulas is $\mathbf{S}$. Let TS be a stubborn set generator and $(\underline{\mathbf{W}},\underline{\mathbf{E}},\mathbf{T})$ the corresponding reduced state space such that the following hold:

(1) $\underline{\mathbf{W}}$ is finite

(2) For every $\underline{s} \in \underline{\mathbf{W}}$, either
 (a) $TS(\underline{s})$ contains no enabled visible transitions, or
 (b) $vis(\Phi) \subseteq TS(\underline{s})$

(3) For every $\underline{s} \in \underline{\mathbf{W}}$, if there is an enabled invisible transition, then $TS(\underline{s})$ contains an enabled invisible transition

(4) Every elementary cycle of $(\underline{\mathbf{W}},\underline{\mathbf{E}},\mathbf{T})$ contains at least one state $\underline{s}$ such that $vis(\Phi) \subseteq TS(\underline{s})$.

Claim: $\varphi \in \Phi$ is valid at $\mathbf{s}_0$ in $(\mathbf{W},\mathbf{E},\mathbf{T})$ if and only if φ is valid at $\mathbf{s}_0$ in $(\underline{\mathbf{W}},\underline{\mathbf{E}},\mathbf{T})$. □

Proof It is sufficient to show that $(\mathbf{W},\mathbf{E},\mathbf{T})$ contains an execution sequence σ which starts at $\mathbf{s}_0$ and satisfies $\neg\varphi$ if and only if $(\underline{\mathbf{W}},\underline{\mathbf{E}},\mathbf{T})$ contains an execution sequence $\underline{\sigma}$ with the same properties. The "if"-part is obvious, as $\underline{\mathbf{W}} \subseteq \mathbf{W}$ and $\underline{\mathbf{E}} \subseteq \mathbf{E}$. To prove the "only if"-part we construct from σ an execution sequence $\underline{\sigma}$ such that $\underline{\sigma}$ starts at $\mathbf{s}_0$ and the sequences of the occurrences of visible transitions in σ and $\underline{\sigma}$ are the same. Then the claim follows from the facts that σ satisfies $\neg\varphi$ and φ is stuttering-invariant. We consider three cases. Let $\underline{s}_i$, σ_i, ρ_i, $\underline{\sigma}_i$, $j(i)$, $d_{1,i} \ldots d_{j(i),i}$ and t'_i be defined as in Construction 2.7.

(1) σ is stopping. Let n be its length. We apply Construction 2.7 n steps on σ to get $\underline{\sigma} = \underline{\sigma}_n$. Assume inductively that $\sigma_i = \sigma$ and the sequence of the occurrences of visible transitions in $\underline{\sigma}_i{}^\bullet\rho_i$ is the same as in σ. Because no transition is enabled at s_n (the last state of σ and σ_i), branch (2) of Construction 2.7 cannot be taken. Thus $\sigma_{i+1} = \sigma$ and we conclude that $\sigma_n = \sigma$ and $\rho_n = s_n$. Because of Assumption (2) of the theorem, if $d_{j(i),i}$ is visible, then $vis(\Phi) \subseteq TS(\underline{s}_i)$, thus $d_{1,i} \ldots d_{j(i)-1,i}$ are invisible. As a result, the sequence of the occurrences of visible transitions in $\underline{\sigma}_{i+1}{}^\bullet\rho_{i+1}$ is the same as in $\underline{\sigma}_i{}^\bullet\rho_i$. The same is obviously true if $d_{j(i),i}$ is invisible. We conclude that the sequence of the occurrences of visible transitions in $\underline{\sigma}_n = \underline{\sigma}_n{}^\bullet\rho_n$ is the same as in σ.

(2) σ is infinite and contains a finite number of visible transition occurrences. Let $|\underline{W}|$ be the number of states in the reduced state space (finite by Assumption (1)), and let m be the length of the prefix of σ up to and including the last occurrence of a visible transition. Let $\sigma' = s_0 -t_1\rightarrow \ldots -t_n\rightarrow s_n$ be a prefix of σ of length $n = (m+1)|\underline{W}|$. We apply Construction 2.7 on σ' n times to generate $\underline{\sigma}_1, \ldots, \underline{\sigma}_n$.

Consider first $0 \leq i < n$ such that ρ_i contains the occurrence of a visible transition. As above, Assumption (2) of the theorem guarantees that if $d_{j(i),i}$ is visible, then $d_{1,i} \ldots d_{j(i)-1,i}$ are invisible. This implies that branch (1) of the construction must be taken. As a result, the sequence of the occurrences of visible transitions in $\underline{\sigma}_{i+1}{}^\bullet\rho_{i+1}$ is the same as in $\underline{\sigma}_i{}^\bullet\rho_i$ and thus the same as in σ. The same is obviously true if $d_{j(i),i}$ is invisible.

Suppose still that ρ_i contains the occurrence of a visible transition $d_{v(i),i}$. Remember that by Construction 2.7 $d_{x,i} = d_{x,i+1}$ for $x = 1, \ldots, j(i)-1$. Furthermore, the construction can be continued until all transition occurrences in ρ_i are consumed. Assume for a while that none of $d_{1,i}, \ldots, d_{v(i),i}$ is consumed during the next $|\underline{W}|$ construction steps, that is, $j(i), j(i+1), \ldots, j(i+|\underline{W}|-1) > v(i)$ and $d_{1,i} = \ldots = d_{1,w}$ and $\ldots$ and $d_{v(i),i} = \ldots = d_{v(i),w}$, where $w = i+|\underline{W}|$. Consider the $|\underline{W}|+1$ states $\underline{s}_i, \ldots, \underline{s}_w$. They must contain an elementary circuit of $(\underline{W},\underline{E},T)$. By Assumption (4) of the theorem there is $h \in \{i, \ldots, i+|\underline{W}|-1\}$ such that $vis(\Phi) \subseteq TS(\underline{s}_h)$. Because $d_{v(i),h} = d_{v(i),i} \in vis(\Phi)$, by Construction 2.7 $j(h) \leq v(i)$, a contradiction. We conclude that there is $0 \leq M \leq m|\underline{W}|$ such that the occurrences of visible transitions in σ' are consumed by exactly M construction steps. That is, ρ_0, $\ldots, \rho_{M-1}$ contain an occurrence of a visible transition, but ρ_M does not. Thus the sequence of the occurrences of visible transitions in $\underline{\sigma}_M$ is the same as in σ. Anticipating Case (3) below, note that so far we have not used the assumption about the length of σ'.

Consider (at last!) $M \leq i < n$ such that ρ_i does not contain an occurrence of a visible transition. Because $i < n$, ρ_i contains a transition occurrence, thus there is an enabled invisible transition at $\underline{s}_i$. By Assumption (3), Construction 2.7 can be continued so that $d_{j(i),i}$ is invisible. If we do so, the sequence of the occurrences of visible transitions in $\underline{\sigma}_i$ is the same as in σ when $M \leq i \leq n$. Because $M \leq n-|\underline{W}|$, at least the last $|\underline{W}|$ transition occurrences in $\underline{\sigma}_n$ correspond to invisible transitions. $\underline{\sigma}_n$ has thus a prefix $\underline{\sigma}_e = \underline{\sigma}_v{}^\bullet\underline{\mu}$ such that $M \leq v < e \leq n$, the sequence of the occurrences of visible transitions in $\underline{\sigma}_v$ is the same as in σ, $\underline{\mu}$ does not contain occurrences of visible transitions, and $\underline{s}_e = \underline{s}_v$. We define $\underline{\sigma} = \underline{\sigma}_v{}^\bullet\underline{\mu}{}^\bullet\underline{\mu}{}^\bullet\underline{\mu}{}^\bullet \ldots$.

(3) σ contains an infinite number of visible transition occurrences. Let σ_i be a prefix of length i of σ. We saw in Case (2) that there is $0 \leq M(i) \leq i|\underline{\mathbf{W}}|$ such that all occurrences of visible transitions in σ_i are consumed by $M(i)$ steps of Construction 2.7. Let us denote the resulting sequence by $\underline{\sigma}_i$. By Case (2) the sequence of the occurrences of visible transitions in $\underline{\sigma}_i$ is the same as in σ_i. We apply the reasoning of the proof of König's Lemma to show that there is an infinite execution sequence $\underline{\sigma}$ such that the sequence of the occurrences of visible transitions in it is the same as in σ. Let $\Omega_0 = \{0, 1, \ldots\}$ and $\underline{s}_0 = \mathbf{s}_0$. Because $\mathbf{T}$ is finite there is a transition t_1 and a state $\underline{s}_1$ such that $\underline{\sigma}_i$ is of the form $\underline{s}_0 -t_1\rightarrow \underline{s}_1 \ldots$ for an infinite set $\Omega_1 \subseteq \Omega_0$ of different values of i. Similarly, there are t_2 and $\underline{s}_2$ such that $\underline{\sigma}_i$ is of the form $\underline{s}_0 -t_1\rightarrow \underline{s}_1 -t_2\rightarrow \underline{s}_2 \ldots$ for an infinite set $\Omega_2 \subseteq \Omega_1$ of different values of i. The same reasoning may be continued to produce t_3, $\underline{s}_3$ and Ω_3 and so on. $\underline{\sigma}$ is the resulting infinite execution sequence $\underline{s}_0 -t_1\rightarrow \underline{s}_1 -t_2\rightarrow \ldots$. Let t_v be a visible transition occurring at position v in σ. By Case (2) t_v is consumed by at most $v|\underline{\mathbf{W}}|$ construction steps. Thus the position of t_v in $\underline{\sigma}_v, \underline{\sigma}_{v+1}, \ldots$ and, consequently, in $\underline{\sigma}$ is at most $v|\underline{\mathbf{W}}|$. Thus $\underline{\sigma}$ contains the same visible transition occurrences and in the same order as σ. □

3. IMPLEMENTATION OF THE METHOD

To implement the LTL preserving stubborn set method it is necessary to find a practical way of generating reduced state spaces such that Assumptions (1), ..., (4) of Theorem 2.8 hold. Then the validity of the formulas in Φ can be checked using the reduced state space instead of the ordinary one. The use of the stubborn set method is thus transparent to the model checking phase.

Assumption (1) is certainly satisfied if the system under analysis has a finite number of states because $\underline{\mathbf{W}} \subseteq \mathbf{W}$. If $\mathbf{W}$ may be infinite then the LTL formula validity decision problem is generally undecidable and it is not surprising if an analysis method does not work. In some cases, though, $\underline{\mathbf{W}}$ may be finite even if $\mathbf{W}$ is not. An example of this is the simple system consisting of a producer and a consumer connected by an unbounded fifo buffer. The stubborn set method may alternate the buffer read and write operations (they accord with each other) in such a way that items do not accumulate in the buffer, if the state predicates of Φ do not test the content of the buffer.

In the following sections we show how Assumptions (2), (3) and (4) can be ensured.

3.1 Constructing Stubborn Sets

To show how Assumptions (2) and (3) may be ensured it is necessary to discuss the problem of finding "good" stubborn sets. From a v/t-system and its state s we derive a finite directed graph Γ_s using the definition of semistubborn sets (Definition 2.2) as follows. The vertices of Γ_s correspond one to one to the transitions of the system. The vertex corresponding to $t \in \mathbf{T}$ is denoted by $vx(t)$. We say that a vertex is enabled if the corresponding transition is. Consider an enabled transition t. If t belongs to a semistubborn set T, then (2) of Definition 2.2 implies that every transition t' such that $\neg\, t \leftrightarrow t'$ belongs to T. This is represented in Γ_s by defining that $(vx(t), vx(t'))$ is an edge for every t' with the above property. Now, let t be disabled. Assume $V \subseteq \mathbf{V}$ is chosen such that $\neg\, en(s,t,V)$. We require that $\forall\, t' \in wrup(t,V)$: $t' \in T$ and represent it again by defining that $(vx(t), vx(t'))$ is an edge. Γ_s is now ready. Let *reachable*(t) be the set of transitions such that the corresponding vertices are reachable from $vx(t)$ in Γ_s. A set of transitions T is semistubborn if (but not necessarily only if) $\forall\, t \in T$: $reachable(t) \subseteq T$. If t_e is enabled, $reachable(t_e)$ is stubborn.

On the basis of the above a stubborn set can be constructed by finding an enabled transition t_e and computing $reachable(t_e)$. However, such stubborn sets may be much larger than necessary. If t_1 and t_2 are enabled, $t_2 \in reachable(t_1)$ but $t_1 \notin reachable(t_2)$, then $reachable(t_2)$ is stubborn and a proper subset of $reachable(t_1)$. Obviously $reachable(t_2)$ is better than $reachable(t_1)$. What we want to find is a stubborn set such that all the corresponding enabled vertices are reachable from each other. Such a set of vertices has an associated maximal strongly connected component (strong component, for short) C of Γ_s such that it contains all the enabled vertices of the set and no other strong component containing enabled vertices is reachable from C. $reachable(c)$ for any $c \in C$ corresponds to the stubborn set we are after. However, in practice it is sufficient to find C as we do not need the disabled transitions of a stubborn set. Strong components can be found in time linear in the size of Γ_s using Tarjan's algorithm [Aho & 74]. Tarjan's algorithm can be adapted to distinguish a strong component with the above required property and then terminate without worsening its time complexity. See [Valmari 88a or 88b] for details.

In the derivation of the above algorithm we did not specify how the $V \subseteq \mathbf{V}$ such that $\neg\, en(s,t,V)$ should be chosen when t is disabled. The above algorithm produces stubborn sets independent of how the choice is made, but the choice may affect the sizes of the stubborn sets. In this respect the stubborn sets produced by the algorithm are not necessarily optimal. ([Valmari 88b] discusses a different, typically quadratic stubborn set search algorithm which is capable of optimising the choice.) After fixing the choices of V, the above algorithm produces stubborn sets which are minimal with respect to set inclusion for the given choices.

The above algorithm (call it $\mathcal{A}$) can be used as a building block of an algorithm finding stubborn sets satisfying Assumptions (2) and (3) of Theorem 2.8. Consider a state with an enabled transition. In the first stage the algorithm tries to find a stubborn set not containing enabled visible transitions. This is done by running $\mathcal{A}$ starting at every enabled invisible vertex not visited so far until the required set is found or the enabled invisible vertices are exhausted. Note that during this stage a strong component search need not enter the vertices which have already been investigated during an earlier search, if the information that a visible enabled vertex is reachable from the current vertex of the search is carefully backward propagated and stored with the vertices. If successful, the first stage produces a stubborn set satisfying Assumptions (2a) and (3).

If the first stage fails then T_u, the union of $reachable(t)$ of every visible t, is computed by performing an ordinary graph traversal. Note that the union of stubborn sets is stubborn and T_u satisfies Assumption (2b). If there are no enabled invisible transitions or if T_u contains one also Assumption (3) is satisfied. Otherwise a third stage is needed. It consists of running $\mathcal{A}$ starting at an enabled invisible transition such that the search does not enter any of the vertices corresponding to T_u. The required stubborn set is the union of T_u and the result of the third stage.

3.2 Ensuring Assumption (4)

A depth-first search can be used to show the presence or absence of elementary cycles in a directed graph as follows. A vertex is called *new* if it has not been encountered by the search as yet, *unready* if it has been encountered but not backtracked from, and *ready* if it has been backtracked from. A basic property of the depth-first search is that the unready vertices form a path starting at the root of the search and ending at the current vertex. The current vertex is always unready. Furthermore, as long as no cycles have been found, only ready vertices can be reached from a ready vertex. A cycle is found when the search finds an unready vertex.

Now assume that the vertices of the graph are either *red* or *green* and we want to eliminate *red cycles*, that is, cycles consisting of only red vertices. Furthermore, we can paint a red vertex green but we do not want to do so unnecessarily often. We apply the depth-first search almost as above, with the exception that when a green vertex is entered or the current vertex is painted green the search does not proceed but backtracks to the previous (usually red) vertex. When there are no unready red vertices the search jumps to some green vertex. It is not important which unready green vertex is selected. The following properties are invariant during the search:

- Only ready red vertices and unready or ready green vertices can be reached from a ready red vertex via red vertices.
- If the current vertex is red, the unready red vertices form a path ending at the current vertex. If the current vertex is green, there are no unready red vertices.
- The current vertex does not have known unready red immediate successors. The other unready red vertices have exactly one known unready red immediate successor each. ("Known" means that the edge to the immediate successor has been investigated.)
- There are no cycles consisting of ready red vertices.

Initially all the vertices but the current vertex are new and the current vertex is unready, so the invariants hold. Assume that the search is at an unready red vertex. If the search finds a green vertex of any status it does not proceed to it, thus the invariants remain valid. If it finds a ready red vertex it does not proceed to it and the invariants remain valid. No cycle has been found because the current vertex is unready and thus, by the first invariant, it is not reachable from the found vertex. If the search proceeds to a new red vertex it changes it to an unready red vertex and the invariants remain valid. If the search finds an unready red vertex it does not proceed to it. A red cycle has been found and the third invariant is violated. The invariant is re-established by painting the current vertex green and backtracking from it. When the search backtracks from an unready red vertex the status of the vertex changes to ready. The invariants remain valid; the first by the third and the fourth by the first. The third invariant prevents the backtracked vertex from being its own immediate successor.

The search is continued at a green vertex only when there are no unready red vertices. Thus the invariants remain valid when the search moves to a green vertex independent of to which unready green vertex it moves. When the search enters an unready red vertex from a green vertex the invariants remain valid. When the search is finished the fourth invariant implies that there are no red cycles.

Assumption (4) of Theorem 2.8 is guaranteed by applying the above red vertex — green vertex algorithm. The reduced state space is generated in the order required by the algorithm. State $\underline{s}$ is green if $vis(\Phi) \subseteq TS(\underline{s})$ and red otherwise. $\underline{s}$ is painted green by computing a new stubborn set $TS'(\underline{s})$ containing $vis(\Phi)$ as shown in Section 3.1. $TS(\underline{s}) \cup TS'(\underline{s})$ is then used as the new stubborn set of $\underline{s}$. It satisfies Assumptions (2b) and (3). The new stubborn set may introduce new immediate successors to $\underline{s}$, but this causes no problem, because $\underline{s}$ can be considered unready and the order in which green vertices are handled is not important.

4. EXAMPLE

To demonstrate the power of the LTL preserving stubborn set method we consider a version of the resource allocator system specified in [Pnueli 86]. Our system consists of a resource allocator and $n \geq 2$ customers which communicate via $2n$ Boolean variables r_i and g_i, initially **F**. The behaviour of customer i is shown below.

1:	r_i := **T**	/* t_{i1} */
2:	**when** $g_i \Rightarrow$ **goto** 3	/* t_{i2} */
3:	r_i := **F**	/* t_{i3} */
4:	**when** $\neg g_i \Rightarrow$ **goto** 1	/* t_{i4} */

The customer may use the resource when it is in state 3. The resource allocator behaves as follows.

1:	**when** $r_i \Rightarrow g_i$:= **T**; **goto** $2i$	/* d_{i1} */
$2i$:	**when** $\neg r_i \Rightarrow g_i$:= **F**; **goto** 1	/* d_{i2} */

The system works basically so that the i:th customer may request the resource by moving from 1 to 2, the resource is granted to it when the allocator moves from 1 to $2i$, the customer takes the resource by moving to 3 and releases it by moving to 4, and finally both the customer and the allocator move back to their initial states. The customers may proceed concurrently but two customers cannot be simultaneously in state 3.

Let us compute the size of the ordinary state space of the system. The states of the variables r_i and g_i can be determined from the states of the customers and the allocator, so they need not be considered. When the resource allocator is in state 1 the customers may be in state 1, 2 or 4 independent of each other and the allocator. This amounts to 3^n states. When the allocator is in state $2i$ the i:th customer is in state 2, 3 or 4 and the others are in state 1, 2 or 4, corresponding to $n \cdot 3^n$ states. This makes a total of $(n+1) \cdot 3^n$ states.

4.1 Example System as a V/T-System

The resource allocator system can be seen as a variable/transition system:

$\mathbf{V} = \{A, C_1, \ldots, C_n, r_1, \ldots, r_n, g_1, \ldots, g_n\}$

A is the state of the allocator

C_i is the state of the i:th customer

$\mathbf{T} = \{t_{11}, t_{12}, t_{13}, t_{14}, t_{21}, \ldots, t_{24}, \ldots, t_{n1}, \ldots, t_{n4}, d_{11}, d_{12}, d_{21}, d_{22}, \ldots, d_{n1}, d_{n2}\}$

$type(A) = \{1, 21, 22, \ldots, 2n\}$
$type(C_i) = \{1,2,3,4\}$
$type(r_i) = type(g_i) = \{\mathbf{F}, \mathbf{T}\}$

$\mathbf{s}_0(A) = 1 \wedge \forall i: \mathbf{s}_0(C_i) = 1 \wedge \mathbf{s}_0(r_i) = \mathbf{s}_0(g_i) = \mathbf{F}$

next is too large to be listed here in full. It can be determined from the program code. For instance, $next(s,t_{12})$ is defined if $s(C_1) = 2 \wedge s(g_1) = \mathbf{T}$. If $next(s,t_{12}) = s'$ (i.e. $s - t_{12} \rightarrow s'$) then $s'(C_1) = 3 \wedge \forall v \neq C_1: s'(v) = s(v)$.

We need an upper approximation to the set $\{(t,t') \in \mathbf{T}\times\mathbf{T} \mid \neg\, t \leftrightarrow t'\}$. If the sets of variables referred to by t and t' are disjoint, then $t \leftrightarrow t'$, because then the occurrence of t does not directly affect the environment of t' and vice versa. Let ref(v) denote the set of transitions referring to a variable v. We conclude that the union of ref$(v)^2$ for every variable v of the system is an upper approximation to the set. We continue by investigating how the approximation can be improved.

We have ref(A) = $\{d_{11},d_{12},d_{21},d_{22},\ldots,d_{n1},d_{n2}\}$. Because of the control structure of the resource allocator, transitions of the form d_{i2} are never enabled simultaneously with any other transition in ref(A). Thus the left hand side of the implication in the definition of "$\leftrightarrow$" is never satisfied and the implication is always true, if $t = d_{i2}$ and $t' \in$ ref$(A)-\{d_{i2}\}$. Consequently, the corresponding transition pairs (t,t') and (t',t) can be eliminated. A transition seldom accords with itself, so we choose not to try to eliminate the pairs (d_{i2},d_{i2}). $\neg\, d_{i1} \leftrightarrow d_{j1}$ holds for $1 \le i,j \le n$, because d_{i1} can disable d_{j1}. As a result, the pairs (d_{i1},d_{j1}) remain. So we have eliminated all pairs except (d_{i2},d_{i2}) and (d_{i1},d_{j1}), where $1 \le i,j \le n$. By similar argument all pairs except $\{(t,t) \mid t \in \text{ref}(C_i)\}$ can be eliminated from the sets ref$(C_i)^2$, as ref$(C_i) = \{t_{i1},t_{i2},t_{i3},t_{i4}\}$.

As ref$(r_i) = \{t_{i1},t_{i3},d_{i1},d_{i2}\}$, we have investigated ref$(r_i)^2$ except the pairs (t_{ij},d_{ik}) and (d_{ik},t_{ij}), where $j \in \{1,3\}$ and $k \in \{1,2\}$. Consider the states s enabling both t_{i1} and d_{i1}. We have $s(r_i) = \mathbf{T}$ and $s(C_i) = s(A) = 1$. When t_{i1} or d_{i1} occurs the only variables whose values may be changed are r_i, g_i, C_i and A, so we investigate them only. If $s -t_{i1}\rightarrow s_1$ and $s -d_{i1}\rightarrow s'$ then $s_1(r_i) = s'(r_i) = \mathbf{T}$, $s_1(g_i) = s(g_i)$, $s'(g_i) = \mathbf{T}$, $s_1(C_i) = 2$, $s'(C_i) = 1 = s_1(A)$ and $s'(A) = 2i$. Therefore $en(s_1,d_{i1})$ and $en(s',t_{i1})$. Let $s_1 -d_{i1}\rightarrow s'_1$ and $s' -t_{i1}\rightarrow s''_1$. By computing the values of r_i, g_i, C_i and A in s'_1 and s''_1 we see that $s'_1 = s''_1$. Thus $t_{i1} \leftrightarrow d_{i1}$. By similar argument it can be shown that $t_{i3} \leftrightarrow d_{i2}$. Only the pairs (t_{i1},d_{i2}), (t_{i3},d_{i1}) and their inverses and the pairs of the form (t,t) were not eliminated from ref$(r_i)^2$. Investigating ref(g_i) in the similar way leaves as the total only the following pairs left:

$$(t,t),\ (d_{i1},d_{j1}),\ (t_{i1},d_{i2}),\ (t_{i2},d_{i2}),\ (t_{i3},d_{i1}),\ (t_{i4},d_{i1}),\ \text{where } 1 \le i,j \le n \text{ and } t \in \mathbf{T}.$$

Also the predicates $en(s,t,V)$ and the sets $wrup(t,V)$ for $t \in \mathbf{T}$ and for some $V \subseteq \mathbf{V}$ are needed by the stubborn set method. Consider t_{12}, for example. $en(s,t_{12})$ holds iff $s(C_1) = 2 \wedge s(g_1) = \mathbf{T}$. Therefore $en(s,t_{12},\{C_1\}) \Leftrightarrow s(C_1) = 2$ and $en(s,t_{12},\{g_1\}) \Leftrightarrow s(g_1) = \mathbf{T}$. We may choose $wrup(t_{12},\{C_1\}) = \{t_{11}\}$, because it is the only transition whose occurrence can assign 2 to C_1. Similarly $wrup(t_{12},\{g_1\}) = \{d_{11}\}$. Following the same principles we can evaluate $en(s,t,\{v\})$ and define $wrup(t,\{v\})$ for every $t \in \mathbf{T}$ and $v \in \mathbf{V}$. All the so defined $wrup$ sets contain exactly one transition, excluding the sets $wrup(d_{i1},\{A\})$ which are all equal to $\{d_{12},d_{22},\ldots,d_{n2}\}$.

4.2 Reduced State Space of the Example System

Now we construct a reduced state space of the system. We want to know whether the system guarantees that the resource cannot be used simultaneously by two customers (it does), and whether a customer which has requested a resource eventually uses the resource (not true). Because of the symmetry of the system the first requirement can be specified by the LTL formula $\Box(\,(C_1 \ne 3) \vee (C_2 \ne 3)\,)$, where C_1 and C_2 are the states of customers 1 and 2, respectively. The second requirement can be encoded as $\Box(\,(C_1 = 2) \Rightarrow \Diamond(C_1 = 3)\,)$. The transitions which can modify the truth values of the state predicates $C_1 \ne 3 \vee C_2 \ne 3$, $C_1 = 2$ and $C_1 = 3$ are t_{11}, t_{12}, t_{13}, t_{22} and t_{23}. Thus we choose $vis(\Phi) = \{t_{11}, t_{12}, t_{13}, t_{22}, t_{23}\}$.

In the initial state $\mathbf{s}_0$ the transitions t_{i1} and no other transitions are enabled. If we want to build a stubborn set $TS(\mathbf{s}_0)$ around t_{i1} then we have to take d_{i2} into the set because $\neg\, t_{i1} \leftrightarrow d_{i2}$. d_{i2} is disabled, thus we have to find $V \subseteq \mathbf{V}$ such that $\neg\, en(\mathbf{s}_0,d_{i2},V)$ and include $wrup(d_{i2},V)$ to the set. We can choose $V = \{A\}$ and $wrup(d_{i2},\{A\}) = \{d_{i1}\}$. $\neg\, en(\mathbf{s}_0,d_{i1},\{r_i\})$ holds and $wrup(d_{i1},\{r_i\}) = \{t_{i1}\}$, but t_{i1} is already in the set. We can thus stop with the set $TS(\mathbf{s}_0) = \{t_{i1},d_{i1},d_{i2}\}$. It is stubborn and contains exactly one enabled transition, namely t_{i1}. To satisfy Assumption (3) of Theorem 2.8 it is reasonable to choose the i so that $i > 1$, and the algorithm in Chapter 3 indeed does so. So only one transition is fired in $\mathbf{s}_0$, namely t_{i1}. Call the resulting new state s'.

By applying the above reasoning again one can see that if $j \neq i$, $\{t_{j1},d_{j1},d_{j2}\}$ is stubborn in s'. In an attempt to avoid visible transitions, the algorithm in Chapter 3 chooses $\{t_{j1},d_{j1},d_{j2}\}$ for some $j \geq 2$ and so on until the state $s_{12\bullet}$ such that $s_{12\bullet}(A) = s_{12\bullet}(C_1) = 1$ and $s_{12\bullet}(C_i) = 2$ for $2 \leq i \leq n$ is reached. At this stage we have generated n states.

The enabled transitions at $s_{12\bullet}$ are t_{11} and d_{i1} for $i \geq 2$. Assumption (3) forces us to include at least one of d_{i1}, $i \geq 2$, into $TS(s_{12\bullet})$. Because $\neg\, d_{i1} \leftrightarrow d_{j1}$ we conclude $\forall\, i \geq 1$: $d_{i1} \in TS(s_{12\bullet})$. Intuitively, this reflects the fact that it is essential which customer gets the resource. d_{11} is disabled and, as before, takes us to t_{11}. That is, also C_1 is given the chance to take the resource. So we have to take all enabled transitions into $TS(s_{12\bullet})$. In essence, the algorithm gives all the other customers the chance to take the resource before t_{11} occurs, because the occurrence of t_{11} modifies the value of the state predicate $C_1 = 2$ in Φ. The algorithm tries to find out what can happen before the state predicate value is modified.

Let d_{i1}, $i > 2$ occur in $s_{12\bullet}$. The system moves to a state where $C_i = 2$ and $A = 2i$. As before, the stubborn set method concentrates on the invisible transitions and fires t_{i2}, t_{i3}, d_{i2}, t_{i4} and t_{i1} returning to $s_{12\bullet}$. For exactly one i the state immediately before the occurrence of t_{i1} has already been generated. So we have generated a total of $5(n-2)-1$ new states. (Strictly speaking, the already generated state may correspond to $i = 2$, but this case is similar and treated below, so there is no nett error in the state count.)

Now consider the occurrence of t_{11} in $s_{12\bullet}$ resulting in the state $s_{22\bullet}$, where $\forall\, i$: $s_{22\bullet}(C_i) = 2 \wedge s_{22\bullet}(A) = 1$. Transitions d_{i1}, $i \geq 1$ are enabled and they do not accord with each other, so we have to consider all of them. After the occurrence of d_{i1}, $i > 2$, we continue as above by firing t_{i2}, t_{i3}, d_{i2}, t_{i4} and t_{i1} in sequence and return to $s_{22\bullet}$. With $i \leq 2$ we have the problem that t_{11}, t_{12}, t_{13}, t_{22} and t_{23} are visible, thus the above reasoning does not immediately apply. But it turns out that when t_{12}, t_{13}, t_{22} or t_{23} occurs, it is the only enabled transition, there is thus no problem. When t_{11} is due to occur the state proves to be $s_{12\bullet}$, which has already been investigated. We have generated $5n$ new states.

We still have not investigated the occurrence of d_{21} in $s_{12\bullet}$. The resulting pattern resembles the case of d_{i1}, $i > 2$, occurring in $s_{12\bullet}$, but the fact that t_{22} and t_{23} are visible has to be taken into account. Its effect is that also t_{11} has to investigated when t_{22} or t_{23} occurs. But the two states resulting from the occurrence of t_{11} have already been generated as indirect successors of $s_{22\bullet}$. 5 new states emerge.

The total number of states generated is thus $11n-6$. Every elementary cycle of the reduced state space contains either $s_{12\bullet}$ or $s_{22\bullet}$, thus Assumption (4) of Theorem 2.8 is satisfied without further action.

The validity of the formulas $\Box((C_1 \neq 3) \vee (C_2 \neq 3))$ and $\Box((C_1 = 2) \Rightarrow \Diamond(C_1 = 3))$ can now be checked from the reduced state space. The former is true, the latter is not. The reduced state space is linear in the number of customers, while the ordinary state space is exponential.

5. CONCLUDING REMARKS

We showed how to generate reduced state spaces such that the truth values of LTL formulas are preserved, provided that the formulas are given before the reduced state space generation commences and they do not contain the "next state" operator. In the example in Chapter 4 the reduction of the size of the state space is from exponential to linear in the size of the system. This is a very good result. However, it is currently not known how well the LTL preserving stubborn set method performs on the average. One may expect that the size of the reduced state space increases when more and more variables are referred to by the formulas to be preserved.

By the time of the writing of this paper the LTL preserving stubborn set method has not been implemented. However, a related stubborn set state space reduction method has been implemented into a tool called *Toras* [Wheeler & 90]. Toras is being developed in Telecom Australia Research Laboratories. Among other features, it supports an as yet unpublished version of the stubborn set method which preserves the *failure semantics* [Brookes & 84] of systems. The method differs from the one presented in this paper in that it does not need Assumptions (1) and (4) of Theorem 2.8. To give an example of the performance of Toras, a certain version of the n dining philosophers system has 3^n-1 states, and the basic stubborn set method reduces the number to $3n^2-3n+2$ states. For the 100 philosopher system ($\approx 10^{47}$ states) Toras generated the predicted 29 702 states in 20 minutes CPU time on a Sun 3/60 [Wheeler & 90].

It is necessary to implement the LTL preserving stubborn set method to get practical results about its performance. For the method to be useful there must be a tool also for the model checking. [Lichtenstein & 85] discuss the principles of a suitable model checking algorithm. An interesting open theoretical question is how the method may be modified to handle various fairness assumptions.

ACKNOWLEDGEMENTS

This work has been supported by the Technology Development Centre of Finland (TEKES).

REFERENCES

[Aho & 74] Aho, A. V., Hopcroft, J. E. & Ullman, J. D.: *The Design and Analysis of Computer Algorithms.* Addison-Wesley, Reading, Massachusetts 1974, 470 p.

[Brookes & 84] Brookes, S. D., Hoare, C. A. R. & Roscoe, A. W.: *A Theory of Communicating Sequential Processes.* Journal of the ACM 31 (3) 1984, pp. 560–599.

[Clarke & 86] Clarke, E. M., Emerson, E. A. & Sistla, A. P.: *Automatic Verification of Finite-State Concurrent Systems using Temporal Logic Specifications.* ACM Transactions on Programming Languages and Systems 8 (2) 1986 pp. 244–263.

[Lamport 83] Lamport, L.: *What Good is Temporal Logic?* Information Processing '83, North-Holland pp. 657–668.

[Lichtenstein & 85] Lichtenstein, O. & Pnueli, A.: *Checking that Finite State Concurrent Programs Satisfy their Linear Specification.* Proceedings of the Twelfth ACM Symposium on the Principles of Programming Languages, January 1985 pp. 97–107.

[Overman 81] Overman, W. T.: *Verification of Concurrent Systems: Function and Timing*. Ph.D. Dissertation, University of California Los Angeles 1981, 174 p.

[Pnueli 86] Pnueli, A.: *Applications of Temporal Logic to the Specification and Verification of Reactive Systems: A Survey of Current Trends*. In: Current Trends in Concurrency, Lecture Notes in Computer Science 224, Springer 1986 pp. 510–584.

[Valmari 88a] Valmari, A.: *Error Detection by Reduced Reachability Graph Generation*. Proceedings of the Ninth European Workshop on Application and Theory of Petri Nets, Venice, Italy 1988 pp. 95–112.

[Valmari 88b] Valmari, A.: *Heuristics for Lazy State Generation Speeds up Analysis of Concurrent Systems*. Proceedings of the Finnish Artificial Intelligence Symposium STeP-88, Helsinki 1988 Vol. 2 pp. 640–650.

[Valmari 88c] Valmari, A.: *State Space Generation: Efficiency and Practicality*. Ph.D. Thesis, Tampere University of Technology Publications 55, 1988, 169 p.

[Valmari 89a] Valmari, A.: *Eliminating Redundant Interleavings during Concurrent Program Verification*. Proceedings of Parallel Architectures and Languages Europe '89 Vol 2, Lecture Notes in Computer Science 366, Springer 1989 pp. 89–103.

[Valmari 89b] Valmari, A.: *Stubborn Sets for Reduced State Space Generation*. Proceedings of the Tenth International Conference on Application and Theory of Petri Nets, Bonn, FRG 1989 Vol. 2 pp. 1–22. A revised version to appear in Advances in Petri Nets 90, Lecture Notes in Computer Science, Springer.

[Wheeler & 90] Wheeler, G. R., Valmari, A. & Billington, J.: *Baby Toras Eats Philosophers but Thinks about Solitaire*. Proceedings of the Fifth Australian Software Engineering Conference, Sydney, NSW, Australia, 1990 pp. 283–288.

DIMACS Series in Discrete Mathematics
and Theoretical Computer Science
Volume 3, 1991

PAPETRI : Environment for the Analysis of PETRI nets

G. BERTHELOT (*), C. JOHNEN (**), L. PETRUCCI (***)

(*) CEDRIC-IIE, 18 Allée Jean Rostand, BP 77, 91002 EVRY CEDEX

(**) L.R.I., U.A. au CNRS 410, Bât. 490, Université Paris-Sud, 91405 ORSAY CEDEX

(***) MASI, tour 65, Université Pierre et Marie Curie, 4 place Jussieu, 75252 PARIS CEDEX 05

Work supported by ESPRIT project DEMON (BRA 3148).

Abstract

In this paper, we present PAPETRI, a general and integrated environment for editing and analysing Petri nets. PAPETRI allows us to work with different classes of nets. Several analysis tools are available for each of these classes.

The kernel of PAPETRI is the graphical and interactive editor, PETRIX. Five analysis tools are available : a graphical simulator of Petri nets, two constructors of covering graphs, a generator of semi-flows (COMBAG), a method using rewriting techniques (PETRIREVE), an analyser and simulator of algebraic nets (VAERA). Each of these tools, as well as PETRIX, is detailed hereafter.

The analysis of a protocol to access a one-way part of a railway with PAPETRI tools will be our running example.

Keywords : Petri nets, coloured nets, algebraic nets, nets analysis environments.

1/ PAPETRI : overview

Petri nets[1] and their extensions are special classes of transitions systems where states are specified by tuples and transitions by a couple of tuples representing respectively precondition and modification of the state.

The goal of PAPETRI (in French, *Poste pour l'Analyse des réseaux de PETRI*) is to provide a friendly editing environment and to afford a great deal of analysis tools for several classes of Petri nets. These tools are not only based on nets classical methods but also on original approaches using rewriting techniques and abstract data types.

A particular care was accorded, in the conception of PAPETRI, on one hand to the flexibility of edition and on the other hand to the easy integration of new tools. The actual functionnalities are given hereafter, for each class of nets :

- *Petri nets* : simulation, construction of covering graphs, computation of a generative family of semi-flows, analysis with rewriting techniques;
- *Coloured nets* [Jen81] : construction of covering graphs, semi-flows calculus;
- *Algebraic nets* [Vau85] : syntactic verification, simulation, skeleton analysis (Petri net), normed net analysis (algebraic net).

---

[1] Basic notions and definitions on Petri nets can be found in [Rei85].

This list will be enlarged : the edition of nets may be extended to other classes, and other tools can be integrated in PAPETRI.

The main function of PAPETRI is PETRIX, a graphical and interactive environment for the edition of nets. The characteristics of PETRIX are detailed in section 2. The commands which may be used are called throughout a set of unfolding menus and dialog windows. They also allow us to call analysis tools for the class of the edited net.

Five analysis tools are presently integrated in PAPETRI :

• *Simulator* : graphical simulation of token game, described in section 2;

• *Covering graphs* : a first procedure builds the classical Karp and Miller covering graph and allows to check some net properties. A second one computes the minimal covering graph. These two tools are presented in section 4;

• *COMBAG* : semi-flows computation, introduced in section 5;

• *PETRIREVE* : analysis of Petri nets using rewriting techniques, described in section 6;

• *VAERA* : tool for syntactic verification, simulation and analysis of algebraic nets, detailed in section 7.

In section 3, we introduce a protocol to access a one-way part of a railway to exemplify the various possibilities of analysis. Thus, we validate the access conditions.

2/ PETRIX : the editor-simulator

The graphical and interactive editor of nets, PETRIX, uses the XWINDOWS environment on UNIX workstations.

PETRIX allows to draw a net, using the mouse and unfolding menus. It is possible to create, modify, drag and suppress the basic objects of a Petri net that are places, transitions, arcs and tokens. An arc joining a place and a transition together may be created as a broken arrow between the two objects. A token must be created inside a place but may be moved into another one. So, the graph obtained is a correct Petri net. Each of the objects may be named. It is possible to create texts which are either free (comments, title of the net, ...), or linked to another object. The shape and the size of graphical objects may be changed.

Several nets can be edited at the same time, each of them being displayed in separate windows. Operations such as cut (or copy)/paste allow the user to insert a part of a net into another net.

The size of edited nets is only limited by the available memory (several thousands of objects for 1Mb).

The simulation of an ordinary net allows to play the token game using the usual enabling rules. Evolutions can be viewed, as the marking of each place is modified in the window. The token game can be performed according to three modes : manual, semi-automatic or automatic. In manual mode, the simulator selects the enabled transitions and the user chooses the one he wants to occur. In semi-automatic mode, the evolution is done as long as only one transition can occur. Otherwise, all enabled transitions are highlighted and the user chooses the one to occur. In automatic mode, choices are randomly made. The simulation stops when no transition can occur. The user can stop the simulation whenever he wants, in any mode.

3/ Introducing an example

In order to exemplify the usefullness and the functionning of the various tools, we have chosen to study a simple problem of synchronization for the access to a critical ressource : we model the entering and exiting protocols on a one-way part of a railway.

Let us assume that the railway between WEST_TERMINUS and EAST_TERMINUS is composed of a one-way section between WEST_JUNCTION and EAST_JUNCTION (see figure hereafter). This problem is derived from the classical readers/writers problem, and it can be complicated a lot by the adjunction of one or more shunts on the one-way section.

The driving rules the trains must observe on the one-way part of the railway are the following :

- the one-way part of the railway must not be used by two trains going in opposite ways;

- the one-way section can be used by one or several trains going in the same way;

- there is no priviledged direction for driving;

- we suppose that only three trains can be on the one-way section at the same time (due to a minimal distance between trains).

Let us model this example using ordinary Petri nets. We are going to detail the net obtained (see figure 1) :

• Places : Place *Departure West_Terminus* (resp. *Departure East_Terminus*) contains the trains which have not yet started their trip, departing from West_Terminus (resp. East_Terminus). Place *Part W->E* (resp. *Part E->W*) contains the trains on the one-way part of the railway and going from West_Junction to East_Junction (resp. from East_Junction to West_Junction). Place *Arrival East_Terminus* (resp. *Arrival West_Terminus*) contains the trains which have finished their trip and are arrived. *MUTEX* manages the mutual exclusion on the one-way section between the trains going from West_Junction to East_Junction and those going in the opposite way. Finally, *Maxnumtrains* contains as many tokens as trains can enter the one-way part of the railway. This place ensures that not more than three trains are on the section.

• Transitions and functionning : transition *First W* (resp. *First E*) occurs when the first train enters the one-way part of the railway at West_Junction (resp. East_Junction). The other trains will fire transition *Enter West_Junction* (resp. *Enter East_Junction*). They will get out of this section of the railway when transition *Exit East_Junction* (resp. *Exit West_Junction*) will occur, except the last train which will fire transition *Last E* (resp. *Last W*). Let us point out that the net is perfectly symmetrical.

• Initial marking : we suppose that there are six trains to depart from West_Terminus and also six trains to depart from East_Terminus. There is one token in place *MUTEX* to ensure the mutual exclusion. This token will be used when the first train will enter the one-way section and will be put back in its place when the last train will exit. At the beginning, no train circulates, so three trains may enter the section, and we put three tokens in place *Maxnumtrains*.

figure 1

Access protocol to a one-way section (Petri net)

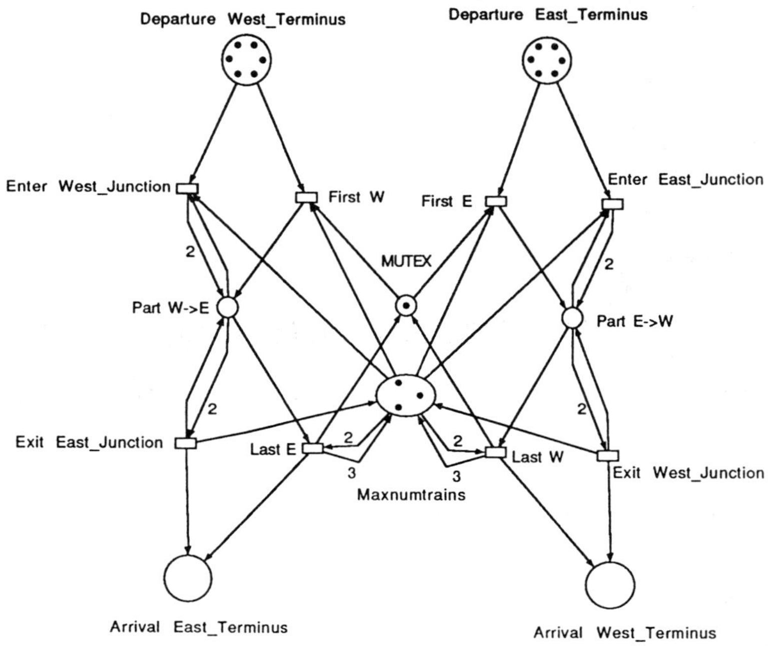

figure 2

Access protocol to a one-way section (coloured net)

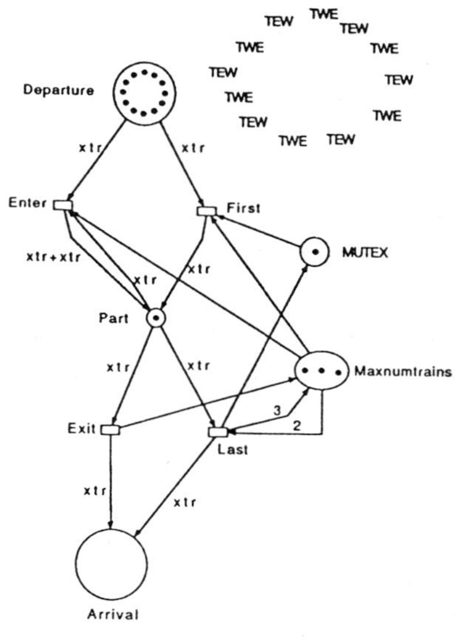

4/ Covering graphs

The more classical technique for the analysis of Petri nets consists in the construction of the *covering graph* : all the reachable markings are listed; when the set of reachable markings is infinite, a shortened list leads to build a bounded graph [KM69]. Important properties (boundedness of places, quasi-live transitions, regularity of the language) are decidable by procedures which analyze the graph.

A procedure to build this graph is available for places/transitions nets and coloured nets in PAPETRI. Once the graph is computed, an analyser of covering graphs provides diagnosis functions for the net as well as a visualisation of the graph. These functions are obtained through a menu.

The analysis functions concern the following properties : regularity of the language, existence of repetitive sequences of transitions, list of unbounded places and quasi-live transitions.

The analyser can compute the covering graph terminal connected components, which gives more information concerning liveness of transitions, home states, deadlocks of the net.

The mutual exclusion between places can be tested. The users specifies the places he wants to test the mutual exclusion of, then the analyser tests it on all nodes labels. Moreover, the analyser indicates whether these places can be empty all together.

The user can specify a sequence of transitions to occur, starting from the initial marking, and the analyser indicates the node obtained if the sequence labels a path in the graph. In the case of a bounded net, this function allows a behavioural study.

The graph may be scanned : the analyser lists all nodes labels.

The construction of a covering graph is expensive as well in time as in space. These difficulties may be reduced by the use of the minimal covering graph [Fin89] : are only kept the nodes with uncomparable labels. The number of nodes may thus be drastically decreased (several orders of magnitude). The construction of such a graph, both for places/transitions nets and coloured nets, was integrated in PETRIX [Mor89]. For the moment, the diagnosis functions are not available with this sort of graph.

For coloured nets, it could be possible to reduce furthermore the size of covering graphs, by the use of this technique combined with Jensen's, which are based on symmetries between colours ([HJJJ86]).

The analysis of our example (figure 1) leads to the following results :

- the net is bounded (places *Part*... contain at most three tokens);
- there are 259 reachable markings;
- all the transitions are quasi-live;
- the net has no live transition nor a repetitive sequence of transitions;
- the covering graph has only one connected terminal component consisting in only one node. This node label is the only home state, and it is a deadlock.
- so, the net terminates. The terminal marking is composed of six tokens in places *Arrival East_Terminus* and *Arrival West_Terminus*, one token in place *MUTEX* and three tokens in place *Maxnumtrains*. The twelve trains effectively arrive; three trains can now again circulate on the section, all of them going in the same way.

The study of mutual exclusion implies that places *Part W->E* and *Part E->W* do never contain tokens at the same time, and that both of them can be empty. The trains circulating on the section go in the same way (the section can be empty).

The minimal covering graph has also 259 nodes : all the reachable markings are uncomparable.

The construction of the covering graph allows to analyse the net for only one initial marking. However, a lot of nets model systems with various initial states. In our example, the protocol must be valid whichever the number of trains departing from West_Terminus or East_Terminus is. This goal can be reached by the study of semi-flows.

5/ COMBAG : semi-flows computation

In this section, we present COMBAG [Tre88] : a structural analysis method (i.e. which relies only on edges between places and transitions) which allows the validation of net properties independently of the initial marking. COMBAG is a tool for semi-flows computation in a Petri net. Two kinds of semi-flows are derived from the incidence matrix C :

- a *p-semi-flow* f weights every place such that $f^T.C = 0$. It leads to a net invariant : the weighted sum of tokens in places is constant for any reachable marking;
- a *t-semi-flow* f associates a number of occurrences with every transition such that $C.f = 0$. It denotes the repetitive and stationnary sequences of transitions.

In the following, we will use the term *semi-flow* to denote as well p-semi-flows as t-semi-flows.

The semi-flows over $\mathbb{N}$ (weights are positive integers) allow a finite generative family : every semi-flow over $\mathbb{N}$ can be expressed as a linear combination of the semi-flows in the generative family.

COMBAG calculates a generative family of semi-flows over $\mathbb{N}$, using Farkas' algorithm, the complexity of which is exponential w.r.t. time. To decrease the answering delays, COMBAG proposes, at each step, two optimization techniques. The first one consists in looking for the incidence matrix column that generates less operations, and the other one suppresses unusefull lines.

The use of coloured nets, which shorten Petri nets, leads to more concise nets. The arcs can be valued by variables (in this case, any token may be used, with no constraint as concerns its colour), or by colours (the token must have the same colour as the valuation). Hence, the coloured net of figure 2 represents the same protocol as the net of figure 1 and it contains twice less places and transitions. Trains are denoted as tokens in both nets. In the places/transitions net, the way trains circulate is represented by places; for example, place *Departure East_Terminus* contains only tokens (trains) going from East_Terminus to West_Terminus. In the coloured net, the way the trains circulate is represented by the colour of tokens : TEW, train going from East_Terminus to West_Terminus; TWE, train going from West_Terminus to East_Terminus.

COMBAG provides semi-flows for coloured nets having a finite number of colours. It calculates a generative family for three classes of semi-flows over $\mathbb{Z}$ (weights are signed integers) of a coloured net [VM84] :

- *type 1* : semi-flows f such that $f^T.|C| = 0$. They indicate invariant assertions on the number of tokens in places, without considering their colour. The generative family of this class of semi-flows, calculated by COMBAG, consists in the following :

$$f_1(Departure) = f_1(Part) = f_1(Arrival) = 1 \text{ and } 0 \text{ otherwise}$$

$$f_2(Maxnumtrains) = f_2(Part) = 1 \text{ and } 0 \text{ otherwise}$$

- *type 2* : semi-flows f such that $f^T.C = 0$. They denote invariant assertions on the number of tokens of a colour, valid for every colour (except the distinguished neutral one). As concerns our example, these colours are TEW and TWE. The generative family of this class of semi-flows, calculated by COMBAG, consists in the following :

$$f_1(Departure) = f_1(Part) = f_1(Arrival) = 1 \text{ and } 0 \text{ otherwise}$$

$$f_2(Maxnumtrains) = 1 \text{ and } 0 \text{ otherwise}$$

$$f_3(MUTEX) = 1 \text{ and } 0 \text{ otherwise}$$

The invariant associated with the first semi-flow indicates that the whole number of tokens with colour TEW (resp. TWE) in places *Departure*, *Part* and *Arrival* is always constant. The last two invariants express that places *MUTEX* and *Maxnumtrains* receive only black tokens and not tokens labelled by TEW or TWE. So, they never receive tokens corresponding to trains.

• *type 3* : semi-flows f such that $f^T.\Pi_a(C) = 0$ (where $\Pi_a(C)$ is the projection of the incidence matrix on colour a). They represent invariant assertions on the number of tokens with colour a. This sort of semi-flows can be calculated for each colour (except the black one). As concerns our example, this class of semi-flows for colour TEW (resp. TWE) is identical to the previous one, because there is no invariant particular to colour TEW (resp. TWE).

This technique enables us to find out structural properties of a net, but it does not give any piece of information concerning the net behaviour. In the next section, we introduce an original approach for the analysis of nets w.r.t. a class of initial markings.

6/ PETRIREVE : analysis with rewriting techniques

PETRIREVE [CJ85] analyses Petri nets with validation techniques based on rewriting systems. This technique studies the behaviour of a net independently of the initial martking, and thus validates structural properties of the net (termination).

PETRIREVE builts a set of oriented equations representing the behaviour of the net. Knuth-Bendix's completion transforms these equations into a rewriting system. To do so, PETRIREVE uses system REVE [FG83] - rewriting laboratory - developped by the CRIN (Nancy, France) and the MIT (Cambridge, USA). Studying the obtained rewriting system leads to straightly deduce some net properties (confluence, boundedness, ...). Other properties (invariants, quasi-liveness, reachability, termination, ...) may be validated by testing the completion of the rewriting system to which a specific equation is added.

We conceived a general technique to represent the behaviour of a net by a canonical rewriting system. A marking is represented by a term with operator *state* as header, this operator having as many arguments as there are places in the net. Each argument corresponds to a place : the value of the argument is the number of tokens in the place.

Each transition is represented by an equation. The left term denotes the precondition necessary for the transition to occur. Every term that unifies with the left term denotes a marking for which the transition can occur. The right term represents the state obtained after the occurrence of the transition.

For example, let us consider transition *First W* of figure 1. If places *Departure East_Terminus, MUTEX* and *Maxnumtrains* contain tokens, the transition is enabled. Every term which can be unified with :

state(s(vDepartureEast_Terminus), vDepartureWest_Terminus, vPartE->W, vPartW->E, s(vMUTEX),
s(vMaxnumtrains), vArrivalWest_Terminus, vArrivalEast_Terminus)

represents a marking which satisfies the precondition for the transition *First W* to occur. Let us notice that integers are denoted with two generators : *0*, which is integer zero, and *s* which stands for the successor operator. s(x) is then x + 1. In the case where a place-argument is (or contains) a variable, its name is the name of the place prefixed with letter v. The term giving the number of tokens in place *MUTEX* is of the form s(vMUTEX). Thus, the place cannot be empty (0 cannot be unified with s(x)). The marking obtained after the occurrence of transition *First W* is denoted by a term which can be unified with :

state(vDepartureEast_Terminus, vDepartureWest_Terminus, s(vPartE->W), vPartW->E, vMUTEX,
vMaxnumtrains, vArrivalWest_Terminus, vArrivalEast_Terminus).

PETRIREVE creates other equations in order to rewrite any term representing a deadlock as an identical term where the header is *deadlock*.

The rewriting system obtained after performing Knuth-Bendix's completion must converge to be used for proofs. Moreover, the equations have to be oriented as the occurrences of transitions. These two requirements leaded us to design an order on place-arguments of *state* to ensure termination, if possible, taking into account the orientation induced by transitions.

PETRIREVE automatically constructs the equations related to the net behaviour. These equations are transmitted to REVE which transforms them into a rewriting system with Knuth-Bendix's completion. The completion reduces the two members of an equation into their normal form and orientates the equation. Critical pairs eventually generate new equations. So, on one hand, rules can be redundant and thus suppressed of the system. On the other hand, these new equations may lead to reduction of already existing rules.

Let us analyse our example with PETRIREVE. The completion orientates all the equation-transitions in the way the transitions occur. This result immediatly indicates that the net is bounded for any initial marking.

As the rewriting system is canonical, proofs which usually need a recurrent procedure can be performed. The equation to satisfy is added to the rewriting system. If the system (with this new equation) completion does not build any equation between generators (s, 0), then the equation is satisfied in the initial algebra [HH80]. To verify that the whole number of tokens in places *Departure East_Terminus, Part E->W* and *Arrival West_Terminus* is constant, the user adds to the system the rewriting rule :

inv(state(vDepartureEast_Terminus, vDepartureWest_Terminus, vPartE->W, vPartW->E, vMUTEX,

vMaxnumtrains, VArrivalWest_Terminus, vArrivalEast_Terminus)) ==

vDepartureEast_Terminus + vPartE->W + vArrivalWest_Terminus.

The left member of this equation unifies with the left member of each rule in the rewriting system. Thus, the completion generates as many critical pairs as there are rules in the system. They are as follows :

inv(state(before occuring) == inv(state(after occuring)).

If these equations were generating rules between generators, the invariant would not be coherent with the occurrence of the transition. As the completion did not generate any equation between generators, the invariant is valid. We can as well verify that the whole number of tokens in places *Departure West_Terminus, Part W->E* and *Arrival East_Terminus* is constant.

We deduce, with this technique, the termination of the net for a class of initial markings (and not for only one initial marking). So, the validation of the previous equation proves that if, at the beginning, Part W->E is empty (this section allows at most three trains in the same way), all the trains arrive (whichever their number is).

state(vDepartureEast_Terminus,vDepartureWest_Terminus, 0, 0, 1, 3,vArrivalWest_Terminus,vArrivalEast_Terminus)

==

deadlock(0, 0, 0, 0, 1, 3, vDepartureEast_Terminus + vArrivalWest_Terminus,

vDepartureWest_Terminus + vArrivalEast_Terminus)

PETRIREVE is the first tool to verify the termination of a net for a class of initial states (a net terminates from an initial state if it always reaches a unique terminal state without successors). For the moment, PETRIREVE only analyses Petri nets. It may be extended to other classes of nets. However, the delays of completion will certainly be quite long.

7/ VAERA : Verification, Analysis and Evolution of algebRAic nets

The algebraic nets defined by J. Vautherin [Vau85] add another extent to Petri nets. The main interest of such a sort of high level nets is structuring data by the use of a specification of abstract data types. This is consistent with recent programming methods.

Let us shortly recall the way an algebraic net works. Such a net consists in the association of two components : a specification of abstract data types, which describes the data types used and the operations on them, and a Petri net. The places of the net are sorted, the tokens being constants of the specification, of the same sort as their containing place. The arcs are valued by terms of the specification.

The transitions are enabled according to the following criteria : first of all, the conditions on the amount of tokens in places must be valid, as in places/transitions nets, secondly, conditions between arcs valuations and tokens values must be satisfied. Each arc entering a transition is valued either by a term without variables, or by a single variable. In the first case, a token with this value must be used to fire the transition. In the second case, the variable takes the value of one of the tokens for the occurrence of the transition. Afterwards, if the transition occurs, the term associated with each arc exiting the transition is evaluated (variables occurring in the term have the value given when examining the entering arcs), and a token having this value is created in the corresponding place. Transitions may also have an auxiliary condition (sound equation) associated with them. This equation, the hand-sides of which are functions of the entering variables, must be valid for the transition to occur.

The analysis of algebraic nets is more complex than the analysis of places/transitions nets, due to the powerfulness of abstract data types. Tokens meaning is denoted as well by their presence in places as by their value. Two tokens within a same place may have different interpretations. We will first explain how PETRIX deals with the definition of an algebraic net, and we will detail afterwards the analysis tools we have designed.

When the user wants to design an algebraic net, he describes the Petri net part with PETRIX and uses VAERA for everything related to algebraic specifications. These must have been created by ASSPEGIQUE (in French, ASsistance à la SPEcification algébrIQUE, see [Cho88]), environment for algebraic specification developped at the LRI (Orsay, France), which can be called by VAERA.

Once the net is created, VAERA provides a tool for syntactic verification of the terms in the net and checks the coherence of the sorts used (tokens must be of sort of the place they are in, ...).

It is also possible to use the simulator included in VAERA to observe the behaviour of the net : the user selects, with the mouse, the transition to occur. The tool verifies that the conditions necessary for the transition to occur are satisfied, and if true, fires the transition as described before. As tokens may have different values, two modes of simulation are available : either the user chooses which tokens will be used for the transition to occur, or the tokens are randomly chosen by VAERA.

Let us model our example with an algebraic net (see figure 3). We characterize the trains travelling from West_Terminus to East_Terminus (TWE) and the trains going from East_Terminus to West_Terminus (TEW).

Place *Departure* contains the trains which are still in their departure station. Place *Arrival* contains trains which have finished their trip. Place *Maxnumtrains* represents the number of trains the one-way section can allow. And finally, place *Part* contains the trains circulating on the one-way part of the railway. This place a FIFO file : the trains must arrive with their departure order.

When a train wants to enter the one-way section, there are two possibilities. Either there is no train on the section (the file is empty), and the train can enter it by the occurrence of transition *Entrance1*. Then, it is in the file, and the number of trains which can still be allowed decreases. Or there are already trains on the section (the file is not empty). If this is the case, the train must verify that these trains circulate in the same way as it does, and also that the section can support new trains. If true, transition *Entrance* occurs, the train is pushed into the file, and the number of trains which can still be allowed decreases. When the first train on the section has finished this part of its trip, transition *Exit* occurs, it gets out of the file, and the number of trains which can be allowed increases.

```
SPEC : BOOL
    SORTS : bool
OPERATIONS :
    true : -> bool
    false : -> bool
    not _ : bool -> bool
    _ and _ : bool bool -> bool
    _ or _ : bool bool -> bool
VARIABLES :
    x : bool
AXIOMS :
    not-1 : not true = false
    not-2 : not false = true

    and-1 : true and true = true
    and-2 : x and false = false
    and-3 : false and x = false

    or-1 : false or false = false
    or-2 : true or x = true
    or-3 : x or true = true
```

```
SPEC : TRAIN
    WITH : NAT
    SORTS : file train
OPERATIONS :
    emptyfile : -> file
    push _ _ : train file -> file
    pop _ : file -> file
    top _ : file -> train
    fempty _ : file -> bool
    norm _ : file -> Nat
    TWE : -> train
    TEW : -> train
    norm _ : train -> Nat
    _ sameway _ : train train -> bool
VARIABLES :
    x, y, tr : train
    z : Nat
    f, f1, f2 : file
AXIOMS :
    fempty-1 : fempty emptyfile = true
    fempty-2 : fempty push x f1 = false

    pop-1 : pop push x emptyfile = emptyfile
    pop-2 : pop push x push y f1 = push x pop push y f1

    top-1 : top push x emptyfile = x
    top-2 : top push x push y f1 = top push y f1

    sameway-1 : x sameway x = true
    sameway-2 : x # y => x sameway y = false

    norm-1 : norm emptyfile = 0
    norm-2 : norm push x f1 = s norm f1
    norm-3 : norm pop f1 = p norm f1
    norm-4 : norm tr = 0
```

```
SPEC : NAT
    WITH : BOOL
    SORTS : Nat
OPERATIONS :
    0 : -> Nat
    nul _ : Nat -> bool
    p _ : Nat -> Nat
    s _ : Nat -> Nat
    _ + _ : Nat Nat -> Nat
    _ < _ : Nat Nat -> bool
    norm _ : Nat -> Nat
VARIABLES :
    x, y : Nat
AXIOMS :
    nul-1 : nul 0 = true
    nul-2 : nul s x = false

    p-1 : p s x = x

    plus-1 : x + 0 = x
    plus-2 : 0 + x = x
    plus-3 : s x + y = s (x + y)

    <-1 : s x < 0 = false
    <-2 : 0 < x = true
    <-3 : s x < s y = x < y

    norm-1 : norm 0 = 0
    norm-2 : norm s x = s norm x
    norm-3 : norm p x = p norm x
```

figure 3

Access protocol to a one-way section, algebraic net

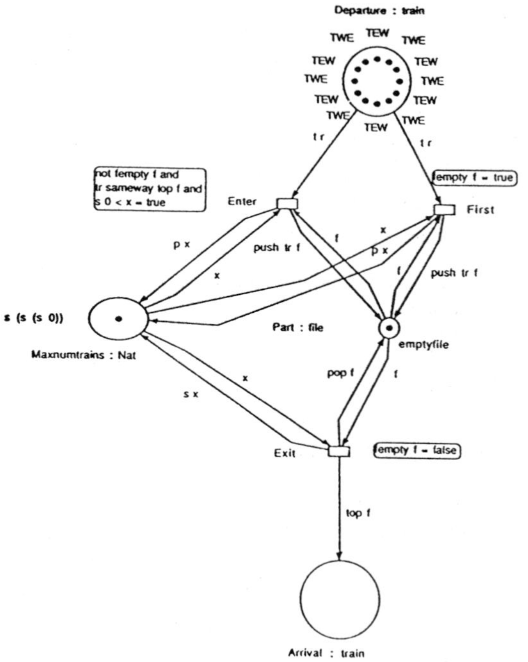

figure 4

skeleton

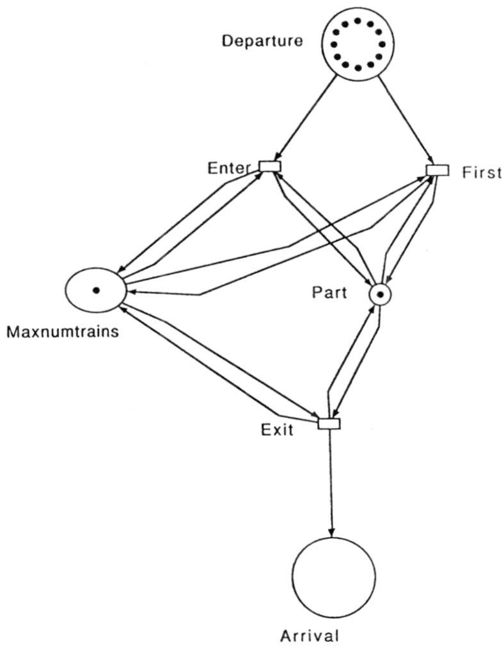

figure 5

normed net

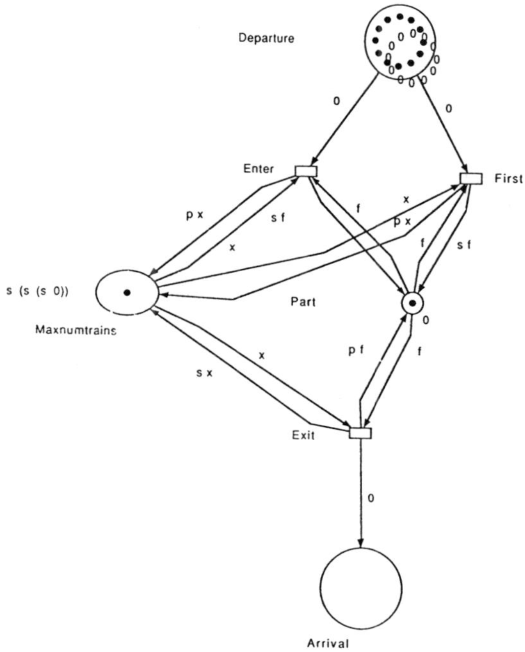

We notice that this net is smaller than the places/transitions net associated to the same problem. The maximum number of trains allowed on the section is not a structural constraint (valuation of arcs), but the value of a token. Moreover, the trains on the one-way part of the railway cannot overtake, and this condition could not be stated with an ordinary net.

Let us now introduce our various analysis tools. It may be interesting, in order to analyse algebraic nets, to yield simpler models by ignoring some pieces of information. These models are the skeleton and the normed net. Indeed, the difficulty of analysis of algebraic nets countervails their expression power.

The skeleton of an algebraic net :

The first model that we will study is the underlying Petri net, named *skeleton* (see [Vau85]). VAERA can automatically generate the skeleton of an algebraic net : auxiliary conditions for transitions to occur, arcs valuations and tokens values are forgotten. Thus, we obtain a net which can be analysed by the places/transitions tools previously described.

Some properties such as the boundedness of a place, the non quasi-liveness of a transition and the termination of a net are true for the algebraic net if they are true for its skeleton.

The study of the example's skeleton (see figure 4) implies the following results : the skeleton is not terminating, all the transitions are quasi-live, and places *Departure*, *Maxnumtrains* and *Part* are bounded by their initial number of tokens. So, the corresponding places in the algebraic net are bounded by the same number.

The normed net :

Another model extracted from the algebraic net is the *normed net* (see [Pet88]). The normed net is a model between the algebraic net (the most structured model) and the skeleton (the less structured model). This net contains less information than the algebraic net, but more than the skeleton. The analysis of the normed net is indeed more complex than the analysis of the skeleton, but its importance leans upon the degree of information it contains. This is the reason why its analysis allows us to detect properties of the algebraic net which are not valid for the skeleton. The main analysis which can be performed concerns the termination of the net.

The normed net is derived from the algebraic net by forgetting the auxiliary conditions for transitions to occur and by changing the valuation of arcs as well as the values of tokens into their norm. The norm maps a term to its "size". For example, as concerns files, function *norm* counts the number of elements in the file.

To generate this net, the user will have to include in his specification the definition of function *norm*. The net obtained is then such that the values of the tokens are unsigned integers. A transition will not occur if this operation leads to create strictly negative tokens.

We have proved [Pet88] that the termination of the normed net implies the termination of the algebraic net. To validate this property, we observe the behaviour of the net transitions. If a transition cannot infinitely occur, we will say that it can be blocked. If all the transitions can be blocked, the net terminates. VAERA decides, in a lot of cases, of the blocking of transitions. It tells the user if the net terminates, and otherwise gives the list of blockable transitions.

When considering our example, we construct the normed net of figure 5. Studying this net confirms its termination. So, the algebraic net terminates.

For the moment, we only have semi-decision procedures. They cannot infirm termination. Hence, to improve the analysis of algebraic nets, we intend to add a tool for the construction of finite reachable markings graphs.

8/ Conclusion and future work

We presented, in this paper, PAPETRI. It offers, in a graphical environment, various analysis tools. Moreover, it deals with different classes of nets : ordinary nets, coloured nets, algebraic nets. The mixture of techniques from various theories allows to study all the aspects of a system and to fully validate it. This large set of possibilities is shortly described with a simple but significant problem.

As concerns the examples we studied, the answering delays are reasonable w.r.t. the calculus complexity (a few minutes for the longer operations).

PAPETRI is an open system which may easily receive other tools and may be extended to other classes of nets. The next foreseen extensions deal with the construction of reachable markings graphs for algebraic nets and verification of home states.

Bibliography

[Cho88] Choppy C. : *ASSPEGIQUE user's manual*, rapport de recherche L.R.I. n°452, Orsay, 1988.

[CJ85] Choppy C., Johnen C. : *PETRIREVE : proving Petri net properties with rewriting systems*, LNCS vol.202, rewriting techniques and applications, Jouannaud J.P. Ed., Springer Verlag, 1985.

[FG83] Forgaard R., Guttag J.V. : REVE : *a term rewriting system generator with failure resistant Knuth Bendix*, Proc. of an NSF Workshop on the rewrite rule laboratory, 1983.

[Fin89] Finkel. A. : *A minimal coverability graph for Petri nets*, rapport de recherche L.R.I., Orsay, 1989.

[Jen81] Jensen K. : *Coloured Petri nets and the invariants method*, T.C.S., 14-3, pp. 317-336, 1981.

[KM69] Karp R.M., Miller R.E. : *Parrallel program schemata*, JCSS,4, pp. 147-195, 1969.

[HH80] Huet G., Hullot J.J. : *Proofs by induction in equational theories with constructors*, JCSS, 25, pp. 239-266, 1982.

[HJJJ86] Huber P., Jensen M., Jensen K., Jepsen O. : *Reachability trees for High-Level Petri nets*, T.C.S., 45, 1986.

[Mor89] Moreau J. M. : *Graphe de couverture minimal dans les réseaux de Petri*, rapport de stage de D.E.A., Université d'Orléans, 1989.

[Pet88] Petrucci L. : *Etude d'un environnement d'exécution de réseaux de Petri algébriques*, rapport de stage de D.E.A., Université de Paris-Sud, Orsay, 1988.

[Rei85] Reisig W. : *Petri Nets* , Springer Verlag, 1985.

[Tre88] Treves N. : *A comparative study of different techniques of semi-flows computation in higher-level nets*, Proc. IX European Workshop on Applications and Theory of Petri nets, 1988.

[Vau85] Vautherin J. : *Un modèle algébrique, basé sur les réseaux de Petri, pour l'étude des systèmes parallèles*, Thèse de doctorat d'ingénieur, Université Paris-Sud, Orsay, 1985.

[VM84] Vautherin J., Memmi G. : *Computation of flows for unary-Predicate/Transition nets*, LNCS 188, G. Rozenberg ed., Springer Verlag, pp. 307-327, 1984.

DIMACS Series in Discrete Mathematics
and Theoretical Computer Science
Volume 3, 1991

Compositional Minimization of Finite State Systems*

Susanne Graf †   Bernhard Steffen ‡

Abstract

In this paper we describe a method for the obtention of the minimal transition system, representing a communicating system given by a set of parallel processes, avoiding the complexity of the non minimal transition system. We consider minimization with respect to observational equivalence, but the method may be adapted to any other equivalence.

An interesting method to achieve this goal is to proceed by stepwise composition and minimization of the components of the system. However, if no precautions are taken, the intermediate state graphs generated by this method may contain a lot of transitions which are impossible in the whole context. We give here a variant of this method which allows to avoid these impossible transitions by taking into account at each composition step a guess of the interface behaviour of the context. This "interface specification" must be provided by the user. The method is based on a *reduction operator* for the composition of a subsystem with its interface specification, which is similar to the parallel operator but introduces *undefinedness* predicates whereever the interface "cuts off" a transition. The parallel operator is defined in a way that these undefinedness predicates disappear again in the full context if and only if the corresponding transition is in fact impossible in the whole system.

The *efficiency* of the method depends in fact on the accuracy with which the designer is able to approximate the possible sequences of the context, but its *correctness* does not. The proof of the correctness of the method is based on a preorder relation similar to the one defined by Walker.

1 Motivation

Many tools for the automatic analysis or verification of finite state concurrent systems are based on the construction of the global state graph of the system under consideration (cf. [CES83,FSS83,CPS89,CPSb89]). Thus they often fail because of the *state explosion problem*: the state space of a system potentially increases geometrically in the number of its

*This work has been partially supported by ESPRIT Basic Research Action 'Spec'
†IMAG-LGI, BP 53X, F-38041 Grenoble
‡Department of Computer Science, University of Aarhus, DK-8000 Aarhus C

parallel components. To overcome this problem techniques have been developed in order to avoid the construction of the complete state graph (cf. [BFH90,CLM89,ClSte90,Fer88, Jos87,KuMcM89,Kr89,LaTh88,LaXi90,Pnu85,ShGr90,StGr89,Wa88,Win90,WoLo89]). In this paper we present a method for the compositional minimization of finite state concurrent systems, which is practically motivated by the following observation:

> For the verification of a system it is usually sufficient to consider an abstraction of its global state graph, because numerous computations are irrelevant from the observers point of view. Such abstractions often allow to reduce the state graph drastically by collapsing semantically equivalent states to a single state, which does not affect the observable behaviour. For example, the so obtained "minimization" of a complex communication protocol may be a simple buffer.

Let us refer to the size of the original state space of a system S as its *apparent* complexity, and to the size of the minimized state space as its *real* complexity. The intention of our method is to compositionally construct the minimal system representation and therefore to avoid the apparent complexity. Unfortunately, the straightforward idea to just successively combine and minimize the components of the system is not satisfactory, because "local" minimization does not take context constraints into account and therefore may even lead to subsystems with a higher real complexity than the apparent complexity of the overall system. This is mainly due to the fact that parts need to be considered that can never be reached in the global context. *Partial* or *loose* specifications allow to "cut off" these unreachable parts (see section 4.3). As in [ClSte90,Kr89,LaTh88,ShGr90,Wa88]) we will exploit this feature to take advantage of context information. Furthermore, we will refer to the size of the maximal transition system that is encountered by our method as the *algorithmic* complexity.

Our method is tailored for establishing $\mathcal{P} \models \phi$, i.e. whether $\mathcal{P}$ satisfies the property ϕ, when $\mathcal{P}$ is a system of the form $(p_1 \|_{\mathrm{I}_1} \cdots \|_{\mathrm{I}_{n-1}} p_n)\langle L\rangle$. Here we assume the processes p_i already as to be given as transition systems, I_i to be *interface specifications*, namely supersets of the set of sequences that can be observed at the associated interfaces, $\|$ to represent parallel composition and $\langle L\rangle$ a hiding operator that abstracts from the activities considered as internal by transforming them into the unobservable action τ (see Section 3.2).

The point of our method is the successive construction of *partially defined* transition systems P_i with the following properties:

1. P_i is less specified than $p_1 \| \cdots \| p_i$, i.e. is smaller in the sense of $\preceq$ (see Definition 4.2).
2. P_n is semantically equivalent to the full system $(p_1 \| \cdots \| p_n)\langle L\rangle$, whenever the interface specifications are correct.
3. P_i has the least number of states and transitions in its semantic equivalence class.

Subsequently, the validation of $P_n \models \phi$ completes the proof. Of course, this requires that ϕ is preserved by the semantic equivalence under consideration. In this paper, we are dealing with a refinement of observational equivalence (see Definition 4.1)). However, the method easily adapts to other equivalences as well (see Section 5.3).

Important factor in this approach are the interface specifications, which should be provided by the program designer. However, the correctness of the method does not depend on the correctness of these specifications. They are only used to "guide" the proof. Thus wrong interface specifications will never lead to wrong proofs. They may only prevent a successful verification of a valid statement. It should be noted that the total definedness (see Definition 3.1) of P_n already implies $P_n \approx^d \mathcal{P}$, which is enough to guarantee successful verification of all ($\approx^d$- consistent) properties satisfied by $\mathcal{P}$.

The power of our method is demonstrated by means of a setup that handles the mutually exclusive access of processes to a common resource. In fact, in this example, the apparent complexity is exponential, whereas the algorithmic and the real complexity are linear (see Section 6). This is illustrated by means of numerical results that have been obtained using the Aldebaran verification tool [Fer88]. The method can easily and efficiently be implemented in systems like the Edinburgh Concurrency Workbench [CPS89] or the Aldebaran verification tool [Fer88].

2 Related work

A great effort has been made in order to avoid the construction of the complete state graph, and therefore to avoid the state explosion problem. Roughly, the proposed methods can be split into two categories, the *compositional verification* and the *compositional minimization*. Characteristic for the former category is that the global system need not be considered at all during the verification process, and for the latter that a minimal semantically equivalent representation of the global system is constructed. This minimal representation can subsequently be used for all kinds of verification.

A pure approach to compositional verification has been proposed by Winskel in [Win90], where rules are given to decompose assertions of the form $\mathcal{P} \models \phi$ depending on the syntax of the program $\mathcal{P}$ and the formula ϕ. Unfortunately, the decomposition rules for processes involving the parallel operator are very restricted. Larsen and Xinxin [LaXi90] follow a similar line, however, their decomposition rules are based on an operational semantics of contexts rather than the syntax.

In order to deal with the problems that arise from parallel compositions Pnueli [Pnu85] proposed a "conditional" inference system, where assertions of the form $\phi\mathcal{P}\psi$ can be derived, meaning that the program $\mathcal{P}$ satisfies the property ψ under the condition that its environment satisfies ϕ. This inference system has been used by Shurek and Grumberg in [ShGr90], where a semi-automatic modular verification method is presented, which, like ours, is based on "guesses" for context specifications. However, in contrast to our method it requires a separate proof of the correctness of these guesses.

Josko [Jos87] also presented a method, where the assumptions on the environment of a component are expressed by a formula, which must be proved in a separate step. The main disadvantage of his method is that the algorithm is exponential in the size of the assumptions about the environment.

A method of the second category was proposed by Halbwachs et al. in [BFH90]. It constructs directly a transition system minimized with respect to bisimulation by successive refinement of a single state. In this method, which has been tailored for Lustre [CHPP87], symbolic computation is needed in order to keep the expressions small which, in general, may grow exponentially.

Another approach of this category was presented by Clarke et al. [CLM89]. They exploit the knowledge about the alphabet of interest in order to abstract and minimize the systems components. Using $\langle L \rangle$ operations together with an elementary rule for distributing them over the parallel operator (see Proposition 3.3) our method covers this approach.

Krumm [Kr89] considers minimization on the fly and reduction with respect to an interface specification that must be provided by the system designer. Whereas the former is limited, because the complete state graph need to be traversed, the latter requires a separate proof of the correctness of the interface specification. This contrasts with our method, where such a proof is unnecessary.

Larsen and Thomsen [LaTh88], and Walker [Wa88] use partial specifications in order to take context constraints into account. Our method is an elaboration of theirs. It uses a more appropriate preorder allowing to define a strategy for (semi-)automatic proofs where the required user support is kept to a minimum.

The methods proposed in [WoLo89,StGr89,KuMcM89,BCG86] are tailored to verify properties of classes of systems that are systematically built from large numbers of similar processes. These methods are somewhat orthogonal to ours. This suggests to consider a combination of both types of methods.

3 Representation of Processes

In this section, we establish our framework in which processes (systems) are labelled transition systems extended by an undefinedness predicate with parallel composition and hiding defined on them. The extension provides a notion of partial definedness, which naturally leads to a specification-implementation relation between processes (see section 4.3).

3.1 Extended Transition Systems

An *extended finite state transition system* T is a quadruple $(S, \mathcal{A} \cup \{\tau\}, \longrightarrow, \uparrow)$ where

1. S is a finite set of *processes* or *states*;
2. $\mathcal{A}$ is a finite *alphabet* of *observable actions*, and τ represents an internal or unobservable action not in $\mathcal{A}$;
3. $\longrightarrow$ is a mapping associating with each $a \in \mathcal{A} \cup \{\tau\}$, a *transition relation* $\xrightarrow{a} \subseteq S \times S$;
4. $\uparrow \subseteq S \times 2^{\mathcal{A} \cup \{\tau\}}$ is a predicate expressing *guarded undefinedness*. Given $(p, L) \in \uparrow$ we write $p\uparrow a$ for $a \in L$.

Typically S is a set of program states, and the relationship $p \xrightarrow{a} q$ indicates that p can evolve to q under the observation of a. Finally, $p\uparrow a$ expresses that an a-transition would allow p to enter an undefined state. We say that p is *a-undefined* in this case. Thus, transition systems involving the undefinedness predicate are only *partially defined* or *specified*. It is this notion of partial specification together with its induced partial order which provides the framework for proving our method correct.

Processes are rooted transition systems, i.e. pairs consisting of a transition system and a designated start state. Given a transition system $T = (S, \mathcal{A} \cup \{\tau\}, \longrightarrow, \uparrow)$, we identify (as usual) a state $p \in S$ with the process $((S_p, \mathcal{A}_p \cup \{\tau\}, \longrightarrow_p, \uparrow_p), p)$, where

- S_p is the set of states that are reachable from p in T,
- $\mathcal{A}_p = \mathcal{A}$ and
- $\longrightarrow_p$ and $\uparrow_p$ are $\longrightarrow$ and $\uparrow$ restricted to S_p, respectively.

In future, obvious indices will be dropped. The following property characterizes the subset of "standard" transition systems.

Definition 3.1 *A process is* totally defined *if its undefinedness predicate* $\uparrow$ *is empty.*

3.2 Critical Processes

We now introduce a binary parallel operator $\|$ and unary hiding operators $\langle L \rangle$, where L is a set of observable actions. Intuitively, $p\|q$ is the parallel composition of p and q with synchronization of the actions common to both of their alphabets and interleaving of the others, and $p\langle L \rangle$ is the process in which only the actions in L are observable. The transition relations for the resulting processes are defined by:

1. $$\frac{p \xrightarrow{a} p'}{p\langle L\rangle \xrightarrow{a} p'\langle L\rangle} \quad a \in L$$

2. $$\frac{p \xrightarrow{a} p'}{p\langle L\rangle \xrightarrow{\tau} p'\langle L\rangle} \quad a \notin L$$

3. $$\frac{p \xrightarrow{a} p'}{p\|q \xrightarrow{a} p'\|q} \quad (\, a \notin \mathcal{A}_q \,)$$

4. $$\frac{q \xrightarrow{a} q'}{p\|q \xrightarrow{a} p\|q'} \quad (\, a \notin \mathcal{A}_p \,)$$

5. $$\frac{p \xrightarrow{a} p' \quad q \xrightarrow{a} q'}{p\|q \xrightarrow{a} p'\|q'} \quad a \neq \tau$$

their undefinedness predicates by:

6. $$\frac{p\!\uparrow a}{p\langle L\rangle\!\uparrow a} \quad a \in L$$

7. $$\frac{p\!\uparrow a}{p\langle L\rangle\!\uparrow \tau} \quad a \notin L$$

8. $$\frac{p\!\uparrow a}{(p\|q)\!\uparrow a} \quad (\, a \notin \mathcal{A}_q \text{ or } q \xrightarrow{a} q')$$

9. $$\frac{q\!\uparrow a}{(p\|q)\!\uparrow a} \quad (\, a \notin \mathcal{A}_p \text{ or } p \xrightarrow{a} p')$$

10. $$\frac{p\!\uparrow a \quad q\!\uparrow a}{(p\|q)\!\uparrow a}$$

and their alphabets by $\mathcal{A}_{p\langle L\rangle} = \mathcal{A}_p \setminus L$ and $\mathcal{A}_{p\|q} = \mathcal{A}_p \cup \mathcal{A}_q$. Finally $S_{p\langle L\rangle} = \{p'\langle L\rangle | p' \in S_p\}$ and $S_{p\|q}$ is the set of pairs $p'\|q'$ that are reachable in $p\|q$.

Thus $p\!\uparrow a$ ($q\!\uparrow a$) implies $(p\|q)\!\uparrow a$, whenever q (p) does not preempt the execution of a, i.e. whenever $a \notin \mathcal{A}_q$ or $q \xrightarrow{a} q'$ ($a \notin \mathcal{A}_p$ or $p \xrightarrow{a} p'$). Remember that $\tau \notin \mathcal{A}_p$ for any p, thus $\xrightarrow{\tau}$ is defined by the clauses 2, 3 and 4, and $\uparrow \tau$ by clauses 7, 8 and 9. The exact meaning of this definition will become clear in Section 5.2, where we introduce the reduction operator. We have:

Proposition 3.2 $\|$ *is associative and commutative.*

Thus processes of the form $(p_1\|\cdots\|p_n)\langle L\rangle$ are well-defined. Our method will concentrate on this form[1].

Proposition 3.3 $\forall p,q\,\forall L.\ (p\|q)\langle L\rangle \;=\; (p\langle L\cup\mathcal{A}_q\rangle \,\|\, q\langle L\cup\mathcal{A}_p\rangle)\langle L\cap\mathcal{A}_p\cap\mathcal{A}_q\rangle$

This proposition is particularly important, because it allows to localize global hiding informations. In fact, this localization is the essence of the construction of the 'interface processes' in [CLM89].

4 Equivalence and Partial Order

In this section, we define a semantics of extended labelled transition systems in terms of observational equivalence (cf. [Mi80]) and establish a specification-implementation relation in terms of a preorder, which is compatible with this semantics. This preorder plays a key role in the correctness proof of our method.

4.1 Guarded Undefinedness

The $\longrightarrow$ relation does not distinguish between observable and unobservable actions. In order to reflect that τ is internal, and hence not visible, we define the *weak* transition relation $\Rightarrow$ and the *weak* undefinedness predicate $\Uparrow$ for arbitrary $p,q\in S$ and $a\in\mathcal{A}$ as the least relation defined by:

1. $p \xrightarrow{\tau}{}^{*} \xrightarrow{a} \xrightarrow{\tau}{}^{*} q$ implies $p \stackrel{a}{\Longrightarrow} q$
2. $p \xrightarrow{\tau}{}^{*} q$ implies $p \stackrel{\varepsilon}{\Longrightarrow} q$
3. $q{\uparrow}a \;\wedge\; p \stackrel{\varepsilon}{\Longrightarrow} q$ implies $p{\Uparrow}a$.
4. $q{\uparrow}\tau \;\wedge\; p \stackrel{\varepsilon}{\Longrightarrow} q$ implies $p{\Uparrow}\varepsilon$.
5. $q{\Uparrow}\varepsilon \;\wedge\; p \stackrel{a}{\Longrightarrow} q$ implies $p{\Uparrow}a$.
6. $p{\Uparrow}\varepsilon$ implies $p{\Uparrow}a$.

As usual, the effect of weakening is to swallow the invisible τ-actions.

[1]This form is called *standard concurrent form* in CCS ([Mi80,Mi89]).

4.2 Semantic Equivalence

Our notion of semantics is defined by means of the following equivalence relation[2]:

Definition 4.1 $\approx^d$ *is the union of all relations* $R \subseteq S \times S$ *satisfying that* pRq *implies for all* $a \in \mathcal{A}$*:*

1. $p \Uparrow a$ *if and only if* $q \Uparrow a$
2. $p \stackrel{a}{\Rightarrow} p'$ *implies* $\exists q'.\ q \stackrel{a}{\Rightarrow} q' \wedge p'Rq'$
3. $q \stackrel{a}{\Rightarrow} q'$ *implies* $\exists p'.\ p \stackrel{a}{\Rightarrow} p' \wedge p'Rq'$

Note that $\approx^d$ coincides with the well-known observational equivalence $\approx$ (cf. [Mi80, Mi89]) if the first of the three defining requirements is dropped.

4.3 The Specification - Implementation Relation

The following preorder between processes is the basis of the framework in which we establish the correctness of our method:

Definition 4.2 $\preceq$ *is the union of all relations* R *satisfying that* pRq *implies for all* $a \in \mathcal{A}$*:*

1. $p \stackrel{a}{\Rightarrow} p'$ *implies* $\exists q'.\ q \stackrel{a}{\Rightarrow} q' \wedge p'Rq'$
2. $\neg p \Uparrow a$ *implies* $(\neg q \Uparrow a$ *and* $q \stackrel{a}{\Rightarrow} q'$ *implies* $\exists p'.\ p \stackrel{a}{\Rightarrow} p' \wedge p'Rq')$

$\preceq$ is a variant of the divergence preorder $\lessapprox$ (cf. [Wa88]) in which a-divergence does not require the potential of an a-move. Our modification serves for a different intend. We do not want to cover divergence, i.e. the potential of an infinite internal computation, but (guarded) undefinedness. This establishes $\preceq$ as a specification-implementation relation: a partial specification p is met by an implementation q iff $p \preceq q$, i.e. in contrast to [ClSte90,Wa88] we do not require an implementation to be able to pass these guards. This modification enhences the practicality of preorders as specification-implementation relations. In fact, similar definitions of preorders already appeared in [Ste89,Sti87], but have not been investigated as specification-implementation relations.

Observational equivalence $\approx$, divergence preorder $\lessapprox$, and our preorder $\preceq$ induce slightly different semantics on processes. However, it turns out that $\approx^d$ is a refinement of all of them:

Proposition 4.3 *If* $p \approx^d q$ *then* $p \approx q$ *and* $p \lessapprox q$ *and* $p \preceq q$*.*

Furthermore, on totally defined processes $\approx^d$ and $\preceq$ and $\approx$ all coincide. Finally we have the following monotonicity properties:

Proposition 4.4 *For all processes* p, p' *and* q*, and all sets of actions* L*, we have:.*

[2] A similar definition has been given in [ClSte90].

1. $p \preceq p'$ implies $p\|q \preceq p'\|q$
2. $p \approx^d p'$ implies $p\|q \approx^d p'\|q$
3. $p \preceq p'$ implies $p\langle L\rangle \preceq p'\langle L\rangle$
4. $p \approx^d p'$ implies $p\langle L\rangle \approx^d p'\langle L\rangle$

5 The Reduction Method

5.1 Interface Specifications

In this section we introduce our notion of *interface specification* together with a notion of *correctness*, which guarantees the success of our method. These notions concentrate on the set of sequences that may pass the interface. Thus the *exact* specification of the interface between p and q is the *language* of $(p\|q)\langle \mathcal{A}_p \cap \mathcal{A}_q\rangle$, i.e. its set of observable sequences. Denoting the language of a process p by $\mathcal{L}(p)$ we have:

Proposition 5.1 $\forall p, q.\ \mathcal{L}((p\|q)\langle \mathcal{A}_p \cap \mathcal{A}_q\rangle) = \mathcal{L}(p\langle \mathcal{A}_q\rangle) \cap \mathcal{L}(q\langle \mathcal{A}_p\rangle)$

We are going to use interface specifications in order to express context constraints. Thus interface specifications are correct or safe if the corresponding exact interface specification is more constraint. This motivates the following definition:

Definition 5.2 *Given two processes p and q we define:*

1. *A totally defined process I is an* interface specification *for p iff $\mathcal{A}_I \subseteq \mathcal{A}_p$, and it is an interface specification for p and q if it is an interface specification for both p and q.*
2. *An interface specification I for p and q is called* correct *for p and q iff $\mathcal{L}((p\|q)\langle \mathcal{A}_p \cap \mathcal{A}_q\rangle) \subseteq \mathcal{L}(I)$.*

The set of all interface specifications for p is denoted by $\mathcal{I}(p)$, and the set of all correct interface specifications for p and q is denoted by $\mathcal{I}(p,q)$.

Proposition 5.4 will show that these language-based definitions are adequate for our purpose.

5.2 The Reduction Operator

For a process p and an interface specification $\mathtt{I} \in \mathcal{I}(p)$ the reduction $\Pi_{\mathtt{I}}(p)$ of p with respect to $\mathtt{I}$ is essentially the projection of $p\|\mathtt{I}$ onto its first component; we define $\Pi_{\mathtt{I}}(p) = ((S, \rightarrow, \mathcal{A} \cup \{\tau\}, \uparrow), p)$, where:

- $S = \{q \in S_p \mid \exists i' \in S_{\mathtt{I}}.\ q\|i' \in S_{p\|\mathtt{I}}\}$
- $\mathcal{A}_p = \mathcal{A}$

- $\forall q, q' \in S\ \forall a \in \mathcal{A} \cup \{\tau\}.\ q \xrightarrow{a} q'$ iff $\exists i, i' \in S_{\mathtt{I}}.\ q \| i \xrightarrow{a}_{p\|\mathtt{I}} q' \| i'$
- $\forall q \in S\ q \uparrow \tau$ iff $q \uparrow \tau$ in the transition system of p
- $\forall q \in S\ \forall a \in \mathcal{A}.\ q \uparrow a$ iff one of the following conditions holds:
  - $q \uparrow a$ in the transition system of p
  - $\exists q' \in S_p.\ q \xrightarrow{a}_p q'$ and $\neg \exists q' \in S.\ q \xrightarrow{a} q'$

The only difference between $\Pi_{\mathtt{I}}(p)$ and the projection of $p \| \mathtt{I}$ onto p are the undefinedness predicates: $\Pi_{\mathtt{I}}(p)$ inherits all undefinedness predicates from p, and new ones are introduced where transitions of p have been "cut" away by $\mathtt{I}$. The point of the reduction operator is that for correct interface specifications this second kind of undefinedness disappears again in the full context. Remember, if an a-transition of p has been replaced by $\uparrow a$, this predicate disappears again in $\Pi_{\mathtt{I}}(p) \| q$ exactly if q in the corresponding state preempts the execution of an a-transition. This means that $\uparrow a$ are used as 'error' predicates, indicating in the full context where an interface specification is badly defined.

It is possible to show that $\Pi_{\mathtt{I}}(p)$ can be constructed in time proportional to the product of the number of transitions of p and $\mathtt{I}$. We have:

Proposition 5.3 $\forall p\ \forall \mathtt{I} \in \mathcal{I}(p).\ \Pi_{\mathtt{I}}(p) \preceq p$

Now we establish the promised independency of the reduction operator of the specific representation of the language specifying the interface.

Proposition 5.4

1. $\forall p\ \forall \mathtt{I}, \mathtt{I}' \in \mathcal{I}(p).\ \mathcal{L}(\mathtt{I}) = \mathcal{L}(\mathtt{I}')$ implies $\Pi_{\mathtt{I}}(p) = \Pi_{\mathtt{I}'}(p)$
2. $\forall p\ \forall \mathtt{I}, \mathtt{I}' \in \mathcal{I}(p).\ \mathcal{L}(\mathtt{I}) \subseteq \mathcal{L}(\mathtt{I}')$ implies $\Pi_{\mathtt{I}}(p) \preceq \Pi_{\mathtt{I}'}(p)$
3. $\forall p\ \forall \mathtt{I} \in \mathcal{I}(p).\ \mathcal{L}(p) \subseteq \mathcal{L}(\mathtt{I})$ implies $\Pi_{\mathtt{I}}(p) = p$

The correctness of our method is based on the following theorem:

Theorem 5.5 *Let p and q be processes and $\mathtt{I}$ an interface specification for p and q. Then $\mathtt{I} \in \mathcal{I}(p, q)$ implies that $p \| q$ is isomorphic to $\Pi_{\mathtt{I}}(p) \| q$, i.e. there exists a bijection $\iota : S_{p\|q} \to S_{\Pi_{\mathtt{I}}(p)\|q}$, such that:*

1. $\forall p', p'' \in S_{p\|q}\ \forall a \in \mathcal{A}_{p\|q} \cup \{\tau\}.\ p' \xrightarrow{a}_{p\|q} p''$ iff $\iota(p') \xrightarrow{a}_{\Pi_{\mathtt{I}}(p)\|q} \iota(p'')$
2. $\forall p' \in S_{p\|q}\ \forall a \in \mathcal{A}_{p\|q} \cup \{\tau\}.\ p' \uparrow a$ iff $\iota(p') \uparrow_{\Pi_{\mathtt{I}}(p)\|q} (a)$

This result illustrates the generality of the reduction operator. It can be used for the verification of all properties, which are preserved by isomorphism. In particular, we have the following corollary concerning the semantic equivalence we are focusing on here:

Corollary 5.6 $\forall q, p\ \forall \mathtt{I}.\ \mathtt{I} \in \mathcal{I}(p, q)$ implies $p \| q \approx^d \Pi_{\mathtt{I}}(p) \| q$

5.3 The Method

In this section, we show how the reduction operator can be used for the compositional minimization of processes of the form $\mathcal{P} = (p_1\| \cdots \|p_n)\langle L\rangle$. This form is of particular interest, because it is responsible for the state explosion problem and therefore characterizes the processes that are critical during analysis and verification[3]. Our method works by successive construction of minimal extended transition systems for components of $\mathcal{P}$. The point of this construction is that it exploits interface specifications expressing context constraints for the component under investigation. Thus, the *effect* of the method depends on the information, which has been be provided by the designer of the system. However, the *correctness* of the method does not depend on the correctness of these interface specifications. They are only used to "guide" the proof; and more precise interface specifications may allow more reductions. Thus, wrong interface specifications will never lead to wrong proofs. They may only prevent a successful verification of a valid statement.

The method expects $\mathcal{P}$ as to be annotated with interface specifications that describe the interface between the right hand process and the left hand process of the parallel operator they are attached to:

$$\mathcal{P} \;=\; (p_1\|_{\mathtt{I}_1} p_2\|_{\mathtt{I}_2} \cdots \|_{\mathtt{I}_{n-1}} p_n)\langle L\rangle$$

where $\mathtt{I}_i$ is an interface specification for $p_1\|\cdots\|p_i$ and $p_{i+1}\|\cdots\|p_n$. We proceed by successive construction of reductions P_i for the prefix processes of $\mathcal{P}$:

$$\underbrace{\underbrace{\underbrace{(p_1\|_{\mathtt{I}_1}}_{P_1} p_2}_{P_2}\|_{\mathtt{I}_2} \cdots \|_{\mathtt{I}_{n-1}} p_n)\langle L\rangle}_{P_n}$$

where the P_i are defined as follows:

- $P_1 = \mathcal{M}(\Pi_{\mathtt{I}_1}(\mathcal{M}(p_1\langle \mathcal{A}_{\mathtt{I}_1} \cup L\rangle)))$
- $P_i = \mathcal{M}(\Pi_{\mathtt{I}_i}(\mathcal{M}((P_{i-1}\|p_i)\langle \mathcal{A}_{\mathtt{I}_i} \cup L\rangle)))$ for $2 \le i \le n-1$
- $P_n = \mathcal{M}((P_{n-1}\|p_n)\langle L\rangle)$

and $\mathcal{M}$ is a function that minimizes extended transition systems up to $\approx^d$.

The goal of this method is to avoid unnecessarily large intermediate transition systems during the construction of the minimal transition system representing the semantic equivalence class of $\mathcal{P}$. Thus it is important to minimize all the intermediate constructions as it is done above. Our method does not depend on a particular semantic equivalence. Other equivalences can be dealt with just by changing the minimization function accordingly (cf. Theorem 5.5). Of course, in order to prove the correctness of the method, this also requires to adapt the preorder definition.

[3]Of course, the method can be applied in a structured way to each of the p_i. Thus it is possible to analyze complex structures by successive construction of increasingly large transition systems.

Independently of the correctness of the interface specification, we obtain just by means of Propositions 4.4 and 5.3:

Proposition 5.7 *Let* $1 \leq i \leq n$ *and* Q *denote the parallel composition of all processes with index greater than* i. *Then we have:* $(P_i \| Q)\langle L\rangle \preceq \mathcal{P}$,

This is already enough to guarantee the correctness of our method, i.e. that the success of a subsequent validation of a ($\approx^d$-consistent) property for P_n proves this property for $\mathcal{P}$. The correctness of the interface specifications comes into play in order to guarantee the success of the method. The following theorem is an immediate consequence of Corollary 5.6:

Theorem 5.8 *For* $1 \leq j \leq n$ *let* Q_j *denote the parallel composition of all processes with index greater than* j. *Then we have for* $1 \leq i \leq n$:

$$\forall j \leq i.\ \mathtt{I}_j \in \mathcal{I}(p_1\| \cdots \|p_j, Q_j) \text{ implies } (P_i\|Q_i)\langle L\rangle \approx^d \mathcal{P}$$

The situation for totally defined overall systems is particularly simple, as can be inferred from:

Corollary 5.9 *Whenever* P_n *is totally defined, we have:* $P_n \approx^d \mathcal{P}$.

6 An Application

Finally, we demonstrate our method by means of a setup that is intended to ensure round robin access of n processes P_i to a common resource R. Figure 1 illustrates this setup for $n = 4$. The idea is to pass a "token" via the communication channels tk_i in round robin manner and to allow access to R only for the process that currently possesses the token. This process then sends its request via ps_i to the resource R, which responds by transmitting the object requested. The corresponding transmission line is modelled by a buffer B_i. This is motivated by thinking of large objects whose transmission cannot be modelled just by an atomic "handshake" communication.

Let us now assume that we want to prove that the access is modelled as intended. For this purpose we can hide everything but the actions corresponding to the transmission of the token, and subsequently prove that the resulting process is equivalent to the process $Spec(n)$ that just iteratively executes the sequence $tk_1, \cdots, tk_n$, i.e. it is enough to show for

$$System(n) =_{def} (R\|P_1\|B_1\| \cdots \|P_n\|B_n)\langle\{tk_1, .., tk_n\}\rangle$$

that $System(n) \approx^d Spec(n)$.

It is easy to see that the apparent complexity of $System(n)$ is exponential in n, whereas its real complexity is linear. In fact, it is also possible to obtain an algorithmic complexity that is linear in n. This can be achieved by processing the system according to the structure indicated below, where the $\mathtt{I}_i$ denote the exact interface specifications:

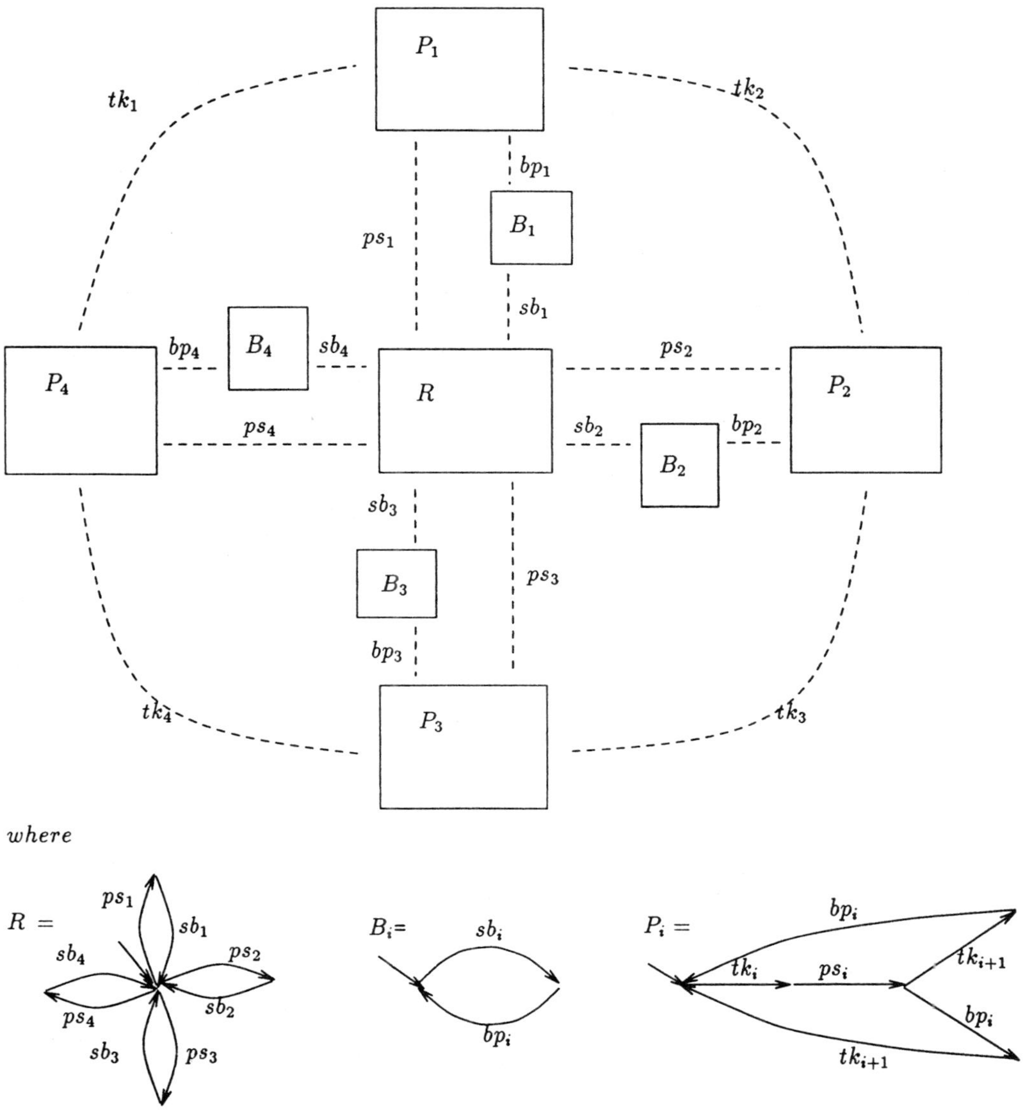

Figure 1: Round Robin Access

$$(\overbrace{R \,\|\, P_1 \| B_1} \,\|_{I_1}\, \overbrace{P_2 \| B_2} \,\|_{I_2} \cdots \|_{I_{n-1}}\, \overbrace{P_n \| B_n})\langle\{tk_1, .., tk_n\}\rangle$$

The table below summarizes a numerical investigation of the efficiency of our method by means of the Aldebaran verification tool [Fer88]. It displays the size of the global state graph (its apparent complexity), the size of the maximal transition system constructed during stepwise minimization when exploiting the exact interface specifications (the algorithmic complexity), and finally the size of the minimized global state graph (its real complexity).

n	apparent complexity		algorithmic complexity		real complexity	
	states	trans.	states	trans.	states	trans.
4	144	368	20	29	4	4
5	361	1101	24	35	5	5
6	865	3073	28	41	6	6
7	2017	8177	32	47	7	7

It is worth mentioning that the method which works by stepwise composition and minimization of components encounters transition systems that are even larger than the global state graph:

n	states	trans.
4	96	243
5	324	927
6	972	3024
7	2916	9801

This stresses the importance of interface specifications for automatic proof techniques. It is our opinion that a software designer should always provide these specifications as part of the implementation. We believe that besides enabling automatic verification, this requirement also leads to a transparent and well structured programming[4].

7 Conclusions and Future Work

We have presented a method for the compositional minimization of finite state systems, which is intended to avoid the state explosion problem. This method can be used to support the verification of any property that is consistent with $\approx^d$. Even better, the choice of $\approx^d$ only affects the minimization step. Thus our method can be adapted to other semantic equivalences simply by modifying the minimization operator $\mathcal{M}$.

In this paper, we proposed a left to right strategy for the minimization of a highly parallel process. Of course, the correctness of our method does not depend on this particular choice.

[4]This reminds for the situation for *while*-programs, where automatic verification depends on *loop invariants* that also need to be provided by the programmer.

The *effect* of our method depends on interface specifications, which we assume as to be given by the program designer. However, the *correctness* of the method does not depend on the correctness of these interface specifications. Wrong interface specifications will never lead to wrong proofs. They may only prevent a successful verification of a valid property. This is very important, because it allows the designer to simply "guess" interface specifications, while maintaining the reliability of a successful verification.

Another way to obtain interface specifications is by exploiting the property we are going to verify. This is what Clarke et al. [CLM89] had in mind. However, their approach only exploits the alphabet of the property under consideration. A refined treatment of property constraints using our notion of interface specification is under investigation.

Finally, it should be mentioned that our method can easily be implemented. In fact, an implementation in the Edinburgh Concurrency Workbench [CPS89] and in the Aldebaran verification tool [Fer88] is planned.

Acknowledgements

We would like to thank Rance Cleaveland and Ernst–Rüdiger Olderog for helpful discussions.

References

[BKP84] H. Barringer, R. Kuiper, A. Pnueli. *Now you may compose Temporal Logic Specifications*, Symposium on Theory of Computing, 1984

[BFH90] A. Bouajjani, J.-C. Fernandez, N. Halbwachs. *Minimal Model Generation*, this volume

[BCG86] M.C.Browne, E.M. Clarke, O. Grumberg. *Reasoning about networks with many identical finite state processes*, in ACM Symposium on Principle of Distributed Computing, 1986

[CES83] E.M. Clarke, E.A. Emerson, E. Sistla. *Automatic Verification of Finite State Concurrent Systems using Temporal Logic Specification: A Practical Approach*, POPL 1983

[CLM89] E.M. Clarke, D.E. Long, K.L. McMillan. *Compositional Model Checker*, LICS, 1989

[CPS89] R. Cleaveland, J. G. Parrow and B. Steffen. *The Concurrency Workbench*, Proceeding of the Workshop on Automatic Verification Methods for Finite State Systems, Grenoble, France, 1989, LNCS 407

[CPSb89] R. Cleaveland, J. G. Parrow and B. Steffen. *A Semantics based Verification Tool for Finite State Systems*, in the proceedings of the Ninth International Symposium on Protocol Specification, Testing, and Verification; North Holland, 1989

[ClSte90] R. Cleaveland, and B. Steffen. *When is "Partial" Complete? A Logic-Based Proof Technique using Partial Specifications*, in Proceedings of LICS'90, 1990

[Fer88] Fernandez, J.-C. *Aldébaran: Un Système de Vérification par Réduction de Processus Communicants*, Ph.D. Thesis, Université de Grenoble, 1988

[FSS83] J.C. Fernandez, J.Ph. Schwartz, J.Sifakis. *An Example of Specification and Verification in Cesar*, Proceedings of 'The Analysis of Concurrent Systems', 1983, LNCS 207

[CHPP87] Caspi P., Halbwachs N., Pilaud N., Plaice J. *LUSTRE, a declarative language for programming synchronous systems*, Proceedings of 14th POPL, Munich, 1987

[Jos87] B. Josko. *MCTL - An extension of CTL for modular verification of concurrent systems*, Workshop on Temporal Logic in Specification 1987, LNCS 398

[KuMcM89] R.P. Kurshan, K. McMillan. *A Structural Induction Theorem for Processes*, in ACM Symposium on Principles of Distributes Computing, 1989

[Kr89] H.Krumm. *Projections of the Reachability Graph and Environment Models, two approaches to facilitate the functional analysis of Systems of cooperating finite state machines*, Proceedings of the Workshop on Automatic Verification of Finite State Systems, Grenoble 89, LNCS 407.

[LaMi87] Larsen, K.G., and R. Milner. *Verifying a Protocol Using Relativized Bisimulation*, in *Proceedings of ICALP 87*, Lecture Notes in Computer Science 267. Springer-Verlag, Berlin.

[LaTh88] Larsen, K.G., and B. Thomsen. *Compositional Proofs by Partial Specification of Processes*, in Proceedings LICS'88, 1988

[LaXi90] K.G. Larsen, L. Xinxin. *Compositionality through an Operational Semantics of Contexts*, in Proceedings ICALP'90, LNCS, 1990

[Mi80] R. Milner. *A Calculus for Communicating Systems*, LNCS 92, 1980

[Mi89] R. Milner. *Communication and Concurrency*, Prentice Hall, 1989

[Old90] E.-R. Olderog. *Nets, Terms and Formulas: Three Views of Concurrent Processes*, Habilitationsschrift, Universität Kiel, to appear in Tracts in Theoretical Computer Science, Cambridge University Press

[Pnu85] A.Pnueli. *In Transition from Global to Modular Temporal Reasoning about Programs*, in Logics and Models for Concurrent Systems, Nato ASI Series F, Vol. 13, Springer Verlag

[ShGr90] G. Shurek, O. Grumberg. *The Modular Framework of Computer-aided Verification: Motivation, Solutions And Evaluation Criteria*, this volume

[StGr89] Z. Stadler, O. Grumberg. *Network Grammars, Communication Behaviours and Automatic Verification*, in Proceeding of the Workshop on Automatic Verification Methods for Finite State Systems, Grenoble, France, 1989, LNCS 407

[Ste89] B. Steffen. *Characteristic Formulae*, in Proceedings ICALP 1989

[Sti87] C. Stirling. *Modal Logics for Communicating Systems*, TCS 49, pp. 311-347, 1987

[StiWa89] C. Stirling and D. J. Walker. *Local Model Checking in the Modal Mu-Calculus*, in Proceedings CAAP 1989

[Wa88] D.J. Walker. *Bisimulation and Divergence in CCS*, in Proceedings LICS 1988

[Win90] G. Winskel. *Compositional Checking of Validity on Finite State Processes*, Workshop on Theories of Communication, CONCUR, 1990

[WoLo89] P. Wolper, V. Lovinfosse. *Verifying Properties of Large sets of Processes with Network Invariants*, in Proceeding of the Workshop on Automatic Verification Methods for Finite State Systems, Grenoble, France, 1989, LNCS 407

DIMACS Series in Discrete Mathematics
and Theoretical Computer Science
Volume 3, 1991

Verifying Temporal Properties of Sequential Machines Without Building their State Diagrams

Olivier Coudert Jean Christophe Madre Christian Berthet

Bull Research Center, PC 62 A13,
68, Route de Versailles
78430 Louveciennes FRANCE

Abstract

This paper presents the algorithm we have developed for proving that a finite state machine holds some properties expressed in temporal logic. This algorithm does not require the building of the state-transition graph nor the transition relation of the machine, so it overcomes the limits of the methods that have been proposed in the past. The verification algorithm presented here is based on Boolean function manipulations, which are represented by typed decision graphs. Thanks to this canonical representation, all the operations used in the algorithm have a polynomial complexity, expect for one called the computation of the "critical term". The paper proposes techniques that reduce the computational cost of this operation.

1 Introduction

Within the community of people working on the verification of sequential machines, the word "verification" has received two quite different meanings. For some people, to verify a machine is to prove that it holds some properties, such as "liveness" or "safety" properties [3] [4] [6] [7] [13]. For the others, to verify a machine is to prove that it is correct with respect to its behavioral specification, which is a program written in some high level hardware description language. This problem comes down to proving the behavioral equivalence between two machines [12]. Though both kinds of verification are needed to design complex circuits, these two verification problems have until recently been studied independently and the techniques that have been proposed to deal with them had few relations one with another.

The behavioral equivalence between two machines $\mathcal{M}_1$ and $\mathcal{M}_2$ can be proved by traversing the state-transition graph of the product machine $\mathcal{M}_1 \times \mathcal{M}_2$ [12]. The first techniques that have been proposed were based on a double enumeration of the states and the input patterns of the machines [11], so that in many cases the time needed to traverse the state-transition graph of the machine $\mathcal{M}_1 \times \mathcal{M}_2$ grows exponentially with the number of inputs of $\mathcal{M}_1$ and $\mathcal{M}_2$.

More recently we have presented [8] [9] a proof procedure that manipulates sets of states and inputs. The basic concepts that underlie this proof procedure are (1) to represent sets either by their *characteristic functions* or by *vectors of Boolean functions* [9] and (2) that the operations on these Boolean functions can be efficiently performed by denoting them with typed decision graphs [1]. This procedure can handle machines that could not be treated with the enumeration based methods referenced above.

It has been shown [3] [6] that these basic concepts can also be used to develop a procedure for proving that a sequential machine holds some temporal properties without building its state-transition graph. Such procedures could overcome the limits of the methods that have been developed in the past, which all required the partial or total building of this graph. However the procedure described in [3] [6] requires the construction of the *transition relation* of the sequential machine, which is not feasible for most of complex machines.

1052-1798/91 $1.00 + $.25 per page

This paper presents the proof procedure that we have independently developed to check that a machine holds a temporal property. This technique, which does not use the transition relation of the machine, is based on specific operators called "restrict" and "expand". The paper is divided into 4 parts. Part 2 defines the model of machines and the temporal formulas that the proof system handles. Part 3 gives the proof algorithm used by the system and shows that it uses the standard Boolean operations in addition with the computation of a specific term. Part 4 is dedicated to the computation of this term that we call the "critical term" because its evaluation is the bottleneck of the algorithm. Part 5 gives some experimental results and discusses them.

2 Definitions

This section describes the inputs of the verification system presented in this paper. First it defines the model of sequential machines that the system handles. Then it gives the syntax and the semantics of the temporal formulas to be verified.

2.1 Model of Sequential Machines

The sequential machine $\mathcal{M}$ that must be verified is an uncompletely specified deterministic Moore Machine. This machine is defined by the 6-tuple $(n, m, r, \omega, \delta, \text{Init})$, where: the state space of the machine $\mathcal{M}$ is $\{0, 1\}^n$, its input space is $\{0, 1\}^m$, and its output space is $\{0, 1\}^r$; ω is the partial output function from $\{0, 1\}^n \times \{0, 1\}^m$ into $\{0, 1\}_\perp{}^r$; δ is the partial transition function from $\{0, 1\}^n \times \{0, 1\}^m$ into $\{0, 1\}^n{}_\perp$; Init is the characteristic function of the set of initial states of $\mathcal{M}$.

In this description, all the Boolean functions are denoted by typed decision graphs (TDG), which are a very compact graphical canonical form [1]. These Boolean functions are automatically computed from the behavioral description of the machine using symbolic execution [2]. The transition function δ is denoted by a couple (Cns, F). Cns denotes the domain where the transition function is defined, so it is a Boolean function from $\{0, 1\}^n \times \{0, 1\}^m$ into $\{0, 1\}$. It is built out of constraints given by the designer and additional constraints computed during the symbolic execution [2]. F is a vectorial function $[f_1 \ldots f_n]$, where each f_j is a Boolean function from $\{0, 1\}^n \times \{0, 1\}^m$ into $\{0, 1\}$. The partial function δ is then defined by:

$$\delta = \lambda s.\lambda p.(\textbf{if } \text{Cns}(s, p) \textbf{ then } F(s, p) \textbf{ else } \perp)$$

For the sake of simplicity, we assume in the following that ω is completely specified. We also assume that $(\forall s\ \exists p\ \text{Cns}(s, p))$ is a tautology, which means that any state has at least one successor. This is not a restriction, because temporal formula have no meaning on states that have no successor. If $(\forall s\ \exists p\ \text{Cns}(s, p))$ is not a tautology, then $\lambda s.(\forall p\ \neg\ \text{Cns}(s, p))$ is the characteristic function of the set of states that have no successor.

2.2 State Formulas in Computation Tree Logic

The temporal formulas handled by the verification system are the *state formulas* of the computation tree logic CTL [7]. This logic is a formalism specially developed to express properties about the states and the computation paths of finite state systems. The different kinds of state formulas and their meanings with respect to a state of the machine $\mathcal{M} = (m, n, r, \omega, \delta, \text{Init})$ are the following:

1. $r_1, r_2, \ldots, r_n, o_1, o_2, \ldots, o_r$ are state formulas. For any state s of the machine, $s \models r_j$ if and only if the j-th component of s is 1. For any state s of the machine, $s \models o_j$ if and only if the j-th component of $\omega(s)$ is 1.
2. If f and g are state formulas then so are $(\neg f)$, $(f \wedge g)$, $(f \vee g)$, $(f \Leftrightarrow g)$, and $(f \Rightarrow g)$. The logical connectors have their usual meanings, for instance $s \models (f \wedge g)$ if and only if $s \models f$ and $s \models g$.
3. If f is a state formula, so are the formulas $EX(f)$ and $AX(f)$. For any state s of $\mathcal{M}$, $s \models EX(f)$ iff there exists at least one input pattern p such that $\delta(s, p) \models f$; by definition, $AX(f)$ is $\neg\ (EX(\neg f))$.

4. If f and g are state formulas, so are the formulas $E[f\ U\ g]$ and $A[f\ U\ g]$. For any state s of the machine $\mathcal{M}$, s $\models E[f\ U\ g]$ if and only if there exists at least one path $(s_0, s_1, ...)$ with $s_0 = s$, and: $\exists i\ ((s_i \models g) \wedge (\forall j\ (0 \leq j < i \Rightarrow s_j \models f)))$.
   For any state s of the machine $\mathcal{M}$, s $\models A[f\ U\ g]$ if and only if for all paths $(s_0, s_1, ...)$ such that $s_0 = s$, we have: $\exists i\ ((s_i \models g) \wedge (\forall j\ (0 \leq j < i \Rightarrow s_j \models f)))$.

Abbreviations have been added to make the CTL formulas more legible: $EF(f) =_{def} E[True\ U f]$; $AF(f) =_{def} A[True\ U f]$; $EG(f) =_{def} \neg (AF(\neg f))$; $AG(f) =_{def} \neg (EF(\neg f))$, i. e. f holds on any state of every path. The reader can refer to [7] to find examples of properties of finite state systems expressed in CTL.

3 Verification Algorithm

It has been shown that "safety properties" of finite state systems [4] can be checked with the forward symbolic traversal procedure presented in [9]. The machine $\mathcal{M}$ holds a safety property f if and only if some sub-part $\mathcal{M}_f$ of the machine $\mathcal{M}$ has an observable behavior that is equivalent to the one of the most general model of the formula f, which is a finite state automaton $\mathcal{A}_f$. This automaton can be automatically built from the formula f. In this case the proof is made by traversing the state diagram of the machine $\mathcal{M}_f \times \mathcal{A}_f$ without building it [10]. For instance, to prove that the machine $\mathcal{M}$ satisfies the formula $f = AG\ (r_i \Rightarrow AX\ (AG\ (o_j)))$ comes down to proving that the observable behavior of $\mathcal{M}$ at its j-th output is equivalent to the behavior of the automaton $\mathcal{A}_f$ given in Figure 1.

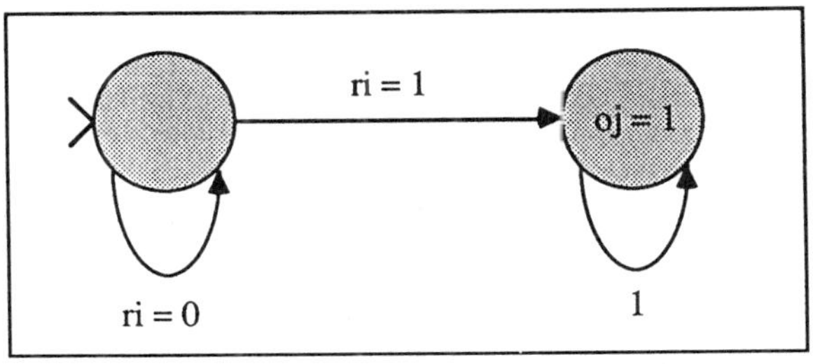

Figure 1. The automaton $\mathcal{A}_f$ associated with the formula $f = AG\ (r_i \Rightarrow AX\ (AG\ (o_j)))$.

This part presents a more general proof algorithm that can handle all the state formulas described in the previous section. This algorithm does not require the building of the state-transition graph of the machine $\mathcal{M}$ under verification. The algorithm takes as inputs a machine $\mathcal{M}$ described by its 6-tuple $(m, n, r, \omega, (\text{Cns}, \text{F}), \text{Init})$ and the formula f to be verified. It recursively computes the set of states $\mathcal{F}$ that satisfy the formula f from the sets of states that satisfy the sub-formulas of f. At each step there are only four basic cases to consider that correspond to the four kinds of formulas given in Section 2.2. Once the set of states $\mathcal{F}$ that satisfy the whole formula f is obtained, to check whether s $\models f$ for some state s of $\mathcal{M}$ comes down to checking whether s belongs to $\mathcal{F}$.

The verification algorithm manipulates sets represented by their characteristic functions, and functions are denoted by their TDG's. We will make no distinction between a function and its TDG, or between a set and the TDG of its characteristic function. The quantifiers that will be used in the sequel are handled using the following identities: if x is a propositional variable, $(\exists x\ f(x))$ is equivalent to $(f(0) \vee f(1))$, and $(\forall x\ f(x))$ is equivalent to $(f(0) \wedge f(1))$. For any n-tuple $x = [x_1 \dots x_n]$ of propositional variables, $(Qx\ f(x)) =_{def} (Qx_1 \dots Qx_n\ f(x_1, ..., x_n))$, where Q is either the existential or universal quantifier. The size of a TDG f is its number of vertices, and it will be noted $|\ f\ |$. The size of the TDG of a function is relative to a total ordering of its variables [1] [5].

The formulas of type (1) and (2) are trivial to treat. For instance, if $f = (f_1 \wedge f_2)$, and $\mathcal{F}_1$ and $\mathcal{F}_2$ are the characteristic functions of the sets of states that satisfy f_1 and f_2 respectively, then $\mathcal{F}$ is $\lambda s.(\mathcal{F}_1(s) \wedge \mathcal{F}_2(s))$.

The cost of computing the TDG of $\mathcal{F}$ is in $O(|\mathcal{F}_1| \times |\mathcal{F}_2|)$ [1] [5]. As soon as $\mathcal{F}$ is computed, it is trivial to determine whether $s \models f$, since this is equivalent to compute $\mathcal{F}(s)$, which is done in $O(n)$, and then to test whether $\mathcal{F}(s) = 1$. In the same way, $M \models f$ if and only if $\lambda s.(\text{Init}(s) \Rightarrow \mathcal{F}(s)) = 1$, which can be tested in $O(|\text{Init}| \times |\mathcal{F}|)$.

3.1 *EX* and *AX* Formulas

The sets of states that satisfy the formulas of types *EX* and *AX* are computed in a single step. Let f be a formula and $\mathcal{F}$ be the set of states that satisfy f. The set of states $\mathcal{EX}$ of the machine $\mathcal{M}$ that satisfy the temporal formula *EX(f)* is $\{s \,/\, \exists p\ \ \delta(s, p) \in \mathcal{F}\}$. Its characteristic function is:

$$\mathcal{EX} = \lambda s.(\exists p\ \ \text{Cns}(s, p) \wedge \mathcal{F}(F(s, p))). \tag{1}$$

By definition, $AX(f) = \neg EX(\neg f)$, so the characteristic function $\mathcal{AX}$ of the set of states of the machine $\mathcal{M}$ that satisfy *AX(f)* is:

$$\mathcal{AX} = \lambda s.(\neg (\exists p\ \ \text{Cns}(s, p) \wedge \neg \mathcal{F}(F(s, p)))). \tag{2}$$

3.2 *EU* and *AU* Formulas

The sets of states that satisfy the formulas of types *EU* and *AU* are computed using iterating algorithms [3]. Let f and g be two formulas and $\mathcal{F}$ and $\mathcal{G}$ be the sets of states that satisfy f and g respectively. The set $\mathcal{EU}$ of states that satisfy the formula *E[f U g]* is the limit of the converging sequence $(\mathcal{E}_j)$ of sets defined by: $\mathcal{E}_0 = \mathcal{G}$, and $\mathcal{E}_{k+1} = \mathcal{E}_k \cup \{s \,/\, (s \in \mathcal{F}) \wedge (\exists p\ \ \delta(s, p) \in \mathcal{E}_k)\}$. The characteristic functions of these sets are the following:

$$\begin{aligned} \mathcal{E}_0 &= \mathcal{G}, \text{ and} \\ \mathcal{E}_{k+1} &= \lambda s.(\mathcal{E}_k(s) \vee (\mathcal{F}(s) \wedge (\exists p\ \ \text{Cns}(s, p) \wedge \mathcal{E}_k(F(s, p))))). \end{aligned} \tag{3}$$

In the same way the set $\mathcal{AU}$ of states that satisfy the formula *A[f U g]* is the limit of the converging sequence $(\mathcal{A}_j)$ defined by: $\mathcal{A}_0 = \mathcal{G}$, and $\mathcal{A}_{k+1} = \mathcal{A}_k \cup \{s \,/\, (s \in \mathcal{F}) \wedge (\forall p\ \ (\delta(s, p) \neq \bot) \Rightarrow (\delta(s, p) \in \mathcal{A}_k))\}$. The characteristic functions of these sets are the following:

$$\begin{aligned} \mathcal{A}_0 &= \mathcal{G}, \text{ and} \\ \mathcal{A}_{k+1} &= \lambda s.(\mathcal{A}_k(s) \vee (\mathcal{F}(s) \wedge (\forall p\ \ \text{Cns}(s, p) \Rightarrow \mathcal{A}_k(F(s, p))))). \end{aligned} \tag{4}$$

3.3 The Critical Term

In the formulas (1) to (4) given in the preceding sections two terms appear that actually are the same one, and that has the form $(\exists p\ \text{Cns}(s, p) \wedge \chi(F(s, p)))$. Since the formula $(\forall x\ f)$ is equivalent to $(\neg (\exists x\ \ \neg f))$, the equation (4) can be rewritten into:

$$\mathcal{A}_{k+1} = \lambda s.(\mathcal{A}_k(s) \vee (\mathcal{F}(s) \wedge \neg (\exists p\ \ \text{Cns}(s, p) \wedge \neg \mathcal{A}_k(F(s, p))))). \tag{4'}$$

From the computational point of view this means that the four basic cases of CTL formulas can be treated with the standard Boolean operations (negation, conjunction, and disjunction), the test of equivalence (to test whether the sequences $\mathcal{E}_k$ and $\mathcal{A}_k$ have converged), in addition with the evaluation of the *critical term* $(\exists p\ \text{Cns}(s, p) \wedge \chi(F(s, p)))$, where χ is a TDG denoting a Boolean function from $\{0, 1\}^n$ into $\{0, 1\}$. Next section explains why we call this term the "critical term", and discusses how it can be computed.

4 Computing the Critical Term

This section shows that computing the critical term is the bottleneck in the verification algorithm. Then it explains why the critical term can be easily computed when the TDG of the transition relation of the machine can be build [3] [6]. At last, it presents the techniques we have developed to perform this computation when it is not possible to build this TDG, which happens for most of complex circuits.

4.1 Is the Computation of the Critical Term a Difficult Problem?

Since typed decision graphs are canonical, testing the equivalence of two TDG's is in O(1). All the standard Boolean operations on TDG's are performed in polynomial time [1] [5]. This means that the computational cost of the formal verification of CTL formulas depends on the evaluation of the critical term.

The complexity of the critical term computation can be studied using the composition problem. This problem is to compute the TDG of $\lambda x.g(f_1(x), ..., f_n(x))$, from the TDG's of the Boolean functions $f_1, ..., f_n$, and g. The three following theorems show that computing the critical term is the most difficult operation that appears in the algorithm presented in Section 3.

Theorem 1. *Composition is O(1)-reducible to the computation of the critical term.*

Proof. It is immediate that composition is reducible in O(1) to the evaluation of the critical term $(\exists p\ Cns(s, p) \wedge \chi(F(s, p)))$, where $Cns = 1$, $\chi = g$, and $\lambda s.\lambda p.F = [\lambda s.\lambda p.f_1(s)\ ...\ \lambda s.\lambda p.f_n(s)]$. ❑

Theorem 2. *Composition is NP-hard.*

Proof. Let $C = \bigwedge_j c_j$ be a 3-conjunctive normal form made of n clauses c_j. Each clause c_j is a disjunction $(q_{1j} \vee q_{2j} \vee q_{3j})$ of three literals, so the TDG's of these clauses are built in O(n). Let g be the function $\lambda[y_1 ... y_n].(\bigwedge_j y_j)$, whose TDG is a n-vine built in O(n). Since C is satisfiable if and only if the TDG of $g(c_1, ..., c_n)$ is not equal to 0, which can be tested in O(1), composition is NP-hard. ❑

The two preceding theorems show that the computation of the critical term is NP-hard. By using the hypothesis that the order of the variables cannot be modified, we obtain the more precise following theorem.

Theorem 3. *Composition is a no polynomial problem if the ordering of the variables is fixed. More precisely, there exist some TDG's $f_1, ..., f_n$ and g, using O(n) variables, with $\Sigma_j |f_j| + |g| = O(n)$, and such that $|g(f_1, ..., f_n)| = O(2^n)$.*

Proof. We consider 2n variables $y_1 < ... < y_{2n}$, and the function $g = \lambda[x_1 ... x_n].(\bigwedge_j x_j)$. Its TDG is a n-vine of size n. We define n functions $f_j = \lambda[y_1 ... y_{2n}].(y_j \Leftrightarrow y_{2n-j+1})$. The TDG's of the functions f_j have all two vertices, so $\Sigma_j |f_j| = 2n$. Thus $|g| + \Sigma_j |f_j| = O(n)$. The composition $g(f_1, ..., f_n)$ is the function $\lambda y.(\bigwedge_j f_j(y))$, that is $\lambda[y_1 ... y_{2n}].(\bigwedge_j (y_j \Leftrightarrow y_{2n-j+1}))$, but the TDG of the term $((y_1 \Leftrightarrow y_{2n}) \wedge (y_2 \Leftrightarrow y_{2n-1}) \wedge ... \wedge (y_n \Leftrightarrow y_{n+1}))$ with the order $y_1 < ... < y_{2n}$ has a size in $O(2^n)$. ❑

This last theorem shows that computing the critical term on TDG's is at least exponential in the worst case. However, one can object that the TDG of the term $((y_1 \Leftrightarrow y_{2n}) \wedge (y_2 \Leftrightarrow y_{2n-1}) \wedge ... \wedge (y_n \Leftrightarrow y_{n+1}))$ is in O(n) with the variable ordering $y_1 < y_{2n} < y_2 < y_{2n-1} < ... < y_n < y_{n+1}$. We may think that composition can be computed more efficiently if we are able to find a variable ordering that minimizes the sizes of the manipulated TDG's. The following remarks show that the problem is not such simple. First, finding such a variable ordering is itself a NP-hard problem. Second, there exist some functions of n variables whose TDG has a no polynomial size with respect to n, whatever the variable ordering. The last remark shows that the composition, and so the computation of the critical term, is *certainly* no polynomial even if an oracle would provide dynamically a "good order".

4.2 Using the Transition Relation to Compute the Critical Term

The transition relation Δ of the machine $\mathcal{M}$ is a subset of $\{0, 1\}^n \times \{0, 1\}^n$. For any states s and s', (s, s') belongs to Δ if and only if $(\exists p\ (s' = \delta(s, p)))$. This means that $\Delta =_{def} \lambda s.\lambda s'(\exists p\ Cns(s, p) \wedge (s' \equiv F(s, p)))$. Using the transition relation Δ, the critical term $(\exists p\ Cns(s, p) \wedge \chi(F(s, p)))$ can be rewritten into:

$$(\exists s'\ \Delta(s, s') \wedge \chi(s')).$$

Given the TDG's of Δ and χ, the TDG of $(\Delta(s, s') \wedge \chi(s'))$ can be built in $O(|\Delta| \times |\chi|)$, and the critical term is obtained by eliminating from this TDG the n atoms associated with s'. Note that this process is exponential: we can prove that there exist some functions f such that $|\exists x\, f| = O(|f|^{\wedge}(n2^n / 2n+1))$, where $n = |x|$; in particular, for these functions, we have $|\exists x\, f| > 2^{\sqrt{|f|}}$. But experience shows it remains polynomial on practical examples.

Assuming that the TDG of Δ can be built, this technique is very efficient [6], because it uses only the standard logical operators, which have relatively low computational costs. But building Δ requires the computation of the TDG of $(Cns(s, p) \wedge (\wedge_j\, (s_j' \equiv f_j(s, p))))$, which is done in $O(|Cns| \times 2^n \times \Pi_j |f_j|)$. Experience shows that for complex machines, it is not possible to build the TDG of the transition relation Δ, so others techniques are needed.

4.3 Using the "Restrict" and the "Expand" Operators to Compute the Critical term

The term $(\exists p\, Cns(s, p) \wedge \chi(F(s, p)))$ can be computed in two steps. The first step consists in building the TDG of the formula $(Cns(s, p) \wedge \chi(F(s, p)))$, and the second step in eliminating from this TDG the atoms associated with the input pattern p. The term $\chi(F(s, p))$ is the composition of χ with the vectorial function $F = [f_1 \dots f_n]$, whose computation is exponential in the worst case, as shown in Section 4.1. More precisely, it can be easily shown, using a proof similar to the one of Theorem 3, that the computation of the TDG $\chi(f_1, \dots, f_n)$ from the TDG's of $f_1, \dots, f_n$ and χ is at least in $O(|\chi| \times \Pi_j |f_j|)$.

Section 4.3.1 shows that it is not necessary to build explicitly the TDG of $\chi \circ F$ to compute the term $(\exists p\, Cns(s, p) \wedge \chi(F(s, p)))$. It presents the "expand" operation that avoids this construction. Section 4.3.2 presents the "restrict" operator that further reduces the computational cost of the composition by reducing the sizes of the TDG's f_j used in the term $\chi(F(s, p))$.

4.3.1 The "Expand" Operation

The idea that underlies the "expand" operation is to express the term $\chi(F(s, p))$ as a sum of functions whose TDG's have less vertices than the TDG of $\chi(F(s, p))$. Using these functions noted $h_1, \dots, h_k$, the critical term can be rewritten into $(\exists p\, Cns(s, p) \wedge (\vee_j\, h_j(s, p)))$, that can be further transformed into $(\exists p \vee_j (Cns(s, p) \wedge h_j(s, p)))$ by distributing the conjunction over the disjunction. The existential quantifier commutes with the disjunction, so that the critical term finally becomes:

$$(\vee_j\, (\exists p\, Cns(s, p) \wedge h_j(s, p))).$$

The final form of this term expresses that its computation can be decomposed into a sequence of computation of simpler terms.

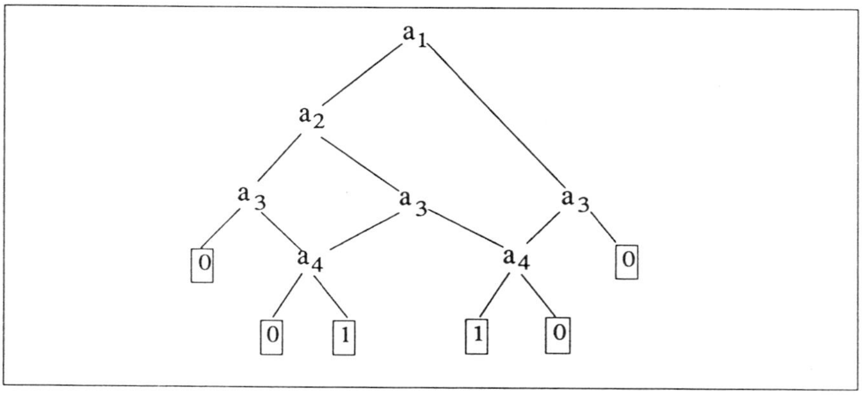

Figure 2. Application of the Expand Function on a BDD.

Consider the functions $f_1, ..., f_n$, and g. The computation of a list of Boolean functions $h_1, h_2, ..., h_k$ whose sum is equal to the function $g(f_1, ..., f_n)$ is made by the "expand" operation. For clarity's sake, we present here the "expand" operation on the binary decision diagrams (BDD) [5].

Each path in the BDD of the function g starting from the root and leading to the leaf 1 defines a *cube* c_j of the function g, and the function g is equal to the sum of its cubes. This means that we could take as functions h_j the functions $c_j(f_1, ..., f_n)$. However this would lead to perform many redundant computations. Consider for instance the BDD drawn in Figure 2. In this BDD there are four paths from the root to 1 that define the four cubes $c_1 = (\neg a_1 \wedge \neg a_2 \wedge a_3 \wedge a_4)$, $c_2 = (\neg a_1 \wedge a_2 \wedge \neg a_3 \wedge a_4)$, $c_3 = (\neg a_1 \wedge a_2 \wedge a_3 \wedge \neg a_4)$, and $c_4 = (a_1 \wedge \neg a_3 \wedge a_4)$. The four functions that would be computed are $h_1 = (\neg f_1 \wedge \neg f_2 \wedge f_3 \wedge f_4)$, $h_2 = (\neg f_1 \wedge f_2 \wedge \neg f_3 \wedge f_4)$, $h_3 = (\neg f_1 \wedge f_2 \wedge f_3 \wedge \neg f_4)$, $h_4 = (f_1 \wedge \neg f_3 \wedge f_4)$.

It is obvious that during the computation of the functions h_1, h_2, h_3, and h_4, the product $(\neg f_1 \wedge f_2)$ is computed twice, as well as the product $(\neg f_3 \wedge f_4)$. Some of these redundant computations can be eliminated by storing partial results in the vertices of the graph that are shared. This is done by the function "expand" that performs a top-down traversal of the BDD of g, and stores in each vertex v of g, the BDD of the function $C_v(f_1, ..., f_n)$, where C_v is the sum of all the cubes represented by the paths starting from the root of the BDD of g and leading to v. Each time the top-down traversal reaches a leaf equal to 1, the function "expand" produces one of the functions h_j.

For the BDD drawn in Figure 2, the partial functions stored in the vertices are: (1) for the vertex (1), $(\neg f_1)$ for (2), $(\neg f_1 \wedge \neg f_2)$ for (3), $(\neg f_1 \wedge f_2)$ for (4), $((\neg f_1 \wedge \neg f_2 \wedge f_3) \vee (\neg f_1 \wedge f_2 \wedge \neg f_3))$ for (5), (f_1) for (6), and $((\neg f_1 \wedge f_2 \wedge f_3) \vee (f_1 \wedge \neg f_3))$ for (7), and the "expand" operation produces two functions:

$$h_1 = (f_4 \wedge ((\neg f_1 \wedge \neg f_2 \wedge f_3) \vee (\neg f_1 \wedge f_2 \wedge \neg f_3))), \text{ and}$$

$$h_2 = (\neg f_4 \wedge ((\neg f_1 \wedge f_2 \wedge f_3) \vee (f_1 \wedge \neg f_3))).$$

Experience shows that the TDG's of the functions $h_1, ..., h_k$ generated by the "expand" operation are smaller than the TDG of the function $\chi(F(s, p))$. The time needed to compute each of these functions directly depends on the sizes of the TDG's of the functions $\delta_j(s, p)$. Next section presents a new Boolean operator that can be used to reduce the sizes of the TDG's used in the term $\chi(F(s, p))$.

4.3.2 The "Restrict" Operator

This section presents the Boolean operator "restrict" [8], noted "$\Downarrow$", that gives a means to reduce the sizes of the TDG's of the functional vector F(s, p) used in the formula $\chi(F(s, p))$. This directly reduces the computational cost of the critical term.

The idea that led to the "$\Downarrow$" operator is made clear by the following remark. In the equation (3) given in Section 3.2:

$$\mathcal{E}_{k+1} = \lambda s.(\mathcal{E}_k(s) \vee (\mathcal{F}(s) \wedge (\exists p \ \mathrm{Cns}(s, p) \wedge \mathcal{E}_k(F(s, p))))), \tag{3}$$

as soon as the term $\mathcal{E}_k(s)$ is equal to 1 or the term $\mathcal{F}(s)$ is equal to 0, the value of $\mathcal{E}_{k+1}(s)$ does not depend on the value of $(\exists p \ \mathrm{Cns}(s, p) \wedge \mathcal{E}_k(F(s, p)))$. In the same way, when Cns(s, p) is equal to 0, then so is $(\mathrm{Cns}(s, p) \wedge \mathcal{E}_k(F(s, p)))$. This means that in the term $\mathcal{E}_k(F(s, p))$, the vectorial function F can be replaced with its restriction $F|_D$ to the domain D, where D is the set whose characteristic function is $\lambda s.\lambda p.(\neg \mathcal{E}_k(s) \wedge \mathcal{F}(s) \wedge \mathrm{Cns}(s, p))$.

The same remark holds for the equations (1), (2), and (4'). For these equations the characteristic functions of the domains D to which the vectorial function F can be restricted are the following:

$$D = \mathrm{Cns} \quad \text{for (1) and (2)},$$

$$D = \lambda s.\lambda p.(\neg \mathcal{E}_k(s) \wedge \mathcal{F}(s) \wedge \mathrm{Cns}(s, p)) \quad \text{for (3)},$$

$$D = \lambda s.\lambda p.(\neg \mathcal{A}_k(s) \wedge \mathcal{F}(s) \wedge \mathrm{Cns}(s, p)) \quad \text{for (4')}.$$

The "restrict" operator has two arguments f and D that are Boolean functions. Its definition [8] on Shannon's canonical form is given in Figure 3. In this figure, we note f.root the atom that occurs at the top of Shannon's canonical form of f, and (f/¬a, f/a) the Shannon's decomposition of f with respect to the atom a. We have the identity: $f = (\neg a \wedge f/\neg a) \vee (a \wedge f/a)$.

```
function restrict(f, D);
if D = 0 then error;
if D = 1 or f = 0 or f = 1 then return f;
let a = D.root in {
    if D/¬a = 0  then return restrict(f/a, D/a);
    if D/a  = 0  then return restrict(f/¬a, D/¬a);
    if f/¬a = f/a then return restrict(f, D/¬a ∨ D/a);
    return ((¬a ∧ restrict(f/¬a, D/¬a)) ∨ (a ∧ restrict(f/a, D/a)));
    }
```

Figure 3. Semantics of the Restrict Operator.

The main properties of "⇓" are expressed by the following theorems. Their proofs are made by induction on Shannon's canonical form of f and D, and by case analysis.

Theorem 4. *Let f and $D \neq 0$ be two Boolean functions. Then $D(x) = 1$ implies that $(f \Downarrow D)(x) = f(x)$.*

Theorem 5. *Let f and D be two Boolean functions, with $D \neq 0$. If $(D \Rightarrow f)$, then $(f \Downarrow D) = 1$. If $(D \Rightarrow \neg f)$, then $(f \Downarrow D) = 0$.*

Theorem 6. *Let f and D be two Boolean functions, with $D \neq 0$. Amongst the functions g such that $(D \Rightarrow (g \equiv f))$ is a tautology, there is a unique function whose Shannon's canonical form has a minimal number of vertices, and this function is $(f \Downarrow D)$.*

Corollary. *Let f and D be two Boolean functions, with $D \neq 0$. Shannon's canonical form of $(f \Downarrow D)$ has less or the same number of vertices than Shannon's canonical form of f.*

Remark. Each time the case (D/¬a = 0) or (D/a = 0) is used to recursively compute Shannon's canonical form of $(f \Downarrow D)$, one branch of Shannon's canonical form of f is deleted. This means that the sooner a "0" occurs on a branch of Shannon's canonical form of D, the larger the number of vertices of Shannon's canonical form of f that are deleted is. This expresses that the smaller the set D is, the greater the reduction is [8].

The operator "⇓" defined above on Shannon's canonical form can also be defined on TDG's. Passing from the canonical tree representation to the canonical graph representation gives rises to the following problem. It can happen that during the application of the operator "⇓" to the TDG's of the function f and of D, some vertex that is shared in the TDG of the function f gets restricted several times, with different vertices from D. This can lead to create a TDG that has more vertices than the one of f. This means that Theorems 4 and 5 hold for this operator on TDG's, but not Theorem 6. To assure that the application of the operator "⇓" on the TGD's of f and D does not return a larger TDG than the one of f, it is sufficient to compare the sizes of the TDG's of f and of $(f \Downarrow D)$, and to return the smallest TDG, which is done in $O(\max(|f|, |D|))$. The function that returns the TDG of $(f \Downarrow D)$ uses Theorem 5 to speed up the computation, and uses a cache to avoid redundant computations. The complexity of the operator "⇓" is $O(|f| \times |D|)$.

4.4 Discussion of the Method

The computation of the critical term depends mainly on the cost of the composition. The "expand" operation can be basically seen as a method to compute $\chi(f_1, ..., f_n)$, whose computational cost is at least

$O(|\chi| \times \Pi_j |f_j|)$. Theorem 4 assures that, for the domains D associated with the equations given in Section 4.3.2, the application of the operator "$\Downarrow$" on each component of the vectorial function $[f_1 \dots f_n]$ is valid. Applying the "$\Downarrow$" operator on F with the domain D is made in $O(|F| \times |D|)$. The TGD's of the functions $(f_1 \Downarrow D), \dots, (f_n \Downarrow D)$ have less vertices than the ones of the functions $f_1, \dots, f_n$. Assuming that each TDG f_j has a size s, and that each TDG $(f_j \Downarrow D)$ has a size bound by $(s \times R)$, where $0 \leq R \leq 1$ is the reduction ratio, the composition $\chi \circ (F \Downarrow D)$ is made with a gain of complexity of at least $(1 / R)^n$. This means that the quadratic operation "$\Downarrow$" can reduce exponentially the computational cost of the critical term.

5 Experimental Results - Discussion

The verification algorithms given above have been written in LISP, and the CPU times given here have been obtained on a BULL DPX5000 mini computer with 32 megabytes of main memory.

CLM is a part of an interface board designed at BULL. It is made of 33 state registers, and has 14 inputs. The TDG's that represent its vector of transition functions has more than 10000 vertices and the largest of them has more than 2500 vertices. This machine has about 377000 valid states out of the 2^{33} possible states. The CPU time needed to symbolically traverse the state diagram of this machine using the procedure given in [9] is 1227 seconds. We did not succeed in computing the TDG of the transition relation of this machine, so the method of [3] and [6] cannot be applied. The property to be proved valid required the computation of only one fixed point that was obtained in 38 steps that took 4000 seconds of CPU time. The composition $\chi(F(s, p))$ of the critical term stopped to be computable using the standard composition algorithm after a very small number of steps. The "restrict" operation was very useful since it reduced the TDG's of $f_1, \dots, f_{33}$ used in the "expand" operation in such a way that during the iteration, none of them had more than 186 vertices. The largest TDG representing one of the sets in the sequence had 352 vertices.

The sequential circuit *Sync* is a synchronizer that has 21 state registers and 4 inputs. The property to be verified was that the value of the output *OK* is equal to 1 in any valid state of the machine. This property is expressed by the formula Init $\models$ *AG(OK)*. The CPU time needed to prove this formula valid was 4160 seconds and the fixed point was found in 9 steps. The "restrict" operator was not very useful since it did not reduce the sizes of the TDG's of the transition function. The largest TDG found in the sequence of sets had more than 2000 vertices.

The property *AG(OK)* is a safety property. It can be checked by traversing the state diagram of the machine *Sync* using the symbolic traversal procedure given in [9]. At each step during the traversal that starts from the initial state Init of the machine, it is sufficient to check that the output *OK* is equal to 1. Since the traversal stops when all the valid states of the machine have been reached, the proof is complete. Verifying the property in this manner takes 140 seconds, and the traversal is done in 20 steps. This example seems to show that whenever the formula to be proved is a safety property, it is better to use this procedure rather than the general procedure described in this paper.

In fact, the technique proposed here cannot give to the symbolic backward traversal procedure the performance of the forward symbolic traversal procedure [9]. The symbolic backward traversal procedure is essentially based on the computation of the reverse image of the vectorial function δ on a set χ. The symbolic forward traversal procedure is based on the computation of the image of the vectorial function δ on a set χ. Given a machine specified by the 6-tuple $(n, m, r, \omega, \delta, \text{Init})$, we can show that the reverse image symbolic computation is *intrinsically* more difficult than the image symbolic computation. It is easy to show that image symbolic computation is NP-hard —for instance, composition is linearly reducible to image symbolic computation—, and that it is $O(n)$-reducible to reverse image symbolic computation. The problem is that reverse image symbolic computation is not polynomially reducible to image symbolic computation, because reducing the former to the latter requires to inverse the vectorial function δ, which is a NP-hard problem —image symbolic computation itself is $O(1)$-reducible to vectorial Boolean function inversion—.

Conclusion

This paper has presented a proof algorithm that automatically checks whether some uncompletely defined deterministic Moore machine $\mathcal{M}$ holds some property expressed in the CTL formalism. This algorithm does not require the building of the state-transition graph of the machine $\mathcal{M}$, so it overcomes the limits of the previous methods based on this construction. This means that sequential machine with a very large number of states and transitions can be handle with this method.

Moreover, this proof algorithm does not require the building of the TDG of the transition relation of the machine, which is too large for many practical examples. The proof algorithm is based on the algorithms that were initially developed for proving the equivalence between two machines, in addition with a procedure that essentially computes the reverse image of a vectorial Boolean function. This procedure is the bottleneck of the proof algorithm.

References

[1] J. P. Billon, "Perfect Normal Forms for Discrete Functions", *BULL Research Report N°87019*, 1987.

[2] J. P. Billon, J. C. Madre, "Original Concepts of PRIAM, an Industrial Tool for Efficient Formal Verification of Combinational Circuits", in *The Fusion of Hardware Design and Verification*, G. J. Milne Ed., North Holland, 1988.

[3] S. Bose, A. Fisher, "Automatic Verification of Synchronous Circuits Using Symbolic Logic Simulation and Temporal Logic", in *Proc. of the IFIP Int. Workshop, Applied Formal Methods for Correct VLSI Design*, Leuven, Nov. 1989.

[4] A. Bouajjani, J. C. Fernandez, N. Halbwachs, "An Executable Temporal Logic for Expressing Safety Properties", July 1990.

[5] R.E. Bryant, "Graph-based Algorithms for Boolean Functions Manipulation", *IEEE Transaction on Computers*, Vol C35 N°8, 1986.

[6] S. Burch, E. M. Clarke, K. L. McMillan, "Sequential Circuit Verification Using Symbolic Model Checking", in *Proc. of Design Automation Conference (DAC)*, Orlando FL, USA, July 1990.

[7] E. M. Clarke, O. Grumbreg, "Research on Automatic Verification of Finite-State Concurrent Systems", *Annual Revue Computing Science*, vol. 2, pp 269-290, 1987.

[8] O. Coudert, C. Berthet, J. C. Madre, "Verification of Synchronous Sequential Machines Based on Symbolic Execution", in *Lecture Notes in Computer Science: Automatic Verification Methods for Finite State Systems*, Volume 407, J. Sifakis Editor, Springer-Verlag, pp 365-373, June 1989.

[9] O. Coudert, C. Berthet, J. C. Madre, "Verification of Sequential Machines Using Boolean Functional Vectors", in *Proc. of the IFIP Int. Workshop, Applied Formal Methods for Correct VLSI Design*, Leuven, November 1989.

[10] O. Coudert, C. Berthet, J. C. Madre, "Formal Boolean Manipulations for the Verification of Sequential Machines", in *Proc. of the First European Design Automation Conference (EDAC)*, Glasgow, March 1990.

[11] G. J. Holtzman, "Algorithms for Automated Protocol Validation", in *Lecture Notes in Computer Science: Automatic Verification Methods for Finite State Systems*, Volume 407, J. Sifakis Editor, Springer-Verlag, June 1989.

[12] Z. Kovahi, *Switching and Finite Automata Theory*, McGraw- Hill Book Edition, 1978.

[13] J. P. Queille, J. Sifakis, "Fairness and Related Properties in Transition Systems", *Acta Informatica*, pp 195-220, 1983.

DIMACS Series in Discrete Mathematics
and Theoretical Computer Science
Volume 3, 1991

Minimal Model Generation *

A. Bouajjani , J-C. Fernandez , N. Halbwachs
IMAG/LGI (U.A. CNRS 398)
B.P. 53, 38041 - Grenoble, France

Abstract

This paper adresses the problem of generating a minimal state graph from a program, without building first the whole state graph. The minimality is considered here with respect to bisimulation. A generation algorithm is presented and illustrated.

1 Introduction

Model generation consists of building a state graph from a program, a formula or any comprehensive expression of a transition system. It is used in program verification ("model checking" [6,11]) and compiling (scanner and parser generation [1], control structure synthesis [2,5],...). A crucial problem with model generation is the size of the graph, which can be prohibitive. This size can be large not only because of the intrinsic complexity of the model, but also because the graph contains a lot of states which are in some sense equivalent. Some solutions have been given to this problem, by applying reduction algorithms [8,9,10]. However, these algorithms can only be applied once the graph has been entirely generated. It is often the case that a tremendous amount of time and memory is necessary to generate a graph, which afterward reduces to a very simple one. It even happens that an infinite model reduces to a finite one. So, it would be interesting to reduce the graph during the generation, on one hand to improve the performances of model generation, and on the other hand, to allow finite model generation from infinite systems. This paper presents and illustrates an algorithm performing this task, when the equivalence considered on states is a bisimulation.

After fixing some terminology and notations (section 2), the algorithm is presented (section 3) and illustrated on a simple example (section 4).

2 Definitions and notations

Let $\mathcal{S} = (Q, \rightarrow, q_{init})$ be a transition system, where Q is a set of states, $\rightarrow \subseteq Q \times Q$ is a transition relation, and q_{init} is the initial state. Let $\sim$ be an equivalence relation on Q. Our problem is to explicitely build the quotient of the set of reachable states from q_{init}, by the coarsest bisimulation compatible with $\sim$. Of course, this is only possible if this quotient has finitely many elements. Moreover, the method presented here only works if the quotient of Q by the coarsest bisimulation is finite (notice that Q itself can be infinite). The basic idea is to progressively build a partition of Q, by distinguishing two parts of Q only when their respective elements clearly don't bisimulate each other. Henceforth, we shall consider partitions instead of equivalence relations.

Let ρ be a partition of the set of states Q. The following notations will be used:

*This work was partially supported by ESPRIT Basic Research Action "SPEC"

For any state $q \in Q$, $post_\rho(q)$ is the set of classes in ρ immediately reachable from q

$$post_\rho(q) = \{X \in \rho \mid \exists q' \in X \text{ such that } q \to q'\}$$

An equivalence relation, noted $\overset{\rho}{\sim}$, is associated with ρ as follows:

$$q_1 \overset{\rho}{\sim} q_2 \iff post_\rho(q_1) = post_\rho(q_2)$$

A subset X of Q is said to be *stable with respect to* ρ if and only if it is included in some equivalence class of $\overset{\rho}{\sim}$. The partition ρ is said to be stable if and only if all of its classes are stable with respect to itself. In other words, a partition is stable if and only if it is the set of classes of a bisimulation.

A *refinement* of a partition ρ is a partition ρ' such that: $\forall X \in \rho', \exists Y \in \rho$ such that $X \subseteq Y$

The *reduction* of a transition system $\mathcal{S}$ with respect to a stable partition ρ is the transition system $(\rho, \rightsquigarrow, [q_{init}]_\rho)$, where

$[q_{init}]_\rho$ is the class of the initial state in ρ

$X \rightsquigarrow Y \iff \exists q \in X,\, q' \in Y$ such that $q \to q'$

With the above terminology, given an initial partition ρ of Q, we are looking for the reduction of $\mathcal{S}$ with respect to the least stable refinement of ρ.

3 Algorithm

The algorithm consists of progressively refining the partition ρ. At each step, two subsets of classes will be distinguished:

- The set R of *reachable classes*, i.e. the classes containing at least one element which has been found reachable from q_{init}.
- The set S of *stable classes*, i.e. the reachable classes which have been found to belong to $\overset{\rho}{\sim}$.

The algorithm is the following:

$R = \{[q_{init}]_\rho\}; S = \phi;$
while $R \neq S$ do
  choose X in $R - S$;
  let $N = X/\overset{\rho}{\sim}$;
  if $N = \{X\}$ then
    $S := S \cup \{X\}; R := R \cup \{post_\rho(q) \mid q \in N\};$
  else
    $R := R - \{X\};$
    if $\exists Y \in N$ such that $q_{init} \in Y$ then $R := R \cup \{Y\};$
    $S := S - \{Y \in S \mid X \in post_\rho(Y)\};$
    $\rho := (\rho - \{X\}) \cup N;$
  fi
od

Proof : Let *Rea* be the set of reachable states, that is the least subset X of Q containing q_{init}, and such that

$$(q \in X \wedge q \rightarrow q') \Longrightarrow q' \in X$$

Then,

(*i*) $X \in S \Longrightarrow X \in \overset{\rho}{\sim}$
since a subset X is only put into S if $X = X/\overset{\rho}{\sim}$ (line 6), and as soon as a refinement of ρ can involve a refinement of $X/\overset{\rho}{\sim}$, X is extracted from S (line 10).

(*ii*) $X \in R \Longrightarrow X \cap Rea \neq \phi$
since a subset X is only put into R if either it contains q_{init} (line 9), or it contains successor states of a stable subset belonging to R (line 6).

(*iii*) When $R = S$, all the reachable classes are in R: If $X \in S$, all the classes directly reachable from X have been put in R (line 6).

(*iv*) So, when $R = S$, R defines a stable partition of *Rea*.

(*v*) The finiteness of the set of classes insures that the algorithm terminates.

Splitting a class : Line 4 of the algorithm splits a reachable class X into a partition $N = X/\overset{\rho}{\sim}$, whose elements are stable with respect to the current ρ. Let us detail the computation of this partition. Let *pre* denote the precondition function $\lambda Y.\{q \in Q \mid \exists q' \in Y \text{ such that } q \rightarrow q'\}$. Then,

$$N = \{X \cap \bigcap_{Y \in \rho} Z_Y \mid Z_Y \in \{pre(Y), Q - pre(Y)\}\}$$

Instead of considering such an exponential number of intersections, most of which are generally empty, we propose to compute N as follows:

```
N = {X};
for each Y ∈ ρ do
  M := φ;
  for each W in N do
    let W1 = W ∩ pre(Y);
    if W1 = W or W1 = φ then M := M ∪ {W}
    else M := M ∪ {W1, W − W1};
  od;
  N := M;
od
```

4 Example

Let us consider the following program, which could be a boolean abstraction of a more realistic program:

```
x := true; y := false; read(a);
loop
  write(x or y);
  z := a; read(a);
  w := x; x := not y; y := w or z;
end;
```

We want to examine all the possible input/output behaviours of this program. So, we consider it as a transition system, whose states are the values of the variables when the output is written. Now, since we are only interested in the output, we may consider as equivalent all the states which produce the same output. So, we start with the initial partition:

$$\{(a,w,x,y,z) \mid x \vee y = true\}, \{(a,w,x,y,z) \mid x \vee y = false\}$$

In the following, classes are represented by their characteristic formulas. The initial partition will be noted:

$$\rho : \quad C_1 = \{x \vee y\} \quad C_2 = \{\neg x \wedge \neg y\}$$

Standard rules of weakest precondition provide the precondition of a class X, with respect to the body of the loop:

$$pre(X) = X[w \vee z/y][\neg y/x][x/w] \downarrow a[a/z]$$

where $X \downarrow a = \exists a_0 X[a_0/a] = X[false/a] \vee X[true/a]$

So, $pre(C_1) = x \vee \neg y \vee a$, $pre(C_2) = \neg x \wedge y \wedge \neg a$

The successive partitions built by the algorithm are illustrated on figure 1.

Step 1: The only reachable class is C_1, since x is initially true. For splitting it, we compute:

$$C_1 \wedge pre(C_1) = (x \vee y) \wedge (x \vee \neg y \vee a) = x \vee (y \wedge a)$$

$$C_1 \wedge \neg pre(C_1) = \neg x \wedge y \wedge \neg a$$

$$C_1 \wedge pre(C_1) \wedge pre(C_2) = false$$

$$C_1 \wedge \neg pre(C_1) \wedge pre(C_2) = C_1 \wedge \neg pre(C_1)$$

So, C_1 is split into:

$$C_{11} = \{x \vee (y \wedge a)\} \quad C_{12} = \{\neg x \wedge y \wedge \neg a\}$$

and only C_{11} is reachable. We have: $pre(C_{11}) = x \vee \neg y \vee a$, $pre(C_{12}) = y \wedge (x \vee a)$

Step 2: For splitting C_{11}, we compute:

$$C_{11} \wedge pre(C_{11}) = (x \vee (y \wedge a)) \wedge (\neg y \vee x \vee a) = C_{11}$$

$$C_{11} \wedge pre(C_{11}) \wedge pre(C_{12}) = (x \vee (y \wedge a)) \wedge (y \wedge (x \vee a)) = y \wedge (x \vee a)$$

$$C_{11} \wedge pre(C_{11}) \wedge \neg pre(C_{12}) = x \wedge \neg y$$

$$C_{11} \wedge pre(C_{11}) \wedge pre(C_{12}) \wedge pre(C_2) = false$$

$$C_{11} \wedge pre(C_{11}) \wedge \neg pre(C_{12}) \wedge pre(C_2) = false$$

So, C_{11} is split into:

$$C_{111} = y \wedge (x \vee a) \quad C_{112} = x \wedge \neg y$$

and only C_{112} is reachable. We have: $pre(C_{111}) = x \vee a$, $pre(C_{112}) = \neg x \wedge \neg y \wedge \neg a$

Step 3: When splitting C_{112}, we get only:

$$C_{112} \wedge pre(C_{111}) \wedge \neg pre(C_{112}) \wedge \neg pre(C_{12}) \wedge \neg pre(C_2) = x \wedge \neg y = C_{112}$$

So C_{112} is stable, and leads to C_{111}. So, C_{111} is reachable.

Step 4: C_{111} is also found stable since:

$$C_{111} = C_{111} \wedge pre(C_{111}) \wedge \neg pre(C_{112}) \wedge pre(C_{12}) \wedge \neg pre(C_2)$$

It leads to itself and to C_{12}, which is found reachable.

Step 5: C_{12} is stable, and leads to C_2, since:

$$C_{12} = C_{12} \wedge \neg pre(C_{111}) \wedge \neg pre(C_{112}) \wedge \neg pre(C_{12}) \wedge pre(C_2)$$

Step 6: C_2 is split into:

$$C_{21} = C_2 \wedge pre(C_{111}) = \neg x \wedge \neg y \wedge a$$

$$C_{22} = C_2 \wedge pre(C_{112}) = \neg x \wedge \neg y \wedge \neg a$$

So C_{12} is removed from stable classes. We have: $pre(C_{21}) = pre(C_{22}) = \neg x \wedge y \wedge \neg a$

Step 7: C_{12} is again found stable, since:

$$C_{12} \wedge pre(C_{21}) \wedge pre(C_{22}) = C_{12}$$

It leads to C_{21} and C_{22}.

Step 8 and 9: From

$$C_{21} = C_{21} \wedge pre(C_{111})$$

$$C_{22} = C_{22} \wedge pre(C_{112})$$

we get that C_{21} and C_{22} are stable, and respectively lead to C_{111} and C_{112}.

We get a graph with 5 vertices (Fig. 2), instead of 16, which would be produced by standard generation (Fig. 3).

5 Conclusion

We have presented an algorithm combining generation and reduction methods. In our opinion, this algorithm is interesting for program verification: a state graph with several thousands (or even infinitely many) states may be reduced to one with a few number of states by considering an equivalence relation.

Of course, one must be capable to compute the function *pre* and intersections of classes, and to decide the inclusion of classes. Such a symbolic computation is achievable in the boolean case, with reasonable average cost [3,7].

Applying our algorithm to program verification appears very close to formal proof (in the Floyd/Hoare sense) or to what is now called "symbolic model checking" [4]. Concerning other applications, the algorithm is being implemented in the new version of the LUSTRE compiler.

We have not presented complexity measures. A comparison with classical reduction methods is difficult, mainly because the complexity of these methods is evaluated as a function of the size of the initial graph, whereas the cost of our method obviously depends on the size of the reduced graph.

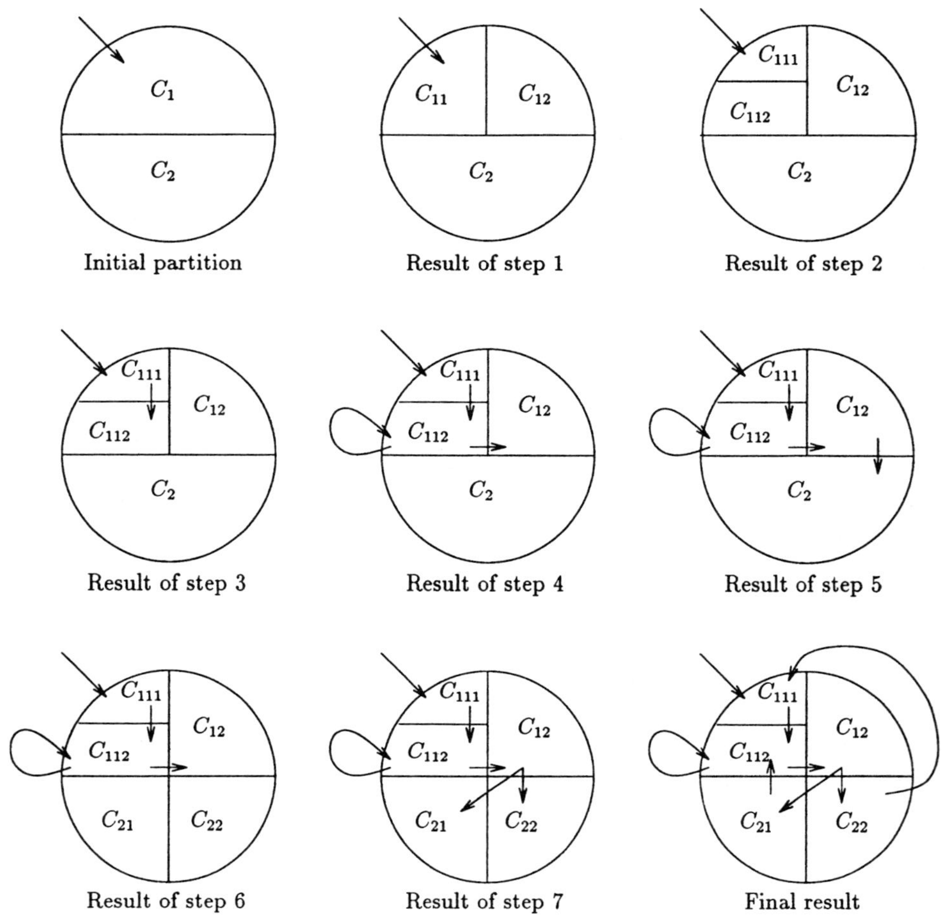

Figure 1: The successive partitions built by the algorithm

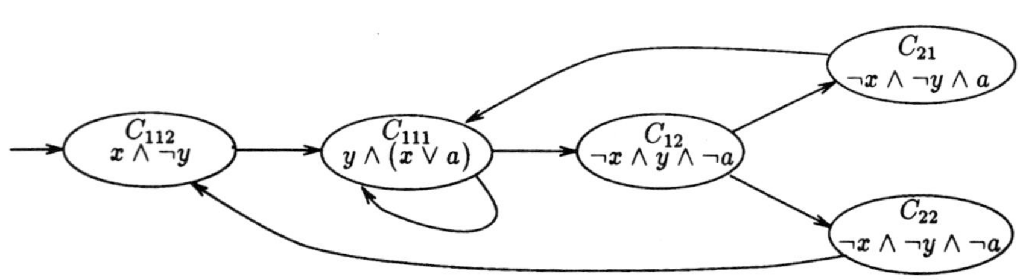

Figure 2: The reduced graph of the example

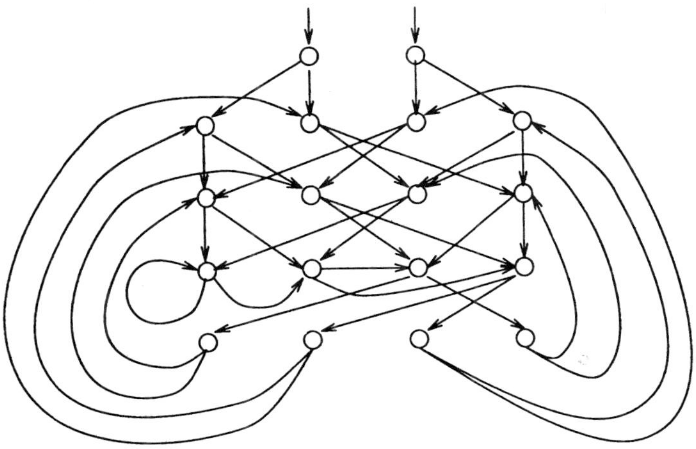

Figure 3: The complete graph of the example

References

[1] A. Aho, R. Sethi, J. Ullman. *Compilers: Principles, Techniques and Tools.* Addison-Wesley, 1986.

[2] G. Berry and G. Gonthier. *The synchronous programming language Esterel, design, semantics, implementation.* Tech. Report 327, INRIA, 1985. To appear in Science of Computer Programming.

[3] R. E. Bryant. Graph-based algorithms for boolean function manipulation. *IEEE Transactions on Computers*, C-35(8), 1986.

[4] J.R. Burch, E.M. Clarke, K.L. McMillan, D.L. Dill, J. Hwang. *Symbolic Model Checking: 10^{20} states and beyond.* Technical Report, Carnegie Mellon University, 1989.

[5] P. Caspi, D. Pilaud, N. Halbwachs, J. Plaice. Lustre: a declarative language for programming synchronous systems. In *14th POPL*, january 1987.

[6] E. Clarke, E.A. Emerson, A.P. Sistla. Automatic verification of finite state concurrent systems using temporal logic. In *10th. Annual Symp. on Principles of Programming Languages*, 1983.

[7] O. Coudert, C. Berthet, J. C. Madre. Verification of synchronous sequential machines based on symbolic execution. In *International Workshop on Automatic Verification Methods for Finite State Systems, LNCS 407*, Springer Verlag, 1989.

[8] J. C. Fernandez. An implementation of an efficient algorithm for bisimulation equivalence. *Science of Computer Programming*, 13(2-3), May 1990.

[9] P. Kanellakis and S. Smolka. CCS expressions, finite state processes and three problems of equivalence. In *Proceedings ACM Symp. on Principles of Distributed Computing*, 1983.

[10] R. Paige and R. Tarjan. Three partition refinement algorithms. *SIAM J. Comput.*, 16(6), 1987.

[11] J.L. Richier, C. Rodriguez, J. Sifakis, J. Voiron. Verification in Xesar of the sliding window protocol. In *17th International Workshop on Protocol Specification Testing and Verification*, 1987.

DIMACS Series in Discrete Mathematics
and Theoretical Computer Science
Volume 3, 1991

Verifying Liveness Properties By Verifying Safety Properties (Extended Abstract)

Jerry R. Burch
School of Computer Science
Carnegie Mellon University
Pittsburgh, PA 15213

Abstract

Conventional techniques for automatically verifying liveness properties of circuits involve explicitly modeling infinite behaviors with either infinite paths through a Kripke structure or with strings in an ω-regular language. This paper describes how *timed trace structures* [2, 3] can be used to convert liveness properties (including unbounded liveness properties such as strong fairness) to safety properties. Such properties can then be modeled and verified using only finite traces. No new algorithms are needed. All that is required is a new interpretation of what behaviors are represented by the finite traces. A mapping is defined between timed trace structures and *complete trace structures* [5], which contain infinite traces, to show that this new interpretation makes sense. The method is demonstrated on a fair mutual exclusion circuit.

1 Introduction

The primary goal of research in automated circuit verification is to devise algorithms that check whether circuits enjoy certain desirable properties. It is common to classify these properties into *safety* and *liveness* properties. Informally, safety properties state that nothing bad happens, while liveness properties state that something good happens.

In formalisms that explicitly model time (such as timed trace structures, described below), it is also possible to express *bounded liveness* properties. Bounded liveness properties state that something good happens within a bounded amount of time. This is in contrast to *unbounded liveness* properties, which state that something good happens eventually without specifying a time bound. A bounded liveness property can be viewed as a form of safety property; conventional methods used to automatically verify safety properties are also adequate for bounded liveness properties.

There are many systems described in the literature that can automatically verify unbounded liveness properties. All of these explicitly model behaviors as being infinite: most often as infinite paths through a Kripke structure or as strings in an ω-regular language. This paper describes a method for converting unbounded liveness properties to bounded liveness properties, and thereby to safety properties. This makes it possible to verify unbounded liveness properties (such as strong fairness) without explicitly modeling infinite behaviors. Instead, methods based on prefix-closed trace structures of finite traces are shown to be adequate. This result adds new insight concerning Black's proof [1] that delay-insensitive fair mutual exclusion cannot be represented using conventional finite traces.

The method merely requires a new interpretation of the behaviors represented by timed trace structures. It was not necessary to develop new algorithms or modify the automatic verifier in any way. The method is demonstrated using a simple buffer circuit as an example. In addition, a mutual exclusion circuit is used to show how strong fairness properties can be modeled and verified.

Using this method a mapping can be defined from timed trace structures on finite traces to complete traces structures [5] on infinite traces. Conformation between trace structures is preserved under this mapping. Thus, the mapping can be used to check whether a specification properly captures the desired liveness properties.

2 Timed Trace Theory

We can only give a short overview of timed trace theory in the space of an extended abstract; the interested reader may refer to [2] and [3]. The formalism is based on Dill's trace theory [5, 6]. We begin by describing trace theory, and then show how it can be extended to include timing information.

In trace theory based verification digital circuits and their specifications are modeled by *trace structures,* which are ordered 4-tuples of the form $\mathcal{T} = (I, O, S, F)$. The set I is the set of input wire names of the circuit; O is the set of output wire names. The set $A = I \cup O$ is called the *alphabet* of $\mathcal{T}$. The sets S (the *success set*) and F (the *failure set*) are regular sets of finite strings, called *traces,* over the alphabet A. Each trace models a possible behavior of the circuit by having each symbol in the trace represent a transition on the corresponding wire.

The S set and the F set are used to give partial specifications. A partial specification of a device describes requirements for the proper use of the device, and specifies the behavior of the device given that those requirements are satisfied. In a trace structure, the set S describes the behaviors of a device when it is used properly. The set F describes behaviors resulting from improper use. For example, in asynchronous circuits a gate is typically modeled so that the F set contains all behaviors that cause a hazard and the S set contains all behaviors that do not cause a hazard.

The set $P = S \cup F$ is the set of all *possible traces.* The S and P sets must be prefix-closed, and P must be non-empty. For some non-deterministic devices (such as the vending machine example in [8], which is also discussed in [5]) a given trace can be both a success and a failure, so the S and F sets need not be disjoint. Also, no circuit can control its inputs, which is modeled by requiring that $PI \subseteq P$ (if A and B are sets of strings or symbols, then $AB = \{ab : a \in A \text{ and } b \in B\}$). This is called the *receptiveness* requirement.

As an example, consider how a buffer might be modeled by a trace structure $\mathcal{T} = (I, O, S, F)$. Let b be the name of the input wire of the buffer, and let x be the name of the output wire. Then $I = \{b\}$ and $O = \{x\}$. Assume that both the input and the output of the buffer are initially low. We define the set of successful behaviors of the buffer to be those behaviors in which the environment does not cause a hazard. Thus, the set S of successful behaviors is equal to $(bx)^*(b+\epsilon)$. If the environment does cause a hazard, then we do not restrict the ensuing behavior of the circuit, so $F = (bx)^*bb(b+x)^*$.

Consider two variations of the above buffer. The first buffer always eventually fires when it is put in a firable state. The second buffer sometimes might never fire. Both of these buffers would be modeled by exactly the same trace structure. This shows that liveness properties cannot be modeled with this form of trace theory.

A composition operation (denoted by "$\circ$") can be defined on trace structures. Let $\mathcal{T}_0 = (I_0, O_0, S_0, F_0)$ and $\mathcal{T}_1 = (I_1, O_1, S_1, F_1)$ be trace structures. The composition of $\mathcal{T}_0$ and $\mathcal{T}_1$ is defined when O_0 and O_1 are disjoint. The set of outputs of the composition is $O_0 \cup O_1$, the set of inputs is $(I_0 \cup I_1) - (O_0 \cup O_1)$. Let $A_0 = I_0 \cup O_0$ and $A_1 = I_1 \cup O_1$. If x is a trace in $(A_0 \cup A_1)^*$, define the projection $x\,|_{A_0}$ to be the trace formed from x by removing all symbols not in A_0. The S and F sets of the composition are given by

$$\begin{aligned} S &= \{x \in (A_0 \cup A_1)^* : x\,|_{A_0} \in S_0 \wedge x\,|_{A_1} \in S_1\}, \\ F &= \{x \in (A_0 \cup A_1)^* : (x\,|_{A_0} \in F_0 \wedge x\,|_{A_1} \in F_1) \vee \\ &\quad (x\,|_{A_0} \in F_0 \wedge x\,|_{A_1} \in S_1) \vee \\ &\quad (x\,|_{A_0} \in S_0 \wedge x\,|_{A_1} \in F_1)\}. \end{aligned}$$

The operation **hide** is also defined on trace structures. The trace structure denoted by **hide**$(D)[(I, O, S, F)]$ (where D is a subset of O) is equal to $(I, O-D, S\,|_{A'}, F\,|_{A'})$, where $A' = I \cup (O - D)$ and the projection operation is extended to sets of traces in the normal manner.

A trace structure is said to be *failure free* when its F is empty. If the composition of several trace structures is failure free, then the components have been composed in such a way that each of their environmental requirements has been satisfied. The trace theory verifier can efficiently check whether the composition of a set of trace structures is failure free. The S and F sets of trace structures are represented in the verifier with deterministic finite automata. Checking if the composition of two trace structures is failure free is done in time linear in the product of the sizes of these automata. If a composition is not failure free, then usually only a small fraction of the states of the composition need to be explored before a short error trace can be given.

Let $\mathcal{T}_0$ and $\mathcal{T}_1$ be trace structures. Then, $\mathcal{T}_0$ *conforms to* $\mathcal{T}_1$ (written $\mathcal{T}_0 \preceq \mathcal{T}_1$) if $I_0 = I_1$ and $O_0 = O_1$ and for any trace structure $\mathcal{T}$, the composition of $\mathcal{T}$ and $\mathcal{T}_1$ being failure free implies that the composition of $\mathcal{T}$ and $\mathcal{T}_0$ is failure free. The idea captured here is that if a circuit works correctly with $\mathcal{T}_1$ as a component, then it works correctly with $\mathcal{T}_1$ replaced by $\mathcal{T}_0$ (up to safety properties). If $\mathcal{T}_0 \preceq \mathcal{T}_1$ and $\mathcal{T}_1 \preceq \mathcal{T}_0$, then $\mathcal{T}_0$ and $\mathcal{T}_1$ are *conformation equivalent*. This means that $\mathcal{T}_0$ and $\mathcal{T}_1$ are functionally interchangeable. There exists a canonical form for trace structures such that two canonical trace structures are conformational equivalent if and only if they are equal.

A trace structure $\mathcal{T}_1$ can also be used to represent a specification. A circuit $\mathcal{T}_0$ satisfies the specification $\mathcal{T}_1$ if $\mathcal{T}_0$ conforms to $\mathcal{T}_1$. Checking that a circuit satisfies a specification would be very expensive if it required checking the failure freedom of compositions with all possible trace structures. However, there exists an operation on trace structures, called *mirroring*, that makes checking conformation practical. If $\mathcal{T} = (I, O, S, F)$ is a canonical trace structure, then $\mathcal{T}^M = (O, I, S, A^* - (S \cup F))$ is its mirror. If $\mathcal{T}'$ has the

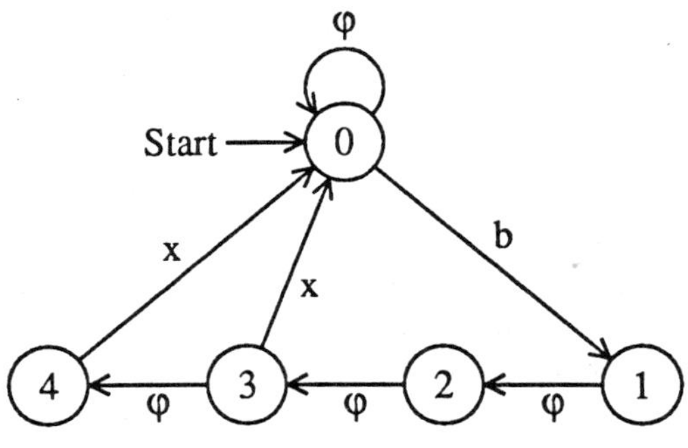

Figure 1: Automata that accepts the S set of a 3/2 rule buffer.

same inputs and outputs as $\mathcal{T}$, then the composition of $\mathcal{T}'$ and $\mathcal{T}^M$ is failure free if and only if $\mathcal{T}'$ conforms $\mathcal{T}$. Thus, checking if a circuit satisfies a specification only requires checking if the composition of the circuit with the mirror of the specification is failure free. This is done in time linear in the product of the sizes of the automata representing the circuit and the specification. If there is an error in the circuit, then usually on a small fraction of the state space needs to be explored before a failure is found. When a failure is found, a short trace of the transitions that led to the failure is output to the user.

2.1 Adding Timing Information

In standard trace theory, only speed-independent circuits can be represented. However, trace theory can be extended to allow the representation of a large class of timing models for different circuit components. We call the extended theory *timed trace theory.* The extension allows traces to contain additional symbols not corresponding to physical wires. The presence of such a symbol in a trace is interpreted as representing the passage of some fixed amount of time. (Later in the paper, we describe an alternative interpretation that allows for the modeling of unbounded liveness properties.) More formally, we allow trace structures of the form (I, O, V, S, F), where V is a set of symbols (disjoint from I and O), and where S and F are sets of traces over the alphabet $A = I \cup O \cup V$. The elements of V do not correspond to any physical wires, so they are called *virtual wire names.* In order to give an intuitive explanation of the intended meaning of such structures, we will consider the case in which V contains a single element, call it φ.

The presence of a φ in a trace indicates the passage of a unit of time, call it τ. The trace $\varphi\varphi x\varphi$ represents a single behavior in which a transition occurs on wire x at time $T_0 + 2\tau$, where T_0 is the time at which the behavior described by the trace began.

Figure 1 is an automata describing the S set of a buffer with input b, output x, and virtual wire φ. Notice that since S is prefix-closed, all of the states in this automata are accepting states. The F set of this buffer is equal to $(SI - S)A^*$. Interpreted in discrete time, this buffer clearly has a minimum delay of 2τ and a maximum delay of 3τ.

Consider a circuit formed using the output of a pulse generator as the input to a buffer as in Figure 2. The pulse generator has a period of 8τ and a 50% duty cycle. Its S set is described by the automata in Figure 3 and its F set is empty. The buffer is the same as

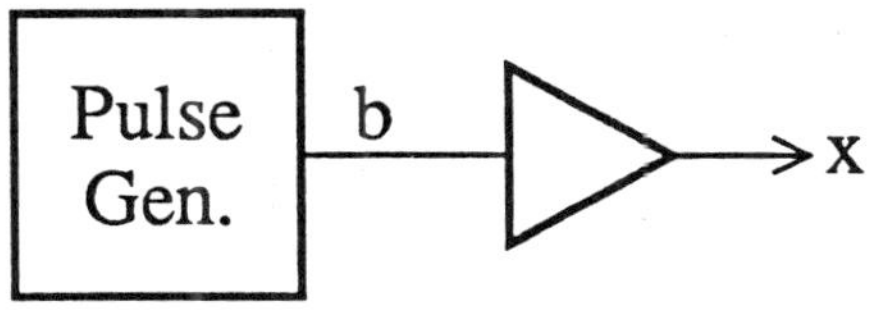

Figure 2: Example circuit for demonstrating effects of timing assumptions.

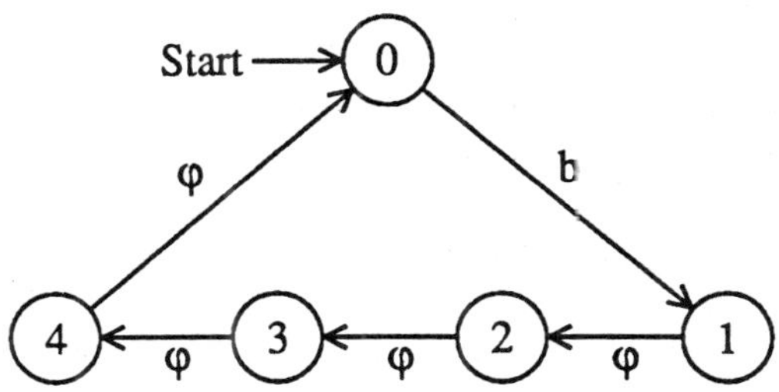

Figure 3: Automata that accepts the S set of a pulse generator.

that represented in Figure 1. The trace structure representing the resulting circuit has an empty F set and its S set is given by the automata in Figure 4. Since the maximum delay of the buffer is 3τ and the pulse generator waits 4τ between outputting transitions, there are never two consecutive b transitions without an x transition in between. This fact could not be represented with the speed-independent timing model.

It is clear how to modify the buffer in Figure 1 to have a minimum delay of $m\tau$ and a maximum delay of $n\tau$ for any non-negative integers m and n such that $m \leq n$. The buffer can also be made to have an unbounded maximum delay by removing state 4 and adding a transition on φ from state 3 back to state 3.

The trace theory verifier has been extended to include virtual wires in this way. Checking that a circuit satisfies a specification has the same complexity as before. Examples

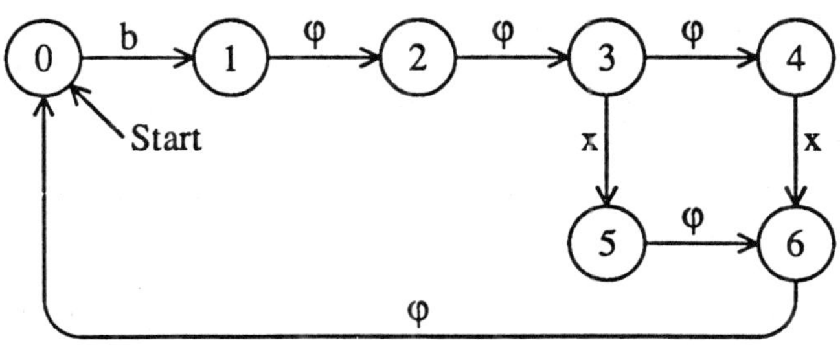

Figure 4: Automata representing the behavior of the example circuit in Figure 2.

of using the verifier to check the correctness of circuits can be found in [2] and [3].

3 Converting Liveness to Safety

Let trace structure $\mathcal{T}$ model a buffer with input b, output x, and virtual wire φ_s that has a minimum delay of zero and a maximum delay of one. Normally, the presence of a φ_s in a trace in $\mathcal{T}$ indicates the passage of a unit of time. For the purpose of converting liveness properties to safety properties, we propose a different interpretation. A transition on b can be seen as placing an obligation on the buffer to eventually toggle x in response. We interpret the nth φ_s in a trace to mean that the buffer has met all such obligations that the environment placed on it before the $(n-1)$th φ_s in the trace. In this interpretation, there need not be a constant, or even bounded, amount of time between each φ_s.

As an example of how this interpretation can be used to verify liveness properties, consider verifying that two buffers in series conform to a single buffer. Let trace structure $\mathcal{T}_1$ model a buffer with input b and output c, and let $\mathcal{T}_2$ model a buffer with input c and output x. Both trace structures have a single virtual wire φ_i, with a minimum delay of zero and a maximum delay of one. In order to show that $\mathcal{T}_1 \circ \mathcal{T}_2$ conforms to the specification, we compose $\mathcal{T}_1 \circ \mathcal{T}_2$ with a constraining trace structure $\mathcal{C}$. The trace structure $\mathcal{C}$ has no inputs or outputs, but has virtual wires φ_i and φ_s. The sole purpose of $\mathcal{C}$ is to constrain the transitions on the virtual wires so that there must be at least three transitions of φ_i between any two transitions of φ_s. Three is the minimum necessary to assure that the two buffers in series satisfy their obligation to transition x before a transition of φ_s. The automatic verifier can be used to show that

$$\mathbf{hide}(\{c, \varphi_i\})[\mathcal{T}_1 \circ \mathcal{T}_2 \circ \mathcal{C}] \preceq \mathcal{T}.$$

If $\mathcal{C}$ is chosen badly, then the verifier may incorrectly report that the circuit does not conform to the specification. The user is required to find a $\mathcal{C}$ that allows the verification to go through. Regardless of what $\mathcal{C}$ is used, however, the verifier will never report that a circuit conforms to a specification when in fact it does not. In this sense, the verification method is *conservative.*

4 Mapping to Complete Traces

In the previous section, we described a specification for a live buffer in terms of a timed trace structure. But it may not be intuitively clear that the timed trace structure captures the properties of a live buffer. This section describes a way of addressing that problem.

In [5], Dill describes a version of trace theory, called complete trace theory, where liveness properties are modeled using infinite traces. Dill describes a complete trace structure that is a specification for a live buffer, as a special case of a live gate. It is easy to give an intuitive argument that this complete trace structure accurately captures the properties of a live buffer. If it could be shown that the timed trace specification for a live buffer was in some way equivalent to the complete trace specification, that would be strong evidence that the timed trace specification is correct.

One way to do this is to define a mapping Φ from timed trace structures to complete trace structures such that

$$\mathcal{T}_1 \preceq \mathcal{T}_2 \Rightarrow \Phi(\mathcal{T}_1) \preceq \Phi(\mathcal{T}_2),$$

and such that the timed trace specification of the live buffer is mapped by Φ to the complete trace specification. As an example of why the implication is in only one direction, let $\mathcal{T}_1$ be a buffer with a maximum delay of one and let $\mathcal{T}_2$ be a buffer with a maximum delay of two. Both $\mathcal{T}_1$ and $\mathcal{T}_2$ have a minimum delay of zero, and have one virtual wire φ. In this case $\Phi(\mathcal{T}_1) = \Phi(\mathcal{T}_2)$, so $\Phi(\mathcal{T}_2) \preceq \Phi(\mathcal{T}_1)$. However $\mathcal{T}_2$ clearly does not conform to $\mathcal{T}_1$.

The definition of such a mapping Φ, and proofs of its properties, are included in the full version of the paper [4].

5 Fair Mutual Exclusion Example

The verification example above for buffers only deals with a simple kind of liveness property. In this section to use the verifier to check a strong fairness property, which is a more difficult kind of liveness property.

The specification for a fair mutual exclusion element has inputs *ai* and *bi* and outputs *ao* and *bo*. It requires that any transition on *ai* be followed by a transition on *ao* with a maximum delay of one, as measured against the virtual wire φ_s. A similar requirement is placed on *bi* and *bo*. Again, this does not represent a bounded response time. The response time is unbounded since there is not assumed to be a bound on the amount of time between transitions of φ_s. However, this specification can not be satisfied by any implementation unless it is assumed that the environment will always eventually release the mutual exclusion element. This assumption is represented by introducing another virtual wire φ_r. After a request is granted by the mutual exclusion element, the environment must release it with a maximum delay of one, as measured against φ_r. Otherwise, it is a failure. This does not actually bound the time for which a channel can hold the mutual exclusion element. It simply means that the environment must constrain φ_r so that the mutual exclusion element is released before there are to many φ_r transitions. Let $\mathcal{T}$ be the trace structure representing this specification. Also, let $\mathcal{T}_1$ represent a buffer with input *ai* and output *ao* and a maximum delay of one, as measured against φ_s. Finally, let $\mathcal{T}_2$ represent an inverter with input *bo* and output *bi* and a maximum delay of one, as measured against φ_r. The automatic verifier can be used to show that

$$\mathbf{hide}(\{bi, bo, \varphi_r\})[\mathcal{T} \circ \mathcal{T}_2] \preceq \mathcal{T}_1.$$

This is evidence that the formal specification captures the desired informal specification for a fair mutual exclusion element.

The verifier was used to verify the speed-independent fair mutual exclusion circuit described in [2]. This involves modeling each component of the circuit as having a maximum delay of one relative to a virtual wire φ_i, and then constraining φ_s relative φ_i and φ_r.

It is straightforward to define a complete trace structure that is a specification for fair mutual exclusion. The mapping Φ described above maps the timed trace theory specification for fair mutual exclusion to such a complete trace specification. This shows that the notion of fair mutual exclusion is accurately modeled by the timed trace theory specification.

In [1] Black shows that conventional finite traces cannot model delay-insensitive fair mutual exclusion. His arguments can be used to show that conventional finite traces cannot model speed-independent fair mutual exclusion if there is no bound on the number requests granted to one channel before a request is granted to the other channel, as in the specification above. Black proposed that infinite traces be used to model fair mutual exclusion. We have shown how trace theory can be modified to model fair mutual exclusion without introducing infinite traces.

6 Conclusions

We have shown how liveness properties, such as unbounded fairness, can be converted to safety properties. This makes possible the automatic verification of liveness properties using finite traces. No new algorithms are needed, just a reinterpretation of the meaning of timed traces. We showed this new interpretation makes sense by defining a mapping Φ from timed trace structure to complete trace structures that have infinite traces. This mapping preserves conformation, and maps specifications in timed trace theory to specifications in complete trace theory that are more obviously correct.

Unfortunately, tests indicate that the method presented here may not be efficient in practice. The same fair mutual exclusion circuit that was verified in this paper was verified more quickly using CTL in [2]. Verifying the circuit with timed traces required that a much larger state space be explored than was necessary using CTL. Thus, the results in this paper may be of more theoretical than practical interest. However, there may be areas of practical application. For example, in machine aided verification techniques like those in [9], a circuit is proved correct by proving it satisfies an invariant. Previously, only safety properties could be verified in this way. But the method described here might be adapted to allow verification techniques based on invariants to also verify unbounded liveness properties.

References

[1] David L. Black. On the existence of fair delay-insensitive arbiters: Trace theory and its limitations. *Distributed Computing*, 1(4):205–225, 1986.

[2] Jerry R. Burch. Combining CTL, trace theory and timing models. In Joseph Sifakis, editor, *Automatic Verification Methods for Finite State Systems, International Workshop, Grenoble, France*, volume 407 of *Lecture Notes in Computer Science*. Springer-Verlag, June 1989.

[3] Jerry R. Burch. Modeling timing assumptions with trace theory. In *IEEE International Conference on Computer Design*, October 1989.

[4] Jerry R. Burch. Verifying liveness properties by verifying safety properties. In Robert Kurshan and Edmund M. Clarke, editors, *Computer-Aided Verification, Proceedings of the 1990 Workshop*, volume 3 of *DIMACS Series in Discrete Mathematics and Theoretical Computer Science*. American Mathematical Society, 1991. To appear.

[5] David L. Dill. *Trace Theory for Automatic Hierarchical Verification of Speed-Independent Circuits*. PhD thesis, Carnegie Mellon University, Pittsburgh, PA 15213, 1988. Also appeared as [7].

[6] David L. Dill. Trace theory for automatic hierarchical verification of speed-independent circuits. In Jonathan Allen and F. Thomson Leighton, editor, *Advanced Research in VLSI: Proceedings of the Fifth MIT Conference*. MIT Press, 1988.

[7] David L. Dill. *Trace Theory for Automatic Hierarchical Verification of Speed-Independent Circuits*. ACM Distinguished Dissertations. MIT Press, 1989.

[8] Robin Milner. *A Calculus of Communicating Systems*, volume 92 of *Lecture Notes in Computer Science*. Springer-Verlag, 1980.

[9] Jorgen Staunstrup, Stephen J. Garland, and John V. Guttag. Localized verification of circuit descriptions. In Joseph Sifakis, editor, *Automatic Verification Methods for Finite State Systems, International Workshop, Grenoble, France*, volume 407 of *Lecture Notes in Computer Science*. Springer-Verlag, June 1989.

DIMACS Series in Discrete Mathematics
and Theoretical Computer Science
Volume 3, 1991

Extension of the Karp and Miller Procedure to Lotos Specifications[1]

Michel Barbeau, Gregor V. Bochmann
Université de Montréal, Département d'IRO
C.P. 128, Succ. "A", Montréal, Canada, H3C 3J7

Abstract

In a companion paper [Barb 90a,b], we proposed a Place/Transition-net (P/T-net) semantics for a subset of Lotos. This subset is such that Lotos specifications can be translated into finite structure Petri nets. It is therefore possible to apply P/T-net verification techniques since they require finite structure. It this paper we demonstrate that it is possible to apply P/T-net verification methods without an a priori construction of the P/T-net associated to the Lotos specification to be analysed. In particular we consider a well known reachability analysis technique for P/T-nets, namely, the Karp and Miller procedure.

1 Introduction

The goal of this work is to investigate verification techniques for basic Lotos specifications [Bolo 87a]. Basic Lotos is presented in § 4. It can be demonstrated that this language has the computational power of Turing machines. Therefore, generally non-trivial properties are undecidable.

Our verification method is based on Petri nets. Petri net verification techniques are transfered to Lotos. So far, two transfer approaches have been proposed. A first approach consists of translating Lotos specifications into Petri nets and evaluating the properties on the equivalent Petri net models. Presently, there are at least two tools based on this approach ([March 89] and [Gara 90]).

We found that the Lotos to Petri nets translation step is, in many cases, as complex, in terms of time and space, as the verification step itself. In this paper, we propose a second approach which involves no translation from one formalism to another. We adapt Place/Transition-net (P/T-net) verification algorithms to Lotos. We consider a well known P/T-net reachability analysis technique, the Karp and Miller procedure [Karp 69].

The main difficulty in these approaches is to make shure that the P/T-net, that models the Lotos specification, has a finite structure. That is, a bounded number of places and

[1]This work has been funded by the Natural Sciences and Engineering Research Council of Canada, the Centre de recherche informatique de Montréal and Bell Northen Research.
1052-1798/91 $1.00 + $.25 per page

transitions. We model Lotos local process states as Petri net places. We put syntactical constraints so that the number of local process states is bounded. Moreover, the number of alternatives to make a transition, from a given local state, is also bounded. Consequently, the Petri net has a finitre structure. However, these constraints do not bound the number of identical parallel processes. Therefore, finite structure Petri nets does not necessarily mean finite state systems. Consequently, the "Lotos" that can be analyzed with our method is computationaly more powerful than the "Lotos" that can be verified with finite state/transition system based methods (e.g. [Bolo 87b]).

Petri nets can represent, with finite structures, infinite state systems. This means that classical finite state system reachability analysis [Boch 78] is not applicable. Karp and Miller trees and graphs are finite and partial representations of, in general infinite, Petri net reachability trees and graphs. P/T-nets are introduced in § 2. Reachability analysis for P/T-nets is discussed in § 3. Modelling of Lotos by P/T-nets is presented in § 5. The extension of Petri net reachability analysis to Lotos is discussed in § 6.

2 P/T-nets

We slightly deviate from the usual notation for P/T-nets [Pete 81]. We represent a P/T-net as a tuple (P, T, Act, M_0) where:

- P is a set of places $\{p_1, ..., p_n\}$,
- $T \subseteq \mathcal{N}^P \times Act \times \mathcal{N}^P$, is a transition relation,
- Act is a set of transition labels, and
- $M_0 \in \mathcal{N}^P$, is the initial marking.

A P/T-net has a **finite structure** if the sets P, T and Act are finite.

$\mathcal{N}$ is the set of non-negative integers. $\mathcal{N}^P$ denotes the set of multi-sets over the set P. A multi-set is a set that can contain multiple instances of the same element. An element $t = (X, a, Y) \in T$ is also denoted as $X - a \rightarrow Y$. Its **preset** $pre(t)$ is X, its **postset** $post(t)$ is Y and **action** $act(t)$ is a. The multi-sets X and Y are also called, respectively, the input and output places of t. We denote as $pre(t)(p)$ $(post(t)(p))$ the number of instances of the element p in the preset (postset) of t.

A Petri net marking is also a multi-set. We denote by $M(p_i)$ the number of instances of the element p_i in the multi-set M. A marking M is also denoted as a n-tuple $(M(p_1), ..., M(p_n))$.

Instances of the element p_i are also called tokens inside place p_i. The operators $\leq$, $+$ and $-$ denote respectively multi-set inclusion, summation and difference. The symbol ω denotes infinity and we have the following usual convention:

$$(\forall n \in \mathcal{N})[n + \omega = \omega + \omega = \omega - n = \omega \wedge n < \omega]$$

$pre(t)(p)$ is the number of tokens that place p must contain to enable transition t. A transition $t \in T$ is **enabled** in marking M if $pre(t) \leq M$. This is denoted as $M(t >$. An enabled transition can be **fired** and the successor marking M' is defined as:

$$M' = M - pre(t) + post(t)$$

this is represented as $M(t > M'$.

A **tree** is a directed graph such that one vertex, called the root, has no edge directed into it, every other vertex has exactly one edge directed into it and is reachable from the root. If there is an edge from x to y then y is a successor of x. Vertices without successors are called leafs. A **labelled tree** is a tree where vertices have labels. If x and y are distinct vertices of a tree and there is a path from x to y then we write $x \prec y$.

Given a P/T-net $N = (P, T, Act, M_0)$, we define the **reachability tree** (RT) as the labelled tree $RT(N) = (V, E, v_0)$ where:

1. V is a set of vertices labelled with markings of N,
2. $v_0 \in V$, is the root labelled with M_0,
3. $E \subseteq V \times T \times V$, is a set of directed edges, and
4. for all $v \in V, t \in T$, if $label(v)(t > M'$ then there is a vertex $v' \in V$, labelled with M' and an edge $(v, t, v') \in E$.

The **reachability graph** (RG) of a P/T-net $N = (P, T, Act, M_0)$ is a graph $RG(N) = (RS, E, M_0)$ that can be derived from $RT(N)$ by merging nodes with identical labels. In the RG, we speak indistinctly of a vertex or of the marking that labels the vertex. The set RS of markings is called the reachability set.

A P/T-net is illustrated in Fig. 1. In the graphical representation, places are shown as circles, transitions as bars and tokens as dots inside places. There is a directed edge from place p_i (transition t_j) to transition t_j (place p_i) *iff* $p_i \in pre(t_j)$ $(p_i \in post(t_j))$. If $pre(t)(p) > 1$ $(post(t)(p) > 1)$ we may label the corresponding edge with the value of $pre(t)(p)$ $(post(t)(p))$. The initial marking of this particular net can be denoted as the multi-set $\{p_1, p_2\}$ or as the 6-tuple $(1, 1, 0, 0, 0, 0)$.

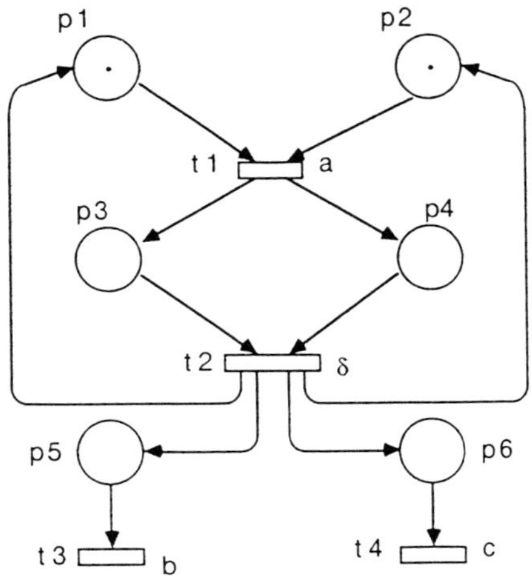

Figure 1: A P/T-net

3 Reachability Analysis for P/T-nets

With respect to Holzmann's classification [Holz 89], the Karp and Miller tree construction procedure is a *stack search strategy*. It is a depth-first technique that minimizes memory usage at the expense of run time. The whole state space does not have to be maintained in memory, only the path starting from the initial marking to the current marking. In general, Petri nets are not finite state systems. The Karp and Miller tree is also called the **coverability tree** because for every reachable marking M of a Petri net, there exists a marking M' in the Karp and Miller tree such that $M \leq M'$.

Definition 1 *The* **coverability tree** *(CT) associated to a P/T-net* $N = (P, T, Act, M_0)$ *is a tree, where vertices are labelled with markings of* N *and edges are labelled with elements in* T. *The CT can be recursively defined as follows:*

1. *The root is labelled with* M_0*;*

2. *Let* x *denote a vertex with label* M,

   (a) *if there exists a vertex* y *such that* $y \prec x$ *and label*$(y) = M$, *then* x *is a leaf;*

   (b) *else, the successors of* x *are in one-to-one correspondence with the elements of the set:*

$$S = \{(t, M') : t \in T \wedge M(t > M'\}$$

- $(1,1,0,0,0,0)$
  - $t_1 \longrightarrow (0,0,1,1,0,0)$
    - $t_2 \longrightarrow (1,1,0,0,\omega,\omega)$
      - $t_1 \longrightarrow (0,0,1,1,\omega,\omega)$
        - $t_2 \longrightarrow (1,1,0,0,\omega,\omega)$
        - $t_3 \longrightarrow (0,0,1,1,\omega,\omega)$
        - $t_4 \longrightarrow (0,0,1,1,\omega,\omega)$
      - $t_3 \longrightarrow (1,1,0,0,\omega,\omega)$
      - $t_4 \longrightarrow (1,1,0,0,\omega,\omega)$

Figure 2: Coverability Tree

Let $(t, M') \in S$, we create the successor vertex z and the edge (x,t,z). The label of z is determined as follows: For $i = 1, ..., n$ (n is the number of places),

i. *if there exists a vertex y such that $y \prec z$, $label(y) \leq M'$ and $label(y)(p_i) < M'(p_i)$ then $label(z)(p_i) = \omega$;*

ii. *else, $label(z)(p_i) = M'(p_i)$;*

Example 1: For the net of Fig. 1, the CT is shown in Fig. 2. The CT has 9 vertices and the maximal path depth is 4. This means that no more than 4 vertices have to be maintained in memory. No deadlock is indicated.

If the whole CT is maintained in memory, the **coverability graph** (CG) can be obtained simply by merging vertices with identical labels. In that case, more memory is required but full connectivity information is stored and can be used to analyse loops.

Example 2: For the net of Fig. 1, the CG is shown in Fig. 3.

With the CT and the CG, the following six problems [Fink 90] become decidable:

1. *Termination* Is the reachability tree finite?

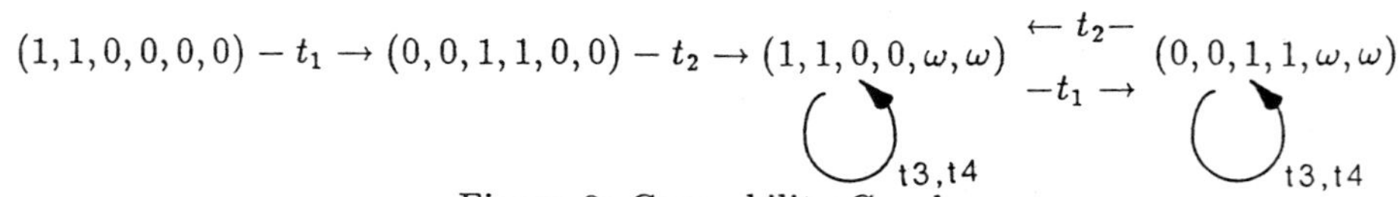

Figure 3: Coverability Graph

2. *Finiteness* Is the reachability set finite?

3. *Coverability* Given a marking M, is there a reachable marking M' such that $M \leq M'$?

4. *Quasi-liveness* Given an action a, is there a reachable marking M such that a is executed from M ?

5. *Boundedness* Is the number of tokens in a given place bounded?

6. *Regularity* Is the Petri net language recognizable by a finite state automaton?

The language of a Petri net is the set of transition sequences starting from the initial marking. If the language is regular the Petri net can be simulated by a finite state automaton. The Karp and Miller tree and graph constructions do not necessarily detect every deadlock in a nonfinite state system.

In general the CT and the CG obtained from Definition 1 are not the minimal CT and CG. Finkel [Fink 90] has proposed an algorithm for computing the **minimal coverability tree** (MCT) and the **minimal coverability graph** (MCG), the ones that have the minimal number of vertices. One of the advantages of this reduction algorithm is that it computes the MCT directly and it is not necessary to go through the construction of a nonminimal CG. There are other graph reduction operators (e.g. R and V in [Rea 87]) that are also not devoid of qualities (see [Barb 89]) but can be computed only after the initial construction of the global transition system. Every above problem, decidable with the nonminimal CT and CG, is decidable as well with Finkel's MCT and MCG construction procedure. Finkel's procedure is therefore an interesting solution to reduce the impact of the state space explosion problem. However, if the system is not finite state some deadlocks may be indicated in the nonminimal CT but not in the MCT.

4 Lotos

A basic Lotos behavior expression is formed out of the following terms:

Inaction	**stop**
Action prefix	$a; B$
Choice	$B_1[]B_2$
Process instantiation	$p[g_1, ..., g_n]$
Pure interleaving	$B_1\|\|\|B_2$
General parallel composition	$B_1\|[g_1, ..., g_n]\|B_2$
Successful termination	**exit**
Sequential composition	$B_1 >> B_2$
Disabling	$B_1[> B_2$
Hiding	**hide** $g_1, ..., g_n$ **in** B_1

where B, $B1$ and $B2$ are behavior expressions. The semantics of Lotos is given in [Bolo 87a]. It uses transition trees to model execution of programs. More precisely, given a starting behavior expression B_0, a **transition tree** $TT = (N, E, n_o)$ is associated such that:

- N, is a set of nodes labelled with behavior expressions,
- $n_o \in N$, is the root labelled with B_0,
- $E \subseteq N \times Act \times N$, is a transition relation, and
- Act denotes a set of user-defined actions together with the successful termination action δ and the internal action i.

An element $(n, a, n') \in E$ is defined if $label(n) = B$, $label(n') = B'$, and B may perform the action a and transforms into B' (also denoted as $B - a \rightarrow B'$). The $B - a \rightarrow B'$ relation is defined by means of a set of inference rules presented in the Reference.

The **transition graph** (TG), of a starting behavior expression B_0, is a graph $TG = (RS, E, n_0)$ that can be derived from the TT by merging nodes with identical labels. In the TG, we speak indistinctly of a vertex or of the behavior expression labeling that vertex.

Proposition 1 *The language under consideration has the computational power of Turing machines*[2].

[2]Proofs of propositions and theorems are given in [Barb 90a]

Since this subset of Lotos has the computational power of Turing machines, nontrivial properties are generally undecidable. P/T-nets (with finite structures) do not have the computational power of Turing machines as Lotos does. In the rest of this section we define a subset of Lotos, PLotos, that can be modelled by finite structure P/T-nets and, conversely, into which P/T-nets can be simulated. The mapping from PLotos to P/T-nets is introduced in the next section whereas P/T-nets simulation in PLotos is discussed in [Barb 90a,b].

We assume that PLotos specifications satisfy the following constraints:

1. *Guarded recursive processes.* A process instantiation term is guarded if it is in the scope of a prefixing operator ";" or the in right sub-expression B_2 of a sequential composition $B_1 >> B_2$ or of a disabling $B_1[> B_2$.

2. *No combination of recursion and general parallel composition.* The general parallel operator "$|[g_1, ..., g_n]|$" is not allowed on recursive paths.

3. *Tail recursion.* The process in which $B_1 >> B_2$ (or $B_1[> B_2$) is defined may not be called from sub-expression B_1. And, in B_1 combination of recursion and parallelism is not allowed.

4. *"Noexit" functionality in pure interleaving.* Operands B_1 and B_2 in parallel composition $B_1|||B_2$ must have the *noexit* functionality.

Paths are defined by choice alternatives and can be illustrated as a tree with behavior expressions labelling nodes, as in the following example:

process $p[a, b, c]$: **noexit** :=

  $(a; stop|||b; stop)$

  []

  $c; p[a, b, c]$

endproc

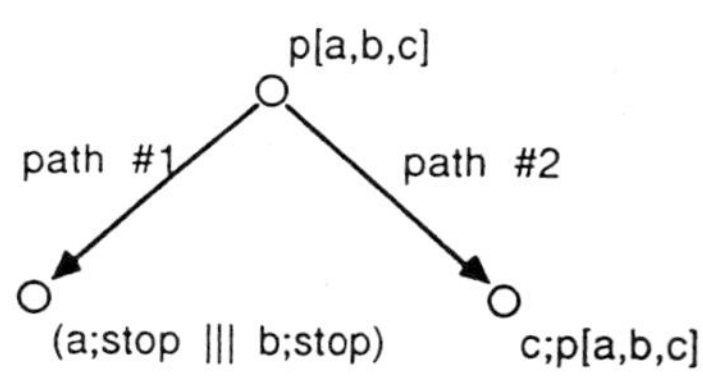

The path number 1 is not recursive whereas the path number 2 is because the process p is recursively called at node labelled "$c; p[a, b, c]$". A process that does not contain parallel operator "|||" and "$|[g_1, ..., g_n]|$" on recursive paths can be modelled by a finite state system. If the operator "$|[g_1, ..., g_n]|$" is disallowed on recursive paths, whereas "|||" is allowed with functionality *noexit* operands, the system is not finite state but can still be represented by a finite structure P/T-net. The **functionality** of a behavior B is equal to *exit iff* it terminates with the successful termination action δ, otherwise it is equal to *noexit*, [Bolo 87a].

PLotos specifications are rewritten into simpler forms. Non-recursive paths are expanded, that is, process definitions are substituted for process calls. Then we distinguish every parallel composition $B_1|[g_1,...,g_n]|B_2$ by labelling the operator with an unique value k. This is represented as $|[g_1,...,g_n]|_k$. For example,

process $p_1[a,b,c]$: **noexit** :=
$(p_2[a,b]|||p_2[a,b])$
$[]$
$c; p_1[a,b,c]$
endproc
process $p_2[a,b]$: **noexit** :=
$a; b; stop|[a]|a; stop$
endproc

is rewritten as:

process $p_1[a,b,c]$: **noexit** :=
$((a; b; stop\ |[a]|_1\ a; stop)$
$|||(a; b; stop\ |[a]|_2\ a; stop))$
$[]$
$c; p_1[a,b,c]$
endproc

5 Modelling of Lotos with P/T-nets

The mapping from Lotos to P/T-nets is based on the work of Olderog [Olde 87] for CCSP. In general, a Lotos behavior expression B represents the composition of several concurrent components. The expression B is explicitly decomposed into its parallel components that become tokens when this behavior is activated. Parallel components and states of parallel components are respectively modelled by Petri net tokens and places. The place in which a token is contained denotes the component state. Lotos gates are modelled by Petri net transitions. Tokens, contained in transition input places, represent components synchronized on this gate. Tokens deposited into output places represent the successor components after the transition has occurred. Several tokens, contained in the same place, represent several identical components. This models unbounded process instantiation with finite structure P/T-nets.

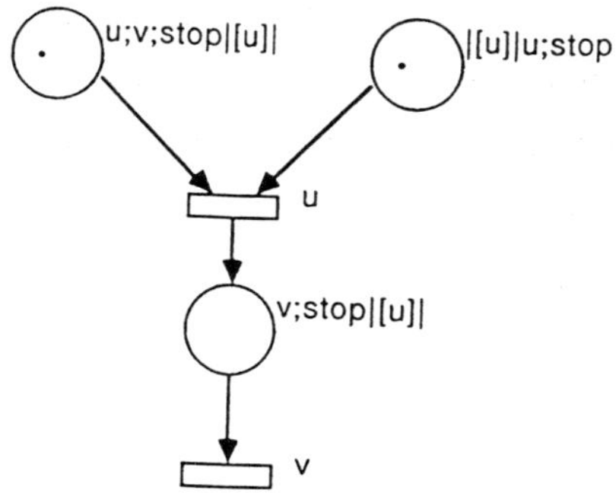

Figure 4: Example

For example, the Lotos expression $u; v; stop|[u]|u; stop$ represents two concurrent components. The first component executes actions u and v and then stops. The second component executes action u and becomes inactive. Both components are coupled on gate u and are therefore dependent on each other with respect to the occurrence of u. The decomposition of $u; v; stop|[u]|u; stop$ into components is denoted as the set $\{u; v; stop|[u]|, |[u]|u; stop\}$. In this syntax, we represent explicitly the fact that components are coupled on gate u by concatenating the symbol $|[u]|$ to the right of $u; v; stop$ and to the left of $u; stop$.

Fig. 4 depicts the corresponding P/T-net model. Places modelling states of components are labelled with the corresponding expression components. Transitions are labelled with gate names. The "stop" expression represents inaction and does not appear in the P/T-net. In our construction, edges from place to transition are always one valued (i.e. $\forall t, p.pre(t)(p)$ equals 0 or 1) and every place has a distinct label. **We unambiguously denote a place by its label.** The above "u" labelled transition is represented as the triple:

$$\{u; v; stop|[u]|, |[u]|u; stop\} - u \rightarrow \{v; stop|[u]|\}$$

To derive such triples, we define i) a function decomposing PLotos expressions, and ii) a system of inference rules. The head of each rule matches a term of the form:

$$\{p_1, ..., p_m\} - a \rightarrow \{q_1, ..., q_n\}$$

A rule can be used to infer, as a function of place label structures, a transition with preset $\{p_1, ..., p_m\}$, action a and postset $\{q_1, ..., q_n\}$. For instance the rule:

if $M_1 - a \rightarrow M_1'$ and $a \notin \{S, \delta\}$
then $M_1.|[S]|_k - a \rightarrow M_1'.|[S]|_k$

has been used to infer the transition:

$$\{v; stop|[u]|\} - v \rightarrow \{\}$$

We substituted $\{v; stop\}$, u and v to respectively M_1, S and a. M_1' is empty because the decomposition of "stop" is defined as the empty set. We first introduce the decomposition function in §5.1, then we present in §5.2 the inference rules.

5.1 Decomposition Function

The decomposition function is denoted as *dec*. Its domain is the set of well-formed PLotos behavior-expressions. Its range is the set of all possible multi-sets of place labels.

Let B_1, B_2 denote syntactically correct PLotos behavior expressions, a denote an action name and $S = g_1, ..., g_n$ a list of synchronization gates,

$$\begin{array}{lrl}
(d1) & dec(stop) & := \{\} \\
(d2) & dec(a; B_1) & := \{a; B_1\} \\
(d3) & dec(B_1[]B_2) & := \{B_1[]B_2\} \\
(d4) & dec(p[g_1, ..., g_n]) & := dec(B_p[g_1/h_1, ..., g_n/h_n]) \\
(d5) & dec(B_1|||B_2) & := dec(B_1) + dec(B_2) \\
(d6) & dec(B_1|[S]|_k B_2) & := dec(B_1).|[S]|_k + |[S]|_k.dec(B_2) \\
(d7) & dec(B_1 >> B_2) & := \{B_1 >> B_2\} \\
(d8) & dec(B_1[> B_2) & := \{B_1[> B_2\} \\
(d9) & dec(hide\ S\ in\ B_1) & := hide\ S\ in.dec(B_1) \\
(d10) & dec(exit) & := \{exit\}
\end{array}$$

where

- B_p represents the body of process definition p,
- in (d4), $g_1, ..., g_n$ is a list of actual gates,
- $h_1, ..., h_n$ is a list of formal gates,
- $[g_1/h_1, ..., g_n/h_n]$ is the relabelling postfix operator, gate h_i becomes gate g_i ($i = 1, ..., n$), and
- the expression $dec(B_1).|[S]|_k$ denotes $\{x|[S]|_k : x \in dec(B_1)\}$, similarly for $|[S]|_k.dec(B_2)$ and the expression $hide\ S\ in.dec(B_1)$ denotes $\{hide\ S\ in\ x : x \in dec(B_1)\}$.

The *dec* function is deterministic, taking into account operator precedences. The restriction to guarded recursive processes is required to stop recursion in the *dec* function. The relabelling operator is not user accessible and exists for the semantic description of process instantiation. In Lotos, relabelling is dynamic. Gates are renamed at the execution time. We show in [Barb 90a] that for injective relabelling operators, static and dynamic relabelling are equivalent. For the sake of simplicity, hereafter we consider solely injective relabellings and perform static renaming.

5.2 Inference Rules

This section presents the inference rules of the mapping from PLotos to P/T-nets. The P/T-net $N = (P, T, Act, M_0)$ associated to a PLotos behavior B is such that:

1. $M_0 = dec(B)$, $(\forall p)[M_0(p) > 0 \Rightarrow p \in P]$,
2. if $X \subseteq P$ and $X - a \rightarrow Y$ then $(\forall p)[Y(p) > 0 \Rightarrow p \in P]$, $(X, a, Y) \in T$, $a \in Act$, and
3. only the elements that can be obtained from items 1 or 2 are in P, T and Act.

The transition instances are inferred from the rules bellow.

For all PLotos expressions B_1, B_1', B_2, B_2', action name a, list $S = g_1, ..., g_n$ of synchronization gates and place multi-sets M_1, M_2, M_1', M_2':

(r1) $\{a; B_1\} - a \rightarrow dec(B_1)$
(r2) if $B_1 - a \rightarrow B_1'$
then $\{B_1 [] B_2\} - a \rightarrow dec(B_1')$
(r3) if $B_2 - a \rightarrow B_2'$
then $\{B_1 [] B_2\} - a \rightarrow dec(B_2')$
(r4) if $M_1 - a \rightarrow M_1'$ and $a \notin \{S, \delta\}$
then $M_1.|[S]|_k - a \rightarrow M_1'.|[S]|_k$
(r5) if $M_2 - a \rightarrow M_2'$ and $a \notin \{S, \delta\}$
then $|[S]|_k.M_2 - a \rightarrow |[S]|_k.M_2'$
(r6) if $M_1 - a \rightarrow M_1'$ and $M_2 - a \rightarrow M_2'$ and $a \in \{S, \delta\}$
then $M_1.|[S]|_k + |[S]|_k.M_2 - a \rightarrow M_1'.|[S]|_k + |[S]|_k.M_2'$
(r7) if $B_1 - a \rightarrow B_1'$ and $a \neq \delta$
then $\{B_1 >> B_2\} - a \rightarrow \{B_1' >> B_2\}$
(r8) if $B_1 - \delta \rightarrow B_1'$
then $\{B_1 >> B_2\} - \delta \rightarrow dec(B_2)$

(r9) if $B_1 - a \rightarrow B_1'$ and $a \neq \delta$
then $\{B_1[> B_2\} - a \rightarrow \{B_1'[> B_2\}$
(r10) if $B_1 - \delta \rightarrow B_1'$
then $\{B_1[> B_2\} - \delta \rightarrow dec(B_1)$
(r11) if $B_2 - a \rightarrow B_2'$
then $\{B_1[> B_2\} - a \rightarrow dec(B_2')$
(r12) if $M_1 - a \rightarrow M_1'$ and $a \notin \{S\}$
then $hide\ S\ in.M_1 - a \rightarrow hide\ S\ in.M_1'$
(r13) if $M_1 - a \rightarrow M_1'$ and $a \in \{S\}$
then $hide\ S\ in.M_1 - \mathrm{i} \rightarrow hide\ S\ in.M_1'$
(r14) $\{exit\} - \delta \rightarrow \{stop\}$

In the "if part" of inference rules (r2), (r3), (r7) (r8), (r9), (r10) and (r11) behavior B_1 (B_2) makes a transition to behavior B_1' (B_2') on action a or δ in accordance with the original Lotos semantics in [Bolo 87a].

Theorem 1 *The Petri net semantics of Lotos is consistent with the standard Lotos semantics. That is, for all PLotos behavior expression B and marking M with $dec(B) := M$:*

- $(\exists B')(\exists a)B - a \rightarrow B' \Rightarrow (\exists M')(\exists t)M(t > M' \wedge act(t) = a \wedge dec(B') := M'$
- $(\exists M')(\exists t)M(t > M' \Rightarrow (\exists B')B - act(t) \rightarrow B' \wedge dec(B') := M'$

An interesting property of our P/T-net semantics for Lotos is:

Corollary 1 *Let B be a PLotos behavior expression, with transition graph TG, and let $N = (P, T, Act, M_0)$ be the associated P/T-net with reachability graph $RG(N)$. The dec function is a homomorphism from TG reachability set to $RG(N)$ reachability set.*

The *dec* function is a graph homomorphism because it identifies equivalent Lotos behavior expressions $B_1|||B_2$ with $B_2|||B_1$, and $B_1|||(B_2|||B_3)$ with $(B_1|||B_2)|||B_3$. These equivalences are in accordance with the commutativity and the associativity laws in [ISO 88]. Solely syntactic nature information is lost, "*dec*" preserves all semantic properties. This can be illustrated by the following commutative diagram:

$$\begin{array}{ccc} B & -dec \rightarrow & M \\ | & & | \\ a & & a \\ \downarrow & & \downarrow \\ B' & -dec \rightarrow & M' \end{array}$$

Example 3: Consider the following specification[3]:

specification $p1[a,b,c]$: **exit** :=
$p2[a,b]|[a]|p2[a,c]$
where
process $p2[x,y]$: **exit** :=
$x; exit >> (y; stop|||p2[x,y])$
endproc
endspec

The P/T-net derived from this specification is identical to the net depicted in Fig. 1 with:

$p_1 = a; exit >> (b; stop|||p2[a,b])|[a]|$ $p_2 = |[a]|a; exit >> (c; stop|||p2[a,c])$
$p_3 = exit >> (b; stop|||p2[a,b])|[a]|$ $p_4 = |[a]|exit >> (c; stop|||p2[a,c])$
$p_5 = b; stop|[a]|$ $p_6 = |[a]|c; stop$

6 Coverability Graphs for Lotos

Given a Lotos specification, it is possible to construct an equivalent P/T-net model by successive applications of the above inference rules. This P/T-net then becomes the input of the reachability analysis algorithm to evaluate the properties. In the worst case, the P/T-net can have more vertices and edges than the coverability graph. In our approach we skip the intermediate Lotos to P/T-nets translation step. We derive the coverability graph directly from the Lotos specification then properties are evaluated.

The syntax of Lotos coverability graphs slightly deviates from the usual syntax for Karp and Miller graphs. Markings are multi-sets of Lotos behavior expression components. We label the root of the graph with the decomposition of the Lotos expression that represents the initial behavior. For example, the decomposition of the initial behavior in Example 3 yields a state represented as the following box:

$1/a; exit >> (b; stop|||p2[a,b])|[a]|$
$1/|[a]|a; exit >> (c; stop|||p2[a,c])$

Every line in the box defines the number of instances of one behavior expression component type in the current state. In case there is an infinite number of occurrences, the expression component is paired with the ω symbol.

[3]For the sake of simplicity, labelling of the $|[a]|$ operator is omitted in this example.

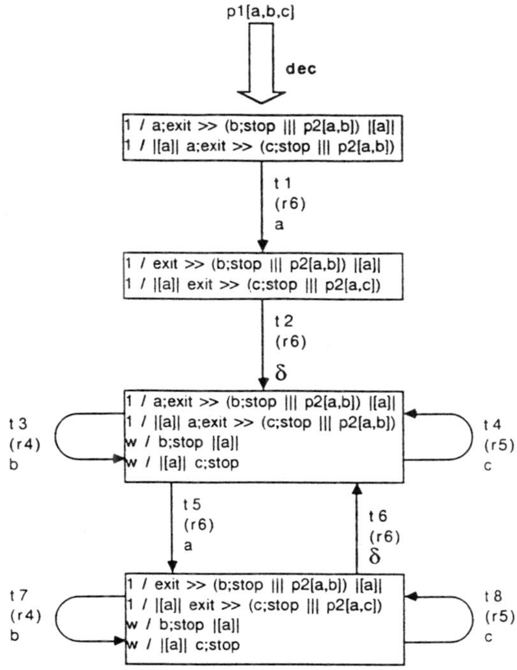

Figure 5: Coverability graph of Example 3

We go from one marking to another by application of the inference rules. An inference rule is applicable from one marking if a finite subset of the expression component multi-set matches the preset of the transition in the head of the rule. The successor state is obtained by removing this preset from the current state and adding the postset defined by the transition (reformulation of the usual P/T-net firing rule). Every edge is labelled with the number of the inference rule which has been applied to derive the transition and the action name of the transition. The coverability graph of Example 3 is illustrated in Fig. 5.

The six problems stated in § 3 can be solved as follows:

1. *Termination* The reachability tree is infinite if there is at least one circuit in the CG.

2. *Finiteness* The reachability set is infinite if the CG contains one marking and one process p paired with ω.

3. *Coverability* Given a marking M, there exists a reachable marking M' such that $M \leq M'$, if there exists in the CG a marking M'' with $M \leq M''$.

4. *Quasi-liveness* The action a is quasi-live if there exists an edge in the CG labelled with a.

5. *Boundedness* Instantiation of process p is unbounded if there exists a marking M in which $dec(p)$ is paired with ω.

6. *Regularity* The language is regular if every elementary circuit of the CG is labelled by a sequence of transitions $t_1, t_2, ..., t_n$ such that for every place p:

$$post(t_1)(p) - pre(t_1)(p) + post(t_2)(p) - pre(t_2)(p) + \cdots post(t_n)(p) - pre(t_n)(p) \geq 0$$

In general, conclusions can be easily drawn from visual inspection of the CT and the CG. For instance, the language of Example 3 is not regular since:

$$post(t_3)(b; stop|[a]|) - pre(t_3)(b; stop|[a]|) = -1 < 0$$

7 Conclusion

We have presented a reasonable subset of Lotos that can be verified using Petri net reachability analysis techniques. Our method does not require explicit translation from Lotos to Petri nets. Analysis is performed in the Lotos world to which the Karp and Miller procedure is extended. To cope with state space explosion, MCT and MCG can be computed to solve the aforementioned six problems. We experimented the MCT construction procedure and obtained satisfactory results. The MCTs were several times less complex than the nonminimal CTs.

8 References

[Barb 90a] Barbeau, M., Bochmann, G. V. *Deriving Analysable Petri Nets from Lotos Specifications*, Research report no. 707, Dept. d'IRO, Université de Montréal, 1990.

[Barb 90b] Barbeau, M., Bochmann, G. V. *Verification of Lotos Specifications: A Petri Net Based Approach*, Proc. of Canadian Conference on Electrical and Computer Engineering, Ottawa, 1990.

[Barb 89] Barbeau, M., Bochmann, G. V. *Experiences with Automated Verification Tools: Application to Discrete Event Systems*, Proc. of Workshop on Automatic Verification Methods for Finite State Systems, Grenoble, 1989.

[Boch 78] Bochmann, G. V. *Finite State Description of Communication Protocols*, Computer Network 2, (361-372), 1978.

[Bolo 87a] Bolognesi, T., Brinksma, E. *Introduction to the ISO Specification Language Lotos*, Computer Networks and ISDN, Vol. 14, No. 1, (25-59), 1987.

[Bolo 87b] Bolognesi, T., Smolka, S. A. *Fundamental Results for the Verification of Observational Equivalence: A Survey*, Proc. of PSTV VII, Zurich, 1987.

[Fink 90] Finkel, A. *A Minimal Coverability Graph for Petri Nets*, Proc. 11th Int. Conf. on Application and Theory of Petri Nets, Paris, 1990.

[Gara 89] Garavel, H., Najm, E. *Tilt: From Lotos to Labelled Transition Systems*, in P. H. J. van Eijk, C. A. Vissers and M. Diaz (Eds.): The Formal Description Technique Lotos, North-Holland, 1989.

[Holz 89] Holzmann, G. J. [1989]. *Algorithms for Automated Protocol Validation*, Proc. of Workshop on Automatic Verification Methods for Finite State Systems, Grenoble 1989.

[ISO 88] ISO *Lotos - A Formal Description Technique Based on the Temporal Ordering of Observational Behavior*, ISO 8807, E. Brinksma (Ed.), 1988.

[Karp 69] Karp, R. M., Miller, R. E. *Parallel Program Schemata*, J. Computer and System Sciences 3, (147-195), 1969.

[Marc 89] Marchena, S., Leon, G. *Transformation from Lotos Specs to Galileo Nets*, in K. J. Turner (Ed.): Formal Description Techniques, North-Holland, 1989.

[Olde 87] Olderog, E.-R. *Operational Petri Net Semantics for CCSP*, LNCS 266, Springer-Verlag, 1987.

[Pete 81] Peterson, J. L. *Petri Net Theory and the Modelling of Systems*, Prentice Hall, 1981.

[Rea 87] Rea, K, Johnston, R. deB. *Automated Analysis of Discrete Communication Behavior*, IEEE TSE, Vol. SE-13, No. 10, 1987.

DIMACS Series in Discrete Mathematics
and Theoretical Computer Science
Volume 3, 1991

Formal Verification of Digital Circuits Using Symbolic Ternary System Models*

Randal E. Bryant
School of Computer Science
Carnegie Mellon University
Pittsburgh, PA 15213 USA

Carl-Johan H. Seger
Department of Computer Science
University of British Columbia
Vancouver, B.C. V6T 1W5
Canada

Abstract

Ternary system modeling involves extending the traditional set of binary values $\{0, 1\}$ with a third value X indicating an unknown or indeterminate condition. By making this extension, we can model a wider range of circuit phenomena. We can also efficiently verify sequential circuits in which the effect of a given operation depends on only a subset of the total system state.

This paper presents a formal methodology for verifying synchronous digital circuits using a ternary system model. The desired behavior of the circuit is expressed as assertions in a notation using a combination of Boolean expressions and temporal logic operators. An assertion is verified by translating it into a sequence of patterns and checks for a ternary symbolic simulator. This methodology has been used to verify a number of full scale circuit designs.

1. Introduction

1.1. Ternary Modeling

Most formal models for hardware verification assume that every signal always has a well-defined, discrete value. For example, a *binary* model assumes that each signal must be either 0 or 1. In this paper we present a methodology for formal verification in which a third value

*This research was supported by the Defense Advanced Research Projects Agency, ARPA Order Number 4976, and by the National Science Foundation, under grant number MIP-8913667.

X is added to the set of possible signal values, indicating an unknown or indeterminate logic value. By shifting to a ternary system model, we gain several advantages.

As a first advantage, this extension makes it possible to model an increased range of circuit phenomena. For example, we can deal with circuits in which nondigital voltages are generated in the course of normal circuit operation. This occurs frequently when modeling circuits at the switch-level [6], due to (generally transient) short circuits or charge sharing. We can also deal with circuits in which indeterminate behavior occurs due either to timing hazards or to circuit oscillation. In all of these cases, the modeling algorithm expresses this uncertainty by assigning value X to the offending circuit nodes, indicating that the actual digital value cannot be determined [8,15].

As a second advantage, we can efficiently verify many aspects of digital circuit behavior by representing the circuit with a ternary system model. We do this by *ternary symbolic simulation*, in which a simulation algorithm designed to operate on scalar values 0, 1, and X, is extended to operate on a set of symbolic values. Each symbolic value indicates the value of a signal for many different operating conditions, parameterized in terms of a set of symbolic Boolean variables. Since the value X indicates that a signal could be either 0 or 1 (or a non-digital voltage), we can often represent many different operating conditions by the constant value X, rather than with a more complex symbolic value. For example, we can verify that a particular sequence of actions will yield a 1 (or 0) on some node regardless of the initial state by verifying that this value results when starting from an initial state where all nodes are set to X. This requires far less effort than analyzing the effect of the action on all possible initial binary states.

Simulators that support ternary modeling intentionally err on the side of pessimism for the sake of efficiency. That is, they will sometimes produce a value X even where exhaustive case analysis would indicate that the value should be binary (i.e., 0 or 1). For example, most ternary simulators evaluate logic functions in a ternary algebra created by extending the standard Boolean operators. This algebra does not obey the law of excluded middle, because $X +_t \overline{X}^t = X$, where $+_t$ and $^{-t}$ are ternary extensions of Boolean sum and complement, respectively. On the other hand, symbolic simulation avoids this pessimism, because it can resolve the interdependencies among signal values, and compute $a + \overline{a} = \mathbf{1}$ (the Boolean function that always yields 1). By combining the expressive power of symbolic values with the computational efficiency of ternary values, we can trade off precision for ease of computation.

When creating a ternary system model, we impose the following monotonicity requirement: In the presence of X signals as stimuli, no action of the circuit should yield a binary value, unless the same value would occur if any subset of these stimuli had binary values instead. That is, when given incomplete information about the exact circuit state, we should never produce a signal inconsistent with one that would result if more information were available. This monotonicity condition makes it possible to verify properties of circuit operation in the presence of potential sources of indeterminate behavior by representing this indeterminacy with the value X. It also makes it possible to verify properties under a ternary system model as a means of proving properties under a binary system model.

1.2. Contribution of Paper

In earlier work, we demonstrated the utility of ternary modeling for verifying a variety of circuits [1,7]. Our methodology is based on ternary simulation, either scalar or symbolic. With a simulator we can verify assertions specifying a postcondition on the circuit state that would result given some precondition on the previous state and some condition imposed on the input values. By restricting the form of the precondition, postcondition, and input condition to specifying that some of the nodes have particular binary values, we can verify such an assertion by a single simulation sequence. That is, we perform a simulation in which any node that is constrained by the precondition or action is set to its specified value, while all other nodes are set to X. If the resulting node states satisfy those specified by the postcondition, then the assertion is proved. For some classes of circuits, e.g., random access memories, we can verify correct circuit operation for all possible inputs and initial states by simulating only a polynomial number of scalar patterns. For other circuits, we can overcome the combinatorial complexity of considering many different initial states and input actions via symbolic simulation.

This earlier work demonstrated the viability of circuit verification by symbolic simulation, but it fell short in terms of generality, ease of use, and degree of automation. We did not have a formal notation for specifying the desired circuit properties, nor a method to generate simulation patterns directly from the specification. Instead, we derived symbolic simulation patterns by hand and argued informally that these patterns served to verify the desired properties. Furthermore, it was particularly cumbersome to verify operations requiring multiple state transitions, such as occurs in pipelined systems.

In this paper, we correct this shortcoming by presenting a formal state transition model for a ternary system, a formal syntax for expressing desired properties of the system, and an algorithm to decide whether or not the system obeys the specified property. Our state transition system is quite general, and is compatible with a number of circuit modeling techniques.

Our specifications take the form of *symbolic trajectory formulas* mixing Boolean expressions and the temporal *next-time* operator. The Boolean expressions provide a convenient means of describing many different operating conditions in a compact form. For example, we can express the desired behavior of an ALU on all possible inputs. By allowing only the most elementary of temporal operators, the class of properties we can express is relatively restricted, as compared to other temporal logics [10,17]. In particular, we can only reason about circuit operations involving a bounded number of state transitions. Nonetheless, we have found that we can readily express many aspects of synchronous digital systems. It is quite adequate for expressing many of the subtleties of system operation, including clocking conventions and pipelining.

Our decision algorithm is based on ternary symbolic simulation. It tests the validity of an assertion of the form $[A \Longrightarrow C]$, where both A and C are trajectory formulas. That is, it determines whether or not every state sequence satisfying A (the "antecedent") must also satisfy C (the "consequent"). It does this by generating a symbolic simulation sequence corresponding to the antecedent, and testing whether the resulting symbolic state sequence satisfies the consequent.

An important property of our algorithm is that it requires a comparatively small amount of simulation and symbolic manipulation to verify an assertion. The restrictions we impose on the formula syntax guarantee that there is a unique weakest symbolic sequence satisfying the antecedent. Furthermore, the symbolic manipulations involve only variables explicitly mentioned in the assertion. Unlike other symbolic circuit verifiers [3], we do not need to introduce extra variables denoting the initial circuit state or possible primary inputs. Finally, the length of the simulation sequence depends only on the depth of nesting of temporal next-time operators in the assertion.

Some insight into the expressive power of our specification language can be gained by reviewing the different classes of properties examined by Pnueli in his original paper on temporal logic [17]. In particular, our notation is adequate for expressing many *invariant* properties of the system. That is, we formulate an assertion where the antecedent expresses a set of guaranteed constraints on the circuit inputs and state over multiple time intervals, while the consequent expresses some additional desired constraints on the state. These desired constraints generally include the state guarantees of the antecedent shifted to a later time. By verifying such an assertion, we prove that the desired properties of the consequent hold over the entire operation of the circuit as long as the inputs are supplied correctly. On the other hand, properties that require reasoning about *eventuality* cannot be expressed with our notation, unless the desired property is guaranteed to hold within a bounded time.

1.3. Overview of Implementation

By modifying the COSMOS symbolic simulator[5], we have been able to implement the algorithm described in this paper and to verify several full scale circuit designs. COSMOS represents a MOS circuit at the switch level as a network of transistor switches connecting charge storage nodes. The simulator preprocesses the circuit to transform it into a Boolean representation [6]. Using this Boolean representation, the switch-level behavior of the circuit can be computed by simply evaluating a series of Boolean operations, much as one would in a gate-level circuit model. The preprocessor is quite general and accurate. The resulting representation is guaranteed to produce the exact same results as more traditional switch-level simulators.

Simulating a circuit symbolically involves evaluating these Boolean operations over the algebra generated by a set of Boolean variables. That is, each element of the algebra corresponds to a Boolean function over the variables. By representing these functions as Ordered Binary Decision Diagrams [4], complex functions can be represented and manipulated efficiently. Of course, all of the verification properties we wish to decide require solving NP-hard problems. Our approach has a worst case time and space requirement that grows exponentially with the size of the formulas expressing the circuit property to be decided. Nonetheless, we have been successful at avoiding exponential blow-up for many useful cases. Our system model and checking algorithm is not strongly tied to the use of BDDs. In this work, we view our BDD code as simply a package for representing and manipulating Boolean functions symbolically.

COSMOS supports a ternary model of switch-level circuit operation, where X indicates either an unknown or potentially non-digital node voltage. It does this by encoding each node state as a pair of binary values, according to a "dual rail" coding of the circuit state. The excitation function for a node is given by a pair of Boolean expressions specifying how the encoded new state of the node should be computed from the encoded old states [6]. This implementation decision greatly helped the development of our symbolic simulator. When simulating the circuit symbolically, we simply represent the state of each node by pointers to the roots of two BDDs, and compute the new states of a node by evaluating the excitation expressions symbolically.

The following table indicates the performance of our verifier on several different circuits. All CPU times were measured on a DEC 3100 (a 10–20 MIPS machine). We also list the maximum memory requirement of the process, as this is more often the limiting factor in BDD-based symbolic manipulation than is CPU time.

Circuit	Transistors	CPU Time	Memory
64 × 32 bit moving data stack	16,470	1.25 min.	3.1 MByte
64 × 32 bit stationary data stack	15,873	7.5 min.	5.7 MByte
1K static RAM	6,875	3.7 min.	9.5 MByte

The two stack examples demonstrate the abstraction capability of our approach. Both started with the same high level specification—defining the behavior of PUSH, POP, and HOLD operations in an abstract stack. We then define how the circuit represents the abstract machine state by trajectory formulas, examples of which will be shown later. The two different circuits represent the abstract state in totally different fashions. In the moving data stack, bit i (counting from the top of stack) of the abstract machine is stored in cell i of the circuit. In the stationary data stack, bit i is stored in location $d - i$ of a static RAM, where d is the current depth of the stack. As a consequence, the symbolic values manipulated in verifying the stationary data stack are more complex than those for the moving data stack. In both cases, however, the performance was acceptable.

The static RAM example indicates just how efficient symbolic verification can be. We have already demonstrated that this class of circuit can be verified by simulating just $O(N \log N)$ scalar patterns [7]. By exploiting the bit-level parallelism of the logical instructions in a conventional machine, we were able to simulate 32 of these patterns at a time. Even so, the symbolic verification outperforms the scalar verification by a factor of 4. Furthermore, creating the assertions for the symbolic verification was far more straightforward than was generating the patterns for the scalar verification. For the scalar verification, we had to consider details of the memory's row and column addressing structure to avoid errors caused by the pessimistic modeling of X. No such tuning was required for the symbolic verification.

We are now applying our verifier to more complex circuits such as pipelined data paths and simple microprocessors. We have found the expressive power of our notation and the performance of the verifier acceptable in most of the cases we have considered.

1.4. Related Work

Our approach to verification relates most closely to the symbolic model checking algorithms devised by Bose and Fisher (BF) [2,3], and by Burch, Clarke, McMillan, and Dill (BCMD) [9]. In fact, all of these approaches are implemented using the same Boolean manipulation code! Furthermore, Bose and Fisher implemented their checker by extending COSMOS. Despite these internal similarities, however, there are significant differences in the capabilities and complexities of the algorithms. In particular, our method is the most restricted in terms of the class of properties that can be verified. Both the BF and BCMD algorithms can decide a class of formulas consisting of a complete branching time, propositional temporal logic under a binary system model. Our method can only be used to verify properties of bounded state sequences. What we loose in expressive power, however, we make up for in computational efficiency. The computational effort required by our model checker is considerably less than theirs. One can view the combined effect of these research projects as providing a spectrum of checking-based verifiers that trade off between expressiveness and performance.

In the BF algorithm, the underlying circuit is assumed to be synchronous and deterministic. Nondeterministic sequential behavior arises due only to the different signals that may be applied to the inputs (expressed as existentially quantified Boolean variables). Their algorithm requires creating an explicit representation of the next state function for every state variable in the system. They create this representation by symbolic simulation. That is, the user identifies each place state is stored in the system, either as charge on a node, or as a pair of complementary values within a static memory element. They then symbolically simulate a single system cycle, where each state variable and each input signal is represented by a distinct Boolean variable. Once they have obtained this Boolean representation of the next state behavior, the validity of a temporal formula can be derived by (rather extensive) symbolic Boolean manipulation. This process of extracting the explicit next state function can be quite costly. In contrast, our method represents the next state function implicitly as a combination of circuit structure and simulation algorithm. We only compute the next state behavior for the particular patterns required to verify a given assertion. These patterns involve far fewer variables than is required by Bose and Fisher's functional extraction.

In the BCMD algorithm, the underlying system can be nondeterministic, and the user can even impose fairness constraints on the model. To support this great expressive power, they require an explicit representation of the next state relation for the entire system. That is, for each state variable in the system they have one Boolean variable representing its "old" value and one representing its "new" value. The next state relation is represented as a Boolean function of all of these variables, where the function yields 1 when the old and new state are related, and 0 otherwise. Generating this characteristic function is even more difficult than generating representations of the individual next state functions for each state variable.

Most other automated approaches to sequential circuit verification are based on testing state machine equivalence [11,13]. Such methods are useful for comparing two different (but hopefully equivalent) representations of the system, such as one at a register-transfer level and one at a gate level. However, they do not work well for verifying the correctness of incompletely specified systems, nor for reasoning about systems that employ methods, such as pipelining, that shift the sequencing of activities in time. Furthermore, most of these methods assume that the system starts in some known initial state. In actual circuits, the initial state usually cannot be predicted.

Other researchers have suggested symbolic simulation as a means of circuit verification [12,18]. None of this work has presented a clear methodology for sequential circuit verification, however.

2. Ternary System

Let $\mathcal{B} = \{0, 1\}$ be the set of the binary values and let $\mathcal{T} = \{0, 1, X\}$. The value X is introduced to denote an "unknown", or "don't care" value.

Define the partial order $\sqsubseteq$ on $\mathcal{T}$ as follows: $a \sqsubseteq a$ for all $a \in \mathcal{T}$, $X \sqsubseteq 0$, and $X \sqsubseteq 1$. The partial ordering orders values by their "information content." That is, X indicates an absence of information while 0 and 1 represent specific, fully-defined values.

We say that ternary values a and b are *compatible*, denoted $a \sim b$, when there is some value $c \in \mathcal{T}$ such that $a \sqsubseteq c$ and $b \sqsubseteq c$. In other words, two scalar values are compatible unless one is 0 and the other is 1.

The *meet*, denoted $\sqcap$, of two ternary values a and b is defined as the largest element $c \in \mathcal{T}$ in the partial order such that $c \sqsubseteq a$ and $c \sqsubseteq b$. For scalar values, if $a = b$, then $a \sqcap b = a$, while if $a \neq b$ then $a \sqcap b = X$. Furthermore, given two compatible ternary values a and b, the *join* between them, denoted $a \sqcup b$, is defined to be the smallest element $c \in \mathcal{T}$ in the partial order such that $a \sqsubseteq c$ and $b \sqsubseteq c$.

It is convenient to define an algebra over $\mathcal{T}$ with operators $\sqcup$, $\sqcap$, as well as the operators $\cdot_t$, $+_t$, and $^{-t}$, where the latter are defined in Table 1. These operations are simply extensions of the corresponding Boolean operations $\cdot$ (product), $+$ (sum), and $^{-}$ (complement).

Let $\mathcal{T}^n$, $n \geq 1$, denote the set of all possible vectors of ternary values of length n, i.e., $\{\langle a_1, \ldots, a_n \rangle | a_i \in \mathcal{T}, 1 \leq i \leq n\}$. The partial order $\sqsubseteq$ is extended to $\mathcal{T}^n$ pointwise: $\vec{a} \sqsubseteq \vec{b}$

$\cdot_t$	a: 0	1	X
b: 0	0	0	0
1	0	1	X
X	0	X	X

$+_t$	a: 0	1	X
b: 0	0	1	X
1	1	1	1
X	X	1	X

a	$\overline{a}^t$
0	1
1	0
X	X

Table 1: Ternary Extensions of $\cdot$, $+$, and $^-$.

iff $a_i \sqsubseteq b_i$ for $1 \leq i \leq n$. Similarly, if $\vec{a} \in \mathcal{T}^n$ and $\vec{b} \in \mathcal{T}^n$, we say $\vec{a} \sim \vec{b}$ iff there exists a vector $\vec{c} \in \mathcal{T}^n$ such that $\vec{a} \sqsubseteq \vec{c}$ and $\vec{b} \sqsubseteq \vec{c}$. In other words, two vectors of scalar ternary values are not compatible when one vector has a 1 in some position and the other vector has a 0 in the same position. Finally, the operations $\sqcup$ and $\sqcap$ are extended pointwise. Note that $\vec{a} \sqcup \vec{b}$ is defined only when $\vec{a} \sim \vec{b}$.

A ternary function, $f: \mathcal{T}^n \rightarrow \mathcal{T}$, is said to be *monotone* when for any $\vec{a} \in \mathcal{T}^n$ and $\vec{b} \in \mathcal{T}^n$ we have

$$\vec{a} \sqsubseteq \vec{b} \Rightarrow f(\vec{a}) \sqsubseteq f(\vec{b})$$

This definition is extended pointwise to vector functions, $\vec{f}: \mathcal{T}^n \rightarrow \mathcal{T}^m$.

The above monotonicity definition is consistent with our use of information content. If a function is monotone, we cannot "gain" any information by reducing the information content of the arguments to the function. In other words, changing some signals from binary values to X will either have no effect on the output values, or it will change some binary values to X.

To express the behavior of a circuit operating over time, we must reason about *sequences* of states. Conceptually, we will consider the state sequences to be infinite, although the properties we will express can always be determined from some bounded length prefix of the sequence. Define a the set $\mathcal{S}^n$ to consist of all sequences $[\vec{a}_0, \vec{a}_1, \ldots]$ where each $a_i \in \mathcal{T}^n$. The relations $\sqsubseteq$ and $\sim$ are extended from vectors to sequences pointwise. That is, two sequences $[\vec{a}_0, \vec{a}_1, \ldots]$ and $[\vec{b}_0, \vec{b}_1, \ldots]$ are ordered (compatible) if and only if each pair $\vec{a}_i$ and $\vec{b}_i$ is ordered (compatible), for all $i \geq 0$.

For vector $\vec{a}$ and sequence S, the expression $\vec{a}S$ denotes the sequence consisting the vector $\vec{a}$ followed by the vectors in S.

3. Circuit Model

The underlying model of a circuit we use is quite simple, as well as general. A circuit $\mathcal{C}$ is a triple $(\mathcal{N}, \vec{Y}, \mathcal{V})$, where

$\mathcal{N}$: is a set of nodes. Let $s = |\mathcal{N}|$.

$\vec{Y}$: is a vector of excitation functions.

$\mathcal{V}$: is a set of symbolic Boolean variables with which parameterized properties of the circuit are to be expressed.

In the mathematical presentation we will refer to the nodes as $n_1, n_2, \ldots, n_s$, whereas in our examples we often will use more descriptive names. In logic gate circuits the nodes correspond to the primary inputs and the outputs of the gates. In switch-level circuits the nodes correspond to electrical nodes in the circuit.

The excitation functions are defined in a non-traditional way. We view them as expressing "constraints" on the values the nodes can take on one time unit later given the current values on the nodes. By constraint we mean specific binary values, whereas the value X indicates that no constraint is imposed. Since the value of an input is controlled by the external environment, the circuit itself does not impose any constraint on the value; hence the excitation of an "input node" is X. More formally, if node n_i corresponds to an input to the circuit then $Y_{n_i}(\vec{a}) = X$ for every $\vec{a} \in \mathcal{T}^s$. Nodes that do not correspond to inputs are called *function nodes*. For a function node n_i the excitation function is a monotone ternary function $Y_{n_i}: \mathcal{T}^s \to \mathcal{T}$ determined by the circuit topology and functionality. For example, if the current input to a unit delay inverter is 0, then the output of the inverter one time unit later is constrained to be 1.

It should be pointed out that the "time unit" referred to above is the smallest period of time that is distinguishable in the circuit model. The minimum delay in any individual component of the circuit can be significantly larger. Thus we are not limited to unit delay circuit models. For example, by using the transformation technique described in [19], both nominal delay and bounded delay circuit models can be used.

The excitation function for a function node is often a ternary extension of a binary excitation function. Note that such an extension can be done in many ways. However, we require that the extended function agrees with the binary on binary inputs, and that the ternary function obtained is monotone. These requirements simply ensure that we do not "loose" information by extending a binary excitation function to the ternary domain. In other cases, like in switch-level models, the excitation functions derived are already ternary, since even binary signals can generate nondigital voltages on some nodes.

3.1. Circuit Trajectories

State sequences are useful when reasoning about circuit behaviors. However, not all state sequences represent possible behaviors of a circuit. The excitation functions generally restrict the possible state sequences significantly. We formalize this property by introducing the concept of a circuit trajectory. Given a circuit $\mathcal{C}$ and an arbitrary sequence $[\vec{a}_0, \vec{a}_1, \ldots] \in \mathcal{S}^s$

we say that the sequence is a *circuit trajectory* if and only if

$$\vec{Y}(\vec{a}_i) \sqsubseteq \vec{a}_{i+1} \text{ for } i \geq 0.$$

The set of all trajectories of circuit C is denoted $S(C)$. The above rule for trajectories is consistent with our definition of an excitation function, i.e., a function computing a constraint on the possible value of a node one time unit later. Thus if the current excitation of a node is binary, say a, then the node must take on the value a in the next state in a valid trajectory. On the other hand, if the excitation is X, then the node value is not constrained.

On first reading, it may seem strange to define a circuit model where a node with excitation X may spontaneously change to either 0 or 1. This is our way of capturing the property that a circuit may exhibit a variety of different behaviors due to variations in the initial state, the primary input values, and the outcome of marginal operating conditions. An input node may change to 0 or 1 on every step, reflecting the fact that its value is determined solely by the operating environment. An uninitialized internal node may change to 0 or 1 to reflect the fact that it could have had either of these as initial values. An internal node that was set to X due to marginal operating conditions can change to 0 or 1 to reflect that the actual node state could resolve in either direction.

To illustrate trajectories, consider a unit delay inverter. Assume the circuit contains two nodes: the inverter input n_1 and the inverter output n_2. The excitation functions are $\vec{Y}(\vec{a}) = \langle X, \overline{a_1}^t \rangle$. It is easy to accept that, for example, $\langle 0,1\rangle, \langle 0,1\rangle, \langle 0,1\rangle, \ldots$ and $\langle 0,1\rangle, \langle 1,1\rangle, \langle 1,0\rangle, \langle 1,0\rangle, \ldots$ are valid circuit trajectories. Less intuitive is that a sequence like $\langle X,0\rangle, \langle X,1\rangle, \langle X,0\rangle, \langle X,1\rangle, \ldots$ is also a valid circuit trajectory. Our methodology verifies properties for all circuit trajectories that satisfies some conditions, including degenerate trajectories as this one. Consequently it will perform a stronger verification than absolutely necessary. Again, this pessimism stems from our use of X as a completely unknown value.

4. Specification Language

Our specification language describes a property of the circuit as an *assertion* of the form $[A \Longrightarrow C]$, where both A and C are *symbolic trajectory formulas* expressing constraints on the circuit trajectory.

4.1. Symbolic Expressions

Before we can define our language, we need to introduce some notation and definitions. If $\mathcal{V}$ is a set of symbolic Boolean variables then an *interpretation*, ϕ, is a function $\phi\colon \mathcal{V} \rightarrow \mathcal{B}$ assigning a binary value to each variable. Let Φ be the set of all possible interpretations, i.e., $\Phi = \{\phi\colon \mathcal{V} \rightarrow \mathcal{B}\}$. A *domain constraint*, $\mathcal{D} \subseteq \Phi$, defines a restriction on the values assigned to the variables. We will denote such domain constraints by Boolean expressions.

That is, let E be a Boolean expression over elements of $\mathcal{V}$.[1] This expression defines a Boolean function $e\colon \Phi \to \mathcal{B}$ and thus denotes the domain constraint $\mathcal{D} = \{\phi | e(\phi) = 1\}$. The set of all interpretations Φ is denoted by the Boolean function $\mathbf{1}$, defined as yielding 1 for all interpretations. Expressing domain constraints by Boolean expressions allows us to compactly specify many different circuit operating conditions with a single formula.

4.2. Symbolic Trajectory Formulas

A trajectory formula expresses a set of constraints on a circuit trajectory. When the formula contains Boolean expressions, each interpretation of the variables yields a different set of constraints.

A *step-level* symbolic trajectory formula is defined recursively as:

1. **Constants:** TRUE is a trajectory formula.
2. **Atomic propositions:** for $n_i \in \mathcal{N}$ the following are trajectory formulas:
   (a) $(n_i = 1)$
   (b) $(n_i = 0)$.
3. **Conjunction:** $(F_1 \wedge F_2)$ is a trajectory formula if F_1 and F_2 are trajectory formulas.
4. **Domain restriction:** $(E \to F)$ is a trajectory formula if E is a Boolean expression over $\mathcal{V}$ and F is a trajectory formula.
5. **Next time:** $(\mathbf{X}_s F)$ is a trajectory formula if F is a trajectory formula.

We say that a formula is *instantaneous* when it does not contain any next time operator $\mathbf{X}_s$. For convenience, we often drop parentheses when the intended precedence is clear.

The truth of a formula F is defined relative to a circuit, an interpretation ϕ of the variables in $\mathcal{V}$, and a circuit trajectory. The truth of F, written $\mathcal{C}, \phi, S \models F$, is defined recursively. In the following, assume that both S and $\vec{a}S$ are trajectories of $\mathcal{C}$.

1. $\mathcal{C}, \phi, S \models$ TRUE holds trivially.
2. (a) $\mathcal{C}, \phi, \vec{a}S \models (n_i = 1)$ iff $a_i = 1$.
   (b) $\mathcal{C}, \phi, \vec{a}S \models (n_i = 0)$ iff $a_i = 0$.
3. $\mathcal{C}, \phi, S \models (F_1 \wedge F_2)$ iff $\mathcal{C}, \phi, S \models F_1$ and $\mathcal{C}, \phi, S \models F_2$

[1]For the sake of brevity, we omit a formal syntax of Boolean expressions. Any standard expression syntax suffices.

4. $\mathcal{C}, \phi, S \models (E \rightarrow F)$ iff $e(\phi) = 0$ or $\mathcal{C}, \phi, S \models F$, where e is the Boolean function denoted by the Boolean expression E.

5. $\mathcal{C}, \phi, \vec{a}S \models \mathbf{X}_s F$ iff $\mathcal{C}, \phi, S \models F$.

For an instantaneous formula, its truth can be defined relative to a single state. For instantaneous formula F, the notation $\mathcal{C}, \phi, \vec{a} \models F$ indicates that F holds under interpretation ϕ for state $\vec{a}$. A formal definition of this notation can be derived by a straightforward adaptation of rules 1–4 above.

4.3. Assertions

Our verification methodology entails proving *assertions* about the model structure. These assertions are of the form $[A \Longrightarrow C]$, where the *antecedent* A and the *consequent* C are trajectory formulas. The truth of an assertion is defined relative to a circuit $\mathcal{C}$ and an interpretation ϕ. Unlike a formula, however, an assertion is considered true only if it holds for all trajectories. That is, $\mathcal{C}, \phi \models [A \Longrightarrow C]$, when for every $S \in \mathcal{S}(\mathcal{C})$ we have that $\mathcal{C}, \phi, S \models A$ implies that $\mathcal{C}, \phi, S \models C$. Given a circuit and an assertion, the task of our checking algorithm is to compute the Boolean function expressing the set of interpretations under which the assertion is true. For most verification problems, this should simply be the constant function **1**, i.e., the assertion should hold under all variable interpretations.

4.4. Properties

We have intentionally chosen to introduce only a heavily restricted trajectory formula syntax for our base logic. As discussed previously, we do not support more sophisticated temporal operators such as "Until," "Globally," or "Eventually." Furthermore, we do not even permit such elementary logic operators such as disjunction, negation, or quantification.

By imposing these restrictions, we can guarantee the following key property:

Proposition 1 *For any trajectory formula F, and any interpretation ϕ, one of the following cases must hold:*

1. *There is no trajectory $S \in \mathcal{S}(\mathcal{C})$ for which $\mathcal{C}, \phi, S \models F$, or*

2. *There exists a unique trajectory $S_{F,\phi} \in \mathcal{S}(\mathcal{C})$ such that for every $S \in \mathcal{S}(\mathcal{C})$ we have $\mathcal{C}, \phi, S \models F$ if and only if $S_{F,\phi} \sqsubseteq S$.*

In the first case above, we say that the formula F is not satisfiable under interpretation ϕ. In the second case, we refer to the sequence $S_{F,\phi}$ as the *weakest trajectory* satisfying formula F under interpretation ϕ.

This theorem can be proved by induction on the formula structure. Rather than prove it here, however, we will show in a later section how the checking algorithm can compute a Boolean function describing the set of interpretations for which the formula is satisfiable, as well as a symbolic sequence representing the weakest trajectory for every satisfiable interpretation.

Note that this theorem expresses a very strong property of our logic. It demonstrates the reason why we can verify an assertion by simulating a single symbolic sequence, namely the one encoding the weakest trajectories allowed by the antecedent for every interpretation. It is stronger than the simple monotonicity condition that if $\mathcal{C}, \phi, S \models F$ and $S \sqsubseteq S'$, then $\mathcal{C}, \phi, S' \models F$.

Observe that this property would not hold if our formula syntax permitted disjunction. For example, given a circuit with two nodes n_1 and n_2, the two formulas $n_1 = 1$ and $n_2 = 0$ have weakest trajectories where the first elements are the vectors $\langle 1, X \rangle$ and $\langle X, 0 \rangle$, respectively. Furthermore, the formula $n_1 = 1 \wedge n_2 = 0$ has a weakest trajectory where the first element is the vector $\langle 1, 0 \rangle$. On the other hand, there is no weakest trajectory satisfying $n_1 = 1 \vee n_2 = 0$. Only the vector $\langle X, X \rangle$ is less than both $\langle 1, X \rangle$ and $\langle X, 0 \rangle$, and a trajectory having this as its initial vector satisfies neither $n_1 = 1$ nor $n_2 = 0$.

If our syntax permitted negation, then we would lose even the monotonicity property of our logic. For example, in the circuit described above, the formula $\neg(n_1 = 1)$ would be satisfied by a trajectory having first element $\langle X, X \rangle$, but not by one having first element $\langle 0, X \rangle$.

Disjunction is a useful extension to our language and can be implemented using quantified Boolean variables. Negation, on the other hand, seems to run contrary to the principles of ternary modeling.

5. Extensions

The logic, as described above, is convenient for deriving the underlying theory. Unfortunately, expressing "interesting" assertions about real circuits using only the constructs above is very tedious. Two shortcomings make using the logic cumbersome: the fine granularity of the timing, and the lack of more powerful logical constructs. It is convenient to add extensions that do not add any expressive power, but make it easier to write assertions.

This basic structure of starting with a minimal basic logic and then adding more elaborate structures as extensions also mirrors our current implementation. The implementation consists of two parts. The underlying logic, with some few extensions, is taken care of by our modified version of the COSMOS symbolic switch-level simulator. The syntactic extensions are supported by a front-end written in SCHEME. The user writes SCHEME code that, when evaluated, generates a file of low-level simulation commands which are then evaluated by the simulator.

5.1. Timing Extensions

Our main interest is in verifying synchronous circuits. Hence, we would like to reason about time on a more abstract level than basic time units. For many VLSI circuits it is natural to describe the desired behavior in terms of phase level behavior. Let a *phase* be a period of time during which no inputs (including clocks) change. For simplicity, we will assume that all phases have the same duration. For example, if phases are k basic time units long and we want the instantaneous formula F to hold for the entire phase, then this can be translated into

$$\bigwedge_{j=0}^{k-1}(\,\mathbf{X}_s^j F)$$

where, for $j \geq 0$, $\mathbf{X}_s^j$ denotes j repetitions of the next time operator, and $\mathbf{X}_s^0 F$ is defined as F. A "next phase" operator $\mathbf{X}_p$ can then be defined simply as $\mathbf{X}_s^k$. Thus a formula in a logic based on the next-phase operator can be translated directly into a formula in the logic based on the next-step operator. In our current implementation, the logic we support only allows the next-phase operator. The transformation into the next-step logic is taken care of by the COSMOS simulator so that event-scheduling can be used to achieve good performance.

If we are using a nominal or bounded delay circuit model, the basic time step corresponds directly with some exact amount of time. Hence, in these cases it is easy to define the length of a phase. At other times we are not interested in the exact timing of the circuit. For example, if we do not have delay values available (pre-layout) or if we use a very inexact delay model (say unit-delay) it is somewhat meaningless to say that a phase is k time units long. In these situations we often want to express the behavior in terms of "stable" phases. In other words, a phase may defined to be as long as needed for the circuit to stabilize. Of course, we must also take care of the case where the circuit never stabilizes, but oscillates. The solution to this is to "force" the circuit simulator to stabilize after some reasonably large number of steps. This can be accomplished by running a normal simulation for that many steps and then forcing oscillating nodes to X. We leave the details on how to do this symbolically to the interested reader.

5.2. Logic Extensions

As already alluded to above, it is very tedious to write assertions using only the basic logic. Thus in our SCHEME translator we provide a number of syntactic extensions. In the following we will outline some of these.

An obvious extension to the logic is to allow the user to write $n_i = E$, where $n_i \in \mathcal{N}$ and E is some Boolean expression over $\mathcal{V}$. The meaning of this is simply $[E \rightarrow (n_i = 1)] \wedge [\overline{E} \rightarrow (n_i = 0)]$, where $\overline{E}$ denotes the Boolean complement of E.

Our next addition is to allow the user to specify a domain restriction for a complete assertion, i.e., we allow the user to write $E \rightarrow [A \implies C]$, meaning that the assertion only needs to hold for interpretations $\phi \in \Phi$ such that $e(\phi) = 1$, where e is the Boolean function denoted by E. This is simply a short-hand for the assertion $[(E \rightarrow A) \implies (E \rightarrow C)]$.

The following extension to the logic, dealing with finite integer domains, is perhaps the most useful of them all. Let $\vec{E}$ be a vector of Boolean expressions over $\mathcal{V}$. Given an interpretation $\phi \in \Phi$ one can view $\vec{E}$ as the binary encoding of a number. Denote this number as $|\vec{E}(\phi)|$. Let $\vec{v}$ denote a vector of Boolean expressions over $\mathcal{V}$. What we would like to do is to use $\vec{E}$ to select one of the Boolean expressions in $\vec{v}$. Assume the vector contains m expressions. The *value* of $v_{\vec{E}}$ is defined as:

$$v_{\vec{E}} \;=\; (|\vec{E}| = 0) \cdot v_0 + \ldots + (|\vec{E}| = m-1) \cdot v_{m-1}$$

where if $\vec{E} = \langle E_{p-1}, E_{p-2}, \ldots, E_0 \rangle$, and $\langle i_{p-1}, i_{p-2}, \ldots, i_0 \rangle$ is the binary encoding of i, the expression $(|\vec{E}| = i)$ is the Boolean expression $\prod_{0 \leq j < p}((E_j \cdot i_j) + (\overline{E_j} \cdot \overline{i_j}))$. It is easy to see that given an interpretation $\phi \in \Phi$ this definition captures our intuition of what it means to select one of the expressions.

Finally, we would also like to do a similar selection among circuit nodes. Let $\vec{\mathrm{N}}$ denote a vector of m circuit nodes. The expression $\mathrm{N}_{\vec{E}} = G$, where G is some Boolean expression over $\mathcal{V}$, is simply a shorthand for the trajectory formula

$$[(|\vec{E}| = 0) \rightarrow (\mathrm{N}_0 = G)] \wedge [(|\vec{E}| = 1) \rightarrow (\mathrm{N}_1 = G)] \wedge \ldots \wedge [(|\vec{E}| = m-1) \rightarrow (\mathrm{N}_{m-1} = G)]$$

where $(|\vec{E}| = i)$ is defined as above.

Given that the front-end is embedded in SCHEME, and the user actually writes SCHEME code, it is easy to define new extensions. In fact, by writing SCHEME procedures it becomes very natural to express the circuit assertions in a hierarchical way, improving the readability of—and consequently the confidence in—the assertions.

6. Examples

The extensions described in the previous section provide a phase level timing model and an assertion language with more powerful abstraction capabilities. Given these extensions we claim the logic is both sufficiently powerful and easy to use to express a wide variety of verification tasks. We support this claim with some illustrative examples.

6.1. Moving Data Stack

First we will demonstrate the utility of the temporal notation in expressing phase level circuit timing and pipelining. The example is the nMOS stack circuit described by Mead and Conway [16]. Figure 1 shows the block diagram for the circuit. The circuit operates with a two-phase nonoverlapping clock. The stack command is specified by a pair of signals, OP1 and OP2, which are multiplexed onto a single circuit input `Op`. The control signal is

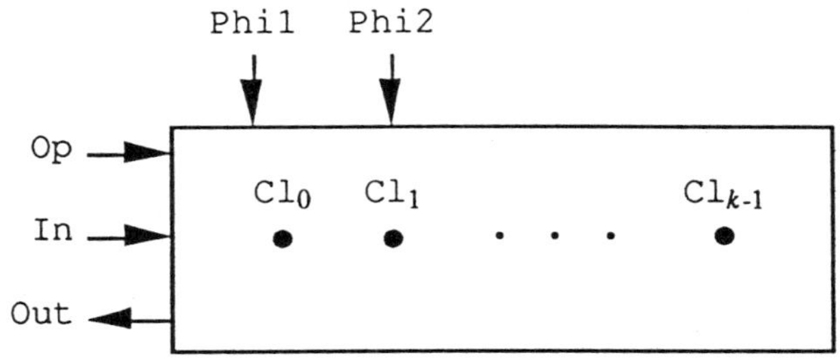

Figure 1: **Block Diagram of Mead and Conway Stack.** Bit i of the stack is stored as charge on node $\texttt{Cl}_i$ (negative logic).

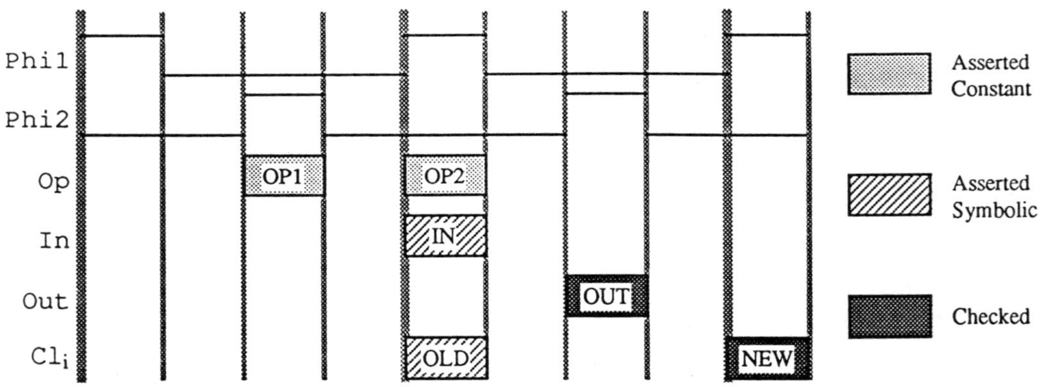

Figure 2: **Timing Diagram Stack.** The two control signals for a given clock cycle are multiplexed on the single line Op.

pipelined half a clock cycle ahead of the data signals. That is, the command for clock cycle t is specified by supplying OP1 on the `Phi2` phase of cycle $t-1$, and OP2 on the `Phi1` phase of cycle t, as illustrated in Figure 2. The input data must be supplied during the `Phi1` phase, and the output is valid during the `Phi2` phase. For each stack cell, we will consider the "state" of the cell to be the value held there during the `Phi1` phase of the cycle.

To illustrate a possible assertion for this circuit, consider checking that a PUSH operation can be carried out correctly. The PUSH operation is performed when OP1 $= 1$ and OP1 $= 0$. Informally, a PUSH operation should put the pushed value into cell 0 and move the contents of cell i to cell $i+1$ for every i less than $k-1$, where k is the number of cells in the circuit. For convenience, let $\vec{\mathtt{Cl}}$ be a vector of the $\mathtt{Cl}_i$ nodes. Let u and v be two Boolean variables, and $\vec{I} = \langle I_{p-1}, \ldots, I_0 \rangle$ be a vector of Boolean variables, where $k \leq 2^p$.

First, we must include in our assertions a specification of how the clocks should be operated. We use the shorthand *Clock1* to denote the trajectory formula

$$(\mathtt{Phi1} = 1) \wedge \ \mathbf{X}_p(\mathtt{Phi1} = 0) \wedge \ \mathbf{X}_p^2(\mathtt{Phi1} = 0) \wedge \ \mathbf{X}_p^3(\mathtt{Phi1} = 0),$$

Clock2 to denote

$$(\mathtt{Phi2} = 0) \wedge \ \mathbf{X}_p(\mathtt{Phi2} = 0) \wedge \ \mathbf{X}_p^2(\mathtt{Phi2} = 1) \wedge \ \mathbf{X}_p^3(\mathtt{Phi2} = 0),$$

and *Cycles* to denote

$$\mathit{Clock1} \wedge \mathit{Clock2} \wedge \ \mathbf{X}_p^4(\mathit{Clock1} \wedge \mathit{Clock2}) \wedge \ \mathbf{X}_p^8[(\mathtt{Phi1} = 1) \wedge \mathtt{Phi2} = 0].$$

The formula *Cycles* defines the operation of the clocks over the time period illustrated in Figure 2.

Since negative logic is used internally, let $\mathit{Stored}(\vec{I}, u)$ be a shorthand for $\mathtt{Cl}_{\vec{I}} = \overline{u}$. Let *Push* denote $\mathbf{X}_p^2(\mathtt{OP} = 1) \wedge \ \mathbf{X}_p^4(\mathtt{OP} = 0)$.

We can now express our assertion as:

$$\begin{aligned}&(\vec{I} < k-1) \rightarrow \\ &\quad \left[\mathit{Cycles} \wedge \mathit{Push} \wedge \ \mathbf{X}_p^4[(\mathtt{In} = v) \wedge \mathit{Stored}(\vec{I}, u)] \implies \ \mathbf{X}_p^8[\mathit{Stored}(0, v) \wedge \mathit{Stored}(\vec{I} + 1, u)]\right].\end{aligned}$$

Note that $(\vec{I} < k-1)$ denotes a Boolean function over $\mathcal{V}$ that, for a given interpretation ϕ, is 1 iff $|\vec{I}(\phi)| < k-1$. Similarly, $\vec{I} + 1$ denotes a vector of Boolean functions such that, for a given interpretation ϕ, $|(\vec{I} + 1)(\phi)| = |\vec{I}(\phi)| + 1$.

In this example, verifying the correct operation of a single stack operation requires reasoning about the behavior of the circuit over parts of 3 clock cycles. Such is often the case in pipelined systems.

6.2. Static RAM

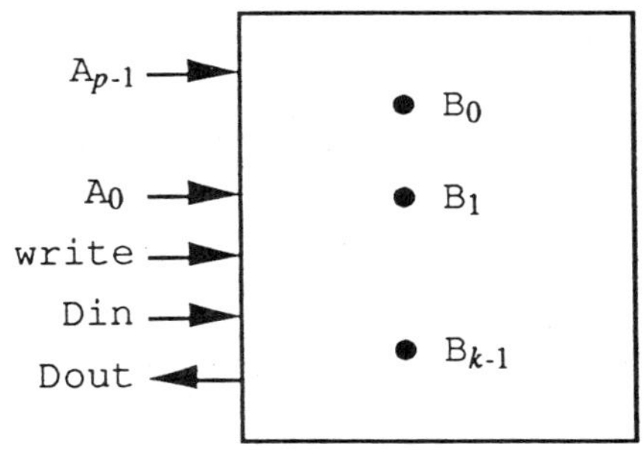

Figure 3: **Block Diagram of Static RAM.** State is stored in location i on node $\mathtt{B}_i$.

Our second example is a static RAM, as illustrated in Fig. 3. The data for each memory location i are stored on node $\mathtt{B}_i$. Within the RAM cell there is a node holding the complementary value. Unlike in the verification described in [7] based on scalar simulation, we do not need to explicitly set or check this complementary value. The phase level timing abstraction handles this automatically. That is, asserting a value on node $\mathtt{B}_i$ for 2 unit steps forces the complementary value onto the other node in the RAM cell.

The only information we need to know about the design of the RAM is the names of the input/output signals, and the names of the storage nodes of each memory cell, as well as the interface timing. For ease of exposition, we express the specification at the *clock cycle* level in terms of a cycle-level next-time operator $\mathbf{X}_c$. Given a specification of the clocking patterns and interface timing for the circuit, we could translate these assertions into phase-level assertions in a similar fashion as we did for the moving data stack.

It is easy to convince oneself that a correct memory must satisfy three properties: 1) that we can write successfully into each cell i, 2) that we can nondestructively read the content of each cell i, and 3) that, unless we are writing into some cell i, the value stored in cell i should not change. Let $\vec{\mathtt{B}}$ be a vector of the $\mathtt{B}_i$ nodes. Furthermore, let u, v, and w be Boolean variables, and $\vec{I} = \langle I_{p-1}, \ldots, I_0 \rangle$ and $\vec{J} = \langle J_{p-1}, \ldots, J_0 \rangle$ be vectors of Boolean variables, where p is the width of the address input. We use the shorthand *Operate*$(\vec{I}, u, w)$ to denote the formula

$$(\mathtt{write} = w) \wedge (\mathtt{Din} = u) \wedge (\mathtt{A}_{p-1} = I_{p-1}) \wedge \cdots \wedge (\mathtt{A}_0 = I_0).$$

Similarly, let *Stored*$(\vec{J}, v)$ denote the formula $(\mathtt{B}_{\vec{I}} = v)$. With this notation, we can express all three conditions mentioned above with a single assertion.

$$\begin{aligned} & \mathit{Stored}(\vec{J}, v) \wedge \mathit{Operate}(\vec{I}, u, w) \\ \Longrightarrow\ & \mathbf{X}_c \ \Big([w \rightarrow \mathit{Stored}(\vec{I}, u)] \ \wedge \\ & \qquad [\overline{w} \cdot (\vec{I} = \vec{J}) \rightarrow (\mathrm{DOUT} = v)] \ \wedge \\ & \qquad [\overline{w} + (\vec{I} \neq \vec{J}) \rightarrow \mathit{Stored}(\vec{J}, v)] \Big) \end{aligned}$$

where $\vec{I} \neq \vec{J}$ denotes the Boolean expression $(I_{p-1} \oplus J_{p-1}) + \cdots + (I_0 \oplus J_0)$, and $\vec{I} = \vec{J}$ denotes the complement of this expression.

6.3. Discussion

As these examples illustrate, our notation works well for expressing the state transition properties of a circuit. For circuits, such as memories and data paths, this is a fairly natural form of specification. That is, one thinks of each operation of the system as updating some portion of the stored system state.

In these examples, we constructed the specification assertion in a hierarchical way, starting with low level timing information and working up to more abstract system operations. This seems to be a fairly convenient way to view the system at different abstraction levels. Because we actually write our specifications as SCHEME programs, we can use the procedural abstraction capabilities of SCHEME to express the specification hierarchically.

7. Symbolic Simulation

As we have shown, symbolic formulas provide a concise means to specify desired properties of the circuit under many different operating conditions. We are now ready to introduce a method of verifying the assertions via symbolic simulation. The key idea is to preserve the symbolic structure of the formulas in the verification algorithm. By doing so, we can replace the need for large amounts of case analysis with algebraic manipulation.

7.1. Symbolic Algebras

In creating a symbolic model, we extend the scalar model defined in terms of the binary and ternary domains $\mathcal{B}$ and $\mathcal{T}$, to one defined in terms of binary- and ternary-valued functions over the variables $\mathcal{V}$. Define the symbolic domain $\mathcal{B}(\mathcal{V})$ (respectively, $\mathcal{T}(\mathcal{V})$) as denoting the set of functions mapping an interpretations in Φ to $\mathcal{B}$ (resp., $\mathcal{T}$). More formally

$$\mathcal{B}(\mathcal{V}) = \{f : \Phi \rightarrow \mathcal{B}\}$$

and

$$\mathcal{T}(\mathcal{V}) = \{f : \Phi \rightarrow \mathcal{T}\}.$$

We then extend the operations defined over scalar values to create a symbolic algebra.

We can also extend the vector and sequence algebra defined over scalar values to their counterparts defined over symbolic values. That is, define the vector domain $\mathcal{T}(\mathcal{V})^n$ as

$$\mathcal{T}(\mathcal{V})^n \;=\; \{\langle a_1, \ldots, a_n\rangle | a_i \in \mathcal{T}(\mathcal{V})\}.$$

In implementing a symbolic simulator, we in effect extend the excitation function $\vec{Y}$ to the symbolic domain as $\vec{Y}: \mathcal{T}(\mathcal{V})^s \to \mathcal{T}(\mathcal{V})^s$. For $\vec{a} \in \mathcal{T}(\mathcal{V})^n$, let $\vec{a}(\phi) \in \mathcal{T}^n$ denote the vector with each element i equal to $a_i(\phi)$. In this way, we can view the symbolic vector $\vec{a} \in \mathcal{T}(\mathcal{V})^n$ either as a vector of symbolic elements, or as a symbolic value which for a given interpretation yields a scalar vector.

We extend most operations from scalar to symbolic domains in a uniform way. Consider an operation $op: \mathcal{D}_1 \times \mathcal{D}_2 \to \mathcal{D}_3$, defined over vectors, single elements, or a combination of the two. Its symbolic counterpart $op: \mathcal{D}_1(\mathcal{V}) \times \mathcal{D}_2(\mathcal{V}) \to \mathcal{D}_3(\mathcal{V})$ is defined such that for all $a \in \mathcal{D}_1(\mathcal{V})$ and $b \in \mathcal{D}_2(\mathcal{V})$, we have $(a \; op \; b)(\phi) = a(\phi) \; op \; b(\phi)$. We use this method to extend the ternary algebraic operations $\cdot_t$, $+_t$, and $^{-t}$, as well as the operation $\sqcap$.

When extending a relation R symbolically, we define the result to be a function specifying the interpretations under which its arguments are related. That is, given a binary relation $R \subseteq \mathcal{D}_1 \times \mathcal{D}_2$, define $R: \mathcal{D}_1(\mathcal{V}) \times \mathcal{D}_2(\mathcal{V}) \to \mathcal{B}(\mathcal{V})$ as $(a \; R \; b)(\phi) = 1$ if and only $a(\phi) \; R \; b(\phi)$. We use this method to define operations $\sim$ and $\sqsubseteq$ over both single elements and vectors.

7.2. Special Operations

We require one operation that is extended to vectors in a nonstandard way. Define the infix operator $?: \mathcal{B} \times \mathcal{T} \to \mathcal{T}$ as $a \; ? \; b$ equals b if a is 1, and equals X otherwise. When extending this operation to vectors, only the second argument is vector-valued. That is the operation $?: \mathcal{B} \times \mathcal{T}^n \to \mathcal{T}^n$ is defined as $(a \; ? \; \vec{b})_i = a \; ? \; b_i$. This operation is then extended symbolically in the manner described above.

As a final operation, we define a variant of the join operation that is defined even when for some $\phi \in \Phi$, we have $\vec{a}(\phi) \not\sim \vec{b}(\phi)$. When using this operation, we will separately keep track of the conditions under which the arguments are compatible. Define the operation $\overset{*}{\sqcup}: \mathcal{T}(\mathcal{V})^n \times \mathcal{T}(\mathcal{V})^n \to \mathcal{T}(\mathcal{V})^n$ as

$$(\vec{a} \overset{*}{\sqcup} \vec{b})(\phi) = \begin{cases} \vec{a}(\phi) \sqcup \vec{b}(\phi), & \vec{a}(\phi) \sim \vec{b}(\phi) \\ \vec{X}, & \text{otherwise} \end{cases}$$

where $\vec{X}$ denotes a vector with all elements equal to X.

7.3. Translating Instantaneous Formulas to Symbolic Vectors

Given the above definitions, we first give a procedure that for an instantaneous formula F derives a Boolean function OK_F and a vector $\vec{a}_F$. The function OK_F can be viewed as a "domain" function of F in the sense that $OK_F(\phi) = 1$ iff under this interpretation F can hold true in some state. Furthermore, we will also show that $\vec{a}_F$ is "weakest" in the sense that if $OK_F(\phi) = 1$, then $\vec{a}_F(\phi) \sqsubseteq \vec{b}$ for every state vector $\vec{b} \in \mathcal{T}^s$ for which F holds under this interpretation. The function OK_F and the vector $\vec{a}_F$ are defined recursively as:

1. If F is TRUE then $OK_F = 1$, and $\vec{a}_F = \langle X, \ldots, X\rangle$.

2. (a) If F is $(n_i = 1)$ then $OK_F = 1$, and $\vec{a}_F = \langle X, \ldots, X, 1, X, \ldots, X\rangle$, where the 1 is in position i.

   (b) If F is $(n_i = 0)$ then $OK_F = 1$, and $\vec{a}_F = \langle X, \ldots, X, 0, X, \ldots, X\rangle$, where the 0 is in position i.

3. If F is $(F_1 \wedge F_2)$ then $OK_F = OK_{F_1} \cdot OK_{F_2} \cdot (\vec{a}_{F_1} \sim \vec{a}_{F_2})$, and $\vec{a}_F = \vec{a}_{F_1} \overset{*}{\sqcup} \vec{a}_{F_2}$.

4. If F is $(E \to F_1)$ then $OK_F = \overline{e} + OK_{F_1}$, and $\vec{a}_F = e \mathbin{?} \vec{a}_{F_1}$, where e is the Boolean function denoted by the expression E.

Proposition 2 *Given a circuit $\mathcal{C}$, let F be an instantaneous formula and OK_F and $\vec{a}_F$ be derived as above. Then $OK_F(\phi) = 1$ iff there exists some state $\vec{b} \in \mathcal{T}^s$ such that $\mathcal{C}, \phi, \vec{b} \models F$. Furthermore, if $OK_F(\phi) = 1$, then $\mathcal{C}, \phi, \vec{b} \models F$ iff $\vec{a}_F(\phi) \sqsubseteq \vec{b}$.*

This proposition can be proved by induction on the formula structure.

7.4. Checking Assertions

Our first step in verifying an assertion is to rewrite the antecedent and consequent into a normal form where all next-time operators are collected together. It is easy to show that a trajectory formula F can be rewritten into $F_0 \wedge \mathbf{X}_s F_1 \wedge \mathbf{X}_s^2 F_2 \wedge \ldots \wedge \mathbf{X}_s^{k-1} F_{k-1}$, for some $k \geq 1$, where each F_i is instantaneous. Note that some of the F_i's might be the trivial formula TRUE. Note also that such a sequence can be extended by appending $\mathbf{X}_s^i$TRUE for $i \geq k$. Hence, without any loss of generality, we will henceforth assume that the antecedent and the consequent in an assertion are trajectory formulas in normal form containing the same number of terms.

Given an assertion $[A \Longrightarrow C]$ of the form

$$\left[A_0 \wedge \mathbf{X}_s A_1 \wedge \ldots \wedge \mathbf{X}_s^{k-1} A_{k-1} \Longrightarrow C_0 \wedge \mathbf{X}_s C_1 \wedge \ldots \wedge \mathbf{X}_s^{k-1} C_{k-1}\right]$$

define a sequence of symbolic ternary vectors $\vec{x}_0, \ldots, \vec{x}_{k-1}$ as follows:

$$\vec{x}_i = \begin{cases} \vec{a}_{A_0}, & i = 0 \\ \vec{Y}(\vec{x}_{i-1}) \overset{*}{\sqcup} \vec{a}_{A_i}, & i > 0. \end{cases}$$

Define the Boolean function $OK_A = \prod_{0 \leq i < k} OK_{A_i}$, where $\prod$ denotes Boolean product. This function yields 0 for those interpretations for which the antecedent contains some internal inconsistency. For example, the formula $A = (n_i = a) \wedge (n_i = b)$ would have $OK_A = \overline{a \oplus b}$, because this formula cannot be satisfied when $\phi(a) \neq \phi(b)$. Define the Boolean function $\mathit{Traj} = \prod_{1 \leq i < k} [\vec{Y}(\vec{x}_{i-1}) \sim \vec{a}_{A_i}]$. This function yields 0 for those interpretations where an incompatibility arises in the trajectory.

We can show that A is satisfiable under some interpretation ϕ if and only if $OK_A(\phi) \cdot Traj(\phi) = 1$. Furthermore, we can extend the sequence $\vec{x}_0, \ldots, \vec{x}_{k-1}$ to be an infinite sequence by defining $\vec{x}_i = \vec{Y}(\vec{x}_{i-1})$ for all $i \geq k$. It can then be shown that for interpretation ϕ the sequence $\vec{x}_0(\phi), \vec{x}_1(\phi), \ldots$ is the weakest trajectory satisfying A under interpretation ϕ. This construction then provides a proof of Proposition 1. This demonstrates how our symbolic simulator can set up the weakest allowable conditions allowed by the antecedent under all possible interpretations.

To check the consequent, define the Boolean function $OK_C = \prod_{0 \leq i < k} OK_{C_i}$. This function yields 0 for those interpretations for which the consequent contains some internal inconsistency. Finally, define the Boolean function $Check = \prod_{0 \leq i < k} [\vec{a}_{C_i} \sqsubseteq \vec{x}_i]$. This function yields 0 for those interpretations where some trajectory satisfying the antecedent may violate the consequent.

Now define $OK_{[A \implies C]}$ as:

$$\overline{OK_A} + \overline{Traj} + (OK_C \cdot Check)$$

Informally, this equation states that the assertion is true under those interpretations for which the antecedent is unsatisfiable (due either to internal inconsistencies or to an incompatibility in the trajectory), as well as those for which the consequent holds (i.e, it is both internally consistent and is satisfied.)

The main result of this paper is captured in the following theorem:

Theorem 1 *Given a circuit $\mathcal{C}$ and an assertion $[A \implies C]$ let $OK_{[A \implies C]} \in \mathcal{B}(\mathcal{V})$ be derived as above. Then*

$$\mathcal{C}, \phi \models [A \implies C] \quad \textit{if and only if} \quad OK_{[A \implies C]}(\phi) = 1.$$

Hence, determining whether a circuit satisfies $[A \implies C]$ is reduced to determining whether $OK_{[A \implies C]} = 1$.

7.5. Summary of Checking Algorithm

In practice, we encode the ternary values as pairs of Boolean values. Thus we express the ternary functions over $\mathcal{V}$ as pairs of Boolean functions over $\mathcal{V}$. It is straightforward to implement all of the operations defined above using such an encoding. Thus, to verify an assertion $[A \implies C]$ we would proceed as follows: First convert A and C into normal forms containing the same number of terms. Now compute the sequence of states $\vec{x}_i$ as defined above. Although the above presentation used $k-1$ distinct vectors, the implementation need only retain the most recent vector. In fact, the vector can be viewed as the current circuit state. Hence, step i in this verification would entail: 1) run the simulator one "step" to derive $\vec{Y}(\vec{x}_{i-1})$, 2) compute OK_{A_i} and $\vec{a}_{A_i}$, 3) compute $(\vec{Y}(\vec{x}_{i-1}) \sim \vec{a}_{A_i})$, 4) compute $\vec{x}_i$,

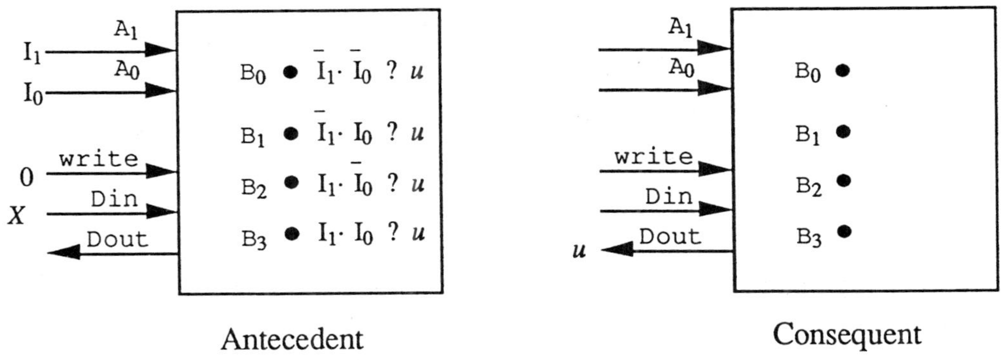

Figure 4: **Verification of 4-bit RAM Read Operation.**

5) compute $\vec{a}_{C_i}$, and 6) compute $(\vec{a}_{C_i} \sqsubseteq \vec{x}_i)$. As we go along, we also maintain Boolean functions representing the 4 constituent components of the expression for $OK_{[A \Longrightarrow C]}$ shown above. In the end we simply combine these together and check that the derived function is the constant function **1**. If the test fails, the derived function provides diagnostic information indicating which cases encounter difficulties.

7.6. Example

Consider a 4-bit memory of the structure illustrated in Figure 3. In verifying the correctness of a READ operation, one aspect of the desired behavior is that after reading with the address inputs set to i, the valued stored in memory location i should appear on the output. This can be expressed by the assertion (in our extended syntax):

$$\left[\mathit{Stored}(\vec{I}, u) \wedge (\texttt{write} = 0) \wedge (\vec{\texttt{A}} = \vec{I}) \implies \mathbf{X}_c(\texttt{Dout} = u)\right]$$

The antecedent is converted into the simulation pattern outlined on the left hand side of Figure 4. That is, the address inputs are set to Boolean variables encoding the possible addresses. The cells are set to more complex symbolic values. For each possible interpretation of the address variables, we see that one cell is set to value u, while the other three cells are set to X. Verifying the assertion involves simulating this single pattern and then checking that the resulting value on Dout equals u.

This example illustrates how our methodology combines symbolic and ternary values when encoding the different possible circuit states. In a binary model of the circuit, there are 16 (in general 2^k) different initial binary states of the memory. These cases can be covered by 8 (in general $2k$) different initial states in a ternary model—each having one cell set to 0 or 1 and the rest set to X. These different cases can then be encoded by a single symbolic case in terms of 3 (in general $1 + \log k$) Boolean variables. As the size of the memory k grows, the reduction in complexity we realize becomes dramatic.

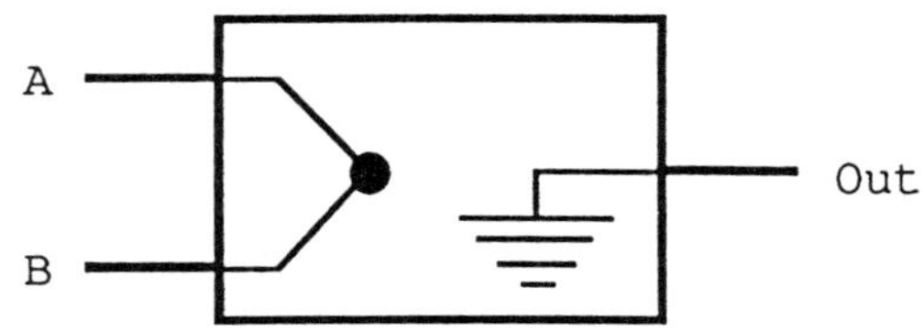

Figure 5: **Pseudo XOR Circuit.** This circuit satisfies the assertion for an XOR circuit, because the antecedent trajectory fails whenever the output should be nonzero.

8. Areas for Future Investigation

The checker described here provides a useful tool for reasoning about digital circuits. However, some questions remain as to exactly how it can and should be used. In particular, we must still devise a comprehensive theory on how to prove that a circuit fully realizes its specification.

To illustrate one subtlety of this task, consider trying to show that a circuit with inputs A and B and output Out implements the exclusive-or function. Intuitively, it seems that it would be sufficient to prove that circuit obeys the assertion $[(\mathtt{A} = a) \wedge (\mathtt{B} = b) \implies \mathbf{X}_s(\mathtt{Out} = a \oplus b)]$. Unfortunately, this is not the case. For example, this assertion is obeyed by the rather useless circuit of Figure 5, where the two inputs are tied together, and the output is always 0. Any interpretation of the variables ϕ for which $\phi(a) \neq \phi(b)$ will cause the antecedent trajectory to fail, because inputs A and B are electrically equivalent. The only interpretations for which the trajectory succeeds are ones for which the output should be 0, in which case the consequent is also satisfied.

Any checking based purely on testing implications is prone to this sort of "false positive" error. Problems of this sort have been encountered by people using other systems for hardware verification such as HOL [14] and EMC [10]. We believe that shortcomings of this sort can be corrected by more careful attention to the cases where an implication succeeds due a failure of its antecedent. In all of the verification examples we have performed, we make sure that the specification is formulated in such a way that the antecedent trajectory should never fail. We check this as part of the verification process.

9. Conclusions

In terms of mathematical sophistication, the problem solved by our algorithm is far less ambitious than what is attempted by full-fledged temporal logic model checkers. However, we believe that our language is rich enough to be able to describe many important properties of a circuit and to provide a direct path by which such properties may be automatically verified. By keeping the goals of our verifier simple, we obtain an algorithm that is capable of dealing with much larger circuits.

One interesting property of our algorithm, in fact, is that its computational complexity is relatively insensitive to the circuit size. That is, the complexity is determined largely by the complexity of the assertion to be verified, measured in terms of the number of symbolic variables, and the depth of nesting of next time operators. We have found that in many circuits, properties can be expressed in terms of a surprisingly small number of variables. For example, our formulas providing a complete specification of of a k-bit static RAM involve only $2 + 2\log k$ variables. Thus, we can perform the verification in polynomial time irrespective of the heuristic efficiency of the Boolean manipulator.

References

[1] D. L. Beatty, R. E. Bryant, and C.-J. H. Seger, "Synchronous Circuit Verification by Symbolic Simulation: An Illustration," *Sixth MIT Conference on Advanced Research in VLSI*, 1990.

[2] S. Bose, and A. L. Fisher, "Verifying Pipelined Hardware Using Symbolic Logic Simulation," *International Conference on Computer Design*, IEEE, 1989.

[3] S. Bose, and A. L. Fisher, "Automatic Verification of Synchronous Circuits using Symbolic Logic Simulation and Temporal Logic," *IMEC-IFIP International Workshop on Applied Formal Methods for Correct VLSI Design*, 1989, pp. 759–764.

[4] R. E. Bryant, "Graph-Based Algorithms for Boolean Function Manipulation", *IEEE Transactions on Computers*, Vol. C-35, No. 8 (August, 1986), 677–691.

[5] R. E. Bryant, D. Beatty, K. Brace, K. Cho, and T. Sheffler, "COSMOS: a Compiled Simulator for MOS Circuits," *24th Design Automation Conference*, 1987, 9–16.

[6] R. E. Bryant, "Boolean Analysis of MOS Circuits," *IEEE Transactions on Computer-Aided Design of Integrated Circuits and Systems*, Vol. CAD-6, No. 4 (July, 1987), 634–649.

[7] R. E. Bryant, "Formal Verification of Memory Circuits by Switch-Level Simulation," To appear in *IEEE Transactions on Computer-Aided Design of Integrated Circuits and Systems*, 1990.

[8] J. A. Brzozowski, and M. Yoeli. "On a Ternary Model of Gate Networks." *IEEE Transactions on Computers C-28*, 3 (March 1979), 178–183.

[9] J. R. Burch, E. M. Clarke, K. L. McMillan, and D. L. Dill, "Sequential Circuit Verification Using Symbolic Model Checking," *27th Design Automation Conference*, 1990.

[10] E. M. Clarke, E. A. Emerson, and A. P. Sistla, "Automatic Verification of Finite-State Concurrent Systems Using Temporal Logic Specifications," *ACM Transactions on Programming Languages*, Vol. 8, No. 2 (April, 1986), pp. 244–263.

[11] O. Coudert, C. Berthet, and J. C. Madre, "Verification of Sequential Machines using Boolean Functional Vectors," *IMEC-IFIP International Workshop on Applied Formal Methods for Correct VLSI Design*, 1989, pp. 111–128.

[12] J. A. Darringer, "The Application of Program Verification Techniques to Hardware Verification," *16th Design Automation Conference*, 1979, 375–381.

[13] S. Devadas, H.-K. T. Ma, and A. R. Newton, "On the Verification of Sequential Machines at Differing Levels of Abstraction," *24th Design Automation Conference*, 1987, 271–276.

[14] M. Gordon, "Why higher-order logic is a good formalism for specifying and verifying hardware," *Formal Aspects of VLSI Design*, G. Milne and P. A. Subrahmanyam, *eds.*, North-Holland, 1986, pp. 153–177.

[15] J. S. Jephson, R. P. McQuarrie, and R. E. Vogelsberg, "A Three-Level Design Verification System," *IBM Systems Journal* Vol. 8, No. 3 (1969), 178–188.

[16] C. A. Mead, and L. Conway, *Introduction to VLSI Systems*, Addison-Wesley, 1980.

[17] A. Pnueli, "The Temporal Logic of Programs," *18th Symposium on the Foundations of Computer Science*, IEEE, 1977, pp. 46–56.

[18] D. S. Reeves, and M. J. Irwin, "Fast Methods for Switch-Level Verification of MOS Circuits", *IEEE Transactions on CAD/IC*, Vol. CAD-6, No. 5 (Sept., 1987), pp. 766–779.

[19] C-J. Seger, and R. E. Bryant, "Modeling of Circuit Delays in Symbolic Simulation", *IMEC-IFIP International Workshop on Applied Formal Methods for Correct VLSI Design*, 1989, pp. 625–639.

DIMACS Series in Discrete Mathematics
and Theoretical Computer Science
Volume 3, 1991

An Algebra for Delay-Insensitive Circuits

Mark B. Josephs
Programming Research Group
Oxford University Computing Laboratory
11 Keble Road, Oxford OX1 3QD, U.K.
Phone: +44-865-272574
E-mail: mark%prg.oxford.ac.uk@nss.cs.ucl.ac.uk

Jan Tijmen Udding
Department of Computer Science
Washington University
Campus Box 1045, St. Louis, MO 63130, U.S.A.
Phone: 314-889-6110
E-mail: jtu@cs.wustl.edu

September 20, 1990

A novel process algebra is presented; algebraic expressions specify delay-insensitive circuits in terms of voltage-level transitions on wires. The approach appears to have several advantages over traditional state-graph and production-rule based methods. The wealth of algebraic laws makes it possible to specify circuits concisely and facilitates the verification of designs. Individual components can be composed into circuits in which signals along internal wires are hidden from the environment.

1052-1798/91 $1.00 + $.25 per page

0 Introduction

A circuit is connected to its environment by a number of wires. If the circuit functions correctly irrespective of the propagation delays in these wires, the circuit is called *delay-insensitive.* Delay-insensitive circuits are attractive because they can be designed in a modular way; indeed no timing constraints have to be satisfied in connecting such circuits together. As a result of the latest Turing Award Lecture [14], the design of delay-insensitive circuits has drawn renewed interest.

The design of delay-insensitive circuits is made difficult by the need to consider situations in which a signal (voltage-level transition) has been transmitted at one end of a wire but has not yet been received at the other end. The algebraic notation presented in this paper may be helpful in the following ways:

1. The functional behavior of primitive delay-insensitive components can be captured by algebraic expressions.

2. All possible behaviors of the circuit that results when such components are connected together can be determined by symbolic manipulation.

3. The algebra facilitates the precise specification of the circuit that the designer has to build, including obligations to be met by the environment.

4. The algebra supports verification of the design against the specification.

The algebra is based upon Hoare's CSP notation [8]. It adapts the theory of asynchronous processes [9, 10] to the special case of delay-insensitive circuits. The possibilities of *transmission interference* and *computation interference*, characterized by Udding [15, 16], are faithfully modeled in the algebra; the designer is able to reason about these errors, and so avoid them. Underpinning the algebra is a denotational semantics similar to those given in [2, 10]; the semantics is compatible with the failures-divergences model of CSP [1, 8].

Our approach complements that taken by Martin [11, 12] to the design of delay-insensitive circuits. Martin's approach, however, is more general in that he makes use of components that are not delay-insensitive, namely his *isochronic forks.* We have discovered that it is possible to understand many of Martin's circuits by treating the isochronic fork, together with one of the gates to which it connects, as a single primitive delay-insensitive component. (Ebergen [5], on the other hand, has investigated how components that are sensitive to delay can be synchronized to form delay-insensitive circuits.)

The remainder of this paper provides a step-by-step introduction to the algebra. We also prove, as a case study, some of Martin's nontrivial circuit designs to be algebraically correct. Similar verifications have been done by Dill [4, 3]. His verifications, however, are performed at the semantics level rather than at the syntactic/algebraic level. Algebraic calculations are arduous but humanly feasible, as well as mechanizable, and seem to avoid a state explosion.

1 Basic Notions and Operators

A process is a mathematical model at a certain level of abstraction of the way in which a delay-insensitive circuit interacts with its environment. Typical processes are P and Q. A circuit receives signals from its environment on its input wires and sends signals to its environment on its output wires. Thus with each process are associated an alphabet of input wires and an alphabet of output wires. These alphabets are finite and disjoint. Typical names for input wires are a and b; typical names for output wires are c and d. The time taken by a signal to traverse a wire is indeterminate.

In the remainder of this section and section 2, we consider processes with a particular alphabet I of input wires and a particular alphabet O of output wires.

The process P is considered to be "just as good" as the process Q ($P \sqsubseteq Q$) if no environment, which is simply another process, can when interacting with P determine that it is not interacting with Q. (This is the refinement ordering of CSP [1, 8], also known as the *must* ordering [7].) Two processes are considered to be equal when they are just as good as each other.

The refinement ordering is intimately connected with nondeterministic choice. The process $P \sqcap Q$ is allowed to behave either as P or as Q. It reflects the designer's freedom to implement such a process by either P or Q. (Thus $\sqcap$ is obviously commutative, associative and idempotent.) Now $P \sqcap Q = Q$ exactly when $P \sqsubseteq Q$.

A wire cannot accommodate two signals at the same time; they might interfere with one another in an undesirable way. This and any other error are modeled by the process $\perp$ (Bottom or Chaos). The environment must ensure that a process never gets into such a state. The process $\perp$ is considered to be so undesirable that any other process must be an improvement on it:

Law 0. $P \sqsubseteq \bot$

It follows that $\sqcap$ has $\bot$ as a null element.

We shall mostly be concerned with recursively-defined processes. The meaning of the recursion $\mu X.F(X)$ is the least fixpoint of F. Its successive approximations are $\bot$, $F(\bot)$, $F(F(\bot))$, *etc.* All the operators that we shall use to define processes are continuous (and therefore monotonic); all except for recursion are distributive (with respect to nondeterministic choice).

In earlier approaches to an algebra for delay-insensitive circuits, *cf.* [15, 2, 5], a particular input signal is allowed only when this is explicitly indicated, and otherwise is assumed to lead to interference. This is in contrast with the algebra presented here: an input need not result in interference even though the possibility of such an input has not been made explicit in the algebraic expression. This follows the approach taken in [10] and is more convenient in algebraic manipulation, even though at first it may appear less natural.

Thus we write $a?; P$ to denote a process that must wait for a signal to arrive on $a \in I$ before it can behave like P. It is quite permissible for the environment to send a signal along any other input wire b. Such a signal is ignored at least until a signal is sent along a (or a second signal is sent along b causing interference).

A process that waits for input on a and then for input on b before being able to do anything is actually just waiting for inputs on both a and b, their order being immaterial:

Law 1. $a?; b?; P = b?; a?; P$

Complementary to input-prefixing $a?; P$ is output-prefixing $c!; P$, where $c \in O$. This is a process that outputs on c and then behaves like P. The environment may send a signal on any input wire even before it receives the signal on c; whether or not it can do so safely depends on P.

Two outputs by a process on the same wire, one after the other, is unsafe because of the danger of the two signals interfering with one another before they reach the environment. Also, since any output of a process may arrive at the environment an arbitrary time later, the order in which outputs are sent is immaterial. Therefore, we have the following two laws:

Law 2. $c!; c!; P = \bot$

Law 3. $c!; d!; P = d!; c!; P$

Example 0 Law 2. allows us to prove that $c!; \bot = \bot$. This should not be surprising: the process $c!; \bot$ behaves like $\bot$ after it has output on c; a wholly undesirable state results even before the output has reached the environment.

$$c!; \bot$$
$$= \quad \{ \text{ Law 2. } \}$$
$$c!; c!; c!; \bot$$
$$= \quad \{ \text{ Law 2. } \}$$
$$\bot$$

■

Finally, as in CSP, prefixing is distributive (with respect to nondeterministic choice). For both input and output-prefixing we have the law

Law 4. $x; (P \sqcap Q) = (x; P) \sqcap (x; Q)$

Example 1 We are now in a position to specify a number of elementary delay-insensitive components, *viz.* the Wire, the Fork, and the C-element.

Consider a circuit with one input wire a and one output wire c. In response to each signal on a, the circuit should produce a signal on c. The precise behavior of this circuit is given by the following algebraic expression:

$$\mu X.\ a?; c!; X$$

which we shall refer to as the process W because it can be readily implemented by a wire. Now unfolding the recursion, we have that $W = a?; c!; W$. As in CSP, this equation itself uniquely defines W because its right hand side is guarded.

Next consider the process, with one input wire a and two output wires c and d, defined by the equation $F = a?; c!; d!; F$. This models the behavior of a fork.

Finally, the Muller C-element repeatedly waits for inputs on wires a and b before outputting on c. It is defined by $C = a?; b?; c!; C$.

When we introduce the *after* operator in the next section, it will become clear that the above expressions do indeed correctly specify the components. ■

A more general form of input-prefixing is input-guarded choice. Such a choice allows a process to take different actions depending upon the input received. The choice is made between a number of alternatives of the form $a? \to P$. For S a finite set of alternatives, the guarded choice $[S]$ selects one of them. An alternative $a? \to P$ can be selected only if a signal has been received on a. The choice cannot be postponed indefinitely once one or more alternatives become selectable.

Choice with only one alternative is no real choice at all:

Law 5. $[a? \to P] = a?; P$

The environment cannot safely send a second signal along an input wire until the first signal has been acknowledged. Thus the result of sending two signals on a to the process $a?; a?; P$ is $\perp$ rather than P. The process is as useless as a choice with no alternative:

Law 6. $a?; a?; P = a?; [\,] = [\,]$

If two alternatives are guarded on a, then either may be chosen after input has been supplied on a. Indeed, the designer has the freedom to implement only one of the two. This is captured in the following law, where the symbol $\Box$ separates the various alternatives:

Law 7.
$$\begin{aligned} & [a? \to P \ \Box\ a? \to Q \ \Box\ S] \\ = \ & [a? \to (P \sqcap Q) \ \Box\ S] \\ = \ & [a? \to P \ \Box\ S] \sqcap [a? \to Q \ \Box\ S] \end{aligned}$$

Example 2 With the above law we can prove the following absorption theorem. An alternative guarded on a is absorbed by $a? \to \perp$:

$$[a? \to \perp \ \Box\ a? \to P \ \Box\ S]$$

$= \quad$ { combining alternatives using Law 7. }

$$[a? \to (\perp \sqcap P) \ \Box\ S]$$

$= \quad$ { $\perp$ is the null element of $\sqcap$ }

$$[a? \to \perp \ \Box\ S]$$

■

Until a process acknowledges receipt of an input signal, a second signal on the same wire can result in interference. So, for S_0 and S_1 sets of alternatives, we have

Law 8. $[a? \to [S_0] \;\square\; S_1] = [a? \to [a? \to \bot \;\square\; S_0] \;\square\; S_1]$

Example 3 As a matter of fact, we can replace $\bot$ in Law 8. by any process P.

$$\begin{aligned} & [a? \to [a? \to \bot \;\square\; S_0] \;\square\; S_1] \\ = \quad & \{ \text{ absorption law derived in Example 2 } \} \\ & [a? \to [a? \to \bot \;\square\; a? \to P \;\square\; S_0] \;\square\; S_1] \\ = \quad & \{ \text{ Law 8. } \} \\ & [a? \to [a? \to P \;\square\; S_0] \;\square\; S_1] \end{aligned}$$

■

Indeed, if it is unsafe for the environment to send a signal along a particular input wire, it remains unsafe at least until an output has been received. Therefore, we also have the following absorption law.

Law 9. $[a? \to \bot \;\square\; b? \to [a? \to P \;\square\; S_0] \;\square\; S_1] \;=\; [a? \to \bot \;\square\; b? \to [S_0] \;\square\; S_1]$

Example 4 With the input-guarded choice we can model somewhat more interesting delay-insensitive components, such as the Merge, the Selector and the Decision-Wait element.

The Merge is a circuit with two input wires a and b and one output wire c. In response to a signal on either a or b, it will output on c:

$$M = [a? \to c!; M \;\square\; b? \to c!; M]$$

We shall discover, in the next section, that this definition implies that it is unsafe for the environment to supply input on both a and b before receiving an output on c.

The Selector is a circuit with one input wire a and two output wires c and d. Upon reception of an input it outputs on one of the two wires:

$$S = [a? \rightarrow c!; S \;\square\; a? \rightarrow d!; S]$$

Actually, in this case there is no need to use guarded choice. By Laws 5. and 7., an equivalent formulation is $S = a?; ((c!; S) \sqcap (d!; S))$.

Finally, we can define the Decision-Wait element (2×1 in this case). It expects one input change in its row and one input change in its column. It produces as output the single entry which is indicated by the two changing inputs – there are two entries in this case:

$$\begin{aligned} DW \;=\; & [r0? \rightarrow [r1? \rightarrow \bot \;\square\; c? \rightarrow e0!; DW] \\ & \square\;\; r1? \rightarrow [r0? \rightarrow \bot \;\square\; c? \rightarrow e1!; DW]\,] \end{aligned}$$

(A C-element can be viewed as a 1×1 Decision-Wait element.) ■

2 More Advanced Constructs

In the last section we provided enough operators to allow us to specify many interesting delay-insensitive components from which larger circuits might be constructed. To better understand these specifications we need to be able to determine how a circuit will behave after some signals have been exchanged with its environment. Before we can do this, it turns out that we need a more general form of guarded choice which allows for *skip* guards as well as input guards.

Recall that a guarded choice $[S]$ consists of a set S of alternatives of the form $a? \rightarrow P$. We shall now allow S to include also alternatives of the form $skip \rightarrow P$. Such an alternative can be selected whether or not any input is supplied. (As in CSP, if no input is supplied, it must eventually be selected.)

The laws of the last section remain valid, but in addition we have several laws involving *skip* guards. (These are also laws in occam [13].) As before, a choice with one alternative is no real choice at all:

Law 10. $[skip \rightarrow P] = P$

Nondeterministic choice can be regarded as a special case of guarded choice:

Law 11. $[skip \rightarrow P \;\square\; skip \rightarrow Q] = P \sqcap Q$

The selection of a *skip*-guarded alternative is an internal (unobservable) action of a process. This gives rise to the following three laws. In the first, a nondeterministic choice arises after input has been supplied on a because the signal may arrive before the selection of a *skip*-guarded alternative:

Law 12.
$$[a? \to P \;\square\; skip \to [a? \to Q \;\square\; S_0] \;\square\; S_1]$$
$$= [skip \to [a? \to (P \sqcap Q) \;\square\; S_0] \;\square\; S_1]$$

The second law states that a nested *skip*-guarded alternative can be selected in preference to any other alternative:

Law 13. $[skip \to [skip \to P \;\square\; S_0] \;\square\; S_1] = [skip \to P \;\square\; S_0 \;\square\; S_1]$

The third law is a convexity property of guarded choice:

Law 14. $[skip \to [S_0] \;\square\; skip \to [S_0 \;\square\; S_1] \;\square\; S_2] = [skip \to [S_0] \;\square\; S_1 \;\square\; S_2]$

Example 5 Two *skip*-guarded alternatives can be combined together:

$[skip \to P \;\square\; skip \to Q \;\square\; S]$

$=$ { nesting the *skip* guards using Law 13. }

$[skip \to [skip \to P \;\square\; skip \to Q] \;\square\; S]$

$=$ { $\sqcap$ as guarded choice, Law 11. }

$[skip \to (P \sqcap Q) \;\square\; S]$

■

Example 6 With impunity we can extend the set of alternatives in a guarded choice with that guarded choice itself as a new *skip*-guarded alternative:

$[S]$

$=$ { one choice is no choice, Law 10. }

$[skip \to [S]\,]$

$=$ { convexity, Law 14. }

$[skip \rightarrow [S] \;\square\; S]$

■

Example 7 A *skip*-guarded alternative can always be chosen, and so no other alternative need be offered, *i.e.*, $P \sqsubseteq [skip \rightarrow P \;\square\; S]$.

$P \sqcap [skip \rightarrow P \;\square\; S]$

$=$ { $\sqcap$ as guarded choice, Law 11. }

$[skip \rightarrow P \;\square\; skip \rightarrow [skip \rightarrow P \;\square\; S]\,]$

$=$ { unnesting the *skip* guards, Law 13. }

$[skip \rightarrow P \;\square\; S]$

In particular, $[skip \rightarrow \bot \;\square\; S] = \bot$ by Law 0. ■

Example 8 As another example of interference between two consecutive outputs on a wire, we have

$c!; [skip \rightarrow c!; P \;\square\; S]$

$\sqsupseteq$ { Example 7 and monotonicity of prefixing }

$c!; c!; P$

$=$ { interference between outputs, Law 2. }

$\bot$

which means that $c!; [skip \rightarrow c!; P \;\square\; S] = \bot$, by Law 0. ■

We are now able to define how a process behaves after the environment has sent input to it or received output from it. We consider the after-input case first.

The process $P/a?$ behaves like P after its environment has sent it a signal on $a \in I$. Notice that this does not mean that P has received this input yet; the signal may still be on its way. Indeed it is impossible to tell whether P has received the signal until some acknowledging signal has been received from P.

The first two laws for after-input are concerned with undesirable behavior. A process which has entered an unsafe state remains unsafe. Also, sending two signals in a row on an input wire may cause interference and is therefore unsafe.

Law 15. $\bot/a? = \bot$

Law 16. $P/a?/a? = \bot$

The order in which signals are sent does not determine the order in which they are received, and so

Law 17. $P/a?/b? = P/b?/a?$

An output-prefixed process can do nothing but output, even when sent input:

Law 18. $(c!; P)/a? = c!; (P/a?)$

After-input (on a) distributes through the alternatives in a guarded choice, except for those alternatives guarded on a. Those become *skip*-guarded. Furthermore, an extra alternative is required to indicate that interference can result after a second input on a.

Law 19. $[S]/a? = [a? \to \bot \,\Box\, S']$,
where S' is formed by substituting for each alternative $A \in S$ the new alternative $A/a?$, defined by

$$\begin{aligned}&(\mathit{skip} \to P)/a? = \mathit{skip} \to (P/a?)\\ &(a? \to P)/a? = \mathit{skip} \to P\\ &(b? \to P)/a? = b? \to (P/a?), \text{ for } b \neq a.\end{aligned}$$

As a consequence of Laws 5. and 19., we have that

$$(a?; P)/a? = [a? \to \bot \,\Box\, \mathit{skip} \to P]$$

and

$$(b?; P)/a? = [a? \to \bot \,\Box\, b? \to (P/a?)].$$

Example 9 When $\bot$ is guarded on a, the environment must not supply input on a.

$[a? \to \bot \,\Box\, S]/a?$

$=$ { after through guarded choice, Law 19.,
S' being some set of alternatives derived from S }

$[a? \to \bot \;\square\; skip \to \bot \;\square\; S']$

$= \quad \{$ unguarded $\bot$, Example 7 $\}$

$\bot$

■

The following law allows us to expand the set of alternatives in a guarded choice to make the behavior after a particular input explicit:

Law 20. $[S] = [a? \to [S]/a? \;\square\; S]$

Example 10 It follows that after-input is distributive:

$(P/a?) \sqcap (Q/a?)$

$= \quad \{ \sqcap$ as guarded choice, Law 11. $\}$

$[skip \to (P/a?) \;\square\; skip \to (Q/a?)]$

$= \quad \{$ adding an a-guarded alternative with Law 20. $\}$

$[a? \to [skip \to (P/a?) \;\square\; skip \to (Q/a?)]/a?$

$\square\; skip \to (P/a?) \;\square\; skip \to (Q/a?)]$

$= \quad \{$ interference between inputs using Laws 16. and 19. and Example 7 $\}$

$[a? \to \bot \;\square\; skip \to (P/a?) \;\square\; skip \to (Q/a?)]$

$= \quad \{$ after through guarded choice, Law 19. $\}$

$[skip \to P \;\square\; skip \to Q]/a?$

$= \quad \{ \sqcap$ as guarded choice, Law 11. $\}$

$(P \sqcap Q)/a?$

■

Example 11 Here is a case in which *skip* can be eliminated:

$[a? \to \bot \;\square\; skip \to [S]]$

$=$ { adding an a-guarded alternative with Law 20. }

$[a? \to \bot \;\Box\; skip \to [a? \to ([S]/a?) \;\Box\; S]]$

$=$ { postponing alternative until after *skip*, Law 12. }

$[skip \to [a? \to (([S]/a?) \sqcap \bot) \;\Box\; S]]$

$=$ { one choice is no choice and $\bot$ is null element of $\sqcap$ }

$[a? \to \bot \;\Box\; S]$

■

Example 12 If $[S]/a? = \bot$, then $[a? \to \bot \;\Box\; S] = [S]$:

$[a? \to \bot \;\Box\; S]$

$=$ { $[S]/a? = \bot$ }

$[a? \to [S]/a? \;\Box\; S]$

$=$ { removing a-guarded alternative with Law 20. }

$[S]$

■

The following law is a useful generalization of Example 12:

Law 21. If $[S]/a? = \bot$, then $[a? \to P \;\Box\; S] = [S]$.

The process $P/c!$ behaves as P after the environment has received a signal on $c \in O$. It does not make sense to define after-output for just any output, but only for those that are actually possible.

Definition 0 We define $out(P)$, the set of wires on which P can output initially (and so for which after-output is defined), as follows:

$$out(\bot) = O$$

$$out(c!; P) = \begin{cases} O & \text{if } c \in out(P) \\ \{c\} \cup out(P) & \text{otherwise} \end{cases}$$

$$out([\,]) = \emptyset$$

$$out([a? \to P \;\Box\; S]) = out([S])$$

$$out([skip \to P \;\Box\; S]) = out(P) \cup out([S])$$

$$out(\mu X.F(X)) = (\bigcap n : n \geq 0 : out(F^n(\bot)))$$

■

It follows from this (by structural induction) that $out(P) \subseteq out(P/a?)$. Also note that another way to define $out(P)$ is that $c \in out(P)$ if and only if $c!; P = \bot$.

We can now examine the laws for after-output. In Laws 24.-27. it is understood that expressions involving after-output are well-defined, *i.e.*, $c \in out(d!; P)$, $c \in out([S])$, $c, d \in out(P)$ and $c \in out(P)$, respectively.

Law 22. $\bot/c! = \bot$

Law 23. $(c!; P)/c! = \begin{cases} \bot & \text{if } c \in out(P) \\ P & \text{otherwise} \end{cases}$

Law 24. For $c \neq d$, $(d!; P)/c! = \begin{cases} \bot & \text{if } d \in out(P) \\ d!; (P/c!) & \text{otherwise} \end{cases}$

Law 25. $[S]/c! = [S']$,
where S' consists of an alternative $skip \to (P/c!)$ for each alternative in S of the form $skip \to P$ for which $c \in out(P)$.

Law 26. $P/c!/d! = P/d!/c!$

Law 27. $P/c!/a? \subseteq P/a?/c!$

Law 27. may be understood as follows. It is assumed that P can output on c. Even when the environment receives the output on c after having supplied the input on a, P might have output on c before receiving that input. Thus the possible behaviors of $P/a?/c!$ include those of $P/c!/a?$. As an example, consider $P = [skip \to c!; [\,] \;\Box\; a? \to \bot]$. Then, $P/c!/a? = a?; \bot$, but $P/a?/c! = \bot$.

Example 13 Now we can determine how components such as the C-element and the Merge behave after they have interacted with their environment. For the C-element we derive

$(a?;b?;c!;C)/a?$

$=$ { after through input-prefixing, corollary to Law 19. }

$[a? \to \bot \;\square\; \mathit{skip} \to b?;c!;C]$

$=$ { one choice is no choice and eliminating *skip* as in Example 11 }

$[a? \to \bot \;\square\; b? \to c!;C]$

Hence, a C-element that has been sent one input on a must not be sent another (until a signal on c has been received). Before it will output on c, however, it has to input on b. This is exactly how we want a C-element to behave after being sent a signal on a. We can now also compute $C/a?/b?$. A little calculation shows that the result is $[a? \to \bot \;\square\; b? \to \bot \;\square\; \mathit{skip} \to c!;C]$, as desired. Taking it one step further, we can show that $C = C/a?/b?/c!$.

A more interesting example of the possibility of interference is seen in the specification of Merge. Once a signal has been sent on either input wire, both input wires become unsafe to use. In this case we compute

$M/a?$

$=$ { definition of M }

$[a? \to c!;M \;\square\; b? \to c!;M]/a?$

$=$ { after through choice and prefixing, Laws 18. and 19. }

$[a? \to \bot \;\square\; \mathit{skip} \to c!;M \;\square\; b? \to c!;(M/a?)]$

$=$ { $M/a?$ is of the form $[\mathit{skip} \to c!;P \;\square\; S]$, Example 8 }

$[a? \to \bot \;\square\; b? \to \bot \;\square\; \mathit{skip} \to c!;M]$

This shows that signals on both a and b should be withheld until the Merge outputs on c. ■

3 Composition

In this section we define a parallel composition operator. With it we can determine the overall behavior of a circuit from the individual behavior of its components.

It is understood that if the output wire of one component has the same name as the input wire of another, then these wires are supposed to be joined together; any signals transmitted along such a connection are hidden from the environment. The parallel composition operator is fundamental to a hierarchical approach to circuit design. It permits an initial specification to be decomposed into a number of components operating in parallel, and each of these components can be designed independently of the rest.

The simplicity of the laws enjoyed by parallel composition is one of the main attractions of our algebra. Indeed, in [15] certain restrictions had to be placed on processes before their composition could even be considered; and in [2] the fixed-point definition of parallel composition was rather unwieldy.

Parallel composition is denoted by the infix binary operator $\|$. All the operators we have met so far do not affect the input and output alphabets of their operands; so, for example, in the nondeterministic choice $P \sqcap Q$, we insist that the input alphabet of P is the same as that of Q, and declare that it is the same as that of $P \sqcap Q$. In the parallel composition $P \parallel Q$, however, the input alphabet of P should be disjoint from that of Q; likewise, the output alphabet of P should be disjoint from that of Q. (These rules prohibit fan-in and fan-out of wires; the explicit use of Merges and Forks is required.) The input alphabet of $P \parallel Q$ then consists of those input wires of each process P and Q which are not output wires of the other. Similarly, the output alphabet of $P \parallel Q$ consists of those output wires of each process which are not input wires of the other.

Parallel composition is commutative. It is also associative, provided we ensure that a wire named in the alphabets of any two processes being composed is not in the alphabets of a third process. If one process in a parallel composition is in an undesirable state, then the overall state is undesirable:

Law 28. $P \parallel \bot = \bot$

When an output-prefixed process $c!;P$ is composed with another process Q, the output is transmitted along c. Depending on whether or not c is in the input alphabet of Q, the signal on c is sent to Q or to the environment:

Law 29. $$(c!;P) \parallel Q = \begin{cases} P \parallel (Q/c?) & \text{if } c \text{ is in the input alphabet of } Q \\ c!;(P \parallel Q) & \text{otherwise} \end{cases}$$

It remains only to consider parallel composition of guarded choices. The following law specifies the alternatives in the resulting guarded choice.

Law 30. $[S_0] \parallel [S_1] = [S]$,
where S is formed from the alternatives in S_0 and S_1 in the following way. For each alternative in S_0 of the form $skip \rightarrow P$, we have $skip \rightarrow (P \parallel [S_1])$ in S. For each alternative in S_0 of the form $a? \rightarrow P$ with a not in the output alphabet of $[S_1]$, we have $a? \rightarrow (P \parallel [S_1])$ in S. The alternatives in S_1 contribute to the alternatives in S in a similar way.

Example 14 If one component is able to send a signal that it is unsafe for the other to receive, then their parallel composition is $\perp$.

$$(a!; P) \parallel [a? \rightarrow \perp \;\Box\; S]$$

$=$ { internal communication on a, Law 29. }

$$P \parallel [a? \rightarrow \perp \;\Box\; S]/a?$$

$=$ { Example 9 and $\perp$ null element of parallel composition, Law 28. }

$$\perp$$

■

Example 15 We compute a number of simple compositions in this example. Although the resulting behaviors are well-known, it has never previously been possible to give a straightforward algebraic derivation.

Consider first connecting two wires W_0 and W_1 together. Let $W_0 = a?; b!; W_0$ and $W_1 = b?; c!; W_1$. Then, in their parallel composition, signals on b are hidden from the environment.

$$W_0 \parallel W_1$$

$=$ { definitions of W_0 and W_1 }

$$(a?; b!; W_0) \parallel (b?; c!; W_1)$$

$=$ { one choice is no choice and parallel composition through guarded choice, Law 30., using that b is internal }

$$a?; ((b!; W_0) \parallel (b?; c!; W_1))$$

$=$ { internal communication on b, Law 29. }

$$a?; (W_0 \parallel (b?; c!; W_1)/b?)$$

$=$ { after through prefixing }

$a?;(W_0 \parallel [b? \to \bot \,\Box\, skip \to c!;W_1])$

$=$ { substituting for W_0 and applying Law 30., parallel composition through guarded choice, using that b is internal }

$$a?;\ [\ \ a? \to ((b!;W_0) \parallel [b? \to \bot \,\Box\, skip \to c!;W_1]) \\ \Box\ skip \to (W_0 \parallel (c!;W_1))\ ]$$

$=$ { one choice is no choice and absorption as in Example 3 }

$a?;[skip \to (W_0 \parallel (c!;W_1))]$

$=$ { one choice is no choice and external communication with Law 29. }

$a?;c!;(W_0 \parallel W_1)$

Since this recursion is guarded, we conclude that $W_0 \parallel W_1 = W$.

A more interesting example is the composition of a "one-hot" C-element and a Fork in the following way. The C-element is specified by $C = a?;b?;c!;C$ and the Fork by $F = c?;a!;d!;F$. This is a circuit involving feedback.

$C/a? \parallel F$

$=$ { Example 13 and definition of F }

$[a? \to \bot \,\Box\, b? \to c!;C] \parallel (c?;a!;d!;F)$

$=$ { one choice is no choice and parallel composition through guarded choice, using that a and c are internal }

$b?;((c!;C) \parallel (c?;a!;d!;F))$

$=$ { internal communication on c }

$b?;(C \parallel (c?;a!;d!;F)/c?)$

$=$ { definition of C and after through prefixing }

$b?;((a?;b?;c!;C) \parallel [c? \to \bot \,\Box\, skip \to a!;d!;F])$

$=$ { one choice is no choice, parallel composition through guarded choice, using that a and c are internal, and definition of C }

$b?;[skip \to (C \parallel (a!;d!;F))]$

= { one choice is no choice, internal communication on a and external communication on d }

$b?; d!; (C/a? \parallel F)$

By uniqueness of guarded recursion, this combination of C-element and Fork behaves just like a wire. Although the Fork signals on a and d "in parallel", this did not lead to a doubling of the number of states which we had to analyze. We could deal with a entirely before d was pulled out of the parallel composition. This technique can be more generally applied and that is why these algebraic manipulations do not lead to a state explosion. ■

4 A Small Case Study

In [12] Martin presents a number of designs of circuits that are delay-insensitive but for their use of Isochronic Forks. An Isochronic Fork has one input wire and two output wires and operates in a way similar to the Fork of Example 1, except that

> the difference in delays between the two branches of the fork is shorter than the delays in the operators to which the fork is an input.

Thus the environment needs to register the arrival of only one output signal before it can safely send another signal to the fork. The Isochronic Fork clearly does not connect components in a delay-insensitive way. Fortunately, one output of an Isochronic Fork is usually fed into an AND-element and this combination as a whole, which we call an Isochronic AND-element (Figure 0), is insensitive to delay. Treating the Isochronic Fork and an AND-element to which it connects as one delay-insensitive component enables us to verify many of Martin's designs. (This treatment suffices for the verification of many designs but not all, because the property of Isochronic Forks stated above is actually slightly stronger than the property that we choose to model.) In particular, our algebra clarifies the relationship between his Q-element and D-element, and we show in this section that a single proof suffices for the verification of his designs for both these components. First, we give the specification of an Isochronic And-element, which we deduce from the assumed behavior of an Isochronic Fork.

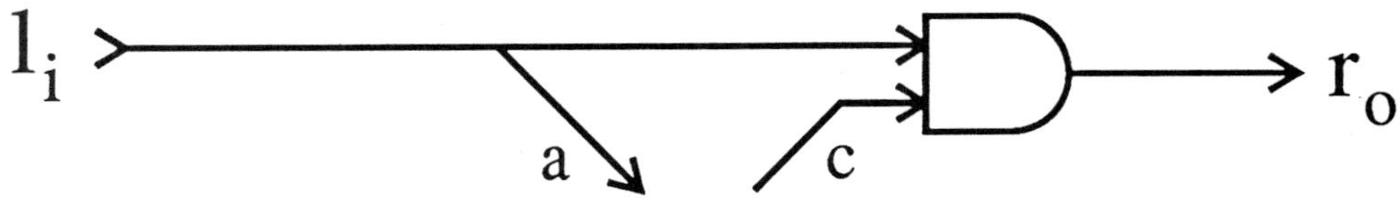

Figure 0: The delay-insensitive Isochronic And-element. (The internal connection of fork to AND-element is sensitive to delay, but the external connections are all delay-insensitive.)

As usual, we assume that all wires connecting a component to its environment are initially low. Signals on a wire between a component and its environment are either up-going or down-going transitions. The problem with the AND-element in delay-insensitive circuit design is that once both of its input wires are high, we can never allow both of them to go low again. The AND-element would output only once and, hence, acknowledge only one of the two down-going inputs. There is no way that the environment could tell that the signal on the other input wire had been received. When the AND-element is combined with an Isochronic Fork, however, it is possible to allow both its input wires to go low again. One branch of the fork connects to one input wire of the AND-element. Any signal that reaches the AND-element on that wire is effectively acknowledged by the corresponding signal on the other branch of the fork. We treat the Isochronic AND-element as a primitive delay-insensitive component and do not model the Isochronic Fork on its own.

The behavior of an Isochronic AND-element (IND for short) can be specified by four mutually-recursive equations. Each defines a quiescent state of the component, *i.e.*, a state in which nothing will be output until further input is supplied. The processes IND° and $\mathrm{IND}^{\bullet}$ may be thought of as "one-hot" Isochronic AND-elements, one input wire being high in each case. IOR may be thought of as an Isochronic OR-element.

Definition 1 An Isochronic AND-element is defined as follows, where input to the Isochronic Fork is on wire li, output from the Isochronic Fork to the environment is on wire a, input from the environment to the AND-element is on wire c and output from the AND-element to the environment is on wire ro.

$$\begin{array}{lcl} \text{IND} & = & [li? \to a!; \text{IND}^\bullet \ \Box \ c? \to \text{IND}^\circ] \\ \text{IND}^\bullet & = & [li? \to a!; \text{IND} \ \Box \ c? \to ro!; \text{IOR}] \\ \text{IND}^\circ & = & [li? \to a!; ro!; \text{IOR} \ \Box \ c? \to \bot] \\ \text{IOR} & = & [li? \to a!; ro!; \text{IND}^\circ \ \Box \ c? \to ro!; \text{IND}^\bullet] \end{array}$$

■

In checking that this definition is reasonable, one discovers, for example, that $\text{IND}/li?/a!/c?/li? = \bot$. (One input wire to an AND-element going low while the other goes high may cause a glitch.)

The following lemmas will be useful later. In some of the proofs we abbreviate certain processes to P and Q and certain sets of alternatives to S_0 and S_1 when their exact expression is no longer of interest.

Lemma 0 $(ro!; \text{IOR})/li? = \bot$.

Proof

$(ro!; \text{IOR})/li?$

$=$ { After through prefixing and guarded choice }

$ro!; [li? \to \bot \ \Box \ skip \to a!; ro!; \text{IND}^\circ \ \Box \ c? \to (ro!; \text{IND}^\bullet)/li?]$

$=$ { Example 8, using that $a!; ro!; P = ro!; a!; P$ }

$\bot$

■

Lemma 1 $(ro!; \text{IOR})/c? = \bot$.

Proof

$(ro!; \text{IOR})/c?$

$=$ { After through prefixing and guarded choice }

$ro!; [li? \to (a!; ro!; \text{IND}^\circ)/c? \ \Box \ c? \to \bot \ \Box \ skip \to ro!; \text{IND}^\bullet]$

$=$ { Example 8 }

$\perp$

■

Lemma 2 $\mathrm{IND}^{\bullet}/c? = ro!;\mathrm{IOR}$.

Proof

$\mathrm{IND}^{\bullet}/c?$

$=$ { After through guarded choice }

$[li? \rightarrow (a!;\mathrm{IND})/c? \;\square\; c? \rightarrow \perp \;\square\; skip \rightarrow ro!;\mathrm{IOR}]$

$=$ { on account of Lemma 1 and Example 7 we have $[li? \rightarrow P \;\square\; skip \rightarrow ro!;IOR]/c? = \perp$. Apply Example 12 }

$[li? \rightarrow (a!;\mathrm{IND})/c? \;\square\; skip \rightarrow ro!;\mathrm{IOR}]$

$=$ { Law 21., using Lemma 0 }

$[skip \rightarrow ro!;\mathrm{IOR}]$

$=$ { one choice is no choice }

$ro!;\mathrm{IOR}$

■

Lemma 3 $\mathrm{IND}/c? = \mathrm{IND}^{\circ}$.

Proof

$\mathrm{IND}/c?$

$=$ { After through guarded choice and prefixing }

$[li? \rightarrow a!;(\mathrm{IND}^{\bullet}/c?) \;\square\; c? \rightarrow \perp \;\square\; skip \rightarrow \mathrm{IND}^{\circ}]$

$=$ { Lemma 2 and Example 6 }

IND°

■

The D and Q-Elements

We are now ready to consider Martin's D and Q-elements [12, pp. 30-33]. Each provides an interface between a left environment and a right environment: they communicate with the left environment by inputting on li and outputting on lo, and with the right environment by inputting on ri and outputting on ro. Both use a four-cycle signaling scheme to implement a one-place buffer. The left environment fills the buffer by signaling on li (which is acknowledged on lo), and this cycle has to be repeated (return-to-zero); the signal ro empties the buffer (which the right environment acknowledges on ri), and this cycle also has to be repeated. Four mutually-recursive equations serve to define both the D and Q-elements.

Definition 2

$$\begin{array}{rcl} D &=& [li? \to lo!; Q \;\square\; ri? \to \bot] \\ Q &=& [li? \to ro!; \tilde{D} \;\square\; ri? \to \bot] \\ \tilde{D} &=& [ri? \to ro!; \tilde{Q} \;\square\; li? \to \bot] \\ \tilde{Q} &=& [ri? \to lo!; D \;\square\; li? \to \bot] \end{array}$$

(Note that $\tilde{D}$ and $\tilde{Q}$ are just D and Q, respectively, with left and right switched over. Notice also that in any state exactly one of the two inputs is expected.) ∎

Martin's designs for each element involve essentially two IND's, a C-element, and a Fork. These can be verified as follows.

Straightforward calculation shows that the C-element $C = a?; b?; x!; C$ composed with the Fork $F = x?; c!; d!; F$ yields the following C-element with two outputs

$$\mathrm{CF} = a?; b?; c!; d!; \mathrm{CF},$$

which we will use in the verification. As a matter of fact, we start out with $\mathrm{CF}/b?$ the specification of which is easily computed to be

$$\mathrm{CF}/b? = [b? \to \bot \;\square\; a? \to c!; d!; \mathrm{CF}].$$

It is also easy to see that

$$\mathrm{CF}/b?/a? = [a? \to \bot \;\square\; b? \to \bot \;\square\; skip \to c!; d!; \mathrm{CF}].$$

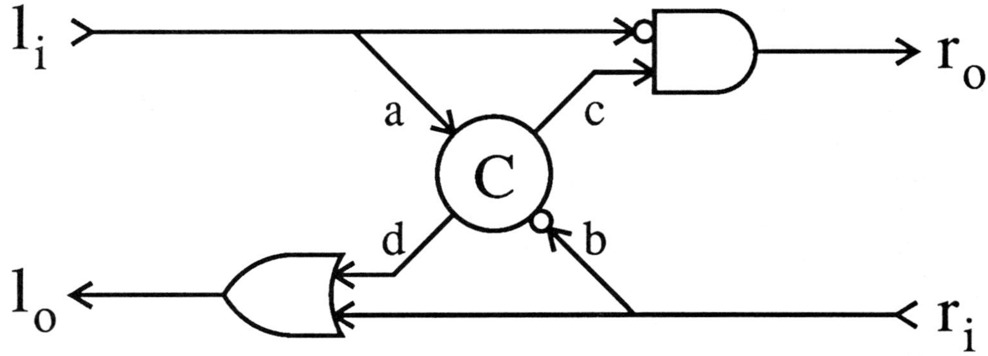

Figure 1: The composition of an IND$^\bullet$, an $\overline{\text{IOR}}$, a C-element, and a Fork

Martin's design for a D-element is depicted in Figure 1. It uses an IND$^\bullet$ and an IOR. We need to rename the wires of the latter, so that li becomes ri, ro becomes lo, a becomes b and c becomes d; we shall denote such a change by an overbar. The specification of IND$^\bullet$ and its derivatives are given in Definition 1. The specification of $\overline{\text{IOR}}$ and its derivatives are, according to the above renaming,

$$
\begin{array}{lcl}
\overline{\text{IND}} & = & [ri? \to b!; \overline{\text{IND}^\bullet} \ \square\ d? \to \overline{\text{IND}^\circ}] \\
\overline{\text{IND}^\bullet} & = & [ri? \to b!; \overline{\text{IND}} \ \square\ d? \to lo!; \overline{\text{IOR}}] \\
\overline{\text{IND}^\circ} & = & [ri? \to b!; lo!; \overline{\text{IOR}} \ \square\ d? \to \bot] \\
\overline{\text{IOR}} & = & [ri? \to b!; lo!; \overline{\text{IND}^\circ} \ \square\ d? \to lo!; \overline{\text{IND}^\bullet}]
\end{array}
$$

We now compute $\text{CF}/b? \parallel \text{IND}^\bullet \parallel \overline{\text{IOR}}$ and show that it implements the D-element.

$$\text{CF}/b? \parallel \text{IND}^\bullet \parallel \overline{\text{IOR}}$$

$=$ { parallel composition through guarded choice, using the specifications of $\text{CF}/b?$, IND$^\bullet$, and $\overline{\text{IOR}}$, and the fact that a, b, c and d are internal wires }

$$[li? \to (\text{CF}/b? \parallel (a!; \text{IND}) \parallel \overline{\text{IOR}}) \ \square\ ri? \to (\text{CF}/b? \parallel \text{IND}^\bullet \parallel (b!; lo!; \text{IND}^\circ))]$$

$=$ { Example 14, using that $\text{CF}/b?$ is of the form $[b? \to \bot \ \square\ S]$ }

$$[li? \to (\text{CF}/b? \parallel (a!; \text{IND}) \parallel \overline{\text{IOR}}) \ \square\ ri? \to \bot]$$

$=$ { internal communication on a }

$[li? \rightarrow (\mathrm{CF}/b?/a? \parallel \mathrm{IND} \parallel \overline{\mathrm{IOR}}) \;\square\; ri? \rightarrow \bot]$

$=$ { parallel composition through guarded choice, because a, b, c, and d are internal }

$[\;\; li? \rightarrow [skip \rightarrow ((c!; d!; \mathrm{CF}) \parallel \mathrm{IND} \parallel \overline{\mathrm{IOR}}) \;\square\; li? \rightarrow P \;\square\; ri? \rightarrow Q]$
$\square\;\; ri? \rightarrow \bot]$

$=$ { Law 9, Example 3, and one choice is no choice }

$[li? \rightarrow ((c!; d!; \mathrm{CF}) \parallel \mathrm{IND} \parallel \overline{\mathrm{IOR}}) \;\square\; ri? \rightarrow \bot]$

$=$ { internal communication on c and d and Lemma 3 }

$[li? \rightarrow (\mathrm{CF} \parallel \mathrm{IND}^{\circ} \parallel \overline{\mathrm{IOR}}/d?) \;\square\; ri? \rightarrow \bot]$

$=$ { after through guarded choice }

$[\;\; li? \rightarrow (\mathrm{CF} \parallel \mathrm{IND}^{\circ} \parallel [ri? \rightarrow P \;\square\; d? \rightarrow \bot \;\square\; skip \rightarrow lo!; \overline{\mathrm{IND}}^{\bullet}])$
$\square\;\; ri? \rightarrow \bot]$

$=$ { parallel composition through guarded choice, because a, b, c, and d are internal }

$[\;\; li? \rightarrow [li? \rightarrow Q \;\square\; ri? \rightarrow R \;\square\; skip \rightarrow (\mathrm{CF} \parallel \mathrm{IND}^{\circ} \parallel (lo!; \overline{\mathrm{IND}}^{\bullet}))]$
$\square\;\; ri? \rightarrow \bot]$

$=$ { Law 9, Example 3 and one choice is no choice }

$[li? \rightarrow (\mathrm{CF} \parallel \mathrm{IND}^{\circ} \parallel (lo!; \overline{\mathrm{IND}}^{\bullet})) \;\square\; ri? \rightarrow \bot]$

$=$ { external communication on lo }

$[li? \rightarrow lo!; (\mathrm{CF} \parallel \mathrm{IND}^{\circ} \parallel \overline{\mathrm{IND}}^{\bullet}) \;\square\; ri? \rightarrow \bot]$

Now if we look at the definition of the D-element we see that the beginning is there. Left to prove is that Q is implemented by $\mathrm{CF} \parallel \mathrm{IND}^{\circ} \parallel \overline{\mathrm{IND}}^{\bullet}$, which is Martin's design for a Q-element! We derive

$\mathrm{CF} \parallel \mathrm{IND}^{\circ} \parallel \overline{\mathrm{IND}}^{\bullet}$

$=$ { parallel composition through guarded choice, because a, b, c, and d are internal }

$[li? \rightarrow (\mathrm{CF} \parallel (a!; ro!; \mathrm{IOR}) \parallel \overline{\mathrm{IND}}^{\bullet}) \;\square\; ri? \rightarrow (\mathrm{CF} \parallel \mathrm{IND}^{\circ} \parallel (b!; \overline{\mathrm{IND}}))]$

$\subseteq$ { guarded choice is monotonic with respect to refinement, internal communication on a and external communication on ro }

$$[li? \rightarrow ro!; (\text{CF}/a? \parallel \text{IOR} \parallel \overline{\text{IND}^{\bullet}}) \ \Box \ ri? \rightarrow \bot]$$

(Note that we have indeed weakened the specification: the implementation is capable of more than that required by the specification.)

Now we use symmetry. We interchange a and b, c and d, li and ri, and lo and ro. For the two Isochronic elements this means that the top and bottom elements are interchanged. The C-element, however, is unaffected on account of Laws 1. and 3. Hence, we have also proved that

$$\text{CF}/a? \parallel \overline{\text{IND}^{\bullet}} \parallel \text{IOR} \ = \ [ri? \rightarrow ro!; (\text{CF} \parallel \overline{\text{IND}^{\circ}} \parallel \text{IND}^{\bullet}) \ \Box \ li? \rightarrow \bot]$$

and

$$\text{CF} \parallel \overline{\text{IND}^{\circ}} \parallel \text{IND}^{\bullet} \ \subseteq \ [ri? \rightarrow lo!; (\text{CF}/b? \parallel \overline{\text{IOR}} \parallel \text{IND}^{\bullet}) \ \Box \ li? \rightarrow \bot]$$

Using the commutativity and associativity of parallel composition, the unique solution of guarded recursions and the monotonicity of recursion, we conclude that

$$\text{CF}/b? \parallel \text{IND}^{\bullet} \parallel \overline{\text{IOR}} \ \subseteq \ D$$

$$\text{CF} \parallel \text{IND}^{\circ} \parallel \overline{\text{IND}^{\bullet}} \ \subseteq \ Q$$

That is, through algebraic manipulation we have verified Martin's designs for the D and Q-elements.

5 Conclusion

An algebraic approach has been taken to the specification and verification of delay-insensitive circuits. It has not been necessary to express explicitly all the states that such a circuit can enter; instead the possibility of them arising can be deduced using algebraic laws. This has led to concise specifications and short proofs. Another simplifying factor has been that, following [15], we do not distinguish between high and low-going transitions; this exposes many symmetries that would not otherwise be apparent. The main advantage of our approach is the ease with which we can compute the parallel composition of components. We have worked through many examples in which we used algebraic laws either to prove further laws or to investigate the behavior of specified circuits. As a case study, we verified some of Martin's designs, bringing to light interesting facts about his Isochronic Forks, D-elements and Q-elements.

We have postulated a large number of laws in this paper. It is possible to give a semantics for the algebraic expressions, compatible with the failures-divergence model of CSP [1, 8], so as to prove the laws correct with respect to that semantics. This will establish the soundness of the algebra. A topic of future research is that of the completeness of the algebra. This means that any two expressions with the same meaning can algebraically be transformed into one another. Finally, we would like to answer the question what processes with a delay-insensitive semantics, cf. [2, 10], can be denoted by a term in the algebra. Since we allow processes to be defined recursively, we have left the realm of regular languages. However, what exactly can and cannot be expressed in the algebra requires further study.

In the case study of the previous section and in the earlier examples the algebra has merely been used for post hoc verification. Given a specification and a proposed design we have verified the correctness of the design. We have chosen to show how expressions can be manipulated in the algebra rather than to show how (complicated) expressions can be written as the parallel composition of (simpler) expressions. Preliminary research indicates, however, that these expressions do indeed suggest ways in which they can be decomposed. For example, the specifications of the D-element and the Q-element suggest a decomposition into a couple of Toggles and Merges. This has also been observed by Ebergen [6]. We believe that the algebra is a powerful tool for both design and verification, although only the latter has been addressed in this paper.

Acknowledgements

We are most grateful to Tony Hoare and Tom Verhoeff for their interest and encouragement. The hospitality of the Department of Computer Science at Washington University helped make it possible for us to collaborate over this research. The work was partially funded by the Science and Engineering Research Council of Great Britain and the ESPRIT Basic Research Action CONCUR.

References

[1] S. D. Brookes and A. W. Roscoe. An improved failures model for communicating sequential processes. In G. Winskel, editor, *Proceedings of the NSF-SERC Seminar on Concurrency*, number 197 in Lecture Notes in Computer Science, pages 281–305. Springer-Verlag, 1985.

[2] W. Chen, J. T. Udding, and T. Verhoeff. Networks of communicating processes and their (de)-composition. In J. L. A. van de Snepscheut, editor, *The Mathematics of Program Construction*, number 375 in Lecture Notes in Computer Science, pages 174–196. Springer-Verlag, 1989.

[3] D. L. Dill. *Trace Theory for Automatic Hierarchical Verification of Speed-Independent Circuits.* PhD thesis, C.S. Dept., Carnegie Mellon Univ., Pittsburgh, PA, Feb. 1988.

[4] D. L. Dill and E. M. Clarke. Automatic verification of asynchronous circuits using temporal logic. In H. Fuchs, editor, *1985 Chapel Hill Conference on Very Large Scale Integration*, pages 127–143. Computer Science Press, 1985.

[5] J. C. Ebergen. *Translating Programs into Delay-Insensitive Circuits.* PhD thesis, Dept. of Math. and C.S., Eindhoven Univ. of Technology, 1987.

[6] J. C. Ebergen. A heuristic for the design of speed-independent circuits. Personal Memorandum, 1989.

[7] M. Hennessy. *Algebraic Theory of Processes.* Series in Foundations of Computing. The MIT Press, Cambridge, Mass., 1988.

[8] C. A. R. Hoare. *Communicating Sequential Processes.* Prentice-Hall, 1985.

[9] He Jifeng, M. B. Josephs, and C. A. R. Hoare. A theory of synchrony and asynchrony. In *Proceedings IFIP Working Conference on Programming Concepts and Methods*, to appear, 1990.

[10] M. B. Josephs, C. A. R. Hoare, and He Jifeng. A theory of asynchronous processes. *J. ACM*, (submitted), 1989.

[11] A. J. Martin. Compiling communicating processes into delay-insensitive VLSI circuits. *Distributed Computing*, 1(4):226–234, 1986.

[12] A. J. Martin. Programming in VLSI: From communicating processes to delay-insensitive circuits. Technical Report Caltech-CS-TR-89-1, Caltech Computer Science, 1989.

[13] A. W. Roscoe and C. A. R. Hoare. The laws of occam programming. *Theoretical Computer Science*, 60(2):177–229, 1988.

[14] I. E. Sutherland. Micropipelines. *Comm. ACM*, 32(6):720–738, 1989. Turing Award Lecture.

[15] J. T. Udding. *Classification and Composition of Delay-Insensitive Circuits.* PhD thesis, Dept. of Math. and C.S., Eindhoven Univ. of Technology, 1984.

[16] J. T. Udding. A formal model for defining and classifying delay-insensitive circuits. *Distributed Computing*, 1(4):197–204, 1986.

DIMACS Series in Discrete Mathematics
and Theoretical Computer Science
Volume 3, 1991

Synthesizing Processes and Schedulers from Temporal Specifications*

Howard Wong-Toi and David L. Dill
Department of Computer Science
Stanford University

Abstract

We examine two closely related problems: synthesizing processes to satisfy temporal specifications of reactive systems, and the synthesis of a scheduler to interact with and control a group of processes in order to meet a specification. Processes communicate through shared and distributed variables, either synchronously or asynchronously. In the finite state case, processes and specifications are arbitrary ω-regular languages, and both synthesis problems are solvable in doubly exponential time and space. The framework we present is flexible enough to incorporate dense real time into the model of concurrency, thereby allowing us to study the synthesis of real-time processes and schedulers. Real-time implementability and scheduling are also doubly exponential.

1 Introduction

A specification for a reactive process defines the desired ongoing behavior of the process along with its environment (the other processes with which it is connected). If the implementer of a process does not have control over the behavior of its environment, it may not be possible to meet the specification. For example, if a specification says that "x is always 1", and the environment can assign x the value 2, the specification is not implementable — the process can set x to 1, but cannot prevent it from taking other values.

The fundamental problem is to derive an implementation that always satisfies the specification in the face of adversarial behavior by the environment. Intuitively, an implementation is a winning strategy in the game against the environment. When the specification can be modeled as a finite automaton on infinite words, or a formula from a logic interpreted over infinite words, the synthesis problem is decidable. The decision procedures consist of reductions to the Church *solvability problem*, [Chu63, BL69] which can be solved using automata on infinite trees [Rab72].

Our processes have three different types of variables: shared, read-only and "distributed" (variables that can be read by other processes but written only by one). We solve the problem of process synthesis under both synchronous and asynchronous paral-

*Supported by NSF under grant MIP-8858807.

lel execution[1]. These results extend, unify, and simplify previous work [PR89a, PR89b, ALW89]. Next, we solve the problem of synthesizing a *scheduler*, which controls a collection of processes so as to satisfy a specification. For example, a critical region between processors could be enforced by a scheduling strategy instead of explicit synchronization within the processes. We also discuss ways to solve the same problem when there are additional constraints on the scheduler (such as processor availability). Finally, we show how these methods can also be applied to meet general real-time requirements.

For arbitrary untimed ω-regular processes and specifications the implementability problem is doubly exponential in the length of a temporal logical formula for the specification. Scheduling is doubly exponential in the number of processes as well, but only singly exponential in their size. If there is any implementation, the algorithm produces one of at most doubly exponential size. Real-time implementability and scheduling are also doubly exponential.

A more detailed presentation of this material appears in [WD91].

2 Background

In the last two years, several people have discovered the relevance of the solvability problem for determining the implementability of a linear temporal specification. In 1988, Dill considered it as a well-formedness condition on models of asynchronous circuits, but did not consider synthesis *per se* [Dil89]. Later, Pnueli and Rosner, [PR89a, PR89b], solved the problem of synthesizing concurrent reactive processes (in both synchronous and asynchronous models of concurrency) from a linear temporal logic specification. This solution was applicable to systems with only *distributed variables*. Shared variables were not considered.

Abadi, Lamport and Wolper did model shared variables, but their *computers* had the ability to observe *every* change made to the state, including all changes made by the environment [ALW89]. This is inappropriate for asynchronous parallelism, since in reality an implementation can base its actions only on the state changes that it can observe by reading variables. It does not have full observability of the entire system. Our implementations use only information directly inferable from their own actions.

We simplify considerably Pnueli and Rosner's solution for the asynchronous case. Their original solution involved transformations on several different representations of the specification. In our case, all transformations are performed on automata. The key simplifying idea is that the reductions used in solving implementability can be derived from projections that are used in the definitions of parallel composition.

3 Processes

We choose an action-based model of execution as opposed to a state-based approach because a process is always completely aware of its actions, but often ignorant of the

[1]By synchronous parallelism, we mean that processes run in "lock-step"; by asynchronous parallelism, we mean that processes run at arbitrary speeds. Elsewhere, these terms are used incompatibly to mean unbuffered and buffered communication between processes.

global state of the system. Our model of a single execution of a process is an infinite sequence of *sets* of actions (the reading and writing of variables). A process is assumed to be acting in an environment of other agents, with the only communication being indirectly through the values of the variables.

Suppose that $\mathcal{P}$ is an infinite set of primitive process names. Then every process P has a *type* describing what basic components it has and the variables it can access. Its type is a quadruple $\mathcal{T}_P = (Procs, \mathcal{D}_P, \mathcal{S}_P, \mathcal{E}_P)$, where $Procs$ is a subset of $\mathcal{P}$, $\mathcal{D}_P$ is a set of (owned) *distributed variables* which it can write and other processes can only read, $\mathcal{S}_P$ is a set of *shared variables* which may be written and read by P and other processes, and $\mathcal{E}_P$ is a set of *external* variables, which are read-only. We omit the subscripts when no confusion arises and abbreviate $\mathcal{D} \cup \mathcal{S} \cup \mathcal{E}$ by $\mathcal{V}$. $Procs$ is used to record which primitive process performs each action; this seems necessary for dealing with shared variables[2]. We assume all variables range over a common domain, *Dom*. When discussing finite-state processes, we assume *Dom* is finite.

The process P has a set of *primitive actions*, A_P, consisting of *reads* of the form $read(P_i, x, v)$ and *writes* of the form $write(P_i, x, v)$, where P_i is a primitive agent name in $\mathcal{P}$ and v is some value in the domain of the variable x. A *composite action* is a set of primitive actions. These correspond to primitive actions occurring *simultaneously*. The empty set of actions represents nothing happening and is called *skip*.

A *state*, s, is a function assigning a value to every variable in $\mathcal{V}$. An initial condition is a function mapping from $\mathcal{D}$ to $\mathcal{V}$. Let σ_i be the ith element of σ. An *action run* of P corresponds to a trace of actions that P may perform, in which P's distributed variables change only if P writes to them, but shared and external variables can change arbitrarily. Formally, an action run $a = a_1, a_2, \ldots$ of type $(Proc, \mathcal{D}, \mathcal{S}, \mathcal{E})$ is an infinite sequence of composite actions of P, satisfying the following consistency condition: there exists an infinite sequence of states $s = s_1, s_2, \ldots$ such that

(*read-values*) if $read(-, x, v)$ occurs in a_i, then x must have the value v in s_i,

(*write-values*) for every $x \in \mathcal{D}$, if a $write(-, x, v)$ occurs in a_i, x's value in s_{i+1} must be v' where some $write(-, x, v')$ occurs in a_i, and, if no $write(-, x, v)$ occurs in a_i, then $s_i(x) = s_{i+1}(x)$, and,

(*type*) whenever $write(-, x, v)$ occurs in a, x is in $\mathcal{D} \cup \mathcal{S}$.

Intuitively, if several writes to x occur simultaneously, x arbitrarily takes the value of one of them. An action run is consistent with an initial condition I if there is a satisfying sequence of states starting with state s_1 which agrees with I on $\mathcal{D}$.

The behavior of a process P is a triple $(\mathcal{T}_P, I_P, Runs_P)$, where $\mathcal{T}_P$ is its type, I_P its initial condition and $Runs_P$ (usually denoted $R(P)$) a set of action runs of type $\mathcal{T}_P$ consistent with its I. We sometimes use P as an abbreviation for $R(P)$.

3.1 Synchronous parallelism

The behaviors of two processes can be combined to yield the behavior that results when they run synchronously (in "lock step"). The runs of $P = P_1 \parallel P_2$ are essentially those

[2]The usual way of dealing with shared variables is have two labels for actions: π if performed by the process or ϵ if by the environment[BKP84]. But for scheduling, it helps to record exactly which component executed each action.

runs of the correct type which look like runs to both P_1 and P_2. P is defined only when $Proc_1$ and $Proc_2$ are disjoint, $\mathcal{D}_1$ and $\mathcal{D}_2 \cup \mathcal{S}_2$ are disjoint, and $\mathcal{D}_2$ and $\mathcal{D}_1 \cup \mathcal{S}_1$ are disjoint. Let del_Q be the projection on action runs that deletes all atomic actions made by agents in Q. Formally, $P = (\mathcal{T}_P, I_P, R(P))$ is defined as follows:

$\mathcal{T}_P = (Proc, \mathcal{D}, \mathcal{S}, \mathcal{E})$ where $Proc = Proc_1 \cup Proc_2$, $\mathcal{D} = \mathcal{D}_1 \cup \mathcal{D}_2$, $\mathcal{S} = \mathcal{S}_1 \cup \mathcal{S}_2$ and $\mathcal{E} = (\mathcal{E}_1 \cup \mathcal{E}_2) - (\mathcal{D} \cup \mathcal{S}) = (\mathcal{E}_1 - (\mathcal{D}_2 \cup \mathcal{S}_2)) \cup (\mathcal{E}_2 - (\mathcal{D}_1 \cup \mathcal{S}_1))$.

I_P is the extension of both I_{P_1} and I_{P_2}, i.e. I_P has domain D and agrees with I_{P_1} on D_1 and I_{P_2} on D_2.

$$\begin{aligned} R(P) &= del_{P_2}{}^{-1}(R(P_1)) \cap del_{P_1}{}^{-1}(R(P_2)) \\ &= \{r \mid r \text{ is a consistent run of type } \mathcal{T}_P,\ del_{P_2}(r) \in R(P_1) \text{ and } del_{P_1}(r) \in R(P_2)\}. \end{aligned}$$

Because del_{P_2} removes all P_2 actions, it deletes everything unobservable to P_1, leaving exactly what P_1 observes. Thus $del_{P_2}(r)$ is in $R(P_1)$ precisely when P_1's view of the global system run r is a run of P_1.

3.2 Generalized parallelism

A run is in the parallel composition of P_1 and P_2 when it is observed as a run by each process. Process observability is represented by functions f_{P_i} mapping consistent runs of type $\mathcal{T}_{P_1||P_2}$ to sets of consistent runs of type $\mathcal{T}_{P_i}$[3]. In synchronous parallelism, we had del_{P_2} for f_{P_1} and del_{P_1} for f_{P_2}.

$$\begin{aligned} R(P_1||P_2) &= f_{P_1}{}^{-1}(R(P_1)) \cap f_{P_2}{}^{-1}(R(P_2)) \\ &= \{r \mid r \text{ is a consistent run of type } \mathcal{T}_{P_1||P_2},\ f_{P_1}(r) \subseteq R(P_1) \text{ and } f_{P_2}(r) \subseteq R(P_2)\} \end{aligned}$$

3.3 Asynchronous parallelism

We often want the behaviors of processes that run in parallel *asynchronously*. Inserting an arbitrary (finite) number of *skip* actions between the non-*skip* actions of a run of a process allows other processes to perform as many or as few actions as they like between the process's non-trivial actions. The *delskip* operation removes the *skip* actions of P from an action run. The *stuttering closure* of an action run a is $stut(a) = \{a' \mid delskip(a') = delskip(a)\} = delskip^{-1}(delskip(a))$, see [ALW89]. This definition is extended to sets of action runs. For convenience, we use $R_{AS}(P)$ to refer to $stut(R(P))$, the asynchronous runs, or asynchronous closure of P. Asynchronous parallelism, $||_{AS}$, is defined as synchronous parallelism on the asynchronous closure of the processes. That is, $R(P_1 ||_{AS} P_2) = del_{P_2}{}^{-1}(R_{AS}(P_1)) \cap del_{P_1}{}^{-1}(R_{AS}(P_2))$. This is a special case of generalized parallelism, with $delskip^{-1} \circ delskip \circ del_{P_2}$ for f_{P_1}.

4 Implementability

A reactive system consists of processes which communicate among one another and that are intended to run forever. Suppose P is a reactive process where the only other agent is a completely unpredictable environment over which it has no control. The interaction between the implementation and the environment can be viewed as a game. Each player

[3]It may be that P_i's view of a run is not uniquely determined.

takes a turn extending a computation. The implementation wins the game if the resulting computation satisfies the specification. Moves in the game are actions, the *reads* and *writes* of variables. If the process always wins the game, it is an implementation of the specification.

A *strategy* is a special kind of process which decides its next *move* from its history of actions. Its moves are *read* requests of *individual* variables, and *write* requests of individual variables *with* the values to be written. Notice that a strategy is restricted to accessing only one variable at a time. Formally, a strategy S of type $(\{S\}, \mathcal{D}, \mathcal{S}, \mathcal{E})$ is a function $g_S : A_S^* \rightarrow Moves_S$, where $Moves_S = \{read(var) \mid var \in \mathcal{S} \cup \mathcal{E}\} \cup \{write(var, value) \mid var \in \mathcal{D} \cup \mathcal{S}, value \in Dom\}$. Because the adversarial environment can choose to return any value in response to a *read* action by the strategy, $R(S)$ is the set of *all* sequences of actions $a = a_1, a_2, \ldots$ such that

1. $a_{i+1} = read(S, var, value)$ for some $value \in Dom$ whenever $g_S(a_1, a_2, \ldots, a_i) = read(var)$, and
2. $a_{i+1} = write(S, var, value)$ whenever $g_S(a_1, a_2, \ldots, a_i) = write(var, value)$.

The strategy plays against an environment E of type $(\{E\}, \mathcal{E}, \mathcal{S}, \mathcal{D})$ whose actions runs are simply all the action runs for its type. It is an implementation of the specification *Spec* if every run of the combined system is in the specification, i.e. $R(S \parallel E) \subseteq Spec$. In this case it is a winning strategy for the game.

4.1 The Read-Write Game

Most of the games in this paper are games of partial information. The implementation cannot directly observe the entire system; the only information it knows about the environment has been inferred from reading the environment's variables. A solution to the implementability problem proceeds in two steps: reducing the game of partial information to one of complete information, and then solving the game of complete information, the *read-write game*. Runs in this game are traces of strategy actions only, and therefore reflect exactly what the strategy observes. A read-write game is stated as: Given a set of action runs W of type $(\{S\}, \mathcal{D}, \mathcal{S}, \mathcal{E})$, is there a strategy of the same type whose action runs are all in W, i.e. $R(S) \subseteq W$?

If the problem is finite state (the behaviors of the processes and the specification can be represented as ω-regular sets), the read-write game is decidable. This result follows almost immediately from the decidability of the Church-Büchi solvability problem, [BL69, Rab72].

Theorem 1 *If A is a nondeterministic Büchi automaton (NBA) defining Spec, a set of action runs for S over k variables, and A has n states, then the read-write game for Spec is decidable in $2^{O[(kn)^2 \cdot log(kn)]}$ time. If there is a winning strategy for Spec, the algorithm produces one of size $2^{O[(kn)^2 \cdot log(kn)]}$.*

Proof: The solution is a variant on the algorithm for "synchronous implementability" in [PR89a]. We construct from A a tree automaton that accepts exactly the trees that correspond to winning strategies. Hence to solve the game, we need only test for emptiness of the tree automaton [EJ88]. □

4.2 Cores

It is usual for a process to have only partial observability of the entire system; it cannot observe all the environment's actions. Suppose the environment owns the variables x_1 and x_2. While the process is reading x_1, the environment may be changing the value of x_2. In an asynchronous system the environment may even change x_2 an arbitrary number of times without the process noticing. Suppose two global system runs, r_1 and r_2, both look like the action run r to the implementation. The implementation only observes r and cannot distinguish whether r_1 or r_2 actually occurred. Thus a strategy or implementation should only generate action runs for which *all* corresponding global runs lie in the specification.

We assume the strategy has observability function f_S. The core of a global specification *Spec*, *core*(*Spec*), is the set of (local) strategy action runs whose inverse images under f_S are all in the specification, i.e. $core(Spec) = \{a \mid f_S^{-1}(a) \subseteq Spec\}$.

Lemma 1 *Let Spec a global specification and S be a strategy, interacting with the environment under observability function f_S. The following are equivalent:*
1) S is an implementation of Spec,
2) $R(S \parallel E) \subseteq Spec$,
3) $f_S^{-1}(R(S)) \subseteq Spec$,
4) $R(S) \subseteq core(Spec)$,
5) S wins the read-write game over core(Spec). □

Therefore implementability can be constructively reduced to the read-write game, provided we have an algorithm for finding the core of a specification. Since $core(Spec) = \{r \mid {f_S}^{-1}(r) \subseteq Spec\}$, its complement is $\{r \mid {f_S}^{-1}(r) \cap \overline{Spec} \neq \emptyset\}$. Thus a run r is in the complement of the core if it has a preimage under f_S in $\overline{Spec}$. Hence $core(Spec) = \overline{f_S(\overline{Spec})}$. The core can be found by complementation and substitution.

4.3 Synchronous Implementability

Until now, we have not considered any particular representation of the specification. We want to express ω-regular properties of variables ranging over *finite non-boolean* domains. Rather than defining a temporal logic with finite-valued variables, for convenience we express properties in *propositional* temporal logic (PTL). We encode each bit-value of a variable as a separate proposition.

Lemma 2 *Let ϕ be a PTL formula for Spec. Then there is an algorithm to find a deterministic Rabin automaton A accepting the $core_S(Spec)$.*

Proof: We proceed in three steps. From ϕ construct a NBA B for $\overline{Spec}$, ([WVS83]). Find a NBA B' for $\overline{core_S(Spec)}$, (apply the homomorphism $f_s = del_E$). Complement and determinize B' to get the desired Rabin automaton A ([EJ89]). □

Theorem 2 *Let ϕ be a PTL formula of length n_0 for the property Spec. There is a $2^{O[(k\cdot|Dom|)^3\cdot 2^{(3n_0)}]}$ algorithm for testing the synchronous implementability of Spec. If Spec is implementable, it yields a $2^{O[(k\cdot|Dom|)^3\cdot 2^{(3n_0)}]}$ folded representation of a strategy for Spec.*

Proof: Because of Lemma 1, we can follow the algorithm of Lemma 2, then solve the read-write game. The complexity result follows from analysis of the tableau procedure of [WVS83], the simultaneous complementation and determinization algorithm of [EJ89] and the test for emptiness of [EJ88]. □

4.4 Asynchronous Implementability

The analogous results from the last section all hold; the observability function is merely changed from del_E to $f_{AS} = delskip^{-1} \circ delskip \circ del_E$. Performing the substitution for *delskip* is similar to way ϵ-transitions are removed from finite state automata.

Theorem 3 *Let ϕ be a PTL formula of length n_0, defining the set Spec . The* asynchronous *implementability of Spec is solvable with complexity* $2^{O[(k\cdot|Dom|)^3\cdot 2^{(3n_0)}]}$. □

5 Scheduling

The ideas in the previous section can be used to find a *scheduler*, which is a process designed to provide external coordination for other processes. Such a situation may arise in the fields of distributed computing, networking, hardware design and operating systems. For example, we may be given information about the processes to be run and asked to schedule them to ensure mutual exclusion of their critical sections or guarantee some form of liveness. The scheduling problem is stated as follows: For processes $P_1, \ldots, P_n$ and specification *Spec*, is there a strategy S such that $P_1 \parallel \cdots \parallel P_n \parallel S$ satisfies *Spec*?

Let $P_1 \parallel \cdots \parallel P_n$ be collectively called P. We assume $P \parallel S$ is a system with no external variables. A truly external environment can always be modeled as another primitive process within P. By definition, $\overline{R(P \parallel S)} = f_P^{-1}(R(P)) \cap f_S^{-1}(R(S))$. Thus, $P \parallel S$ satisfies *Spec* iff $f_S^{-1}(R(S)) \subseteq \overline{f_P^{-1}(R(P))} \cup Spec$. This problem is essentially implementability with a modified specification, $Spec' = \overline{f_P^{-1}(R(P))} \cup Spec$, (see Lemma 1). Intuitively, *Spec'* asserts that if the environment behaves like the processes we need to schedule, then the original specification *Spec* must hold.

5.1 The Finite State Case

For both synchronous and asynchronous parallelism, the processes can be effectively composed and *Spec'* computed, which implies that the scheduling problem is decidable. The complexity is the same as that for implementability except for an extra multiplicative factor of p^n for the composition of the processes.

Theorem 4 (Scheduling) *Assume the system has k variables over a finite domain Dom. Let $\{P_1, \ldots, P_n\}$ be a set of processes, where each P_i is defined by a NBA A_i of size $\leq p$. Let Spec, a property of action runs with agents $P_1, \ldots, P_n$ and S, be defined by a PTL formula of length n_0. Then there is a* $2^{O[(k\cdot|Dom|\cdot p^n)^3\cdot 2^{(3n_0)}]}$ *algorithm to solve the scheduling problem, giving a* $2^{O[(k\cdot|Dom|\cdot p^n)^3\cdot 2^{(3n_0)}]}$ *state scheduler, if any scheduler exists.* □

6 Transformations on specifications

Most specifications are only expected to be met when the processes or the environment behave in a certain way. For example, a mutual exclusion program may only guarantee progress for each process requesting entry to its critical section if every process has a terminating critical section, $Term\_Crit \Rightarrow Prog$. An environmental assumption can be handled simply by adding it as an antecedent to the specification.

Additional requirements on the scheduler can be added to specifications. For example, suppose we have 10 processes to schedule, but only 5 processors available. Then we must disallow any scheduler that permits more than 5 processes to execute at any time. To handle this, the scheduler needs substantial control over the processes. One way to achieve this control is to add a new boolean *signal variable* $\{go_i\}$ for each process P_i. These new variables are owned by the scheduler. The process constraints C_P would assert that no P_i ever makes a non-*skip* action when the signal go_i is false. We then add to the scheduler the constraint "no more than 5 go_i signals are true at any time".

A constraint on the scheduler C_{sched} can be enforced by changing the specification $Spec$ to $C_{sched} \cap Spec$. The same technique can be used for more general resource constraints, for example, "processes 5 and 6 can never run on processors A and B at the same time."

7 Real-Time Processes and Specifications

Synchronous parallelism requires all processes to advance at the same time, while asynchronous parallelism is completely speed independent. It is often the case that processes are not interacting synchronously, but nor are they completely asynchronous — something is known about timing in the system. Knowledge of the relative timing of events may make it easier to implement a specification.

We consider *dense time*, where events may occur at arbitrarily close times; time is interpreted over the nonnegative real numbers, **R**. The execution of concurrent real-time processes is modeled by *timed traces* [AD90], which are infinite sequences together with times at which each event occurs. We use the *timed Büchi automata (TBA)* of [AD90] to model specifications and processes.

7.1 Real-Time Implementability

The methodology for solving the implementability and scheduling problems for untimed processes and specifications also works when timing is introduced. Using the techniques of section 6, timing assumptions on the environment and timing requirements on the implementation may all be written into the specification. Thus a strategy has no timing information itself and may be synthesized from the core of the specification, exactly as above for the untimed case. To solve real-time implementability then, we must show how to derive the core from a timed specification.

Timed processes are the natural extension of untimed processes, in that runs are now *timed* traces. Let $Untime$ be the function that removes the timing information from a timed trace. The observability function for P_1 in $P_1 \parallel_t P_2$ is given by $f_{P_1} = Untime \circ delskip^{-1} \circ delskip \circ del_{P_2}$. A strategy S is a timed implementation of $Spec$ iff $Untime^{-1}(S) \parallel_t Untime^{-1}(E) \subseteq Spec$. It is clear that S implements $Spec$ if it wins the read-write game over $core_t(Spec) = \{r \mid {f_S}^{-1}(r) \subseteq Spec\}$.

Lemma 3 *Given a TBA* $\overline{A' = \langle \Sigma, S, s_0, F, C, \delta \rangle}$ *for* $\overline{Spec}$, *there is an* $O(|A|^3)$ *algorithm to find a NBA B for* $\overline{core(Spec)}$. *B has size* $O(|C|! \cdot |\Sigma| \cdot |S|^2 \cdot 2^d)$, *where d is the number* *of bits in the binary encoding of A''s timing constants.* □

Because TBAs are not closed under complementation, we allow only specifications given as *deterministic* timed automata (DTA), which have at most one run for any timed trace. However DTA with Büchi acceptance conditions are not as expressive as DTA with Muller, Streett and Rabin acceptance condition. All these deterministic automata can be complemented, yielding TBA with polynomially many states[4]. Timing conditions on the implementation and the environment can also be included as antecedents to the original specification. An implementation of $Spec' = \overline{T_{imp} \cap T_e} \cup Spec$ implements $Spec$ under the timing assumptions T_{imp} and T_e.

Theorem 5 (Real-time Implementability) *Let A_{imp} and A_e be TBA for timing requirements T_{imp} and T_e on the implementation and the environment respectively. Let A_{Spec} be a DTA[4] for the specification. Then the real-time implementability problem is solvable in time exponential in the number of variables and number of states of the automata but doubly exponential in the number of bits to encode the timing constants.* □

7.2 Real-time Scheduling

The problem of scheduling a collection of activities to meet hard real-time deadlines has been studied extensively [CS88]. Previous work in this area considers scheduling a set of tasks (which may or may not be known in advance) to meet completion deadlines, usually on a single processor. These problems are usually at least NP-hard.

Here we are scheduling *reactive* processes which communicate and coordinate among themselves. This problem is not easily modelled as a collection of tasks to be scheduled. Furthermore, the expressiveness of ω-regular languages enables us to handle a far wider range of timing properties than the simple meeting of completion deadlines. For example, restrictions can be enforced on the ordering of completion times and the differences between completion times. Our specifications include, for example, "if y is set to 1 sufficiently often, it will eventually always be reset to 0 within 2 seconds".

We deal only with static scheduling, where the characteristics of the processes are known beforehand. Specifications must be given as *deterministic* TBA, but the processes to be scheduled and the timing constraints on the scheduler and the environment may be arbitrary TBA. Real-time scheduling then reduces to real-time implementability over the property "if the environment behaves like the processes, and all components in the system satisfy their timing constraints, then the specification is met".

Theorem 6 (Real-time Scheduling) *Let $\{P_1, \ldots, P_n\}$ be a set of processes with each P_i defined by the TBA A_i of size $\leq p$. Let Spec be given by a DTBA A_{Spec} of size n_{Spec}, and let T_{imp} and T_e be timing constraints on the implementation and the environment defined by TBA A_{imp} and A_e of sizes n_{imp} and n_e respectively. Let c be the total number of clocks and d the number of bits in the encoding of all the timing constants. The complexity of real-time scheduling is $2^{O([c!\cdot p^n \cdot k \cdot |Dom|(n_{imp} \cdot n_e \cdot n_{Spec})^2 \cdot 2^d]^3)}$.* □

---

[4]The number of acceptance pairs of a Rabin automaton must be logarithmic in the number of states.

8 Conclusion

The idea of finding strategies for games is not only applicable to the specific problem of synthesizing an implementation from a temporal specification, but can also be applied to other problems such as scheduling. Although the doubly exponential algorithms presented here may prove to be useful for very small systems, easier special cases of the problem will have to be discovered for real practicality. The problem of synthesizing a scheduler to meet a specification closely resembles the *supervisory control problem* in the study of discrete event systems, which historically has fallen in the domain of control theory. The relations between these two problems should be studied in greater detail.

Acknowledgements

We thank Amir Pnueli for a helpful discussion, and Elizabeth Wolf for careful reading of a draft of the paper.

References

[AD90] R. Alur and D. Dill, "Automata for modeling real-time systems", ICALP 1990.

[ALW89] M. Abadi, L. Lamport, P. Wolper, "Realizable and unrealizable specifications of reactive systems", *International Colloquium on Automata, Languages, and Programming,* Lecture Notes in Computer Science, 1989, Springer-Verlag

[BKP84] H. Barringer, R. Kuiper and A. Pnueli, "Now you can compose temporal logic specifications", *Proceedings of the ACM Symposium on Theory of Computing*, 1984, pp. 51–63.

[BL69] J. R. Büchi and L. H. Landweber, "Solving sequential conditions by finite-state strategies", Transactions of the American Mathematical Society, 138, 1969, pp. 295–311.

[CS88] S-C. Cheng and J. A. Stankovic, "Scheduling algorithms for hard real-time systems – a brief survey", in *Hard Real-Time Systems*, IEEE Press, 1988, pp. 150–173.

[Chu63] A. Church, "Logic, arithmetic, and automata", in *Proceedings of the International Congress of Mathematicians, 1962*, Institut Mittag-Leffler, 1963, pp. 23–35.

[Dil89] D.L. Dill, "Trace Theory for Automatic Hierarchical Verification of Speed-Independent Circuits", MIT Press, 1989.

[EJ88] E.A. Emerson, C.S. Jutla, "The complexity of tree automata and logics of programs", *Proc. of the 29th IEEE Symp. on Foundations of Computer Science*, 1988, pp. 328–337.

[EJ89] E.A. Emerson, C.S. Jutla, "On simultaneously determinizing and complementing ω-automata", *Pro. of the Symp. on Logic in Computer Science*, 1989, pp. 333–342.

[PR89a] A. Pnueli, R. Rosner, "On the synthesis of a reactive module", *Proc. 16th ACM Symp. Principles of Programming Languages*, 1989, pp. 179–190.

[PR89b] A. Pnueli, R. Rosner, "On the synthesis of an asynchronous reactive module", *International Colloquium on Automata, Languages, and Programming*, Lecture Notes in Computer Science Vol 372, 1989, Springer-Verlag

[Rab72] M. O. Rabin, "Automata on infinite objects and church's problem", *Regional Conference Series in Mathematics*, Vol. 13, American Mathematical Society, 1972.

[WVS83] P. Wolper, M.Y. Vardi, A.P. Sistla, "Reasoning about infinite computation paths", *Proc. of the 24th IEEE Symp. on Foundations of Computer Science*, 1983, pp. 185–194.

[WD91] H. Wong-Toi and D.L. Dill, "Synthesizing processes and schedulers from temporal specifications", *Computer-Aided Verification (Proc. CAV90 Workshop)*, DIMACS Series in Discrete Mathematics and Theoretical Computer Science Vol. 3 (American Mathematical Society, 1991).

DIMACS Series in Discrete Mathematics
and Theoretical Computer Science
Volume 3, 1991

Verification of Multiprocessor Cache Protocol using Simulation Relations and Higher-Order Logic

Paul Loewenstein*(Mitsubishi Electronics America) and
David L. Dill†(Stanford University)

November 14, 1990

Abstract

We present a formal verification method for concurrent systems. The technique is to show a correspondence between state machines representing an implementation and specification behaviour. The correspondence is called a *simulation relation*, and is particularly well-suited for theorem-provers. Since the method does not rely on enumerating all the states, it can be applied to systems with an infinite or unknown number of states. This substantially expands the class of hardware designs that can be formally verified. The method is illustrated by proving the correctness of a particularly subtle example which is likely to be of increasing importance: a directory-based multiprocessor cache protocol. The proof is carried out using the HOL ("higher-order logic") theorem-prover.

1 Introduction

It has been shown that formal verification of certain types of non-trivial hardware designs using theorem-provers is possible. Indeed, several complete (albeit simple) processors have been formally verified [4, 3, 10, 16] using theorem-proving. But before theorem-proving techniques can be applied to most hardware designs, the problem of modelling and verifying concurrent behaviour must be solved. Unforseen interactions among concurrent systems are one of the major source of bugs in hardware designs today. Significantly, architectural features that depend on concurrent interaction have been omitted from almost all formally verified processor designs. Such features include caches, pipelining, interrupts, and page mapping. Concurrency is crucial for high performance, so its use will continue to increase. In particular, multiprocessor architectures are certain to be a major source of complicated concurrent hardware designs. This paper describes a successful attempt to verify a high-level concurrent hardware design.

The cache coherence protocol of our example is *directory-based*. The main idea behind directory-based cache coherence protocols is to maintain a partial model of cache contents in a central directory (or directories). When, for example, an address is written for the first time in one cache, the directory can send "invalidate" messages to the specific caches which have data present at the

*Work performed at Stanford University while visiting scientist for National Semiconductor Corporation.

†Supported by NSF grant number MIP-8858807 and DARPA contract number N00014-87-K-0828.

same address. The advantage is that cache invalidations do not have to be broadcast, eliminating a major bottleneck in large multi-processors.

The protocol involves the interaction of several concurrent agents, namely the individual cache controllers and the directory. In general, models of concurrent systems are fundamentally different from, say models of functional programs. It is not sufficient to model a system as an input/output mapping on data-values or memory states. A model must also capture the *sequence* of states during a computation, because the interactions with other systems depends crucially on the intermediate states of the computation.

We use a well-known method for verifying concurrent systems based on *simulation relations.* The implementation (system behaviour) and specification (desired behaviour) are both represented as state graphs. A simulation relation establishes a correspondence between implementation states and specification states. For the example of the paper, we use *successive refinement.* We transform a higher-level description into an implementation in several steps. A simulation relation is used to show that the automaton after the step implements the automaton before the step.

Previously, there have been proposals for logics and methods for reasoning about concurrent systems too numerous to mention. However, one promising class of methods is to establish a correspondence between two automata (the correspondence can be a single- or multi-valued function, or a relation) [13, 1, 12, 18, 11, 15]. Although some impressive examples have been done by hand these methods have generally not been combined with automatic theorem-proving (the one other example of which we are aware is due to Nipkow [17]).

The most successful methods for verifying concurrent hardware are *state enumeration methods.* In essence, these methods *automatically* find all of the reachable states in a state graph representing the behaviour of a circuit For example, in the *model-checking* approach, the state graph is compared with *temporal logic* formulas, which make assertions about the execution paths through the system [2, 7]. Other techniques compare formal languages of finite automata [5, 6].

Of course, the major limitation of state enumeration methods is that they potentially need to inspect all of the states of the system. They all suffer from the *state explosion problem* — the number of states grows exponentially with the size of the system being verified. They are also generally not good for systems that are partially described. For example, in our description of a cache system, we never specify the number of processes, size of memory, size of data values, or many other parameters of the system. Hence, we have verified the correctness of all the systems that can be obtained by giving values to these parameters. A state enumeration approach would require not only that values be assigned to these parameters, but that the values be very small integers. In some case, one can show that a parameterized design is correct for all parameter values by verifying a few designs with particular values for the parameters. However, these techniques are either highly restricted or it requires human insight to extend the proof from special cases to the general case.

Theorem proving and state enumeration are complementary methods — each is appropriate for a particular class of problems. Eventually, we hope that a combination of both approaches will make hardware design verification economically feasible (and routine).

We carry out proofs using the HOL theorem-prover ("Higher-Order Logic") [9, 8]. HOL uses a very expressive logic, in which variables can range over functions as well as values (unlike first-order logic). Our approach is to express everything in HOL: indeed, the implementation and specification state graphs are given as logical formulas. There are several advantages to using

HOL. The generality of HOL allowed us to prove theorems not only about hardware, but about mathematics, including the soundness of the method itself. Also, the HOL type system allows proofs about objects without knowing their types. This results in generic results that can easily be reused.

Notation

For concreteness, we have included some of the theorems and formulas used to describe state graphs. For brevity, we have omitted most of them (for a computer file containing everything, including the proofs, please contact one of the authors). We have modified the notation somewhat to improve readability for readers who are not very familiar with HOL (for a detailed description of the HOL logic, see [8]).

HOL logic reads very much like ordinary predicate logic with a few differences. Variables and constants can range over functions or values, depending on their type. Types are left implicit in this paper; they are explained in the text unless they are obvious ¿from context. Function application is indicated by juxtaposition: $f\,x$ denotes function f applied to x. Often multiple argument functions are *curried*; $g\,x\,y$ denotes a function g applied to x, the result (again a function) is then applied to y. Informally one can consider g to be merely a multiple argument function. The (x, y) notation is used for pairing (cartesian product). When pairs are used as function arguments, as in $f(x, y)$, the notation appears conventional. Sometimes, (as in this paper), currying and pairing are mixed, as in $g(u, v)(x, y)$. Logical implication is indicated by the $\supset$ operator.

Proven theorems are indicated by the symbol $\vdash$. These have been derived using the mechanical theorem prover, HOL[9], and should be correct apart from errors in typographic transcription. Defined constants are printed in sans serif typeface.

$\mathsf{Let}\, x = y \,\mathsf{in}\, z[x]$ is syntactic sugar for a term semantically identical to $(\lambda x.z[x])y$, which reduces to $z[y]$. It saves repeating a long term y which may have multiple occurrences in $z[y]$.

We have adopted some special conventions for describing sequences and automata. Variables denoting sequences, or functions of time, are emboldened ($\boldsymbol{e}$), as opposed to other variables (e). When referring to next-state relations, unprimed variables (s) refer to the current state, and primed variables (s') to the next state. The initial state predicate of an automaton Name, is always $\mathsf{Name\_P}$. The next-state relation is $\mathsf{Name\_N}$ and any invariant is $\mathsf{Name\_I}$.

In the remainder of the paper, the next section outlines the proof strategy. The third section describes the state-graph model of behaviour and the definition of simulation relations. The fourth section outlines our proof strategy, which transforms a high-level description of the system through several intermediate representations to the final implementation. Then we describe a model for asynchronous communication. We then develop a specification for a multi-port memory without a cache. After that, part of the proof is presented in more detail. The final section discusses the appropriate use of the verification method and future work.

2 The Automaton Model

We regard a digital system as a concurrent process with externally visible variables (corresponding to wires, or bundles of wires) and invisible internal state. A *trace* is an infinite externally-visible execution history of the system. The *behaviour* of a system is the set of all possible traces that it can exhibit. This set can be related to a state graph, consisting of of a set of states, a subset of those representing possible initial states, and a next-state relation.

Each state is denoted by a pair (e, s), where e is the visible (or external) part of the state and s is the hidden, or internal, part. A state graph is defined by a pair of predicates, (P, N): P is a unary predicate that defines the the initial states, and N is a binary next-state relation: $N(e,s)(e',s')$ holds when (e', s') is a possible successor state to (e, s).

A binary predicate (Labelled State Automaton) $\mathsf{LSA}(P,N)e$ is defined in HOL. It states that e is a trace of the automaton defined by P (initial states) and N (next state relation). In more detail, e is a trace of the automaton if there exists a sequence of hidden states, s, such that the first state satisfies P, and N holds between every state and the next state in the sequence:

$$\mathsf{LSA}(P,N)e = (\exists s.P(e\,0, s\,0) \wedge (\forall t.N(e\,t, s\,t)(e(t+1), s(t+1)))) \tag{1}$$

2.1 Relating two automata

One automaton *implements* another if every trace of the first is an trace of the second. A powerful method for showing that this relation holds is to find a *simulation relation* between them. A simulation relation is a three-way relation $R\,e\,s_1\,s_2$, where (e, s_1) is a state of the implementation graph and (e, s_2) is a state of the specification graph (the visible components must be the same in both states). It must satisfy two properties: first, every initial state of the implementation is related by R to some initial state of the specification; and, second, whenever $R\,e\,s_1\,s_2$ and $N_1(e,s_1)(e',s_1')$ both hold, there exists some specification state (e', s_2') such that $N_2(e,s_2)(e',s_2')$ and $R\,e'\,s_1'\,s_2'$ both hold.

To show that one automaton is an implementation of another, it is sufficient to produce a simulation relation from the first to the second. This fact is proven as a theorem in HOL[1]

$$\begin{array}{l}\vdash \forall P_1 N_1 P_2 N_2. \\ \quad (\exists R. \\ \quad\quad (\forall e s_1.P_1(e,s_1) \supset (\exists s_2.P_2(e,s_2) \wedge R\,e\,s_1\,s_2))\wedge \\ \quad\quad (\forall e e' s_1 s_1' s_2.R\,e\,s_1\,s_2 \wedge N_1(e,s_1)(e',s_1') \supset \\ \quad\quad\quad (\exists s_2'.R\,e'\,s_1'\,s_2' \wedge N_2(e,s_2)(e',s_2')))) \supset \\ \quad (\forall e.\,\mathsf{LSA}(P_1,N_1)e \supset \mathsf{LSA}(P_2,N_2)e)\end{array} \tag{2}$$

The types of e and s_1 and s_2 are left open; the theorem therefore holds for finite, countable and uncountable-state automata.

This is a soundness result for our verification method: to prove that one automaton implements another, it is sufficient to supply a simulation relation. The converse would be that whenever one graph is an implementation of another, there exists a simulation relation. This property (completeness) does not hold. Nevertheless, simulation relations have been adequate for our purposes. Abadi and Lamport have shown that a similar method can be made complete if additional variables are added to the implementation to record the past and predict the future (history[2] and prophecy variables), a result that is equally applicable to our method, should the addition of variables prove necessary [1].

[1]This theorem was derived from the corresponding theorem for a labelled *transition* automaton presented with proof in [14].

[2]Our simulation relation approach does not need history variables.

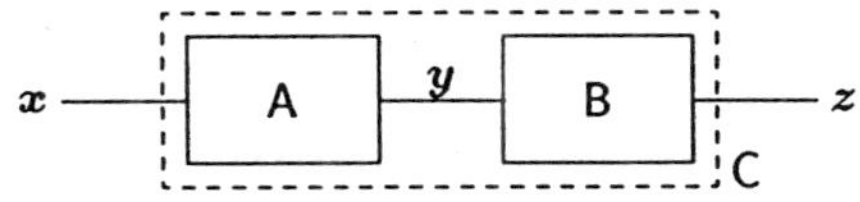

Figure 1: Composition of automata

An *invariant* is a predicate on individual states. An invariant is satisfied by a state graph if every state that is reachable from an initial state satisfies the invariant. An important method for simplifying proofs is to prove strong invariants before proving more complicated properties. Once an invariant has been proven on a state graph, the invariant can be incorporated into the next-state relation of original state graph without changing the behaviour of the system (this just makes explicit a property that is already implicitly true). This transformation often substantially reduces the amount of case analysis that must be used in proofs of additional properties of the state graph. The correctness of this transformation (although obvious) has been proven in HOL by specialisation of theorem 2.

2.2 Composition and Hiding

Figure 1 shows a parallel composition of two circuits, A and B. Given definitions of the automata for A and B, we can write the initial state predicate of C as

$$\mathsf{C\_P}((x,z),s_1,s_2,y) = \mathsf{A\_P}((x,y),s_1) \wedge \mathsf{B\_P}((y,z),s_2), \tag{3}$$

where A_P and B_P are the initial state predicates of the individual machines. Note that y has moved from being a component of the *label* of the individual machine state, to being a component of the *hidden state* of the composed machine (composition without hiding can be defined be putting y with x and z, if desired). The next-state of C relation is defined similarly.

$$\mathsf{C\_N}((x,z),s_1,s_2,y)((x',z'),s_1',s_2',y') = \mathsf{A\_N}((x,y),s_1)((x',y'),s_1') \wedge \mathsf{B\_N}((y,z),s_2)((y',z'),s_2') \tag{4}$$

Although it is possible to prove a general schema as a theorem, we found it simpler to merely prove each composition individually by expanding the LSA definition.

3 Roadmap of the Specification and Proof

This section gives a very high-level overview of the proof. The next few sections discuss the various steps in greater detail.

A preliminary task was to define a common master-slave communication channel MS, which is used at every interface where data is read or written. In the figures, communication delay is represented by circled arrows, with the arrow pointing from master to slave.

The actual proof refines an specification of memory behavior, which we call a *multi-port memory* to a memory system consisting of a cache for each processor, a directory, and a main memory (which is also a multi-port memory). The first step is to refine the multi-port memory into what we call a *cached multi-port memory*. The cached multi-port memory is not directly implementable as specified — it exists only to simplify the proof. The second part of the proof is to refine a component of the cached multi-port memory into a composition of two smaller components. Finally, the directory-based system is formed by merging several of the components into a single directory. This "step" is just an application of composition and hiding. We now discuss these steps in more detail.

3.1 Multi-port Memory

We first define the behaviour of an multi-port memory with no cache, MP_Mem. MP_Mem serves two purposes: MP_Mem is a part of the implementation because a cached memory is just a multi-port memory with a cache in front; MP_Mem is also the specification of the *desired* behaviour of the cached memory (a correctly functioning cached memory behaves just like an uncached memory).

The per-port components of the multi-port memory are modelled as functions from a port address, q, to a value. The type of q is left open so that the proofs are valid for any number of ports. This is possible because none of the consistency proof uses induction; the only induction is over time in the derivation of theorem 2.

3.2 Refining the Multi-Port Memory to a Cached Multi-port memory

We first refine the (uncached) multi-port memory to a simple cached multi-port memory. This involves defining the cached memory and proving that it implements the multi-port memory.

The cached multi-port memory consists of three types of components: a multi-port memory for primary storage, a cache Port_Cache (there is one for each port) and unit EXA which goes between the memory and port caches, whose primary purpose is to enforce mutually-exclusive access by the port caches to the multi-port memory.

Port_Cache looks very much like a single processor cache, except that it has a "window", s_d, on part of its hidden state. This window is used by EXA both to obtain information about the internal state of the Port_Caches, and to constrain the behaviour of the Port_Caches. It is very important to note that we have provided no explicit mechanism for this two-way transfer of information between Port_Cache and EXA (it is therefore not indicated on figure 2 as a circled arrow). This is an example of a generally useful trick: partitioning the logical description of a design in ways that are not physically implementable. In this case, it makes the proof much easier, and no harm is done because the next step will refine the unimplementable description into an implementable one by repartitioning the system. EXA uses s_d to decide when it can grant a cache exclusive access, and also to block access directly.

It is then proved, by the use of a simulation relation, that the cached multi-port memory implements the uncached multi-port memory.

3.3 Splitting the port cache

The second part of the proof refines the Port_Cache into a composition of two halves: Cache, which is the final (per-port) cache specification, and Port_Dir, which will eventually end up being part of the directory. The two halves communicate via two MS communication channels, one of which carries directory-to-cache requests, and the other carries cache-to-directory requests. It is then proved, by using a simulation relation, that Port_Dir coupled to Cache implements Port_Cache. Despite the

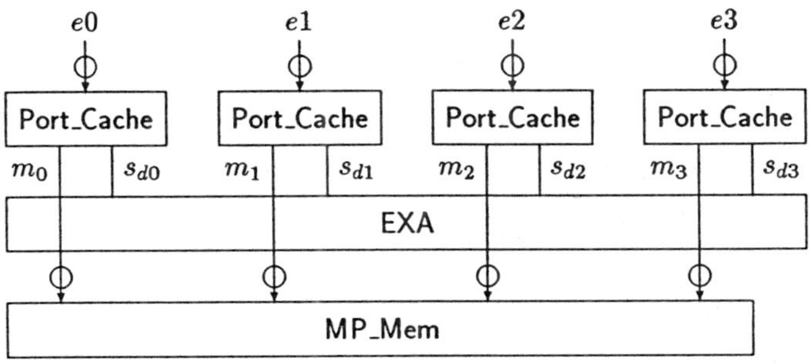

Figure 2: Modified caches with exclusive access control for a multi-port memory

fact that this can be carried out on one port in isolation, it is by far the largest part of the total multiport correctness proof. The final system is shown in figure 3.

The directory consists of the EXA and the collection of Port_Dir, which is regarded as a single unit. The division between EXA and Port_Dir in the intermediate representation was only a convenient fiction which may have nothing to do with the actual implementation of the directory. The s_d state is now completely enclosed in the directory, so the communication between the different components is entirely via master-slave channels. So, if the individual components are implementable (as we believe they are), the entire system is also implementable.

At this point, we cease refining the specification. If we wanted to carry it further, MP_Mem could be implemented with a memory request arbiter and a single port memory, in which case the arbiter would also become part of the directory. One could also use another level of caching at each processor. However, proving these transformations is relatively routine compared with the steps we are focussing on in this paper.

4 Modelling communication

All communication in our systems occurs over master-slave channels. Every communication consists of several steps. Initially, each end of the channel is in an *acknowledge* state. The master places a *request value* on its end of the channel (basically, it sends a message). After an arbitrary delay, this value appears at the slave end of the channel. After another delay, the slave responds by placing an *acknowledge value* on its end of the channel, which is eventually relayed back to the master. At this point, the communication cycle has completed (both ends of the channel are in the acknowledge state) and a new cycle can begin.

The states of the channel are the combinations of values on the two ends of the channel, as shown in figure 4, along with an additional bit of internal state, b, which is needed to distinguish the states labelled RA (when a request is in transit from the master to the slave) and RA' (when an acknowledge is in transit from the slave back to the master). The additional b is true during RA', false during RA, and undefined otherwise.

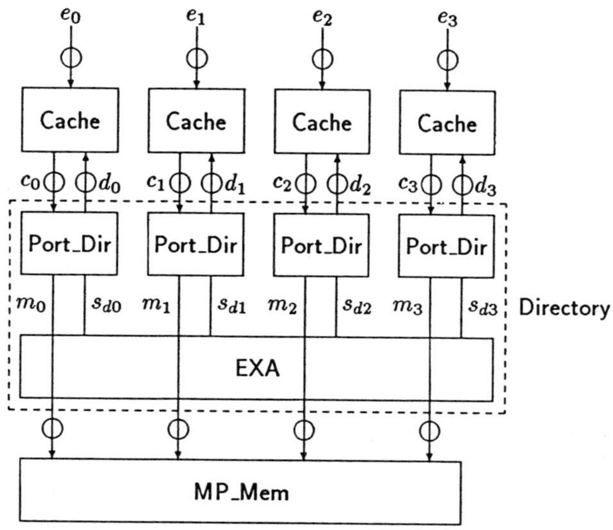

Figure 3: Configuration after splitting the caches

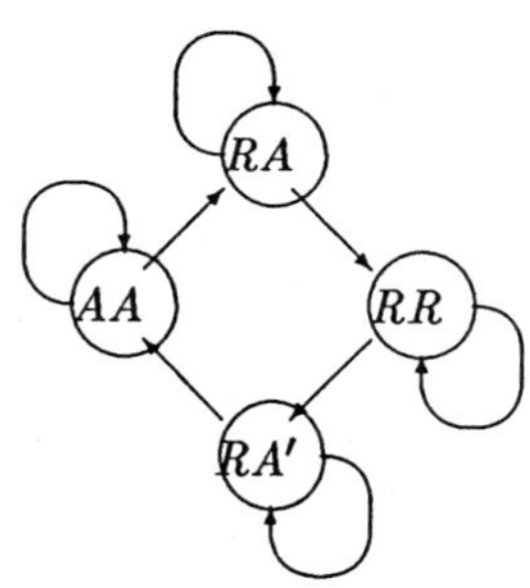

Figure 4: States and transitions of master-slave channel

Representation in HOL

The definition of master-slave communication in HOL takes a predicate defining the acknowledge values as a parameter, so that the same definition can be use regardless of the set of values. So $A\,x$ says "x is an acknowledge value", while $\neg A\,x$ says "x is a request value." The state is of the form $((m, s), b)$ where m is the value on the master end of the channel, s is the value on the slave end, and b is the internal bit mentioned above.

The initial state predicate of MS master-slave communication channel states that initially both ports have the same acknowledge value:

$$\mathsf{MS\_P}\,A\,((m, s), b) = A\,m \wedge (s = m) \tag{5}$$

and the next-state relation, MS_N, defines the state-graph in figure 4.

$$\begin{array}{l}\mathsf{MS\_N}\,A((m, s), b)((m', s'), b') = \\ (A\,m \wedge A\,s \supset (s' = s) \wedge ((m' = m) \vee \neg A\,m' \wedge \neg b'))\wedge \\ (\neg A\,m \wedge A\,s \wedge \neg b \supset (m' = m) \wedge ((s' = s) \wedge \neg b' \vee (s' = m)))\wedge \\ (\neg A\,m \wedge \neg A\,s \supset (m' = m) \wedge ((s' = s) \vee A\,s' \wedge b'))\wedge \\ (\neg A\,m \wedge A\,s \wedge b \supset (s' = s) \wedge ((m' = m) \wedge b' \vee (m' = s)))\end{array} \tag{6}$$

We prove the invariant, MS_I, which indicates state "AR" is never seen, and that in the AA and RR states, the values on the two ports are identical:

$$\mathsf{MS\_I}\,A((m, s), b) = (Am \supset A\,s) \wedge ((A\,m = A\,s) \supset (m = s)) \tag{7}$$

5 Modelling a Multi-Port Memory

A multiport memory is modelled with internal state as a function from address to data. This function is updated and referred to atomically from multiple read-write ports. The specification is separated into two parts. One specifies communication interfaces; the other enforces internal consistency.

5.1 Memory port communication

There are two types of request: RR a and WR $a\,b$ for reading and writing an address, a, and two corresponding acknowledgments:, RA b and WA.

The predicate defining the acknowledge values is:

$$\mathsf{Ack}\,m = (\exists b.m = \mathsf{RA}\,b) \vee (m = \mathsf{WA}) \tag{8}$$

and the allowed acknowledge transitions are defined by:

$$\begin{array}{l}\mathsf{Port\_N}\,m\,m' = \\ ((\mathsf{Ack}\,m' = \mathsf{Ack}\,m) \supset (m' = m))\wedge \\ (\mathsf{Ack}\,m' \supset \\ \;((\exists a.m = \mathsf{RR}\,a) \supset (\exists d.m' = \mathsf{RA}\,d))\wedge \\ \;((\exists ad.m = \mathsf{WR}\,a\,d) \supset (m' = \mathsf{WA})))\end{array} \tag{9}$$

5.2 Definition of a multiport memory

The visible state is modelled as a function, e, from the *port address*, q, to a "memory port" value, $e\,q$. The types of the port address, memory address and data are left open. The memory hidden state, s, is modelled as a function from address to data. The address and data types can be replaced with, for example, bit-vectors of some length, without re-running the proofs.

Initially all ports are in an acknowlege state:

$$\mathsf{MP\_Mem\_P}(e, s) = (\forall q.\, \mathsf{Ack}(e\,q)) \tag{10}$$

The next state relation has three components:

- All ports satisfy the memory request-acknowledge protocol.
- All read acknowledge transitions return the value stored in the memory.
- If there is a write acknowledge transition on some port, then the internal state of the memory is updated with the data value on some (not necessarily the same) port with a write acknowledge transition, otherwise the memory state is unchanged.

which in HOL is defined as:

$$\begin{array}{l}\mathsf{MP\_Mem\_N}(e, s)(e', s') = \\ (\forall q.\, \mathsf{Port\_N}(e\,q)(e'\,q)) \wedge \\ (\forall a. \\ \quad (\forall q.(e\,q = \mathsf{RR}\,a) \wedge \mathsf{Ack}(e'\,q) \supset (e'\,q = \mathsf{RA}(s\,a))) \wedge \\ \quad ((\exists q.(\exists b.e\,q = \mathsf{WR}\,a\,b) \wedge \mathsf{Ack}(e'\,q)) \to (\exists q.(e\,q = \mathsf{WR}\,a(s'\,a)) \wedge \mathsf{Ack}(e'\,q)) | (s'\,a = s\,a)))\end{array} \tag{11}$$

5.3 Loosening of specification to allow for port delay

This specification is sufficient for the specification of the multiport cache-memory system. It is however rather restrictive as a general specification of a multiport memory. As defined, all memory operations, whether read or write, occur simultaneously with the request-to-acknowledge transitions on the ports. In practice there will be an unknown delay between the issuance of a request and its execution, and between the execution of a request and its acknowledgement.

The specification is therefore loosened to allow the memory state to be updated, or data to be read, at any time between the issuance of a request and its acknowledgement. We do this by adding an acknowledgement delay to each port. This adds extra hidden state to store the reply between request execution and its acknowledgement at the port.

On this loosened specification, MS communication delays can be added to the ports and the system remains a loosened multiport memory. This can be demonstrated by simulation relation. This result formally justifies the communication delays on the e_i ports and the MP_Mem ports in figure 3.

6 Refining a Multiport Memory to a Cached Multiport Memory

In this section we define an initial specification for a multi-port cache, and prove that it is consistent when assembled with a multiport memory as in figure 2.

The multi-port cache specification is composed from two types of component, a *per-port* component, Port_Cache, which superficially resembles a single-port cache, and an exclusive access control, EXA, which prevents two caches having write access to the same address. Exclusive access inevitably involves communication between the ports, and therefore cannot be expressed independently of the multiport structure.

6.1 The per-port component

Port_Cache (figure 2) has three visible state components, e and m, representing the processor and memory ports, and s_d, which provides a "window" to part of the internal state. Both the e and m ports operate on the same protocol as the memory port in section 5.1, the cache acting as a "slave" on the processor port, and as "master" on the memory port.

The internal memory of a cache is typically implemented as some sort of associative memory. We do not want to bind a particular implementation of the associative memory, so we choose a very abstract model: the state is a function s_c from addresses to Exclusive(b) ("dirty" data), Non_Ex(b) ("clean" data) or Absent.

The exclusive access control uses a window on the internal state (drawn as the s_{dn} connections on figure 3). This should not be thought of as a physical connection. It is merely an abstraction in an intermediate specification. $s_d\, a$ can take 3 values:

- Ab, indicating that the address is absent in the cache.
- Nx, indicating that the address is not exclusive (i.e. either non-exclusive or absent) in the cache.
- Ex, indicating that the address may be exclusive in the cache.

The reason for this organisation is to minimise directory memory requirements, since the s_{dn} (one for each port) are destined to become internal state of the directory (section 3.3). The directory memory would normally be implemented as a single associative memory which keeps track of the status of each address in each cache. In practice it could be excessively costly to keep track of *all* the addresses in *all* the caches. However, it is not necessary to keep track of which caches have read-only access to an address. It is only necessary to know which addresses are write-accessed in what cache, and to have a means of forcing other caches to relinquish access to that address (which may involve some wasted commands).

Initially, the ports are in acknowledge states, the cache is empty, and s_d does not indicate the presence of exclusive data.

$$\text{Port\_Cache\_P}((p, s_d, m), s_c) = \text{Ack}\, p \wedge \text{Ack}\, m \wedge (\forall a. s_c\, a = \text{Absent}) \wedge (\forall a. \neg(s_d\, a = \text{Ex})) \qquad (12)$$

The next-state relation, Port_Cache_N, has a port and a per-address statement. The English version is rather imprecise to avoid excessive verbosity:

- The processor and memory ports obey the memory protocol.
- For each address:
  - The cache may acknowledge a WR $a\, b$ from the processor only if the cache has exclusive access to a.

- If the cache receives a $\mathsf{RR}(a)$ request from the processor, it can only respond if there is data present for address a.
- If $s_c\, a = \mathsf{Exclusive}\, y$ then
  - * If there is a $\mathsf{RR}\, a$ request from the processor, then any acknowledgement has to be $\mathsf{RA}\, y$.
  - * Any write request to the memory for address a must be with either data y, or, if a write acknowledge is being issued to the processor for a $\mathsf{WR}\, a\, b$ request, b.
  - * If a write request is being issued to memory, then $s_c\, a$ is updated to non-exclusive or absent, otherwise it remains exclusive.
- If $s_c\, a = \mathsf{Non\_Ex}\, y$ then
  - * No memory write operations are allowed for address a.
  - * Acknowledgement of processor $\mathsf{RR}\, a$ requests is $\mathsf{RA}\, y$.
  - * $s_c\, a$ can remain unchanged, or be updated to Absent or $\mathsf{Exclusive}\, y$
- If $s_c\, a = \mathsf{Absent}$ then
  - * No memory write operations are allowed for address a.
  - * Either $s_c\, a$ remains unchanged, or, upon a memory read acknowledge on a, can be updated to non-exclusive or exclusive with the data read from memory.
- If $s_c\, a$ is updated to exclusive, then $s_d\, a$ is updated to Ex.
- $s_d\, a$ is updated to Ab, only if $s_c\, a$ is updated to Absent.

We have used non-determinism to avoid over-specifying. For instance there is nothing ensuring that a memory read occurs in the event of a read miss. This could involve a sequence of events such as writebacks to make free space. These details are irrelevant to the consistency of the cache system; however they would be needed for a later liveness proof, and are allowed by this specification.

We then demonstrate invariant Port_Cache_I:

$$\begin{array}{l}\mathsf{Port\_Cache\_I}((p, s_d, m), s) = \\ (\forall a. \\ \quad (\forall b b'.(m = \mathsf{WR}\, a\, b) \wedge (s_c\, a = \mathsf{Non\_Ex}\, b') \supset (b = b'))\wedge \\ \quad ((\exists b. s_c\, a = \mathsf{Exclusive}\, b) \supset (s_d\, a = \mathsf{Ex})) \wedge ((s_d\, a = \mathsf{Ab}) \supset (s_c\, a = \mathsf{Absent})))\end{array} \tag{13}$$

which confirms that data being written back is consistent with any non-exclusive copy in the cache, as well a trivially provable relation between s_c and s_d.

6.2 Exclusive Access Control

In order to maintain consistency we must ensure that if a cache has exclusive access to an address, then that address must be absent in all other caches. It is the function of EXA to enforce this. Also while an address is exclusive in some cache, or data is being written back to memory at that address, then no other cache can issue a memory request for that address, lest it receive "stale" data from the memory.

The EXA next-state relation restricts the states that the Port_Caches can enter:

- s_d for port q and address a can only be updated to Ex, if the s_d for all other ports and address a become Ab, and no other port issues, or has a pending memory request for address a.
- If data is being written back to address a for port q, then no other cache may issue a memory request, or have a pending request for address a.

And formally it is given by:

$$\begin{array}{l}\mathsf{EXA\_N}(m, s_d)(m', s'_d) = \\ (\forall aqr. \\ \quad ((s'_d\, q\, a = \mathsf{Ex}) \supset (q = r) \vee (s'_d\, r\, a = \mathsf{Ab}) \wedge ((\mathsf{Addr}(m'\, r) = a) \supset \mathsf{Ack}(m'\, r))) \wedge \\ \quad ((\exists b. m'\, q = \mathsf{WR}\, a\, b) \supset (q = r) \vee ((\mathsf{Addr}(m'\, r) = a) \supset \mathsf{Ack}(m'\, r))))\end{array} \tag{14}$$

It has no invisible state component. For defining EXA, the behaviour, we use a version of LSA with no hidden state component.

6.3 Proving Invariants of the Cached Memory

EXA is assembled with multiple Port_Caches to form a complete multiport cache, MP_Cache.

$$\mathsf{MP\_Cache}\, p\, m = (\exists s_d. \mathsf{EXA}\, m\, s_d \wedge (\forall q. \mathsf{Port\_Cache}(p\, q)(s_d\, q)(m\, q))) \tag{15}$$

The number of ports, not necessarily finite, is determined by the type of q. s_d is now a *hidden* state component of MP_Cache. All state components derived ¿from Port_Cache are now functions of q, the port address. The Port_Cache_I invariant is included. We also prove an additional invariant of the multiport cache:

- If s_d indicates exclusive access for address a on a port then, for all other ports, s_d indicates address a is absent, and there can be no memory for address a.
- If there is a writeback in progress to address a from a cache, then for all other caches, address a is absent, and there is no memory transaction for address a.

Or, in HOL:

$$\begin{array}{l}\mathsf{MP\_Cache\_I}((p, m), s, s_d) = \\ (\forall aqr. \\ \quad ((s_d\, q\, a = \mathsf{Ex}) \supset (q = r) \vee (s_d\, r\, a = \mathsf{Ab}) \wedge ((\mathsf{Addr}(m\, r) = a) \supset \mathsf{Ack}(m\, r))) \wedge \\ \quad ((\exists b. m\, q = \mathsf{WR}\, a\, b) \supset (q = r) \vee (s_c\, r\, a = \mathsf{Absent}) \wedge ((\mathsf{Addr}(m\, r) = a) \supset \mathsf{Ack}(m\, r))))\end{array} \tag{16}$$

After composing MP_Cache with the loosened version of MP_Mem (section 5.3), an additional invariant MPCM_I is demonstrated:

- If a memory read is being acknowledged, then the returned value agrees with the value in the memory (no writebacks from other caches are possible during a memory read).
- During a memory write acknowledgement, the contents of memory agrees with the datum written (Only one cache can writeback for a given address at a time).

- If there is non-exclusive data in a cache, then either that data is in memory, or it is in the process of being written back by that cache.

Which is given in HOL by:

$$\begin{array}{l} \mathsf{MPCM\_I}(e, x, s_c, s_d, y, s) = \\ (\forall a q. \\ \quad ((x\,q = \mathsf{RR}\,a) \wedge \mathsf{Ack}(y\,q) \supset (y\,q = \mathsf{RA}(s\,a))) \wedge \\ \quad (\forall b.(x\,q = \mathsf{WR}\,a\,b) \wedge \mathsf{Ack}(y\,q) \supset (y\,q = \mathsf{WA}) \wedge (s\,a = b)) \wedge \\ \quad (\forall b.(s_c\,q\,a = \mathsf{Non\_Ex}\,b) \supset (x\,q = \mathsf{WR}\,a\,b) \vee (s\,a = b))) \end{array} \tag{17}$$

where y is the extra state which stores the memory operation before its acknowledgement by the loosened multiport memory.

6.4 Simulation Relation Consistency Proof

We now determine how the MPCM state models the state in MP_Mem. If we use s_1 to denote the cached memory internal state, and s_2 to denote the specification memory state, we can define a suitable simulation relation, for each address a:

- If there is exclusive data, b, for a in some cache, then $s_2\,a = b$.
- If no cache has exclusive access to a, but some cache is writing back data b to address a, then $s_2\,a = b$.
- If no cache has exclusive access or is writing back to a, then $s_2\,a = s_1\,a$.

Formally this is defined as:

$$\begin{array}{l} \mathsf{MPCM\_R}\,e(x, s_c, s_d, y, s_1)s_2 = \\ (\forall a. \\ \quad (\forall b.(\exists q.s_c\,q\,a = \mathsf{Exclusive}\,b) \supset (s_2\,a = b)) \wedge \\ \quad ((\forall q b.\neg(s_c\,q\,a = \mathsf{Exclusive}\,b)) \supset \\ \quad\ (\forall b.(\exists q.x\,q = \mathsf{WR}\,a\,b) \supset (s_2\,a = b)) \wedge \\ \quad\ ((\forall q b.\neg(x\,q = \mathsf{WR}\,a\,b)) \supset (s_2\,a = s_1\,a)))) \end{array} \tag{18}$$

Demonstration that MPCM_R is a simulation relation is straightforward, if somewhat tedious. We can then use theorem 2 to complete the proof that the caching scheme in figure 2 implements a multiport memory, as outlined in section 3.2.

7 Splitting Port_Cache

Although we now have a consistent specification for a multiport cache, it is not very useful as it is not partitioned in a form suitable for implementation. We therefore proceed to split Port_Cache into two halves, Cache and Port_Dir, connected by two request-acknowledge communication channels, one for cache to directory requests, and one for requests to the cache ¿from the directory. Cache is the cache local to each processor. Port_Dir is destined to be part of the directory, and maintains a partial record of the status of the cache memory. This partial record is the implementation of the "miraculous" s_d in section 3.2.

Because only a single Port_Cache need be considered for this splitting transformation, the cache splitting proof does not concern itself with the multiport nature of the entire cache system.

The formal definitions of Cache and Port_Dir are too long to present in this paper. We shall therefore forego further formal detail and outline the transformations and proof.

7.1 Cache to Directory requests

A similar request-acknowledge protocol is used for cache-directory requests as for memory read-write requests.

Cache-to-directory requests are of six types. Each expresses a desired state change in the cache, and there is one type of request for each of the six possible changes of status between Absent, Non_Ex, and Exclusive. Data fields include the address, and for "downgrades" from Exclusive, a data field containing the dirty data.

There are three types of acknowledgements, acknowledge with data for "upgrades" from Absent, acknowledge without data, and refusal.

7.2 Directory to Cache requests

Since the directory only ever needs to issue downgrade requests, directory-to-cache requests come in only two flavours. Each one says what the desired state in the cache is, independently of the previous state. They are "Request absent" and "Request non-exclusive".

Since the cache cannot refuse directory-to-cache requests, there are only two forms of reply: acknowledge with (exclusive) data, and acknowledge with no data.

7.3 Definition of Cache

Cache is similar to Port_Cache, except that it interacts with the c and d cache-directory communication ports instead of the memory port, and has no s_d port (see figure 3). The processor port is also blocked from any operation incompatible with the result of an pending downgrade request. A key feature is the blocking of directory-to-cache requests when there is an outstanding cache-to-directory request pending for the same address. This means that cache-to-directory requests are always consistent with the cache state. For liveness, however, directory-to-cache requests must always be allowed to complete. The directory ensures completion by always replying to a blocking cache-to-directory request, even if there is a pending directory-to-cache request for the same address. Since it is not always possible to honour a cache-to-directory request, the directory can reply with a refusal. Once the cache receives the directory's reply, it can reply to the directory.

7.4 Elimination of cache-to-directory delay

We can show, by a simulation relation, that the Cache with a MS delay on the c port implements a cache with no MS delay. From now on we can safely ignore the c delay and concentrate on the directory-to-cache, or d delay. This simplification reduces enormously subsequent case analysis, since the case analysis can treat one communication channel at a time, rather than having to deal with all possible combinations of communication channel state. It is allowing cache-to-directory requests to block directory-to-cache requests that makes this simplification possible.

7.5 Definition of Port_Dir

Port_Dir provides the interface between the c and d cache-directory communication ports, the memory port, m, and s_d, now an independently maintained partial copy of the Cache state, s_c. Port_Dir has no hidden state. Its next-state relation is specified by:

- The ports obey their appropriate protocols.
- For each address, a:
  - During pending directory-to-cache requests for a, refuse cache-to-directory upgrade requests for a.
  - Data supplied to caches (for upgrades from Absent) is obtained by reading memory.
  - A memory WR a b is issued if and only if either:
    - A cache-to-directory request for a with (dirty) data is acknowledged, or
    - An acknowledgement for directory-to-cache request for a is received with (dirty) data.
  - If $s_d\, a =$ Ex then:
    - Upon acknowledgement of a "request absent" (issued on the c port or received on the d port), then $s_d\, a$ is updated to Ab or Nx, else
    - Upon acknowledgement of a "request non-exclusive" (on either c or d ports), then $s_d\, a$ is updated to Nx, else
    - $s_d\, a$ remains unchanged.
  - If $s_d\, a =$ Nx then:
    - Only "request absent" directory-to-cache requests can be issued for address a.
    - Upon acknowledgement of a "request absent", then $s_d\, a$ is updated to Ab or Nx, else
    - Upon acknowledgement of a "request exclusive" (on the c port), then $s_d\, a$ is updated to Ex, else
    - $s_d\, a$ remains unchanged.
  - If $s_d\, a =$ Ab then:
    - No directory-to-cache requests are issued for address a.
    - Upon acknowledgement of a "request exclusive" (on the c port), then $s_d\, a$ is updated to Ex, else
    - Upon acknowledgement of a "request non-exclusive" (on the c port), then $s_d\, a$ is updated to Nx, else
    - $s_d\, a$ is updated to Nx or remains unchanged.

7.6 Invariants of the Port_Dir-Cache assembly

Port_Dir is composed with Cache with a MS communication delay in the cache-to-directory (d) channel. The communication ports are hidden. An invariant is proven for each address a:

- If an acknowledgement with data is in progress on the d channel, then $s_d\, a = \mathsf{Ex}$.
- If a "request absent" is being acknowledged on the d channel, then $s_c\, a = \mathsf{Absent}$.
- If a "request non-exclusive" is being acknowledged on the d channel, then $s_c\, a \neq \mathsf{Exclusive}\, b$.
- Any data in a d channel acknowledgement agrees with any non-exclusive data in Cache.
- If $s_c\, a = \mathsf{Exclusive}\, b$ then $s_d\, a = \mathsf{Ex}$.
- If $s_d\, a = \mathsf{Ab}$ then $s_c\, a = \mathsf{Absent}$.

7.7 Simulation relation to Port_Cache

It is this last simulation relation proof that is by far the most tedious. Proof is largely by case analysis on the state-transitions of MS, with case splitting at various times on the cache contents and the type of cache-to-directory request. The simulation relation is, however, extremely simple. Defining s_{c1} to be the state of Cache, and s_{c2} to be the state of Port_Cache the relation is for each address a;

- If there is a request for a being acknowledged with data b on the d channel, then $s_{c2}\, a = \mathsf{Exclusive}\, b$, otherwise
- $s_{c2}\, a = s_{c1}\, a$

Once this proof is complete, the MS communication delay can be reintroduced on Cache c port, using the delay elimination result in section 7.4.

8 Conclusion

Before this project we thought that one could perform verification by proceeding top-down from a single specification, deciding on specifications of components and proving that the assembly of components conforms to the specification. Then repeat the process on the components, thus following a "verification tree" all the way to the primitive components.

Things are not that simple. For example, modelling the directory state as a memory *per cache* requires *merging*, not division, as the proof proceeds down the abstraction hierarchy. The design hierarchy becomes a much more dynamic concept, different compositions being used to derive different properties.

The principle inspiration in the proofs is the finding of the simulation relation, and choosing invariants that are sufficiently strong. The rest of the proof is tedious and boring. To make this a routine way to approach engineering design we need a much improved user interface, much better interactive response time, and better automatic proof procedures than HOL currently offers. The re-running (without interaction) of the proofs in this paper takes about 20 hours on an unloaded Sun 3/60.

The automaton theory used in this paper was derived in a general-purpose theorem prover. The theory required to reason about other aspects of the design (such as Temporal logic for deriving liveness properties), can also be derived in the same environment, thus allowing more complete proof inside a single environment.

By using a suitable powerful logic can prove the correctness of systems with an arbitrarily large state-space, provided that the system is structured to allow a tractable case analysis.

References

[1] Martín Abadi and Leslie Lamport. *The Existence of Refinement Mappings.* SRC Report 29, Digital Equipment Corporation, 1988.

[2] Michael C. Browne, Edmund M. Clarke, David L. Dill, and Bud Mishra. Automatic verification of sequential circuits using temporal logic. *IEEE Transactions on Computers*, C-35(12):1035–1044, December 1986.

[3] Avra Cohn. Correctness properties of the Viper block model: the second level. In G. Birtwistle and P. A. Subrahmanyam, editors, *Current Trends in Hardware Verification and Automated Theorem Proving*, pages 1–91, Springer-Verlag, 1989.

[4] Avra. J. Cohn. A proof of correctness of the Viper microprocessor: the first level. In G. Birtwistle and P. A. Subrahmanyam, editors, *VLSI Specification, Verification and Synthesis*, Kluwer Academic Publishers, 1988.

[5] Srinivas Devadas, Hi Keung Ma, and A. Richard Newton. *On the Verification of Sequential Machines at Differing Levels of Abstraction.* Memorandum UCB/ERL M86/93, University of California, Berkeley, 1986.

[6] David L. Dill. *Trace Theory for Automatic Hierarchical Verification of Speed-independent Circuits.* MIT Press, 1989.

[7] Masahirio Fujita, Hedehiko Tanaka, and Tohru Moto-oka. Verification with prolog and temporal logic. In T. Uehara and M. Barbacci, editors, *IFIP Sixth Computer Hardware Description Lanquages and their applications*, pages 103–114, North Holland Publishing Company, 1983.

[8] Mike Gordon. *HOL: A Machine Oriented Formulation of Higher-Order Logic.* Technical Report 68, University of Cambridge Computer Laboratory, 1985.

[9] Mike Gordon. HOL: a proof generating system for higher-order logic. In G. Birtwistle and P. A. Subrahmanyam, editors, *VLSI Specification, Verification and Synthesis*, Kluwer Academic Publishers, 1988.

[10] W. A. Hunt, Jr. The mechanical verification of a microprocessor design. In D. Borrione, editor, *From HDL Descriptions to Guaranteed Correct Circuit Designs*, North Holland, 1987.

[11] Nils Klarlund and Fred B. Schneider. *Verifying Safety Properties using Non-Deterministic Infinite-State Automata.* Technical Report TR 89-1037, Cornell University Computer Science Department, 1989.

[12] R. P. Kurshan. Analysis of discrete event coordination. In J.W. de Bakker, W.-P. De Roever, and G. Rozenberg, editors, *Stepwise Refinement of Distributed Systems*, Springer-Verlag, 1990.

[13] Simon S. Lam and A. Udaya Shankar. Protocol verification via projections. *IEEE transactions on software engineering*, SE-10(2):137–151, July 1984.

[14] Paul Loewenstein. Reasoning about state-machines in higher-order logic. In M. Leeser and G. Brown, editors, *Hardware Specification, Verification and Synthesis: Mathematical Aspects*, Springer Verlag, 1990.

[15] Nancy A. Lynch and Mark R. Tuttle. *Hierarchical correctness proofs for distributed algorithms.* Technical Report TR-387, MIT Laboratory for Computer Science, 1987.

[16] Paliath Narendran and Jonathan Stillman. Formal verification of the sobel image processing chip. In G. Birtwistle and P. A. Subrahmanyam, editors, *Current Trends in Hardware Verification and Automated Theorem Proving*, pages 92–107, Springer-Verlag, 1989.

[17] Tobias Nipkow. Formal verification of data type refinement. In J.W. de Bakker, W.-P. De Roever, and G. Rozenberg, editors, *Stepwise Refinement of Distributed Systems*, Springer-Verlag, 1990.

[18] A. P. Sistla. *A complete proof system for proving correctness of non-deterministic safety specifications.* Technical Report TC-0060-8-89-378, GTE Laboratories, 1989.

DIMACS Series in Discrete Mathematics
and Theoretical Computer Science
Volume 3, 1991

Memory Efficient Algorithms for the Verification of Temporal Properties

C. COURCOUBETIS, M. VARDI, P. WOLPER AND M. YANNAKAKIS

ABSTRACT. This paper addresses the problem of designing memory efficient algorithms for the verification of temporal properties of finite-state programs. Both the programs and their desired temporal properties are modeled as automata on infinite words (Büchi automata). Verification is then reduced to checking the emptiness of the automaton resulting from the product of the program and the property. This problem is usually solved by computing the strongly connected components of the graph representing the product automaton. Here, we present algorithms which solve the emptiness problem without explicitly constructing the strongly connected components of the product graph. By allowing the algorithms to err with a small probability, we can implement them with a randomly accessed memory of size $O(n)$ bits, where n is the number of states of the graph, instead of $O(n \log n)$ bits which the presently known algorithms require.

1. Introduction

Reachability analysis is one of the most successful strategies for analyzing and validating computer protocols. It was first proposed by West [**19**], and further studied by many researchers (cf. [**11, 13**]. Reachability analysis is applied to a protocol by systematically exercising all the protocol transitions. Such analysis can detect syntactical errors such as *static deadlock*, *unspecified reception*, or *unexercised code.* The simplicity of the strategy lends itself to easy implementation. Indeed, automated reachability analyses detected errors in published standards such as the X.21 (cf. [**20**]). The approach is less successful when it comes to *protocol verification*, i.e., verifying that the given protocol achieves its functional specification. This limitation is due to the fact that a functional specification cannot be directly checked by reachability analysis. To apply reachability analysis to such a task, one first has to manually translate the functional specification to a property of the protocol state graph. While this can be done for some specific specifications (cf. [**14**]), it is not a general approach.

A general approach to protocol verification is to use a *theorem-prover* for an appropriate logic. Early systems used to focus on input/output behavior of

---

The work of the first and third authors is supported by the European Community ESPRIT BRA project SPEC (3096).

protocols rather than on ongoing behavior (cf. [**15**]), but systems that are based on temporal logic overcame this shortcoming (cf. [**7**]). Unfortunately, theorem-proving systems are semi-automated at best, and their success at dealing with real-life protocols is not as impressive as that of reachability analysis (cf. [**6**]).

A new approach that emerged in the 1980's is the so-called *model-checking* approach [**5, 4, 10, 12**]. Model checking is based on the idea that verifying a propositional temporal logic property of a finite-state program amounts to evaluating that formula on the program viewed as a temporal interpretation. The algorithms for doing this are quite efficient, since their time complexity is a linear function of the size of the program. As was shown later in the automata-theoretic approach of [**16, 17, 21**], model checking can be viewed as an augmented reachability analysis; the model-checking algorithm uses the temporal logic specification to guide the search of the protocol state space in order to verify that the protocol satisfies its functional specification. Model checking thus seems to solve one of the limits of reachability analysis: the inability to automatically verify functional specifications.

Model checking suffers, however, from the same fundamental problem plaguing the reachability-analysis approach: the ability to explore only limited-size state spaces. This problem, called the *state-explosion* problem, is the most basic limitation of both approaches. It has been the subject of extensive research both in the context of reachability analysis (cf. [**11, 13**]) and in the context of model checking (cf. [**4**]). A recent development [**8**] has substantially pushed back the state-explosion limit for reachability analysis. The main idea behind this development is that, at the price of possibly missing part of the state space, the amount of randomly accessed memory necessary for exploring a state space of a given size could be substantially reduced (essentially from $O(n \log(n))$ to $O(n)$ for a graph with n states). The essence of the method is the use of hashing without collision detection.

In this paper, we show that model checking can also benefit from a similar reduction in the required random memory. This result is obtained by a combination of techniques. We approach model checking from the automata-theoretic perspective of [**16, 17, 21**]. This has the advantage of essentially reducing model checking to reachability analysis, though on a state space that is the cross product of the original state space with the state space of an automaton describing the functional specification. It is then possible to adapt techniques inspired by those of [**8**] to solve this problem. However, while Holtzmann's technique is suitable for searching for "bad" states in the state space, model checking involves searching for "bad" cycles. We thus had to develop some special purpose algorithms that are presented here.

This paper is organized as follows. We first review some background on model checking using the automata-theoretic approach and define the corresponding graph-theoretic problem. Then, we discuss the requirements that algorithms for solving this problem have to satisfy. Next we present our solutions. Finally, we present some extensions and some final remarks.

2. Temporal Logic Verification using Büchi Automata

The model-checking problem we consider is the following. Given a program P described as a product of finite-state transition systems P_i and a temporal logic formula f, check that all infinite computations of P satisfy f. To solve this problem, we use the following steps:

(1) Build the finite-automaton on infinite words for the *negation* of the formula f (one uses the negation of the formula as this yields a more efficient algorithm). The resulting automaton is $A_{\neg f}$.
(2) Take the product of the program $P = \prod P_i$ and the automaton $A_{\neg f}$.
(3) Check if the product automaton is non-empty.

The approach we have just outlined has one major advantage over other model-checking approaches: it does not build the entire state graph for the program (the product of the P_i) before checking that it satisfies the temporal property f. Indeed, the product $\prod P_i \times A_{\neg f}$ can be computed in one pass. This can lead to more efficiency for various reasons. In the first place, the product of P and $A_{\neg f}$ only accepts sequences that do not satisfy the requirement. One expects few of these (none if the program is correct). It is thus possible that the product of P and $A_{\neg f}$ will have fewer reachable states than P. Furthermore, when building $P \times A_{\neg f}$, it is not necessary to store the whole state-graph. It is sufficient to keep just enough information for checking that condition 3 above is satisfied. This is exactly what the algorithms we present in Section 3 will do. The advantages of reducing model checking to a reachability problem are also investigated in [9], but only for pure safety properties. In that case, it is sufficient to check that some states are simply reachable and the algorithms we develop in this paper are not needed.

To be able to describe our algorithms, we need more details about Büchi automata and how to check their emptiness. A *Büchi automaton* is a tuple $A = (\Sigma, S, \rho, S_0, F)$, where

- Σ is an alphabet,
- S is a set of states,
- $\rho : S \times \Sigma \rightarrow 2^S$ is a nondeterministic transition function,
- $S_0 \subseteq S$ is a set of starting states, and
- $F \subseteq S$ is a set of designated states.

A *run* of A over an infinite word $w = a_1 a_2 \ldots$, is an infinite sequence $s_0, s_1, \ldots$, where $s_0 \in S_0$ and $s_i \in \rho(s_{i-1}, a_i)$, for all $i \geq 1$. A run $s_0, s_1, \ldots$ is *accepting* if there is some designated state that repeats infinitely often, i.e., for some $s \in F$ there are infinitely many i's such that $s_i = s$. The infinite word w is *accepted* by A if there is an accepting run of A over w. The set of denumerable words accepted by A is denoted $L(A)$.

From the definition of Büchi automata, it is relatively easy to see that a Büchi automaton is nonempty iff it has some state $f \in F$ that is reachable from the initial state and reachable from itself (in one or more steps) [**18**]. In graph theoretic terms, this means that the graph representing the automaton has a reachable cycle that contains at least one state in F. In what follows, we will give a memory-efficient algorithm to solve this problem.

To formalize our verification approach, we define a program P as being a finite-state transition system consisting of

- a state space V,
- a nondeterministic transition function $\sigma : V \times \Sigma \rightarrow 2^V$ (Σ is the alphabet common to the program and the automaton for the property $A_{\neg f}$) and
- a set of starting states $V_0 \subseteq V$.

The accepting runs of P are defined by viewing P as a restricted type of Büchi automaton in which the set of designated states is the whole set of states V.

According to the definitions above, if $A_{\neg f} = (\Sigma, S, \rho, S_0, F)$, the product $P \times A_{\neg f}$ is a Büchi automaton with

- state set $V \times S$,
- transition function $\tau : V \times S \rightarrow 2^{V \times S}$ defined by $(v_2, s_2) \in \tau((v_1, s_1), a)$ iff $v_2 \in \sigma(v_1, a)$ and $s_2 \in \rho(s_1, a)$,
- and set of designated states $V \times F$.

This product automaton accepts all runs which are possible behaviors of P (accepted by the automaton P) and violate the formula f (are accepted by the automaton $A_{\neg f}$). Hence we have reduced the problem of proving that the program P satisfies the formula f to the problem of checking the emptiness of the Büchi automaton $P \times A_{\neg f}$.

It is interesting to note that the product automaton $P \times A_{\neg f}$ has the Büchi type of acceptance condition because the acceptance condition for P is the trivial one. In the case in which the program P is modeled as an arbitrary Büchi automaton, the problem of checking the emptiness of $P \times A_{\neg f}$ is different and will be examined in Section 4.

3. Verification Algorithms

3.1. Requirements on the Algorithms. We characterize the memory requirements of any verification algorithm as follows. We consider the data structures used by the algorithm. The total amount of space used by these data structures corresponds to the total space requirements of the algorithm. The above space can be divided into memory that is *randomly accessed* and into memory that is *sequentially accessed*. For example, for implementing a hash table we need randomly accessed memory, while a stack can be implemented with sequentially accessed memory.

As correctly pointed out in [**8**], the bottleneck in the performance of most verification algorithms is directly related to the amount of the randomly accessed memory these algorithms require, and is due to the significant amount of paging involved during the execution of the algorithm. Holzmann observed that there is a tremendous speed-up for an algorithm implemented so that its randomly accessed memory requirements do not exceed the main memory available in the system (since sequentially accessed memory can be implemented in secondary storage).

The basic problem that Holzmann considered is how to perform reachability analysis by using the least amount of randomly accessed memory. For a graph with n states, his scheme involves a depth-first search in the graph, where the information about the states visited is stored in a bit-array of size m as follows. When a new state is generated, its name is hashed into an address in the array; if the bit of the corresponding location is on, then the algorithm considers that the above state has already been visited; if the bit is off then it sets the bit and adds the state on the stack used by the depth-first search. Since there is no collision detection it follows that the above search is partial; there is always a possibility that a state will be missed.

The key assumption behind this method, see **[8]**, is that in general one can choose the value of m large enough and construct a hash function so that the number of collisions becomes arbitrarily small. Furthermore, since the limiting factor in reachability analysis is usually the space required by the computation rather than the time required to do the computation, one could significantly reduce the probability of error by running the algorithm a few times with different hash functions. Indeed, Holzmann claims that, for most practical applications, choosing a hash table of size $m = O(n)$ together with appropriate hash functions is sufficient for the effect of collisions to become insignificant. Is this really so?

To answer this question, let us consider the memory requirements of the general reachability problem defined as follows. We assume that the states of the graph G have names from a name space U. In many applications (for example protocols), $|U|$ is many orders of magnitude larger than the number n of reachable states of G. In this case, complete reachability analysis (no missed states whatever the input graph) appears to require $O(n \log |U|)$ bits of randomly accessed memory, and probably can not be done with less memory (unless the names of the reachable states of G are not randomly selected from U). Indeed, representing each state with less than $\log |U|$ bits amounts to mapping the state space U to a smaller state space. Now, for any such mapping there will always be subsets of U on which it is not one-to-one and hence on which complete reachability will not be guaranteed.

The situation is different is one analyses the problem from a probabilistic point of view. Consider all possible mappings from the set $S = \{1, \ldots, n\}$ into the set $\{1, \ldots, m\}$. There are m^n such mappings of which $m!/(m-n)!$ are one-to-one. Thus, if one assumes that the mapping implemented by a hash function is randomly selected, the probability that it is one-to-one (no collisions) is $m!/((m-n)!m^n$ which for $n << m$ can be approximated by $e^{-n^2/m}$. This implies that in the case of a name space U and a graph with n reachable states, we can do partial reachability (with arbitrarily small probability of missing reachable states) by using $O(n \log n)$ bits of randomly accessed memory (instead of $O(n \log |U|)$ bits for complete reachability) as follows. First hash the n reachable states into a set $1, \dots, m$ with an arbitrarily small probability of collision. As we have just seen, this is possible if we take $m = O(n^2)$. Then, do complete reachability using the set $1, \dots, m$ as the name space for the states.

Holtzmann's technique goes one step further and only uses one bit of randomly accessed memory per reachable state. This is equivalent to assuming that there exists a hash function mapping U into $1, \dots, m$, $m = O(n)$, with a small probability of collisions. As the analysis above shows, this is not possible if we just assume that the hash function is random. It can however be possible if the state space U is only of size $O(n)$ or if the set of reachable states has a particular structure that can be used by the hash function. In these cases, the gain in randomly accessed memory, size $O(n)$ instead of size $O(n \log |U|)$ is quite significant for large state spaces.

However, this gain in memory use is only obtained for straightforward reachability analysis. To verify general temporal properties we have to check nonemptiness of the product automaton. One way to accomplish this is to construct the strongly connected component of the product automaton state graph and then to check whether one of the strongly connected component contains an accepting state. Unfortunately, we cannot apply Holtzmann's method to the standard algorithm for constructing the strongly connected components of the graph [**2**]. Indeed, although in that algorithm the states of the components are stored in a stack, it requires access to information (depth-first and low-link number) about states randomly placed in the stack, which implies the need of at least $O(n \log n)$ bits of randomly accessed memory. Hence, given a fixed amount of memory, the size of the problems we could efficiently analyze with the above algorithm is substantially smaller than the size of the problem that can be analyzed with the technique of [**8**].

From the previous discussion the following problem emerges. Assuming that reachability analysis in graphs of size n can be efficiently done with randomly accessed memory of size $O(n)$, can we solve the emptiness problem for Büchi automata using only randomly accessed memory of size $O(n)$? The answer to this problem is positive and the corresponding algorithms are described in the following section.

3.2. The Algorithms. In this section we provide algorithms for the following problem.

PROBLEM 1 (nonemptiness of Büchi automata). *Given directed graph G, start node s_0, distinguished set of accepting nodes F, determine whether there is a member of F which is reachable from s_0 and belongs to a cycle, or equivalently, to a nontrivial strong component.*

We make the following representation assumptions. The graph G is given by a *successor* function: a function that takes a node as argument and returns an ordered list of its successors. The set F is specified by a membership routine. We assume that we have a function h mapping one-to-one every node to an integer in the range $1, \ldots, m$.

Algorithm A:
The algorithm consists of two depth-first-searches (DFS's). The two searches can be performed one after the other, or can be done together in an interleaved fashion. It is simpler to describe first the noninterleaved execution. The purpose of the first DFS is to (1) determine the members of F that are reachable from s_0, and (2) order them according to last visit (i.e., in postorder) as $f_1, \ldots, f_k$.[1] The second DFS explores the graph using this ordering; it does not perform k searches but only one. In more detail, the main data structures are as follows: a stack S (to hold the path of DFS from root to current node), a (FIFO) queue Q to hold the reachable members of F in postorder and a bit-array M indexed by the hash values $1, \ldots, m$ for the "marked" bit (whether the node has been visited). The two passes share the same structures S and M.

The first DFS is as follows:

1. Initialize: $S := [s_0]$, $M := 0$, $Q := \emptyset$.
2. Loop: while $S \neq \emptyset$ do

```
        begin
            v := top(S);
            if M[h(w)] = 1 for all w ∈ succ(v)
              then begin
                    pop v from S;
                    if v ∈ F insert v into Q
                    end
              else begin
                    let w be the first member of succ(v) with M[h(w)] = 0;
                    M[h(w)] := 1;
                    push w into S
              end
        end
```

The second DFS is as follows:

1. Initialize: $S := \emptyset$, $M := 0$.
2. Loop: while $Q \neq \emptyset$ do

```
        begin
            f := head(Q);
            remove f from Q;
            push f into S;
```

[1] f_1 is the first postorder reachable accepting state and f_k is the last

```
        while S ≠ ∅ do
              begin
                    v := top(S);
                    if f ∈ succ(v) then halt and return "YES";
                    if M[h(w)] = 1 for all w ∈ succ(v)
                      then pop v from S
                      else begin
                             let w be the first member of succ(v) with M[h(w)] = 0
                             M[h(w)] := 1;
                             push w into S
                           end
              end
end
```

The correctness of the algorithm is based on the following claims.

LEMMA 1. *Let $f_1, \ldots, f_k$ be the members of Q after the first DFS, i.e., the members of F that are reachable from s_0 in postorder (f_1 is the first member of F to be reached in postorder, f_k the last). If for some pair f_i, f_j with $i < j$ there is a path from f_i to f_j, then node f_i belongs to a nontrivial strong component.*

Proof: Suppose that there is a path from f_i to f_j. If no node on this path was marked before f_i, then the DFS would have reached f_j from f_i, so f_j would have come before f_i in the postorder. Thus, some node p on the path was marked before f_i. If p comes before f_i in the postorder, then f_j also should come before f_i in the postorder. Since p was marked before f_i, but comes after f_i in the postorder, it must be an ancestor of f_i. Thus, f_i can reach an ancestor and therefore belongs to a nontrivial strong component. □

THEOREM 1. *If the second DFS halts and returns "YES", then some reachable node of F belongs to a nontrivial strongly connected component. Conversely, suppose that some reachable node of F belongs to a nontrivial strongly connected component. Then the second DFS will return "YES".*

Proof: The first part is clear: suppose the second DFS returns "YES" while processing node f_j of Q. Then, it is building a tree with root f_j and discovers a back edge to the root f_j, and therefore f_j is obviously in a cycle. For the converse, let f_j be a reachable member of F that belongs to a nontrivial strongly connected component and has the smallest index j among all such members. Consider a path p from f_j to itself. We claim that no node of p is reachable from a f_i with a smaller i. For, if some node was reachable, then f_i would also reach f_j, which by Lemma 1 contradicts the choice of f_j. Therefore, no node of the path p is marked when we push f_j into S in the second DFS, and thus we will find a back edge to the root f_j. □

Note that the creation of both S and Q and access to them in both searches are sequential. Hence, both can be stored in secondary memory as needed.

So far we analyzed the algorithm under the assumption that the hash function f is perfect. One of the main features of our algorithm is its behavior in the presence of hash collisions. In that case, although the algorithm might erroneously conclude (due to collisions) that the Büchi automaton does not accept any word, it will never mistakenly conclude that the automaton accepts some word. In terms of the underlying verification problem, this means that our algorithm might miss some errors, but will never falsely claim that the protocol is incorrect. Thus, the algorithm should be viewed more as a systematic debugging tools rather than as a verification tool.

The following alternative does away with the queue Q by using instead a second stack S_2 and bit array M_2. We do the second DFS on top of the first in an interleaved way. The details are as follows.

Algorithm B:

```
1. Initialize: S1 := [s0], S2 := ∅, M1 := M2 := 0.
2. while S1 ≠ ∅ do
   begin
           x := top(S1);
           if there is a y in succ(x) with M1[h(y)] = 0
           then begin
                   let y be the first such member of succ(x);
                   M1[h(y)] := 1;
                   push y into S1
                   end
           else begin
                 pop x from S1;
                 if x ∈ F then begin
                    push x into S2;
                    while S2 ≠ ∅ do begin
                            v := top(S2);
                            if x ∈ succ(v) then halt and return "YES";
                            if M2[h(w)] = 1 for all w ∈ succ(v)
                              then pop v from S2
                              else begin
                                let w be the first member of succ(v) with M2[h(w)] = 0;
                                M2[h(w)] := 1;
                                push w into S2
                            end
                 end
              end
        end
end
```

The above algorithm requires twice as much space as that required by the first algorithm. Its advantage is that if the automaton is found to be nonempty, an accepted word can be extracted from the stacks S_1 and S_2. In verification terms, this means that, if the protocol is found to be incorrect by the algorithm, a sample incorrect path can be produced. This is essential for debugging to be possible.

4. Extensions and Concluding Remarks

An extension of the verification problem described in Section 2 is the verification of programs with liveness conditions, see [**1**]. In this case the program is given in terms of components, each having it own liveness conditions. Each such component is modeled as a Büchi automaton. Hence, the product $P \times A_{\neg f}$ corresponds to an automaton whose transition table G is the product of the corresponding transition tables and its acceptance condition is given in terms of a set of sets of designated states $\{F_1, \ldots, F_k\}$. A run is accepting if it repeats some state from each of these sets infinitely often. Clearly, checking the emptiness of $P \times A_{\neg f}$ is equivalent with checking for the existence of a strongly connected component in the product transition table which is reachable from the initial state and intersects all these sets. Let S be the state space of the product transition table. We can construct a Büchi automaton B with $k|S|$ states, such that the emptiness of B is equivalent with the emptiness of $P \times A_{\neg f}$ (see for instance [**17**]).

- The graph of B consists of k copies of G with the transitions modified as follows. Consider the k copies $G_1, \ldots, G_k$ of G. For $i = 1, \ldots, k$, replace the transitions from every state $f \in F_i$ of G_i by similar transitions to the states in $G_{(i \bmod k)+1}$.
- The initial states of B are those of one copy of G, say G_1.
- The accepting states of B are the states F_i of the copy G_i of G, for some arbitrary i. For instance we can take $F_1 \subset G_1$.

Hence, if we apply the algorithms of the previous section to B, we can do verification with $O(k|S|)$ bits of randomly accessed memory.

Another remark is the following. In many applications it is reasonable to assume that the predecessor function of the graph is given as well. In this case one can use the algorithm in Section 6.7 in [**3**] for constructing the strongly connected components of the graph G by using randomly accessed memory of size $O(n)$. Let G_r be the directed graph corresponding to G by reversing its edges. This algorithm performs first a DFS on G and numbers the states in order of completion of the recursive calls (in postorder). This can be implemented by pushing the states in a stack according to their postorder visit by the DFS; this stack can use sequentially accessed memory. Then the algorithm performs a DFS on G_r (by using the predecessor function of G) starting with the state with the highest postorder sequence number (top of stack). This DFS on G_r must be restricted to the states reached during the first DFS, and uses a hashing mechanism for marking the states already visited. If the search does not reach all states, the algorithm starts the next DFS on G_r from the highest-numbered state which has not been already visited by the previous DFS. This can be easily done by poping the postorder stack until a state which has not been visited (the corresponding bit in the hash table is zero) is found. Since each tree in the resulting spanning forest is a strongly connected component, one can easily check for the properties of each such component while it is being generated.

REFERENCES

1. S. Aggarwal, C. Courcoubetis, and P. Wolper. Adding liveness properties to coupled finite-state machines. *ACM Transactions on Programming Languages and Systems*, 12(2):303–339, 1990.
2. Alfred V. Aho, John E. Hopcroft, and Jeffrey D. Ullman. *The Design and Analysis of Computer Algorithms.* Addison Wesley, Reading, 1974.
3. Alfred V. Aho, John E. Hopcroft, and Jeffrey D. Ullman. *Data Structures and Algorithms.* Addison Wesley, Reading, 1982.
4. E. M. Clarke and O. Grümberg. Avoiding the state explosion problem in temporal logic model-checking algorithms. In *Proc. 6th ACM Symposium on Principles of Distributed Computing*, pages 294–303, Vancouver, British Columbia, August 1987.
5. E.M. Clarke, E.A. Emerson, and A.P. Sistla. Automatic verification of finite-state concurrent systems using temporal logic specifications. *ACM Transactions on Programming Languages and Systems*, 8(2):244–263, January 1986.
6. R. Grotz, C. Jard, and C. Lassudrie. Attacking a complex distributed systems from different sides: an experience with complementary validation tools. In *Proc. 4th Work. Protocol Specification, Testing, and Verification*, pages 3–17. North-Holland, 1984.
7. B.T. Hailpern. Tools for verifying network protocols. In K. Apt, editor, *Logic and Models of Concurrent Systems, NATO ISI Series*, pages 57–76. Springer-Verlag, 1985.
8. G. Holzmann. An improved protocol reachability analysis technique. *Software Practice and Experience*, pages 137–161, February 1988.
9. C. Jard and T. Jeron. On-line model-checking for finite linear temporal logic specifications. In *Automatic Verification Methods for Finite State Systems, Proc. Int. Workshop, Grenoble*, volume 407, pages 189–196, Grenoble, June 1989. Lecture Notes in Computer Science, Springer-Verlag.
10. O. Lichtenstein and A. Pnueli. Checking that finite state concurrent programs satisfy their linear specification. In *Proceedings of the Twelfth ACM Symposium on Principles of Programming Languages*, pages 97–107, New Orleans, January 1985.
11. M.T. Liu. Protocol engineering. *Advances in Computing*, 29:79–195, 1989.
12. J.P. Quielle and J. Sifakis. Specification and verification of concurrent systems in cesar. In *Proc. 5th Int'l Symp. on Programming*, volume 137, pages 337–351. Springer-Verlag, Lecture Notes in Computer Science, 1981.
13. H. Rudin. Network protocols and tools to help produce them. *Annual Review of Computer Science*, 2:291–316, 1987.
14. H. Rudin and C.H. West. A validation technique for tightly-coupled protocols. *IEEE Transactions on Computers*, C-312:630–636, 1982.
15. C.A. Sunshine. Experience with automated protocol verification. In *Proceedings of the International Conference on Communication*, pages 1306–1310, 1983.
16. M. Vardi. Unified verification theory. In B. Banieqbal, H. Barringer, and A. Pnueli, editors, *Proc. Temporal Logic in Specification*, volume 398, pages 202–212. Lecture Notes in Computer Science, Springer-Verlag, 1989.
17. M.Y. Vardi and P. Wolper. An automata-theoretic approach to automatic program veri-

fication. In *Proc. Symp. on Logic in Computer Science*, pages 322–331, Cambridge, june 1986.
18. M.Y. Vardi and P. Wolper. Reasoning about infinite computation paths. IBM Research Report RJ6209, 1988.
19. C.H. West. Generalized technique for communication protocol validation. *IBM J. of Res. and Devel.*, 22:393–404, 1978.
20. C.H. West and P. Zafiropulo. Automated validation of a communication protocol: the ccitt x.21 recommendation. *IBM Journal of Research and Development*, 22:60–71, 1978.
21. P. Wolper. On the relation of programs and computations to models of temporal logic. In B. Banieqbal, H. Barringer, and A. Pnueli, editors, *Proc. Temporal Logic in Specification*, volume 398, pages 75–123. Lecture Notes in Computer Science, Springer-Verlag, 1989.

36 Dedalou Street, P.O. Box 1385, 71110 Iraklio, Crete, Greece

E-mail: courcou@ariadne.uucp

IBM Almaden, Department K55/802, 650 Harry Road San Jose, California 95120-6099, U.S.A.

E-mail: vardi@ibm.com

Université de Liège, Institut Montefiore, B28, B-4000 Liège Sart-Tilman, Belgium

E-mail: pw@montefiore.ulg.ac.be

AT&T Bell Labs, 600, Mountain ave, Murray Hill, NJ 07974,U.S.A.

E-mail: mihalis@research.att.com

DIMACS Series in Discrete Mathematics
and Theoretical Computer Science
Volume 3, 1991

Vectorized Model Checking for Computation Tree Logic

HIROMI HIRAISHI, SHINTARO MEKI, AND KIYOHARU HAMAGUCHI

ABSTRACT. The aim of this paper is to show how big model checking problems for Computation Tree Logic (CTL) can be handled by using current powerful vector processors. Although efficient recursive model checking algorithms for CTL, which run in time proportional to both the size of Kripke structures and the length of formulas, have been already proposed [**7**, **2**], their algorithms cannot be vectorized due to recursive procedure calls. In this paper we propose a new model checking algorithm, called a vectorized model checking algorithm, for CTL which is suitable for the execution on vector processors. It can handle more than 1 million state Kripke structure derived from a deterministic sequential machine.

1. Introduction

Recently various kinds of formal methods for automatic verification have been widely studied. Among them, the model checking approach based on a branching time temporal logic called CTL (Computation Tree Logic) [**2**, **3**, **5**, **6**, **7**] is one of the most efficient approaches. In verification of a system which consists of several machines communicating with each other, however, there is a problem so-called a state explosion problem. Although there are several works trying to avoid this problem [**8**, **10**], it seems to be difficult to avoid the problem in general, and there are strong requirements for verification of large systems[**11**].

We mainly aim at clarifying how big machines can be verified based on the model checking algorithm for CTL by using current powerful vector processors. As the first step to this purpose, we challenged to vectorize model checking for CTL on Kripke structures this time. Although the model checking algorithms in [**7**] are efficient and runs in time proportional to both the size of Kripke structures and the length of CTL formulas, they are not suitable for vector processors because it is difficult to vectorize them due to their recursive procedure calls. In order to extract high performance of vector processors, we need an algorithm

1980 *Mathematics Subject Classification* (1985 *Revision*). Primary Primary 03B45; Secondary 68M15.

1052-1798/91 $1.00 + $.25 per page

using repeated uniform operations on array type data. It is easy to develop such a model checking algorithms based on fixpoint calculations of CTL semantics, but the direct implementation of the fixpoint calculations would easily lead to an algorithm whose time complexity is proportional to $|S|^3$ or $|S|^4$, where $|S|$ is the number of states of Kripke structure.

The new model checking algorithm called vectorized model checking algorithm proposed here can be vectorized for executions on vector processors. It runs in time linear to both the size of Kripke structures (i.e. $|S| + |R|$, where $|R|$ is the number of edges of Kripke structure) and the length of CTL formulas. We also implemented the algorithm on a vector processor FACOM VP400E. The analysis of storage requirement of the implementation shows that it can manipulate more than 1 million state Kripke structure derived from a deterministic sequential machine. We also present the result of an experiment which shows the efficiency of our implementation.

This paper is organized as follows: Section 2 summarizes the definition of CTL. Section 3 describes the vectorized model checking algorithm for CTL in conjunction with explanations about vector processors. In Section 4 we explain the implementation of the algorithm on a vector processor FACOM VP400E and show its experimental result. Section 5 concludes this paper with summarizing remaining future problems.

2. Computation Tree Logic

Computation Tree Logic (CTL)[**6**] is a branching time temporal logic. Let AP be a set of atomic propositions. CTL formulas are inductively defined as follows:

- If $p \in AP$, p is a CTL formula.
- If η is a CTL formula, then so are $\neg\eta$, $EX\eta$ and $EG\eta$.
- If η and ξ are CTL formulas, then so are $\eta \vee \xi$ and $E[\eta \mathcal{U} \xi]$.

The semantics of CTL is defined over a Kripke structure $K = (S, R, I)$, where

- S is a non-empty finite set of states.
- $R \subseteq S \times S$ is a total binary relation on S (i.e. for $\forall s \in S$, there exists $s' \in S$ such that $(s, s') \in R$).
- $I : S \rightarrow 2^{AP}$ is an interpretation function which labels each state with a set of atomic propositions true at that state.

An infinite sequence of states $\pi = s_0 s_1 s_2 \ldots$ is called a *path* from s_0 if $(s_i, s_{i+1}) \in R$ for $\forall i \geq 0$. $\pi(i)$ denotes the i-th state of the sequence π (i.e. $\pi(i) = s_i$).

The truth-value of a CTL formula is defined at a state of a Kripke structure and $K, s \models \eta$ denotes that a CTL formula η hold at a state s of a Kripke structure K. If there is no ambiguity, we will omit K and just write as $s \models \eta$. The relation $\models$ is recursively defined as follows:

- $s \models p \, (\in AP)$ iff $p \in I(s)$.

```
DO 10 I = 1,N                DO 10 I = 1,N,K              DO 10 I = 1,N
10 A(I) = B(I) + C(I)        10 A(I) = B(I) + C(I)        10 A(I) = B(L(I))
```

(a) contiguous access (b) constant strided access (c) indirectly addressed access

FIGURE 1. Three types of vector accesses.

```
DO 10 I = 1,N
   IF (A(I) .GT. 0.0) THEN
      B(I) = B(I) + C(I)
   ENDIF
10 CONTINUE
```

(a) DO loop with a conditional statement

```
K = 0
DO 10 I = 1,N
   IF (A(I) .GT. 0.0) THEN
      K = K + 1
      C(K) = B(I)
   ENDIF
10 CONTINUE
```

(b) vector compress function

FIGURE 2. DO loops containing conditional statements

- $s \models \neg\eta$ iff $s \not\models \eta$.
- $s \models \eta \vee \xi$ iff $s \models \eta$ or $s \models \xi$.
- $s \models EX\eta$ iff there exists some next state s' of s (i.e. $(s, s') \in R$) such that $s' \models \eta$.
- $s \models EG\eta$ iff there exists some path π on K starting from the state s such that $\pi(i) \models \eta$ for $\forall i \geq 0$.
- $s \models E[\eta\mathcal{U}\xi]$ iff there exists some path π on K starting from the state s such that $\exists i \geq 0$, $\pi(i) \models \xi$ and $\pi(j) \models \eta$ for $0 \leq \forall j < i$.

In addition to other Boolean operators such as *conjunction* ($\wedge$), we often use the following abbreviations:

- $AX\eta = \neg EX\neg\eta$.
- $A[\eta\mathcal{U}\xi] = \neg(EG(\eta \wedge \neg\xi) \vee E[(\eta \wedge \neg\xi)\mathcal{U}(\neg\eta \wedge \neg\xi)])$.
- $AF\eta = \neg EG\neg\eta$.
- $EF\eta = E[true\,\mathcal{U}\eta]$, where *true* represents Boolean tautology.
- $AG\eta = \neg EF\neg\eta$.

Intuitively, the first letters of temporal operators, A and E, represent universal and existential path quantifier respectively. The second letters of temporal operators represents:

- $X\eta$ represents that η holds at the next state;
- $G\eta$ represents that η holds at every state on the path;
- $F\eta$ represents that η holds at some state on the path;
- $\eta\mathcal{U}\xi$ represents that ξ holds at some state and η holds always before that state on the path.

3. Vectorized Model Checking Algorithm

3.1. Vector processors. Vector processors are supercomputers for large-scale computations. They achieve more than several hundred MFLOPS (Million FLoating-point Operations Per Seconds) by vector instructions which execute uniform operations on array-structured data using pipelined functional units, and they usually have large main memory of several hundred mega bytes. In conjunction with floating-point operations, they also support integer and bit-wise logical operations.

Although the maximum speed of vector processors are very high, the following two points are very important in programming for vector processors to achieve their maximum performance:

Vectorization ratio:: Vectorization ratio is the rate of the operations executed by vector instructions to the whole operations in a program. This ratio should be more than 90% to obtain high performance of vector processors.

Vector length:: Since there are some overheads for setting up vector instructions, the length of operands (vector length) of vector instructions should be large enough; it should be larger than several hundreds to get maximum performance of vector processors.

As for data transmission between the main memory and vector registers, there are load/store pipelines which support basically the following three types of vector accesses (see Figure 1):

- contiguous vector access
- constant strided vector access
- indirectly addressed vector access

Furthermore, vector processors support **DO** loop with conditional statements and *vector compress function* shown in Figure 2. These vectorized functions are very powerful in implementing vectorized model checker for CTL.

3.2. Vectorized model checking algorithm. The vectorized model checking algorithm for CTL runs in bottom-up way by labeling each state with the truth value of each sub-formula contained in a given CTL formula. $Label(\eta, s)$ denotes the truth value of a CTL formula η at a state s in the following.

3.2.1. Case Boolean operations ($\neg\eta$ or $\eta \vee \xi$). We need only to calculate logical negation or disjunction of $Label(\eta, s)$ and $Label(\xi, s)$ for each state s as $Verify\_Not(\neg\eta)$ and $Verify\_Or(\eta \vee \xi)$ in Figure 3.

3.2.2. Case $EX\eta$. First, it assumes that $EX\eta$ is *false* at all the states. For all states where η holds, it labels $EX\eta$ to be *true* at their all predecessor states (see $Verify\_EX(EX\eta)$ in Figure 4.).

3.2.3. Case $E[\eta\mathcal{U}\xi]$. From the definition of $E[\eta\mathcal{U}\xi]$, it holds at the states where ξ is *true* and it also holds at the states which are reachable to such states only through the states where η holds. Therefore, this is a kind of reachability problem.

```
procedure  Verify_Not(¬η)
    for all  s ∈ S  do
        Label(¬η, s) := Not(Label(η, s));
    return;
end of procedure
```

```
procedure  Verify_Or(η ∨ ξ)
    for all  s ∈ S  do
        Label(η ∨ ξ, s) := Or(Label(η, s), Label(ξ, s));
    return;
end of procedure
```

FIGURE 3. Vecorized model checking algorithms for Boolean operators

In the procedure *Verify_EU*($E[\eta \mathcal{U} \xi]$), the sets of states N_1 and N_2 are used to keep track of the states where the truth value of $E[\eta \mathcal{U} \xi]$ is newly determined to be *true*. It first initializes the set N_1 and then it calculates the initial values of $Label(E[\eta \mathcal{U} \xi], s)$ at each state (lines 3 ~ 11) as follows: if ξ holds at the state, it is 1 (i.e. *true*) because $E[\eta \mathcal{U} \xi]$ holds at the state apparently and the state is inserted to N_1; if neither ξ nor η hold at the state, it is 0 (i.e. *false*) because $E[\eta \mathcal{U} \xi]$ does not hold at the state apparently; if η holds but ξ does not, it is temporarily assigned to 2 indicating that it will be determined later by checking the reachability.

Next, for all the states labeled 2, the procedure checks the reachability to the state labeled 1 through the states labeled 2, and the reachable states are labeled 1 and the unreachable states are labeled 0. This step is done as follows (lines 12 ~ 27): For each state where $E[\eta \mathcal{U} \xi]$ is newly determined to be *true*, the labels of its predecessor states are checked, and if they are 2, then they are relabeled as 1 because they are the reachable states and they are added to the set which keeps track of the states newly labeled as 1. This step is repeated until no more states are newly labeled as 1 (lines 12 ~ 23). Finally, the states whose labels are still 2 are labeled as 0 because they are the unreachable states (lines 24 ~ 27).

3.2.4. Case $EG\eta$. From the definition of $EG\eta$, it holds at the states on the loops which are constructed only by the states where η holds. It also holds at the states reachable to such loops through the states where η holds. In order to find out such states, the procedure *Verify_EG*((η) (Figure 4) first constructs a sub-graph logically by extracting all the states where η holds. Then it keeps removing the states from the sub-graph whose out degrees are 0 while such states exist. It is almost clear that $EG\eta$ holds at the states contained in the resulting sub-graph and it does not hold at all other states.

More precisely, after initializing the set of states N_1 to be empty, the procedure labels the states as 1 where η holds; it labels the states as 0 where η does not hold (lines 3 ~ 10). For each state labeled 1, if the labels of its predecessor states

```
procedure Verify_EX(EXη)
 for all s ∈ S do
  Label(EXη, s) := 0;
 for all s' ∈ S do
  if Label(η, s') = 1 then
   for all s such that (s, s') ∈ R do
    Label(EXη, s) := 1;
 return;
end of procedure
```

```
procedure Verify_EU(E[η U ξ])
 N1 := ∅
 for all s ∈ S do
  if Label(ξ, s) = 1 then
   begin
    Label(E[η U ξ], s) := 1;
    N1 := N1 ∪ {s};
   end
  else if Label(η, s) = 0 then
   Label(E[η U ξ], s) := 0;
  else Label(E[η U ξ], s) := 2;
 while N1 ≠ ∅ do
  begin
   N2 := ∅;
   for all s' ∈ N1 do
    for all s such that (s, s') ∈ R do
     if Label(E[η U ξ], s) = 2 then
      begin
       Label(E[η U ξ], s) := 1;
       N2 := N2 ∪ {s};
      end
   N1 := N2;
  end
 for all s ∈ S
  if Label(E[η U ξ], s) = 2 then
   Label(E[η U ξ], s) := 0;
 return;
end of procedure
```

```
procedure Verify_EG(EGη)
 N1 := ∅;
 for all s ∈ S do
  if Label(η, s) = 1 then
   begin
    Label(EGη, s) := 1;
    N1 := N1 ∪ {s};
   end
  else
   Label(EGη, s) := 0;
 if N1 ≠ ∅ then
 begin
  for all s' ∈ N1 do
   for all s such that (s, s') ∈ R do
    if Label(EGη, s) ≥ 1 then
     Label(EGη, s) := Label(EGη, s) + 1;
  N2 := ∅;
  for all s ∈ N1 do
   if Label(EGη, s) = 1 then
    begin
     Lable(EGη, s) := 0;
     N2 := N2 ∪ {s}
    end
  N1 := N2;
 end
 while N1 ≠ ∅ do
 begin
  N2 := ∅;
  for all s' ∈ N1 do
   for all s such that (s, s') ∈ R do
    begin
     if Label(EGη, s) = 2 then
      N1 := N1 ∪ {s};
     Label(EGη, s) := Label(EGη, s) − 1;
    end
  N1 := N2;
 end
 for all s ∈ S do
  if Label(EGη, s) ≥ 2 then
   Label(EGη, s) := 1;
  else
   Label(EGη, s) := 0;
 return;
end of procedure
```

FIGURE 4. Vectorized model checking algorithms for temporal operators

are greater than 0, then they are incremented (lines 13 $\sim$ 16). At this point, the label 0 means that $EG\eta$ does not hold at the state; the label greater than 0 means that η holds at the state and it has its label - 1 successor states where η holds. Next, for each state labeled 1 (i.e. the state which has no successor states where η holds), the label is relabeled to 0 and the state is inserted to the set N_2 which keeps track of the states newly labeled as 0 (lines 18 $\sim$ 23). In lines 27 $\sim$ 37, for each state newly determined that $EG\eta$ does not hold on it, the labels of its predecessor states are decremented. This step is repeated for those states that become to have no successor states until no more such states exist. Finally, the states whose labels are greater than 1 are relabeled to 1 because such states have at least one infinite path on which η always holds; the other states are labeled as 0 (lines 38 $\sim$ 43).

3.3. Time complexity. It is clear that *Verify_Not*$(\neg\eta)$ and *Verify_Or*$(\eta \vee \xi)$ runs in time proportional to $|S|$.

In the case of *Verify_EX*$(EX\eta)$, the lines 6 $\sim$ 7 are executed only $|R|$ times in total by adopting a data structure which assigns a list of its predecessor states to each state. Therefore, the time complexity is $O(|S| + |R|)$.

In the case of *Verify_EU*$(E[\eta \mathcal{U} \xi])$, the *for loop* from the line 3 to the line 11 is executed $|S|$ times. The lines 16 $\sim$ 21 are executed only $|R|$ times in total because it never checks the predecessor states of the same state twice. The last *for loop* (lines 24 $\sim$ 26) is repeated $|S|$ times. Therefore, its time complexity is $O(|S| + |R|)$.

In the case of *Verify_EG*$(EG\eta)$, it is easy to see that the lines 3 $\sim$ 10, 18 $\sim$ 23, and 38 $\sim$ 43 are executed $|S|$ times in total, and the lines 14 $\sim$ 16 and 30 $\sim$ 35 are repeated $|R|$ times in total in the same way as in the case of *Verify_EU*$(E[\eta \mathcal{U} \xi])$. Therefore, its time complexity is $O(|S| + |R|)$.

Since one of these procedures is executed for each sub-formula of a given CTL formula η, the time complexity of the vectorized model checking algorithm is $O((|S| + |R|)|\eta|)$, where $|\eta|$ denotes the length of η.

3.4. Fairness constraints. Fairness constraints can be handled efficiently by labeling *fair states* which have at least one *fair path* [7, 9]. This can be done by first obtaining *fair strongly connected components* and then getting reachable states to the fair strongly connected components.

There is a well known linear time algorithm to get strongly connected components of a directed graph based on the depth first search [1]. It seems to be difficult to vectorize this algorithm. We leave the vectorization of this part as a future problem and decided to use the non-vectorized well known algorithm.

Once the strongly connected components have been obtained, it is easy to vectorize the decision procedure if they are fair or not. The reachability problem can be also vectorized in the same way as model checking of $E[\eta \mathcal{U} \xi]$.

The labeling for fair states should be done once before starting model checking and it should be also done when evaluating EG operator.

4. Vectorized Model Checker

4.1. Implementation. We have implemented the vectorized model checking algorithm on a vector processor FACOM VP400E as a vectorized model checker. In the VP400E, three pipelined vector functional units, each of which consists of 4 pipelined units, can operate in parallel with 7 nano second cycle time. Its peak performance is about 1714 MFLOPS. It has a 256 M byte main memory, in which we can use 200 M bytes as a user area.

The input of the vectorized model checker is a Moore type deterministic sequential machine and CTL formulas to be verified. It creates the corresponding Kripke structure internally from a given Moore type deterministic sequential machine.

Let $M = (X, Z, \Sigma, \delta, \lambda, s_0)$ be a Moore type deterministic sequential machine, where

- X is a finite and nonempty set of binary input signals (atomic propositions);
- Z is a finite and nonempty set of binary output signals (atomic propositions);
- Σ is a finite and nonempty set of states;
- $\delta : 2^X \times \Sigma \rightarrow \Sigma$ is the state transition function;
- $\lambda : \Sigma \rightarrow 2^Z$ is the output function;
- s_0 is the initial state.

Then, the corresponding Kripke structure $K = (S, R, I)$ becomes as follows:

$$\begin{aligned} S &= \{s'_{i,j} | s_i \in \Sigma, j \in 2^X \text{and } \delta(j, s_i) \text{ is defined.}\} \\ R &= \{(s_{i,j}, s_{i',j'}) | s_{i,j}, s_{i',j'} \in S, \delta(j, s_i) = s_{i'}\} \\ I(s'_{i,j}) &= \{x | x \in j\} \cup \{z | z \in \lambda(s_i)\} \end{aligned}$$

Intuitively, there is a one to one correspondence between the transition edges of the Moore type deterministic sequential machine M and the states of the corresponding Kripke structure K. The size of the Kripke structure becomes as follows:

$$\begin{aligned} O(|S|) &= O(|\Sigma| \times 2^{|X|}) \\ O(|R|) &= O(|E| \times 2^{|X|} \\ O(|S| + |R|) &= O((|\Sigma| + |E|) \times 2^{|X|}), \end{aligned}$$

where $|E|$ represents the number of transition edges of M and $O(|E|) = O(|\Sigma| \times 2^{|X|})$.

Almost all parts of the model checker are vectorized except the fairness constraints handling. The most time consuming parts of the vectorized model checker are the calculations of labels of the predecessor states for each state (lines 16 $\sim$ 21 of *Verify_EU*($E[\eta \mathcal{U} \xi]$) and lines 30 $\sim$ 35 of *Verify_EG*($EG\eta$) in Figure 4). The average vector length for these parts becomes the average number of predecessor states of each state, and it is $|R|/|S|$ or $2^{|X|}$. That is, if the machine has 10 input signals, the average vector length becomes 1024 and it is enough large to extract high performance of a vector processor.

In order to represent a transition relation R of a Kripke structure, we use 2 integer arrays Q and R. Since there is a one to one correspondence between the edges of a sequential machine and the states of the corresponding Kripke structure, it is possible to number the states of the Kripke structure so that each state s_i has its predecessor states $s_{Q(i)} \sim s_{R(i)}$. By using this numbering method, we can use the efficient *contiguous access* (see Figure 1) to calculate the labels of predecessor states which is the most time consuming parts of the model checker. The sizes of the arrays Q and R are both $|S|$. These 2 integer arrays Q and R can be easily constructed directly from a given Moore type deterministic sequential machine in proportional time to the size of the corresponding Kripke structures by using one additional integer array of size $|S|$.

As for N_1 and N_2 used in *Verify_EU*($E[\eta\, \mathcal{U} \xi]$) and *Verify_EG*($EG\eta$), we also use integer arrays with the corresponding index variables. Initialization of N_i (i.e. $N_i := \emptyset, i = 1, 2$) can be done by just substituting 0 to the corresponding index variable. Insertion of a state s to N_i (i.e. $N_i := N_i \cup \{s\}$) can be done by just storing the state s at the place in N_i pointed by its corresponding index variable and updating the index variable. As for copying data from N_2 to N_1 (i.e. $N_1 := N_2$), we just exchange the role of N_1 and N_2 instead of copying data actually. The maximum required sizes of the arrays N_1 and N_2 are $|S|$.

In order to store the truth value of each sub-formula at each state, we use 1 bit each. Therefore, the required memory in total for this purpose is $|S| \times |\eta|/8$ bytes. In addition, we use a working integer array of size $|S|$ to store a label for each state in checking temporal operators.

We also use $|S| \times 8$ words to handle fairness constraints.

Note that 1 integer word consists of 4 bytes. Therefore, the total amount of required memory is

$$|S| \times 56 + |S| \times |\eta|/8 \text{ bytes.}$$

For CTL formulas which contains 256 and 1024 operators, the vectorized model checker can manipulate Kripke structures of 2.3 million and 1.1 million states respectively with main memory of 200 M bytes.

4.2. Example. In order to measure the efficiency of the vectorized model checker, we applied it to the verification of two large Kripke structures (SR8 and SR9) corresponding to synchronous shift registers with parallel load and serial output. SR8 consists of 131,072 states and 67,108,864 edges (each state has 512 edges), and SR9 consists of 524,288 states and 536,870,912 edges (each state has 1024 edges) as shown in Table 1. The CTL formulas which give their full specifications contain more than 300 different sub-formulas with no fairness constraints.

The benchmark results without fairness handling are shown in Table 2. Both SR8 and SR9 are verified to be *true* on the vector machine of VP-400E in about

TABLE 1. Kripke structures for benchmark tests

Name	$\|Spec\|$	#States	#Edges
SR8	Total: 306 Boolean: 212 Temporal: 94	131,072	67,108,864
SR9	Total: 368 Boolean: 254 Temporal: 114	524,288	536,870,912

TABLE 2. Results of benchmark tests

Name	Scalar Execution (m sec.)	Vecotr Execution (m sec.)	Acceleration Ratio	Average Vector Length	Used Memory
SR8	132,126	4,998	26.4	512	13 MB
SR9	1,131,039	28,985	39.0	1,024	52 MB

5 seconds and 29 seconds by using 13 MB and 52 MB of memory respectively. This means that our vectorized model checker evaluates about 7 $\sim$ 8 states in a second, which implies that it will be able to verify 1 million state Kripke structure in a couple of minutes. Furthermore, the acceleration ratio obtained by our algorithm is around 26 $\sim$ 39, which is extremely high ratio in non numeric application programs.

We also verified SR8 by using the CTL model checker B1.0 developed by Clarke et al. [5] installed on Sun-3/80. It took about 5,068 seconds to verify SR8. By considering that some benchmark tests show that the scalar unit of FACOM VP-400E is about 6.7 times faster than Sun-3/80, it will still take about 750 seconds to verify SR8 even if we install the CTL model checker B1.0 on VP-400E because their algorithm cannot be vectorized due to recursive calls.

5. Conclusion

We proposed a vectorized model checking algorithm for CTL and implemented the vectorized model checker on a vector processor FACOM VP400E. Almost all parts of the algorithms are vectorized except the parts to obtain strongly connected components for fairness constraints handling.

It can handle about 2.3 million or 1.1 million state Kripke structures derived from a Moore type deterministic sequential machine when the length of a give CTL formula is less than 256 or 1024 respectively.

We also presented examples which show that a CTL formula of 368 different sub-formulas is verified to be true in 29 seconds on a Kripke structure with 524,288 states and 536,870,912 edges.

There are several remaining future problems:

The first is to devise a vectorized algorithm for obtaining strongly connected components of a directed graph, and we need more consideration.

The second is a vectorization of model checking which can handle sequential machines directly [2]. We think this is not so difficult and would like to implement it in the near future.

The third is a vectorization of model checking based on Binary Decision Diagrams (BDD) because the BDD is a powerful technique to reduce necessary amount of memory for model checking dramatically in some cases [4].

Although our current version of the vectorized model checker takes a single Moore type sequential machine as input, it would be interesting to challenge the vectorization of the procedure to create direct product of several concurrent/parallel sequential machines. It would be also interesting to devise vectorized algorithm for model checking without creating direct product explicitly.

Acknowledgments The authors would like to express their appreciations to Prof. E. M. Clarke of CMU for his valuable suggestions. They also would like to express their appreciations to Prof. S. Yajima, Dr. N. Takagi, Mr. N. Ishiura and Mr. H. Ochi of Kyoto University for their precious discussions and comments.

REFERENCES

1. A. V. Aho, J. E. Hopcroft, and J. D. Ullman. *The Design and Analysis of Computer Algorithms.* Addison Wesley, 1974.
2. M. C. Browne. An improved algorithm for the automatic verification of finite state systems using temporal logic. Technical Report CMU-CS-86-156, Carnegie Mellon University, 1986.
3. M. C. Browne, E. M. Clarke, D. L. Dill, and B. Mishra. Automatic verification of sequential circuits using temporal logic. *IEEE Transactions on Computers*, C-35(12):1035–1044, December 1986.
4. J. R. Burch, E. M. Clarke, K. L. McMillan, D. L. Dill, and J. Hwang. Symbolic model checking: 10^{20} states and beyond. In *Proc. Logic in Computer Science*, 1990.
5. E. M. Clarke, S. Bose, M. C. Browne, and O. Grumberg. The design and verification of finite state hardware controllers. Technical Report CMU-CS-87-145, Carnegie Mellon University, July 1987.
6. E. M. Clarke and E. A. Emerson. Synthesis of synchronization skeletons for branching time temporal logic. In *Proc. Workshop on Logic of Programs*, pages 52–71. Springer-Verlag, 1981.
7. E. M. Clarke, E. A. Emerson, and A. P. Sistla. Automatic verification of finite state concurrent systems using temporal logic specifications: A practical approach. Technical Report CMU-CS-83-152, Carnegie Mellon University, 1983.
8. E. M. Clarke and O. Grumberg. Avoiding the state explosion problem in temporal logic model checking algorithms. In *Proc. 6th Annual ACM Symposium of Principles of Distributed Computing*, pages 294–303, August 1987.
9. E. M. Clarke and O. Grumberg. Research on automatic verification of finite-state concurrent systems. Technical Report CMU-CS-87-105, Carnegie Mellon University, January 1987.
10. E. M. Clarke, D. E. Long, and K. L. McMillan. A language for compositional specification and verification of finite state hardware controllers. In *Proc. 9th IFIP Symposium on Computer Hardware Description Languages and their Applications*, pages 281–295, June 1989.
11. S. Graf, J. L. Richier, C. Rodriguez, and J. Voiron. What are the limits of model checking methods for the verification of real life protocols ? In *Proc. Workshop on Automatic Verification Methods for Finite State Systems*, June 1989.

Department of Information Science, Kyoto University, Kyoto 606, Japan
E-mail: hiraishi@kuis.kyoto-u.ac.jp

Department of Information Science, Kyoto University, Kyoto 606, Japan
E-mail: meki@kuis.kyoto-u.ac.jp

Department of Information Science, Kyoto University, Kyoto 606, Japan
E-mail: hama@kuis.kyoto-u.ac.jp

DIMACS Series in Discrete Mathematics
and Theoretical Computer Science
Volume 3, 1991

On Some Implementation of Optimal Simulations

Ryszard Janicki[1]
Department of Computer Science and Systems
McMaster University
Hamilton, Ontario, Canada, L8S 4K1

Maciej Koutny[2]
Computing Laboratory
The University of Newcastle upon Tyne
Newcastle upon Tyne NE1 7RU, U.K.

Abstract

We describe an algorithm for simulating the behaviour of a class of communicating finite state automata in such a way that several interesting properties of the system can be determined without having to construct the entire reachability graph. The paper is an application of the theory developed in [JK89,JK89a].

1) Research partly supported by a grant from NSERC No. OGP 0036539.

2) Research supported by ESPRIT Basic Research Action 3148 (project DEMON).

INTRODUCTION

The process of verification of dynamic properties of a concurrent system generally involves some kind of reasoning about the complete state-space of the system, e.g. proving deadlock-freeness requires showing that it is not possible to reach a state in which no transition is enabled, and proving liveness requires showing that starting from any reachable state it is possible to execute any given action. Reasoning about the complete state-space of concurrent systems has one serious drawback which is a combinatorial explosion of the state-space. Even a simple concurrent system can generate many hundreds or thousands of states. This is further worsened by the fact that the higher the degree of concurrency exhibited by the system is (the degree of concurrency is roughly the number of its sequential sub-systems), the sooner the state-space grows beyond any manageable size. To cope with the combinatorial explosion a number of sophisticated techniques have been invented. Perhaps the most attractive is induction, e.g. [Kel76], which employs invariants to prove that a property is true in all the states of the system. Although such invariants can sometimes be derived automatically [MS82], one usually has to formulate them without any external assistance, which severely limits the applicability of the invariant techniques. An alternative approach is one in which reasoning about the complete state-space is replaced by the analysis of its (reduced) representation [MR87]. Examples of an approach which is aimed at generating reduced state spaces of concurrent systems are [Jen87], [Val89] and [God90]. Another, perhaps the most successful approach from the application point of view, are the temporal logic model checkers of [Bro86, CG87, CES86].

In [JK87,JK89 and JK89a] we investigated a possibility of defining a reachability relation on the system's histories which would be a 'small' subset of the complete reachability relation, and yet provide full information about the relevant dynamic properties of the system.We defined such a reduced reachability relation, called the *optimal simulation*, and proved that it enables the reasoning about many interesting properties of a concurrent system, and at the same time requires a minimal computational effort. The optimal simulation has been defined in a general, trace-based, setting what makes it applicable to a number of models for concurrency, including Petri nets [Pet81,Rei85], and non-trivial subsets of CSP [Hoa85] and CCS [Mil80].
The reachability graphs of concurrent systems can be regarded as *finite representations of reachability relations* (if only such a representation exists). Since the optimal simulation provides the same information about relevant dynamic properties of the system as the full reachability relation [JK89,JK89a], the reachability graph of optimal simulation (i.e. its finite representation) and the full reachability graph may be considered as equivalent. The optimal simulation is always a subset of the full reachability, but of course it does not mean that the reachability graph of optimal simulation is always (much) smaller than the full reachability graph. However, we do claim that in the case of concurrent systems exhibiting high degree of concurrency (i.e. those with many sequential components), the reachability graph of optimal simulation is much smaller than the full reachability graph. Thus it is advantageous to use optimal simulation as a tool to *reduce the size of reachability graphs of concurrent systems*.

Unfortunately, as opposed to the full reachability graph, in the general case it is not clear how to generate the reachability graph for optimal simulation in an efficient way. In some sense this is a negative side-effect of the above mentioned generality of optimal simulation. In this paper we will show how such a graph (as we claim, reduced in the majority of cases) can be constructed in a concrete framework, namely that of finite asynchronous automata [Zie87]. They enjoy clear structural properties making possible an efficient algorithm generating a reachability graph of the optimal simulation. At the same time asynchronous automata can model 1-safe Petri nets, COSY path programs, and non-trivial subsets of CCS and CSP.

We will not prove here that optimal simulation is indeed behaviourally equivalent to the full reachability, nor justify the construction of the optimal simulation relation. These issues have been dealt with in [JK89] which is widely available (Lecture Notes in Computer Science 366), and in [JK89a] which can sent be on request.

We would like to point out that the major methodological difference between our approach to minimise reachability graphs and those developed in [Jen87,Val89,God90] is that we do not try to minimise (or even deal with) the full reachability graph in an explicit way. All what we are trying to do is to build a reachability graph which represents the optimal simulation relation (a subset of full reachability). We then make a claim, based on the general properties of optimal simulation, that in majority of cases, such a graph is much smaller than the original full reachability graph.

Note that our approach is based on the assumption that concurrent behaviours (histories) can be modelled by causal partial orders. We will represent those partial orders by certain equivalence classes of step sequences, generalising the notion of traces of [Maz86].

The paper is organised as follows. In the next section we briefly outline the motivations and basic idea behind the optimal simulation. In Section 2 we introduce the finite asynchronous automata, generalise the concept of trace of [Maz86] and recall the main notions of [JK89a]. The third section contains a formal definition of reachability graphs. The two final sections present an algorithm generating a reachability graph of the optimal simulation.

Throughout the paper we use standard mathematical notation. In particular, if $\sigma = A_1 \ldots A_k$ is a string and A is an element, then $\sigma\hat{}A$ denotes the string $A_1 \ldots A_k A$.

Also, if $I = \{i_1, \ldots, i_k\}$, where $i_1 < \ldots < i_k$, is a set of indices and $S_{i_1}, \ldots, S_{i_k}$ is a sequence of sets, then $\times_{i \in I} S_i$ denotes the Cartesian product $S_{i_1} \times \ldots \times S_{i_k}$.

1 Motivation

We will use Petri nets to illustrate the basic concepts discussed in this section.

Execution paths generated by a concurrent system can be represented by *step sequences* - each step being a finite set of transitions executed simultaneously. Consider the Petri net in Figure 1.1. Its behaviour might be briefly described in the following way: All step sequences must begin with transition a. After that one can simultaneously execute transitions b and c, or execute b followed by c, or execute c followed by b. The net generates three step sequences leading to a deadlocked marking, $\pi_1 = \{a\}\{b,c\}$, $\pi_2 = \{a\}\{b\}\{c\}$ and $\pi_3 = \{a\}\{c\}\{b\}$.

Suppose now that we were asked to find all the deadlocked markings of the net by following as few step sequences as possible and by selecting possibly shortest step sequences. An exhaustive search would include all three step sequences, π_1, π_2 and π_3. However, one may easily observe that all these lead to the same marking. One may also observe that π_1 is shorter a sequence than both π_2

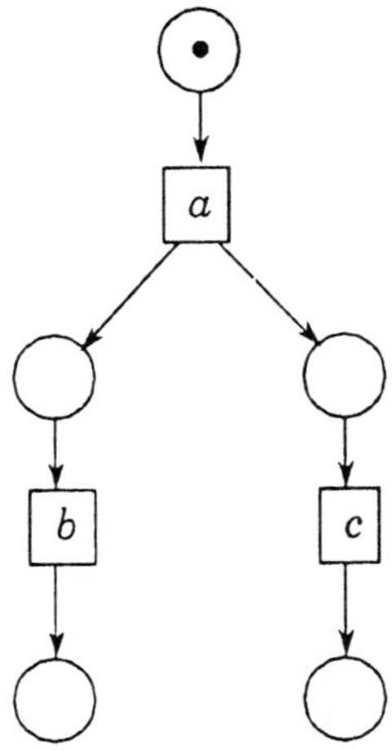

Figure 1.1

and π_3. Hence an efficient search detecting the deadlocked markings of the net should include just one path, π_1.

The above example is an instance of the following general problem: *Is there a way of executing a concurrent which is both expressive and efficient?* By an expressive execution we mean one providing enough information to verify some important properties of the system, e.g. liveness or termination, whereas by an *efficient* execution we mean one which requires minimal computational effort, e.g. by avoiding execution paths providing redundant information

Referring to our example, one may observe that π_1 has a straightforward operational interpretation as it follows the rule: "always choose a maximal set of independent transitions to be executed next", a rule which characterises *maximally concurrent* execution. Employing maximal concurrency is an attractive idea, both conceptually and from the point of view of implementation. Unfortunately, there are cases in which maximally concurrent execution is not as expressive as we

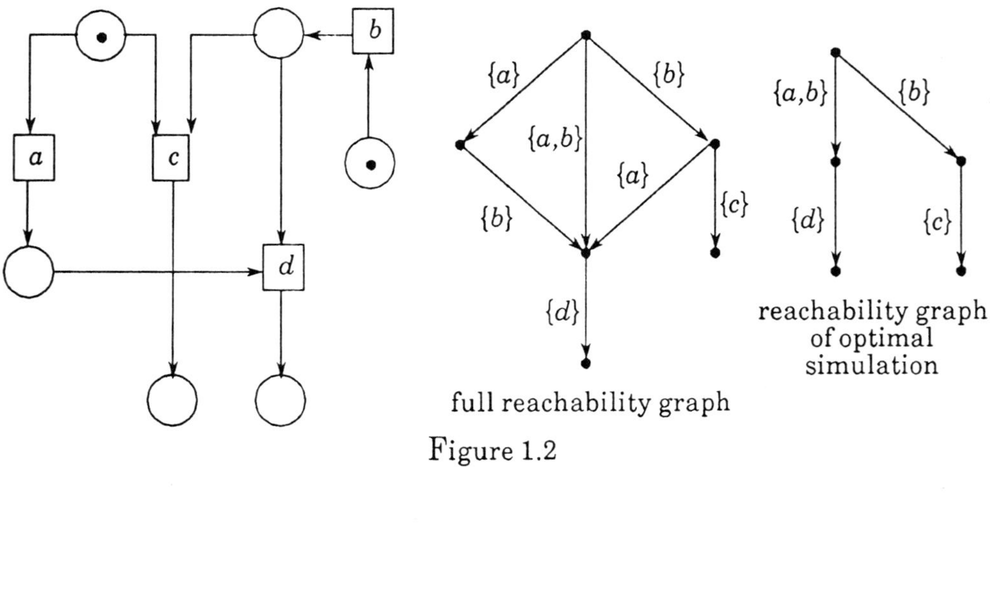

Figure 1.2

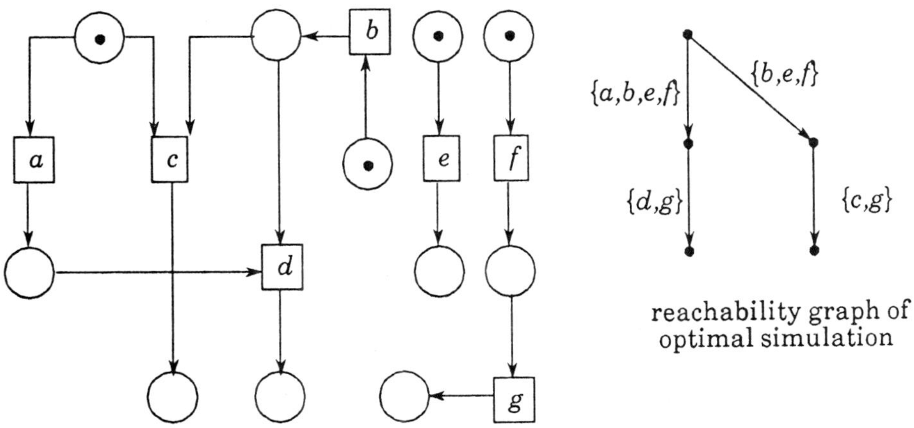

Figure 1.3

would like it to be (see [JLKD86] for necessary and sufficient conditions where it is). To show this we take the net in Figure 1.2. The maximally concurrent execution can find only one deadlocked marking of the net by proceeding along the step sequence $\rho_1 = \{a,b\}\{d\}$. The other deadlocked marking, which might be reached by following $\rho_2 = \{b\}\{c\}$, is left undetected.

In [JK89,JK89a] we have defined, by generalising maximally concurrent execution, the *optimal simulation* which is both expressive and efficient way of executing the net for verification purposes. In the case of net in Figure 1.2, optimal simulation would consist of two step sequences, ρ_1 and ρ_2, together with their prefixes. Figure 1.2 shows both the full reachability graph of the net and the reachability graph of the optimal simulation. The latter one is smaller, but because in this case only two transitions a and b can be fired concurrently, the difference in size is not significant. However, in the case of net in Figure 1.3, the reachability graph of optimal simulation is isomorphic to that in Figure 1.2, while (as one may easily check) the full reachability graph would hardly fit on a single page.

In this paper we will focus on the problem of finding a reachability graph (i.e. a finite representation) of the optimal simulation. There are two reasons why generating such a reachability graph seems to be an important issue. First, if we were able to generate the complete reachability graph of optimal simulation of the net, then it would be in general advantageous to use this graph for verification purposes rather than the (usually much larger) full reachability graph. Second, if the net would generate reachability graph of the size we could not possibly handle, we might use the algorithm(s) developed for the purpose of generating a reachability graph, to implement an efficient simulation technique based on the concept of optimal simulation.

2 Asynchronous Automata and The Optimal Simulation

We have chosen asynchronous automata due to their relative generality. For example, they can model (see [Zie87,JL90,Zie89]) 1-safe labelled Petri nets [Pet81], COSY path programs [LSC81] and asynchronous cellular automata, as well as non-trivial sub-models of CCS [Mil80] and CSP [Hoa85]. Asynchronous automata were introduced by Zielonka [Zie87] to investigate recognisable subsets of partially commutative monoids. Labelled Petri nets can be used as a convenient graphical representation of asynchronous automata (in the same way as Mealy graphs can be used to represent standard finite state automata), and throughout the rest of this paper we will have a Petri net representation for each example of asynchronous automaton.

A *finite (deterministic) asynchronous automaton* (or *FASA*) is a tuple

$M=(P_1,...,P_n,\Delta,F)$

such that for every $i=1,...,n$,

$P_i=(\Sigma_i,S_i,s_{oi})$ is a *process*, where

Σ_i is a finite non-empty set of *events*,

S_i is a finite set of *states* ($S_i \cap S_j = \varnothing$ for $i \neq j$),

$s_{oi} \in S_i$ is the *initial state* of P_i,

Δ is the set of *transition functions*,

$F \subseteq S_1 \times ... \times S_n$ is the set of *final states*.

$S=S_1 \times ... \times S_n$ is the set of *global states*, $\Sigma = \Sigma_1 \cup ... \cup \Sigma_n$ is the set of *events*, and $s_o=(s_{o1},...,s_{on})$ is the *initial state* of M.

For every $a \in \Sigma$, the set $p(a)=\{i \mid a \in \Sigma_i\}$ may be interpreted as the set of processes synchronously performing event a, and the execution of a can be interpreted as resulting from a 'handshake' synchronisation between the processes in $p(a)$.

For every $a \in \Sigma$, the set $\Delta = \{\delta_a \mid a \in \Sigma\}$ contains exactly one partial transition function

$\delta_a : \times_{i \in p(a)} S_i \rightarrow \times_{i \in p(a)} S_i$.

Figure 2.1 shows the relationship between FASA and labelled Petri nets, and Figure 2.2 gives an example of FASA and the corresponding net.

In this paper we will not be concerned with the set F and in what follows each global state will be considered as a final one.

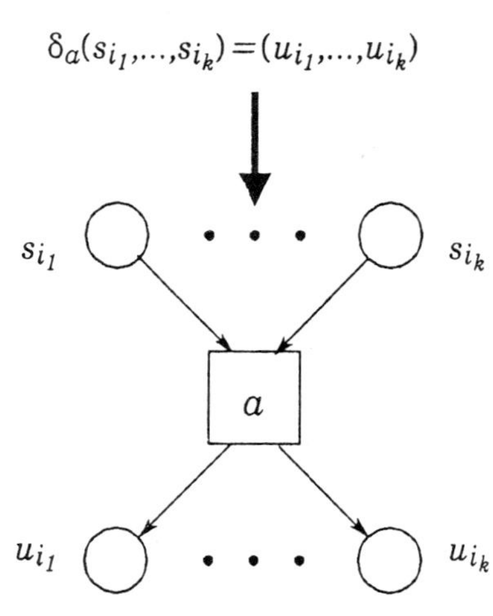

Figure 2.1: Relationship between FASA and labelled Petri nets.

$M=(P_1,P_2,\Delta,F)$, where
$P_1=(\Sigma_1,S_1,s_{o1})$ and $P_2=(\Sigma_2,S_2,s_{o2})$
$\Sigma_1=\{a,c\}$ $\Sigma_2=\{b,c\}$
$S_1=\{s_{o1},s_{11},s_{21}\}$ $S_2=\{s_{o2},s_{12},s_{22}\}$
$\Delta=\{\delta_a,\delta_b,\delta_c\}$ $F=S_1\times S_2$
$\delta_a(s_{o1})=s_{11}$ $\delta_a(s_{11})=s_{21}$
$\delta_b(s_{o2})=s_{12}$ $\delta_b(s_{12})=s_{22}$
$\delta_c(s_{11},s_{12})=(s_{o1},s_{o2})$ $\delta_c(s_{21},s_{22})=(s_{o1},s_{o2})$

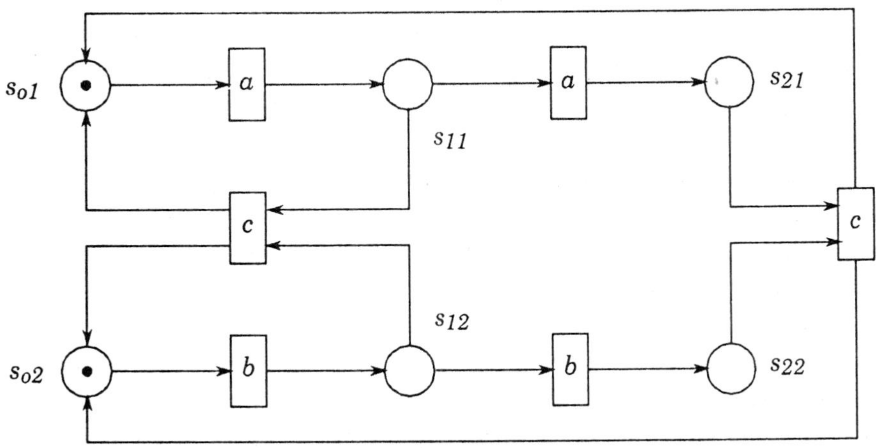

Figure 2.2: FASA and a corresponding Petri net.

The *independency* relation, $ind\subseteq\Sigma\times\Sigma$, is defined as $(a,b)\in ind \Leftrightarrow p(a)\cap p(b)=\varnothing$.
If $(a,b)\in ind$ then a and b are interpreted as *independent* events which, if enabled, can be executed simultaneously.
We define Ind to be a set of *steps* which are non-empty sets of mutually independent events, i.e. $Ind=\{A\subseteq\Sigma \mid A\neq\varnothing \wedge \forall a,b\in A.\ a=b \vee (a,b)\in ind\}$. For every $A\in Ind$, $p(A)=\bigcup_{a\in A} p(a)$.

For every non-empty set $I\subseteq\{1,...,n\}$, let $\Pi_I : S_1\times...\times S_n \rightarrow \times_{i\in I} S_i$ be the standard projection.
For every $A\in Ind$ we define a transition relation $\Rightarrow_A\subseteq S\times S$ as follows:
Let $s,r\in S$. Then $s\Rightarrow_A r$ if
(1) $\delta_a(\Pi_{p(a)}(s))=\Pi_{p(a)}(r)$ for all $a\in A$.
(2) $\Pi_I(s)=\Pi_I(r)$ for $I=\{1,...,n\}-p(A)$.

A *step sequence* of M, $\sigma\in Steps$, is a string $\sigma=A_1...A_k$ such that $k\geq 0$, $A_i\in Ind$ $(i=1,...,k)$, and there are global states $r_0,r_1,...,r_k\in S$ such that $r_0=s_o$ and $r_{i-1}\Rightarrow_{A_i} r_k$ for $i=1,...,k$.
The empty step sequence $(k=0)$ will be denoted by λ. We also denote $state_\sigma=s_k$ (we can do this because FASA is *deterministic*).
For the example of Figure 2.2 we have

$\sigma_1 = \{a,b\}\{b\}\{a\}\{c\} \in Steps$, $\sigma_2 = \{a\}\{b\}\{c\}\{a,b\} \in Steps$

$state_\lambda = (s_{o1}, s_{o2}) = state_{\sigma_1}$ and $state_{\sigma_1} = (s_{11}, s_{12})$.

The semantics of M in terms of steps sequences can be refined by identifying those step sequences which are essentially different realisations of the same underlying partial order history of the FASA.

Let $\simeq$ be the least equivalence relation on step sequences which contains all pairs of non-empty step sequences (σ,ω) such that

$\sigma = A_1 \ldots A_{k-1} A_k A_{k+1} \ldots A_l$ and $\omega = A_1 \ldots A_{k-1} A_{k1} A_{k2} A_{k+1} \ldots A_l$,

where $1 \leq k \leq l$, $A_{k1} \cap A_{k2} = \varnothing$ and $A_{k1} \cup A_{k2} = A_k$. The equivalence class of $\simeq$ containing step sequence σ will be denoted by $[\sigma]$.

Henceforth each equivalence class H of $\simeq$ will be called a *history* of M, $H \in Hist$.

For the FASA of Figure 2.2 we have $[\{a,b\}\{b\}] = \{\{a\}\{b\}\{b\}, \{b\}\{a\}\{b\}, \{a,b\}\{b\}, \{b\}\{a,b\}, \{b\}\{b\}\{a\}\}$.

Step sequences belonging to a history H can be regarded as different realisations of an underlying concurrent history which itself may be represented by a partially ordered set of event occurrences. This partial order is the intersection of all the partial orders induced by the step sequences in that equivalence class. This is illustrated in Figure 2.3(b) for a history $H = [\{a\}\{b\}\{c\}]$ of the FASA (and net) of Figure 2.3(a). Note that step sequence $\sigma = \{a\}\{b,c\}\{a\}$ induces a partial order in which the first occurrence of a precedes the occurrences of b and c, and the occurrences of b and c are unordered and they both precede the second occurrence of a. Every partial order induced by step sequences has the following property: its disorder relation is transitive. Such orders are sometimes called *stratified partial orders*.

The concepts of $\simeq$ and a history $H \in Hist$ are natural generalisations of similar concepts from the theory of *partial commutative monoids* [CF69,Maz86,Zie87]. If we restricted step sequences to just sequences (interleavings) in the definition of *Hist* then we would get exactly the classical notion of Mazurkiewicz traces [Maz86] (the name 'trace' is sometimes used [Hoa85] to mean 'sequence', so we write 'Mazurkiewicz trace' to avoid any confusion). Our representation of a history as a partial order is a natural generalisation of the result of [Szp30] on the representation of partial orders by the set of their linearisations. We represent partial orders by the set of their *stratifications*. The basic advantage of the approach in [Maz86] is that a causal partial order may be represented by just one interleaving. In our case, every history may be represented by just one step sequence. In particular, we may choose the shortest one, i.e. maximally concurrent. This idea will be developed further when the definition of *canonical* step sequence - a very fundamental concept of our approach - will be given.

Let for every $H \in Hist$, $enabled(H) = \{A \in Ind \mid \exists \sigma \in H.\ \sigma\hat{}A \in Steps\}$.

PROPOSITION 2.1

Let $H \in Hist$, $\sigma,\omega \in H$, and $A \in enabled(H)$.

(1) $mar_\sigma = mar_\omega$.

(2) $\sigma\hat{}A \in Steps$.

(3) $\sigma\hat{}A \simeq \omega\hat{}A$. □

The above proposition makes possible the following two definitions.

For every $H \in Hist$ we denote by $state_H$ the global state satisfying $state_H = state_\sigma$, for all $\sigma \in H$.

$M=(P_1,P_2,\Delta,F)$, where

$P_1=(\{a,d\},S_1,s_{o1})$ and $P_2=(\{b,c,d\},S_2,s_{o2})$

$S_1=\{s_{o1},s_{11},s_{21}\}$ $S_2=\{s_{o2},s_{12},s_{22},s_{32}\}$

$\delta_a(s_{o1})=s_{11}$ $\delta_b(s_{o2})=s_{22}$

$\delta_c(s_{22})=(s_{32})$ $\delta_d(s_{o1},s_{o2})=(s_{21},s_{12})$

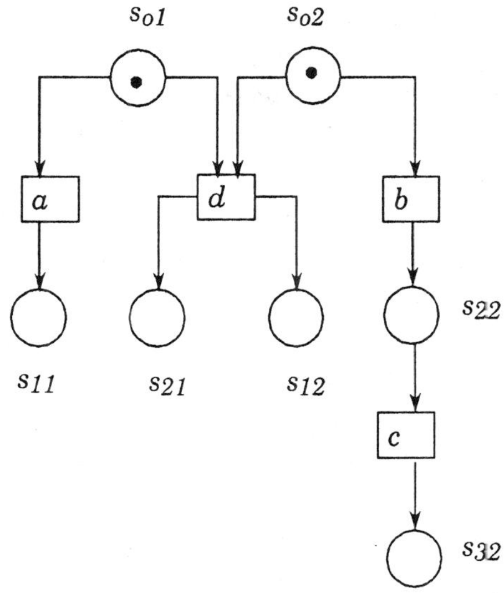

(a)

(b)

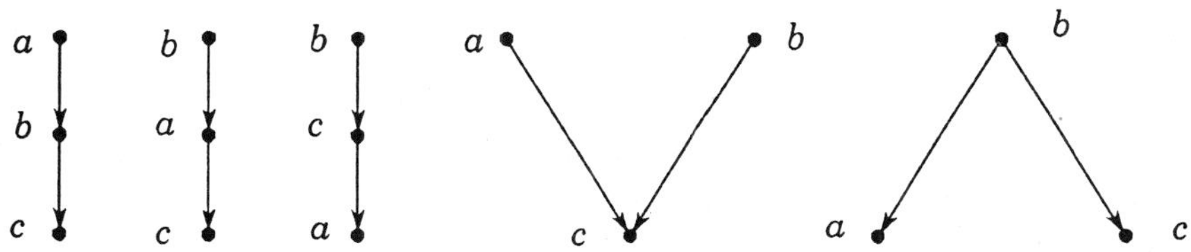

partial orders induced by step sequences belonging to history
$H=\{\{a\}\{b\}\{c\}, \{b\}\{a\}\{c\}, \{b\}\{c\}\{a\}, \{a,b\}\{c\}, \{b\}\{a,c\}\}$

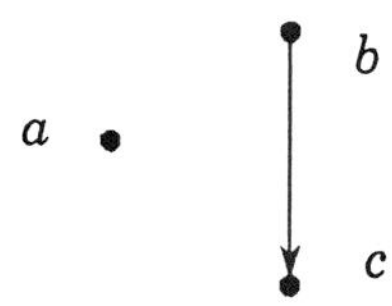

partial order underlying history H

Figure 2.3

For every $H\in Hist$ and every $A\in enabled(H)$ we denote by $H\circ[A]$ the history such that $\sigma^\frown A\in H\circ[A]$, for all $\sigma\in H$.

Proposition 2.2

If $H,J\in Hist$ and $state_H=state_J$ then $enabled(H)=enabled(J)$. □

In [JK89,JK89a] we introduced the notion of a *simulation* of the FASA, or net, which is a kind of reachability relation on the histories of the FASA. In this paper we will deal only with two simulations, the *full* and *optimal* ones.

The full simulation, *FULL*, is defined in the following way.

$FULL=\{(G,H)\in Hist\times Hist \mid \exists A\in enabled(G).\ H=G\circ[A]\}$.

The full simulation represents the complete dynamic behaviour of M. Its advantage is relatively straightforward definition and natural interpretation, its disadvantage is the size of its reachability graph. Even for small FASA's the graph grows beyond any manageable size, making the formal verification of the behavioural properties extremely difficult. It was our goal in [JK89,JK89a] to find possibly smallest simulation which could be used for the verification of the relevant system properties. A solution we proposed there was the *optimal* simulation, *OPT*. There are three reasons why *OPT* can be regarded as the optimal simulation:

(1) There are a number of behavioural properties which are common to *FULL* and *OPT*. For example, *FULL* and *OPT* generate the same sets of deadlocked states. It is also possible to verify liveness using *OPT*. Indeed, we claimed in [JK89a] that *FULL* and *OPT* essentially capture *the same behavioural properties of the FASA (concurrent system, in general)*.
(2) *OPT* involves a minimal set of histories, i.e. each proper subset of *OPT* is less expressive than *FULL*, and it may not be used, e.g., to verify the deadlock-freeness.
(3) The information about the FASA is generated in *OPT* using the shortest step sequences. For instance, each deadlocked marking will be generated by following the shortest step sequence leading to it.

Moreover, there is no other simulation which would satisfy (1)-(3) above.
OPT is defined in the following way.

Let $\zeta \subseteq Ind \times Ind$ comprise all pairs of steps (A,B) such that for every $b \in B$ there is $a \in A$ satisfying $(a,b) \notin ind$. The relation ζ may be interpreted as *forward dependency* on steps.
A step sequence $\sigma = A_1 \ldots A_k$ is *canonical*, $\sigma \in Steps_{can}$, if $k \geq 0$ and $(A_{i-1},A_i) \in \zeta$ for $i=2,\ldots,k$. Intuitively, in a canonical step sequence the execution of events is never delayed (no event can be moved from A_i to A_{i-1}).
The definition of *OPT* is based on the following result.

THEOREM 2.3 [CF69,JLKD86]

Every history contains exactly one canonical step sequence. □

The canonical step sequence contained in a history H will be denoted by $can(H)$. We also introduce an auxiliary reachability relation, *CAN*.

$$CAN = \{(G,H) \in Hist \times Hist \mid \exists A \in enabled(G).\ can(H) = can(G)\,\hat{}\,A\}.$$

Clearly, $CAN \subseteq FULL$.

For every $H \in Hist$, let $maxenabled(H)$ comprise all steps in $enabled(H)$ which are *maximal* w.r.t. set inclusion, i.e. $maxenabled(H) = \{A \in enabled(H) \mid \forall B \in enabled(H).\ A \subseteq B \Rightarrow A = B\}$.
Let $Hist_{maxend} \subseteq Hist$ comprise histories whose canonical step sequences *end with a maximal step*, i.e. $Hist_{maxend} = \{[A_1 \ldots A_k] \mid k \geq 1 \wedge A_1 \ldots A_k \in Steps_{can} \wedge A_k \in maxenabled([A_1 \ldots A_{k-1}])\}$.
Note that $Hist_{maxend}$ is in general a proper subset of $Hist$. For example, if we take the FASA of Figure 2.4 then $[\{a\}] \in Hist - Hist_{maxend}$ since $\{a,c\} \in enabled([\lambda])$.

Finally, the *optimal* simulation, *OPT*, is defined as the smallest subset of *CAN* such that for every $H \in Hist_{maxend}$ there are $H_1,\ldots,H_m$ satisfying $H_1 = [\lambda]$, $H_m = H$, and $(H_i,H_{i+1}) \in OPT$, for $i = 1,\ldots,m-1$.
We also denote $Hist_{opt} = \{[\lambda]\} \cup \{G \mid (G,H) \in OPT\} \cup \{H \mid (G,H) \in OPT\}$.

For the FASA of Figure 2.4 we have the following:

$$\zeta = \{(A,A),(B,B),(C,C),(D,D),(B,A),(B,C),(B,D),(A,B),(C,B),(D,A),(D,B),(D,C)\}$$

$$Steps_{can} = \{\lambda,A,B,C,D\}$$

$M=(P_1,P_2,\Delta,F)$, where

$P_1=(\{a,b\},S_1,s_{o1})$ and $P_2=(\{b,c\},S_2,s_{o2})$

$S_1=\{s_{o1},s_{11},s_{21}\}$ $S_2=\{s_{o2},s_{12},s_{22}\}$

$\delta_a(s_{o1})=s_{11}$ $\delta_c(s_{o2})=s_{22}$

$\delta_b(s_{o1},s_{o2})=(s_{21},s_{12})$

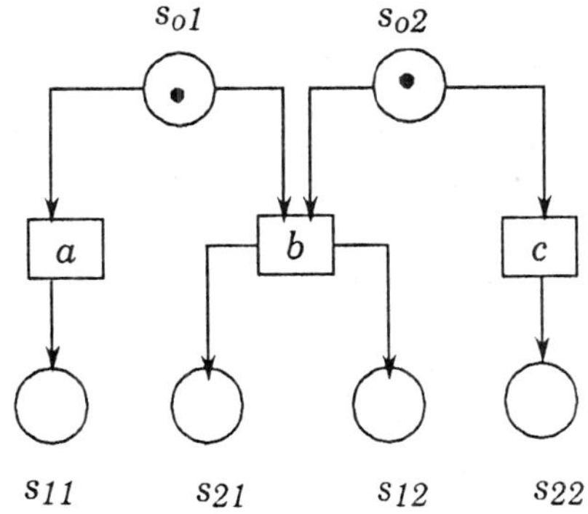

Figure 2.4: FASA illustrating the definition of *OPT*

$CAN=\{([\lambda],[A]),([\lambda],[B]),([\lambda],[C]),([\lambda],[D])\}$

$maxenabled([\lambda])=\{B,D\}$ and $Hist_{maxend}=\{[B],[D]\}$

$OPT=\{([\lambda],[B]),([\lambda],[D])\}$

where $A=\{a\}$, $B=\{b\}$, $C=\{c\}$ and $D=\{a,c\}$.

In this paper we will investigate the implementation aspects of the optimal simulation. In [JK89a] we demonstrated that, under certain 'finiteness' assumptions which are satisfied for the asynchronous automata, there always exists a finite reachability graph of *OPT*. In the rest of this paper we will outline the idea behind an efficient algorithm generating relatively small reachability graph of the optimal simulation.

3 Reachability Graphs

A *reachability graph* is a labelled graph $RG=(V,Arcs,v_{init})$, where V is a non-empty set of vertices with $v_{init}\in V$ being the initial vertex, and $Arcs\subseteq V\times Ind\times V$ is the set of labelled arcs, such that the following hold.

$M=(P_1,P_2,\Delta,F)$

$P_1=(\{a,b\},\{s_{o1},s_{11}\},s_{o1})$

$P_2=(\{a,c\},\{s_{o2},s_{12}\},s_{o2})$

$\delta_a(s_{o1},s_{o2})=(s_{11},s_{12})$

$\delta_b(s_{o1})=s_{o1}$ $\delta_c(s_{o2})=s_{o2}$

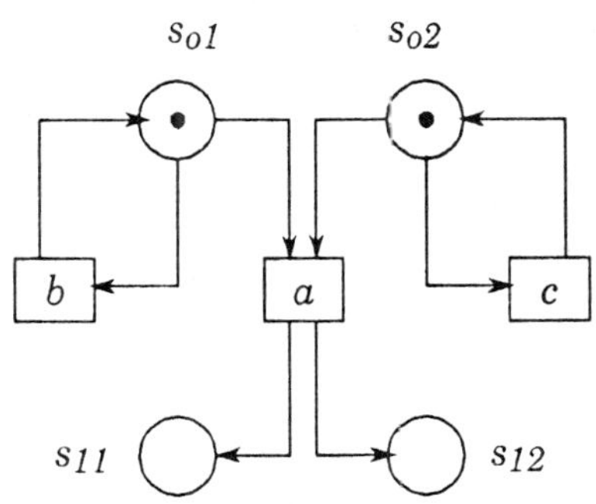

Figure 3.1: FASA for which Proposition 2.2 does not hold.

(1) If (p,A,q) and (p,A,r) are arcs then $q=r$.

(2) If (v_1,A_1,v_2), (v_2,A_2,v_3),...,(v_k,A_k,v_{k+1}) are arcs such that $v_1=v_{init}$ then $A_1...A_k \in Steps$.

The reachability graph RG generates a relation on the histories of M, $Rel(RG) \subseteq Hist \times Hist$, which comprises all pairs of histories (G,H) for which there are arcs (v_1,A_1v_2), (v_2,A_2,v_3),...,(v_k,A_k,v_{k+1}) such that $v_1=v_{init}$, $G=[A_1...A_{k-1}]$, and $H=[A_1...A_k]$.
Clearly, $Rel(RG) \subseteq FULL$.

A reachability graph generating the full semantics, RG_{FULL}, can be defined in the following way.

(1) $V=\{state_H \mid H \in Hist\}$.

(2) $Arcs=\{(state_H,A,state_{H\circ[A]}) \mid H \in Hist \wedge A \in enabled(H)\}$.

(3) $v_{init}=s_o$.

It is not difficult, using Proposition 2.1 and 2.2, to show that RG_{FULL} is indeed a reachability graph of the full simulation, i.e. $Rel(RG_{FULL})=FULL$.

At this point it is not obvious what might be a finite reachability graph of the optimal simulation. The reason is that Proposition 2.2 does not hold for OPT. That is, even if $G,H \in Hist_{opt}$ satisfy

$state_G=state_H$ and $(H,H\circ[A]) \in OPT$

then it does not necessarily follow that $(G,G\circ[A]) \in OPT$. Take, for example, the FASA shown in Figure 3.1. Let $H=[\{b\}]$ and $G=[\{c\}]$. We have $([\lambda],G) \in OPT$, $([\lambda],H) \in OPT$, and $state_G=state_H$. On the other hand, $(H,H\circ[\{b\}]) \in OPT$ while $(G,G\circ[\{b\}]) \notin OPT$ since $([\{c\}],[\{b,c\}]) \notin CAN$. Hence the construction used to derive RG_{FULL} in which histories leading to the same state were assigned the same node of the graph would not work. In Section 5 we will show how to strengthen the condition $state_G=state_H$ in order to guarantee $(H,H\circ[A]) \in OPT \Leftrightarrow (G,G\circ[A]) \in OPT$.

4 A Characterisation of Steps in The Optimal Simulation

In this section we prove a technical result giving a characterisation of steps involved in optimal simulation which will be used in the subsequent construction of its reachability graph.

For every $a \in \Sigma$, let $dom(a)=\{s \in S \mid \delta_a(\Pi_{p(a)}(s))$ *is defined*$\}$.

Let $\nabla \subseteq CAN$ comprise all pairs of histories (G,H) such that there is a non-empty set $I \subseteq \{1,...,n\}$ satisfying the following three conditions (below $A \in Ind$ satisfies $H=G\circ[A]$).

(4.1) $p(A) \cap I=\varnothing$.

(4.2) There is $C \in enabled(G)$ such that $p(C) \subseteq I$.

(4.3) If $b \in \Sigma$ is such that $p(b) \cap I \neq \varnothing$ and $\Pi_{I \cap p(b)}(state_G) \in \Pi_{I \cap p(b)}(dom(b))$ then $p(b) \subseteq I$.

The conditions (4.1)-(4.3) are illustrated in Figure 4.1 using a Petri net representation. The idea behind ∇ is that for all continuations of H in CAN, the processes P_i, for all $i \in I$, are 'frozen'. Hence, by (4.2), no continuation of H in CAN can yield a step which is maximal. To prove this formally, we need an auxiliary lemma.

Lemma 4.1

Let $(G,H) \in \nabla$.

(1) $H \notin Hist_{maxend}$.

(2) If $(H,J) \in CAN$ then $(H,J) \in \nabla$.

Proof

(1) Follows from (4.1) and (4.2).

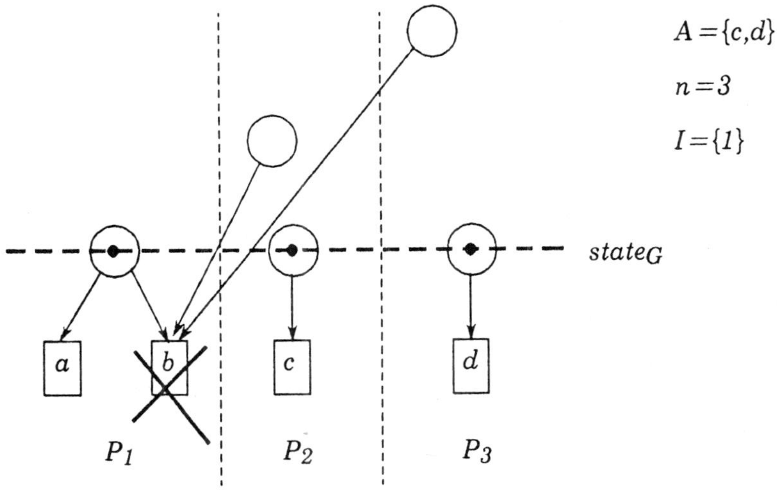

Figure 4.1: b is excluded by (4.3)

(2) Let $I \subseteq \{1,...,n\}$ and A be such that (4.1)-(4.3) hold for (G,H), and let B be such that $J = H \circ [B]$. We will show that I and B satisfy (4.1)-(4.3) for (H,J).

We first observe that from (4.1) it follows that

(4.4) $\quad \Pi_I(state_G) = \Pi_I(state_H)$.

Hence (4.2) and (4.3) hold for (H,J).

Suppose $b \in B$ and $p(b) \cap I \neq \emptyset$. From (4.4) and (4.3) it follows that $p(b) \subseteq I$. This, however, contradicts (4.1) and $(A,B) \in \zeta$. Hence (4.1) holds for (H,J). □

For every $H \in Hist$, let Γ_H be the set of all $A \in Ind$ such that $(H, H \circ [A]) \in CAN\text{-}\nabla$.

THEOREM 4.2

If $(H, H \circ [A]) \in OPT$ then $A \in \Gamma_H$.

PROOF

Suppose $A \notin \Gamma_H$, i.e. $(H, H \circ [A]) \in \nabla$. From Lemma 4.1 it follows that $H \circ [A] \notin Hist_{maxend}$ and, furthermore,there are no $A_1,...,A_l$ such that

$(H \circ [A], H \circ [A] \circ [A_1]) \in CAN$

$(H \circ [A] \circ [A_1], H \circ [A] \circ [A_1] \circ [A_2]) \in CAN$

...

$(H \circ [A] \circ [A_1] \circ ... \circ [A_{l-1}], H \circ [A] \circ [A_1] \circ [A_2] \circ ... \circ [A_{l-1}] \circ [A_l]) \in CAN$

and $H \circ [A] \circ [A_1] \circ [A_2] \circ ... \circ [A_{l-1}] \circ [A_l] \in Hist_{maxend}$.

Hence $(H, H \circ [A]) \notin OPT$, a contradiction. □

Consequently, in generating OPT instead of looking at $enabled(H)$ as the potential next steps for a history $H \in Hist_{opt}$, we can restrict ourselves to the (usually much smaller) set Γ_H.

It is not difficult to check whether $A \in enabled(H)$ belongs to Γ_H.

Suppose $can(H) = A_1...A_k$ and $A \in enabled(H)$.

We first check whether $(H, H \circ [A]) \in CAN$, by taking $L = p(A_k)$ and checking if $L \cap p(a) \neq \emptyset$ for all $a \in A$. All these can be done in $card(A_k) + card(A)$ steps.

To check whether $(H,H\circ[A])\in\nabla$ we proceed as follows.

$I:=\{1,\dots,n\}-p(A)$
WHILE there is $b\in\Sigma$ *such that* $p(b)\cap I\neq\varnothing$ *and* $p(b)-I\neq\varnothing$
  and $\Pi_{I\cap p(b)}(state_G)\in\Pi_{I\cap p(b)}(dom(b))$ *DO*
 BEGIN
  $I:=I-p(b)$
 END

One can show that $(H,H\circ[A])\in\nabla$ if and only if the final I satisfies (4.1)-(4.3). The whole process takes no more than n steps as the sequence of the I's generated during the execution of the loop is a strictly decreasing one.

5 Reachability Graph of The Optimal Simulation

The set Γ_H can be used to identify those histories involved in the optimal simulation which have identical continuations in OPT.
Let $\sim\subseteq Hist\times Hist$ be defined by $G\sim H :\Leftrightarrow (state_G=state_H \wedge \Gamma_G=\Gamma_H)$.

THEOREM 5.1
If $G,H\in Hist_{opt}$ and $G\sim H$ then for all $A\in Ind$ the following hold.
(1) $(G,G\circ[A])\in OPT \Leftrightarrow (H,H\circ[A])\in OPT$.
(2) If $(G,G\circ[A])\in OPT$ then $G\circ[A]\sim H\circ[A]$.

PROOF
(1) Suppose $(G,G\circ[A])\in OPT$. By Theorem 4.2, $A\in\Gamma_G$. Hence, by $\Gamma_G=\Gamma_H$, $A\in\Gamma_H$.
Thus $(H,H\circ[A])\in CAN$.
From $(G,G\circ[A])\in OPT$ it follows that $G\circ[A]\in Hist_{maxend}$, or there are $A_1,\dots,A_l$ such that
$(G\circ[A],G\circ[A]\circ[A_1])\in CAN$
$(G\circ[A]\circ[A_1],G\circ[A]\circ[A_1]\circ[A_2])\in CAN$
...
$(G\circ[A]\circ[A_1]\circ\dots\circ[A_{l-1}],G\circ[A]\circ[A_1]\circ[A_2]\circ\dots\circ[A_{l-1}]\circ[A_l])\in CAN$
and $G\circ[A]\circ[A_1]\circ[A_2]\circ\dots\circ[A_{l-1}]\circ[A_l]\in Hist_{maxend}$.
Thus from $(H,H\circ[A])\in CAN$ it follows that
$(H\circ[A],H\circ[A]\circ[A_1])\in CAN$
$(H\circ[A]\circ[A_1],H\circ[A]\circ[A_1]\circ[A_2])\in CAN$
...
$(H\circ[A]\circ[A_1]\circ\dots\circ[A_{l-1}],H\circ[A]\circ[A_1]\circ[A_2]\circ\dots\circ[A_{l-1}]\circ[A_l])\in CAN$
and $H\circ[A]\circ[A_1]\circ[A_2]\circ\dots\circ[A_{l-1}]\circ[A_l]\in Hist_{maxend}$.
Hence $(H,H\circ[A])\in OPT$.

(2) By (1), $H\circ[A]\in Hist$. Clearly, $state_{G\circ[A]}=state_{H\circ[A]}$. Furthermore, $\Gamma_{G\circ[A]}=\Gamma_{H\circ[A]}$ follows from $can(G\circ[A])=can(G)\hat{}A$ and $can(H\circ[A])=can(H)\hat{}A$. □

We now may define a reachability graph of the optimal semantics, RG_{OPT}, in the following way.
(1) $V=\{(state_H,\Gamma_H)\mid H\in Hist_{opt}\}$.

(2) $Arcs = \{((state_H,\Gamma_H),A,(state_{H\circ[A]},\Gamma_{H\circ[A]})) \mid (H,H\circ[A])\in OPT\}$.

(3) $v_{init} = (s_o,\Gamma_{[\lambda]})$.

THEOREM 5.2

RG_{OPT} is a finite reachability graph of the optimal simulation.

PROOF

$Rel(RG_{OPT}) = OPT$ follows from Theorem 5.1.

The finiteness of RG_{OPT} follows from the finiteness of the FASA. □

Generating reachability graph RG_{FULL} is usually done in a loop which checks the already generated nodes and steps 'enabled' at those nodes (nodes are labelled with states). If there exists a node labelled $state_H$ and a step A which have not yet been tried, the algorithm generates a global state $s = state_{H\circ[A]}$. It then adds a new node labelled s and an arc to the graph if s has not yet been a label; otherwise it draws an arc to the node labelled by state s.
Our algorithm generating RG_{OPT} will follow in principle the same pattern. There is, however, one essential difference. An arc cannot be accepted as belonging to RG_{OPT} before another arc, labelled with a maximal step, is found which can be reached from that arc. Consequently, we first generate an auxiliary graph G_{aux} and then prune those arcs from which an arc labelled with a maximal step cannot be reached, obtaining a graph isomorphic to RG_{OPT}.

We first show how to generate the auxiliary graph G_{aux}.
Let $G_{aux} = (V,Arcs)$, where $V \subseteq Steps_{can} \times 2^{Ind}$ and $Arcs \subseteq V \times Ind \times V$, be a directed graph generated in the following way (below for every $v = (\sigma,\psi) \in V$ we denote $\sigma(v) = \sigma$ and $\psi(v) = \psi$).

```
V := {(λ,Γ[λ])};
Arcs := ∅;
WHILE not all v∈V have ψ(v) empty DO
    BEGIN
        Find any v∈V and A∈ψ(v)
        IF there is w∈V such that [σ(w)]~[σ(v)^A]
            THEN Arcs := Arcs∪{(v,A,w)}
            ELSE
                BEGIN
                    w := (σ(v)^A, Γ[σ(v)^A])
                    V := V∪{w}
                    Arcs := Arcs∪{(v,A,w)}
                END
        ψ(v) := ψ(v)-{A}
    END
```

The resulting graph $G = (V,Arcs)$ contains a graph isomorphic to RG_{OPT}. To obtain it we proceed as follows.

Let $Arcs_o$ be the set of all $(v,A,w) \in Arcs$ such that there are

$(v_1,A_1,v_2),(v_2,A_2,v_3),\ldots,(v_l,A_l,v_{l+1}),\ldots,(v_{l+k},A_{l+k},v_{l+k+1})$

satisfying the following

(1) $v_1 = (\lambda, \Gamma_{[\lambda]})$

(2) $k \geq 1$

(3) $(v_l, A_l, v_{l+1}) = (v, A, w)$

(4) $A_{l+k} \in maxenabled([A_1 \ldots A_{l+k-1}])$.

Then $RG_{opt} = (V_o, Arcs_o, v_{init})$, where $v_{init} = (\lambda, \varnothing)$ and V_o is the set comprising v_{init} and all the end-points of the arcs in $Arcs_o$.

It is not difficult to see that RG_{opt} is isomorphic to RG_{OPT}. Hence $Rel(RG_{opt}) = OPT$.

To illustrate the process of generating G_{aux} we take the FASA of Figure 3.1. In Figure 5.1 we show a possible sequence of steps of the algorithm generating G_{aux}, where $A = \{a\}$, $B = \{b\}$, $C = \{c\}$ and $D = \{b,c\}$. We note that in this case $G_{aux} = RG_{opt}$.

As another example we consider the FASA M_k $(k \geq 2)$ of Figure 5.2(a).The graph G_{aux} is shown in Figure 5.2(a), where $A = \{a\}$, $B_i = \{b_i\}$, $C_i = \{c_i\}$ and $D_i = \{b_i, c_i\}$. Also in this case $G_{aux} = RG_{opt}$, so no unnecessary arcs and nodes are generated. The reachability graph of the optimal simulation for M_k has $3k$ vertices and $3k+5$ arcs which compares favourably with the reachability graph of RG_{FULL} with k^2 vertices and $3k^2+1$ arcs .

Concluding Remarks

In this paper we outlined the idea behind an algorithm generating reachability graph of optimal simulation - a way of executing a system directly generalising the maximally concurrent execution [JK89]. Together with results obtained in [JK89] and [JK89a], this gives a strong indication that the graph RG_{OPT} would in general case be much smaller than the reachability graph of the full execution mode, RG_{FULL}. Furthermore, as our experience has shown, the greater the degree of concurrency the system exhibits, the more can be gained by switching from RG_{FULL} to RG_{OPT}. There are, however, situations where RG_{OPT} is bigger than RG_{FULL}. For example, the reachability graph RG_{OPT} generated in Figure 5.1 is bigger than RG_{FULL}. However, the situation is less paradoxical than it might look at the first glance. Let MAX be the maximally concurrent simulation [JLKD86,JK89], and let RG_{MAX} be its reachability graph. (MAX is the maximal subset of CAN such that $(G,H) \in MAX$ implies $G,H \in Hist_{maxend}$.) It is not difficult to see that $RG_{MAX} \subseteq RG_{FULL}$ and $RG_{MAX} \subseteq RG_{OPT}$, as well as $MAX \subseteq OPT \subseteq FULL$. Furthermore, the relation OPT is a minimal subset of $FULL$ containing the same behavioural information as $FULL$ and, intuitively, OPT is only 'slightly' bigger than MAX. As the result, the reachability graph of OPT, RG_{OPT}, is only 'slightly' bigger than the reachability graph of MAX, RG_{MAX}. On the other hand, $RG_{MAX} \subseteq RG_{FULL}$, and the difference between RG_{MAX} and RG_{FULL} strongly depends on the degree of concurrency exhibited by the system. If the system contains only a few concurrent events (as that of Figure 3.1) then the difference between RG_{MAX} and RG_{FULL} is rather small, and in such a case RG_{OPT} might be bigger than RG_{FULL}. When the FASA contains many concurrent events the difference between RG_{MAX} and RG_{FULL} increases dramatically, while *RG_{OPT} is still only 'slightly' bigger than* RG_{MAX}. If we make the FASA of Figure 3.1 'more concurrent' by replacing events b and c by two 'event cycles', $b_1, b_2, \ldots, b_k$ and $c_1, c_2, \ldots, c_k$ (similar to those in Figure 5.2(a)) the situation will change radically as RG_{OPT} will have $o(k)$ vertices and arcs, whereas RG_{FULL} will have $o(k^2)$ vertices and arcs. Another example may be given by taking the FASA's from Figure 1.2 and 1.3. The latter has

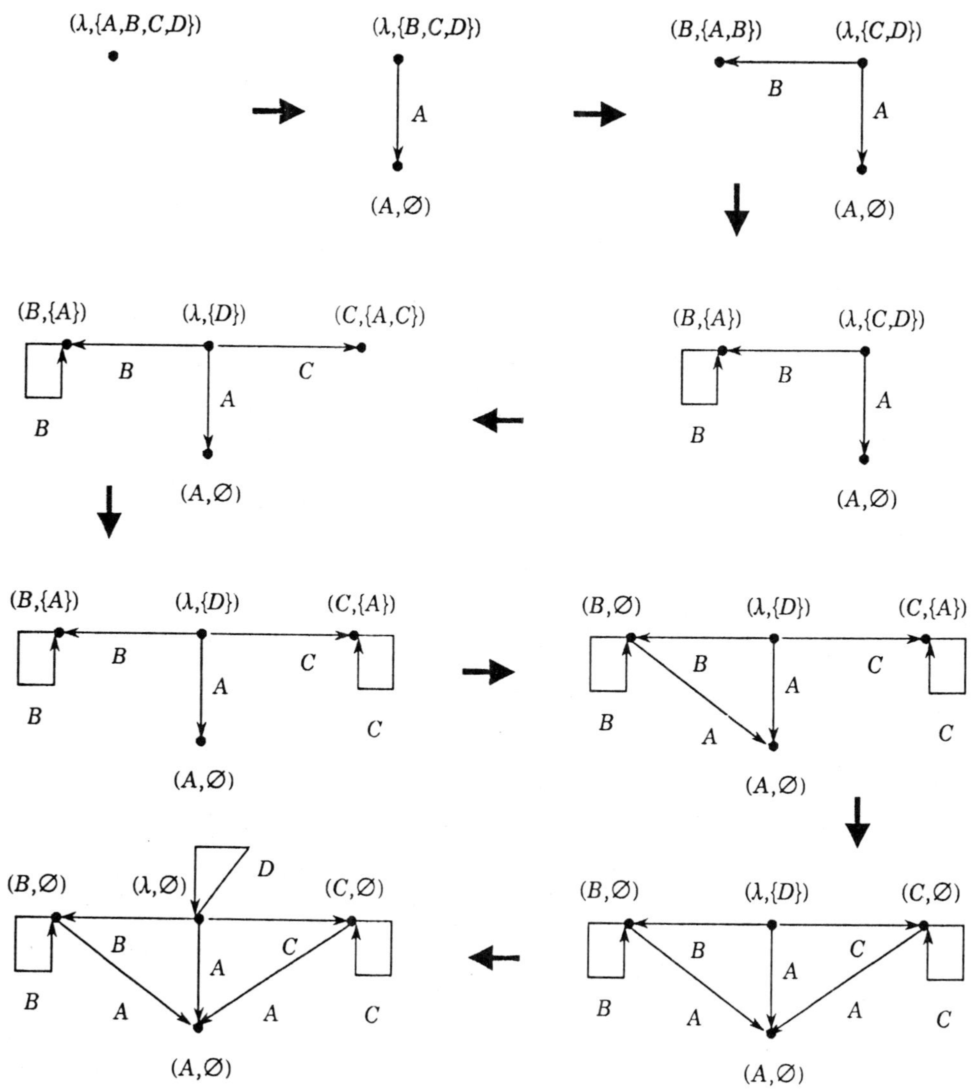

Figure 5.1: Generating graph G_{aux}

been obtained from the former by adding two simple concurrent subsystems. This had no effect on the size of the reachability graph of the optimal simulation, while the size of the full reachability graph has increased significantly. Note also that if RG_{FULL} is loop-free (i.e. $FULL$ is finite) then we always have $RG_{OPT} \subseteq RG_{FULL}$.

A somewhat different kind of problem can be encountered by taking the FASA of Figure 5.2(a). If we replace event cycle $b_1,\dots,b_k$ by $b_1,\dots,b_{k-1}$ then although RG_{OPT} for the modified FASA will be smaller than RG_{FULL}, it will have $o(k^2)$ nodes and arcs. This problem is essentially caused by the generality of OPT. The optimal simulation has been defined in a pure event-based setting. In particular, the lack of any reference to the states of a concurrent system makes OPT applicable to almost all models for non-sequential computation, but sometimes it may lead to less efficient solution as far as the reachability graphs are concerned. Referring to the modified FASA of Figure 5.2(a), by taking into account the particular structure of the states for this FASA, we may further reduce RG_{OPT} to a graph which has $o(k)$ nodes and arcs and is still equivalent to the full reachability graph. What this clearly demonstrates is that for specific system models, e.g. for those which support the notion of state, it is possible to modify RG_{OPT} in a way which takes advantage of some particular properties of that model. RG_{OPT} should not therefore be regarded as a complete blueprint for an efficient reduction of the reachability graphs, but in some cases as particularly suitable starting point for developing algorithms for such a reduction.

$M=(P_1,P_2,\Delta,F)$, where

$P_1=(\{a,b_1,\ldots,b_k\},S_1,s_{o1})$ and $P_2=(\{a,c_1,\ldots,c_k\},S_2,s_{o2})$

$S_1=\{s_{o1},s_{11},\ldots,s_{(k-1)1}\}$ $S_2=\{s_{o2},s_{12},\ldots,s_{(k-1)2}\}$

$\delta_a(s_{o1},s_{o2})=(s_{o1},s_{o2})$

$\delta_{b_k}(s_{(k-1)1})=s_{o1}$ and $\delta_{b_i}(s_{(i-1)1})=s_{i1}$ for $i=1,\ldots,k-1$

$\delta_{c_k}(s_{(k-1)2})=s_{o2}$ and $\delta_{c_i}(s_{(i-1)2})=s_{i2}$ for $i=1,\ldots,k-1$

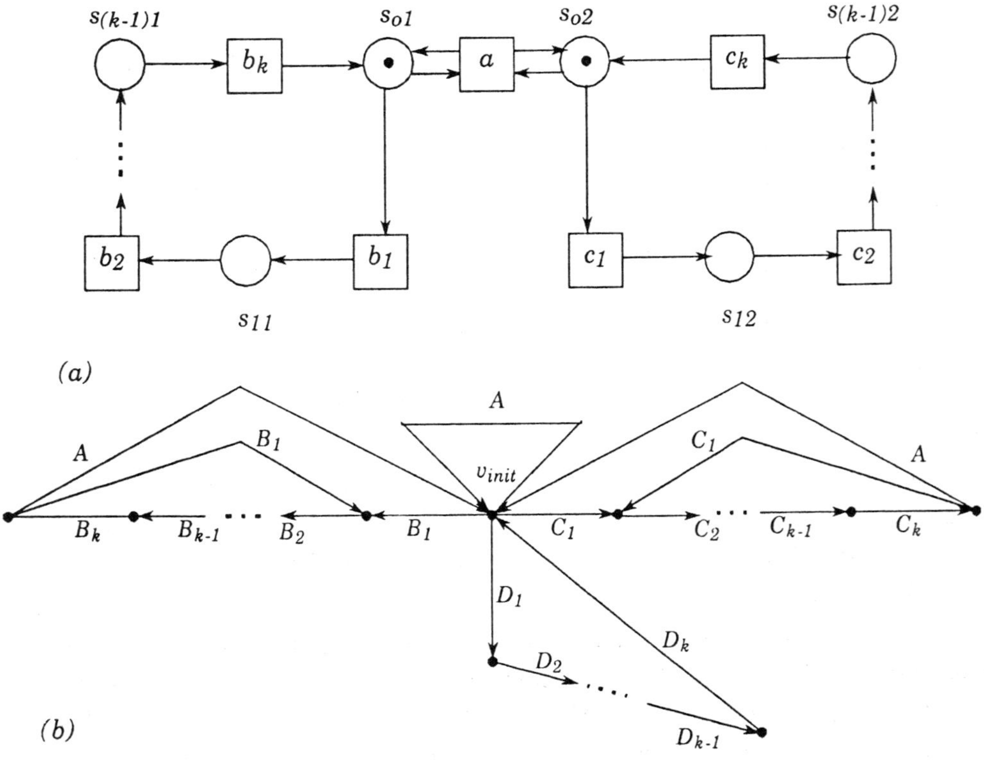

Figure 5.2: Reachability graph of the optimal simulation

Our final comment is that the fusion of our approach with one of the approaches which deal explicitly with reachability graphs, as those of [Val89] and [God90], is likely to lead to highly efficient algorithms for reduced reachability graph generation.

Acknowledgment

We would like to thank Antti Valmari for his comments on the modified FASA of Figure 5.2.

References

[Bro86] Browne M.C., *An Improved Algorithm for the Automatic Verification of Finite State Systems using Temporal Logic*, Proc. Conf. Logic in Computer Science, Cambridge, Mass., 1986, 260-267.

[CF69] Cartier P., Foata D., *Problemes combinatoires de communication et rearrangements*, Lecture Notes in Mathematics 85, Springer 1969.

[CG87] Clarke E.M., Grümberg O., *Research on Automatic Verification of Finite-State Concurrent Systems*, Ann. Rev. Comp. Sci. 2(1987), 269-290.

[CES86] Clarke E.M., Emerson E.A., Sistla A.P., *Automatic Verification of Finite-State Systems using Temporal Logic Specifications*, ACM Transactions on Programming Languages and Systems 8(1986), 244-263

[God90] Godefroid P., *Using Partial Orders to Improve Automatic Verification Methods*, Proc. of CAV'90, this volume.

[Hoa85] Hoare C.A.R., *Communicating Sequential Processes*, Prentice-Hall, 1985.

[JLKD86] Janicki R., Lauer P.E., Koutny M., Devillers R., *Concurrent and Maximally Concurrent Evolution of Non-Sequential Systems*, Theoretical Computer Science 43(1986), 213-238.

[JK87] Janicki R., Koutny M., *On Equivalent Execution Semantics of Concurrent Systems*, Lecture Notes in Computer Science 266, Springer 1987, 89-103.

[JK89] Janicki R., Koutny M., *Towards a Theory of Simulation for Verification of Concurrent Systems*, Lecture Notes in Computer Science 366, Springer 1989, 73-88.

[JK89a] Janicki R., Koutny M., *Optimal Simulation for Verification of Concurrent Systems*, Technical Report No. 89-05, McMaster University,Hamilton, Ontario, 1989.

[JL90] Janicki R., Lauer P. E., *Specification and Analysis of Concurrent Systems; The COSY Approach*, Springer, to appear.

[Jen87] Jensen K., *Coloured Petri Nets*, LNCS 254, Springer 1987, pp. 248-299.

[Kel76] Keller R.M., *Formal Verification of Concurrent Programs*, CACM 19(7), 1976, 371-384.

[LSC81] Lauer P.E., Shields M.W., Cotronis J.Y., *Formal Behavioural Specification of Concurrent Systems without Globality Assumptions*, Lecture Notes in Computer Science 107, Springer 1981, 115-151.

[MS82] Martinez J., Silva M., *A Simple and Fast Algorithm to Obtain All Invariants of a Generalized Petri Net*, Informatik-Fachberichte 52, Springer 1982, 301-310.

[Maz86] Mazurkiewicz A., *Trace Theory*, Lecture Notes in Computer Science 255, Springer 1986, 297-324.

[Mil80] Milner R., *A Calculus of Communicating Systems*, Lecture Notes in Computer Science 92, Springer 1980.

[MR87] Morgan E.T, Razouk R.R., *Interactive State-Space Analysis of Concurrent Systems*, IEEE Transactions on Software Engineering 13(10), 1987.

[Pet81] Peterson J.L., *Petri Net Theory and the Modeling of Systems*, Prentice Hall, 1981.

[Rei85] Reisig W., *Petri Nets*, Springer 1985.

[SM81] Salwicki A., Müldner T., *On Algorithmic Properties of Concurrent Programs*, Lecture Notes in Computer Science 125, Springer 1981, 169-197.

[Szp30] Szpilrajn-Marczewski E., *Sur l'extension de l'ordre partial*, Fundamenta Mathematicae 16 (1930), pp. 386-389.

[Val89] Valmari A., *Stubborn Sets for Reduced State Space Generation*, Proceedings of the 10th International Conference on Application and Theory of Petri Nets, Bonn, June 1989.

[Zie87] Zielonka W., *Notes on Finite Asynchronous Automata*, Informatique Thèorique et Applications 21(2), 1987, pp. 99-135.

[Zie89] Zielonka W., *Safe Executions of Recognizable Trace Languages by Asynchronous Automata*, Lecture Notes in Computer Science 363, Springer 1989.

DIMACS Series in Discrete Mathematics
and Theoretical Computer Science
Volume 3, 1991

TASK-DRIVEN SUPERVISORY CONTROL OF DISCRETE EVENT SYSTEMS

C.H. Golaszewski

R.P. Kurshan

AT&T Bell Laboratories, Murray Hill, NJ 07974

September 1990

Abstract

We extend the supervisory control framework proposed by Ramadge and Wonham to allow the synthesis of supervisors which control a given system to perform an arbitrary ω-regular task specified by a nondeterministic Büchi automaton. A typical application is provided by a communication system with unreliable and unobservable channels, where a sender's message is required to arrive eventually at a receiver. This paper is motivated by the lack of success in applying discrete event system models to the synthesis of communication protocols. We demonstrate the usefulness of our approach with the synthesis of a simple, centralized communication protocol with minimal assumptions on the underlying system; in particular, we do not assume that any feature of the protocol, such as sequence numbers, is already incorporated in the underlying model.

1. Introduction

One of the goals of supervisory control for discrete event systems is the synthesis of supervisors which guarantee that the closed-loop (or supervised) system performs certain prescribed tasks. In this paper a task is assumed to be specified by an ω-regular language defined by a Büchi automaton. The given discrete event system is not constrained to be completely observable, i.e., it may be required to synthesize a supervisor for the system which cannot observe all system events and which may not possess complete state information about the system.

To motivate our study, consider a communication system consisting of a sender A, a receiver B and an unreliable channel C which nondeterministically loses packets. A standard problem is to design a supervisor (communication protocol) that ensures that a packet sent by A is eventually received by B, under the assumption that the channel eventually delivers some message which was transmitted by A. The channel can delay packets arbitrarily long, the loss of a message is unobservable and all components operate asynchronously. Thus, based on the past observations of the system trajectory it cannot be decided whether a packet which was not received was actually lost or will be received at some point in the future. Consequently, a supervisor which is to guarantee

1052-1798/91 $1.00 + $.25 per page

packet delivery must at some point cause retransmission of that message, irrespective of the event of packet loss. The timing of this retransmission is not critical, and for efficiency may be adjusted with consideration of system component delay. Therefore, it is natural to model the supervisory event which causes retransmission as a nondeterministic control action. However, in order for the supervisor to fulfill its required task, it will often be required satisfy certain obligations eventually. These considerations lead in a natural way to the definition of the supervisor as a nondeterministic automaton with acceptance conditions for infinite sequences.

The purpose of this paper is to establish a supervisory control framework to study supervisor synthesis for communication protocols and similar systems. Although many different models have been proposed for the verification of communication protocols, the problem of synthesis has not been addressed adequately. (For a survey of various models see [3]). All of these models assume that the communication protocol to be analyzed is already specified. Presently, the question of how to "conceive" the protocol in the first place has no satisfactory answer. Although our results do not completely resolve this problem, we hope that our research provides a basis for a feasible synthesis procedure.

Models for analysis and synthesis of communication protocols need not be dissimilar. For example, Petri nets, finite state automata and Hoare's paradigm of communicating sequential processes ([5]) have all been used in the supervisory control of discrete event systems. The model we propose is based on the framework by Ramadge and Wonham which is essentially state machine based (RW-Model, [9]). However, our formulation of the supervisor synthesis problem is distinct enough so that we can address control problems which cannot be solved in the standard RW-Model.

The major new contribution of this paper is a synthesis procedure for Büchi supervisors. These can be understood as supervisors that for certain states of the system may choose nondeterministically between several control actions. The choice is not forever arbitrary, however, as we require that eventually some particular sequence of control actions is selected; this ultimate occurrence of sequences of control actions is characterized by automata acceptance conditions. Allowing nondeterminism in the supervisor permits to defer certain timing and sequencing decisions to a lower level of system design, while at the same time providing "place holders" for them during the high-level supervisor synthesis. It was already illustrated with the previous example that such an extension is needed for the synthesis of communication protocols. Lacking this feature, the RW-Model cannot be applied directly to the synthesis of, say, communication protocols as they are conventionally modeled. For example, Cieslak et al. had to include the structure of the alternating bit protocol in the model of the open-loop system in order to derive their solution in [2]. Clearly, a general synthesis procedure should not require much knowledge about the solution. This suggests that the basic RW-Model lacks certain features that are necessary to synthesize a supervisor for this class of systems.

The present investigation is by no means the first attempt to extend the RW-Model so that it can be applied to a larger class of supervisory control problems. For example, recently it was realized by Thistle and Wonham, and Ramadge that one important aspect absent from the basic model is the notion of periodic and asymptotic behaviors of discrete event systems. This led to proposals for extensions of the RW-Model to include infinite sequences ([10,12]). These approaches are very similar in spirit in that they both extend the notion of string controllability of the basic model to the prefix set of (infinite) sequences. The extended model was used to pose and solve several interesting supervisory control problems ([4,10]). However, these extensions do not address the problem of achieving a desired sequential closed-loop behavior which is neither realizable ([12]) nor topologically closed ([10]). Simply stated, this framework applies only to the case of transition structures which are structurally incapable of generating infinite sequences incompatible with the given task. We show in this paper that this is too restrictive in the context of communication protocols. Instead, by defining the supervisor as a nondeterministic L-process, we may place constraints on the eventuality of the closed-loop behavior to exclude arbitrarily long but finite periods of behavior, which, if continued forever, would constitute a failure of the system.

In contrast to the work of Ramadge and Wonham we focus our attention on the supervisor synthesis for arbitrary prescribed sequential behaviors. To this end, we model the sequential behavior of a discrete event process by an L-process G. The control objective is to design a supervisor C for G such that the closed loop system $G \otimes C$ performs a prescribed ω-regular task which is specified by a Büchi automaton. We give conditions for the existence of Büchi supervisors for fully and partially observable systems. In a sequel paper ([7]) we will include results on distributed control and discuss the complexity of synthesis algorithms.

Some of our research overlaps with work done by Brave et al. in [1] and Özveren et al. in [8]; our independently derived results differ from the work of both of these groups, though. Their research is focused on the synthesis of deterministic supervisors which guarantee that the closed-loop system visits a given subset of states infinitely often. In contrast, we use a formal framework for the infinite behavior of discrete event systems, which includes Büchi supervisors and L-processes, to pose and solve supervisory synthesis problems for ω-regular tasks. We hope that one of the eventual benefits of this investigation will be an *a posteriori* explanation of sequence numbers for messages in systems with unreliable and partially observable communication channels by rigorous mathematics.

The organization of the paper is as follows. In Section 2 the discrete event system model is reviewed. Section 3 introduces the control-theoretic framework used in the paper. Section 4 discusses the existence and synthesis of supervisors for fully observable systems. In Section 5 these results are extended to the nondeterministic case, and a mathematical abstraction of sequence numbers is proposed. Finally, in Section 6, the ideas of the paper are applied to an example of a simple communication protocol.

2. L-Processes

In this paper, the logical behavior of discrete event systems is modeled by *L-processes.* The reason is that L-processes are a class of automata with a simple (polynomial) product and the property that the language of the product system is the intersection of the languages of the subsystems. We will review only the basic definitions; for a detailed presentation we refer to [6].

2.1 Boolean Algebras

Let $L = \langle \mathbf{L}, +, *, \sim, 0, 1\rangle$ be a Boolean algebra over the set $\mathbf{L}$ with disjunction and disjunction denoted by $*$ and $+$, respectively, and complement $\sim$, multiplicative identity 1 and additive identity 0. For the purpose of this paper one may think of $\mathbf{L}$ as the power set of a finite set Σ (the *alphabet*), where $1 = \Sigma$, $0 = \emptyset$, $*$ is set intersection, $+$ is set union and $\sim$ is set complementation in Σ.

A Boolean algebra L admits a partial order $\leq$ defined by $x \leq y$ if and only if $x * y = x$. An *atom* of L is a minimal element with respect to this order.

Let $L_1, \ldots, L_k$ be subalgebras of L. Define their product

$$\prod_{i=1}^{k} L_i = \left\{ \sum_{j\in J} x_{1j} * \ldots * x_{kj} | x_{ij} \in L_i \text{ for } i = 1, \ldots, k, j \in J \text{ and } J \text{ is finite} \right\}$$

Note that $\prod L_i$ is a subalgebra of L. $L_1, \ldots, L_k$ are *independent* if

$$0 \neq x_i \in L_i \text{ for } i = 1, \ldots, k \Rightarrow x_1 * \ldots * x_k \neq 0.$$

Subalgebras will be used later to describe the concurrent operation of several discrete event processes.

2.2 L-Matrices and Graphs

Let V be a non-empty set. The map

$$M : V \times V \rightarrow L$$

is an *L-matrix* with state space $V(M)$; the elements of $V(M)$ are the *states* of M.

An L-matrix M is *deterministic* if for all $u, v, w \in V(M)$,

$$v \neq w \Rightarrow M(u,v) * M(u,w) = 0.$$

It is *complete* if for all $v \in V(M)$

$$\sum_{w \in V(M)} M(v,w) = 1.$$

With each L-matrix M we can associate a *graph* G_M with node set equal to $V(M)$. The edges of G_M are the elements $e \in V(M) \times V(M)$ such that $M(e) \neq 0$.

Let M and N be L-matrices over the same Boolean algebra L. Their *tensor product* $M \otimes N$ is defined by

$$(M \otimes N)((v,w),(v',w')) = M(v,v') * N(w,w').$$

Given an L-matrix M, a sequence $s \in L^\omega$ and a sequence of states $w \in V(M)^\omega$, we call w a *state trajectory* of s starting at q if $w_o = q$ and $s_i * M(w_i, w_{i+1}) \neq 0$ for all i.

Finally, for two sets G and H we denote the projection $G \times H \rightarrow G$ by π_G.

2.3 L-Automata

An *L-automaton* is a quadruple

$$\Gamma = (M_\Gamma, I(\Gamma), R(\Gamma), Z(\Gamma)),$$

where M_Γ, the *state transition matrix*, is a complete L-matrix, $\emptyset \neq I(\Gamma) \subseteq V(M_\Gamma)$ are the *initial states*, $U(\Gamma) \subseteq V(M_\Gamma)$ are the *recurring states* (Büchi states), and $Z(\Gamma) \subseteq 2^{V(M_\Gamma)}$ is the *cycle set*. (in [6], $U(\Gamma)$ is replaced by the set of recurring edges $R(\Gamma)$; often, this leads to a more compact representation.)

Denote the set of all sequences of atoms of L by $S(L)^\omega$. A sequence $s \in S(L)^\omega$ is Γ-*cyclic* if for some $N > 0$ and some $C \in Z(\Gamma)$, the corresponding state trajectory w satisfies $w_i \in C$ for all $i > N$; it is Γ-*recurring* if $w_i \in U(\Gamma)$ for infinitely many i. A sequence $s \in S(L)^\omega$ is a *tape* of Γ if it is Γ-cyclic or Γ-recurring. The *behavior* (or language) $\mathcal{L}(\Gamma)$ accepted by Γ is the set of tapes of Γ.

2.4 L-Processes

Let L be a Boolean algebra. An L-process G is a quintuple

$$G = (M_G, S_G, I(G), U(G), Z(G)),$$

where M_G, $I(G)$, $U(G)$, and $Z(G)$ are as in the definition of an L-automaton in the previous section. However, M_G is not required to be complete. Finally, S_G is the nondeterministic *selection function*

$$S_G : V(G) \rightarrow 2^L.$$

We restrict attention to the case where $S_G(q)$ is an atom for every $q \in V(G)$.

A selection $x \in S_G(q)$ enables the state transition $q \to p$ if $x * M_G(q,p) \neq 0$. We also require (see [6]) that

$$\sum_{p \in V(G)} M_G(q,p) \leq \sum_{x \in S_G(q)} x. \tag{2.4.1}$$

If equality holds in (2.4.1) for all $q \in V(G)$, and if $0 \notin S_G(q)$, then G is said to be *lockup-free*. (This is similar to the notion of nonblocking in [10].)

In contrast to the definition of a tape for an L-automaton, an element s of $S(L)^\omega$ is a tape for G if and only if it is not G-cyclic and not G-recurring. If $s \in S(L)^\omega$ is not a tape for G because it is G-cyclic or G-recurring, then s is said to be *excepted* from the language of G. The *behavior* (or language) $\mathcal{L}(G)$ generated by an L-process G is the set of all tapes of G.

2.5 Product Systems

The concurrent operation of several discrete event processes is modeled as follows: each individual component system is represented by an L_i-process G_i over a Boolean algebra L_i. The algebras L_i can be interpreted as subalgebras of a Boolean algebra L which is given by the product $L = \prod_i L_i$. This procedure associates with each L_i-process a corresponding L-process.

Given L-processes $G_1, \ldots, G_k$ define their *synchronous product* to be

$$G = \bigotimes_{i=1}^{k} G_i = (M_G, S_G, I(G), U(G), Z(G)) =$$

$$= \left(\bigotimes_{i=1}^{k} M_{G_i}, \prod_{i=1}^{k} S_{G_i}, I(G_1) \times \ldots \times I(G_k), \bigcup_{i=1}^{k} \pi_{G_i}^{-1} U(G_i), \bigcup_{i=1}^{k} \pi_{G_i}^{-1} Z(G_i) \right),$$

where

$$\left(\prod_{i=1}^{k} S_{G_i} \right) (q_1, \ldots, q_k) = \{ x_1 * \ldots * x_k | x_i \in S_{G_i}(q_i), i = 1, \ldots, k \}.$$

The model described so far applies to the concurrent operation of synchronous discrete event processes, i.e., state transitions occur simultaneously in the component processes. To model partially and completely asynchronous selections we introduce two additional selections called *wait* and *pause*, respectively. A process can select *pause* independently of the selections of the other processes and, for example, use *pause* to self-loop for an undetermined amount of "time." The *wait* selection is used to model the case where the possible state transitions of a process depend on the selections of other processes. Until this process can proceed it remains at the current state (in a self-loop) and selects *wait*.

2.6 Process Homomorphisms and Partial Observations

Consider two alphabets Σ_1 and Σ_2, and let

$$\phi: \Sigma_1 \to \Sigma_2$$

be an arbitrary map. We can extend ϕ to a *language homomorphism* $\Sigma_1^\omega \to \Sigma_2^\omega$ in the usual fashion by setting

$$\phi(s) = \phi(s_1)\phi(s_2)\ldots$$

for all $s \in \Sigma_1^\omega$.

Let G_1 be an L_1-process over the alphabet Σ_1 and G_2 be an L_2-process over the alphabet Σ_2. The pair $\Phi = (\varphi, \phi)$, with $\varphi: V(G_1) \to V(G_2)$ and ϕ a language homomorphism, is said to be a *process homomorphism* if

$$\sigma \in M_1(v, w) \Rightarrow \phi(\sigma) \in M_2(\varphi(v), \varphi(w)). \tag{2.6.1}$$

Now let G be an L-process, not necessarily deterministic. Define a binary relation Ψ on $V(G) \times V(G)$ according to

(1) for all $q \in V(G)$, $(q, q) \in \Psi$, and

(2) $(p, q) \in \Psi$ if there exists $(u, v) \in \Psi$ such that $\sigma \in M(u, p)$ and $\sigma \in M(v, q)$ for some $\sigma \in \Sigma$.

A pair of states $(p, q) \in \Psi$ is said to be *indistinguishable*. Note that in general Ψ is not an equivalence relation.

For two relations Ψ_1 and Ψ_2, a partial order is defined in the usual fashion:

$$\Psi_1 \leq \Psi_2 \quad \Leftrightarrow \quad ((p, q) \in \Psi_1 \Rightarrow (p, q) \in \Psi_2).$$

Partial or incomplete observations of the tapes of G are modeled by letting the language homomorphism ϕ be a projection onto a subset of the alphabet. In this case,

$$\ker(\phi) = \{\sigma \in \Sigma : \phi(\sigma) = 1\}$$

induces a binary relation on $V(G) \times V(G)$ via (2.6.1). The L-process that results from applying the corresponding process projection Φ to G is denoted by G_Φ.

3. Tasks and Supervisors

For a given L-process G with behavior $\mathcal{L}(G)$ we can specify its admissible behavior $\mathcal{L}$. In this paper $\mathcal{L}$ is assumed to be an ω-regular language. Thus a task can be represented by a deterministic L-automaton T. A standard verification problem is to decide if

$$\mathcal{L}(G) \subseteq \mathcal{L}.$$

Suppose this condition fails to hold, but there exists some L-process C such that

$$\mathcal{L}(G \otimes C) \subseteq \mathcal{L}. \tag{3.1}$$

In situations where G is independent from its environment this cannot happen. However, it is often the case that some (but not necessarily all) of the state transitions of G depend on shared or controllable variables which can be accessed by other processes. Thus, the language $L(G)$ can be restricted to a subset by external *control selections.*

We call the L-process C in (3.1), if it exists, a *Büchi supervisor* (or supervisor) for G with respect to the *task* $\mathcal{L}$. Note that we permit a supervisor to be an L-process with nondeterministic selection function. Hence, certain control selections (or sequences of control selections) can be required to occur eventually. For example, in the model of a communication protocol an eventual control action may represent the fact that the protocol will retransmit a window after a time-out has occurred.

As mentioned in the introduction this has the advantage of permitting the implementation of a larger class of sequential behaviors than possible in the RW-Model, namely those that fail to be topologically closed. For example, behaviors which are not closed arise in the modeling of time-outs which are a standard feature in many real-world discrete event systems. Thus, if time-outs are to be included in the discrete event model, then the corresponding behaviors are generally not topologically closed. Furthermore, if the duration of a time-out is assumed to be arbitrary, then the synthesis procedures based on the definitions of the RW-Model do not yield finite state supervisors. In contrast, time-outs can be easily modeled by Büchi supervisors.

Let $h: V(G) \to 2^{V(C)}$ be the map induced by $G \otimes C$, which for each tape in $\mathcal{L}(G \otimes C)$ assigns to the present state q of G the corresponding state(s) $h(q)$ of C. A state transition $M(q,p)$ of G is *disabled*, i.e., prevented from occurring, if

$$S_G(q) * S_C(h(q)) * M(q,p) = 0;$$

otherwise λ is *enabled* by C.

Define an equivalence relation Θ on $V(G) \times V(G)$ according to

$$\Theta = \left\{(q,p) : S_G(q) * S_G(p) * \lambda = 1 \Rightarrow S_C(q') * \lambda = S_C(p') * \lambda \text{ for all } q' \in h(q), p' \in h(p)\right\}$$

Let $B \subseteq V(G)$ be an arbitrary subset. B is *controlled-invariant* if there exists a supervisor C such that every state-trajectory of the process $G \otimes C$, once it enters B, remains indefinitely inside B.

Say that a state $q \in V(G)$ is B*-steerable*, if there exists a supervisor C such that for all $p \in V(G)$, $S_C(h(q)) * M_G(q,p) = 1$, implies that p is either B-steerable or an element of B.

3.1 Modular Supervisors

Complex tasks are usually the result of specifying many simple subtasks simultaneously. In this case it is advantageous to design supervisors for each of the individual tasks rather than trying to deal with the exponentially large state space of their product. It is natural to require that supervisors do not share control variables because this corresponds to the fact that the value of a control variable can be determined by one of the supervisors only.

Now consider two tasks T_1 and T_2 and suppose C_1 and C_2 are two supervisors with

$$\mathcal{L}(G \otimes C_1) = \mathcal{L}_1 \subseteq \mathcal{L}(T_1) \quad \text{and} \quad \mathcal{L}(G \otimes C_2) = \mathcal{L}_2 \subseteq \mathcal{L}(T_2).$$

Their *product* $C = C_1 \otimes C_2$ is just the usual product between L-processes, i.e.,

$$\mathcal{L}(G \otimes C_1 \otimes C_2) = \mathcal{L}_1 \cap \mathcal{L}_2.$$

The problem with the product of supervisors is that the resulting L-process need not be lockup-free even though $G \otimes C_1$ and $G \otimes C_2$ are. A sufficient condition for $G \otimes C_1 \otimes C_2$ to be lockup-free is that G, C_1 and C_2 are independent L-processes, i.e., there exist independent subalgebras L_G, L_1 and L_2 such that (Proposition 4.9 in [6])

$$\begin{aligned} S_G &= \bigcup_{q \in V(G)} S_G(q) \subseteq L_G, \\ S_{C_1} &= \bigcup_{q \in V(C_1)} S_{C_1}(q) \subseteq L_1 \quad \text{and} \\ S_{C_2} &= \bigcup_{q \in V(C_2)} S_{C_2}(q) \subseteq L_2 \end{aligned}$$

Hence, if we prohibit supervisors to share control variables, then they must be independent.

Now suppose we are given an L-process G, a task T and a supervisor C satisfying

$$\mathcal{L}(G \otimes C) \subseteq \mathcal{L}(T),$$

i.e., G is controllable relative to T. Furthermore suppose that for some reason (e.g. distributed implementation) we desire to partition T into subtasks $T_1, \ldots, T_n$ with

$$\mathcal{L}(T) = \mathcal{L}(T_1) \cap \ldots \cap \mathcal{L}(T_n).$$

Since there is a supervisor for T there also exists a supervisor C_i which implements task T_i. However, if there are two tasks T_i and T_j such that any pair of supervisors C_i and C_j, which implement the tasks, is dependent, then it is impossible to realize T through the given subtasks $T_1, \ldots, T_n$.

In certain cases it is possible to eliminate the dependency between supervisors by introducing new variables. For example, let $x \in S_{C_1} \cap S_{C_2}$ be a "shared" control variable and introduce two variables $x_1 \notin S_G \cup S_{C_2}$ and $x_2 \notin S_G \cup S_{C_1}$ (possibly by extending $\mathbf{L}$) and set $x = f(x_1, x_2)$ where f is an appropriately chosen Boolean function. At the present time it is not clear when such an f exists; this question is currently under investigation.

4. Existence of Supervisors: Deterministic Case

Consider an L-process G with deterministic resolution and a task T. Our goal is to derive necessary and sufficient conditions for the existence of a lockup-free supervisor C such that $\mathcal{L}(G \otimes C) \subseteq \mathcal{L}(T)$. Intuitively, such a supervisor C exists if all state trajectories which originate in $I(G)$ and reach a void cycle of uncontrollable events, can be disabled by appropriately selected control actions.

Recall that any tape of $\mathcal{L}(G)$ has to be T-cyclic or T-recurring to be a tape for $\mathcal{L}(T)$. Since the two acceptance conditions are independent of each other, they are investigated separately in the next two sections.

4.1 Largest Invariant Recurrent Set

The case of recurrent sets is similar to the problem considered in [1] and [8] in the context of the RW-Model for supervisory control.

The following algorithm computes the largest invariant recurrent set via a fixpoint iteration:

Algorithm 1:

$K_1 := \pi_G(V(G \otimes T))$;
$H_1 := \pi_G(U(G \otimes T))$;
REPEAT
  $K_{i+1} := \{q \in K_i : q \text{ is } H_i - \text{steerable}\}$;
  $H_{i+1} := H_i \cap K_{i+1}$;
UNTIL $H_{i+1} = H_i$;

Denote the (unique) fixpoint of the algorithm by

$$H(G) = \lim_{i \to \infty} K_i.$$

Since we consider only finite state systems it is clear that the algorithm converges in a finite number of steps.

Proposition 4.1. *The set $H(G)$ is controlled invariant.*

Proof. Every state in $H(G)$ is steerable to some other state in $H(G)$. ◇

Proposition 4.2. *Suppose $I(G) \subseteq H(G)$. Then there exists a lockup-free supervisor C such that $\mathcal{L}(G \otimes C) \subseteq \mathcal{L}(T)$.*

Proof. Follows from control invariance and the fact that the states of every cycle which does not intersect $U(G \otimes T)$ are $U(G \otimes T) \cap H(G)$-steerable. ◇

Proposition 4.3. *Suppose $Z(G \otimes T) = \emptyset$. Then there exists a lockup-free supervisor C such that $\mathcal{L}(G \otimes C) \subseteq \mathcal{L}(T)$ if and only if $I(G) \subseteq H(G)$.*

Proof. (if.) This is the statement of Proposition 4.2.

(only if.) Suppose there is a supervisor such that $\mathcal{L}(G \otimes C) \subseteq \mathcal{L}(T)$. Since $Z(G \otimes T) = \emptyset$ this implies that C can control G in such a way that every state trajectory intersects the set $U(G \otimes T)$ infinitely often, irrespective of the initial state. But now it follows from the construction of the set $H(G)$ that $I(G) \subseteq H(G)$. ◇

The last proposition implies that, in the absence of cycle sets, Algorithm 1 produces the largest set of initial states for which a lockup-free supervisor with the desired properties exists.

4.2 Cycle Sets

We proceed in a similar fashion to derive conditions on the set of initial states with respect to the cycle set.

Algorithm 2:

FOR ALL $K \in Z(G \otimes T)$ DO
  $A(K) :=$ largest controlled invariant set contained in $\pi_G(K)$;
  $H(K) := \{q \in V(G) : q \text{ is } A(K) - \text{steerable }\}$;

The computation of the sets $A(K)$ can be accomplished by a variant of Algorithm 1:

$V_1 := \pi_G(K)$;
$H_1 := \pi_G(K)$;
REPEAT
  $V_{i+1} := \{q \in V_i : q \text{ is } H_i - \text{steerable }\}$;
  $H_{i+1} := H_i \cap V_{i+1}$;
UNTIL $H_{i+1} = H_i$;

For each $K \in Z(G \otimes T)$ set

$$A(K) = \lim_{i \to \infty} V_i,$$

i.e., $A(K)$ is the fixpoint of the above procedure.

The following two propositions summarize the properties of the sets $H(K)$:

Proposition 4.4. *Suppose $I(G) \subseteq \bigcup_{K \in Z(G \otimes T)} H(K)$. Then there exists a lockup-free supervisor C such that $\mathcal{L}(G \otimes C) \subseteq \mathcal{L}(T)$.*

Proof. We can assume without loss that there is only one set $H(K)$. Consider the set $D = H(K) - A(K)$. Since each state in D is $A(K)$-steerable it follows that the states on every cycle in D not contained in $Z(G \otimes T)$ are $A(K)$-steerable. This implies that the required supervisor C exists. ◇

Proposition 4.5. *Suppose $U(G \otimes T) = \emptyset$. Then there exists a lockup-free supervisor C such that $\mathcal{L}(G \otimes C) \subseteq \mathcal{L}(T)$ if and only if $I(G) \subseteq \bigcup_{K \in Z(G \otimes T)} H(K)$.*

Proof. (if.) This is the statement of Proposition 4.4.

(only if.) The argument is identical to the one used for the proof of Proposition 4.3. ◇

4.3 Necessary and Sufficient Conditions

The sufficient conditions of the two previous sections can now be combined to yield necessary and sufficient conditions for the existence of a lockup-free supervisor. Let G be an L-process and T be a task. With the notation from above define

$$B = \bigcup_{K \in Z(G \otimes T)} H(K) \cup H(G).$$

Theorem 4.1. *There exists a lockup-free supervisor C such that $\mathcal{L}(G \otimes C) \subseteq \mathcal{L}(T)$ if and only if $I(G) \subseteq B$.*

Proof.

(if). Follows from the previous propositions.

(only if). Suppose $I(G) \not\subseteq B$. Then there exists a sequence $s \in S(G)^\omega$ starting at an initial state in $I(G) - B$ which, independent of any control actions chosen, eventually reaches a cycle which does not intersect $U(G \otimes T)$ and is not contained in $Z(G \otimes T)$. Hence, there cannot exist a supervisor C with the desired property. ◊

Note that the necessary and sufficient conditions of the theorem can be easily checked by an efficient algorithm.

4.4 Supervisory Synthesis

Based on the above theorem the synthesis of a supervisor C for the task T is straightforward. We already presented algorithms (Algorithm 1, Algorithm 2) which compute the sets $H(K)$ for $K \in Z(G \otimes T)$, and $H(G)$. There is a standard procedure in supervisory control to synthesize a lockup-free controller for a controlled invariant set (see [12]). So, for each of the above sets we can compute a supervisor, say C_i. If we define the supervisor

$$C' = \bigoplus_i C_i,$$

then C' forces G to remain inside the set of "good" states B, provided $I(G) \subseteq B$. Now extend C' to a supervisor C by excepting all states which correspond to void cycles, i.e., C eventually "interrupts" every cycle in $H(G)$ and $H(K)$, $K \in Z(G \otimes T)$, which does not intersect $U(G \otimes T)$ and is not contained in $Z(G \otimes T)$. Then the language of $G \otimes C$ satisfies

$$\mathcal{L}(G \otimes C) \subseteq \mathcal{L}(T).$$

Note, that all of the algorithms have polynomial (in the input size) running time, provided G is deterministic.

5. Existence of Supervisors: Nondeterministic Case

We now consider the case where G is permitted to have nondeterministic resolution; everything else is like in the previous section. In the present setting, L-processes with nondeterministic resolution arise from ones with deterministic resolution and partial observations. As discussed previously, partial observations are modeled by a language homomorphism and give rise to a relation Ψ on the state space of G.

It is clear that Algorithms 1 and 2 can still be used to compute the sets $H(G)$ and $H(K)$. However, the existence of supervisors does no longer follow immediately. For the moment assume that there is a supervisor C with $\mathcal{L}(G \otimes C) \subseteq \mathcal{L}(T)$, which by some unspecified means is capable of exact state observations, independent of Ψ. In this case a necessary and sufficient condition for the existence of an "ordinary" supervisor C' with

$$\mathcal{L}(G \otimes C') = \mathcal{L}(G \otimes C),$$

whose resolution is limited to $\ker(\Psi)$, is that

$$\Psi \leq \Theta. \tag{5.1}$$

This means that whenever two states are indistinguishable the respective control actions must be identical.

We contend that the requirement (5.1) is artificial and too severe for many applications; for example, it is rarely satisfied in communication systems or distributed systems. On the other hand, if (5.1) fails to hold, then clearly a mechanism is needed that allows the supervisor to glean the necessary state information from G. Mathematically, we can always change the system through *cylindrification* and satisfy (5.1). Roughly, this amounts to refining Ψ by adding new selections to transition conditions, i.e., making G_Φ more deterministic. In communication systems cylindrification is achieved through the introduction of sequence numbers or time stamps; more generallly, cylindrification can be interpreted as introducing tags for events. However, to give a generally applicable practical solution appears difficult and remains an open problem for future research.

6. Example: A Simple Communication Protocol

Consider a communication system G consisting of a source SO, a transmitter TR with buffer size M, a channel CH, a receiver RE with buffer size 1, and a destination DE (see Figure 1).

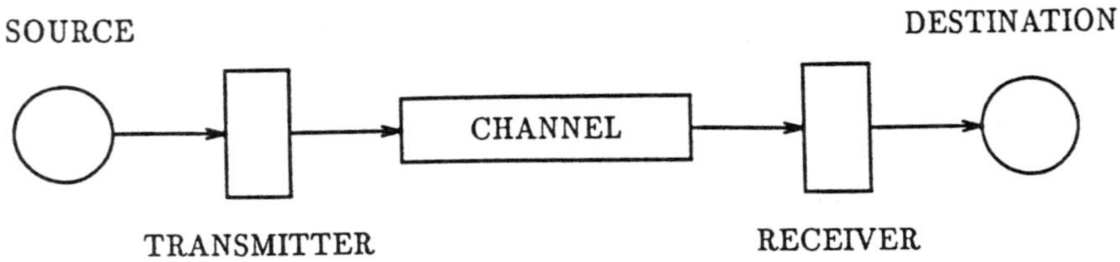

Figure 1. Communication System.

We assume initially that the channel is an arbitrary packet network where packets can arrive in any order; this allows a simple logical model for the channel. However, for the last step of the synthesis (the homomorphic reduction) we require the channel to transmit messages in the order they were sent, i.e., the channel acts as a FCFS queue. Packets can be lost during transmission due to various, not further specified reasons. (One of which could be congestion.) In addition, we assume that the channel will not lose packets forever.

Suppose that the source has an arbitrary number of messages to transmit to the destination. We assume that each message possesses an unique identifier, for example a time stamp of its creation. Thus we can label each message with an integer; we denote the first created message by m_o and the message created after message m_{i-1} by m_i, where $i = 1, 2, \ldots$ and $m_i \neq m_j$ for $i \neq j$.

The task T for the system G is formally defined as:

Task T: If message m_i is created by the source, then exactly one copy of m_i is delivered to the destination. Furthermore, message m_i is delivered before message m_{i+1}.

For the synthesis of the protocol two assumptions are made:

(1) the source creates only a finite number, say $N+1 > M$, of messages; and
(2) there are never more than a fixed number, say B, of copies of a message in the channel.

The assumptions are needed for otherwise the state space of the system would be infinite. However, it will turn out that the the restriction introduced by the first assumption will be eliminated by a homomorphic reduction. The second assumptions is not a severe restriction since it is easy to show by induction that our solution applies to arbitrary values of B.

The individual components of the system are modeled by deterministic, lockup-free L-processes. Let x be an arbitrary event and C be a supervisor. For notational convenience we write cx for the selection of C which enables x. Also, to simplify the example, we allow supervisors to access the same control variable. The interpretation in this case is that a transition is enabled by supervisors C_1 and C_2 if and only if both enable the transition simultaneously.

The individual events are

- $s(i)$: message m_i is created by the sender and loaded into the transmitter buffer;
- $t(i)$: a copy of message m_i is transmitted over the channel;
- $r(i)$: a copy of message m_i is written into the receiver buffer;
- $d(i)$: a copy of message m_i is delivered to the destination;
- $l(i)$: a copy of message m_i is lost during transmission; and
- $cs(i)$, $ct(i)$, $cr(i)$ and $cd(i)$ are the corresponding external selections which can be used to control the system.

As mentioned in Section 2.5, asynchrony is modeled through *wait* and *pause* selections. In the following state transition diagrams, self-loops corresponding to these selections are usually omitted.

Source and destination are the very simple L-processes SO and DE depicted below.

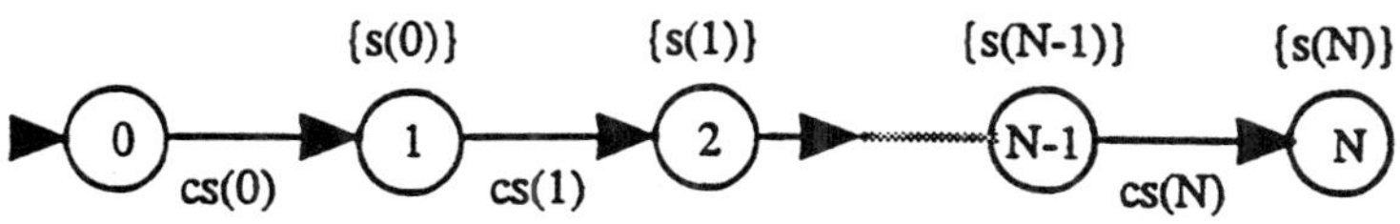

Figure 1. Source SO

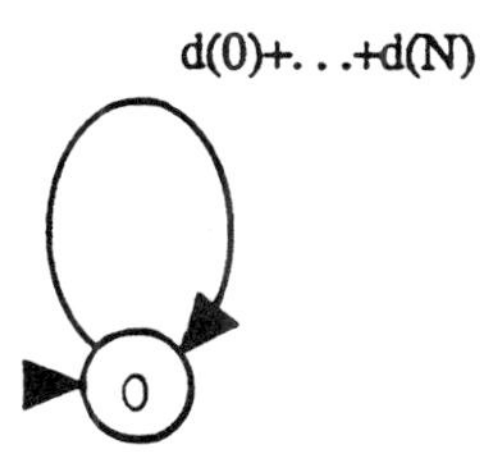

Figure 1. Destination DE

The transmitter TR consists of a buffer holding the last M created messages. The event $t(i)$ corresponds to transmitting a copy of message m_i over the channel. The L-process for TR is shown in Figure 2.

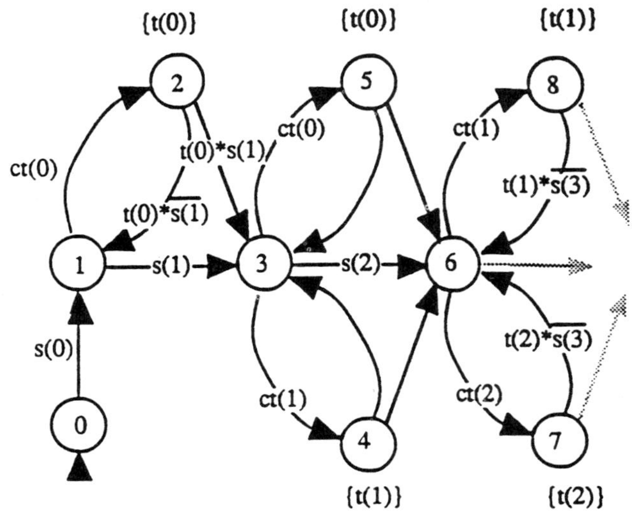

Figure 2. Transmitter TR; $M = 2$

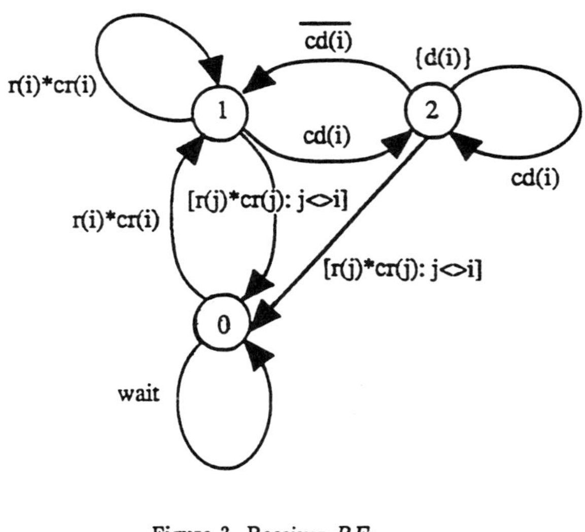

Figure 3. Receiver RE_i

To simplify the state diagrams for the receiver and the channel, they are constructed in a modular fashion. Thus, the L-process for the receiver for message m_i is RE_i which is shown in Figure 3.

The resultant model for the receiver is

$$RE = \bigotimes_{i=0}^{N} RE_i.$$

The model for the channel CH is the product of the channels for the individual messages, i.e., we assume that the interaction of messages during transmission is modeled by the fact that the channel can lose messages:

$$CH = \bigotimes_{i=0}^{N} CH_i,$$

where CH_i is the L-processes depicted in Figure 4.

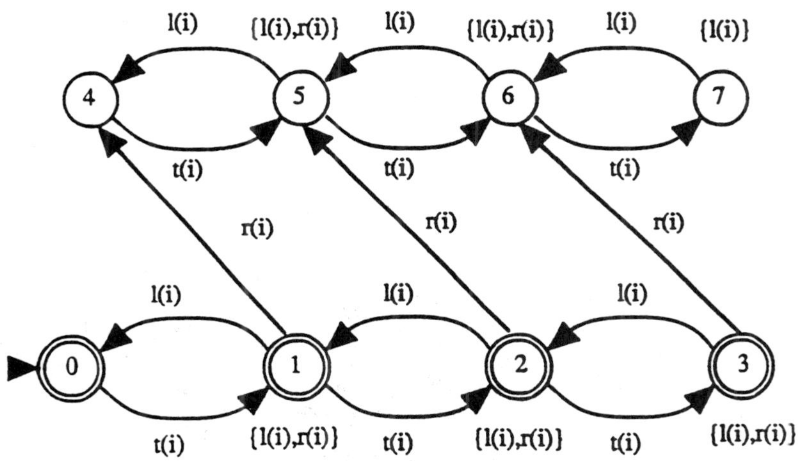

Figure 4. Channel CH_i; $B = 3$

Recall that the event $l(i)$ denotes the loss of message m_i. It is the only event which is unobservable. Also, all state transitions of the channel are uncontrollable.

The subtask for eventual delivery of message m_i is specified through the L-automaton in Figure 5.

The language of the task T_{ME} corresponding to the delivery of all messages satisfies

$$\mathcal{L}(T_{ME}) = \bigcap_{i=0}^{N} \mathcal{L}(T_i).$$

Let C_i be the L-process depicted in Figure 6.

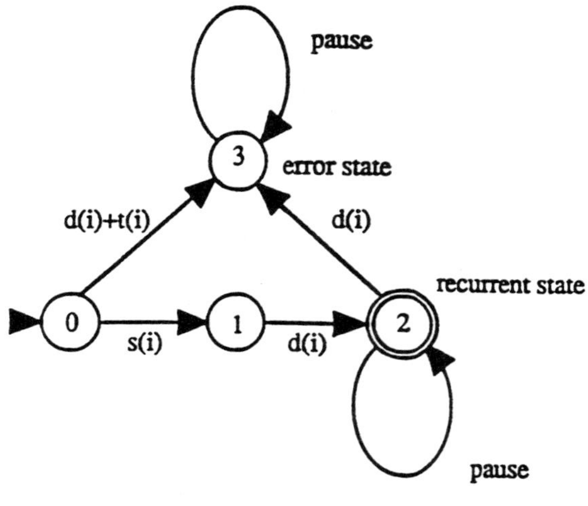

Figure 5. Task T_i

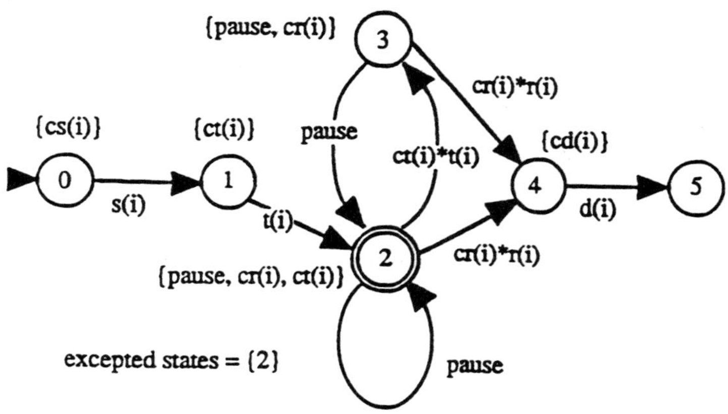

Figure 6. Supervisor C_i

It is easy to verify that C_i is indeed a supervisor for task T_i. Because C_i cannot observe the loss of copies of message m_i it will force retransmission of a message until at least one copy is received.

It follows from

$$\mathcal{L}(G \otimes C_i) \subseteq \mathcal{L}(T_i)$$

and

$$\mathcal{L}(G \otimes (C_o \otimes \ldots \otimes C_N)) \subseteq \bigcap_{i=0}^{N} \mathcal{L}(T_i),$$

that the L-process C_{ME} defined by

$$C_{ME} = \bigotimes_{i=0}^{N} C_i$$

is a supervisor for T_{ME}. Because of the constraints on buffer sizes and message ordering, this supervisor is not a solution to the problem; it permits event sequences that are not in $\mathcal{L}(T)$.

To derive a supervisor that observes the buffer constraints, introduce tasks T_{TB} and T_{RB}. Task T_{TB} is shown in Figure 7.

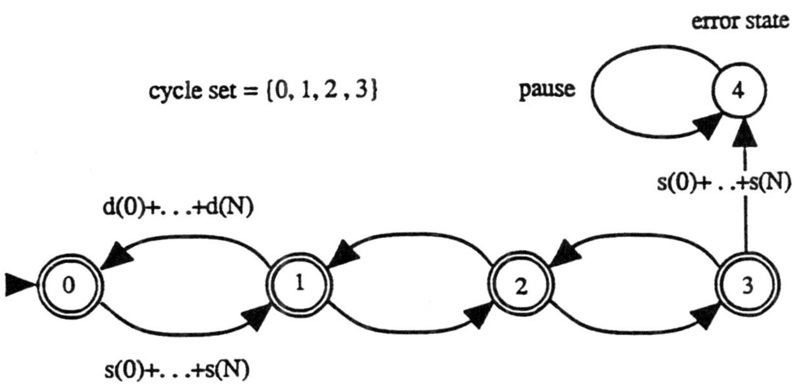

Figure 7. Task T_{TB}; $M = 3$

Task T_{TB} requires that no message is erased from the transmitter buffer before it is delivered. Task T_{RB} is decomposed into $N+1$ subtasks, i.e.,

$$\mathcal{L}(T_{RB}) = \bigcap_{i=0}^{N} \mathcal{L}(T_{RB}^{i})$$

where T_{RB}^{i} is shown in Figure 8.

Task T_{RB} requires that the first received copy of a message is read before the receiver buffer can be overwritten.

The constraint which requires that messages are delivered in the order in which they were created is encoded in task T_{FIFO} depicted in Figure 9.

The L-process in Figure 10 defines a supervisor C_{TB} for T_{TB}.

The supervisor C_{FIFO} for T_{FIFO} is shown in Figure 11.

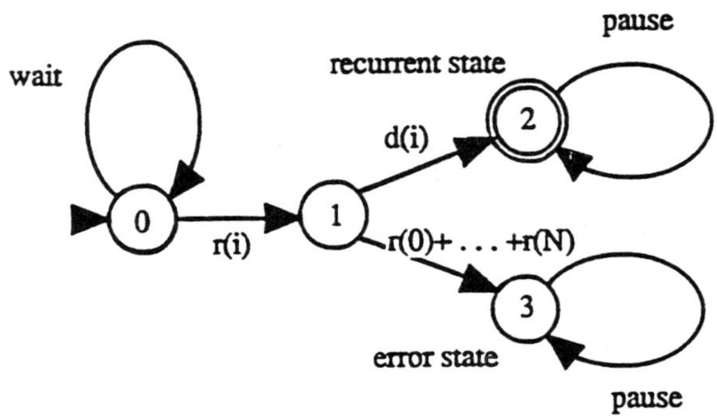

Figure 8. Task T^i_{RB}

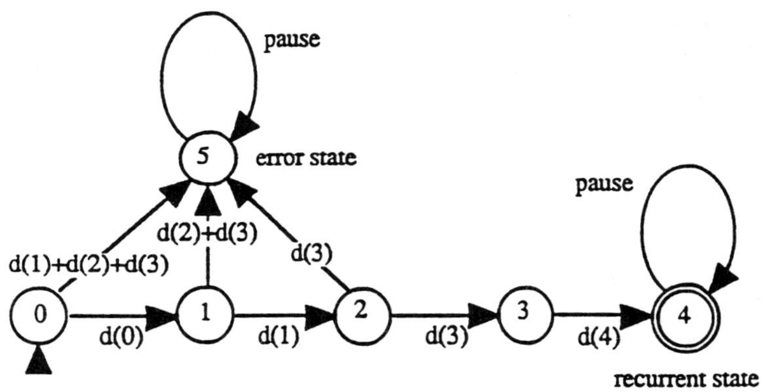

Figure 9. Task T_{FIFO}; $N = 3$

The supervisor C_{RB} for T_{RB} is defined as

$$C_{RB} = \bigotimes_{i=0}^{N} C^i_{RB}$$

where C^i_{RB} is defined in Figure 12.

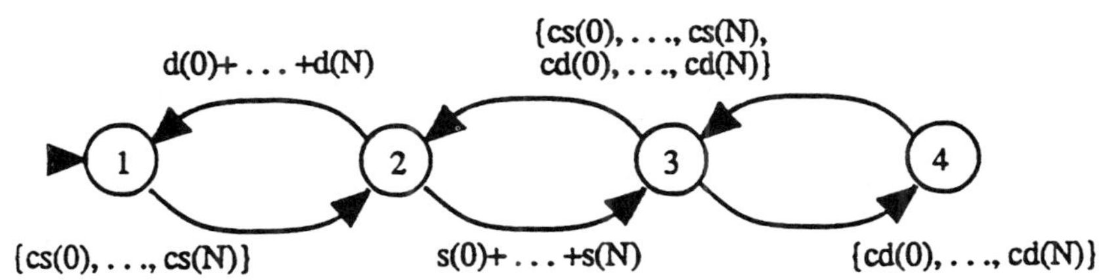

Figure 10. Supervisor C_{TB}; $M = 3$

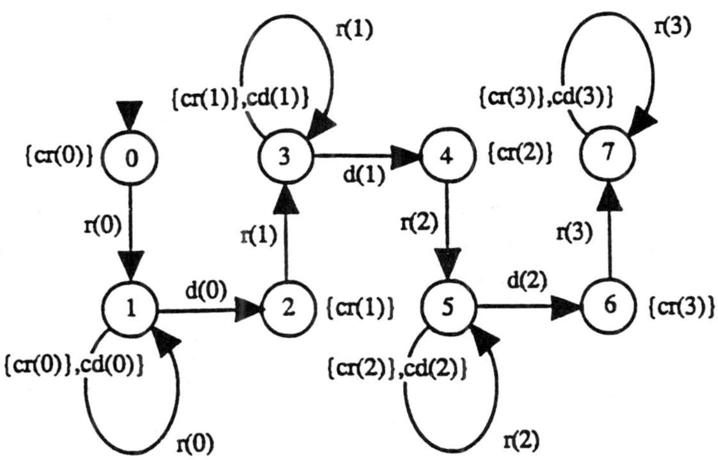

Figure 11. Supervisor C_{FIFO}; $N = 3$

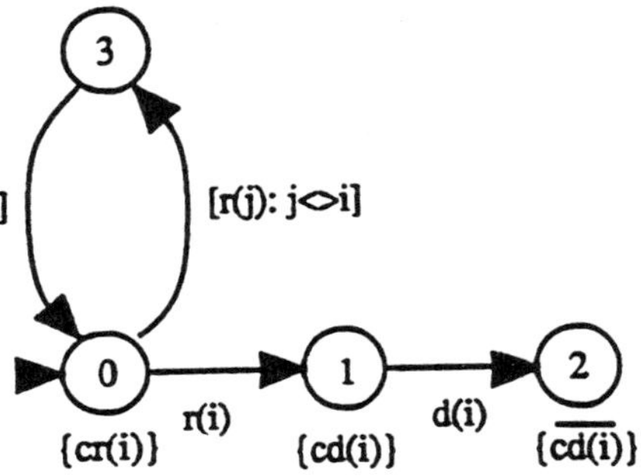

Figure 12. Supervisor C^i_{RB}

It can be verified that the supervisor C defined by

$$C = C_{ME} \otimes C_{TB} \otimes C_{FIFO} \otimes C_{RB}$$

is lockup-free and satisfies

$$\mathcal{L}(G \otimes C) \subseteq \mathcal{L}(T)$$

where

$$\mathcal{L}(T) = \mathcal{L}(T_{ME}) \cap \mathcal{L}(T_{TB}) \cap \mathcal{L}(T_{FIFO}) \cap \mathcal{L}(T_{RB})$$

is the task defined at the beginning of the section.

Although the supervisor C is a solution to the problem, it is not optimal, both in terms of the size of the resultant model and its seeming dependency on N. The construction of C involves an exponential number (in N) of L-automata and L-processes, leading to a state space whose size is exponential in N.

We now assume that the channel is FCFS. This constraint can be added easily to $G \otimes C$ because messages are already transmitted in ascending order. The additional constraint is that message m_i cannot arrive at the receiver before message m_{i+1}. Under this assumption, a closer look at the solution reveals that it depends on the parameter M only, and not on N. Hence we can apply a homomorphic reduction ϕ which maps i to i (mod M); more specifically, for every event $x(i)$

$$\phi(x(i)) = x(i \pmod{M}).$$

Since the reduction is homomorphic, the properties of the initial solution are preserved ([5]). (This is not true without the assumption that the channel is FCFS!) Furthermore, the solution becomes independent of N.

The resulting reduced solution is a centralized version of the standard communication protocol with a transmitter buffer of size M and sequence numbers $0, \ldots, M-1$. The special case $M = 1$ is known as the *Alternating Bit Protocol*. In a second paper we will demonstrate how to synthesize a decentralized solution starting from the centralized solution presented here.

7. Conclusion

A modified supervisory control framework based on the models proposed by Ramadge and Wonham ([9]), and by Kurshan ([5]) was presented. We introduced the notion of a Büchi supervisor to implement arbitrary ω-regular behaviors by a feedback control. In contrast to work along similar lines by other researchers, these behaviors are allowed to violate the conditions of realizability in [12] and topological closure in [10]. We briefly discussed the decomposition of complex tasks into subtasks, a topic which will receive more attention in a future paper. Conditions for the existence of supervisors that lead to feedback implementations of prespecified ω-regular tasks were derived for the case of L-processes with deterministic state transition matrix. Finally, the example of a simple communication system was used to demonstrate that the model can be applied to the synthesis of communication protocols without prior assumptions on the form of the solution.

Future research will address the questions of distributed control, task decomposition and computational aspects. The sum of the results will be used to derive a decentralized version of the *Alternating Bit Protocol*.

References

[1] I. Brave and M. Heymann, "On Stabilization of Discrete-Event Processes", *Proceedings of the 28th Conference on Decision and Control*, Tampa, Florida, 1989.

[2] R. Cieslak, C. Desclaux, A. Fawaz, and P. Varaiya, "Supervisory Control of Discrete Event Processes with Partial Observations", *IEEE Transactions on Automatic Control*, Vol. 33(3), pp. 249-260, March 1988.

[3] V. Freitas, "A Review of Formalisms for the Specification of Computer Communication Protocols", in *Computer Communication Systems*, 1988.

[4] C.H. Golaszewski and P.J. Ramadge, "Mutual Exclusion Problems for Discrete Event Systems with Shared Events", *Proceedings of the 27th Conference on Decision and Control*, pp. 234-239, Austin, Texas, 1988.

[5] Hoare, C.A.R., *Communicating Sequential Processes*, Prentice Hall, Englewood Cliffs, NJ., 1985.

[6] R.P. Kurshan, "Analysis of Discrete Event Coordination", Lecture Notes on Computer Science 430, pp. 414-453, Springer Verlag, 1990.

[7] R.P. Kurshan and C.H. Golaszewski, "Decentralized Supervisory Control of L-Processes", in preparation.

[8] C.M. Özveren and A.S. Willsky, "Output Stabilizability of Discrete Event Dynamic Systems", Discrete-Event Dynamical Systems", *Proceedings of the 28th Conference on Decision and Control*, Tampa, Florida, 1989.

[9] P.J. Ramadge and W.M. Wonham, "Supervisory Control of a Class of Discrete-Event Processes", *SIAM J. Control and Optimization*, Vol. 25(1), pp. 206-230, January 1987.

[10] P.J. Ramadge, "Some Tractable Supervisory Control Problems for Discrete Event Systems Modeled by Büchi Automata", *IEEE Transactions on Automatic Control*, Vol. 34(1), pp. 10-19, January 1989.

[11] P.J. Ramadge, "Observability of Discrete Event Systems", *Proceedings of the 25th Conference on Decision and Control*, Athens, Greece, 1986.

[12] J.G. Thistle and W.M. Wonham, "On the Synthesis of Supervisors Subject to ω-Language Specifications", *22nd Annual Conference on Information Sciences and Systems*, Princeton NJ, pp.440-444, March 1988.

DIMACS Series in Discrete Mathematics
and Theoretical Computer Science
Volume 3, 1991

Finiteness conditions and structural construction of automata for all process algebras

Eric Madelaine ** and Didier Vergamini *

**INRIA
Route des Lucioles, Sophia Antipolis
06565 Valbonne Cedex (France)
email: madelain@mirsa.inria.fr

*CERICS
Rue Albert Einstein, Sophia Antipolis
06565 Valbonne Cedex (France)
email: dvergami@mirsa.inria.fr

Abstract

Finite automata are the basis of many verification methods and tools for process algebras. It is however undecidable in most process algebras whether the semantics of a given term is finite. We give sufficient finiteness conditions derived from the analysis of the operational rules of the algebra operators. From these rules we also generate the functions that compute automata from terms of the algebra. These constructions allow one to use our verification tools for programs written in many process algebras.

Keywords verification, concurrent systems, process algebra, lotos, structural operational semantics, finite automata.

1 Introduction

Verification methods for concurrent systems can be classified in at least three families: theorem proving methods, model-checking, and automata based methods. The first family holds the biggest theoretical power; it may be applied to many sort of undecidable problems and in some sense it can deal with infinite objects. However theorem proving methods have usually a high complexity and there is few hope to make these methods purely automatic. The ECRINS system ([MdSV89]) uses theorem proving methods, together with specialized algorithms, to check for the validity of bisimulation laws in process algebras. The approach is general enough to consider most usual process calculi from the literature; the semantics of the operators is defined in user-defined *calculus description files*, and used by the system to generate specialized behaviour-evaluation algorithms.

We want to apply this parameterized approach for building tools based on automata analysis. The system AUTO ([dSV89], [RS89]) is dedicated to *verification by reduction* of parallel and concurrent programs. AUTO deals only with terms that have finite automata representations. Its main activities are the construction of automata from terms of process algebras and the reduction and comparison of automata along a large family of equivalences. These activities are mostly intertwined, according to congruence properties of the equivalences that allow for reducing subterms of operators before building any global automaton. This approach cuts off partially the space explosion that causes the well-known limitation of such techniques.

The current AUTO system is using a subset of the MEIJE calculus ([Bo85]) as input language. To ensure that terms have finite representations, we use a two layers structure for input terms. In the lower layer, one can write recursive definitions directly encoding automata: recursive variables correspond to states and transitions are specified through action prefixing and non-deterministic choice (see the example 1 in section 2.2); these are *dynamic* operators, for they build the behaviour of components of a system. In the upper layer, one builds networks of automata using *static* operators (asynchronous parallel composition, renamings of signals, and a restriction operator). Finiteness of automata is guaranteed by forbidding occurrences of the parallel and renaming operators inside the recursive definitions. Observational equivalence appears to be a congruence for the MEIJE parallel and restriction operators, so lower layer automata can be reduced before being composed.

Extending the structural construction of automata to parameterized process algebra, we need new finiteness conditions, computable from the very definitional rules of the operators. Here the splitting between static and dynamic operators makes less sense, as many operators can be used both in dynamic and static positions. Moreover there are operators that are asymmetric; recursion on the left argument of the *enable* and *disable* operators of LOTOS may generate infinite structures, whereas recursion on their right arguments can be used safely. We shall deduce from analysis of their rules which operators may in which position accept a recursive variable as one of its arguments, and which operators are preserving finiteness of automata. The rules analysis also provides us functions associated with each operator for building the automata.

In section 2, we give an overview of the concepts from process calculi theory we need in the paper, including a description of the syntax we allow for structural operational semantic rules. In section 3, we discuss finiteness conditions and explain the classification of operators obtained from the analysis of the rules. In section 4 we describe the algorithms for structural construction of the automata, and in the conclusion we describe a prototype system that uses this generic technique and discuss current work.

2 Process Calculi

Process calculi are now a well-accepted generic notion for designing a class of formalisms which share the same definitional principles : CCS [Mi80], SCCS [Mi83], MEIJE [Bo85], TCSP [Bro83], ACP [BK86], BasicLOTOS [BB88] to name a few. We shall assume reader's acquaintance with at least one of these languages and its definitional mechanisms.

Process calculi are based on two main types : *actions* and *processes*. Operators take actions and processes arguments into processes, providing a classical algebraic structure. Operational semantics provides interpretation of closed terms into transition systems, with actions as transition labels. Operators and non-closed expressions are then interpreted as transition system transformers; this semantics is defined through behaviour rules in a structural operational style, with a particular format (see [dSi85], [VG88]). We shall describe our format for rules in section 2.4.

Special operators are action renamings and recursive definitions. They are present in all process calculi.

2.1 Actions

In all process algebras actions are themselves structured. This structure is what allows for synchronization and further communication to be handled in relevant operators rules. Just recall the inverse signals in CCS, which meets in synchronization and produce a hidden action τ. In SCCS and MEIJE there is a full commutative monoid of potential simultaneous actions, again containing a group of invertible signals. Actions structure in ACP is more scarce and parametric, while in BasicLOTOS it is only a set of so-called gate names without structure, but for a distinguished termination action δ.

2.2 Recursive definitions

Most process algebras have some sort of recursion operator. In AUTO and ECRINS we use a common recursive definition mechanism for all process algebras. Here is an example of a recursive definition, written with MEIJE operators:

Example 1

```
let rec {x = a:x + b:x +c:stop
     and y = a:y + b:z
     and z = a:z + c:c:y }     in x//y
```

Such recursive definitions are used in AUTO for building finite automata, perhaps composed later on by other operators of the algebra. One can also build infinite structures, not suitable for analysis in AUTO, such as:

Example 2

```
let rec {many-processes = one-process // many-processes}
    in many-processes
```

2.3 Definition of process algebras

A calculus description contains the concrete syntax and abstract syntax definitions of the calculus operators, together with *structural conditional rules* that gives them an operational semantics. The ECRINS *calculus compiler* uses the first part to produce a scanner, a parser and abstract syntax structures for expressions of the calculus. From the semantics part, the compiler will produce the functions for building and combining automata from terms. We now sketch the allowed syntax formats using examples. Full documentation will be found in [MdSV89] and is recalled shortly below.

```
operator hiding:: Process Label --> Process
syntax   \\   left 4
semantics
   hiding                    p -- a --> p'  &  (a equal s)
                             -----------------------------

                                p\\s -- tau --> p'\\s
   hiding-not              p -- a --> p'  &  not (a equal s)
                           ---------------------------------

                                 p\\s -- a --> p'\\s
end
```

The first line declares the name and the signature of the operator. Then follows the concrete syntax description, along with associativity and priority declarations aimed at the parser generator. Finally the operational rules are listed. The end right upper part of the rules (e.g. `(a equal s)`) is a predicate over actions conditioning the applicability of the rule. Such predicates implement synchronization conditions over the actions of the operator subterms. The plain format of the structural conditional rules will be explained below. Here are two other examples, respectively from MEIJE and BASICLOTOS:

```
operator ticking :: Action Process --> Process
syntax   * right 3
semantics
        ticking       p -- a --> p'
                     -------------------

                     s*p -- a.s --> s*p'
end
```

```
operator disable :: Process Process --> Process
syntax    [> left 4
semantics
        disabling_1        p -- a --> p' & not(a equal d)
                           ------------------------------
                               p[>q -- a --> p'[>q
        disabling_2        p -- a --> p' & (a equal d)
                           ---------------------------
                               p[>q -- d --> p'
        disabling_3           q -- a --> q'
                              ----------------
                              p[>q -- a --> q'
end
```

2.4 The Conditional Rewrite Rules format

The rules must obey the following format:

$$\frac{\{x_j \xrightarrow{u_j} x'_j\}_{\mathcal{J}\subset[1..n]} \quad \& \quad P(\{u_j\}_{\mathcal{J}}, a_1, \ldots, a_m)}{Op(x_1, \ldots, x_n, a_1, \ldots, a_m) \xrightarrow{F(\{u_j\}_{\mathcal{J}}, a_1, \ldots, a_m)} T'(\{x_k\}_{[1..n]-\mathcal{J}}, \{x'_k\}_{\mathcal{J}}, a_1, \ldots, a_m)}$$

Many definitions in this paper rely on the syntax allowed for the various elements of conditional rules. Let us give precise names to these elements.

Definition 1 *We call:*

- premises *the upper part of the rule and* conclusion *its bottom part.*
- subject *the term at the left end part of the conclusion. The head operator Op of the subject has process arguments $x_1, \ldots, x_n$ and action parameters $a_1, \ldots, a_m$. Inside P, F and T' some other actions may appear: they are global constants of the calculus and had to be declared as such previously.*
- formal hypothesis *the part of the premises on the left of the & and* working formal variables *the x_j with $j \in \mathcal{J}$.*
- actions predicate *the part of the premises on the right of the &. P belongs to boolean operators closure of the following basic predicates :* equality, divisibility, set membership.. *over actions terms with synchronization product.*
- resulting action *the label over the arrow of the conclusion. The resulting action F is a function of the formal hypotheses actions (and of the operator action parameters).*
- resulting process *the right end part of the conclusion. The process arguments that appear as formal hypothesis must be transformed as x'_j in the resulting process. This condition does not allow to test a potential future behaviour of a process, without making it explicitly perform its action within the considered rule, therefore saving the possibility of choosing another future. This is a restriction to the format in [VG88]. Beside this condition, the resulting process is built from the action and process variables, and the operators of the calculus (excluding recursive definitions).*

3 Finiteness Conditions

The preceding conditional rules defines in a structural manner the semantics function that computes a *transition system* from a (closed) term of a process algebra. This definition is constructive: you can compute each transition of the transition system by building proof trees which nodes are instances of the rules.

Definition 2 *A term of a process algebra has* transition-system finiteness *(TSF) iff the transition system computed from the term using the operators' rules is strongly bisimular to some finite transition system (i.e. with a finite number of states and a finite number of transitions).*

In many process algebras with recursion, this property is undecidable. Sufficient syntactic conditions to ensure FTS will be given in this section, for any process algebra defined using the conditional rewrite rules from the preceding section. As far as possible, these conditions will be expressed in terms of the semantic rules of the operators. We shall use the following notions:

guarded recursion: It is possible to build infinite proof trees for terms containing recursive definitions. Can we found syntactic conditions to guaranty that all proof trees are finite?

non-growing operators: Making the assumption that the arguments of an operator have finite semantics, is it always true that their composition by this operator is a finite automaton? This property holds for most classical operators, but the rule format allows to define exotic operators that create infinitely many states.

sieves: Some unary operators have the nice property that the resulting automaton has exactly the same states than the argument automaton, only some transitions being transformed, erased, or added. We implement their semantics by *sieves*, that is functions that only modify the transitions of the system. Which operators may be implemented in this way?

switches: Inside a recursive definition, the use of recursive variables should be limited in some way, in order to avoid building infinitely many states, or states with infinitely many transitions. Clearly, parallel composition operators and non-alphabetical renamings should be somehow forbidden inside recursive definitions. At which places (defined as occurrences of operator arguments) is a recursive variable allowed to appear?

3.1 Guarded recursion

In order to avoid divergence in the proof tree construction, we introduce as usual a notion of guarded terms. We define here this notion in a rather abstract way, and we shall give a generic algorithm that computes it in a further section. The definition relies on the fact that if a proof tree is infinite, then either it contains a pattern that occurs infinitely, or the subject of its nodes are strictly growing along the infinite branches. The *guarded* property takes care of infinitely repeated patterns, while growing branches will be addressed later.

Definition 3

- *A proof tree is* unguarded *iff the* subject *of its root is equal to the subject of one of its subtree, or if it has an unguarded subtree.*
- *A term of a process algebra is* unguarded *if it has an unguarded proof tree, or if at least one of its possible reconfigurations is unguarded.*
- *A term of a process algebra is* guarded *if it is not unguarded.*

3.2 Non-growing operators

Growing operators build infinite structures from arguments having finite automata representations. Though no operators of usual process algebras have such a nasty behaviour, we need a syntactical way to ensure that no growing operator is used in a term. In order to obtain infinitely many states in the resulting automaton, one would have to introduce a rule that produces new terms ab infinitum. A natural sufficient condition for ensuring finiteness is to be able to find an order on expressions such that for all rules, the resulting process is not strictly greater than the subject.

It is possible to adapt here many results from the term rewrite system theory, with the difference that we are looking for a non-strict order compatible with our rewrite relation, whereas usual rewrite systems need a strict decreasing order. We give here a simple definition that covers nearly all interesting cases:

We consider families of operators closed under reconfiguration: if an operator belongs to the family, then all operators that occur in the resulting processes of all its rules also belong to the family.

Definition 4 *A family of operators* $\{Op_k\}$ *is* non-growing *iff there exist a simplification ordering* $\lll$ *such that:*

For each rule of each operator, let us denote $Op_k(x_i)$ *the subject of the rule and* $T_{i_j}[x'_i/x_i]$ *its resulting process in which all resulting working variables* x'_i *have been replaced by their corresponding* x_i*, then:*

either $T_{i_j}[x'_i/x_i] \lll Op_k(x_i)$

or $T_{i_j}[x'_i/x_i] = Op_k(x_i)$

where $=$ *is the syntactic equality on terms.*

We say then that all Op_k *are* $\lll$*-non-growing.*

Theorem 1 *Given a family of non-growing operators* $F = \{Op_k\}$*, an operator* $Op_k(x_i)$ *of arity* n*, and* n *terms* T_i *having TSF, then* $Op_k(T_i)$ *has TSF.*

Lemma 1 *Any process expression containing no recursive definition has a finite number of behaviour rules*

This follows trivially from our format for the operators' rules: building the behaviour proof trees, the premise at each node is strictly smaller than the subject of the node, w.r.t. the subterm order. So the depth of the proof tree is bounded by the depth of the expression. Then at each node you can choose among a finite number of rules to apply, the number of proof trees is finite, so the number of behaviours (= successful proof trees) is finite.

Proof of theorem 1:

Consider the set $\mathcal{R}$ of all possible reconfigurations R of $Op_i(T_i)$. The behaviours of each R are computed from instances of the rules of the operators $\{Op_k\}$. As simplification orderings are substitutive, the rules of each R are themselves compatible with the ordering. From lemma 1, each R have only a finite set of rules. Simplification orderings are well-founded, so $\mathcal{R}$ is finite.

Now say the subterms T_i have finite automata A_i. Then we can build the transition system of $Op_k(T_i)$, with states of the form $s =< R, st_{j_1}, ..., st_{j_n} >$, where the st_{j_i} are states from A_k, and initial state $< Op_k(T_i), st_{0_1}, ..., st_{0_n} >$. The number of those states is bounded by $|\mathcal{R}|*|A_1|*...*|A_n|$. The transitions of any such s are computed from the rules of R (finite from lemma 1), and from all combinations of the transitions of the st_{j_i}; they are in finite number.

∎

This proof gives hints toward implementation: A pre-compilation phase can compute the reconfiguration set, and even build the functions that compute the composed transitions. Such a pre-compilation is subsumed in the algorithm in section 4.3.

3.3 Sieves

Definition 5 *A* sifting operator, *or* sieve, *is an operator with exactly one process argument, which rules obey the following conditions: each rule has exactly one working formal variable (the process argument), may have predicates and any form of resulting action, and the resulting process is obtained from the subject of the rule by substituting the working formal variable by the corresponding resulting process variable.*

This definition includes the *renaming*, *restriction*, and *ticking* operators of MEIJE, the *hiding* operator of TCSP and LOTOS.

Sieves are defined in fact as (partial) functions on actions, rather than on automata. More precisely they are functions that map actions to finite sets of actions, for a sieve may have a (finite) number of rules that can apply on a a given action. As such, they can be easily composed. They can also be combined with the automata building functions, leading to efficient implementations where no intermediate automata are built for such operators. This is especially important when the sieve discards some transitions: it is possible to save producing and exploring the corresponding target states.

Another issue is the introduction of sieves inside recursive definitions. We need define here a subclass of sieves such that the language generated by their compositions, modulo some idempotence property, remains finite. Then the states of the generated automaton will be obtained as pairs of a recursive variable and a composition of sieves.

This applies e.g. for alphabetical renamings, hiding, and restriction (for the alphabet of action labels in a term is finite, and the set of all restriction compositions is a finite commutative group).

Of course, it does not apply to ticking or to non-alphabetical renamings, and the MEIJE term: `let rec x= a:y and y = b*x in x` generates infinitely many states `x`, `b*x`, `b*b*x`, etc.

We need here a *non-growing* definition for resulting action functions:

Definition 6 *An operator rule is* action-non-growing *iff its resulting action is an action term built only from the following items: the formal action of the rule, the action parameters of the subject, the constant actions of the calculus, and alphabetical renamings (including renaming a label by an invisible action).*

This definition is trivially fulfilled by all operators of BASICLOTOS and CCS, but not by the ticking operator of MEIJE, nor by non-alphabetical renamings.

Definition 7 *A* non-growing sieve *is a sifting operator which rules are action-non-growing.*

Proposition 1 *Given a finite alphabet of actions, the algebra of all compositions of non-growing sieves has a finite model.*

Proof: Consider the *action-sifting* functions associated to sieves: Given an action (the label of a transition), an action-sifting function returns the (finite) set of possible transformations of the action by the various rules of the sieve that may be applied to this particular action.

It is enough finding a finite model for the compositions of action-to-action transformation: the extension to subsets will keep finiteness. By the way, the set of action-to-action transformations listed in definition 6 is closed under composition (u_p is the formal action of a sieve rule):

- $u_p \longrightarrow u_p$ (the identity) is neutral element for the composition,
- $u_p \longrightarrow a$ with a either an action parameter or constant of the calculus is trivially left absorbing for the composition,
- $u_p \longrightarrow u_p < a_1/b_1, \ldots, a_n/b_n >$ where a_i and b_i are action labels (excluding composed actions) form a finite monoid for the composition.

∎

3.4 Switching operators

The simplest way to generate an arbitrary finite automaton using recursive definition is to write an equation for each state of the automaton that describe the (finite) set of transitions, and of target states, of this particular state. This is achieved usually using recursive equations which right-hand-sides contain finite sums of action prefix operators (this was the case in the preceding version of AUTO).

For the sake of conciseness of the terms and of generality with respect to the now usual operators of process algebras, we want to extend as much as possible the possible language for these recursive equations, while keeping syntactic conditions on this language for TSF. Such extensions have been already published e.g. for Basic Lotos in [Ai86],[GN], allowing for example the following BASICLOTOS-like term, where ":" stands for the action prefixing operator, "[]" for the choice operator, and ">>" for the enabling operator:

Example 3
```
let rec x = a:exit >> y
    and y = b:y [] c:x   in x
```

Definition 8 *Given a family of operators $\mathcal{O}$ and a well founded simplification ordering $\lessdot$ on expressions generated from $\mathcal{O}$,*

An operator Op in $\mathcal{O}$ is a switching operator *(or simply a* switch*) w.r.t. one of its process arguments "p" iff:*

- *All rules in which "p" is a working formal variable verify the following properties: it has no other premise ("p" works alone) and the resulting process is exactly p'.*
- *All rules where the resulting process contains an occurrence of "p" are non-growing for $\lessdot$ and are their resulting processes are themselves switches for "p".*

Remarks:

- This includes non-growing operators with no premise at all.
- The name "*switch*" may refer to the fact that a switching rule selects (at most) one of its process arguments.
- This definition could be extended by allowing rules in which "p" is working to have as resulting processes T such that $T \lessdot Op(..., p', ...)$, with T being also in some sense a switch for p. However, this would complicate to much both the definition and the related proofs, whereas all classical operators fit the restricted definition we have just given.

The *sum* operator of SCCS, the binary *choice* of BASICLOTOS are switches, but also the *delay* operator of MEIJE and the *disabling* operator of LOTOS for its second argument.

Usual prefixing operators are also switches, including the *action prefix* operators of MEIJE and LOTOS, of course, but also the *enabling* operator of LOTOS for its second process argument. The *internal choice* of TCSP is a switch.

Yet the *external choice* operator of TCSP is not a *switching* operator (see its rules in the annex), because for each of its arguments, it has a rule looking as a *switching* rule, and a rule resembling a *sieve* rule. By the way this operator is one we do not want to be involved in a recursive definition:

Example 4
```
let rec x = (tau : x) ext-choice a:y in x
```

This term generates the following sequence of resulting processes:

```
let rec {x = tau:x ext-choice a:x}
        in x ext-choice a:x
let rec {x = tau:x ext-choice a:x}
        in (x ext-choice a:x) ext-choice a:x
let rec {x = tau:x ext-choice a:x}
        in (x ext-choice a:x) ext-choice a:x ext-choice a:x
...
```

Though this specific case could be reduced (to a finite set of terms) by semantical arguments, the finiteness property may no more be guaranteed at a syntactical level. Semantical arguments for finiteness are out of the scope of this paper.

Definition 9 *Given a family of variables V, a term from a process algebra is called a* term suitable for recursion on V *iff either*

1. *it is a variable from V,*
2. *or it does not contain any variable from V and it has a finite automaton semantics,*
3. *or its head operator is a switching operator for some of its arguments, these arguments are subterms suitable for recursion on V, and all other arguments contains no occurrences of variables in V and have finite automaton semantics,*
4. *or its head operator is a non-growing sieve and its argument is suitable for recursion on V.*

Remark: this definition can be extended to handle nested recursive definitions, by adding an item for any recursive declaration "let rec $\{x_i = e_i\}$ in x_0" such that all e_i are suitable for recursion on $V \cup \{x_i\}$. Such an extension preserves the following theorem, though the proof is still more tedious.

Theorem 2 *Let Proc be a recursive definition* "let rec $\{x_i = e_i\}$ in x_0".

If Proc is guarded, and if all expressions e_i are terms suitable for recursion on $\{x_i\}$, then the recursive definition has a finite automaton semantics.

Proof: We define a finite set of *states* by induction on the structure of the term, then we prove that the transition system of the term maps in this state space, with a finite number of transitions from each state. The full proof is in the annex.

This property allows us to guarantee that some recursive definition have a finite semantics. In any process algebra, it permits using any combination of nested recursive definitions, and arbitrary closed terms inside recursive definitions. In the case of MEIJE-SCCS, it naturally includes the classical "well-guarded" condition (sums of action-prefix operators). In the case of BASICLOTOS, it allows the occurrence of recursive variables as second arguments of the *enabling* operator and well-guarded occurrences of recursive variables within the second argument of the *disabling* operator.

3.5 Finite operators

This notion is used to ease the application of definition 9 for simple cases.

Definition 10 Finite *operators are non-growing operators with no process argument.*

Most process algebras have a constant operator without any possible behaviour (the *nil* of CCS, the *stop* of LOTOS). All these are finite operators. Another example is:

```
operator clock : Process
syntax h prefix
semantics
          ----------------------
           h(a) -- a --> h(a)
```

Finite operators are trivially accepted by case 1 of definition 9

3.6 Accessibility

All preceding conditions can be restricted to the *accessible* parts of the term. This is not only an optimization issue: considering only accessible parts allows for rejecting less programs, for any violations of a condition inside a non-accessible part of a term will have no consequence on its semantics.

Accessibility is an *undecidable* property, for it may involve deciding whether some subterms are dead-locked. We only want to give here a reasonable sufficient characterization of *potentially accessible* parts of a term. The main idea is that whenever some equation in a recursive definition is not referenced from an accessible part of the term, anything inside this equation has no influence at all on the semantics of the term. We define here by structural induction whether a sub-term, a recursive variable, or a recursive equation is accessible:

Definition 11

The root of a term is accessible.

If a subterm is accessible and this subterm has the form $Op(arg_i)$*, and if the union of the sets of working formal variables in all the rules of operator Op is* $\{arg_j\}$*, then all the* $\{arg_j\}$ *are accessible (for all operators of classical process algebras,* $\{arg_j\} = \{arg_i\}$*).*

If a recursive definition `let rec xi = ti in T` *is accessible, then all the variables* `xi` *that are free in* `T` *and in the accessible right-hand-sides* `ti` *are accessible, the corresponding recursive equations are accessible, and the right-hand-sides of these equations are accessible.*

If a local definition `let x = t in T` *is accessible, and if* `x` *is free in* `T`*, then the variable* `x` *is accessible, and the sub-term* `t` *is accessible.*

4 Algorithms

We have implemented a new version of AUTO using the preceding results. From the rules of the operators, and from their classification in finite, sifting, switching, and non-growing operators, we derive functions that test the syntactical conditions for finiteness, then build the automaton of a term.

More precisely, given a closed term of a given process algebra, the system goes through the following steps:

1. Computes the accessible part of the term. This is an opportunity to emit diagnostics on potential user errors.
2. Checks whether the accessible part is guarded. If not, rejects the term.
3. Checks whether all accessible recursive definitions are suitable for recursion. If not, rejects the term.
4. Generates the corresponding automaton. This step, as in the original AUTO system, is parameterized by a reduction function (hopefully a congruence) to be applied on each sub-automaton before building their combination.

4.1 Guarded recursion test

Within the framework of AUTO, the only possibility for building an unguarded proof tree is through a recursive declaration: there exist a path in any unguarded proof tree such that the subjects at both ends of the path are identical, and there is an unfolding of (at least) a recursive definition on this path, and all other nodes correspond to rules having at least one premise.

Let us test whether a term of is guarded or not. Informally, we look for a cycle through the immediate dependencies among recursive variables. As such a cycle may appear through several levels of nested recursive declarations, the dependency graph is to be computed for the whole accessible part of the term.

The algorithm builds a directed graph whose vertices are either recursive variables or subterms of the term, and the initial vertices are the accessible recursive variables. There is an arrow between states **T1** and **T2** if the head operator of **T1** has a rule with a premise for its n-th argument, and **T2** is the n-th son of **T1**; there is also an arrow between states **T1** and **T2** if **T1** is a recursive variable, and **T2** is the right-hand-side part of the equation defining **T1**, and between states **T1** and **T2** if **T1** is a recursive definition `let rec xi = ti in T2`

The term is unguarded iff there exists a cycle in this graph.

4.2 Automata Generators and Combinators

We describe now the algorithms used for the structural construction of automata.

We call *tta* (standing for *term to automaton*) the function computing an automaton from a term of a given process algebra. This function builds automata in a bottom-up fashion. A special case raises when a recursive definition is encountered: under the hypotheses of definition 9, the semantics of a recursive definition is a finite automaton (theorem 2). The algorithm *let-rec:tta* checks for the finiteness hypotheses and build the automaton. On all non-growing operators the algorithm *non-growing:tta* builds a resulting automaton from the automata corresponding to the operator parameters. The *tta* algorithms make also a special case for sieves: as the states of `tta`(<sieve>(p)) are included in those of `tta`(p), we propagate sieves as second argument of *tta* and use them each time a transition is generated. This implies that we can compose sieves and that we apply them as deep as possible inside the computations. Thus, *tta* takes two arguments:

- a term which is a non-growing operator applied on several subterms,
- a composed sieve (inherited from the recursive calls).

The basic *tta* algorithms are generic. They are adapted to each algebra through a specific set of functions derived from the operational semantics definition of the operators. The following sections describe the two algorithms *non-growing:tta* and *letrec:tta*, and the form of the specialized functions.

We use the object-oriented convention *type:name* for naming a generic algorithm and its specialized versions (called *methods*) depending on the type of its first argument.

4.3 The residual algorithm for any kind of non-growing operator

We use the same generic algorithm for all non-growing operator but for sieves. The algorithm for sieves is straightforward: $\mathit{sieve{:}tta}(\langle\text{sieve}\rangle_1(p), \langle\text{sieve}\rangle_2) = \mathit{tta}(p, \langle\text{sieve}\rangle_1 \circ \langle\text{sieve}\rangle_2)$

For other non-growing operators, we have to compose the argument automata. In a first step we recursively compute the automata denoted by all subterms. The most difficult point in this recursive application is building the composed sieve to give as second argument for each recursive call: if we want to optimize the size of the subsystems, we need to be able to compute how a composed sieve is modified when going through an operator. For the moment, we only treat simple cases such as the classical sum operator of CCS for which we transmit the inherited sieve without modification. Other operators transmit the trivial identity sieve.

The second step of the algorithm is a classical residual exploration. Each state of the global system is a structure containing the states of each subsystem plus the operator combining them. The initial state of the global system is composed by the operator and the initial states of the subsystems. The unexplored state list is initialized to this initial state. Then the algorithm is a loop taking one state in the unexplored state list, computing all the transitions of this state using the specific functions associated to its operator and the sieve. This computation may produced some states that are not yet discovered and thus added to the unexplored state list. The computation finishes when the unexplored state list is empty. As we only use non-growing operators, the number of reconfigurations of a given operator is finite, and the termination of the algorithm stems from theorem 2.

More formally, we sketch in figure 1 the residual algorithm that computes **tta**($Op(t_1, \ldots, t_n$, <actions parameters>), <sieve>) where <sieve> is the inherited composed sieve, and t_i are the process arguments of the operator Op. In this figure, we call transition a couple of an action and a resulting state.

```
Algorithm for
non-growing:tta(Op(t1,...,tn,<actions parameters>),<sieve>)

∀i, Ti = tta(ti, mk-sieve(Op, <sieve>, <actions parameters>))
state0 = ⟨Op, initial(T1), ..., initial(Tn)⟩
STATES = {state0}
UNEXPLORED = {state0}
TRANSITIONS = ∅
while (UNEXPLORED ≠ ∅) loop
  take s in UNEXPLORED
  UNEXPLORED = UNEXPLORED − {s}
  transitions = ∅
  foreach ⟨action, state⟩ ∈ mk-trans(s, <actions parameters>) do
    action = <sieve>(action)
    if action then
      transitions = transitions ∪ {⟨action, state⟩}
      if state ∉ STATES then
        STATES = STATES ∪ {state}
        UNEXPLORED = UNEXPLORED ∪ {state}
  TRANSITIONS = TRANSITIONS ∪ {⟨s, transitions⟩}
return ⟨STATES, state0, TRANSITIONS⟩
```

Figure 1: The residual algorithm for non-growing operators

The methods *mk-trans* and *mk-sieve* are specific to each operator. For instance, in CCS the method `parallel:mk-sieve` for the parallel operator simply produces the identity sieve. The method `parallel:mk-trans` is sketched in figure 2:
Here the notation $< p'|q >$ stand for the composed state where the first sub-automaton has advanced by one step, and the second sub-automaton is still in its original state.

foreachmove $< p \xrightarrow{a} p' >$ means that we repeat the same group of operations for each transition of p.

4.4 The residual algorithm for recursive declaration

We sketch in figure 3 the algorithm dedicated to recursive declarations.

The *scan* methods uses the classification of operators in *variables*, *switches*, and *non-growing* operators. The method `variable:scan` fails, i.e. indicates that there is a non-guarded variable

Algorithm for `parallel:mk-trans(p|q,<actions parameters>)`

transitions = $\emptyset$

foreachmove $< p \xrightarrow{a} p' >$
  transitions = transitions $\cup < a, p'|q >$
  foreachmove $< q \xrightarrow{b} q' >$
  if a is inverse of b then transitions = transitions $\cup < \tau, p'|q' >$
foreachmove $< q \xrightarrow{b} q' >$
    transitions = transitions $\cup < b, p|q' >$
return transitions

Figure 2: `parallel:mk-trans`

Algorithm for `let-rec:tta`(let rec $\{x_i = t_i\}_{i=1,n}$ in x_k, <sieve>)

Equations = $\{\langle x_i, t_i, \text{<sieve>}\rangle\}_{i=1,n}$
state$_0$ = $\langle x_k, t_k, \text{<sieve>}\rangle$
STATES = {state$_0$}
UNEXPLORED = {state$_0$}
TRANSITIONS = $\emptyset$
while (UNEXPLORED $\neq \emptyset$) loop
  take $s = \langle x, t, \text{<sieve>}\rangle$ in UNEXPLORED
  UNEXPLORED = UNEXPLORED $- \{s\}$
  transition = $\emptyset$
  scan(t, <sieve>)
  TRANSITIONS = TRANSITIONS $\cup \{\langle s, \text{transitions}\rangle\}$
return $\langle$STATES, state$_0$, TRANSITIONS$\rangle$

Figure 3: The residual algorithm for recursive definitions

in the term. The methods `<sieve>:scan` recursively call `scan` on its process argument after modifying its sieve argument. Last, the methods `<switch>:scan` call recursively `scan` on each allowed subterm, producing some transitions, leading to some subterms: when these subterms do not correspond to some variable in the set Equations, the function `check` generates new equations to be added to the set. The function `add-transition` add a transition to the list *transitions*, and add the new states to the STATES and UNEXPLORED lists. Figure 4 sketches the *scan* method for some simple switches and for sieves.

5 Conclusion

The AUTO system we currently distribute is using specific hand-coded algorithms for the operators of the MEIJE0 calculus (stop, prefixing, sum, parallel, restriction, renaming). These algorithms were carefully optimized in order to avoid building parts of product automata that were to be deleted by some restriction operators.

We have built a new prototype of the AUTO system using the generic algorithms of this paper. Tests have been made both for the MEIJE0 calculus and for BASICLOTOS (the prototype has been presented in [MV89]). The MEIJE0 operators are correctly classified by our definitions: the prototype accepts stricty more MEIJE0 programs than the preceeding AUTO system. Some other MEIJE-SCCS operators (see [dSi85]) can be added easily to this syntax, including *ticking*,

operator	parameters	scan algorithm
`sequence`	(<action> : p, <sieve>)	action = <sieve>(action) if action then state = check(p, <sieve>) add-transition(action, state)
`sum`	($p + q$, <sieve>)	scan(p, <sieve>) scan(q, <sieve>)
SCCS `delay`	($delay(p)$, <sieve>)	action = <sieve>(tau-action) if action then state = check(p, <sieve>) add-transition(action, state)
<sieve>	(<sieve>$_1$(p), <sieve>$_2$)	scan(p, <sieve>$_2$ o <sieve>$_1$)

Figure 4: `scan algorithms`

interleaving and the *synchronized product* as non-growing operators. The results are good also for BASICLOTOS operators: the usual finiteness conditions are correctly deduced from the rules. Some limitations of our conditions are listed in the annex. We also obtained efficiency mesurements: this version appears to have the same order of performances than the old version. Moreover, it should be clear that in many cases it allows to build a smaller number of automata (for sieves never require to copy an automaton) and to apply sieves on smaller automata. No optimizations have been done in the first prototype, so the new version is potentially much more efficient than the specialized MEIJE0 version.

The set of programs accepted by the generic version is of course larger than in the former system, and programs may be written in a much more permissive way: for example parallel compositions may be done in many different ways using various operators and nested recursive definitions are allowed.

The congruence properties of some equivalences versus MEIJE composition operators are also to be generalized. It is very important for space efficiency reasons to apply reductions as deep as possible in the term, in order to create and compose smaller automata. We plan to have the ECRINS system proving congruence laws for various equivalences and various operators, so that the congruence properties can be automatically used in AUTO during the automata construction.

6 Annexes

Not all the operators mentionned in this paper are widely known. Let us give the rules of some of them.

In [Bo85] is defined the MEIJE-Sccs calculus, with operators from both Boudol's MEIJE calculus and Milner's SCCS calculus. The operators *hiding, ticking, clock* that occur in the paper are taken from the MEIJE-SCCS calculus. Let us give other examples:

```
operator delta:: Process --> Process
syntax delta
semantics
  delayed           ---------------------------
                    delta(p) -- tau --> delta(p)
  undelayed                p -- a --> p'
                        --------------------
                        delta(p) -- a --> p'
end
operator DELTA:: Process --> Process
syntax DELTA
semantics
  DELTA_1                  p -- a --> p'
                      -----------------------------------
                      DELTA(p) -- a --> delta (DELTA(p'))
end
operator desynchro:: Process --> Process
syntax desynchro
semantics
  desynchronised         ------------------------------------
                         desynchro(p) -- tau --> desynchro(p)
  undesynchronised               p -- a --> p'
                         -----------------------------------
                         desynchro(p) -- a --> desynchro(p')
end
```

The *delta* operator is a switch.

The *DELTA* operator does not match our conditions for non-growing operators, though it could be managed by had-hoc methods.

The *desynchro* operator is a degenerated form of sieve, that does not match our conditions. An extension of the sieves definition towards this type of operators would be possible.

The *external-choice* of TCSP, as mentionned in section 8 have both switching and sifting rules. This prevent us from using it in recursive definitions:

```
operator external-choice :: Process Process --> Process
syntax ext-choice left 6
semantics
p -- a -->p' & (not (a equal tau))      p -- a -->p' & (a equal tau)
----------------------------------    -----------------------------------
    p ext-choice q -- a -->p'         p ext-choice q -- a --> p' ext-choice q
q -- a -->q' & (not (a equal tau))      q -- a -->q' & (a equal tau)
----------------------------------    ------------------------------------
    p ext-choice q -- a -->q'         p ext-choice q -- a --> p ext-choice q'
end
```

Proof of theorem 2: We define a finite set of *states* by induction on the structure of the term, then we prove that the transition system of the term maps in this state space, with a finite number of transitions from each state.

Let us call *state* a pair $St = < P, S >$ where P is a process expression and S is a (composition of) sieve.

We build recursively the set $\mathcal{S}$ of states by:

St1: The *initial state* $St_0 =< e_0, I >$, where I is the identity sieve, is in $\mathcal{S}$.

St2: Given a state $< P, S >$ in $\mathcal{S}$, with $P = x_i$, then $< e_i, S >$ is in $\mathcal{S}$.

St3: Given a state $< P, S >$ in $\mathcal{S}$, with P having TSF, then for all states st of the automaton of P, extending $\lll$ such that st is smaller than all operators, then $< st, S >$ is in $\mathcal{S}$,

St4: Given a state $< P, S >$ in $\mathcal{S}$, with $P = Op(p_i, q_j)$, Op is a switching operator for the x_i arguments corresponding to the p_i, and the q_j have TSF, let $\{r_i\}$ be the rules in which the x_i are working, $\{r_j\}$ the other rules of Op, $\{rp_i\}$ and $\{rq_j\}$ their respective resulting processes, then:

St4.1: for all r_i, $< p_i, S >$ is in $\mathcal{S}$,

St4.2: for each r_j, call q_g the subset of $\{q_j\}$ working in r_j, introduce new constants c_{g_k} of type process, one for each state of each automaton of q_g, stating that these new constants are smaller than all operators w.r.t. $\lll$, then for all terms $q' \lll Op(p_i, q_j)[c_{g_k}/q_g]$ for some combination of the c_{g_k}, $< q', S >$ is in $\mathcal{S}$ (if the rule is an axiom, then take $q' \lll Op(p_i, q_j)$).

St5: Given a state $< P, S >$ in $\mathcal{S}$, with $P =$ <sieve>(P'), then $St' =< P', S \circ$ <sieve> $>$ is in $\mathcal{S}$.

The set of process expressions in $\mathcal{S}$ is finite, because they all are $\lll$ to one of the e_i (cases 1 and 2 introduce some e_i, cases 3 and 4 introduce expressions $\lll$ to some expression in $\mathcal{S}$, case 5 introduces subterms of expressions in $\mathcal{S}$, and $\lll$ is compatible with the subterm ordering). The set of sieves compositions in $\mathcal{S}$ is finite by proposition 1. So $\mathcal{S}$ is finite.

Remark that any process expression can be written $s_1(\ldots(s_k(P))\ldots)$, where all s_i are sieves, and the head operator of P is not a sieve; let us write this decomposition $S(P)$.

Now we prove that all reconfigurations of *Proc* decompose into a $S(P)$ such as $< P, S >$ is in $\mathcal{S}$, and have finitely many transitions. This will achieve the proof.

Let us call subreconfigurations of *Proc* the smaller set R of process expressions containing e_0, closed under *switching argument* relation (if $Op(p_i, q_j) \in \mathcal{R}$ and Op is a switch for the p_i, then the p_i are in $\mathcal{R}$), and closed under the transition relation (if $P' \in \mathcal{R}$ and $P' \xrightarrow{a} P''$ then $P'' \in \mathcal{R}$).

Let us prove by induction that:

$$P' \in \mathcal{R} \Longrightarrow P' \text{ is suitable for recursion on } \{x_i\} \text{ and } P' = P^*(S') \text{ with } < P^*, S' > \in \mathcal{S}$$

Ind0: e_0 verifies our property: it is suitable for recursion by hypotheses of theorem 2, and its decomposition is in $\mathcal{S}$, using construction St1, together with item 5 as many times as the head operator of e_0 is a sieve.

Ind1: For an operator Op that is a switch for p_i, the subterm corresponding to p_i is suitable for recursion by definition 9.3, and from $Op \in \mathcal{R}$ we get $< Op, I > \in \mathcal{S}$, so $< p_i, I > \in \mathcal{S}$ by St4.

Ind2: Consider a transition $P' \xrightarrow{a} P''$, with with $P' \in \mathcal{R}$. By induction hypotheses, P' is suitable for recursion, and its decomposition $< P^*, S' >$ is in $\mathcal{S}$. Trivially, P^* is suitable for recursion. Let us show that $P'' = P^{**}(S'')$ By case analysis on definition 9:

Ind2.1: $P^* = x_i$: then its transitions are those of e_i, that is suitable for recursion, and $< e_i, S' >$ is in $\mathcal{S}$ by St2 leading to the conclusion by application of the induction step on $< e_1, S' >$ (there may be no loop in the proof here, because of the *guarded* hypotheses).

Ind2.2: P^* is closed w.r.t $\{x_i\}$, and has TSF: all its reconfigurations are similarly suitable for recursion (by definition 9.2), and they map to the $< st_i, S' >$ states introduced by ST3.

Ind2.3: $P^* = Op(p_i, q_j)$ where Op is a switch for the x_i arguments corresponding to the p_j; its transitions are built:

Ind2.3.1: either using a rule with one of the p_i working; the resulting process P^{**} is got from a transition of p_i itself, so it verifies our property (apply the induction step on $< p_i, S' >$),

Ind2.3.2: or using a rule in which no p_i are working; by the definition of suitability, all q_j have TSF. Consider the product automaton of the q_j working in this rule (if the rule is an axiom, this is a one state blocked automaton); each state of this compound automaton correspond to a reconfiguration of $Op(p_i, q_j)$ that may involve some of the p_i, but these are unchanged. Those states map to some of the q' of St4.2. Their transitions are those of the compound automaton, plus some new transitions using r_i rules, leading to a p_i' reconfiguration. In the first case, $< q', S' >$ is in $\mathcal{S}$, and we keep suitability (for all subsequent reconfigurations are still switches for the p_i). In the second case, as in Ind2.3.1, we keep our property.

Ind2.4: $P^* =$ <sieve>(p) is not possible by definition of the decomposition.

■

References

[Ai86] G. Ailloux, "*Verification in Ecrins of Lotos Programs*", ESPRIT/SEDOS/C2/N45 (1986)

[BB88] T. Bolognesi, E. Brinksma, "Introduction to the ISO Specification Language LOTOS", in *The Formal Description Technique LOTOS*, North-Holland, 1988

[BK86] J.A. Bergstra, J.W. Klop, "Process Algebra: Specification and Verification in Bisimulation Semantics", *CWI Monographs*, North-Holland, 1986

[BS87] T. Bolognesi, S. A. Smolka, "Fundamental Results for the Verification of Observational Equivalence: a Survey", proc. of the IFIP 7th Internaional Symposium on Protocol Specification, Testing, and Verification, North-Holland, 1987

[Bo85] G. Boudol, "Notes on Algebraic Calculi of Processes", *Logics and Models of Concurrent Systems*, NATO ASI series F13, K.Apt ed., 1985

[Bro83] S. Brookes, "*A Model for Communicating Sequential Processes*", PhD Thesis, University of Oxford, 1983

[dSi85] R. De Simone, "Higher-Level Synchronising Devices in Meije-Sccs", *Theoretical Computer Science* 37, p245-267, 1985

[dSV89] R. De Simone, D. Vergamini, *"Aboard AUTO"*, Technical Report INRIA RT111, 1989

[GN] H. Garavel, E. Najm, "TILT: From LOTOS to Labelled Transition Systems", in *The Formal Description Technique LOTOS*, North-Holland, 1988

[MV89] E. Madelaine, D. Vergamini, "AUTO, a verification tool for distributed systems using reduction of automata", in proceedings of *Forte'89 conference*, Vancouver, North-holland, 1989

[MdSV89] E. Madelaine, R. de Simone et D. Vergamini, *"ECRINS, A Proof Laboratory for Process Calculi, User Manual"*, to appear INRIA, 1990

[Mi80] R. Milner, *"A Calculus for Communicating Systems"*, Lectures Notes in Comput. Sci. 92, 1980

[Mi83] R. Milner, "Calculi for Synchrony and Asynchrony", *Theoretical Computer Science* 25, p267-310, 1983

[RS89] V. Roy, R. De Simone, *"AUTO - AUTOGRAPH"*, submitted to Workshop on Computer-Aided Verification, Princeton, N.J., 1990

[VG88] F.W. Vaandrager, J.F. Groote, "Structured operational semantics and bisimulation as acongruence" CWI report CS-R8845, 1988

[Ve89] D. Vergamini, "Verification of Distributed Systems: an Experiment", in *Formal Properties of Finite Automata and Applications*, LNCS 386, 1990

DIMACS Series in Discrete Mathematics
and Theoretical Computer Science
Volume 3, 1991

A Computational Theory and Implementation of Sequential Hardware Equivalence*

Carl Pixley
Microelectronics and Computer Technology Corporation (MCC)
Computer Aided Design Program

September 27, 1990

Abstract

A common design problem is to replace one sequential design by another. Sequential design replacement may be required in order to meet performance (speed or area) goals in mapping from one technology to another or within a given technology. In practice, the resulting sequential design must be checked against the original, whether the redesign is done by-hand or with synthesis tools. Functional equivalence is often imperfectly checked by partial simulation of the sequential behavior of the redesigned part by itself or in its intended environment.

As an alternative to partial simulation, a theory of synchronous hardware equivalence based upon a gate-level model of behavior is presented. Efficient algorithms based upon predicate logic over boolean domains are employed to decide the equivalence of two designs from their netlist descriptions. These algorithms are based upon an efficient encoding of the state transition diagram that does not require explicit enumeration of all possible states of the designs. These algorithms also do not require any knowledge of state encodings. Algorithms are presented to calculate minimal input sequences that force original and replacement designs to behave identically. The algorithms are illustrated by examples.

*This report is a revision of MCC Technical Report: CAD 448-89 (Q)

It will be obvious that this theory can be extended to encompass a more general theory in which propositions about the behaviors of sequential designs can be decided automatically. To illustrate the power of this approach, an algorithm is presented that computes the set of states for which a formula of branching time temporal logic holds.

1 Introduction

1.1 A Model of Synchronous Design

Synchronous designs are modeled at the gate level in terms of combinational elements and primitive storage elements. A primitive storage element (PSE) is a device that transports its input to its output on a clock event and holds the output value until the next clock event. An example of a primative storage element is a simple D flip-flop (without enable or reset). Fortunately, most real storage devices, such as D flip-flops with enable, reset and both Q and Qn outputs, can be modeled as a network of these primative storage elements and combinational logic.

A synchronous hardware design is defined to be an interconnection of purely combinational elements and primitive storage devices. Each interconnection (i.e., net) is required to have exactly one driver (design input or output of a combinational or storage device). We require that there are no loops of purely combinational elements.

Given a design, D, let $\vec{in}$, $\vec{q}$, and $nx\vec{t}\text{-}q$ be vectors of boolean variables representing, respectively, the inputs, current states of the PSEs, and next-states of the PSEs of the design. Given a netlist and device-models for the design, we compute a predicate (i.e., a boolean valued function of several boolean variables),

$$transition(\vec{q}, \vec{in}, nx\vec{t}\text{-}q)$$

that has value 1 if and only if given inputs $\vec{in}$ and current values $\vec{q}$ of the registers, $nx\vec{t}\text{-}q$ are the values of the registers after one clock event. The predicate, *transition*, is just an encoding of the state transition graph of the design. We employ Ordered Binary Decision Diagrams (obbds) [1, 2] as an efficient way to represent and calculate predicates. The importance of obbds is that the size of an obdd representation of transition is not related to the

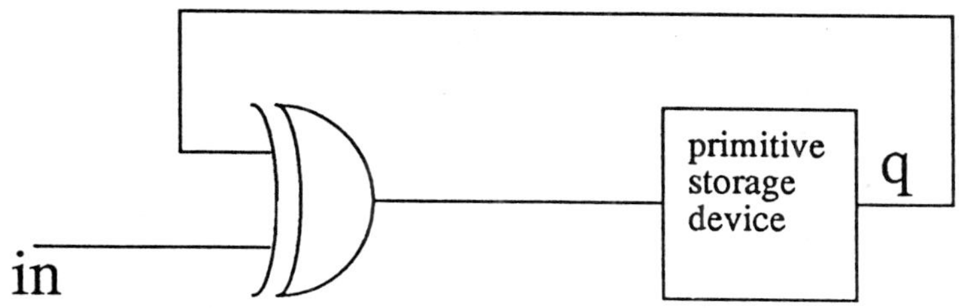

transition(q, in, next-q) = in&(q xor next-q) or ~in&(q xnor next-q)
= ~(in xor q xor next-q)

	q	in	next-q
extension = {	(1,	1,	0),
	(1,	0,	1),
	(0,	1,	1),
	(0,	0,	0) }

Figure 1

size of the state transition graph. The design output functions play no role in the definition of *transition*.

For example, given the design in Figure 1, $transition(q, in, next\text{-}q)$ is represented by the formula,

$$in \& (q \ xor \ next\text{-}q) \ or \ \neg in \& (q \ xnor \ next\text{-}q)$$

The designs are modeled as Mealy machines. Output functions, $Out(\vec{in}, \vec{q})$, are derived from the netlists. We treat Moore machines, in which each output is a combinational function of the PSE's alone, as a special case. A **model** of design D consists of (1) declarations of boolean variables $\vec{in}$, $\vec{q}$, and $\vec{nxt}\text{-}q$, (2) the *transition* function, and (3) the $\vec{Out}$ functions. In our theory, the model totally characterizes the behavior of D.

Sometimes it will be convenient to describe a predicate, such as *transition*, as a set of tuples (in this case, three-tuples: $\vec{q}$, $\vec{in}$, $\vec{nxt}\text{-}q$). What we mean is the set of tuples of vectors of booleans, $(\vec{q}, \vec{in}, \vec{nxt}\text{-}q)$, such that $transition(\vec{q}, \vec{in}, \vec{nxt}\text{-}q) = 1$. As a slight variation of usual mathematical parlance, the set of tuples is called the **extension** of the predicate and the predicate is called the **characteristic function** of the set of tuples.

Sometimes it is necessary to constrain the inputs by a predicate,

$$valid\text{-}inputs(\vec{in})$$

It is trivial to relativize the theory to account for input constraints.

1.2 Comparing Synchronous Design Pairs

Suppose we are given two designs, D_0 and D_1, with the same inputs, $\vec{in}$, and outputs, $\vec{out}$. Suppose vectors of boolean variables, $\vec{q_0}$ and $\vec{q_1}$, represent the values of the storage devices of the two designs. We consider a single design, consisting of disjoint copies of D_0 and D_1 except that corresponding inputs are identified. In addition, we define a new predicate, $Equivalent\text{-}Outputs(\vec{q_0}, \vec{q_1})$ that is TRUE if and only if for each pair of corresponding outputs, Out_0 and Out_1, the following is true. For all inputs, $\vec{in}$, $Out_0(\vec{in}, \vec{q_0}) = Out_1(\vec{in}, \vec{q_1})$.

We say that a pair of states, $(\vec{q_0}, \vec{q_1})$, are **equivalent** if and only if for each pair of states, $(\vec{q_0}', \vec{q_1}')$, reachable from $(\vec{q_0}, \vec{q_1})$ by any finite sequence of inputs (the same for both designs), $Equivalent\text{-}Outputs(\vec{q_0}', \vec{q_1}') = 1$. Therefore, a

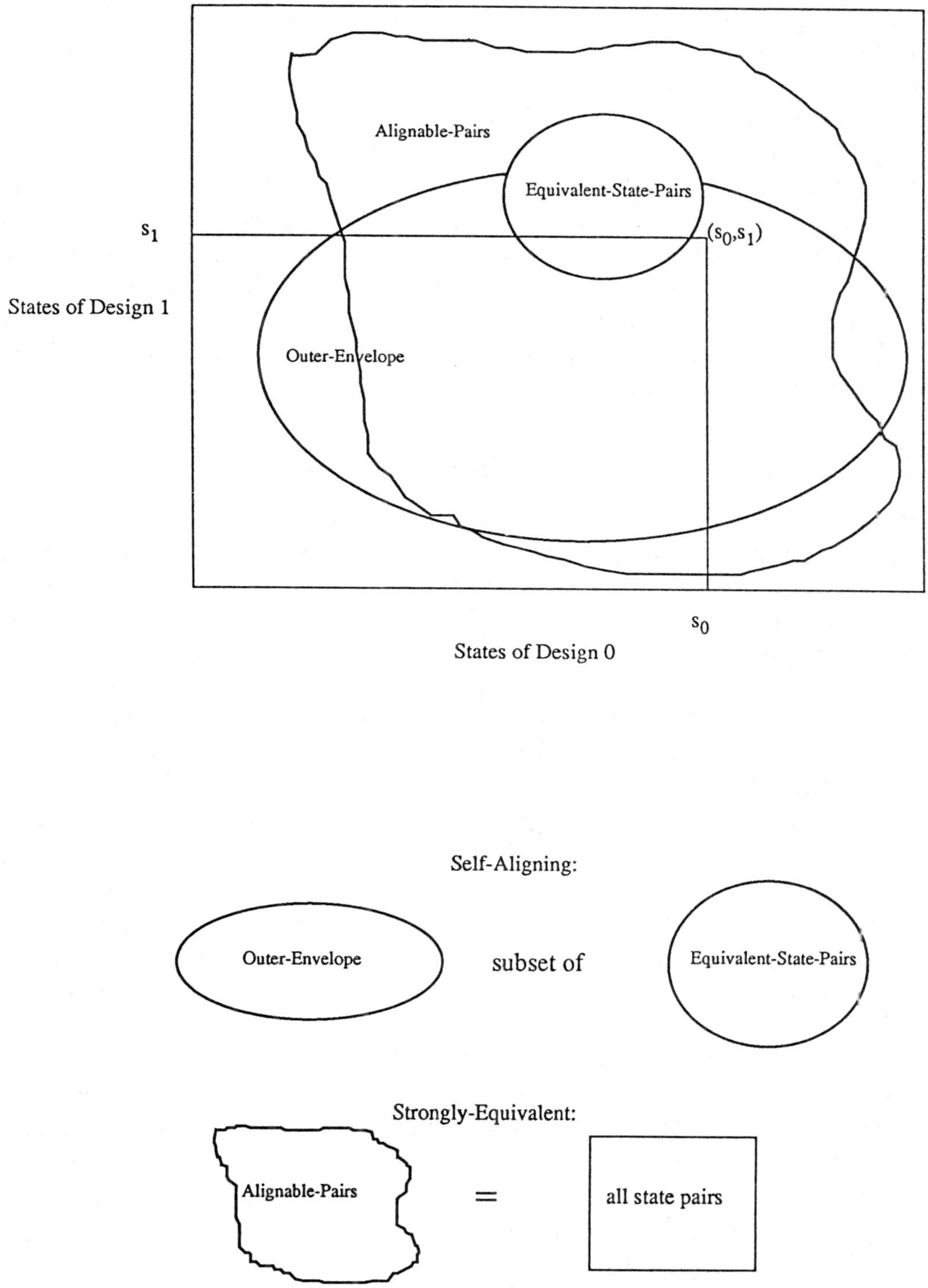
Alignable-Pairs
Equivalent-State-Pairs
s_1
(s_0,s_1)
States of Design 1
Outer-Envelope
s_0
States of Design 0
Self-Aligning:
Outer-Envelope
subset of
Equivalent-State-Pairs
Strongly-Equivalent:
Alignable-Pairs
=
all state pairs

Figure 2

pair, $(\vec{q_0}, \vec{q_1})$, is equivalent if and only if whenever design D_0 is initialized to state q_0 and D_1 is initialized to q_1, the designs will produce the same outputs for any finite sequence, no matter how long, of identical inputs.

Let *equivalent-state-pairs*$(\vec{q_0}, \vec{q_1})$ be the predicate having value 1 if and only if $(\vec{q_0}, \vec{q_1})$ is an equivalent pair. We give an algorithm to compute equivalent-state-pairs. If equivalent-state-pairs has value FALSE, then there is no pair of states, $(\vec{q_0}, \vec{q_1})$, having the property that all states reachable from it generate equivalent output functions. Having equivalent-state-pairs, it is trivial to decide whether machine D_0, initialized to $\vec{q_0}$, will have the same output behavior as machine D_1, initialized to $\vec{q_1}$, for all possible sequences of valid inputs starting from states $(\vec{q_0}, \vec{q_1})$. Figure 2, illustrates several of the concepts discussed in this paper.

1.3 Comparison to Symbolic Simulation

We digress to point out a relationship between symbolic simulation [4] and the present theory. Suppose we are given two designs, D_0 and D_1, which, for the sake of simplicity, are Moore machines. Let p_0 be a pair of specific boolean values, $(\vec{q_0}, \vec{q_1})$, assigned to the storage devices of the two machines, we can symbolically simulate the two machine for any number of cycles and check whether their output functions on the nth cycle are the same boolean function of the first n input variables. If for some cycle there is a discrepancy between some pair of corresponding outputs, then p_0 is not an equivalent pair.

On the other hand, there exists a number of cycles, N, such that if each pair of corresponding outputs agrees (symbolically) for all cycles less than N, then p_0 is an equivalent pair and the machines will produce the same outputs for *all* possible sequences of inputs of any length. What is the number N? Consider the set of all pairs of states, p, reachable from p_0 by some sequence of inputs. For each such pair, p, let n_p be the length of the shortest sequence of inputs to get from p_0 to p. The number N, is the minimum of all the numbers n_p.

If the design is symbolically simulated for N cycles, every pair of states will be visited. Therefore, N is a sufficiently large number to guarantee correctness if the output functions agree for all sequences of N or fewer cycles of symbolic inputs. However, how does one calculate N in practice? We know of no convenient way to find an acceptable lower bound for N by analyzing the physical structure of the circuit (unless there are no feedback loops among registers).

However, with the *transition* function one can find N as follows. For each i, let A_0 be the set containing only the pair, $(\vec{q_0}, \vec{q_1})$. Let A_{i+1} be the union of A_i and all states reachable from A_i by one transition (we will show how to compute this using *transition* later). The sequence, $A_0, A_1, \ldots$, is increasing and since the set of all state pairs is finite, for some i, $A_i = A_{i+1} = \ldots$. The number N is the least number, i, such that $A_i = A_{i+1}$.

This algorithm illustrates the general technique we use to compute equivalent-state-pairs and many other interesting predicates.

1.4 Initialization and Alignability

Correct operation of sequential hardware depends upon the ability to return to a good state from an arbitrary state. A state, S is defined to be a **reset state** if there is some sequence of inputs (called a homing or reset sequence) that will transform an arbitrary state into S. Quite often hardware is designed to have a short reset sequence that will drive the design into a specific reset state. Though we do not discuss it here, the techniques of this paper can be used to discover reset states and reset sequences.

If a designer knows a reset sequence for an original design, she may use the same reset sequence on the set of all pairs of states (of the original design and its replacement) to discover if the set of resulting pairs is a subset of the equivalent-state-pairs. If so, the same reset sequence for the original design resets the new design as well. If not, there is some pair of states, $(\vec{q_0}, \vec{q_1})$, such that the reset sequence will result in a pair of states that are not equivalent.

However, in many cases in which a part of a larger design is being replaced, a designer may not know a reset sequence for the part of the original design. Given that equivalent-state-pairs is not the empty set (i.e., there exists some pair of equivalent states), it is desirable to ask how to get from an arbitrary initial pair of states to an equivalent pair. That is to say, what sequence of inputs will force the replacement design to behave like the original?

A state pair, $(\vec{q_0}, \vec{q_1})$, is **alignable** if and only if there is some sequence of inputs to the two designs such that when applied to $(\vec{q_0}, \vec{q_1})$, the resulting state is in the equivalent-state-pairs. We show how to compute the predicate *alignable-pairs*$(\vec{q_0}, \vec{q_1})$. We prove that if the set of alignable pairs is the set of all possible pairs, then there exists a *single* sequence of inputs that will align *all* pairs of states. We show how to compute the exact conditions on the minimum sequence of inputs that will align every pair of states.

We illustrate by counterexample that if the set of alignable state pairs is not the whole space of pairs then it is not, in general, true that there exists one sequence of inputs that will align every alignable pair of states.

We define a pair of designs to be **strongly equivalent** if every pair of states is alignable. We show that strongly-equivalent is an equivalence relation on the set of all hardware models which are strongly equivalent to themselves. Finally, we prove a theorem that allows a part of a design to be replaced by another part without affecting the behavior of the larger part.

1.5 Self-Comparison

It is useful to compare a sequential design, D to itself. If we compare a machine to itself (by creating distinct logic variables, $\vec{q_A}$ and $\vec{q_B}$, for the storage elements, $\vec{q}$, of D) then the equivalent-state-pairs represents pairs of equivalent states. Of course, for any assignment $\vec{q}$ of values to storage elements, the pair $(\vec{q_A}, \vec{q_B})$ will be represented in the equivalent-state-pairs. It is easy to compute a representation, $Different(\vec{q_A}, \vec{q_B})$, of assignments that differ for at least one storage element q. The predicate *Different&equivalent-state-pairs* represents all equivalent states that are not identical.

If D has an initial reset state, *Reset*, then we can compute the set, *Reachable-Pairs*, of all pairs of the two copies of D reachable from *Reset*. Let *Non-Minimal* be

equivalent-state-pairs&Different&Reachable-Pairs

The following is then an obvious theorem.

Theorem 1 *Non-Minimal has value FALSE if and only if design D does not have two different, but equivalent, states reachable from Reset.*

It is also useful to discover if a design is strongly equivalent to itself. Designs that are not strongly equivalent to themselves allow at least two modes of operation that have distinct output behaviors and which cannot be reset to equivalent states by any input sequence.

1.6 Related Research

In the classical theory of finite state machines [5], two fsms are defined to be equivalent provided they "accept" the same input sequences. A fsm "accepts" an valid input sequence if, starting from a preferred initial state, the sequence leads to one of a set of designated final output states. Because, the theory presented here deals with comparison of two gate-level models (as defined by interconnections of logic gates and primative storage elements), there is no notion of initial state, final state, or valid input to a state (though the notion of equivalence can be modified to reflect input constraints).

The present theory separates the behavior of the hardware from of the intentions of the designer. Therefore, "state" is any one of the 2^n assignments of boolean values to the n many primative storage elements of a design and hardware is modeled in terms of its sequential input/output behavior from any initial state. This viewpoint is dictated by the fact that the initial state of a machine, when it is powered up, cannot be reliably predicted.

Also, the present theory is primarily concerned with equivalence of two hardware designs, rather than comparison of a design to a "specification," though the latter can be accomplished with similar computational techniques. This viewpoint is dictated by the desire to understand when one design can safely be replaced by another, equivalent, design. If *seq* is a "homing" or "reset" sequence intended to drive any state of design A to a preferred initial state, we can test whether *seq* is an aligning sequence for the pair of designs, A and B. If so, *seq* is also a homing sequence for B. In any case, if A is equivalent to B, any aligning sequence for the pair, followed by *seq*, will be a homing sequence for B.

Coudert, Berthet, and Madre [3] present a verification method based upon characteristic functions of sets of tuples of boolean values and quantifiers. Starting from a pair of initial states they compute the set of reachable states until either a discrepancy between output functions is recognized or all reachable state pairs have been examined. By contrast, the method presented here, extracts the set of all pairs (possibly empty) for which the two machines have the same input-output behaviors. Surprisingly, our experiments indicate that our algorithm generally converges faster. However, their more incremental approach purports to be able to handle larger designs.

Devadas et al. [8] describe a method for comparing the state transition graphs of two finite automata (derived either from gate, register transfer or

ISP-like specifications) with boolean simulation. In contrast to the theory presented here, their comparison algorithm presupposes that the machines start in a valid initial state. The efficiency of their algorithm derives from their ability to extract "don't care" information from the RTL or logic-level description which allows them to greatly reduce the number of cases to simulate. In contrast, it is a remarkable property of the bdd representation of the state transition relation employed in the current work, that only dependencies among logic variables which affect the characteristic function are explicitly represented.

Symbolic simulation has been shown by Randal E. Bryant and his students [4] to be a effective method for checking that a hardware implementation responds correctly to a sequence of symbolic inputs. This approach can avoid the combinational explosion that would normally occur in boolean simulation when evaluating circuit operation over many combinations of input and initial state. Symbolic simulation can effectively compare the output behaviors of two designs with the same (finite) sequence of symbolic inputs but cannot, by itself, establish the equivalence of output behaviors for all possible sequences of inputs of arbitrary length as in the present theory. On the other hand, symbolic simulation can complement the present approach by helping to explain to designers why two designs with initial state assignments which are thought to be equivalent, fail to have the same output behaviors.

Bose and Fisher [10] describe a temporal logic model checker for Mealy machines. Their model is specifically adapted to deterministic finite state machines (i.e., the next-state is a function of the current state and inputs). The present work describes model checking in a more general setting and gives a detailed algorithm for evaluating temporal logic formulas.

It is well-known that the sizes of ordered binary decision diagrams is sensitive to the ordering of boolean variables employed. For combinational circuits, orderings of circuit inputs derived from the circuit interconnections are described in [6] and [7]. In the present work, there is the added complication of ordering both circuit inputs and PSE outputs (representing both state and next-state variables). In the present work this is accomplished by a rather simple depth-first heuristic. However, optimal ordering of variables is a continuing subject of investigation.

None of the above works present a theory of equivalent hardware design involving the notions of alignability, self-comparison, or a definition of design equivalence.

2 Theory and Algorithms

Recall that for any design, boolean variables, $\vec{in}$, are associated with the inputs to the design. In addition, boolean variables, $\vec{q}$ and $\vec{nxt} - q$, are associated to the storage elements of the designs. This section gives algorithms for computing useful predicates, such as transition, equivalent-state-pairs, outer-envelope, and alignable-pairs, involving these variables.

From [1] and [2], we can calculate an ordered binary decision diagram (obbd) for any combination of logic functions: or $\vee$, and $\&$, xor $\wedge$, xnor $\equiv$, not $\neg$, and so forth. The meaning of $\forall x, foo(x)$ is $foo(0)\&foo(1)$. The meaning of $\exists x, foo(x)$ is $foo(0) \vee foo(1)$. There are efficient ways of calculating existential and universal quantification of boolean functions.

In our pseudo-code, := is the assignment operator and == is a test for the equality of predicates. We use the "loop until" construct in the usual way. Generally, *char* and *nxt_char* are used as variable names to indicate the characteristic function of a set of tuples and the next characteristic function, respectively.

2.1 Calculating *Transition* and *Equivalent-Outputs*

Using the device model descriptions, we replace each storage device by an equivalent interconnection of combinational and sequential elements. For the input to each primitive storage element, q, express the input to q as a boolean function, *in-q* of variables, $\vec{q}$ and $\vec{in}$. For each variable, *nxt-q*, let *nxt-predicate*$(\vec{q}, \vec{in}, \vec{nxt}\text{-}q)$ be the predicate, $nxt\text{-}q \equiv in\text{-}q$. The transition predicate, *transition*$(\vec{q}, \vec{in}, \vec{nxt}\text{-}q)$, is then the conjunction of all the *nxt-predicate* predicates.

For corresponding outputs, Out_0 and Out_1, of the two designs, express the output as a function, $Out\text{-}fun_0(\vec{in}, \vec{q_0})$ and $Out\text{-}fun_1(\vec{in}, \vec{q_1})$ of the inputs and q-values. Let $Out(\vec{q_0}, \vec{q_1})$ be the predicate $\forall \vec{in}, (Out_0(\vec{q_0}, \vec{in}) \equiv Out_1(\vec{q_1}, \vec{in}))$. *Equivalent-Outputs*$(\vec{q_0}, \vec{q_1})$ is the conjunction of all *Out*'s.

2.2 Calculating *Equivalent-State-Pairs*

Let A_0 be the set of state pairs, $(\vec{q_0}, \vec{q_1})$, for which *Equivalent-Outputs* $\equiv 1$, i.e., corresponding outputs of the two machines agree. In general, let state pair, $(\vec{q_0}, \vec{q_1})$ belong to A_{i+1} if and only if *Equivalent-Outputs*$(\vec{q_0}, \vec{q_1}) \equiv 1$

and the set of all states immediately reachable from $(\vec{q_0}, \vec{q_1})$ are in A_i. A simple induction argument shows that a state-pair, $(\vec{q_0}, \vec{q_1})$, belongs to A_i if and only if $Equivalent\text{-}Outputs(\vec{q_0}, \vec{q_1}) \equiv 1$ and for any sequence of inputs, $in_i, in_{i-1}, \ldots, in_1$, having length i, all of the states-pairs reached by that set of inputs satisfies Equivalent-Outputs. Therefore, the following is a theorem.

Theorem 2 *For all i, A_{i+1} is a subset of A_i.*

The intersection of all the A_i's is called the *equivalent-state-pairs*. We compute equivalent-state-pairs by the following algorithm. Since the space of all $\vec{q}$ is finite, the theorem guarantees that the loop must terminate. The predicate, *q's-are-nxt-q's* has value TRUE if and only if for each storage device, the variables, q and nxt_q, have the same value.

```
char := 1
next_char := Equivalent-Outputs

loop until (char == next_char)
  char := next_char
  correct-next-q's := (exist q's: char&q's-are-next-q's)
  ins-qs-point-to-correct-next-q's
      := (exist next-q's:transition&correct-next-q's)
  qs-always-point-to-correct-next-qs
      := (all ins:ins-qs-point-to-correct-next)
  next_char
      := Equivalent-Outputs&qs-always-point-to-correct-next-qs
end loop

equivalent-state-pairs := char
equivalent-state-pairs-non-empty
       := exist q's:equivalent-state-pairs
```

2.3 Self-Aligning Design and the Outer-Envelope

Sometimes the original design to be replaced is **self-resetting** in the following sense. There exists a number k, such that (if all storage elements are enabled) the state of the design depends only upon the inputs to the last k cycles, no matter what the inputs are. An important case of self-resetting design is pipelined logic circuits.

So given an original self-resetting design, D_0, and replacement, D_1, it may be that no matter which values are assigned to the storage elements, if the machines are clocked a sufficient number of cycles with *any* set of valid inputs, the machines will be in equivalent states. We call such a design pair **self-aligning**. In this section we show how to discover if a pair is self-aligning.

2.3.1 The Outer-Envelope

Let D be any sequential design. We define a set of states, S, to be **invariant under all inputs** if the set of states reachable from S in one clock cycle with arbitrary inputs, is S itself. Let predicate P(state1,state2) be true if and only if for some input vector, $\vec{in}$, there is a transition from state1 to state2.

Definition 1 *We say that state, s_0, points to state, s_1, if $P(s_0, s_1)$ is true.*

Given any set of states, S, let $f(S)$ be the set of states pointed to by some state in S, i.e.,

$$f(S) = \{s'; s \in S \,\&\, P(s, s')\}$$

Theorem 3 *Let A_0 be the set of all states. For all i, let $A_{i+1} = f(A_i)$. Then, for all i, A_{i+1} is a subset of A_i.*

Proof: Clearly, $f(A_0)$ is a subset of A_0.

Suppose, inductively, that A_{i+1}, which is $f(A_i)$, is a subset of A_i. Consider A_{i+2}, which is $f(f(A_i))$. Suppose A_{i+2} is not a subset of A_{i+1}. Let z be an element of A_{i+2} that is not an element of A_{i+1}. Since z is in A_{i+2}, there is a state, w, such that w is in A_{i+1} and such that w points to z. However, by the inductive hypothesis, w is in A_i and since w points to z, z is by definition in A_{i+1} contrary to assumption. Hence, A_{i+2} is a subset of A_{i+1}.

By the principle of mathematical induction, the proof is complete. QED

Definition 2 *The outer-envelope of the state transition graph is the intersection of all A_i's.*

Theorem 4 *Outer-envelope is non-empty.*

Proof: If A is non-empty, $f(A)$ is non-empty, since every state points to at least one other state. **QED**

Theorem 5 *Every set of states that is invariant under all inputs is a subset of the outer envelope.*

Proof: Inductively, every set of states invariant under all inputs is in each A_i. **QED**

If the design in question has input signals to disable all storage devices, then outer-envelope will be the set of all states. In this case, one should calculate outer-envelope with the inputs restricted to always enable the storage elements.

We now give a more formal definition of the notion of being self-aligning

Definition 3 *A pair of designs, D_0 and D_1, is* ***self-aligning*** *if there is a number, k, such that if P is any pair of initial states, and the machines are clocked k cycles with arbitrary but identical inputs, then the resultant pair of states is equivalent.*

Theorem 6 *Given two designs, D_0 and D_1, the pair is self-aligning if and only if the outer-envelope is a subset of the equivalent-state-pairs.*

Proof: Assume that the outer-envelope is a subset of the equivalent-state-pairs. We must show that the pair of designs is self-aligning. Consider the sets A_i of Theorem 3. Let k be the least number i such that $A_i = A_{i+1}$. If any pair of initial states is clocked k times, the resultant pair is in the equivalent-state-pairs and, hence, the resultant pair is equivalent.

Conversely, suppose that the pair of designs is self-aligning. Let k be an integer such that if any pair of initial states is clocked k cycles with arbitrary but identical inputs, then the resultant pair is equivalent. The set of states reachable from some initial pair in i cycles is A_i of Theorem 3. Therefore, A_k is a subset of equivalent-state-pairs and hence the outer-envelope is a subset of the equivalent-state-pairs. **QED**

The following is a program to compute the outer-envelope.

```
char := 0
next_char := 1
loop until (char == next_char)
 char := next_char
 image := exist in's q's:char&transition
 next-char := exist next_q's:image&q's-are-next-q's
end loop
outer-envelope := char
```

2.4 Calculation of *Alignable-Pairs*

Suppose that the set of equivalent state pairs is non-empty, i.e., the predicate, *equivalent-state-pairs*, is not FALSE. Let B_0 be the set of equivalent state pairs. For all $i > 0$, let B_{i+1} be the union of B_i and the set of state-pairs, p, such that for some vector of inputs, $\vec{in}_p$, if $\vec{in}_p$ is applied to p then the resulting state is in B_i. The set, B_i, is the set of state-pairs that can be transformed into equivalent-state-pairs in i or fewer cycles. Since B_i is a subset of B_{i+1}, let the union of all the B_i's be called the **alignable-pairs**. The following algorithm computes the predicate, *alignable-pairs*($\vec{q}$) from *equivalent-state-pairs.*

```
char := 0
next_char := equivalent-state-pairs

loop until (char = next_char)
  char := next_char
  new-align := (exist in's nxt-q's:transition &
                    (exist q's: char & q's-are-next-q's))
  next_char := char or new-align
end loop

alignable-pairs := char
all-pairs-are-alignable := (alignable-pairs xnor 1)
```

If the alignable-pairs is not all of pair-space, i.e., *alignable-pairs*($\vec{q_0}, \vec{q_1}$) does not have value TRUE, then there exists a state pair, $p = (\vec{q_0}, \vec{q_1})$, such that no sequence of inputs will drive the machine into equivalent states from initial states, p. The practical consequence of this information is that design D_1 is unlikely to be a suitable replacement for design D_2. If D_0 were powered up in state $\vec{q_0}$ and D_1 were powered up in state $\vec{q_1}$, then no sequence of inputs would force the machines to have identical behavior.

If p is a state pair in alignable-pairs, then there is some sequence of input vectors, say $align(p)$, that will transform p into the equivalent-state-pairs. There may be a different sequence for each p. The following theorem is, therefore, somewhat surprising.

Theorem 7 *If every state pair is in alignable-pairs, then there is a single sequence of input vectors that will align every pair of states.*

Proof: We show how to construct the aligning sequence. Let P be set of all pairs of states. Let R_0 be the set, $P \setminus$ *equivalent-state-pairs.*

If R_i is not empty, we define R_{i+1} as follows. Let $p(i)$ be an element of R_i. Let $align(p(i))$ be any aligning sequence for $p(i)$ and let $align_{p(i)}$ be the function from P into itself such that $align_{p(i)}(q)$ is the state resulting from applying the aligning sequence, $align(p(i))$ to state pair q.

Since the equivalent-state-pairs is invariant under all inputs, the image of equivalent-state-pairs under $align_{p(i)}$ is contained in equivalent-state-pairs. Let R_{i+1} be the image under $align_p$ of R_i minus equivalent-state-pairs, $align_p(R_i) \setminus$ *equivalent-state-pairs.* The state pair, $align_{p(i)}(p(i))$, is also in the equivalent-state-pairs. Therefore, the number of state-pairs in R_{i+1} is at least one fewer than the number of state-pairs in R_i. Notice that since alignable-pairs is the whole space, every pair in R_{i+1} is alignable.

Let i be any index such that R_i is empty. Letting ++ denote concatenation of sequences, a single aligning sequence for all pairs is $align(p(0))$ ++ $align(p(1))$ ++ ... $align(p(I-1))$. **QED**

Is it always the case that there exists a single aligning sequence for all pairs in alignable-pairs – even if the alignable-pairs is not the set of all state pairs? The answer is "no", as the following theorem shows.

Theorem 8 *There exist two designs such that there is no single sequence that will align every state pair in their alignable-pairs.*

Proof: Design, D_0, has four states numbered 0, 1, 2, 3. Design, D_1, has one state, numbered 3. There are, therefore, four state pairs:

$$\{(0,3),(1,3),(2,3),(3,3)\}$$

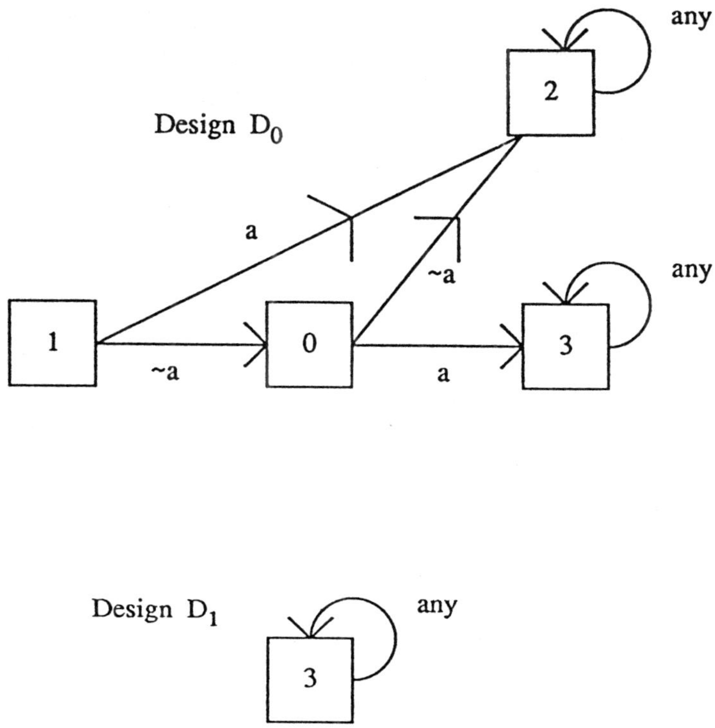

Figure 3

Each design has only one input, a, and two outputs that, for each design, is a binary encoding of the number of the state. Therefore, two states generate the same outputs if and only if their numbers are the same. Figure 3 illustrates the two state machines.

The state transition table for D_0 is the following.

$$\begin{array}{ccc} state & a-input & next-state \\ 1 & \neg a & 0 \\ 1 & a & 2 \\ 2 & any & 2 \\ 0 & a & 3 \\ 0 & \neg a & 2 \\ 3 & any & 3 \end{array} \quad \begin{array}{r} (1) \\ (2) \\ (3) \\ (4) \\ (5) \\ (6) \\ (7) \end{array}$$

The state transition table for D_1 is the following.

$$\begin{array}{ccc} state & a-input & next-state \\ 3 & any & 3 \end{array} \quad \begin{array}{r} (8) \\ (9) \end{array}$$

The equivalent-state-pairs consists of only one state pair:

$$\{(3,3)\}$$

Then alignable-pairs is:

$$\{(1,3),(0,3),(3,3)\}$$

The set of non-alignable state pairs is:

$$\{(2,3)\}$$

Notice that, for the pair $(1,3)$, every aligning sequence begins with $\neg a$ followed by a. However, for the pair $(0,3)$, every aligning sequence starts with a. Therefore, no single aligning sequence can align both $(1,3)$ and $(0,3)$ though they are both alignable. **QED**

We defined a pair of designs, A and B to be **strongly equivalent**, $A \cong B$, if the alignable-pairs is the set of all state pairs.

Theorem 9 *Strong equivalence is a symmetric and transitive relation.*

Proof: Clearly, if $A \cong B$ then $B \cong A$, so $\cong$ is symmetric. To show that $\cong$ is transitive, assume that $A \cong B$ and $A \cong C$. We must show that $A \cong C$. By Theorem 7, it is sufficient to show that there is a single aligning sequence that will drive every state pair of A and C into the equivalent-state-pairs. Let $sequence_{AB}$ be an input sequence that drives every A-B state pair into equivalent-state-pairs. Let $sequence_{BC}$ be an input sequence that drives every B-C state pair into its equivalent-state-pairs. We claim that $sequence_{AB}$ ++ $sequence_{BC}$ drives every A-C pair into its equivalent-state-pairs.

Let (a, c) be any A-C state pair. Let b be any state of design B. Consider the triple, (a, b, c). Applying $sequence_{AB}$ to the triple yields triple, (a_1, b_1, c_1) where a_1 and b_1 produce the same outputs for all input sequences. Applying $sequence_{BC}$ to (a_1, b_1, c_1) produces sequence, (a_2, b_2, c_2) where b_2 and c_2 produce the same output for all input sequences. Since a_2 and b_2 produce the same outputs for all input sequences, a_2 and c_2 must produce the same outputs for all input sequences. **QED**

We say a design, A, is **self-equivalent** if $A \cong A$. The design D_0 in the proof of Theorem 8 is not self equivalent because the pair of states (2,3) cannot be driven into equivalent-state-pairs for any sequence of inputs. There are compelling intuitive reasons for demanding that any acceptable sequential design be self-equivalent. If a design is not self-equivalent then there exist two states of the design such that no sequence of inputs will cause the two states to become equivalent. The following theorem shows that a non-self-equivalent design is isolated from all designs under the strong-equivalence relation.

Theorem 10 *If a design is not self-equivalent then it is not strongly equivalent to any design.*

Proof: If $D_0 \cong D_1$ then, by symmetry, $D_1 \cong D_0$. By transitivity, $D_0 \cong D_0$. **QED**

The following is a corollary to Theorem 8.

Theorem 11 *Strong-equivalence is an equivalence relation on the set of all self-equivalent designs.*

2.5 Calculating all Minimal Aligning Sequences

For this section, we assume that equivalent-state-pairs is non-empty and that the alignable-pairs is the set of all state pairs. Let *iteration-number* be the number of loop iterations required in calculating the alignable-pairs. A lower bound for the number of clock cycles to align from an arbitrary alignable state is (*iteration-number* − 1). It should be pointed out that many state machines are designed to have a very short align.ing sequence. Let $\vec{p}$ represent the pair of current states, $(\vec{q_0}, \vec{q_1})$.

Suppose we are given $transition(\vec{q}, \vec{in}, \vec{nxt-q})$ and $equivalent\text{-}state\text{-}pairs(\vec{q}$. For each i, we need a fresh set of variables, $\vec{in}_{-i}$, that represent the values of the inputs on cycle $-i$. We need a predicate, $in's\text{-}are\text{-}in'_{-i}s(\vec{in}, \vec{in}_{-i})$, that has value TRUE if and only if corresponding $in's$ and $in'_{-i}s$ have the same value. Let $\vec{fresh\text{-}q}$ be a fresh set of variables corresponding to q's. Let *next-q's-are-fresh-q's* be a predicate that has value TRUE if and only if corresponding next-q's and fresh-q's have the same value, and *q's-are-fresh-q's* be a predicate that has value TRUE if and only if corresponding q's and fresh-q's have the same value.

We iterate on i until the set of state pairs alignable by a single input sequence is the set of all state pairs. For each iteration, i, we compute

$$align(\vec{p}, \vec{in}_{-i}, \ldots, \vec{in}_{-1}, \vec{in})$$

which has value TRUE if and only if from initial state $\vec{p}$ and with input sequence, $\vec{in}(-i) \ldots \vec{in}(-1)\ \vec{in}$, the resultant state will be in the equivalent-state-pairs. The characteristic function of the set of state pairs, $\vec{p}$ that can be aligned by a single aligning sequence in i cycles, *alignable-in-i-cycles*, is the following.

$$\exists\, \vec{in}_{-i} \ldots \vec{in}_{-1}\, \vec{in} : \ align$$

```
next-fresh-transition
   := exist next-q's:transition & next-q's-are-fresh-q's
correct-next-envelope
   := exist q's:equivalent-state-pairs & q's-are-next-q's
align := exist next-q's:transition & correct-next-envelope
aligning-sequence := all q's:align
i := 0
loop until (align-sequence ~== 0)
```

```
  i := i-1
  fresh-transition
    := exist in's:next-fresh-transition & in's-are-in-i's
  align := exist q's:align & q's-are-fresh-q's
  align := exist fresh-q's:fresh-transition & align
  aligning-sequence := all q's:align
end loop
```

This program is different from the others we have seen because it requires an unspecified number of fresh variables, $in'_{-i}s$, because we must construct predicates, *in's-are-*$in'_{-i}s$, dependent upon i, and because we must quantify over an increasing set of variables to calculate *char*. In practice, it is wise to set a bound on i and calculate only to that bound. If the minimum length of a aligning sequence is above that bound, we will not discover it.

3 Examples

Our experimental implementation consists of a modification of the binary decision diagram (bdd) routines of Randal Bryant's COSMOS system to implement quantifiers. We also have a program that makes calls to the modified bdd program. The primary purpose of these experiments was to determine if the theory of state machine verification presented in this paper is computationally feasible; what are the sizes of the bdds generated, and how long does it take to calculate a bdd for the various sets of state pairs?

These examples show that, at least for relatively small designs, equivalence of state machines can be decided automatically in relatively little time. The data suggests a direct correlation between the number of boolean variables (inputs and storage devices) and the size of data structures created. We do not know the limits of design complexity for which these techniques are practical.

We have computed *transition*, *equivalent-state-pairs*, *outer-envelope*, *alignable-pairs*, and the minimum aligning sequence for several pairs of designs. It should be pointed out that the comparisons of these designs did not require any knowledge about their intended behaviors. The times given below refer to the time spent by the bdd program only. The examples were run on a Sun 4/110.

The first three pairs of examples were supplied by MCC shareholders. The next three were created by a commercial synthesis tools from existing designs. The rest of the examples come from the Berkeley/MCNC suite of examples which were synthesized with two different state encodings.

design-pair	ins	stores A	stores B	outs	dagsize transition	dagsize cor-env	time
axorb	16	8	9	1	1483	86	4.38 s.
lgc-mvd	10	5	6	3	486	19	2.82 s.
3-1's	1	4	3	1	204	20	1.5 s.
motor	2	2	4	2	69	15	.77 s.
2-mult	3	7	8	1	696	227	4.54 s.
4-mult	3	18	15	1	4394	12216	3.89 m.
bbara a-b	4	4	4	2	192	12	.98 s.
bbara b-o	4	4	10	2	270	38	1.65 s.
tbk a-b	6	5	5	3	1337	77	4.9 s.
tbk b-g	6	5	5	3	1434	81	4.77 s.
bbsse	7	4	4	7	800	44	2.3 s.
bbtas	2	3	3	2	55	17	.84 s.
beecount	3	3	3	4	104	1	.65 s.
cse	7	4	4	7	819	44	2.7 s.
dk14	3	3	3	5	157	19	.98 s.
dk15	3	2	2	5	52	8	.63 s.
dk16	2	5	5	3	534	85	3.12 s.
dk17	2	3	3	3	108	20	.61 s.
dk27	1	3	3	2	57	19	.45 s.
dk512	1	4	4	3	127	43	1.13 s.
donfile	2	5	5	1	416	1	1.46 s.
ex1	9	5	5	19	406	92	2.2 s.
ex2	2	5	5	2	406	92	2.2 s.
ex3	2	4	4	2	177	44	.98 s.
ex4	6	4	4	9	296	42	1.56 s.
ex5	2	4	4	2	205	43	1.2 s.
ex6	5	3	3	8	333	19	1.12 s.
ex7	2	4	4	2	194	44	1.15 s.
keyb	7	5	5	2	2123	88	5.66 s.
kirkman	12	4	4	6	70	44	2.75 s.

design-pair	ins	stores A	stores B	outs	dagsize transition	dagsize cor-env	time
lion	2	2	2	1	42	8	.37 s.
lion9	2	4	4	1	126	34	.71 s.
makrl	5	4	4	16	536	44	1.54 s.
mc	3	2	2	5	63	8	.4 s.
modulo12	1	4	4	1	78	1	.33 s.
opus	5	4	4	6	526	32	1.56 s.
planet	7	6	6	19	9436	188	21.51 s.
planet1	7	6	6	19	9436	188	22.13 s.
s1	8	5	5	6	9332	92	20.2 s.
s1a	8	5	5	6	10850	1	6.24 s.
s8	4	3	3	1	282	1	.4 s.
sand	11	5	5	9	59800	92	2.92 m.
scf	27	7	7	56	18639	388	21.53 s.
shiftreg	1	3	3	1	62	20	.55 s.
sse	7	4	4	7	800	44	2.15 s.
styr	9	5	5	10	6213	92	11.61 s.
tav	4	2	2	4	15	8	.47 s.
tbk	6	5	5	3	1434	81	4.75 s.
train11	2	4	4	1	188	36	.92 s.
train4	2	2	2	1	25	8	.33 s.

4 A New Decision Procedure for Branching-Time Temporal Logic

We demonstrate the applicability of our approach to state machine verification by showing how to model branch time temporal logic and how to calculate the temporal operators in our system. We follow the development in [9]. We do not give complexity measures for these algorithms. We have not implemented these algorithms and do not know their practical value.

We now generalize the notion of state. Let $\vec{q} = (q_1, \ldots, q_n)$ be n many boolean variables that represent **atomic propositions**. A state is an assignment of boolean values to the variables $\vec{q}$. Let $S(\vec{q})$ be any predicate that does not have value FALSE. The predicate, S, represents (i.e., is the characteristic function of) the domain of states in a model.

To model state transitions, we introduce a fresh set of variables, $nxt\text{-}\vec{q} = (nxt\text{-}q_1, \ldots, nxt\text{-}q_n)$. We use the familiar predicate, $q's\text{-}are\text{-}nxt\text{-}q's(\vec{q}, nxt\text{-}\vec{q})$, that has value TRUE if and only if corresponding q's and $nxt\text{-}q$'s have the same value. Let $nxt\text{-}S(nxt\text{-}\vec{q})$ be the predicate

$$\exists \vec{q} : S \& q's\text{-}are\text{-}nxt\text{-}q's$$

The predicate $nxt\text{-}S(nxt\text{-}\vec{q})$ has value TRUE if and only if $nxt\text{-}\vec{q}$ is a state of S.

Let $R(\vec{q}, nxt\text{-}\vec{q})$ represent the state transition relation. So $R(\vec{q}, nxt\text{-}\vec{q})$ has value TRUE if and only if $\vec{q}$ represents an S-state and $next\text{-}\vec{q}$ represents a state of S that is a successor of $\vec{q}$. We require R to be total on S, i.e.,

$$\forall state_0 \in S, \exists state_1 \in S : R(state_0, state_1)$$

In [9], the language CTL is defined as follows.

1. Every atomic proposition, q, is a CTL formula.

2. If f_1 and f_2 are CTL formulae, then so are $\neg f_1$, $f_1 \wedge f_2$, AXf_1, EXf_1, $A[f_1 \cup f_2]$, and $E[f_1 \cup f_2]$.

We now describe what it means for a CTL formula, f, to be **satisfied** by state s_0 of S, i.e., $s_0 \models p$. A **path** is an infinite sequence of states of S, $(s_0, s_1, \ldots)$, such that, for each i, $R(s_i, s_{i+1})$.

$$\begin{array}{lll}
s_0 \models p & iff & p\, is\, assigned\, TRUE\, in\, state\, s_0 \\
s_0 \models \neg f & iff & not(s_0 \models f) \\
s_0 \models f_1 \wedge f_2 & iff & s_0 \models f_1 \, and\, s_0 \models f_2 \\
s_0 \models AXf_1 & iff & for\, all\, states\, t\, such\, that\, R(s_0, t), t \models f_1 \\
s_0 \models EXf_1 & iff & for\, some\, state\, t\, such\, that\, R(s_0, t), t \models f_1 \\
s_0 \models A[f_1 \cup f_2] & iff & for\, all\, paths\, (s_0, s_1, \ldots), \\
 & & \exists i[i \geq 0 \wedge s_i \models f_2 \wedge \forall j[0 \leq j < i \rightarrow s_j \models f_1]] \\
s_0 \models E[f_1 \cup f_2] & iff & for\, some\, path\, (s_0, s_1, \ldots), \\
 & & \exists i[i \geq 0 \wedge s_i \models f_2 \wedge \forall j[0 \leq j < i \rightarrow s_j \models f_1]]
\end{array}$$

For each proposition, P, of CTL, we show how to compute the set of states, $s \in S$, such that $s \models P$, that is, we compute the extension of proposition P. Given a formula of CTL, the algorithm is applied to the parse tree for the formula from the leaves to the root. The leaves correspond to atomic propositions whose bdd representations are known. We show how to compute the representation of a CTL operator applied to propositions whose representations are known.

- If P is a primitive proposition represented by boolean variable q_i, then the set of S-states in which P is satisfied is represented by

$$S\&q_i$$

- If P is the proposition $\neg Q$ and $\hat{Q}(\vec{q})$ represents Q, then P is represented by

$$S\&(\neg\hat{Q})$$

- If P is EXQ and $\hat{Q}(\vec{q})$ represents Q, then to find the representation of P we need to find the representation of all S-states, s, such that for some s' satisfying $\hat{Q}$, $R(s, s')$. Let $next\text{-}\hat{Q}$ be $\exists\vec{q} : q's\text{-}are\text{-}next\text{-}q's\&\hat{Q}$. We see that P is represented by

$$\exists \vec{nxt\text{-}q} : R\&next\text{-}\hat{Q}$$

- If P is AXQ and $\hat{Q}(\vec{q})$ represents Q, then to find the representation of P we apply the standard trick of calculating

$$\neg EX(\neg\hat{Q})$$

- Suppose the proposition is $A[P \cup Q]$ and $\hat{P}$ and $\hat{Q}$ represent P and Q, respectively. There are two ways $A[P \cup Q]$ can *fail* to be satisfied in state, s. The first way is that $\neg P\&\neg Q$ hold in state s or there is a path of states starting at s, each satisfying $P\&\neg Q$ and terminating in a state satisfying $\neg P\&\neg Q$. The second is that s belongs to a loop of states satisfying $P\&\neg Q$ or that there is path of $P\&\neg Q$ states from s to a state on a loop of $P\&\neg Q$ states.

  The following defines the set of states for which there is a $P\&\neg Q$ path to a $\neg P\&\neg Q$.

```
P-and-not-Q := P & (not Q) & S
not-P-and-not-Q := ((not P) & (not Q) & S)
```

```
char := 0
next_char := not-P-and-not-Q
loop until (char == next_char)
 char := next_char
 next-not-P-and-not-Q := exist q's:(char & q's-are-nxt-q's)
 next_char
   := exist next-q's:P-and-not-Q & R & next-not-P-and-not-Q
 next_char := next_char or char
end loop
first-kind-of-bad-states := char
```

Next we calculate the largest set, W, (possibly empty) of $P\&\neg Q$ states having the property that if state s_0 is in W, then for some s_1 in W, $(s_0 R s_1)$. This is just the set of $P\&\neg Q$ states such that there is a $P\&\neg Q$ path to a loop of $P\&\neg Q$ states.

```
char := 0
next_char := P-and-not-Q & S
loop until (char == next_char)
 char := next_char
 bad-next := exist q's:(char & q's-are-next-q's)
 next-char := exist next-q's:(char & R & bad-next)
end loop
second-kind-of-bad-states := char
```

Finally the set of states for which $A[P \cup Q]$ does not hold is given by

```
bad-states
   := first-kind-of-bad-states or second-kind-of-bad-states
good-states := S & (not bad-states)
```

$A[P \cup Q]$ is represented by *good-states*.

- Finally, let P be $E[P \cup Q]$. The following computes the representation for P.

```
P-and-not-Q := P & (not Q) & S
char := 0
```

```
next_char := Q
loop until (char == next_char)
 char := next_char
 good-next := exist q's:char & q's-are-next-q's
 next_char := exist next-q's:P-and-not-Q & R & good-next
 next_char := next_char or char
end loop
```

All CTL formulas have been considered. Therefore, we have shown how to compute, for any CTL formula, a representation the set of states in which the formula holds. This algorithm illustrates the applicability of binary decision diagrams to compute boolean predicates representing states.

5 Acknowledgements

Thanks are due Ramon Acosta for generating the Synopsys test cases to exercise our experimental tool and for reading this manuscript. The author appreciates conversations with Paul Vaughan and David Burgess on map-over equivalence and the continued support of this research by Bill Read, director of the MCC-CAD Systems Design Satellite.

References

[1] R.E. Bryant, Graph-Based Algorithms for Boolean Function Manipulation, *IEEE Transactions on Computers*, Vol. C35 No. 8, August 1986.

[2] J.P. Billon, J.C. Madre, Original Concepts of Priam, An Industrial Tool for Efficient Formal Verification of Combinational Circuits, *IFIP conference on The Fusion of Hardware Design and Verification*, July 3-6, 1988, pp. 481-495.

[3] O. Coudert, C. Berthet, Jean Christolphe Madre, Verification of Sequential Machines Using Boolean Functional Vectors, *Proceedings of the IMEC-IFIP International Workshop on Applied Formal Methods For Correct VLSI Design*, November 13-16, 1989.

[4] R.E. Bryant, Can a Simulator Verify a Circuit?, *Formal Aspects of VLSI Design*, eds. G.J.Milne et al., North-Holland, 1966.

[5] J.E.Hopcroft, J.D.Ullman, *Formal Languages and Their Relation to Automata*, Addison-Wesley, 1969.

[6] M. Fujita, H. Fujisawa, N. Kawato, Evaluation and Improvements of Boolean Comparison Method Based on Binary Decision Diagrams, *ICCAD '88*, pp. 2-5.

[7] S. Malic, A.R.Wang, R.K.Braton, A. Sangiovanni-Vincentelli, Logic Verification using Binary Decision Diagrams in a Logic Synthesis Environment, *ICCAD '88*, pp. 6-9.

[8] S. Devadas, H-K. T. Ma, A.R. Newton, On The Verification of Sequential Machines At Differing Levels of Abstraction, *24th ACM/IEEE Design Automation Conference*, Paper 16.2, pp. 271-276.

[9] E.M. Clarke, E.A. Emerson, A.P. Sistla, Automatic Verification of Finite State Concurrent Systems Using Temporal Logic Specifications: A Practical Approach, *Proceedings of the 10th Symposium on Principles of Programming Languages*, Austin, Texas, Jan. 1983, pp. 117-126.

[10] S.Bose, A.L.Fisher, Automatic Verification of Synchronous Circuits Using Symbolic Logic Simulation and Temporal Logic, *Proceedings of the IMEC-IFIP International Workshop on Applied Formal Methods For Correct VLSI Design*, November 13-16, 1989.

DIMACS Series in Discrete Mathematics
and Theoretical Computer Science
Volume 3, 1991

Using Partial Orders to Improve Automatic Verification Methods

PATRICE GODEFROID

ABSTRACT. In this paper, we present a verification method for concurrent finite-state systems that attempts to avoid the part of the combinatorial explosion due to the modeling of concurrency by interleavings. The behavior of a system is described in terms of partial orders (more precisely in terms of Mazurkiewicz's traces) rather than in terms of interleavings. We introduce the notion of "trace automaton" which generates only one linearization per partial order. Then we show how to use trace automata to prove program correctness.

1. Introduction

Finite-state methods are quite widely used for concurrent program verification. Indeed, they have several advantages: they are simple and easy to understand and they can be fully automated. Unfortunately, these methods also have some serious drawbacks. They are not always applicable and, when they are applicable, they are often limited by combinatorial explosion.

The frustrating fact is that a lot of this combinatorial explosion is unnecessary: it is due to the modeling of concurrency by interleavings. For example, the concurrent composition of two n-state processes having completely independent activities is represented by an n^2-state process.

Of course, it has been recognized for some time that concurrency and nondeterminism are not the same thing. This observation has inspired a fairly large body of work on "partial order" models of concurrency [Lam78] [Maz86] [Pra86] [Win86]. With very few exceptions, work in this area is limited to rather abstract semantical models.

In this paper, we take a very pragmatic point of view towards partial order models. Our goal is to develop verification methods for concurrent finite-state systems that avoid the part of the combinatorial explosion due to the modeling of concurrency by interleaving. We present a framework in which this can be done successfully.

This research is supported by the European Community ESPRIT BRA project SPEC (3096).

To define a verification method, four elements are necessary: a representation of programs, a representation of properties, a semantics according to which we compare programs and properties, and an algorithm for doing this comparison [Wol89].

For representing programs and properties, we chose one-safe place/transition-nets (P/T-nets) [Rei85] [Roz86]. This is a well-known formalism and it fits very well both with the interleaving and the partial order approaches. As semantic model, we use Mazurkiewicz's traces [Maz86]. However, we use this model in such a way that the results of our verification are identical to what one would obtain with an interleaving model of concurrency. Precisely, we verify that the language of firing sequences of the one-safe P/T-net N_I representing the implementation is included in the language of firing sequences of the net N_S representing the specification.

Our verification method works in several steps. First, we build an automaton from the net representing the implementation. However, this automaton does not represent all interleavings. It only represents one interleaving for each Mazurkiewicz trace in the semantics of the implementation. This automaton can be very much smaller than the one representing all interleavings. We then compare this automaton to the net for the specification taking into account the dependency relation of the Mazurkiewicz trace semantics.

2. One-safe P/T-nets

DEFINITION 2.1. *A one-safe place/transition-net (P/T-net), is a quadruple* $N = (S, T, F, M_0)$ *where*

- S *and* T *are finite, disjoint, nonempty sets of respectively* places *(*local states*) and* transitions*;*
- $F \subseteq S \times T \cup T \times S$ *is a* flow relation *such that* $dom(F) \cup cod(F) = S \cup T$ *(no isolated elements); and*
- $M_0 \subseteq S$ *is the* initial marking.

P/T-nets are represented graphically using boxes (or straight lines) to represent transitions, circles to represent places, and arrows to represent the flow relation. In such a representation, circles corresponding to places in the initial marking are marked with dots (tokens).

For each $x \in S \cup T$, the sets

- ${}^{\cdot}x = \{y : (y, x) \in F\}$
- $x^{\cdot} = \{y : (x, y) \in F\}$
- ${}^{\cdot}x^{\cdot} = {}^{\cdot}x \cup x^{\cdot}$

are called respectively the *preset*, the *postset* and the *proximity* of x.

A *marking* M (or *global state*) of such a net N is a subset of S.

DEFINITION 2.2. *Let* M *be a marking of* N. *A transition* $t \in T$ *is* M-firable *iff* $({}^{\cdot}t \subseteq M) \land (t^{\cdot} \cap (M \setminus {}^{\cdot}t) = \emptyset)$.

(Note that a C/E-system [Rei85] (or an Elementary Net system [Roz86]) is more restrictive than a one-safe P/T-net since the requirements for the firing of a transition t in a C/E-system are $({}^{\cdot}t \subseteq M) \wedge (t^{\cdot} \cap M = \emptyset)$.) Each place of each marking will contain at most one token. An M-firable transition $t \in T$ may fire and yield a *successor marking* M' of M which is such that $M' = (M \setminus {}^{\cdot}t) \cup t^{\cdot}$.

If t fires from M to M' we write $M\ [\ t > M'$. A *reachable marking* of N is a marking M such that $\exists t_1, \ldots, t_n \in T$:

$$M_0\ [\ t_1 > M_1\ [\ t_2 > M_2\ [\ \ldots\ [\ t_n > M_n = M.$$

The set of all reachable markings of N will be denoted by $mark(N)$.

The sequence $t_1, t_2, \ldots, t_n$ is called a *firing sequence* of N. The set of all firing sequences of N will be called the *firing sequence language of N*. This language is prefix closed.

DEFINITION 2.3. *A one-safe P/T-net N is called* contact-free *iff for all $M \in mark(N)$ and for all $t \in T$: $({}^{\cdot}t \subseteq M) \Rightarrow (t^{\cdot} \cap (M \setminus {}^{\cdot}t) = \emptyset)$.*

3. Automata

A finite-state deterministic automaton is a quadruple (S, Σ, δ, s_0) where

- S is a finite set of states;
- Σ is an alphabet;
- $\delta : S \times \Sigma \rightarrow S$ is a deterministic transition function; and
- s_0 is the starting state.

The language generated by such an automaton is the set of words $w = a_1 a_2 \ldots a_n$ such that there exist $s_i = \delta(s_{i-1}, a_i)$, for all $i : 1 \leq i \leq n$. The expression $s_0 \overset{w}{\Rightarrow} s_n$ denotes the generation of the word w from the state s_0

An automaton can be represented by a directed graph. The nodes of this graph represent the states of S while the edges represent the transition function and are labeled with elements of Σ.

It is easy to define an automaton that generates the firing sequence language of a given one-safe P/T-net N:

- $S = mark(N)$;
- $\Sigma = T$, the set of transitions of N;
- $\delta(M, t) = M'$ iff $M, M' \in mark(N)$ and $M\ [\ t > M'$;
- $s_0 = M_0$, the initial marking of N.

EXAMPLE 3.1. *The automaton (b) of Figure 1 generates the firing sequences of the one-safe P/T-net (a). Clearly, the states of this automaton correspond to the reachable markings of N.* ∎

The language L accepted by such an automaton will be referred to as the *sequential behavior* of N. The words (sequences) of this language can be viewed as sequential observations of the behavior of N, i.e. observations made by observers

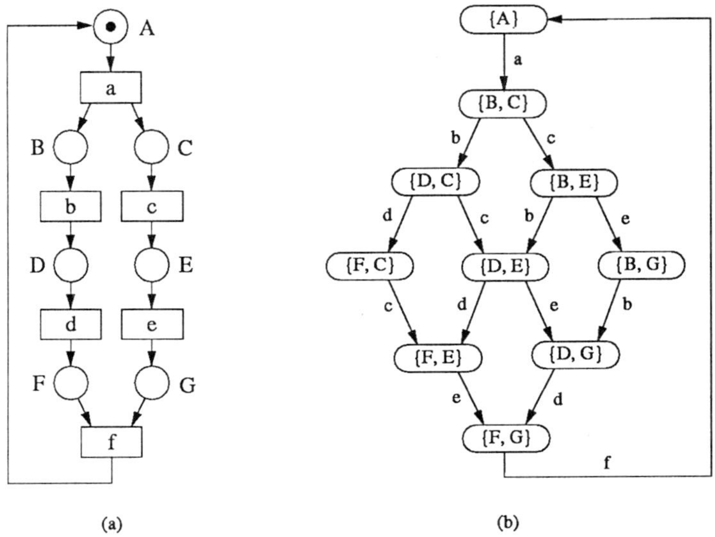

FIGURE 1. Automaton (b) generates the firing sequences of (a)

able to see only a single event occurrence at a time. The ordering of symbols in these words reflects not only the (objective) causal ordering of event occurrences (transitions of the net), but also a (subjective) observational ordering resulting from a specific view of concurrent actions: whenever there are concurrent transitions in the net N, the corresponding automaton introduces an "artificial" nondeterminism whose nondeterministic choices correspond to the possible interleavings of these concurrent transitions. Therefore, the structure of such an automaton alone does not make it possible to decide whether the difference in ordering is caused by a conflict resolution (a nondeterministic decision), or by different observations of concurrency (interleavings). In order to extract the causal ordering of event occurrences, we will use the notion of *trace* [Maz86].

4. Traces

First, we define the notion of *concurrent alphabet.*

DEFINITION 4.1. *A concurrent alphabet is a pair $\Sigma = (A, D)$ where A is a finite set of actions, called the alphabet of Σ, and D is a binary, symmetrical and reflexive, relation in A, called the* dependency *in Σ.*

$D(A) = (A, A^2)$ stands for the concurrent alphabet of total dependency on A, and $I_\Sigma = A^2 \setminus D$ stands for the *independency* in Σ.

DEFINITION 4.2. *Let Σ be a concurrent alphabet; A^* represents the set of all finite sequences (words) of symbols in A, $\cdot$ stands for the concatenation operation, and the empty word is noted ε. We define the relation $\equiv_\Sigma$ as the least congruence in the monoid $[A^*; \cdot, \varepsilon]$ such that $(a, b) \in I_\Sigma \Rightarrow ab \equiv_\Sigma ba$.*

This congruence is an equivalence relation which respects the $\cdot$ and ε operations. The relation $\equiv_\Sigma$ is referred to as the *trace equivalence over* Σ.

DEFINITION 4.3. *Equivalence classes of* $\equiv_\Sigma$ *are called* traces over Σ.

A trace characterized by a word w and a concurrent alphabet Σ is denoted by $[w]_\Sigma$.

Thus a trace over a concurrent alphabet $\Sigma = (A, D)$ represents a set of words defined over A only differing by the order of adjacent symbols which are independent according to D. For instance, if a and b are two symbols of A which are independent according to D, the trace $[ab]_\Sigma$ represents the two words ab and ba. A trace is an equivalence class of words. A *trace language* is a set of traces over a given concurrent alphabet.

Let us return to the one-safe P/T-net N. We define the *dependency* in N as the relation $D_N \subseteq T \times T$ such that:

(1) $$(t_1, t_2) \in D_N \Leftrightarrow {}^{\cdot}t_1 \cap {}^{\cdot}t_2 \neq \emptyset.$$

The complement of D_N is called the *independency* in N. If two independent actions occur next to each other in a firing sequence, the order of their occurrences is irrelevant (since they occur concurrently in this execution). Let $\Sigma_N = (T, D_N)$ be the concurrent alphabet associated with N and let L be the firing sequence language of N. We define the *trace behavior* of N as the set of equivalence classes of L defined by the relation $\equiv_{\Sigma_N}$. These equivalence classes are called *firing traces* of N. Such a class (trace) corresponds to a partial order (i.e. a set of causality relations) and represents all its linearizations (words).

To describe the behavior of a one-safe P/T-net by means of traces rather than sequences, we will need the dependency D_N of N (which can be deduced from the statical structure of N as defined by (1)) and *only one* linearization for each trace. Consider a language L' representing one arbitrary linearization (word) for each possible trace of the net. Let L' be such that

$$L = \bigcup_{w' \in L'} Pref(lin([w']_{\Sigma_N}))$$

where $lin([w]_{\Sigma_N})$ denotes the set of linearizations (words) of the trace (equivalence class) $[w]_{\Sigma_N}$ and $Pref(w)$ denotes the prefixes of w. Clearly, L' is a regular language. So, *the behavior of a one-safe P/T-net is fully characterized by the dependency* D_N *and an automaton which generates exactly* L'. Let us call this automaton a *trace automaton* for N.

To construct such an automaton, we do not need to compute all the reachable markings of N: whenever several independent transitions are firable, we fire only one of these transitions in order to generate only one interleaving (linearization) of these transitions.

In the next section, we present an algorithm to construct a trace automaton for a given contact-free one-safe P/T-net N.

```
1. Initialize: Stack is empty; H is empty;
              enter (M0, ∅, 0) in H;
              push (M0, ∅, 0) into Stack
2. Loop: while Stack ≠ ∅ do
              begin
                pop (M, Sleep, NDinfo) from Stack;
                FT := CHOOSE-AMONGST({tf ∉ Sleep : ·tf ⊆ M}, Sleep, NDinfo)
                for all tf ∈ FT: add tf to succ(M) in H
                for all tf ∈ FT do
                   begin
                     nextM := M − ·tf + tf·;
                     NDinfo := tf->NDinfo; Sleep := tf->Sleep;
                     if nextM is already in H
                       then NDinfo-old := NDinfo associated with nextM in H;
                            if Same-ND-Branch(NDinfo-old,NDinfo)
                              then Sleep := Sleep ∪ succ(nextM);
                                   push (nextM, Sleep, NDinfo) into Stack
                              else Sleep-old := Sleep associated with nextM in H
                                   if Sleep-old ⊈ Sleep
                                     then enter (nextM, Sleep ∩ Sleep-old,
                                                          NDinfo) in H;
                                          push (nextM, Sleep, NDinfo) into Stack
                       else enter (nextM, Sleep, NDinfo) in H;
                            push (nextM, Sleep, NDinfo) into Stack
                   end
              end
```

FIGURE 2. Algorithm

5. Constructing the Trace Automaton

The algorithm presented in Figure 2 is a classical depth-first search of the reachable markings of the net N with some important modifications.

The algorithm uses a *Stack* to hold the configurations that remain to be examined. Each configuration is composed by a marking M and two additional information: a "*Sleep* set" and a "*NDinfo*". A *Sleep* set is a subset of M-firable transitions. A "sleeping" transition will never be fired in the remainder of the search starting from the current marking M. Thus *Sleep* denotes a set of transitions which are firable but which will not be fired. A *NDinfo* is an information which identifies the nondeterministic branch of the current marking M. Different possibilities for solving a nondeterminism lead to different nondeterministic branches. The root of all nondeterministic branches is the initial marking. A (hash) table H is also used to store the states of the automaton that have been explored. As usual, these states will correspond to reachable markings of the net N. A *Sleep* set, a *NDinfo* and a "*succ*" set are associated to each state. The *succ* set contains the transitions leaving that state in the trace automaton. The states reachable from a given state of the trace automaton are obtained by firing, in the net, the transitions of the *succ* set associated with that state.

Since we suppose that N is contact-free, we do not have to check the requirement $t^{\cdot} \cap (M \setminus {}^{\cdot}t) = \emptyset$ before firing a transition t from a marking M. Dealing with contact-free one-safe P/T-nets is not a restriction. Indeed for every one-safe P/T-net there exists an equivalent contact-free one-safe P/T-net (i.e. a net that has the same firing sequence language) [Rei85]. Transitions $t_1, t_2, \ldots, t_n$ are referred to as being *in conflict* iff $({}^{\cdot}t_1 \cap {}^{\cdot}t_2 \cap \ldots \cap {}^{\cdot}t_n) \neq \emptyset$ (their occurrences are mutually exclusive and lead to different nondeterministic branches). A transition t that is not in conflict with another transition is *conflict-free*.

The first modification w.r.t. the classical algorithm is that we do not systematically fire all firable transitions from a given marking: we only choose some of them since we want to construct an automaton that generates only one interleaving whenever several independent transitions are firable. Our choices are motivated by the following two principles:

(1) we want to minimize the number of states of the trace automaton;
(2) we have to consider all possible behaviors of the net: each firing sequence of the net must be represented by some trace generated by the trace automaton.

Remember that each word w generated by the trace automaton defines a trace $[w]_{\Sigma_N}$. This means that all the linearizations of this trace and all the prefixes of these linearizations are firing sequences of the net N.

In order to minimize the number of states to be constructed, we define a priority scheme to choose amongst the firable transitions those that are to be fired. The highest priority (priority 1) is given to conflict-free transitions (these transitions will be fired one by one successively). Next, we give priority (priority 2) to transitions that are in conflict exclusively with firable transitions. When we fire such a transition, we also fire (from the current marking) the transitions that are in conflict with it in order to explore all possible nondeterministic cases (it corresponds to a branching in the trace automaton). All these possible cases lead to different nondeterministic branches (therefore, the $NDinfo$ of the markings obtained after firing these transitions are distinct). Finally, priority 3 is given to firable transitions that are in conflict with at least one nonfirable transition (this is a situation of *confusion* [Rei85]).

When there remain only firable transitions of priority 3, we proceed as follows. Let t be one of these firable transitions. We know that t is in conflict with at least one transition x that is not firable from the current marking. But it is possible that x will become firable because of the evolution of some other tokens not in ${}^{\cdot}t$. In that case, the firing of t could be replaced by the firing of x. In order not to miss this possibility, we consider several cases: we fire t from the current marking M, we fire the firable transitions $t_1, \ldots, t_n$ in conflict with t from M, if any, and then we "give the hand" to other tokens to see if they can make x firable (all these cases correspond to different nondeterministic branches). In the last case, since we have already considered the firing of t and $t_1, \ldots, t_n$ and since we are only interested by the "potential" firing of x, we do not consider $t, t_1, \ldots, t_n$ any longer and we put them in the *Sleep* set associated with the

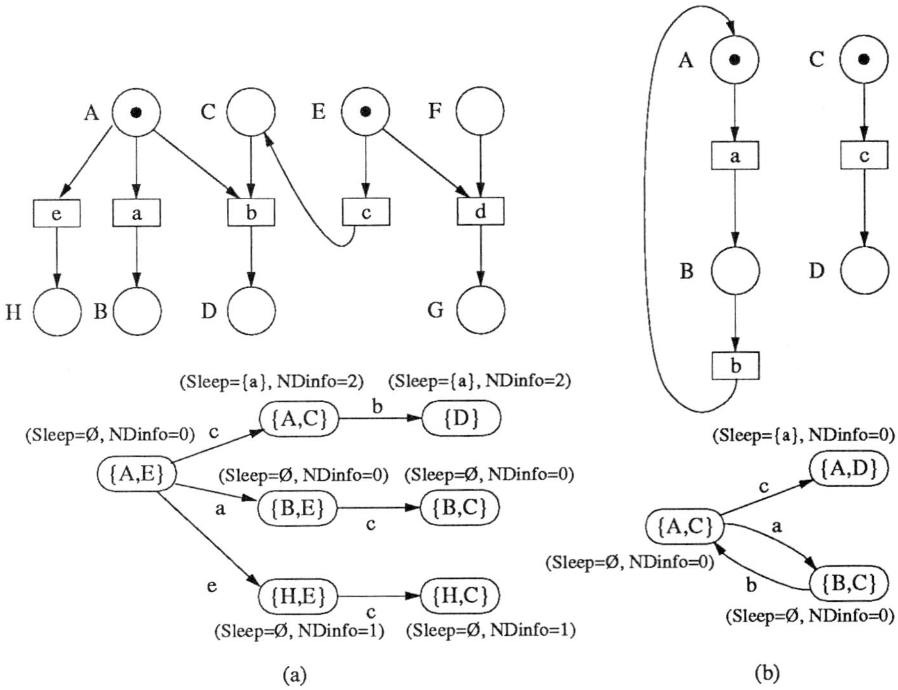

FIGURE 3. Nets and their corresponding trace automaton

current configuration. If there still remain firable transitions of priority 3, we proceed again as described above with the new *Sleep* set, etc. In summary, when there remain only firable transitions of priority 3, all these transitions are fired from the current marking but with different *NDinfo* and *Sleep* sets. An example of such a situation is shown in Figure 3.a: from the current marking all the firable transitions have priority 3. The corresponding trace automaton is given below the net (the value of *Sleep* and *NDinfo* during the construction of the trace automaton is given between parentheses).

The function CHOOSE-AMONGST returns the next transition(s) to be fired according to the priority scheme. Moreover, this function associates two additional information to each chosen transition t_f to be fired: t_f->*Sleep* and t_f->*NDinfo* representing respectively the *Sleep* set and the *NDinfo* to be associated with the new marking obtained after firing t_f.

Another modification w.r.t. the classical depth-first search is that, if the current marking has already been reached, the search does not stop automatically. Indeed, suppose the current marking is M. The search continues and after the firing of some transitions, the current marking becomes M again. In other words some of the tokens of M moved from their place and then returned back to their respective place. But it is possible that other tokens of M have not yet had the possibility of moving. In order to consider all possible behaviors, we have to "give the hand" to these tokens to see which other concurrent transitions could

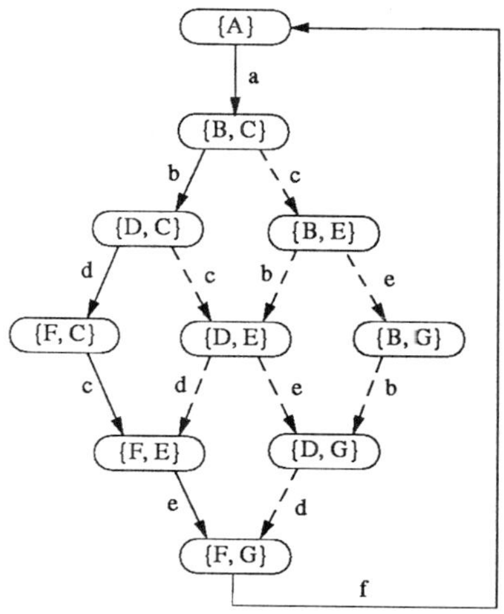

FIGURE 4. Trace automaton for the net (a) of figure 1

be fired (it is a kind of "fairness requirement" on the choices amongst independent firable transitions). This situation is illustrated in Figure 3.b: from the initial marking $\{A, C\}$, assume that we fire a and that next we fire b and go back to $\{A, C\}$; then we have to fire c although the marking $\{A, C\}$ has already been reached. This kind of situation can arise only with markings that can appear several times in the *same* nondeterministic branch of the trace automaton. Now, what happens if the next marking has already been reached in *another* nondeterministic branch ? The search from this next marking may stop if the Sleep set "*Sleep*-old" associated with this marking in the already reached marking table is included in the current Sleep set "*Sleep*". If this requirement is not satisfied, i.e. if there exists some transition t in *Sleep*-old (t is firable) such that t is not in *Sleep*, the search has to continue. Indeed, since t is not in *Sleep*, t (which is firable) ***has to be fired*** eventually and, since t is in *Sleep*-old, we know that t *does not occur* in the remainder of the search previously made starting from the next marking with *Sleep*-old as Sleep set.

EXAMPLE 5.1. *The trace automaton of Figure 4 and the concurrent alphabet*

$$\Sigma_N = D(\{a,b\}) \cup D(\{b,d\}) \cup D(\{d,f\}) \cup D(\{a,f\}) \cup D(\{a,c\}) \cup D(\{c,e\}) \cup D(\{e,f\})$$

(obtained with relation 1) fully characterize the traces of the one-safe P/T-net (a) of Figure 1. The dotted part is not part of the trace automaton. ∎

Of course, there are many possible trace automata corresponding to a given contact-free one-safe P/T-net N. The algorithm described above constructs one of these possible automata. The order of the time complexity for our algorithm is given by the number of constructed transitions times the maximum number of simultaneous firable transitions. We do not claim that the trace automaton constructed by our algorithm is always the minimal one: it is often possible to

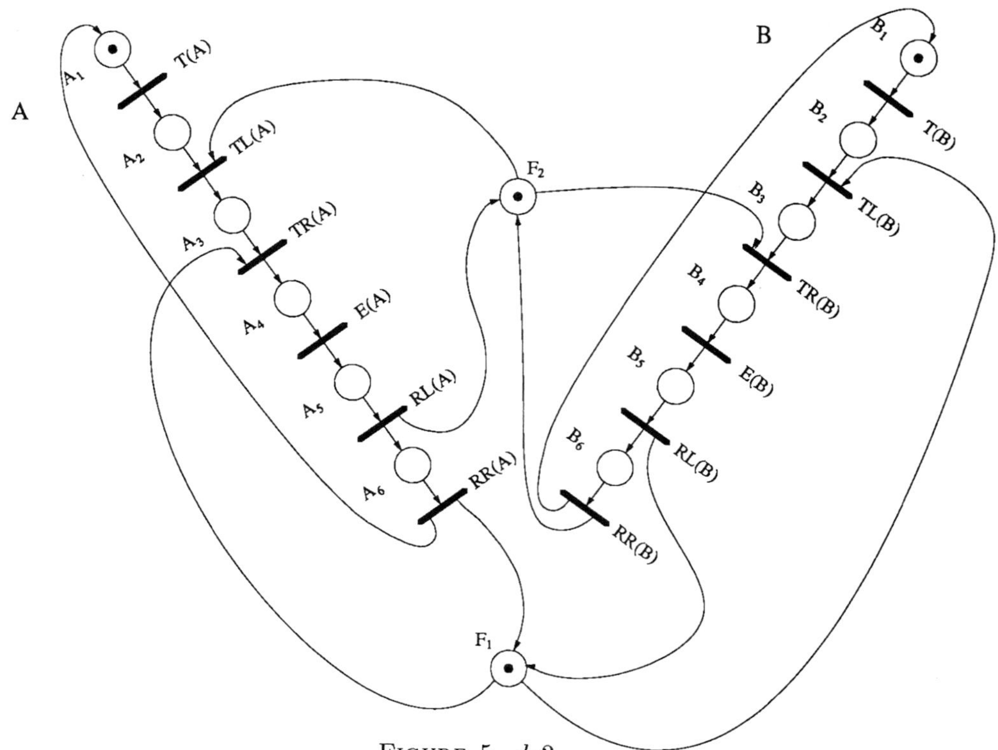

FIGURE 5. *dp*2

further reduce the size of the trace automaton but at the cost of an increased time complexity.

EXAMPLE 5.2. *Let us consider the well-known dining philosophers problem. Figure* 5 *shows a net dp*2 *that represents two philosophers and their adjacent forks. The "classical" automaton, i.e. the one whose states correspond to all the reachable markings, and the trace automaton constructed by our algorithm that correspond to dp*2 *are presented in figure* 6. *The dotted part is not part of the trace automaton.* ∎

The net *dp*2 is susceptible to deadlock: there is no firable transition from the marking $\{A_3, B_3\}$. Note that the state corresponding to the marking $\{A_3, B_3\}$ is a state of the trace automaton. Indeed, remember the main idea behind our algorithm is to choose some of the firable transitions to be fired whenever there are independent firable transitions. Thus, if there is *no* firable transition, we detect it (the deadlock-preserving property of partial order semantics was already pointed out in [Gai88], [Val88] among others).

Table 1 compares the performance of a depth-first search algorithm against the algorithm presented in this section for the nets *dp*2 to *dp*5 (five dining philosophers). The combinatorial explosion both in the number of states and transitions is clearly avoided by using trace automata. We also compare the run time needed

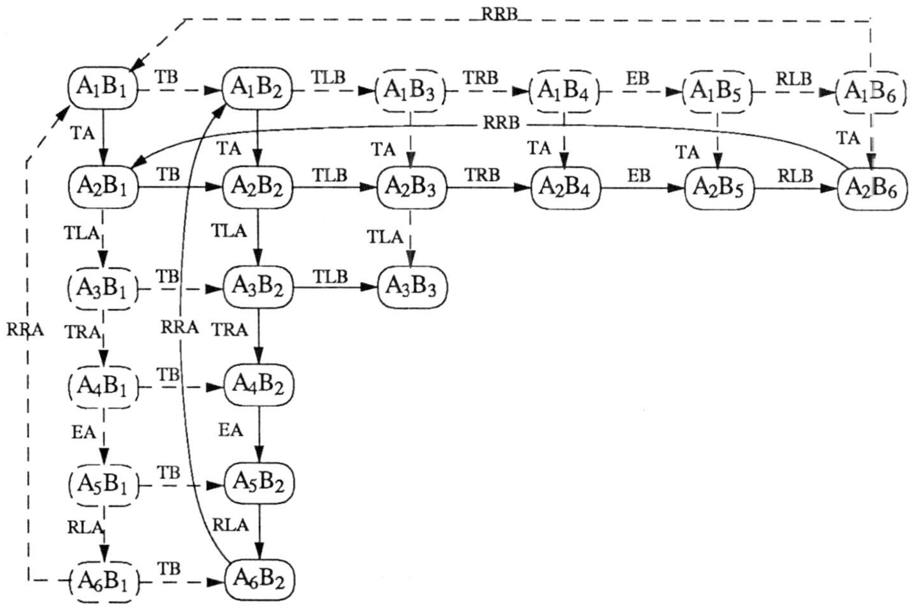

FIGURE 6. Classical automaton and trace automaton for *dp*2

NET	Classical Automaton				Trace Automaton			
	Total Run Time (sec)	States	Trans.	Deadlock	Total Run Time (sec)	States	Trans.	Deadlock
dp2	0.76	21	34	13	0.7	13	14	13
dp3	3.64	99	240	25	1.1	25	28	20
dp4	43.52	465	1508	173	1.98	43	49	27
dp5	692.4	2163	8770	525	3.6	72	83	34

TABLE 1. Classical automaton versus trace automaton

to construct these automata (these results were obtained with a LISP prototype): our algorithm can be much faster than the classical one. Moreover, the deadlock is detected sooner: for *dp*5, the deadlock corresponds to the 34th reached state by our algorithm instead of the 525th one with a classical depth-first search.

Our method for modeling net behaviors has the potential of being *much more efficient, both in time and memory*, than the classical one. In the next sections, we show how to use trace automata to prove program correctness.

6. Verification

Consider a set of processes P and a relation $\leq$ on this set such that $I \leq S$ iff I *refines* S (i.e., I is less nondeterministic than S). Let P be the set of all one-safe P/T-nets and $\leq$ the inclusion relation defined on the languages generated by these nets.

To verify that the implementation N_I effectively meets the specification N_S, the classical method consists in comparing the automata I and S respectively generating the firing sequence language of the nets N_I and N_S. If these automata are deterministic, verifying the language inclusion reduces to "simulating" I by S, i.e. checking that all that I can do can also be done by S. This can be formalized by:

> If I is in a state I_i such that $I_0 \stackrel{w}{\Rightarrow} I_i$ (where I_0 is the initial state of I and w is a sequence generated by I) and if there is a transition $I_i \stackrel{a}{\rightarrow} I_j$, then from the state S_i of S such that $S_0 \stackrel{w}{\Rightarrow} S_i$ (where S_0 is the initial state of S), there must be a transition $S_i \stackrel{a}{\rightarrow} S_j$.

The cost of checking the existence of this simulation increases with the size of the automaton I, i.e. with the total number of reachable markings of the net N_I. Our claim is that this verification can be done at a lower cost, by using a trace automaton for N_I.

Remark. In a nondeterministic automaton, the transition function δ is replaced by a transition relation $\Delta : S \times \Sigma \times S$. A transition $t \in \Sigma$ from a state $s \in S$ may lead to several states. The verification method we have just described is still applicable when the automaton corresponding to the implementation is nondeterministic. This is of great interest because it makes it possible for some of the transitions of N_I to be hidden.

In the next section, we show how to express a criterion equivalent to language inclusion in terms of traces.

7. A Verification Criterion based on Traces

Let L_I (L_S) be the firing sequence language of a given one-safe P/T-net $N_I = (S_I, T_I, F_I, M_{0_I})$ (N_S). Let D_I (D_S) be the dependency in N_I (N_S) as defined previously and $\Sigma_I = (T_I, D_I)$ ($\Sigma_S = (T_S, D_S)$) the concurrent alphabet associated with N_I (N_S). We define $PO([w]_\Sigma)$ as the transitive closure of the relation $\{(a_i, a_j) \ : \ (a_i, a_j) \in D \text{ with } 1 \leq i < j \leq n\}$ if $w = a_1 a_2 \ldots a_n$ and D is the dependency of Σ. $PO([w]_\Sigma)$ represents the set of causality relations corresponding to $[w]_\Sigma$ ($PO([w]_\Sigma)$ is a partial order).

Let I' denote a trace automaton corresponding to N_I and let $L_{I'}$ be the language generated by I'.

THEOREM 7.1.

$$lin([w]_{\Sigma_I}) \subseteq lin([w]_{\Sigma_S}) \Leftrightarrow PO([w]_{\Sigma_I}) \supseteq PO([w]_{\Sigma_S}).$$

Proof: If the causality relations on the elements (letters) of w in Σ_I are more restrictive than (i.e. includes) the causality relations on the elements of w in Σ_S, the trace on Σ_I defined by w will have less linearizations than the trace on Σ_S defined by the same word w (since Σ_S is more "liberal"); and conversely. ∎

This theorem leads us directly to the following consideration: *to verify that the linearizations of a trace $[w']_{\Sigma_I}$ are included in the firing sequence language of a given one-safe P/T-net N_S, one needs to check if the word w' is a firing sequence of N_S and to verify that $PO([w']_{\Sigma_I}) \supseteq PO([w']_{\Sigma_S})$.*

THEOREM 7.2.

$$\left\{ \begin{array}{l} w' \in L_S \\ PO([w']_{\Sigma_I}) \supseteq PO([w']_{\Sigma_S}) \end{array} \right\} \Rightarrow lin([w']_{\Sigma_I}) \subseteq L_S$$

Proof: Since $w' \in L_S$, w' is a firing sequence of S and a linearization of the firing trace $[w']_{\Sigma_S}$ of S: $w' \in lin([w']_{\Sigma_S})$. From $PO([w']_{\Sigma_I}) \supseteq PO([w']_{\Sigma_S})$, we have $lin([w']_{\Sigma_I}) \subseteq lin([w']_{\Sigma_S})$ by Theorem 7.1.1. As we know that $lin([w']_{\Sigma_S}) \subseteq L_S$, the result follows. ∎

The converse of this theorem is valid if we have one more assumption on N_S: N_S must be a *"restricted" one-safe P/T-net.*

DEFINITION 7.1. *A one-safe P/T-net is called "restricted" if the following requirement is satisfied:*

$$\forall x, y \in T : (^{\cdot}x \cap {}^{\cdot}y) \neq \emptyset \Rightarrow ((^{\cdot}x \cap {}^{\cdot}y) \not\subseteq (y^{\cdot} \cap x^{\cdot})). \tag{2}$$

We now show that two words differing only by the order of two adjacent symbols which are dependent according to the dependency D_S of a "restricted" one-safe P/T-net N_S can not be both included in the firing sequence language of N_S. Formally:

LEMMA 7.3. *If $N_S = (S_t, T, F, M_0)$ is a "restricted" one-safe P/T-net, $\forall x, y \in T$:*

$$(x, y) \in D_S \Rightarrow$$

$$\nexists M, M', M'', M''' \subseteq S_t : (M\ [\ x > M'\ [\ y > M'') \wedge (M\ [\ y > M'''\ [\ x > M'').$$

Proof: We know that x and y are dependent according to D_S: ${}^{\cdot}x^{\cdot} \cap {}^{\cdot}y^{\cdot} \neq \emptyset$. This statement is equivalent to:

$$(^{\cdot}x \cap {}^{\cdot}y) \neq \emptyset \vee (^{\cdot}x \cap y^{\cdot}) \neq \emptyset \vee (x^{\cdot} \cap {}^{\cdot}y) \neq \emptyset \vee (x^{\cdot} \cap y^{\cdot}) \neq \emptyset \tag{3}$$

Moreover we have the following requirements on M:

(a) ${}^{\cdot}x \subseteq M$
(b) ${}^{\cdot}y \subseteq M$
(c) ${}^{\cdot}x \subseteq M''' = (M \setminus {}^{\cdot}y) \cup y^{\cdot}$
(d) ${}^{\cdot}y \subseteq M' = (M \setminus {}^{\cdot}x) \cup x^{\cdot}$

We will consider two cases:

- Case (a): ${}^{\cdot}x \cap {}^{\cdot}y = \emptyset$. To satisfy equation(3), there are three possibilities left:
  (1) Assume ${}^{\cdot}x \cap y^{\cdot} \neq \emptyset$. Since y is firable from M, we know that $y^{\cdot} \cap (M \setminus {}^{\cdot}y) = \emptyset$. Since (a), we have: $y^{\cdot} \cap (^{\cdot}x \setminus {}^{\cdot}y) = y^{\cdot} \cap {}^{\cdot}x = \emptyset$. This is in contradiction with our assumption.
  (2) Assume $x^{\cdot} \cap {}^{\cdot}y \neq \emptyset$. This case is similar to the previous one and also yields a contradiction.

(3) Finally, assume: $x^{\cdot} \cap y^{\cdot} \neq \emptyset$. Since y is firable from M', we know that $y^{\cdot} \cap (M' \setminus {}^{\cdot}y) = y^{\cdot} \cap (((M \setminus {}^{\cdot}x) \cup x^{\cdot}) \setminus {}^{\cdot}y) = \emptyset$. Hence $y^{\cdot} \cap (x^{\cdot} \setminus {}^{\cdot}y) = \emptyset$. Since we have already established above that $x^{\cdot} \cap {}^{\cdot}y = \emptyset$, $y^{\cdot} \cap x^{\cdot} = \emptyset$. This is a contradiction.

Thus the assumption ${}^{\cdot}x \cap {}^{\cdot}y = \emptyset$ made in this case is in contradiction with the requirements that have to be satisfied.

- Case (b): ${}^{\cdot}x \cap {}^{\cdot}y \neq \emptyset$. Hence from (a) we get ${}^{\cdot}x \not\subseteq (M \setminus {}^{\cdot}y)$. But we know (c) ${}^{\cdot}x \subseteq ((M \setminus {}^{\cdot}y) \cup y^{\cdot})$. Thus we have $({}^{\cdot}x \setminus (M \setminus {}^{\cdot}y)) \subseteq y^{\cdot}$. Since (a), $({}^{\cdot}x \cap {}^{\cdot}y) \subseteq y^{\cdot}$.
  In the same way, we can deduce $({}^{\cdot}x \cap {}^{\cdot}y) \subseteq x^{\cdot}$.
  From the last two statements we get $({}^{\cdot}x \cap {}^{\cdot}y) \subseteq (x^{\cdot} \cap y^{\cdot})$. This is in contradiction with the property (2) since N_S is a "restricted" one-safe P/T-net.

So we have proved that the requirements (a), (b), (c) and (d) on M are not compatible with our assumptions (2) and (3). The result follows. ∎

Now we are ready to prove the converse of Theorem 7.1.2.

THEOREM 7.4. *If N_S is a "restricted" one-safe P/T-net,*

$$lin([w']_{\Sigma_I}) \subseteq L_S \Rightarrow \begin{cases} w' \in L_S \\ PO([w']_{\Sigma_I}) \supseteq PO([w']_{\Sigma_S}) \end{cases}$$

Proof: If $lin([w']_{\Sigma_I}) \subseteq L_S$, then $w' \in L_S$ (obvious). To show that $PO([w']_{\Sigma_I}) \supseteq PO([w']_{\Sigma_S})$, we will use Theorem 7.1.1. We have to prove that $lin([w']_{\Sigma_I}) \subseteq lin([w']_{\Sigma_S})$. Of course, we know that $w' \in lin([w']_{\Sigma_S})$ ($\subseteq L_S$ by definition of L_S).

Let $w \in lin([w']_{\Sigma_I})$ ($\subseteq L_S$ by assumption) be such that w and w' are only differing by the order of two adjacent symbols x and y (thus these two symbols are independent according to D_I).

Suppose $w \notin lin([w']_{\Sigma_S})$. This means that w does not belong to the equivalence class $[w']_{\Sigma_S}$ and that we have $(x, y) \in D_S$ (x and y are dependent according to D_S). By lemma 7.1.3, we deduce that w and w' cannot both be included in L_S. This is in contradiction with the fact that $lin([w']_{\Sigma_I}) \subseteq L_S$. Hence $w \in lin([w']_{\Sigma_S})$.

Step by step, we can do the same proof for each word of $lin([w']_{\Sigma_I})$ and thus we obtain $lin([w']_{\Sigma_I}) \subseteq lin([w']_{\Sigma_S})$. ∎

Note that this theorem is not valid if N_S is not a "restricted" one-safe P/T-net.

EXAMPLE 7.1. *An example is shown in Figure 7. N_S is not a "restricted" one-safe P/T-net since ${}^{\cdot}x \cap x^{\cdot} = {}^{\cdot}y \cap y^{\cdot} \neq \emptyset$. We clearly see that $lin([xy]_{\Sigma_I}) = \{xy, yx\} \subseteq L_S = \{(x \cup y)^*\}$. But we do not have $PO([w']_{\Sigma_I}) = \emptyset \supseteq PO([w']_{\Sigma_S}) = \{(x, y)\}$ or $\{(y, x)\}$ respectively if $w' = xy$ or $w' = yx$.* ∎

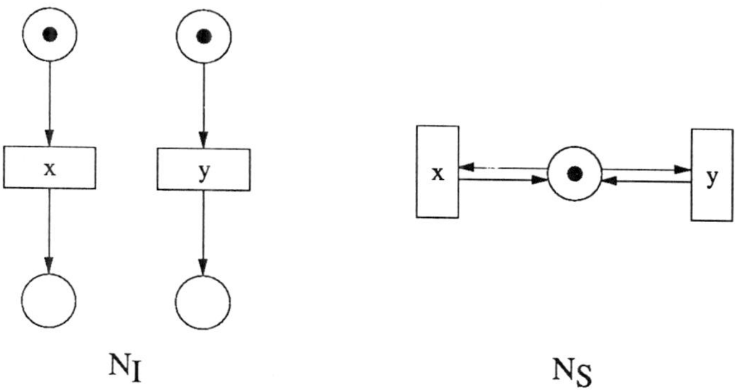

FIGURE 7. N_S is not a "restricted" one-safe P/T-net

Now we generalize Theorem 7.1.2 and Theorem 7.1.4 to the set of traces generated by I'. Remember $L_{I'}$ is the string language generated by the trace automaton I'.

THEOREM 7.5. *Let N_I be a one-safe P/T-net and N_S a "restricted" one-safe P/T-net. We have:*

$$L_I \subseteq L_S \Leftrightarrow \begin{cases} L_{I'} \subseteq L_S \\ \forall w' \in L_{I'} : PO([w']_{\Sigma_I}) \supseteq PO([w']_{\Sigma_S}) \end{cases}$$

Proof: If we apply Theorem 7.1.2 and Theorem 7.1.4 for each $w' \in L_{I'}$ and we use the definitions of L_I and $L_{I'}$, one directly obtains Theorem 7.1.5. ∎

This theorem is of great interest because in most cases verifying the two inclusions of the second member can be done more efficiently than verifying the inclusion between all the linearizations. Indeed, the size of I' can be much smaller than the size of I, if N_I is a concurrent system.

8. The Verification Algorithm

We check that $L_{I'} \subseteq L_S$ as usual (see Section 6) by exploring the reachable states of $I' \times S$. A transition executed by I' must be simulated by S except if this transition is hidden, i.e. if this transition is not in the set T_S of transitions of the net N_S. All the reachable states of $I' \times S$ are stored in a hash table H-new. During the exploration, if the current state has already been stored in H-new, the simulation from this state stops.

Verifying the inclusion relation between the causality relations is done during the checking of the simulation of I' by S. This verification is based on the following observations. Assume the current global state (global marking) is (M_I, M_S) such that M_{I_0} $[\, t_1 > M_{I_1}$ $[\, t_2 > M_{I_2}$ $[\, \ldots$ $[\, t_n > M_I$ and M_{S_0} $[\, t_1 > M_{S_1}$ $[\, t_2 > M_{S_2}$ $[\, \ldots$ $[\, t_n > M_S$ where M_{I_0} and M_{S_0} are respectively the initial marking of N_I and N_S. Let w be the firing sequence $t_1 t_2 \ldots t_n$. Furthermore, assume we have already checked that $PO([w]_{\Sigma_I}) \supseteq PO([w]_{\Sigma_S})$ for w

1. Initialize: *Stack* is empty;
enter (M_{I_0}, M_{S_0}) in H-new;
push (M_{I_0}, M_{S_0}) into *Stack*
2. Loop: while $Stack \neq \emptyset$ do
begin
pop (M_I, M_S) from *Stack*;
for all $t \in succ(M_I)$ in H do
begin
temp:=$\emptyset$;
if $t \in T_S$
then begin
if ${}^{\cdot}t_{/S} \not\subseteq M_S$ then print"$L_I \not\subseteq L_S$";
halt
temp:= $\cup_{token \in {}^{\cdot}t_{/I}} ancestors(token)$;
if $\cup_{token \in {}^{\cdot}t_{/S}}$ direct-ancestor($token$) $\not\subseteq$ temp
then print"$L_I \not\subseteq L_S$";
halt
$M_S := M_S - {}^{\cdot}t_{/S} + t^{\cdot}_{/S}$;
for all $token \in t^{\cdot}_{/S}$: direct-ancestor($token$) := t ;
temp:=temp + t;
end
$M_I := M_I - {}^{\cdot}t_{/I} + t^{\cdot}_{/I}$;
if (M_I, M_S) is not already in H-new
then begin
enter (M_I, M_S) in H-new;
if temp= $\emptyset$ then temp:= $\cup_{token \in {}^{\cdot}t_{/I}} ancestors(token)$;
for all $token \in t^{\cdot}_{/I}$: $ancestors(token)$:= temp;
push (M_I, M_S) into *Stack*;
end
end
end
print"$L_I \subseteq L_S$"

FIGURE 8. Verification algorithm

and all its prefixes. Now suppose the transition t is fired from the current global state. How can we check that $PO([wt]_{\Sigma_I}) \supseteq PO([wt]_{\Sigma_S})$?

We know that the set $PO([wt]_{\Sigma})$ is equivalent to the union of three sets:

(1) $PO([w]_{\Sigma})$;

(2) $\{(t_i, t) : (t_i, t) \in D, 1 \leq i \leq n\}$ where D is the dependency of Σ. This is the set of causality relations formed by the transitions that are dependent with t according to D and that have already occurred. These transitions are the *direct ancestors* of t.

(3) $\{(t_i, t) : (t_i, t_j) \in PO([w]_{\Sigma}) \wedge (t_j, t) \in D$ with $1 \leq i < j \leq n\}$, the transitive closure of (1) and (2). The transitions t_i in (2) and (3) are the *ancestors* of t.

From this observation, we can deduce that if $PO([w]_{\Sigma_I}) \supseteq PO([w]_{\Sigma_S})$ and if the direct ancestors of t in N_S are ancestors of t in N_I, then the other ancestors of t in N_S are ancestors of t in N_I and then $PO([wt]_{\Sigma_I}) \supseteq PO([wt]_{\Sigma_S})$.

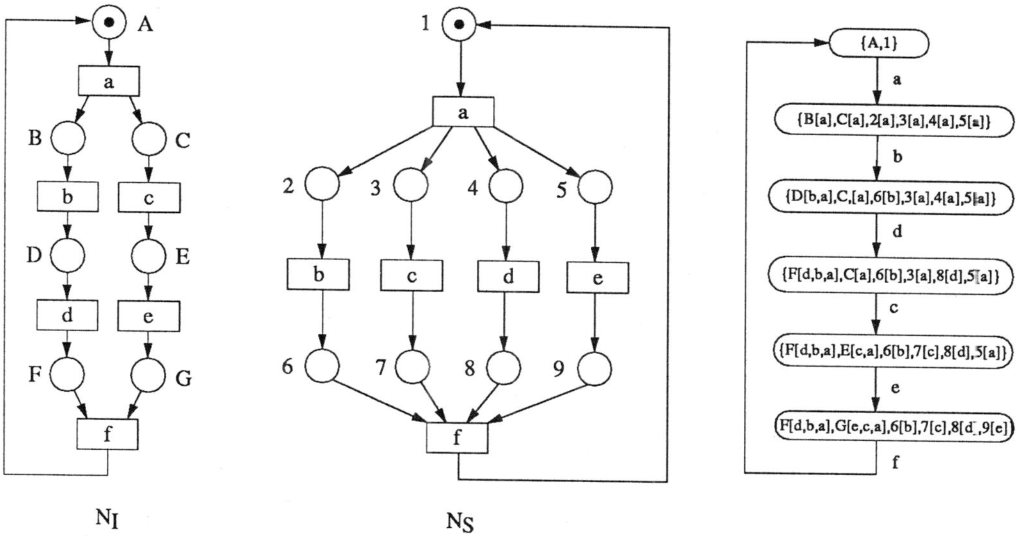

FIGURE 9. Example of verification

We proceed as follows:

- a list of ancestors is associated with each token of N_I;
- a direct ancestor is associated with each token of N_S.

When a transition t is fired both in N_I and N_S (i.e. t is not hidden), we check that the direct ancestors of t in N_S (obtained by taking the union of the direct ancestors associated with the tokens in ${}^{\cdot}t_{/S}$, i.e. the tokens participating in the firing of t in N_S) are ancestors of t in N_I (obtained by taking the union of the ancestors associated with the tokens in ${}^{\cdot}t_{/I}$, i.e. the tokens participating in the firing of t in N_I). Then we add t to the list of ancestors associated with the tokens in ${}^{\cdot}t_{/I}$ to obtain the list of ancestors associated with the tokens in $t^{\cdot}_{/I}$. If a transition t is fired only in N_I (i.e. t is hidden), this transition is not involved in the verification between causality relations and does not need to be stored in some ancestor list. We just have to update the structure of these lists: the union of the ancestor lists associated with the tokens in ${}^{\cdot}t_{/I}$ will be transmitted and associated with the tokens in $t^{\cdot}_{/I}$ (without adding t to this new list).

The main data structures used in the algorithm described in Figure 8 are a *Stack* to hold the global markings that remain to be examined and a hash table *H*-new to store the states of $I' \times S$ that have been explored. From a marking M_I and its associated *succ* set (that has already been established during the construction of the trace automaton) we can obtain the next states of the trace automaton by firing the transitions of *succ* from M_I in N_I. When $L_I \not\subseteq L_S$, a sequence of L_I that is not in L_S can be produced.

EXAMPLE 8.1. *Figure 9 shows two nets N_I and N_S, and the global automaton resulting from the construction of I' and the corresponding simulation by S (the list of ancestors associated with each token of N_I and the direct ancestor*

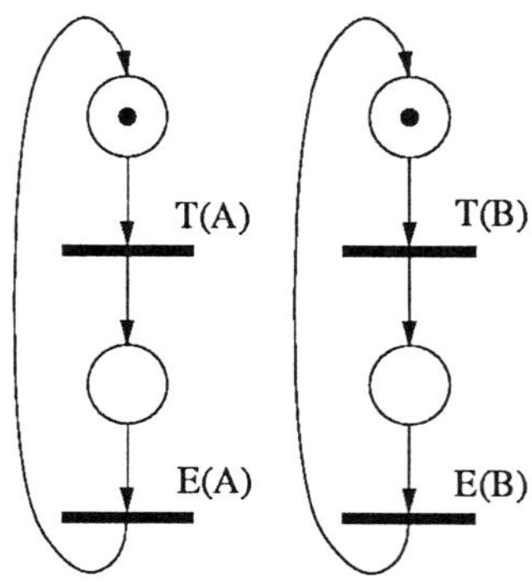

FIGURE 10. Specification for $dp2$

NET	Classical Simulation	Verification with P.O.
	Total Run Time (sec)	Total Run Time (sec)
dp2	0.96	0.74
dp3	5.08	1.18
dp4	60.1	2.36
dp5	940.4	4.58

TABLE 2. Classical verification versus verification with P.O.

associated with each token of N_S are given between brackets with the states of this automaton). One can easily check that $L_{I'} = \{(abdcef)^*\} \subseteq L_S$ and that $PO([(abdcef)^*]_{\Sigma_I}) \supseteq PO([(abdcef)^*]_{\Sigma_S})$. So N_I is indeed an implementation of N_S w.r.t. our criterion. ∎

EXAMPLE 8.2. *Figure* 10 *shows a specification N_S for the two dining philosophers $dp2$ of figure* 5. ∎

Table 2 compares the run time needed to verify that $dp2$ is an implementation of N_S w.r.t. our criterion (theses results were obtained with a LISP prototype). In a similar way, this problem has been extended up to five philosophers and the corresponding results are presented in table 2: our algorithm can be much faster than the classical one. Moreover, from $dp3$ and beyond, the memory required to store the states and the transitions due to the combinatorial explosion in the classical case is larger than the memory required to store ancestor lists with our method (hidden transitions of the implementation are never stored in an ancestor list).

9. Conclusions

Most of the work on partial orders deals with semantic problems or with algebraic properties and remains in a theoretical framework. There are only a few papers related to concurrent system verification using partial order models [Gai88][KP86][KP88][Pen90][PL89][PP90][PW84][Val89]. Our method has the advantages of being simple, fully algorithmic, and of extending directly conventional verification techniques. Moreover, this method has the potential of being much more efficient, both in time and memory, than the classical one. The dining philosophers example emphasizes clearly the power of our method: we avoid the part of the combinatorial explosion due to the modeling of concurrency by interleavings. The verification framework presented in this paper is restricted to nonlabeled P/T-nets and safety properties. Interesting future work would be to broaden the scope of our method to cope with labeled P/T-nets and liveness properties.

Acknowledgements

I would like to thank Professor Pierre Wolper for his enthousiastic supervision, for fruitful discussions and thoughtful hints. I am grateful to Jean-Yves Pirnay and Philippe Simar for helpful comments and contributing to the typesetting of this paper. My thanks are also addressed to Marianne Baudinet, Froduald Kabanza, François Schumacker and Dr. Wojciech Penczek for reading drafts of this paper.

References

[Gai88] H. Gaifman. Modeling concurrency by partial orders and nonlinear transition systems. In *Linear Time, Branching Time and Partial Order in Logics and Models for Concurrency*, LNCS 354, pages 467–488, 1988.

[God89] P. Godefroid. Les modèles ordre partiel du parallélisme (partial order models for concurrency). Undergraduate thesis, Service d'Informatique, Université de Liège, June 1989.

[JK90] R. Janicki and M. Koutny. On some implementation of optimal simulations. To appear in *Proc. Computer-Aided Verification Workshop*, Rutgers, 1990.

[KP86] Y. Kornatzky and S. S. Pinter. A model checker for partial order temporal logic. EE PUB 597, Department of Electrical Enginering, Technion-Israel Institute of Technology, 1986.

[KP88] S. Katz and D. Peled. An efficient verification method for parallel and distributed programs. In *Linear Time, Branching Time and Partial Order in Logics and Models for Concurrency*, LNCS 354, pages 489–507, 1988.

[Lam78] L. Lamport. Time, clocks, and the ordering of events in a distributed system. *Communications of the ACM*, 21(7):558–564, 1978.

[Maz86] A. Mazurkiewicz. Trace theory. In *Petri Nets: Applications and Relationships to Other Models of Concurrency, Advances in Petri Nets 1986, Part II; Proceedings of an Advanced Course*, LNCS 255, pages 279–324, 1986.

[Maz88] A. Mazurkiewicz. Basic notions of trace theory. In *Linear Time, Branching Time and Partial Order in Logics and Models for Concurrency*, LNCS 354, pages 285–363, 1988.

[Pen90] W. Penczek. Proving partial order properties using CCTL. Submitted to *Proc. Concurrency and Compositionality Workshop*, San Miniato, Italy, 1990.

[PL89] D. K. Probst and H. F. Li. Abstract specification, composition and proof of correctness of delay-insensitive circuits and systems. Department of Computer Science, Concordia University, Montreal, Quebec Canada, 1989.

[Pra86] V. Pratt. Modelling concurrency with partial orders. *International Journal of Parallel Programming*, 15(1):33–71, 1986.

[PP90] D. Peled and A. Pnueli. Proving Partial Order Liveness Properties. ICALP, 1990.

[PW84] S. S. Pinter and P. Wolper. A temporal logic for reasoning about partially ordered computations. In *Proc. 3rd ACM Symposium on Principles of Distributed Computing*, pages 28–37. Vancouver, 1984.

[Rei85] W. Reisig. Petri nets: an introduction. EATCS Monographs on Theoretical Computer Science, Springer-Verlag, 1985.

[Roz86] G. Rozenberg. Behaviour of elementary net systems. In *Petri Nets: Applications and Relationships to Other Models of Concurrency, Advances in Petri Nets 1986, Part II; Proceedings of an Advanced Course*, LNCS 254, pages 60–94, 1986.

[Val88] A. Valmari. Error detection by reduced reachability graph detection. In *Proc. 9th International Conference on Application and Theory of Petri Nets*, pages 95–112, Venice, 1988.

[Val89] A. Valmari. Stubborn sets for reduced state space generation. In *Proc. 10th International Conference on Application and Theory of Petri Nets*, vol. 2, pages 1–22, Bonn, 1989.

[Win86] G. Winskel. Event structures. In *Petri Nets: Applications and Relationships to Other Models of Concurrency, Advances in Petri Nets 1986, Part II; Proceedings of an Advanced Course*, LNCS 255, pages 325–392, 1986.

[Wol89] P. Wolper. On the relation of programs and computations to models of temporal logic. In B. Banieqbal, H. Barringer, and A. Pnueli, editors, *Proc. Temporal Logic in Specification*, LNCS 398, pages 75–123, 1989.

[Zie80] W. Zielonka. Proving assertions about parallel programs by means of traces. ICS PAS Report 424, Institute of Computer Science, Polish Academy of Sciences, 1980.

Université de Liège, Institut Montefiore, B28, 4000 Liège Sart-Tilman, Belgium

E-mail: godefroid@montefiore.ulg.ac.be

DIMACS Series in Discrete Mathematics
and Theoretical Computer Science
Volume 3, 1991

A Context Dependent Equivalence Relation between Kripke Structures

BERNHARD JOSKO

ABSTRACT. In [1] Browne, Clarke and Grumberg define a bisimulation relation on Kripke structure and give a characterization of this equivalence relation in temporal logic. We will generalize their results to reactive systems, which are modelled by Kripke structures together with some constraints describing some requirements how the environment has to interact with the module. Our results subsume the result of [1] by using the constraint *true*. Furthermore it answers the questions raised in that paper how the equivalence of Kripke structures with fairness constraints can be characterized.

1. Introduction

In a top down design step of a large system one component may be splitted in several subcomponents (see Fig. 1). Not only the tasks of the subcomponents have to be specified but also the interface between the subcomponents have to be defined. On the one hand the interface has to declare the interconnections of the subcomponents i.e. the inputs and outputs of the components, and on the other hand the protocols for the exchange of data have to be specified too. E.g. if a subsystem consists of two components which are coupled asynchronously together (see Fig. 2), we can use a 4-cycle signalling protocol to send data from one component to the other (see Fig. 3). Module M_1 is responsible for the *Req* signal and it has to guarantee that this signal is set and reset according to the given protocol, and M_2 is responsible for the signal *Ack*. Proceeding in the design process we may define a more concrete representation of module M_2 by splitting this component in several subcomponents together with interface specifications

1980 *Mathematics Subject Classification* (1985 *Revision*). Primary Primary 68Q10, 68Q90; Secondary 03B45.

Key words and phrases. temporal logic, Kripke structure, bisimulation, modular specification, reactive systems, hierarchical design.

The final version of this paper will be submitted for publication elsewhere.

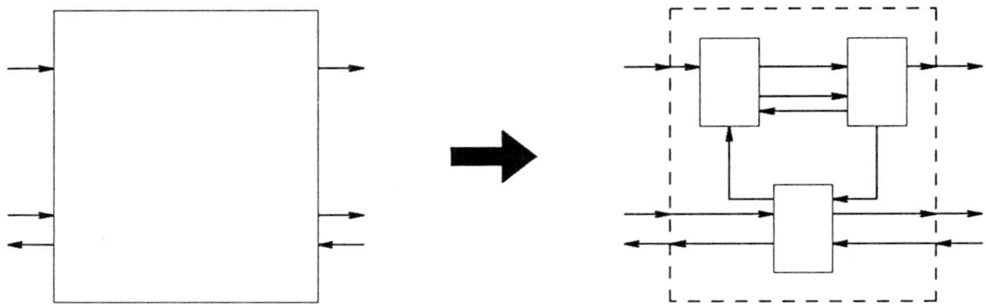

FIGURE 1. Top down design step

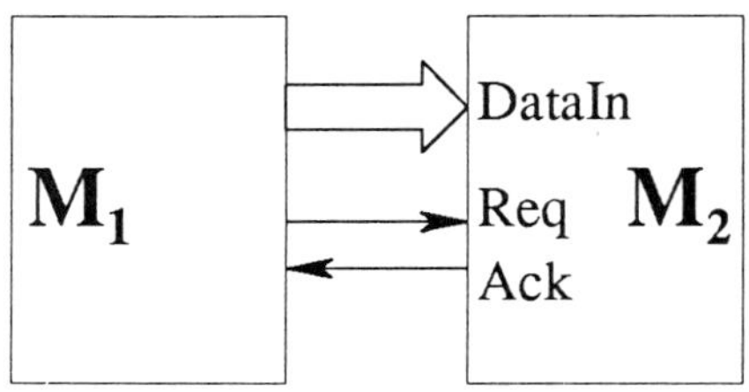

FIGURE 2. asynchrnously coupled components

or by defining an implementation. In such a design step we can use the fact that the environment (i.e. module M_1) acts according to the protocol and hence, it is only required that the module M_2 behaves correctly provided the environment guarantees the given interface constraint. Therefore a module specification is given by a pair $(assm, spec)$ where $assm$ describes some constraints on the environment and $spec$ is the specification of the module, which has to be satisfied by the module provided the environment guarantees $assm$.

In this paper we investigate system design in the framework of temporal logic. We will use the branching time temporal logic CTL* [5] as specification language. In the context of temporal logics an implementation of a system is modelled by a state/transition-graph, also called a Kripke structure. A temporal logic formula is then interpreted in the associated computation tree, which is obtained by unravelling the transition graph. As we are considering open systems (or reactive

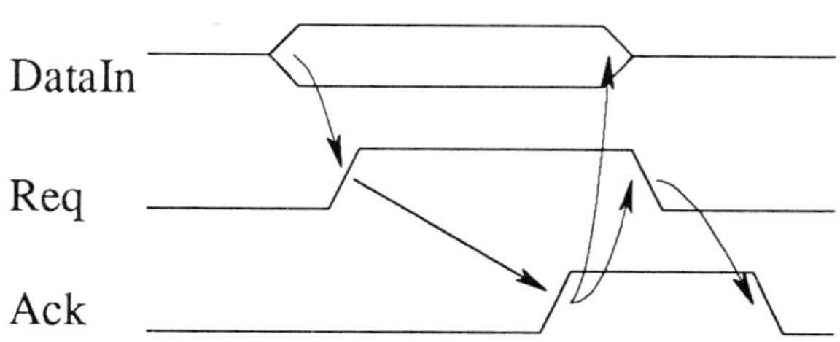

FIGURE 3. 4-cycle signalling protocoll

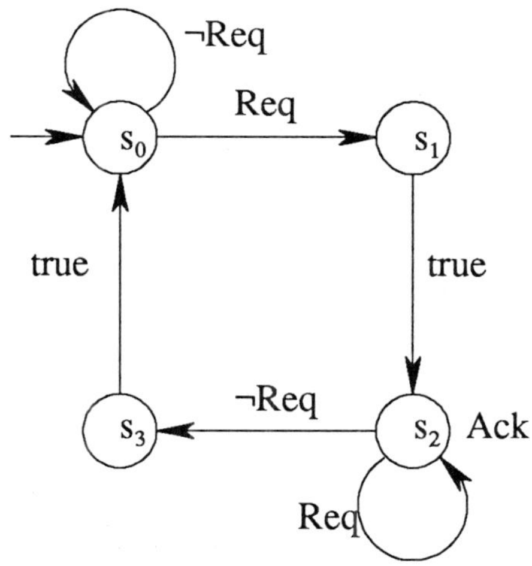

FIGURE 4. Kripke structure $\mathcal{K}_1$

systems [14]), a module will be modelled by a Kripke structure $\mathcal{K}$ together with an interface constraint *assm*. The constraint *assm* restricts the possible paths in the computation trees. Hence to check the validity of a specification *spec* only those paths are considered which satisfy the given assumption *assm*. As the assumptions on the environment reflect a part of the observable behaviour of the environment, these constraints will be given by path formulae. Considering the example in Fig. 2 together with Fig. 3 the environment constraints used by module M_2 can be defined in temporal logic as follows:

$assm_1 := \Box(Ack \wedge \neg Req \;\rightarrow\; [\neg Req\, unless\, \neg Ack \wedge \neg Req])$

A request occurs only if the acknowledge signal is low.

$assm_2 := \Box(Req \;\rightarrow\; [stable(DataIn)\, unless\, Ack])$

Whenever the request signal is high, the input data is stable until an acknowledge is given.

$assm_3 := \Box(Req \;\rightarrow\; [Req\, unless\, Ack \wedge Req])$

Whenever the signal Req is high it remains high unless an acknowledge occurs.

$assm_4 := \Box(Ack \wedge Req \;\rightarrow\; \Diamond \neg Req)$

Whenever an acknowledge is given, the request signal will eventually be reset.

Fig. 4 shows a state/transition graph for M_2 (only the protocol is implemented).

During the design process a module implementation may be replaced by a new one. If a replaced module has already been verified to be correct w.r.t. its specification and one has derived properties of a composed module which

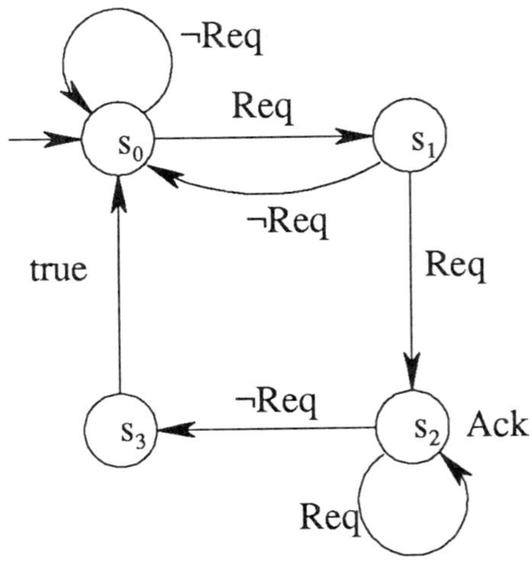

FIGURE 5. Kripke structure $\mathcal{K}_2$

contains that specific module, one is interested in the question whether the derived properties remain valid. Hence one is interested in an equivalence notion on Kripke structures, which guarantees that the replacement of a submodule by an equivalent one does not violate the correctness of derived properties of a composed (sub)system.

In Fig. 5 another implementation for module M_2 is given. In this implementation the request signal is checked after reading the input data. If the request signal is low – this is an violation of the protocol – the input data is ignored. In an environment which guarantees a correct behaviour according to the protocol, this step will never occurs. Thus both implementations, $\mathcal{K}_1$ and $\mathcal{K}_2$, are equivalent with respect to an environment satisfying the constraints.

Two structures will be called equivalent if they cannot be distinguished by any formula. In **[1]** a bisimulation (**[11, 12]**) for *closed* systems, i.e. Kripke structures without constraints on the environment, is given. Two states s and s' are called bisimulation equivalent if they are labelled with the same atomic propositions and for every transition $s \longrightarrow s_1$ in one structure there is a corresponding transition in the other structure leading to a state s_1' which is equivalent to s_1. It is shown that this relation can be characterized by the fact that both structures satisfy the same temporal logic formulae. In this paper we will show how the results of **[1]** can be generalized to *open* systems. We will relativize the correspondence of equivalent steps by requiring that only for those transitions which are on a path satisfying the environment constraints there are corresponding transitions in the other module. We will show that this bisimulation with respect to an assumption *assm* characterizes the structures satisfying the same set of CTL* formulae relatively to the given assumption *assm*, i.e. two structures $\mathcal{K}_1$ and $\mathcal{K}_2$ are bisimulation equivalent w.r.t. *assm* iff the modules $(\mathcal{K}_1, assm)$

and $(\mathcal{K}_2, assm)$ are not distinguishable by any CTL* formula.

For action based transition systems like CCS Larsen has given a notion of bisimulation relatively to some contexts [**8**, **9**], but there is no relation to temporal logic specifications.

2. Basic definitions

A Kripke structure is given by $\mathcal{K} = (S, R, s^0, L)$, where S is a finite set of states, $s^0 \in S$ is the initial state and R is a transition relation $R \subseteq S \times S$, such that for every state s there is some state s' with $(s, s') \in R$, and L is a labelling function which associates with every state $s \in S$ a set of atomic propositions, which are valid in that state.

Dealing with reactive systems one should use labelled transitions, as a transition depends on the input signals. A Kripke structure with inputs is given by $\mathcal{K} = (S, R, s^0, IN, OUT, L)$ where S is a set of states, s^0 is the initial state, IN and OUT are disjoint set of propositions, L is a labelling of states with subsets of OUT, and R is a transition relation with $R \subseteq S \times BExpr(IN) \times S$, such that for every satisfiable boolean expression b and every state s there is some state s' with $(s, b, s') \in R$. IN is a set of propositions on the input signals (usually the input signals themselves) and OUT are propositions on the internal and output signals. The transition labels are constraints on the input signals, restricting a transition to the instances where the actual input signals satisfy the given boolean expression. For every Kripke structure $\mathcal{K}$ with inputs there is a corresponding Kripke structure $\mathcal{K}'$ (without inputs) [**2**]. Therefore we will use the usual notion of Kripke structure in this paper. Fig. 6 resp. Fig. 7 shows the transformed Kripke structures of Fig. 4 resp. Fig. 5.

Given a Kripke structure $\mathcal{K} = (S, R, s^0, L)$, $\pi = (s_i \mid i \in \mathbb{N})$ is a path in $\mathcal{K}$ if $(s_i, s_{i+1}) \in R$ for all $i \in \mathbb{N}$. We will refer to the i-th state s_i also by $\pi(i)$. The suffix of a path starting at state $\pi(i)$ will be denoted by π^i. The (infinite) computation tree of $\mathcal{K}$ is obtained by unravelling the Kripke structure (considered as a graph) starting with the initial state s^0.

As indicated in the introduction a reactive module is given by a pair $(\mathcal{K}, assm)$ where $\mathcal{K}$ is a Kripke structure describing the implementation of the module and $assm$ describes some constraints on the environment. A specification $spec$ of a module specifies the behaviour of the module provided the environment guarantees the given assumption $assm$. Here we will use the branching time logic CTL* as a specification language. The constraints of the environment describe the observable interface specification of the environment, hence this will be given by paths formulae; we will use the linear time logic PTL [**13**] to describe these constraints.

PTL is the linear propositional temporal logic defined by

$$f ::= a \mid \neg f \mid f_1 \wedge f_2 \mid f_1 \vee f_2 \mid Xf \mid [f_1\, U\, f_2] \mid \Box f \mid \Diamond f,$$

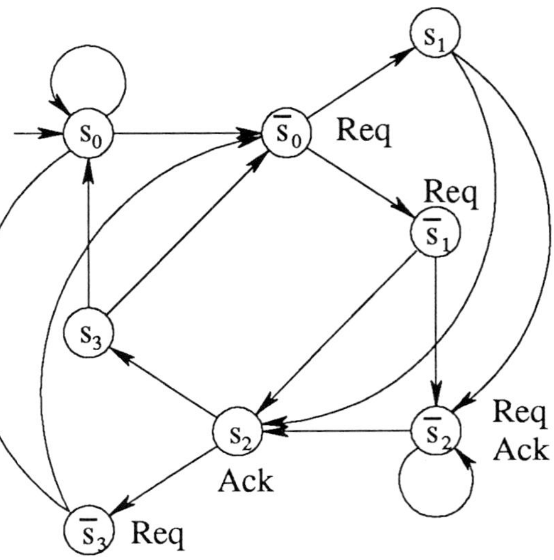

FIGURE 6. Kripke structure $\mathcal{K}_1$ transformed

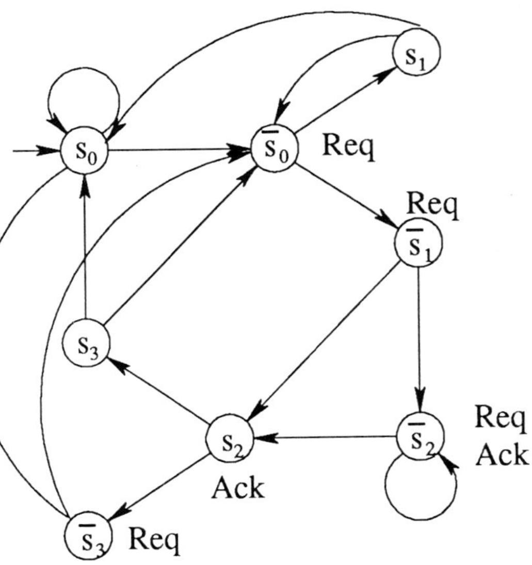

FIGURE 7. Kripke structure $\mathcal{K}_2$ transformed

where a denotes an atomic proposition. $[f_1 \, unless \, f_2]$ will be used as an abbreviation for $\Box f_2 \vee [f_1 \, U \, f_2]$. The validity of a PTL formula f along a path π will be denoted by $\pi \models f$. We also call a path π f-good if π satisfies the formula f. A state s is called f-good iff there is some path π starting at s with is f-good.

CTL and CTL* are branching time temporal logics which are defined by the following rules:

$$\begin{array}{lll} (1) & sf ::= & a \mid \neg sf \mid sf_1 \wedge sf_2 \mid sf_1 \vee sf_2 \mid \forall pf \mid \exists pf \\ (2) & pf ::= & sf \mid \neg pf \mid pf_1 \wedge pf_2 \mid pf_1 \vee pf_2 \mid Xpf \mid [pf_1 \, U \, pf_2] \\ (3) & pf ::= & Xsf \mid [sf_1 \, U \, sf_2] \end{array}$$

The rules in (1) describe the building of state formulae and the rules in (2) and (3) describe the construction of path formulae. CTL is the set of all state formulae defined by (1) and(3) and CTL* is the set of all state formulae defined by (1) and (2). Furthermore we will use the path formula $\Diamond f$ as an abbreviation for $[true \, U \, f]$, and $\Box f$ as an abbreviation for $\neg \Diamond \neg f$. The semantics of CTL and CTL* for closed systems is as usual (cf. [5]).

A module is given by a pair $(\mathcal{K}, assm)$, where $\mathcal{K}$ is a Kripke structure and $assm$ is a PTL formula. The interpretation of a CTL* formula $spec$ in a module $(\mathcal{K}, assm)$ can be described as follows:

(i) Construct the computation tree with root s^0.

(ii) Mark all paths starting at s^0 which satisfy the constraint $assm$.

(iii) Interpret the formula $spec$ as in the usual interpretation of CTL* formulae but with the restriction that all path quantifiers are restricted to the marked paths.

To give an inductive definition of the semantics we need a description of the behaviour of the environment at a specific instance on a computation path. The expected behaviour of the environment at instance i on a path π is determined by the given interface constraint $assm$ and the computation history along π . E.g. if $\Box(Req \rightarrow [Req \, unless \, Req \wedge Ack])$ is a conjunctive of the interface constraint and assume that the signal Req is set to high at instance j (i.e. the proposition Req is valid at instance j) then $[Req \, unless \, Req \wedge Ack]$ is also a part of the expected behaviour of the environment at the next time instance. Using the computation history along a path π we can define a formula $assm_i(\pi, assm)$ which describes the expected behaviour of the environment at instance i on the computation path π . This is done by using a technique which is used in satisfiability procedures for temporal logics (cf. [**10**, 7]). A path formula can be devided into two parts: one describing a property of the actual state and one describing a property of the path starting at the next time instance. To be more precise, every formula is equivalent to a disjunction of such descriptions. Using this representation of a given constraint $assm$ and using the actual state information, we can describe the expected behaviour of the environment at the next time instance.

A temporal logic formula f is in negation normal form ($\neg$-NF) iff a negation occurs only before an atom. Every temporal logic formula f can be transformed into an equivalent formula f' which is in $\neg$-NF. Hence we will assume that all PTL formulae are given in this normal form. For a set of PTL formulae F we can consider closures of F, which are obtain from F by adding for every conjunction its conjunctives, and for every disjunction at least one of its disjunctives. Building closures, $\Box f$ is considered as the conjunction $f \wedge X\Box f$, $\Diamond f$ as the disjunction $f \vee X\Diamond f$, and $[f_1\, U\, f_2]$ is considered as the disjunction $f_2 \vee (f_1 \wedge X[f_1\, U f_2])$.

DEFINITION 1. C is a closure of F (assuming that all formulae of F are in $\neg$-NF) iff

$$
\begin{array}{lll}
(1) & F \subseteq C & \\
(2) & \neg a \in C & \Rightarrow \quad a \notin C \\
(3) & f_1 \wedge f_2 \in C & \Rightarrow \quad f_1,\, f_2 \in C \\
(4) & f_1 \vee f_2 \in C & \Rightarrow \quad f_1 \in C \text{ or } f_2 \in C \\
(5) & [f_1\, U\, f_2] \in C & \Rightarrow \quad f_2 \in C \text{ or } f_1,\, X[f_1\, U\, f_2] \in C \\
(6) & \Box f \in C & \Rightarrow \quad f,\, X\Box f \in C \\
(7) & \Diamond f \in C & \Rightarrow \quad f \in C \text{ or } X\Diamond f \in C
\end{array}
$$

If there exists no closure of F then F is not satisfiable, e.g. if F contains both a and $\neg a$. C is a minimal closure of F if every proper subset of C is not a closure of F. For a finite set F there are only finitely many minimal closures of F.

FACT 1. *Let F be a finite set of PTL formulae, then*

$$\pi \models F \quad \textit{iff} \quad \pi \models \bigvee\{C \mid C \textit{ is a minimal closure of } F\}.$$

For a closure C we define the sets $literal(C)$ and $next(C)$ as follows: $literal(C)$ is the set of all literals contained in C and $next(C)$ is the set of all formulae of C prefixed with the next-operator X.

FACT 2. *Let F be a finite set of PTL formulae and C a closure, then*

(i)

$$\pi \models C \quad \textit{iff} \quad \pi(0) \models literal(C) \;\; \textit{and} \;\; \pi^1 \models \{f \mid Xf \in next(C)\}$$

(ii)

$$\pi \models F \quad \textit{iff} \quad \pi^1 \models \bigvee\{f \mid Xf \in next(C), C \textit{ is a minimal closure of } F \textit{ and } \pi(0) \models literal(C)\}$$

Due to Fact 2, given an assumption $assm$ and a state s we can construct a formula $assm'$ such that $assm$ is valid on a path π starting at s iff $assm'$ is valid on π^1. We will denote that formula $assm'$ by $next\_assm(assm, s)$, i.e.

$$next\_assm(assm, s) := \bigvee\{f \mid Xf \in next(C), C \text{ is a minimal closure of } \{assm\} \text{ and } s \models literal(C)\}$$

This defintion can be extended to all time instances along a path. Given a path π and a formula *assm*, we can determine for every $i \in \mathbb{N}$ a formula $assm_i$ which is valid on π^i iff π satisfies *assm*. We define

$$\begin{aligned} assm_0(f,\pi) &:= f \\ assm_{k+1}(f,\pi) &:= next\_assm(assm_k(f,\pi),\pi(k)) \end{aligned}$$

EXAMPLE. Let *assm* be $\Box(Req \rightarrow [Req\, unless\, Req \wedge Ack])$. The minimal closures of $\{assm\}$ are

$$\begin{aligned} C_1 &= \{\ assm, (Req \rightarrow [Req\, unless\, Req \wedge Ack]), X assm, \neg Req\ \}, \\ C_2 &= \{\ assm, (Req \rightarrow [Req\, unless\, Req \wedge Ack]), X assm, \\ &\quad [Req\, unless\, Req \wedge Ack], Req \wedge Ack, Req, Ack\ \}, \quad \text{and} \\ C_3 &= \{\ assm, (Req \rightarrow [Req\, unless\, Req \wedge Ack]), X assm, \\ &\quad [Req\, unless\, Req \wedge Ack], Req, X[Req\, unless\, Req \wedge Ack]\ \}. \end{aligned}$$

$$\begin{aligned} literal(C_1) &= \{\neg Req\}, & next(C_1) &= \{X assm\}, \\ literal(C_2) &= \{Req, Ack\}, & next(C_2) &= \{X assm\}, \\ literal(C_3) &= \{Req\}, & next(C_3) &= \{X assm, X[Req\, unless\, Ack]\}. \end{aligned}$$

$$next\_assm(s, assm) = \begin{cases} assm & \text{if } s \models \neg Req \vee Ack \\ assm \wedge [Req\, unless\, Req \wedge Ack] & \text{otherwise} \end{cases}$$

Now we can give an inductive definition of the validity of a CTL* formula in a module $(\mathcal{K}, assm)$ using the notion of $next\_assm(assm, s)$ and $assm_i(assm, \pi)$. E.g. $(\mathcal{K}, assm, s) \models \forall[f_1\, U f_2]$ iff for all *assm*-good paths π starting at s it holds: there is some instance k with $(\mathcal{K}, assm_k(assm,\pi), \pi^k) \models f_2$ and for all $j, 0 \leq j < k, (\mathcal{K}, assm_j(assm,\pi), \pi^j) \models f_1$.

DEFINITION 2. Given a Kripke structure $\mathcal{K}$ and a context *assm*, the validity of a formula f at a state s, denoted by $(\mathcal{K}, assm, s) \models f$, is defined as follows:

$(\mathcal{K}, assm, s) \models a$	iff	$a \in L(s)$
$(\mathcal{K}, assm, s) \models \neg sf$	iff	$not(\mathcal{K}, assm, s) \models sf$
$(\mathcal{K}, assm, s) \models sf_1 \wedge sf_2$	iff	$(\mathcal{K}, assm, s) \models sf_1$ and $(\mathcal{K}, assm, s) \models sf_2$
$(\mathcal{K}, assm, s) \models \forall pf$	iff	for all *assm*-good paths π starting at s $(\mathcal{K}, assm, \pi) \models pf$
$(\mathcal{K}, assm, s) \models \exists pf$	iff	there is some *assm*-good path π starting at s with $(\mathcal{K}, assm, \pi) \models pf$
$(\mathcal{K}, assm, \pi) \models sf$	iff	$(\mathcal{K}, assm, \pi(0)) \models sf$
$(\mathcal{K}, assm, \pi) \models \neg pf$	iff	not $(\mathcal{K}, assm, \pi) \models pf$
$(\mathcal{K}, assm, \pi) \models pf_1 \wedge pf_2$	iff	$(\mathcal{K}, assm, \pi) \models pf_1$ and $(\mathcal{K}, assm, \pi) \models pf_2$
$(\mathcal{K}, assm, \pi) \models X pf$	iff	$(\mathcal{K}, assm_1(assm,\pi), \pi^1) \models pf$
$(\mathcal{K}, assm, \pi) \models [pf_1\, U pf_2]$	iff	there is some k with $(\mathcal{K}, assm_k(assm,\pi), \pi^k) \models pf_2$ and for all $j, 0 \leq j < k$: $(\mathcal{K}, assm_j(assm,\pi), \pi^j) \models pf_1$

Furthermore, we say that a module $(\mathcal{K}, assm)$ satisfies a specification f, denoted by $(\mathcal{K}, assm) \models f$, iff $(\mathcal{K}, assm, s^0) \models f$.

3. Relativized equivalence of Kripke structures

Usually two Kripke structures $\mathcal{K}$ and $\mathcal{K}'$ are equivalent if both satisfy the same set of formulae. Dealing with modules embedded in an environment which has to guarantee a behaviour according to an interface specification *assm*, we use a relativized equivalence relation, which demands only that both modules should have the same behaviour in such an environment, i.e. $(\mathcal{K}, assm)$ and $(\mathcal{K}', assm)$ cannot be distinguished by any formula.

DEFINITION 3. Two Kripke structures $\mathcal{K}$ and $\mathcal{K}'$ are equivalent relatively to $assm \in$ PTL, denoted by $\mathcal{K} \equiv_{assm} \mathcal{K}'$, iff for all CTL* formulae f:

$$(\mathcal{K}, assm) \models f \quad \text{iff} \quad (\mathcal{K}', assm) \models f.$$

This (semantical) equivalence relation can be defined by a syntactical relation on the structures (bisimulation) as we will show in the sequel. Furthermore we will characterize the equivalence class of a given Kripke structure by a CTL formula.

Given two Kripke structures $\mathcal{K}$ and $\mathcal{K}'$ a bisimulation between $\mathcal{K}$ and $\mathcal{K}'$ is given by a relation $E \subseteq S \times S'$ with the property that if $(s, s') \in E$ then $L(s) = L'(s')$ and for every successor s_1 of s there should be a corresponding successor s_1' of s' with $(s_1, s_1') \in E$ and vice versa [**1**]. As we consider reactive modules the bisimulation can be restricted to the steps which are consistent with the environment assumption *assm*, i.e. only for those steps $s \to s_1$ which are on a path satisfying *assm* there should be a corresponding step in the other structure. We will define the relativized bisimulation by a chain of approximations considering paths up to depth i.

DEFINITION 4. Given two Kripke structures $\mathcal{K} = (S, s^0, R, L)$ and $\mathcal{K}' = (S', s'^0, R', L')$ with the same set $ATOM$ of atomic propositions and given a PTL formula *assm* we define the relativized equivalence relations $BISIM_i(assm) \subseteq S \times S'$ as follows:

(i) $(s, s') \in BISIM_0(assm)$ iff $L(s) = L'(s')$

(ii) $(s, s') \in BISIM_{i+1}(assm)$ iff

  (a) $L(s) = L'(s')$ and

  (b) for every $next\_assm(assm, s)$-good successor s_1 of s there is a successor s_1' of s' with $(s_1, s_1') \in BISIM_i(next\_assm(assm, s))$

  (c) for every $next\_assm(assm, s')$-good successor s_1' of s' there is a successor s_1 of s with $(s_1, s_1') \in BISIM_i(next\_assm(assm, s))$

(iii) $(s, s') \in BISIM(assm)$ iff $(s, s') \in BISIM_i(assm)$ for all $i \in \mathbb{N}$.

(iv) $(\mathcal{K}, \mathcal{K}') \in BISIM(assm)$ iff $(s^0, s'^0) \in BISIM(assm)$.

Remark. If the labellings of two states are equal, $L(s) = L'(s')$, then we have that $next\_assm(assm, s) = next\_assm(assm, s')$.

EXAMPLE. Consider the Kripke structures of Fig. 6 and Fig. 7. Without consideration of the environment constraints the two structures are different as $\mathcal{K}_1$ can proceed from state s_1 to $\overline{s}_2$ or s_2, whereas in $\mathcal{K}_2$ the next states of s_1 are s_0 or $\overline{s}_0$ as the input signal *Req* is low in state s_1. But as the environment has to guarantee that the signal *Req* remains high unless the response *Ack* occurs – this is expressed by the assumption $assm = \Box(Req \rightarrow [Req\, unless\, Req \wedge Ack])$ –, both structures are equivalent, as under the constraint *assm* the state s_1 in $\mathcal{K}_1$ and the state s_1 in $\mathcal{K}_2$ are not reachable.

The bisimulation $BISIM(assm)$ and the equivalence relation $\equiv_{assm}$ coincide. To prove this fact we first show that two Kripke structures are equivalent relatively to a given constraint *assm* if they are bisimulation equivalent.

THEOREM 1. *If* $(s, s') \in BISIM(assm)$ *then for all* $f \in CTL^*$*:*

$$(\mathcal{K}, assm, s) \models f \quad \text{iff} \quad (\mathcal{K}', assm, s') \models f.$$

To prove this theorem we need some properties of the bisimulation relation $BISIM(assm)$.

FACT 3. *If* $L(s) = L'(s')$ *and if there is no assm-good path starting at* s *and no assm-good path starting at* s', *then* $(s, s') \in BISIM(assm)$.

LEMMA 1. *Let* f *and* f' *be PTL formulae and assume* $\models f \rightarrow f'$, *then* $BISIM(f') \subseteq BISIM(f)$

PROOF. By induction on i we show that

$$BISIM_i(f') \subseteq BISIM_i(f).$$

$i = 0$: $BISIM_0(f)$ is independent of f. Hence $BISIM_0(f) = BISIM_0(f')$.

$i = k+1$: Assume that $(s, s') \in BISIM_{k+1}(f')$. The labelling $L(s)$ of a state can be considered as a formula

$$\bigwedge\{a \mid a \in L(s)\} \wedge \bigwedge\{\neg a \mid a \in ATOM \setminus L(s)\}$$

then we have that $f \wedge L(s)$ is equivalent to $\bigvee\{next(C) \mid C$ is a minimal closure of f and $s \models literal(C)\ \}$ and $f' \wedge L'(s')$ is equivalent to $\bigvee\{next(C) \mid C$ is a minimal closure of f' and $s' \models literal(C)\ \}$. As $L(s) = L'(s')$, and $f \rightarrow f'$, we have that $next\_assm(f, s)$ implies $next\_assm(f', s')$. Now, assume that s_1 is a $next\_assm(f, s)$-good successor of s, then s_1 is also a $next\_assm(f', s)$-good successor of s. Hence there is a successor s_1' of s' with

$$(s_1, s_1') \in BISIM_k(next\_assm(f', s)).$$

By induction hypothesis we can conclude that

$$(s_1, s_1') \in BISIM_k(next\_assm(f, s)).$$

Hence $(s, s') \in BISIM_{k+1}(f)$. $\Box$

LEMMA 2. *Assume that* $(s, s') \in BISIM(assm)$, *then for every assm-good path* $\pi = (s_0, s_1, \ldots)$ *there is an assm-good path* $\pi' = (s'_0, s'_1, \ldots)$ *with* $(s_i, s'_i) \in BISIM(assm_i(assm, \pi))$.

PROOF. For every given formula f there are only finitely many approximations $BISIM_i(f)$ of $BISIM(f)$ as there are only finitely many states. Furthermore the set $\{assm_i(assm, \pi) \mid i \in \mathbb{N}\}$ is finite. Hence let k be the minimal integer such that

$$BISIM_{k+1}(assm_i(assm, \pi)) = BISIM_k(assm_i(assm, \pi)) \qquad \text{for all } i.$$

Therefore

$$BISIM(assm_i(assm, \pi)) = BISIM_k(assm_i(assm, \pi)).$$

By induction on i we show that for every $i \in \mathbb{N}$ there is a state s'_i which is equivalent to s_i. As, by definition of the bisimulation, the labelling of s'_i is the same as the labelling of s_i it is obvious that the resulting path is *assm*-good.

$i = 0$: Let s'_0 be s'. As $assm_0(assm, \pi) = assm$ and $(s_0, s') \in BISIM(assm)$, it is obvious that $(s_0, s'_0) \in BISIM(assm_0(assm, \pi))$.

$i = j + 1$: By induction hypothesis we have that

$$(s_j, s'_j) \in BISIM(assm_j(assm, \pi)).$$

As s_{j+1} is a $next\_assm(assm_j(assm, \pi), s_j)$-good successor of s_j (by Fact 2) there is a successor s'_{j+1} of s'_j with

$$(s_{j+1}, s'_{j+1}) \in BISIM(next\_assm(assm_j(assm, \pi), s_j).$$

As

$$assm_{j+1}(assm, \pi) = next\_assm(assm_j(assm, \pi), s_j)$$

this proves our statement. □

PROOF (THEOREM 1). We prove the following fact:
Let f be a state formula or a path formula, and let $\pi = s, s_1, \ldots$ be an *assm*-good path in $\mathcal{K}$ and $\pi' = s', s'_1, \ldots$ an *assm*-good path in $\mathcal{K}'$ with $(s, s') \in BISIM(assm)$ and $(s_i, s'_i) \in BISIM(assm_i(assm, \pi))$ for all $i \in \mathbb{N}$. Then

$$(\mathcal{K}, assm, s) \models f \quad \text{iff} \quad (\mathcal{K}', assm, s') \models f \quad \text{resp.}$$
$$(\mathcal{K}, assm, \pi) \models f \quad \text{iff} \quad (\mathcal{K}', assm, \pi') \models f$$

This can be proved by induction on the structure of f. Here we show the induction step for $\forall pf$ and for the *until* operator, the other cases are similarily.

$f = \forall pf$: Assume that $(\mathcal{K}, assm, s) \models f$. To prove that $(\mathcal{K}', assm, s')$ satisfies f we show that every *assm*-good path π' starting at s' satisfies pf. Hence, assume that π' is an *assm*-good path starting at s'. As $(s, s') \in BISIM(assm)$ there is a corresponding *assm*-good path π in $\mathcal{K}$ (by Lemma 2). As $\forall pf$ is valid at s in $(\mathcal{K}, assm)$, π satisfies the formula pf. Therefore π' satisfies the formula pf too (as $L'(\pi'(i)) = L(\pi(i))$). Thus $(\mathcal{K}', assm, s') \models f$.

$f = [f_1 \, U f_2]$: Suppose that $(\mathcal{K}, assm, \pi) \models f$. Hence there is a k with

$$\begin{aligned}(\mathcal{K}, assm_k(assm, \pi), \pi^k) &\models f_2 \text{ and} \\ (\mathcal{K}, assm_j(assm, \pi), \pi^j) &\models f_1 \text{ for all } j, 0 \le j < k.\end{aligned}$$

As π and π' are bisimulation equivalent so are π^j and π'^j, for all j. Hence, by induction hypothesis, we have that

$$\begin{aligned}(\mathcal{K}', assm_k(assm, \pi), \pi'^k) &\models f_2 \text{ and} \\ (\mathcal{K}', assm_j(assm, \pi), \pi'^j) &\models f_1 \text{ for all } j, 0 \le j < k.\end{aligned}$$

Thus $(\mathcal{K}', assm, \pi') \models f$. □

To prove the reversion of Theorem 1 we give a characterization of the bisimulation class of a Kripke structure by a CTL formula. Basicly this formula describes the computation tree of the structure. Two states which are bisimulation equivalent have corresponding computation trees w.r.t. the given constraint on the environment. For a state s, let $CT_n(s)$ denote the computation tree of depth n rooted at s. Two trees $CT_n(s)$ and $CT_n(s')$ correspond w.r.t. an assumption $assm$, denoted by $CT_n(s) \equiv_{assm} CT_n(s')$, iff $L(s) = L'(s')$ and for every $next\_assm(s, assm)$-good successor s_1 there is a successor s'_1 of s' such that $CT_{n-1}(s_1)$ and $CT_{n-1}(s'_1)$ corresponds w.r.t. $next\_assm(s, assm)$ and vice versa.

LEMMA 3. *$(s, s') \in BISIM_n(assm)$ iff $CT_j(s) \equiv_{assm} CT_j(s')$ for all $j \le n$.*

PROOF. obvious □

We can describe the computation tree $CT_n(s)$ w.r.t. an assumption $assm$ by a CTL formula. The formula $F_{CT,n}(s, assm)$ is defined by:

$$F_{CT,0}(s, assm) = a_1 \wedge \ldots a_n \wedge \neg b_1 \wedge \ldots \wedge \neg b_m,$$

where $L(s) = \{a_1, \ldots, a_n\}$ and $ATOM \setminus L(s) = \{b_1, \ldots, b_m\}$

$$\begin{aligned}F_{CT,k+1}(s, assm) &= F_{CT,0}(s, assm) \\ &\wedge \bigwedge \{\exists X F_{CT,k}(s', next\_assm(assm, s)) \mid \\ &\qquad s' \text{ is a } next\_assm(assm, s)\text{-good successor of } s\} \\ &\wedge \forall X (\bigvee \{F_{CT,k}(s', next\_assm(assm, s)) \mid \\ &\qquad s' \text{ is a } next\_assm(assm, s)\text{-good successor of } s\})\end{aligned}$$

EXAMPLE. Consider the Kripke structure of Fig. 6 and assume that

$$assm = \Box(Req \rightarrow [Req \, unless \, Req \wedge Ack]).$$

$$
\begin{aligned}
F_{CT,0}(s_0, assm) &= F_{CT,0}(s_1, assm) = F_{CT,0}(s_3, assm) \\
&= \neg Req \wedge \neg Ack \\
F_{CT,0}(\overline{s}_0, assm) &= F_{CT,0}(\overline{s}_1, assm) = F_{CT,0}(\overline{s}_3, assm) \\
&= Req \wedge \neg Ack \\
F_{CT,0}(s_2, assm) &= \neg Req \wedge Ack \\
F_{CT,0}(\overline{s}_2, assm) &= Req \wedge Ack
\end{aligned}
$$

$$
\begin{aligned}
F_{CT,k+1}(s_0, assm) &= (\neg Req \wedge \neg Ack) \wedge \\
&\quad (\exists X F_{CT,k}(s_0, assm) \wedge \exists X F_{CT,k}(\overline{s}_0, assm)) \wedge \\
&\quad \forall X (F_{CT,k}(s_0, assm) \vee F_{CT,k}(\overline{s}_0, assm)) \\
F_{CT,k+1}(\overline{s}_0, assm) &= (Req \wedge \neg Ack) \wedge \\
&\quad (\exists X F_{CT,k}(\overline{s}_1, assm')) \wedge \forall X (F_{CT,k}(\overline{s}_1, assm'))
\end{aligned}
$$

where $assm' = next\_assm(assm, \overline{s}_0) = assm \wedge [Req\, unless\, Req \wedge Ack]$
(Observe that s_1 is not an $assm$-good successor of $\overline{s}_0$.)

$$
\begin{aligned}
F_{CT,k+1}(\overline{s}_1, assm') &= (Req \wedge \neg Ack) \wedge \\
&\quad (\exists X F_{CT,k}(\overline{s}_2, assm')) \wedge \forall X (F_{CT,k}(\overline{s}_2, assm'))
\end{aligned}
$$

(s_2 is not an $assm'$-good successor of $\overline{s}_1$.)
There is no $assm'$-good path starting at s_1, hence:

$$
\begin{aligned}
F_{CT,k+1}(s_1, assm') &= (\neg Req \wedge \neg Ack) \wedge true \wedge \forall X false \\
F_{CT,k+1}(\overline{s}_2, assm') &= (Req \wedge Ack) \wedge \\
&\quad (\exists X F_{CT,k}(s_2, assm') \wedge \exists X F_{CT,k}(\overline{s}_2, assm')) \wedge \\
&\quad \forall X (F_{CT,k}(s_2, assm') \vee F_{CT,k}(\overline{s}_2, assm')) \\
F_{CT,k+1}(s_2, assm') &= (\neg Req \wedge Ack) \wedge \\
&\quad (\exists X F_{CT,k}(s_3, assm) \wedge \exists X F_{CT,k}(\overline{s}_3, assm)) \wedge \\
&\quad \forall X (F_{CT,k}(s_3, assm) \vee F_{CT,k}(\overline{s}_3, assm))
\end{aligned}
$$

$(next\_assm(assm', s_2) = assm.)$

LEMMA 4. *If there is some assm-good path π starting at s, then it holds:*

$$(\mathcal{K}, assm, s) \models F_{CT,n}(s, assm) \qquad \text{for all } n \in \mathbb{N}.$$

PROOF. obvious □

LEMMA 5. *If there is some assm-good path starting at s' and $(\mathcal{K}, assm, s) \models F_{CT,n}(s', assm)$ then $CT_n(s) \equiv_{assm} CT_n(s')$.*

PROOF. Induction on n:
$n = 0$: obvious
$n = k + 1$: By definition of the formula, the labelling of s must be the same as the labelling of s'. Hence

$$next\_assm(assm, s) = next\_assm(assm, s').$$

Now assume that s_1' is some $next\_assm(assm, s')$-good successor of s'. As $(\mathcal{K}, assm, s) \models F_{CT,n}(s', assm)$ there is some $next\_assm(assm, s)$-good successor s_1 of s with

$$(\mathcal{K}, next\_assm(assm, s), s_1) \models F_{CT,n-1}(s_1', next\_assm(assm, s')).$$

Hence, by induction hypothesis,

$$CT_{n-1}(s_1) \equiv_{next\_assm(assm,s')} CT_{n-1}(s_1').$$

Assume that s_1 is some $next\_assm(assm, s)$-good successor of s. As

$$(\mathcal{K}, assm, s) \models F_{CT,n}(s', assm),$$

we have for all $next\_assm(assm, s)$-good successor s_1 of s:

$$(\mathcal{K}, next\_assm(assm, s), s_1) \models \bigvee\{F_{CT,n-1}(s'', next\_assm(assm, s')) \mid s'' \\ next\_assm(assm, s')\text{-good succ. of } s'\}.$$

Hence there is some $next\_assm(assm, s')$-good successor s'' of s' with

$$(\mathcal{K}, next\_assm(assm, s), s_1) \models F_{CT,n-1}(s'', next\_assm(assm, s')).$$

Therefore

$$CT_{n-1}(s_1) \equiv_{next\_assm(assm,s')} CT_{n-1}(s'').$$

□

LEMMA 6. *If s is a state in a Kripke structure $\mathcal{K}$, then there is a CTL formula $F_{BISIM}(\mathcal{K}, s, assm)$ that determines s up to $BISIM(assm)$-equivalence within $\mathcal{K}$, i.e. $F_{BISIM}(\mathcal{K}, s, assm)$ is true relatively to assm at s and at every state of $\mathcal{K}$ that is $BISIM(assm)$-equivalent to s, but false at every state that is not $BISIM(assm)$-equivalent to s.*

PROOF. Consider the computation tree of $\mathcal{K}$ with root s. Let

$$n := |S| \cdot |\{assm_i(assm, \pi) \mid \pi \text{ path rooted at } s,\ i \in \mathbb{N}\}|.$$

Then the computation tree is fully determined by the tree up to the depth $n+1$, i.e. for all states s and s' and every assumption $assm_i(assm, \pi)$ we have that if s and s' are not $BISIM(assm_i(assm, \pi))$-equivalent then $CT_{n+1}(s)$ and $CT_{n+1}(s')$ are not equivalent w.r.t. $assm_i(assm, \pi)$. Hence we choose

$$F_{BISIM}(\mathcal{K}, s, assm) := L(s) \wedge (\exists X true \rightarrow F_{CT,n+1}(s, assm)).$$

If there is no $assm$-good path starting at s then s' is $BISIM(assm)$-equivalent to s if s' has the same labelling as s and there is no $assm$-good path starting at s'. If there is no $assm$-good path starting at s the formula $F_{CT,n+1}(s, assm)$ is equivalent to $false$, hence $(\mathcal{K}, assm, s') \models F_{BISIM}(\mathcal{K}, s, assm)$ if there is no $assm$-good path starting at s' and $L(s') = L(s)$.

Now assume that there is some $assm$-good path starting at s. If s' is a state $BISIM(assm)$-equivalent to s then $CT_{n+1}(s) \equiv_{assm} CT_{n+1}(s')$ and therefore, the formulae describing these computation trees are equivalent. Hence, by Lemma 4, $(\mathcal{K}, assm, s') \models F_{BISIM}(\mathcal{K}, s, assm)$. Furthermore, if s' is not

$BISIM(assm)$-equivalent to s, then the computation trees up to depth $n+1$, $CT_{n+1}(s)$ and $CT_{n+1}(s')$, are not $assm$-equivalent, too. Hence by Lemma 5 $(\mathcal{K}, assm, s')$ does not satisfy the formula $F_{BISIM}(\mathcal{K}, s, assm)$. □

THEOREM 2. *Given a Kripke structure $\mathcal{K}$ with initial state s^0 and an assumption $assm$, then there is a CTL formula $F_{BISIM}(\mathcal{K}, assm)$ that characterizes that structure up to $BISIM(assm)$-equivalence.*

PROOF. Let $\{assm_1, \ldots, assm_m\} = \{\ assm_i(assm, \pi) \mid \pi$ any path in $\mathcal{K}$ starting at s^0, $i \in \mathbb{N}\ \}$. We then define

$$\begin{aligned} G_{BISIM}(\mathcal{K}, s, assm_i) \quad := \quad & \forall\Box(F_{BISIM}(\mathcal{K}, s, assm_i) \quad \rightarrow \\ & \bigwedge\{\exists X F_{BISIM}(\mathcal{K}, s', next\_assm(assm_i, s)) \mid \\ & \qquad s' \ next\_assm(assm_i, s)\text{-good successor of } s\} \quad \wedge \\ & \forall X(\bigvee\{F_{BISIM}(\mathcal{K}, s', next\_assm(assm_i, s)) \mid \\ & \qquad s' \ next\_assm(assm_i, s)\text{-good successor of } s\}) \\ F_{BISIM}(\mathcal{K}, assm) \quad := \quad & F_{BISIM}(\mathcal{K}, s^0, assm) \quad \wedge \\ & \bigwedge\{G_{BISIM}(\mathcal{K}, s, assm_j) \mid s \in S,\ j = 1, \ldots, m\} \end{aligned}$$

If two structures $\mathcal{K}$ and $\mathcal{K}'$ are $BISIM(assm)$ equivalent they satisfy the same formulae (by Theorem 1). Since $(\mathcal{K}, assm, s^0) \models F_{BISIM}(\mathcal{K}, assm)$, that formula is satisfied by $(\mathcal{K}', assm, s'^0)$ too. □

Combining the results of Theorem 1 and Theorem 2 we obtain the following characterizations.

COROLLARY 1. *Given two structures $\mathcal{K}$ and $\mathcal{K}'$ then it holds:*
(i) $(\mathcal{K}, \mathcal{K}') \in BISIM(assm)$ *iff* $\mathcal{K} \equiv_{assm} \mathcal{K}'$
(ii) $(\mathcal{K}, \mathcal{K}') \in BISIM(assm)$ *iff*

$$\forall f \in CTL: \qquad (\mathcal{K}, assm) \models f \quad \text{iff} \quad (\mathcal{K}', assm) \models f.$$

(iii) *If there is some CTL* formula f with $(\mathcal{K}, assm) \models f$ but $(\mathcal{K}', assm) \models \neg f$, then there is already a CTL formula f' which distinguishes both structures.*

4. Conclusion

In this paper we have defined a relativized bisimulation between Kripke structures and we have characterized this relation in temporal logic. Our equivalence relation is a generalization of the relation given in [**1**], the results of [**1**] are subsumed by our results by using the assumption *true*. Our bisimulation is a strong bisimulation, but we can also define a weak bisimulation by weakening the condition of a corresponding step: for every step $s \rightarrow s_1$ on an $assm$-good path there should be a corresponding finite path in the other structure. To give a temporal characterization of the weak bisimulation we have to omit the next operator. This leads to a generalization of the corresponding result of [**1**]. As fairness constraints may be specified as an assumption, we can characterize the equivalence of Kripke structures with fairness constraints, too. This solves a problem raised in [**1**].

If a temporal specification of a system composed of several modules is derived from the specifications of the modules, and the modules are proved correct w.r.t. their specifications, one can replace every module by an equivalent one without loosing the correctness of the derived specification. This can be done as two equivalent modules can not be distinguished by any temporal formula.

Furthermore, the equivalence relation is decidable, as the characteristic formula $F_{BISIM}(\mathcal{K}, assm)$ is constructable and the validity of a formula *spec* in a module $(\mathcal{K}', assm)$ is decidable.

If the given constraints on the environment are safety properties, the definition of an *assm*-good state can be weakened to the requirement that the state has to satisfies only the disjunction of all sets $literal(C)$, where C is a closure of *assm*. Hence this condition can be checked locally.

In the context of the computer architecture design language AADL we use a temporal logic MCTL for module specification [**3, 4, 6**]. MCTL consists of pairs $(assm, spec)$ where *assm* is a restricted PTL formula and *spec* is a CTL formula. By this paper we have an appropriate notion of equivalence for that logic, which can be used in the design process within AADL.

References

1. M. C. Browne, E. M. Clarke, and O. Grumberg, *Characterizing Kripke Structures in Temporal Logic*, Technical Report CMU-CS-87-104, Carnegie Mellon University, Pittsburgh, 1987.
2. M. Browne, *An improved algorithm for the automatic verification of finite state systems using temporal logic*, Proceedings of the Symposium on Logics in Computer Science 1986, 260–266.
3. W. Damm, and G. Döhmen, *AADL: A net based specification method for computer architecture design*, in: de Bakker (Ed.), *Languages for Parallel Architectures: Design, Semantics, and Implementation Models*, Wiley & Sons, Chichester, 1989, 51–109.
4. W. Damm, G. Döhmen, V. Gerstner, and B. Josko, *Modular verification of Petri nets: The temporal logic approach*, REX Workshop on Stepwise Refinement of Distributed Systems: Models, Formalisms, Correctness, Lecture Notes in Computer Science **430**, 1990, 180–207.
5. E. A. Emerson, and J. Y. Halpern, *'Sometimes' and 'not never' revisited: On branching versus linear time temporal logic*, J. Assoc. Comput. Mach. **33** (1986), 151–178.
6. B. Josko, *Verifying the correctness of AADL modules using model checking*, REX Workshop on Stepwise Refinement of Distributed Systems: Models, Formalisms, Correctness, Lecture Notes in Computer Science **430**, 1990, 386–400.
7. F. Kröger, *Temporal Logic of Programs*, EATCS-Monographs, Springer, 1987.
8. K.G. Larsen, *Context-dependent bisimulation between processes*, Ph.D. Thesis, Edingburgh, 1986.
9. K.G. Larsen, and R. Milner, *Verifying a protocol using relativized bisimulation.*, ICALP 87, Lecture Notes in Computer Science **267**, 1987, 126–135.
10. O. Lichtenstein, and A. Pnueli, *Checking that finite state concurrent programs satisfy their linear specification*, 12th ACM Symposium on Principles of Programming Languages, 1985,

97–107.
11. R. Milner, *Calculi for synchrony and asynchrony*, Theoret. Comput. Sci. **25** (1983), 267–310.
12. D. Park, *Concurrency and automata on infinite sequences*, Lecture Notes in Computer Science **104**, 1981, 167–183.
13. A. Pnueli, *The temporal logic of programs*, 18th Annual Symposium on Foundations of Computer Science, 1977, 46–57.
14. A. Pnueli, *Linear and branching structures in the semantics and logics of reactive systems*, ICALP 85, Lecture Notes in Computer Science **194**, 1985, 14–32.

Computer Science Department, University of Oldenburg, W 2900 Oldenburg, Federal Republic of Germany

E-mail: josko@arbi.informatik.uni-oldenburg.de

DIMACS Series in Discrete Mathematics
and Theoretical Computer Science
Volume 3, 1991

THE MODULAR FRAMEWORK OF COMPUTER-AIDED VERIFICATION

Motivation, Solutions, and Evaluation Criteria

Gil Shurek and Orna Grumberg
Computer Science Department, The Technion
Haifa 32000, Israel
E-address: orna@techsel (BITNET), orna@sel.technion.ac.il (CSNET)

1. INTRODUCTION

This paper presents a modular, semi-automatic method for the verification of finite-state programs with temporal logic specifications, based on model checking procedures. The *purpose* of this work, however, goes beyond the search for a technical solution to modular verification. The additional goal is to gain a wider understanding of the modular framework. Most of the introduction is dedicated to a general discussion of this framework - motivation, solutions and evaluation criteria. Previous works in this area are discussed in that context. Moreover, throughout the paper results are related to this discussion.

1.1 Motivation

Temporal logic model checking procedures ([Br86],[CES86], [CPS88],[LP85],[QS81]) have been proven to be a feasible approach to automatic verification of relatively small finite-state systems. As the expressive power of temporal logic makes it adequate for specifying properties of reactive systems and as interesting examples are typically those of concurrent and distributed systems, it seems natural to consider the applicability area of automatic model checking as that of reactive finite-state concurrent systems. This context, however, exposes the major disadvantage of model checkers, which is their global nature. By this we mean that only a complete system can be verified to meet its temporal specifications. This globality, on the one hand, induces the "state explosion problem" and, on the other hand, prevents the realization of the "pre-verified component methodology", presented below.

The *state explosion problem* arises in systems composed of many loosely coupled components. The size of such a system may grow as the product of the sizes of its components. Thus, space requirement might become too large for a model checking procedure to be applicable. Obviously, state explosion affects the time-complexity as well.

The *pre-verified component methodology* refers to the verification of systems containing an already verified component, that avoids redoing the component's proof. An example of a context where it seems natural to apply such a method, is the *superimposition* ([BF88],[CM88],[K87]). There, a superimposed component is designed to apply a certain task (e.g. distributed termination detection) to a component that handles some other computations. The superimposed component, when designed, can only be verified as one-component of a whole system, not yet available. When verifying a specific system, it should be possible to exploit the pre-verification of the superimposed component. As a result, the average verification time-complexity over the set of systems containing this pre-verified component should be reduced.

1.2 Solutions

It seems natural to consider the use of *compositional* reasoning as a solution to the state explosion problem. Exploiting the modular structure of a system, a compositional verification method applies model checking procedures to individual components, to check properties from which the specification of the global system is deduced. The construction of the global system is avoided, and thus space requirements are reduced. Hopefully time-complexity is reduced as well.

The desirability of using pre-verified components raises the need for our method to be *modular* too. By this we mean that it should be possible to apply a verification procedure to an individual component when no specific environment is available.

The difference between compositionality and modularity lies in the amount of information that the verification procedure has about one component when verifying the other. In a compositional approach complete knowledge about the environment may be used, while in the modular approach no information about the actual environment is available. A modular proof is composed of stages, each carried out *separately* in the context of an individual constituent of the compound system. All stages carried out in the context of a given component are referred to as the *proof block* of that component. Using this terminology a pre-verified component is one whose proof block has been carried out. Note that our terminology adapts the terms "compositionality" and "modularity", yielding notions that differ slightly from the traditional ones.

1.3 Proof System Properties and Evaluation Criteria

Following the previous discussion, we will pursue a modular, computer-aided approach. Properties of modular proof systems are considered below, evaluated in view of the desired goals.

- An obvious quality of a global model checker that should be preserved in a modular procedure is the ability to verify arbitrary properties of a distributed system, not being restricted to properties of single components within the compound system.

Modular procedures have no knowledge about the environment. Hence, applying the checking procedure to one component, an *assumption* on the behavior of the environment is often necessary in order to deduce a global property. Incorporating a pre-verified proof block into a proof of a compound system, the assumptions included in that proof block have to be met by the specific environment.

- Since an algorithmic construction of assumptions might not be feasible, our search might result in a *semi-automatic* model checker. In that case, the user is required to supply ("guess") the needed assumptions, while all the other stages of the proof are carried out automatically.
- Unlike a global model checker, an assumption-aided checking procedure that fails to prove its goal does not imply that the compound model fails to satisfy its specification (even in a complete proof system), as the right set of assumptions might not have been tried yet.

In supplying assumptions, the user should consider their strength. By assuming only what is necessary to conduct the proof successfully, one gets *weak assumptions*. These are less specific than strong assumptions in the sense that they are satisfied by a larger set of environments.

- Assuming only partial knowledge about the environment results in a component's proof block, which can be used in proofs of distributed systems that are the composition of this component with various environments. The weaker the assumptions, the larger the set of proofs which can reuse the same proof block. Thus, weaker assumptions further reduce the average time-complexity of the model checking procedure.

- However, weaker assumptions require more insight into the problem. This is undesirable from the point of view of automatic verification, and is referred to as a lower *degree of automation*. We would like the model checker to enable the user to conduct proofs that use stronger assumptions, based on a relatively superficial understanding of the problem. The strength of the assumptions in an actual proof reflects the user's preference with respect to the tradeoff between time-complexity and the degree of automation.
- In some cases (to be indicated in the following sections) weak assumptions reduce time and space requirements directly, i.e. even over a single proof. Here again, the automation-complexity tradeoff is reflected.

1.4 Previous Works

Several works on compositional techniques for reasoning about reactive concurrent systems established the background for our research.

Barringer, Kuiper and Pnueli [BKP84] developed a temporal language which is powerful enough to express the full semantics of their composition operator. Using the adequate inference rule, a formula which is valid in the compound system can be generated from partial specifications proved to be satisfied by the constituent's models. This formula is of a special syntactic structure. In order to prove arbitrary specifications we need a proof system that deduces the implications of the required specification from formulas of this special form. However, a proof system for semantically valid, purely temporal formulas of that type is not provided.

Pnueli [P85] presented the *assume guarantee* (AG) paradigm for a compositional proof system. The primitive formulas in his logic are triples of the form $<\varphi>P<\psi>$ where φ and ψ are temporal formulas, and P is a process. $<\varphi>P<\psi>$ is semantically valid if P is guaranteed to behave according to ψ in every execution in which the environment behaves according to φ. The AG paradigm does not indicate how a model checking procedure is to be implemented. However, since it is general enough and captures many desired properties (e.g., modularity, independence of a specific composition operator, freedom to determine the strength of the assumptions, proof of arbitrary global properties) it inspired other works, including ours.

Josko [J87] adopted the AG paradigm for finite-state systems, to implement a semi automatic model checking procedure. In his algorithm, the environment is modeled by a temporal formula expressing the assumption. This formula is used to restrict the paths in the tree over which the component's specification is interpreted. This method is limited to proving global specifications which are a conjunction of properties of the system's components, a restriction which is not implied by the AG paradigm. Another disadvantage is that even though specifications are limited to a sub-CTL language (which has a linear model checking procedure [CES86]) the overall time complexity of this algorithm is exponential in the size of the assumptions.

Clarke, Long and McMillan [CLM89] introduced the idea of modeling the environment by an *interface process* i.e., to use finite state transition structures as the assumption language. Similarly to Josko's method, the specifications verified by the interface rules are restricted. Only boolean combinations of the component properties can be handled. Another drawback of these rules is that the interface process is equivalent to the environment with respect to the whole specification language (restricted to the component's alphabet), rather than with respect to the verified specification. Considering the interface process as an assumption on the environment, there are no means to weaken the assumption in a way that depends on the given specification. Hence, the use of pre-verified components in this method is limited.

The *fully automated* implementation of the interface rule is an example of a compositional proof system which is not modular (according to our terminology). The verification of one component involves the automatic construction of the interface process from the environment. Thus, the actual environment should be known when the component proof block is carried out. Clearly, this implementation cannot realize the "pre-verified component methodology".

1.5 The Present Work

Based on the previous discussion, this work presents a search for a partially automated modular verification procedure for reactive, finite-state distributed systems. The systems are modeled by finite structures described as state-transition graphs, where asynchronous actions have an interleaving semantics. We assume that specifications are given by propositional temporal logic formulas that are interpreted over the computation tree of the system's transition graph.

As we regard the *Assume – Guarantee* paradigm to be the preferable modular scheme, we begin with an exploration of its underlying mechanism. This is done by presenting a deductive proof system scheme (*AG*) which is an abstraction of that paradigm. In the scheme, a property of the distributed system is deduced in the context of one of its components, considering another global property (which was previously deduced). The deductive process is based on global properties which can be verified in the context of one component, exclusively. Considering a further abstraction (the *AGS*) we claim that, such paradigms essentially manipulate infinite sets of models. The main difficulty in implementing an *AG*-like scheme as a semi-automatic procedure, is identified to be the need to represent these sets so that they can be handled automatically.

We present an *AG*-like proof system scheme, denoted *AGM*, in which a set of models is represented by a single model. The model represents the set of *composed* models which are *smaller* with respect to a *Preorder* with a semantic property, defined over the model domain. The greater model can be viewed as holding partial information about the small ones. The task of combining partial knowledge about the checked system (an assumption) with full information about one of the components is achieved here by applying the composition operator. In this scheme assumptions are expressed by finite state-transition structures, while the specifications are given in a temporal language. The scheme defines complete proof systems, in which soundness depends on conditions that should be satisfied in an actual framework (i.e., an actual model of computation and a temporal language). Implementations of this scheme as semi-automatic procedures are based on an automatic model checker that is not necessarily modular.

We show how a proof system can be derived from the *AGM* scheme. As the model of computation we choose a simplified version of *CCS* [M80] which describes asynchronous communicating systems. The specification language is a sublanguage of CTL^* (CTL^*_{-E}) which is more expressive than *LTL*. In CTL^*_{-E} negation is restricted to the atomic propositions and the **E** ("there exists") path quantifier is eliminated. The suggested preorder can be syntactically identified by an algorithm.

A generalized version of *AGM* (AGM^*) can be realized in this framework. The AGM^* enables simultaneous use of partial information on both components of the distributed system. In AGM^*, verification can be viewed as state-reduction of the global model without construction of the model. The reduction is relative to the property to be verified (as opposed to minimization which guarantees semantic equivalence of models).

Proof systems derived from the *AGM* and the AGM^* schemes can be implemented as semi-automatic procedures for modular verification. Properly used, these procedures enable the use of pre-verified components, and guarantee reduced memory requirement and even improved time complexity, compared to a global model checker.

In Section 2 we describe the *AG* and *AGS* proof system schemes. In Section 3 the *AGM* and its generalized version AGM^* are presented. Section 4 presents the actual proof system derived from these schemes. This section includes an application of that proof system to a program.

2. THE *AG* AND *AGS* SCHEMES

2.1 The *AG* Proof System Scheme

The *AG* deductive proof system scheme presented in this section should be viewed as an abstraction of Pnueli's assume-guarantee paradigm [P85].

Let **L** be a logic used as a reasoning and specification language for distributed systems. Let M be a set of models, containing component models and composed models, for which '$\|$' denotes the composition operation. Each composed model m is accompanied with a fixed decomposition into two component models m_A, m_B, such that $m = m_A \| m_B$. For a component model $m \in M$, $C_A[m]$ denotes the set of all models in M of the form $m_A \| m_B$, for which m_A is m (i.e., $C_A[m] = \{m_A \| m_B \mid m_A = m\}$). $C_B[m]$ is defined similarly.

For $\varphi_1, \varphi_2 \in \mathbf{L}$, the notation $\varphi_1 \models_{C_i[m]} \varphi_2$, $i \in \{A,B\}$ means that every model in $C_i[m]$ which satisfies φ_1, satisfies φ_2 as well. Thus, φ_2 is the consequence of φ_1 in $C_i[m]$. $\vdash_{C_i[m]}$ is a proof system that validates **L** formulas in $C_i[m]$. The soundness of this system is defined as: $\forall \varphi_1, \varphi_2 \in \mathbf{L}\ [\varphi_1 \vdash_{C_i[m]} \varphi_2 \Rightarrow \varphi_1 \models_{C_i[m]} \varphi_2]$. When $\varphi_1 \equiv$ true, the definition results in $\vdash_{C_i[m]} \varphi_2 \Rightarrow \models_{C_i[m]} \varphi_2$ which means that if φ_2 is provable in $\vdash_{C_i[m]}$ then φ_2 is true in the composition of m with *any* environment.

Given the proof systems $\vdash_{C_A[m_1]}$ and $\vdash_{C_B[m_2]}$ we present below a set of *AG* proof systems (one for each model $m_1 \| m_2$). The proof system $\vdash_{AG-m_1\|m_2}$ for the verification of L-formulas in the composed model $m_1 \| m_2$ is defined as follows:

Axioms:

$$\{\varphi \mid \ \vdash_{C_A[m_1]} \varphi\} \cup \{\varphi \mid \ \vdash_{C_B[m_2]} \varphi\}$$

Inference rules:

$$\frac{\vdash_{AG-m_1\|m_2} \varphi \qquad \varphi \vdash_{C_A[m_1]} \psi}{\vdash_{AG-m_1\|m_2} \psi} \qquad\qquad \frac{\vdash_{AG-m_1\|m_2} \varphi \qquad \varphi \vdash_{C_B[m_2]} \psi}{\vdash_{AG-m_1\|m_2} \psi}$$

The soundness of $AG-m_1 \| m_2$ is defined as: $\forall \psi \in \mathbf{L}\ [\ \vdash_{AG-m_1\|m_2} \psi \Rightarrow m_1 \| m_2 \models \psi\ ]$.

Theorem 1: Given sound proof systems $\vdash_{C_A[m_1]}$ and $\vdash_{C_B[m_2]}$, $\vdash_{AG-m_1\|m_2}$ is sound.

The $AG-m_1 \| m_2$ is a scheme for a proof system, capable of verifying properties of the distributed system $m_1 \| m_2$, without being restricted to the behavior of a single component in the context of the global system. Moreover, it can be considered modular according to our terminology. Examining a typical AG proof, its modularity can be observed:

$$\vdash_{C_A[m_1]} \varphi_n\ ,\ \ \varphi_n \vdash_{C_B[m_2]} \varphi_{n-1} \ \ \cdots\ \ \varphi_1 \vdash_{C_A[m_1]} \varphi$$

We are interested in a semi-automatic implementation of the *AG* scheme, where the user is required to supply ("guess") the sequence of assumptions φ_n to φ_1 and the validation in $C_A[m_1]$ and in $C_B[m_2]$ is done automatically.

The m_1 proof-block is the set of proof stages in which $\vdash_{C_A[m_1]}$ is applied. When assumptions are given, the m_1 proof block can be carried out without having to consider any specific "m_2". Given a component model m_i , if the proof-block consisting of all $\vdash_{C_B[\ ]}$ stages can be carried out successfully with respect to $\vdash_{C_B[m_i]}$, then m_1 proof-block can be completed to a full *AG* proof, deducing $m_1 \| m_i \models \varphi$.

In general, the user can determine the strength of the provided assumptions. Thus, in a semi-automatic implementation, appears the previously discussed tradeoff between time complexity and the degree of automation.

A *canonical* proof is defined as a proof sequence in which every φ_i is derived at the earliest possible stage.

Figure 1 is a schematic description of an $AG-m_1 \| m_2$ proof of $\varphi_0 \in \mathbf{L}$. A closed curve represents a set of models, where the set of models that satisfy φ_i is denoted by "φ_i".

The proof sequence (1) $\vdash_{C_A[m_1]} \varphi_1$, (2) $\varphi_1 \vdash_{C_B[m_2]} \varphi_0$ is assumed to be canonical. φ_1 is an *AG* axiom, because it has been proven to be valid in $C_A[m_1]$; this means that the set of models that satisfy φ_1, contains $C_A[m_1]$ ((1) in Figure 1). φ_0 is derivable in $AG-m_1 \| m_2$ because it has been proven to be a consequence of φ_1 in $C_B[m_2]$. This means that the set of models defined by φ_0 contains the *intersection* of $C_B[m_2]$ and the set φ_1 ((2) in Figure 1).

Introducing the concept of *set of models* into *AG*, leads to the *AGS* scheme presented below.

2.2 The *AG*-sets (*AGS*) Proof System Scheme

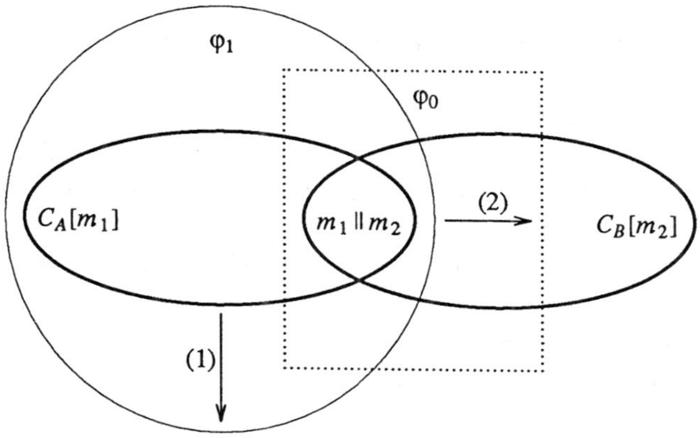

Figure 1 - Proof diagram

In this section we present another proof system scheme, *AG*-sets (*AGS*), which is an abstraction of *AG*, meant to explore the underlying mechanisms of *AG*. In *AGS*, sets of models are used directly as the reasoning language (similar to the *Semantic Model* of [AL89]). However, we still assume temporal specification. Therefore, an inference rule that derives **L** formulas (from a previously "derived" model set) is necessary. Defined below is an *AGS* proof system for $m_1 \| m_2$.

$\mathbf{gs[m_1 \| m_2]}$ is a set of model sets, defined inductively by:
Atoms ("axioms"):

$$\{C \mid C \supseteq C_A[m_1]\} \cup \{C \mid C \supseteq C_B[m_2]\}$$

Closure operations ("inference rules"):

$$\frac{C \in gs[m_1 \| m_2] \qquad C' \supseteq C \cap C_A[m_1]}{C' \in gs[m_1 \| m_2]} \qquad\qquad \frac{C \in gs[m_1 \| m_2] \qquad C' \supseteq C \cap C_B[m_2]}{C' \in gs[m_1 \| m_2]}$$

The inference rule:

$$\frac{\varphi \in \mathbf{L} \qquad C \in gs[m_1 \| m_2] \qquad \models_C \varphi}{\vdash_{AGS-m_1 \| m_2} \varphi}$$

Lemma 1: $C \in gs[m_1 \| m_2]$ iff $m_1 \| m_2 \in C$.

Theorem 2: $AGS\text{-}m_1 \| m_2$ is sound and complete.

Noticing that both *AG* and *AGS* basically manipulate potentially infinite sets of models, leads us to identify the major obstacle to the implementation of *AG*. In *AG*, sets of models are represented either by themselves ($C_A[m_1]$ and $C_B[m_2]$) or by **L**-formulas (i.e., φ defines and represents the set of all models satisfying φ). The lack of uniform representation for these sets forces *AG* implementations (e.g., [J87]) to provide conversion procedures (which are, in general, cumbersome and time-consuming) to enable constructive manipulation of model sets.
In *AGS* we abstract away from the *representation problem.* The *AGS* proof system is theoretic and has no explicit representation for sets. This system is not intended to be implemented directly, but to inspire the definition of other proof system schemes.
In the next section we present an AGS-like proof system, in which sets are represented in a way they can be handled constructively.

3. THE *AG*-MODELS (*AGM*) SCHEMES

3.1 The *AGM* Proof System Scheme

The *AGM* is an *AGS*-like proof system scheme, in which the basic elements handled by the deductive procedure are models (i.e., finite state-transition structures). These basic elements are used to represent the potentially infinite sets of models, discussed in the previous section. The *AGM* proof system is a modular model checking system for finite state distributed system, based on their decomposition into two components.

Let M be a model domain consisting of component models m_i and composed models m, which may differ syntactically. Let Σ be an alphabet and let Σ_m denote the alphabet associated with the model m. Σ_m consists of state labels and/or transition labels. Let $\|$ denote a composition operator that applied to two component models m_i, m_j results in a composed model m, i.e., $m = m_i \| m_j$. Also, let $|_\Sigma$ denote a restriction transformer, that applied to a composed model results in a component model with alphabet set restricted to Σ.

Let **L** be a temporal language interpreted over the domain of models M. Let $\mathbf{L}(\Sigma)$ denote the set of formulas in **L** defined over the set of atomic propositions in Σ. We assume (for the semantics of **L**) that if $\varphi \notin L(\Sigma_m)$ then $m \not\models \varphi$.

Let $\sqsupseteq$ denote a *Preorder* (reflexive and transitive relation) defined over M, which has the following semantic property: $m_A \sqsupseteq m_B$ implies that all **L** formulas that are satisfied in m_A, are satisfied in m_B (i.e., if $m_A \sqsupseteq m_B$ than $\forall \varphi \in \mathbf{L},\ m_A \models \varphi \Rightarrow m_B \models \varphi$).

Comments:

a. Since it is not required that $m_A \sqsupseteq m_B$ iff $\forall \varphi \in \mathbf{L},\ m_A \models \varphi \Rightarrow m_B \models \varphi$, the relation $\sqsupseteq$ is not uniquely defined.

b. The existence of a semantic preorder which is not an equivalence relation even for models that share the same alphabet (Σ), implies that the temporal language ($\mathbf{L}(\Sigma)$) does not contain the negation for every formula.

Let $\vdash_{MC}$ denote a proof system for checking the satisfaction of **L** formulas in M models (i.e., a *model checker).*

The *AGM* definition inherits the structure of the *AGS* definition, where inference rules which do not deduce temporal formulas, are referred to as closure operations. Defined below is an *AGM* proof system for $m_1 \| m_2$.

gm[m₁ || m₂] is a set of models, defined inductively by:
Atoms :

$$\{m' \| m_2 \mid m' \sqsupseteq m_1\} \cup \{m_1 \| m'' \mid m'' \sqsupseteq m_2\}$$

closure operations :

$$\frac{m_1 \| m'' \in gm[m_1 \| m_2] \qquad m' \sqsupseteq (m_1 \| m'')|_{\Sigma_{m_1}}}{m' \| m_2 \in gm[m_1 \| m_2]} \qquad\qquad \frac{m' \| m_2 \in gm[m_1 \| m_2] \qquad m'' \sqsupseteq (m' \| m_2)|_{\Sigma_{m_2}}}{m_1 \| m'' \in gm[m_1 \| m_2]}$$

The inference rule :

$$\frac{m \in gm[m_1 \| m_2] \qquad \varphi \in \mathbf{L}(\Sigma_m) \qquad \vdash_{MC}\ m \models \varphi}{\vdash_{AGM-m_1 \| m_2}\ \varphi}$$

The soundness of the *AGM* scheme depends on the following conditions (regarding the language, the models domain, the composition operator, the restriction transformer and the semantic preorder) :

Soundness Conditions :

1. For all m', m_A, m_B component models in M, if $m_A \sqsupseteq m_B$ then $m_A \| m' \sqsupseteq m_B \| m'$.
2. For all m', m_A, m_B component models in M, if $m_A \| m' \sqsupseteq m_B \| m'$ then $(m_A \| m')|_{\Sigma_{m'}} \sqsupseteq (m_B \| m')|_{\Sigma_{m'}}$.
3. For all m_A, m_B component models in M , $((m_A \| m_B)|_{\Sigma_{m_A}}) \| m_B \sqsupseteq m_A \| m_B$.

When the composition operator is *not commutative* the dual conditions are required as well.

Theorem 3: Given that the soundness conditions are fulfilled then:

3.1 Using a sound proof system $\vdash_{MC}$, $\vdash_{AGM-m_1 \| m_2}$ is sound.

3.2 Using a complete proof system $\vdash_{MC}$, $\vdash_{AGM-m_1 \| m_2}$ is complete.

Proof:

3.1 The condition above ensures that for any model m in $gm[m_1 \| m_2]$, $m \sqsupseteq m_1 \| m_2$. Thus, the specification of $m_1 \| m_2$ can be checked in m .

3.2 Since $m_1 \| m_2 \in gm[m_1 \| m_2]$, one can always choose m in the inference rule to be $m_1 \| m_2$. Thus, any specification satisfied by $m_1 \| m_2$ is proved by *AGM*.

Comments:

a. In the case where component models in M do not differ syntactically from the composed models, the closure operations may be applied recursively to components, handling systems which are composed of more than two components.

b. Similar proof systems can be defined (analogically) for a modular model checking of a distributed system composed of k components, for any pre-defined number k.

3.1.1 Evaluation

The *AGM* scheme is capable of verifying any property of the global distributed system, which is expressible in **L**.

Examining a typical $AGM-m_1 \| m_2$ proof, verifying that $m_1 \| m_2 \models \varphi$, its modularity can be observed:

Stage 1: Verify that $m_A \sqsupseteq m_1$ (concluding: $m_A \| m_2$ is a $gm[m_1 \| m_2]$ atom);
Stage 2: Verify that $m_B \sqsupseteq (m_A \| m_2)|_{\Sigma_{m_2}}$ (concluding: $m_1 \| m_B \in gm[m_1 \| m_2]$);
Stage 3: Verify that $m_C \sqsupseteq (m_1 \| m_B)|_{\Sigma_{m_1}}$ (concluding: $m_C \| m_2 \in gm[m_1 \| m_2]$);
Stage 4: $\vdash_{MC}$ $m_C \| m_2 \models \varphi$.

Two proof blocks can be identified in the above proof sketch: The m_1 proof block is composed of stages 1 and 3, while stages 2 and 4 compose the m_2 proof block. The assumptions are : m_A, m_B, m_C .

Regarding the state explosion problem, this proof system is feasible as long as the size of the structures representing the assumptions is significantly less than the size of the components' models.

Applying the pre-verified components methodology, feasibility depends on the assumptions strength, as discussed in previous sections. As "greater" ($\sqsupseteq$) structures make weaker assumptions, the *AGM* enables the user to weaken the supplied assumptions provided that the knowledge needed for the verification of the temporal specification is preserved.

Assuming that a semi automatic system is implemented, where the user is required to supply the assumptions, the tradeoff between the degree of automation and the average time complexity, arises here, just as in *AG*. In cases where weaker assumptions are expressed by more compact structures, the tradeoff concerns even the time complexity of a single proof.

3.1.2 Resemblance to other Systems

The resemblance between the *AGM* scheme and the *interface rule* [CLM89] is not incidental, since we have adopted here the idea to use finite state-transition structures as the assumption language, and to enable the use of the composition operator itself (rather than its semantics) in the derivation rules. However, being able to control the assumption strength and having the *AGS* proof structure, yields here a proof system scheme that has most of the desired properties discussed in the introduction.

In the *AGM* scheme, a single model of a composed system represents an *AGS* basic element: a set of models. The model represents the set of *composed* models in M which are *smaller* with respect to the semantic relation $\sqsupseteq$ (i.e., m represents a set of models in which every formula satisfied by m is valid). The "greater" model can be viewed as holding partial information about the "small" one. Thus, the preorder between two basic elements of the *AGM* is analogous to the subset relation between *AGS* sets. The task of combining partial knowledge about the checked system with full information about one of the components, which is carried out by the intersection operator in the *AGS* scheme, is achieved here by applying the composition operator itself (aided by the restriction transformer).

3.1.3 Implementation

A computer-aided implementation of the *AGM* proof system is expected to carry out automatically the actions listed below:

a. Models composition.
b. Applying the restriction transformer.
c. Verifying the satisfaction of **L** formulas in *M* models.
d. Verifying the preorder between models.

It is not difficult to imagine implementable definitions of a composition operator and a restriction transformer which make tasks **a.** and **b.** algorithmic. To automate task **c.** , a global model checker can be applied. The only task which is not clearly decidable even for the finite-state case is the verification of the semantic preorder (task **d.**). Any useful implementation of the *AGM* system should suggest a semantic preorder which is syntactically identified by an efficient algorithm.

In section **4.** we suggest an implementation of *AGM* system in which all four tasks are algorithmically handled. As can be seen there, the need for a decidable preorder affects the choice of the temporal specification language.

3.2 The AGM^* Proof System Scheme

Setting stricter soundness conditions, the AGM^* which is a generalized version of *AGM*, can be defined. The AGM^* enables simultaneous use of partial information on both components of the distributed system. Using this scheme it is possible to conduct a proof, along which only the knowledge that is essential for the verification of the final specification is preserved.

$gm^*[m_1 \| m_2]$:

Atoms :

$$\{ m_A \| m_B \mid m_A \sqsupseteq m_1 , m_B \sqsupseteq m_2 \}$$

Closure operation :

$$\frac{\bar{m}, \bar{\bar{m}} \in gm^*[m_1 \| m_2] \qquad m' \sqsupseteq \bar{m}|_{\Sigma_{m_1}} \qquad m'' \sqsupseteq \bar{\bar{m}}|_{\Sigma_{m_2}}}{m' \| m'' \in gm^*[m_1 \| m_2]}$$

The inference rule :

$$m \in gm^*[m_1 \| m_2]$$

$$\varphi \in L(\Sigma_m)$$

$$\vdash_{MC} m \models \varphi$$

$$\overline{\vdash_{AGM^*-m_1 \| m_2} \varphi}$$

Soundness Conditions:

1. For all m', m_A, m_B component models in M, if $m_A \sqsupseteq m_B$ then $m_A \| m' \sqsupseteq m_B \| m'$.
2. For all m_A, m_B, m', m'' component models in M, if $m' \| m'' \sqsupseteq m_A \| m_B$, then $(m' \| m'')|_{\Sigma_{m_A}} \sqsupseteq (m_A \| m_B)|_{\Sigma_{m_A}}$.
3. For all m_A, m_B, component models in M, $((m_A \| m_B)|_{\Sigma_{m_A}}) \| ((m_A \| m_B)|_{\Sigma_{m_B}}) \sqsupseteq m_A \| m_B$.
4. For all m_A, m_B, component models in M, $m_A \sqsupseteq (m_A \| m_B)|_{\Sigma_{m_A}}$.

When the composition operator is not commutative, the dual conditions are required as well.

Theorem 4: Given that the AGM^* soundness conditions are fulfilled then:

4.1 Using a sound proof system $\vdash_{MC}$, $\vdash_{AGM^*-m_1 \| m_2}$ is sound.

4.2 Using a complete proof system $\vdash_{MC}$, $\vdash_{AGM^*-m_1 \| m_2}$ is complete.

4.3 Every *AGM* proof is an AGM^* proof.

The following lemma is a special (and useful) case of the closure operation, which is a consequence of the fourth soundness condition and the reflexivity of $\sqsupseteq$.

Lemma 2: If $m \in gm^*[m_1 \| m_2]$, $m'' \sqsupseteq m|_{\Sigma_{m_2}}$ and $m' \sqsupseteq m_1$ then $m' \| m'' \in gm^*[m_1 \| m_2]$. The dual statement holds as well.

Comments:

a. If the AGM^* soundness conditions are fulfilled, then the *AGM* soundness conditions are guaranteed to hold.

b. It is possible to base the $gm^*[m_1 \| m_2]$ definition on a single atom: $m_1 \| m_2$. All other models currently defined as atoms, can be generated by the closure operation. Using the fourth soundness condition, it is possible to avoid the construction of the global model $m_1 \| m_2$ in this generation procedure.

In the case where "smaller" ($\sqsupseteq$) models can often be represented by smaller structures, the AGM^* enables further reduction in the size of the structures handled by the proof system (in comparison with *AGM*). Thus, further reduction in time and space complexity is achieved.

3.2.1 Resemblance to other Systems

An interesting observation is the resemblance between an AGM^* *atomic proof* (i.e., a proof that does not use the closure operation: $m_A \sqsupseteq m_1$, $m_B \sqsupseteq m_2$, $\vdash_{MC} m_A \| m_B \models \varphi$) and [BKP84] proofs, referred to in the introduction.

Kurshan [Ku90] describes a transformation by which a complex assertion about a complex model may be replaced by a simpler assertion about a simpler model, in such a way that the validity of the latter implies the validity of the former. Moore machines are used as the model of computation, while formal assertions are given as automata on infinite tapes. The verification process can be viewed as a state-minimization of the global model, achieved by applying transformers to individual components. This reduction is relative to the property to be verified.

The verification process resembles the special case of atomic AGM^* proofs, but is more general in that transformation is applied to the assertion as well (the transformed model preserves transformed assertions). The assertion language is limited to the expression of linear-time properties - as opposed to the branching-time language presented in the next section. As in AGM^*, the model of computation is the assumption language. No general means are presented to verify that an assumption is a proper transformation of the original component model. To show that, the assumption has to be supplied with an explicit homomorphism (to be checked). This restriction accomplishes a verification complexity which is linear in the size of the components.

4. ASYNCHRONOUS COMMUNICATING SYSTEMS

4.1 The Model of Computation

Defined below is a simple computation model for distributed systems. The components of these systems are asynchronous processes, that communicate through synchronized actions. This model is similar to the CCS model [M80]. We use here a simplified version, restricted to one application of the composition operation (i.e., every model of a compound system is composed of two process models). In this version, compound models are syntactically guaranteed to be deadlock free.

Let *AcT* be a set of *open actions* that contains:

- The set A of primitive open actions such that $\bar{a} \in A$ whenever $a \in A$ and $\bar{\bar{a}} = a$.
- A set Λ of special actions used for internal transitions that do not require synchronization. Typical elements of this set are denoted λ or λ_i .
- A set E of special actions used to trace the environment non-synchronizing transitions. Typical elements of this set are denoted ε or ε_i .

The set ACTION $\subseteq AcT \times$AcT of *closed actions,* contains synchronized actions of the form $a\bar{a}$, where $a \in A$ and asynchronous transitions of the form $\lambda\, \varepsilon$, $\lambda \in \Lambda$, $\varepsilon \in$ E .

The model domain M consists of *process models* and of *system models*, which are defined as follows:

A *process model m* is a 6-tuple, $m = \langle AC, AP, S, R, s_0, L \rangle$, where

- AC is the subset of *AcT* used in R. AC includes at most one element of Λ and one element of E .
- AP is a set of atomic propositions. *AP* and *AC* are disjoint.
- S is a *finite* set of states.
- $R \subseteq S \times AC \times S$ is the transition relation which must be total in the first argument.

The notation $s_1 \xrightarrow{\alpha} s_2$ is used to indicate that $(s_1, \alpha, s_2) \in R$.

- $s_0 \in S$ is the initial state.
- $L: S \rightarrow \mathbf{P}(AP)$ is a function that labels each state with a set of atomic propositions.

The *system model* definition differs from the above, by the following:

- AC includes at most two elements of Λ and two elements of E .
- $R \subseteq S \times (AC \times AC) \times S$ such that if $(s,\alpha,s') \in R$ then $\alpha \in ACTION$. R must be total in the first argument.

Let $m_1 = \langle AC_1, AP_1, S_1, R_1, s_0^1, L_1 \rangle$ and $m_2 = \langle AC_2, AP_2, S_2, R_2, s_0^2, L_2 \rangle$ be two *process* models. AP_1 , AP_2 , AC_1 , and AC_2 are mutually disjoint.

The system model $m_1 \parallel m_2 = \langle AC,\ AP,\ S,\ R,\ s_0,\ L \rangle$ is defined as follows:

- AC is the subset of $AC_1 \cup AC_2$ used in R.
- AP is the union of AP_1 and AP_2.
- S is the maximal subset of $S_1 \times S_2$, reachable from s_0 by the transitions in R, for which R is total in its first argument.
- $R \subseteq S \times (AC \times AC) \times S$ contains three types of transitions:
  - $(s_1,s_2) \xrightarrow{\lambda_1 \varepsilon_2} (s'_1,s'_2)$ iff $[\ s_1 \xrightarrow{\lambda_1} s'_1$ and $s_2 \xrightarrow{\varepsilon_2} s'_2\ ]$
  - $(s_1,s_2) \xrightarrow{\lambda_2 \varepsilon_1} (s'_1,s'_2)$ iff $[\ s_1 \xrightarrow{\varepsilon_1} s'_1$ and $s_2 \xrightarrow{\lambda_2} s'_2\ ]$
  - $(s_1,s_2) \xrightarrow{a\bar{a}} (s'_1,s'_2)$ iff $[\ s_1 \xrightarrow{a} s'_1$ and $s_2 \xrightarrow{\bar{a}} s'_2\ ]$
- $s_0 = (s_0^1, s_0^2)$
- $L: S_1 \times S_2 \rightarrow \mathbf{P}(AP)$ such that $L((s_1,s_2))$ is the disjoint union of $L_1(s_1)$ and $L_2(s_2)$.

Comments:

a. The defined composition operator is commutative.

b. The totality of R enforces "pruning" on the model's transition graph, eliminating branches that do not end in a nontrivial strongly connected component of that graph. Thus, if a nonempty compound model exists, it will include only infinite paths (i.e., it is deadlock free in the context of reactive systems). We consider such a model to be *syntactically deadlock free*. When a composed system is verified using this type of models, all the deadlocked sub-structures of the computation tree of the system are ignored.

c. The closed action $\lambda\,\varepsilon$ models asynchronous transition provided that every state of a process model has a reflexive ε transition (self-loop), which enables arbitrary interleaving with the environment's λ transitions. Note that adding the ε transition, the computation tree of the process model is an *open computation* tree [P85].

In the sequel, we consider only nonempty models.

The alphabet Σ_m of a (process or system) model m is defined as the union of AC and AP.

Presented below, is a restriction transformer, defined over system models. Let $m = \langle AC,\ AP,\ S,\ R,\ s_0,\ L \rangle$ be a *system model*. Let Σ be $AC_\Sigma \cup AP_\Sigma$ where, $AC_\Sigma \subseteq AcT$, AP_Σ and AcT are disjoint. The Σ- *restricted model* $m_r = m|_\Sigma$ is defined as follows:

- AC_r is the subset of $AC \cap \Sigma$ used in R_r .
- $AP_r = AP \cap \Sigma$
- S_r is the maximal subset of S reachable from s_0^r by R_r, for which R_r is total in its first argument.
- $R_r \subseteq S_r \times AC_r \times S_r$ contains 3 types of transitions:

- ○ $s_1 \xrightarrow{a} s_2$ iff $(s_1, a\bar{a}, s_2) \in R$ and $a \in AC \cap \Sigma$, $\bar{a} \notin AC \cap \Sigma$.
- ○ $s_1 \xrightarrow{\varepsilon_i} s_2$ iff $(s_1, \lambda_j\varepsilon_i, s_2) \in R$ and $\varepsilon_i \in AC \cap \Sigma$, $\lambda_j \notin AC \cap \Sigma$.
- ○ $s_1 \xrightarrow{\lambda_i} s_2$ iff $(s_1, \lambda_i\varepsilon_j, s_2) \in R$ and $\lambda_i \in AC \cap \Sigma$, $\varepsilon_j \notin AC \cap \Sigma$.
- $s_0^r = s_0$
- $L_r: S_r \to P(AP_r)$ such that $L_r(s) = L(s) \cap AP_r$.

The restriction transformer applied to a system model hides all alphabet not in Σ and then eliminates those transitions which remain either with no label or with two labels.
Note that $m|_\Sigma$ is (syntactically) a model of a process restricted to the symbols in Σ.

4.2 The Specification Language

Defined below is a propositional temporal logic which is a sublanguage of CTL^* [CE81]. Let Σ be the disjoint union of AP_Σ and AC_Σ. The language $CTL^*_{-E}(\Sigma)$ is defined as the smallest set of state formulas such that:

(1) If $A \in AP_\Sigma$, then A and $\neg A$ are state formulas.
(2) If φ and ψ are state formulas, then $\varphi \vee \psi$ and $\varphi \wedge \psi$ are state formulas.
(3) If φ is a path formula, then $\mathbf{A}\varphi$ is a state formula.
(4) If φ is a state formula, then φ is a path formula.
(5) If φ and ψ are path formulas, then $\varphi \vee \psi$, $\varphi \wedge \psi$, $\varphi\, \mathbf{U}\, \psi$, $\mathbf{X}\varphi$ and $\mathbf{G}\varphi$ are path formulas.

We use the abbreviation $\mathbf{F}\varphi$ to denote $(\mathbf{true}\mathbf{U}\varphi)$.

The semantics of $CTL^*_{-E}(\Sigma)$ is defined with respect to a (process or system) model $m = \langle AP, AC, S, R, s_0, L\rangle$. A path in m is defined to be a sequence of states, $\pi = s_0, s_1, \cdots$, such that for every $i \geq 0$, $(s_i, \alpha, s_{i+1}) \in R$, for some α. π^i denotes the suffix of π starting at s_i.
$m,s \models \varphi$ denotes that the state formula φ holds at state s in m.
$m,\pi \models \psi$ denotes that the path formula ψ holds along path π in m.

For formulas in $CTL^*_{-E}(\Sigma_m)$, $\models$ is defined as follows (assuming that A is an atomic proposition in AP_{Σ_m}, φ_2 and φ_2 are state formulas, ψ_1 and ψ_2 are path formulas) :

1	$m,s \models A$	iff	$A \in L(s)$
2	$m,s \models \neg A$	iff	$m,s \not\models A$
3	$m,s \models \varphi_1 \vee \varphi_2$	iff	$m,s \models \varphi_1$ or $m,s \models \varphi_2$
4	$m,s \models \varphi_1 \wedge \varphi_2$	iff	$m,s \models \varphi_1$ and $m,s \models \varphi_2$
5	$m,s \models \mathbf{A}\psi_1$	iff	for every path π starting with s, $m,\pi \models \psi_1$
6	$m,\pi \models \varphi_1$	iff	s is the first state of π and $m,s \models \varphi_1$
7	$m,\pi \models \psi_1 \vee \psi_2$	iff	$m,\pi \models \psi_1$ or $m,\pi \models \psi_2$
8	$m,\pi \models \psi_1 \wedge \psi_2$	iff	$m,\pi \models \psi_1$ and $m,\pi \models \psi_2$
9	$m,\pi \models \psi_1 \mathbf{U} \psi_2$	iff	there exists a $k \geq 0$ such that $m,\pi^k \models \psi_2$ and for all $0 \leq j < k$ $m,\pi^j \models \psi_1$
10	$m,\pi \models \mathbf{X}\psi_1$	iff	$m,\pi^1 \models \psi_1$
11	$m,\pi \models \mathbf{G}\psi_1$	iff	$\forall k \geq 0$, $m,\pi^k \models \psi_1$

The satisfaction of a state formula $\varphi \in CTL^*_{-E}(\Sigma)$ in a model m (denoted $m \models \varphi$) is defined by :

(1) If $\varphi \in CTL^*_{-E}(\Sigma_m)$ then $m \models \varphi$ iff $m,s_0 \models \varphi$.

(2) If $\varphi \notin CTL^*_{-E}(\Sigma_m)$ then $m \not\models \varphi$.

Comments:

a. A (process or system) model can be interpreted as a Kripke structure [HC77] by ignoring the transition labels.

b. The path quantifier E ("for some computation path") is eliminated *semantically* from CTL^*_{-E} .

We say that one language is *more expressive* then the other if there exists a formula in the one that does not have an equivalent formula in the other. Let $\succ$ denote "more expressive" and let CTL_{-E} be $CTL \cap CTL^*_{-E}$. Comparing CTL^*_{-E} with CTL^*, CTL and LTL [CG87] we get the following lemma :

Lemma 3: **(a)** $CTL^* \succ CTL^*_{-E}$; **(b)** $CTL^*_{-E} \succ CTL_{-E}$; **(c)** $CTL^*_{-E} \succ LTL$.

Proof:

(a) By comment **b.** above.

(b) There is no CTL_{-E} formula that is equivalent to the CTL^*_{-E} formula $\mathrm{A}F(a \wedge Xa)$. (This formula has been proven to have no CTL equivalent by [EH83].)

(c) There is no LTL formula that is equivalent to the CTL^*_{-E} formula $\mathrm{A}(X\neg a \vee \mathrm{AF}a)$.

We can now choose a model checking procedure $\vdash_{\overline{MC}}$, needed in our modular proof systems (AGM and AGM^*). As CTL^*_{-E} is a sub language of CTL^* there is a model checking algorithm which is exponential in the size of the specification and linear in the size of the Kripke structure (i.e., $O(2^{|\varphi|}(|S|+|R|))$) [EL85]. One may prefer to restrict the specifications to CTL_{-E}, gaining a linear model checking algorithm (i.e., $O(|\varphi|(|S|+|R|))$) [CES86].

4.3 The Preorder

The choice of CTL^*_{-E} as a specification language is motivated by a unique semantic quality: The validity of CTL^*_{-E} formulas is preserved over "*pruned*" computation trees. A description of this property is presented below. In this context, we view the semantics of CTL^*_{-E} as defined over computation trees. The originating transition graphs (if exist) of these trees are omitted.

Let T be a computation tree, and $\varphi \in CTL^*_{-E}$ be a temporal formula. $T,v \models \varphi$ denotes that φ is satisfied at node v in T. $T \models \varphi$ denotes that φ holds at the root of T. A *branching node* in T is a node that has at list two immediate descendants. Eliminating a subtree of T rooted at v, such that v is an immediate descendant of a branching node, is defined as *pruning T at v*.

Theorem 5: If T' is generated by pruning T at nodes which are descendants of v, then

$$\forall \varphi \in CTL^*_{-E} \;\; [\;\; T,v \models \varphi \Rightarrow T',v \models \varphi \;\;].$$

The interesting special case is T', generated from T by arbitrary pruning. Then:

$$\forall \varphi \in CTL^*_{-E} \;\; [\;\; T \models \varphi \Rightarrow T' \models \varphi \;\;].$$

Long [L89] defined a syntactic preorder over Kripke structures, which partially identifies the *pruning relation* of their computation trees. Basically, we adopt his relation as our choice of the required semantic preorder $\sqsupseteq$. Assimilated in our system, the definition is enhanced to encompass more cases for which the semantic property holds, including more cases for which the pruning relation holds. In addition, it is modified to consider transition labels and to verify the containment of the atomic proposition sets.

Let $m = \langle AC,AP,S,R,s_0,L\rangle$ and $m' = \langle AC',AP',S',R',s'_0,L'\rangle$ be either both process models, or both system models, such that $AP \subseteq AP'$. A sequence of relations $\boldsymbol{F}_i$, such that $\forall i \geq 0 \ \boldsymbol{F}_i \subseteq S \times S'$, is defined as follows:

For all $s \in S$, $s' \in S'$

1. $s \, \boldsymbol{F}_0 \, s'$ iff $L(s) = L'(s') \cap AP$.
2. $s \, \boldsymbol{F}_{n+1} \, s'$ iff -
   2.1 $s \, \boldsymbol{F}_n \, s'$.
   2.2 $\forall s'_1 \in S', \forall \alpha \in ACTION \cup AcT \ [\, s' \xrightarrow{\alpha} s'_1 \Rightarrow \exists s_1 \in S \ [\, s \xrightarrow{\alpha} s_1 \ \wedge \ s_1 \, \boldsymbol{F}_n \, s'_1 \,]]$.

$\boldsymbol{F} \subseteq S \times S'$ is the relation defined for m and m' by:

$$\text{For all } s \in S,\ s' \in S', \qquad s \, \boldsymbol{F} \, s' \ \text{ iff } \ \forall i \geq 0 \ \ s \, \boldsymbol{F}_i \, s'.$$

Let $m = \langle AC,AP,S,R,s_0,L\rangle$ and $m' = \langle AC',AP',S',R',s'_0,L'\rangle$. The relation $\geq$ over the model domain M is defined as follows:

$\boldsymbol{m \geq m'}$ iff -

i. m and m' are either both process models, or both system models.

ii. $AP \subseteq AP'$.

iii. $s_0 \, \boldsymbol{F} \, s'_0$.

Theorem 6:

6.1 $\geq$ is a preorder.

6.2 If $m_1 \geq m_2$ then for every Σ, and for every formula $\varphi \in CTL^*_{-E}(\Sigma)$, $m_1 \models \varphi \Rightarrow m_2 \models \varphi$.

There is an algorithm to verify that two models are preordered, which is polynomial-time in their size. Using $\geq$ as the required syntactically-identified semantic preorder, note that "greater" ($\geq$) models are, in general, expressed by smaller structures. The implications of this property were previously discussed.

Comments:

The pruning relation and the syntactic preorder ($\geq$) both capture some, but not all of the cases (m_1,m_2) in which $m_1 \models \varphi \Rightarrow m_2 \models \varphi$ for every φ. However, they are not identical:

a. The syntactic preorder ($\geq$) does not capture cases (m_1,m_2) - for which the pruning relation holds - in which labels of corresponding transitions are not identical. This restriction is required to ensure the satisfaction of the AGM and AGM^* soundness conditions (see Theorem 7 below).

b. The pruning relation does not capture cases (m_1,m_2) - for which $\geq$ holds - in which the sets of state labels are not identical. Defining $\geq$ to hold in these cases is justified by the definition $\varphi \notin CTL^*_{-E}(\Sigma_m) \Rightarrow m \not\models \varphi$.

Theorem 7:

1. For all m', m_A, m_B process models in M, if $m_A \geq m_B$ then $m_A \,||\, m' \geq m_B \,||\, m'$.

2. For all Σ, and for all m_A, m_B *system models* in M, if $m_A \geq m_B$ then $m_A|_\Sigma \geq m_B|_\Sigma$.

3. For all m_A, m_B process models in M: $((m_A \| m_B)|_{\Sigma_{m_A}}) \| ((m_A \| m_B)|_{\Sigma_{m_B}}) \geq m_A \| m_B$.

4 For all m_A, m_B process models in M: $m_A \geq (m_A \| m_B)|_{\Sigma_{m_A}}$.

Since the composition operator is commutative, Theorem 7 implies that the *AGM* (and AGM^*) soundness conditions are satisfied. Establishing that, the presentation of a semi automatic model checking system derived from these schemes, is completed.

4.4 Generalized Models

A helpful tool for generating assumptions is *generalization*. Generalization of a given system or component model produces a model which is greater ($\geq$) than the original. A generalized model captures only a partial set of the properties of the original model and has a smaller set of states. Figure 2 demonstrates an application of the generalization operator to the component model m_1 , yielding the generalized model $(m_1)_G$.

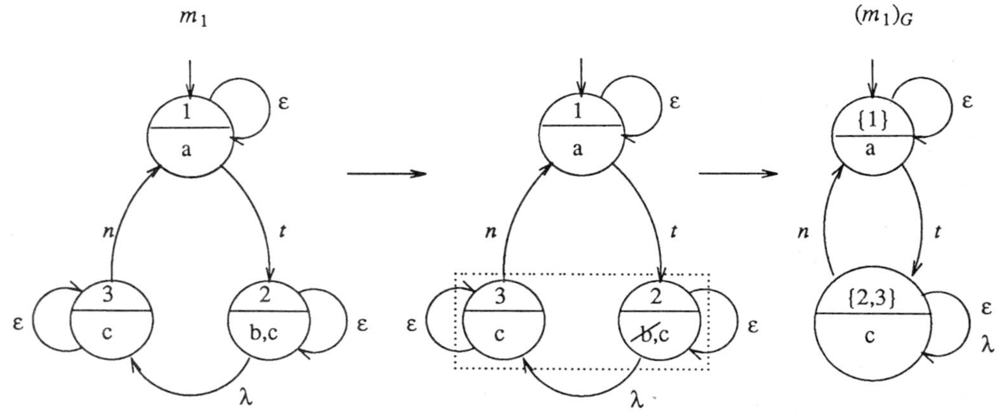

Figure 2 - Generalizing m_1

Presented below, is the generalization operator. Let $m = \langle AC, AP, S, R, s_0, L\rangle$ be a system or a component model. Let Σ be an alphabet and let Pr be a partition of S such that for every $p \in Pr$ and for every $s, s' \in p$, $L(s) \cap \Sigma = L(s') \cap \Sigma$. The generalized model $m_{G(Pr,\,\Sigma)}$ is defined as follows:

- $AC_G = AC$
- $AP_G = AP \cap \Sigma$
- $S_G = Pr$
- R_G : for all $s_i^G, s_j^G \in S_G$ and for all $\alpha \in AcT \cup ACTION$:
  $(s_i^G, \alpha, s_j^G) \in R_G$ iff there exist $s \in s_i^G$, $s' \in s_j^G$ for which $(s, \alpha, s') \in R$.
- s_0^G is the S_G element of which s_0 is a member.
- L_G : $L_G(s_i^G) = L(s) \cap \Sigma$ for some $s \in s_i^G$.

Lemma 4: $m_{G(Pr,\,\Sigma)} \geq m$.

Note that generalized models resemble the *Image* finite state machines described in [LS84].

4.5 An Example

A model of a simple finite-state distributed program is verified below, using the AGM^* proof system. The components' models m_1 and m_2 are presented in Figure 3a . The composed system $m_1 \| m_2$ should be verified to meet the temporal specification $\varphi = \mathbf{AG}((a \wedge \neg f) \rightarrow \mathbf{X}(\neg a \wedge f))$ *without* having to construct its global model (Figure 3b). Note that φ is a property of the compound system which depends on the interaction between the system's components. Before presenting the formal proof sequence we outline the construction of the proof:
Using the minimal state labels and the potential actions in the context of m_1 and m_2 , φ is expressed by the system model $\overline{m} \| \overline{\overline{m}}$ (Figure 3c). A model checker can be applied to verify that $\overline{m} \| \overline{\overline{m}} \models \varphi$. AGM^* is used to verify that $\overline{m} \| \overline{\overline{m}} \geq m_1 \| m_2$ (refer to the proof of Theorem 3.1) without constructing $m_1 \| m_2$.
Determining $\overline{m}$ to be a generalization of m_1 ($\overline{m} = (m_1)_{G(Pr,\,\Sigma)}$, $Pr = \{\{1,4\},\{2,3\}\}$, $\Sigma = \{a\}$) , we conclude that $\overline{\overline{m}}$ is as described in Figure 3c . Unfortunately, $\overline{\overline{m}}$ is *not* greater than m_2 , because $\overline{\overline{m}}$ enables $\overline{n}$ transitions only in a state labeled by f (compare $\overline{\overline{m}}$ with $(m_2)_{G(Pr,\,\Sigma)}$, $Pr = \{\{5,7\},\{6,8\}\}$, $\Sigma = \{f\}$). Hence, this property of $\overline{\overline{m}}$ does not stems from m_2 exclusively, and its derivation should consider properties of m_1 as well. The relevant properties of m_1 and m_2 are captured in m' and m'' (Figure 3c), which are their respective generalizations. Establishing that $\overline{\overline{m}} \geq m' \| m'' \rceil_{\Sigma_{m_2}}$, the proof is completed.

The formal proof sequence is :

- • 1 $m' \geq m_1$
- • 2 $m'' \geq m_2$
- 3 $m' \| m''$ is a $gm^*[m_1 \| m_2]$ atom. (1,2 and the definition)
- • 4 $\overline{\overline{m}} \geq m' \| m'' \rceil_{\Sigma_{m_2}}$
- • 5 $\overline{m} \geq m_1$
- 6 $\overline{m} \| \overline{\overline{m}} \in gm^*[m_1 \| m_2]$ (3,4,5 and lemma 2)
- 7 $\varphi \in CTL^*_{-E}(\Sigma_{\overline{m} \| \overline{\overline{m}}})$
- • 8 $\vdash_{MC} \overline{m} \| \overline{\overline{m}} \models \varphi$
- 9 $\vdash_{AGM^* - m_1 \| m_2} \varphi$ (6,7,8 and the inference rule)

An exhaustive search can show that there is no shorter sequence that ends with the application of a model checker to $\overline{m} \| \overline{\overline{m}}$.

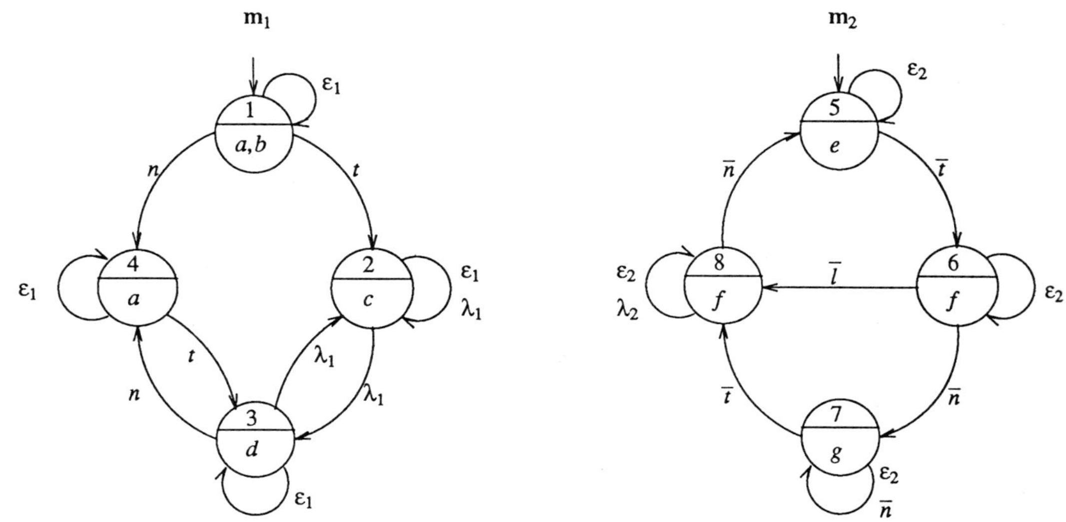

Figure 3a - The components' models

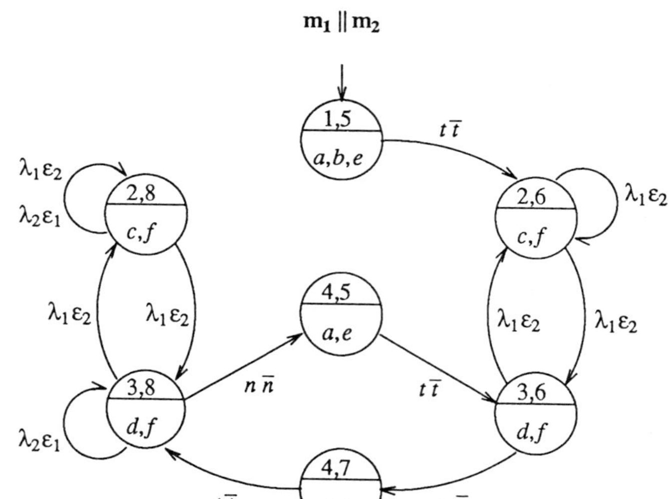

Figure 3b - The global system model

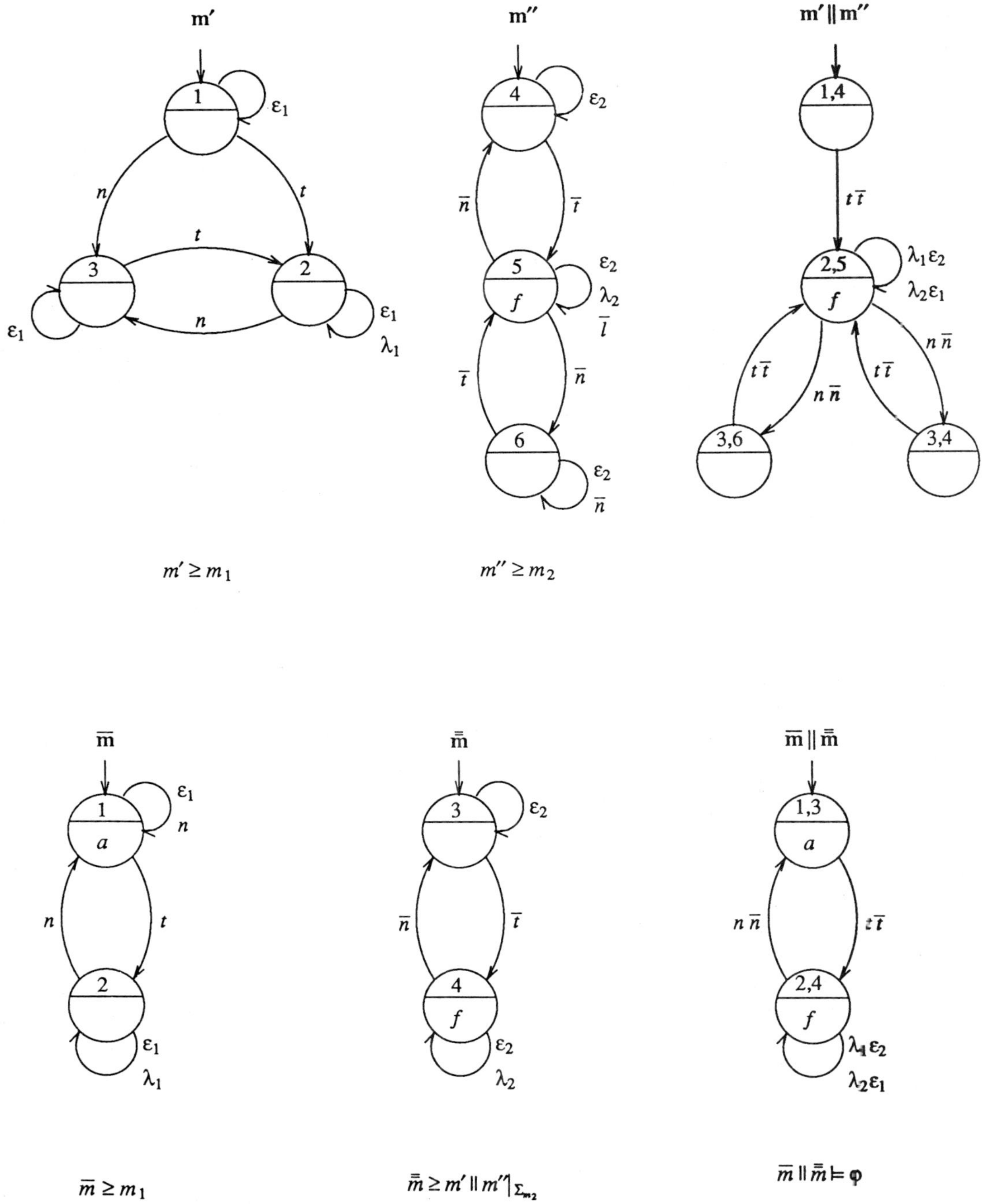

Figure 3c - Assumptions and composed models of the AGM^* proof

REFERENCES

[AL89] M. Abadi and L. Lamport, "Composing Specifications", REX Workshop on Stepwise Refinement of Distributed Systems: Models, Formalisms, Correctness, Mook, May 1989.

[BF88] L. Bouge, and N. Francez, "A Compositional Approach to Superimposition", Proc. 15th ACM POPL Symposium, San Diego, January, 1988.

[BKP84] H. Barringer, R. Kuiper, and A. Pnueli, "Now You May Compose Temporal Logic Specifications", Proceedings of the sixteenth ACM symposium on the Theory of Computing, 1984.

[Br86] M.C. Browne, "An Improved Algorithm for the Automatic Verification of Finite State Systems using Temporal Logic", Proceedings of the 5th Conference on Logic in Computer Science, Cambridge, MA, June 1986, pp. 260-267.

[CE81] E.M. Clarke, and E.A. Emerson, "Synthesis of Synchronization Skeletons for Branching Time Temporal Logic", Proc. of Workshop on Logic of Programs, Yorktown-Heights, 1981.

[CES86] E.M. Clarke, E.A. Emerson, and A.P. Sistla, "Automatic Verification of Finite-State Concurrent Systems using Temporal Logic Specifications", ACM Transactions on Programming Languages and Systems 8, 2, pp. 244-263, 1986.

[CG87] E.M. Clarke, O. Grumberg, "Research on automatic verification of finite-state concurrent systems", Annual Reviews of Computer Science, Vol. 2, 269-290, 1987 (J.F. Traub, editor).

[CLM89] E.M. Clarke, D.E. Long, and K.L. McMillan, "Compositional Model Checker". Proc. of the 4th IEEE Symp. on Logic in Computer Science, Asilomar, June 1989.

[CM88] M. Chandy, and J. Misra, "Parallel Programs Design", Addison-Wesly, 1988.

[CPS88] R. Cleavaland, J. Parrow, and B. Steffen, "The Concurrency Workbench," University of Edinburgh Report, 1988.

[EH83] E.A. Emerson, and J.Y. Halpern, " "Sometimes" and "Not Never" revisited: on Branching versus Linear Time", 10th symposium on Principle of programming Languages, 1983.

[EL85] E.A. Emerson, and C. Lei, "Modalities for Model Checking: Branching Time Strikes Back", 12th Symposium on Principles of Programming Languages, New Orleans, La., January 1985.

[HC77] G.E. Hughes, and M.J. Creswell, "An introduction to Modal Logic", London: Methuen, 1977.

[J87] B. Josko, "MCTL - An Extension of CTL for Modular Verification of Concurrent Systems". Workshop on Temporal Logic, (H. Barringer, ed.) University of Manchester, April 1987, LNCS 398, pp. 165-187.

[K87] S. Katz, "A Superimposition Control Construct for Distributed Systems", TR MCC-STP, MCC, Austin, August 1987.

[Ku90] R.P. Kurshan, "Analysis of Discrete Event Coordination", LNCS 430 (1990) Springer-Verlag, Stepwise Refinment of Distributed Systems, pp. 414-453.

[L89] D.E. Long, Private communication.

[LP85] O. Lichtenstein, and A. Pnueli, "Checking that Finite State Concurrent Programs Satisfy Their Linear Specification", Conference Record of the 12th Annual ACM Symposium on Principles of Programming Languages, New Orleans, La., January 1985.

[LS84] S. S. Lam and A. U. Shankar, "Protocol Verification Via Projections", IEEE Transactions on Software Engineering, SE-10(4), pp. 325-342, July 1984.

[M80] R. Milner, A Calculus of Communicating Systems, Springer Lecture Notes on Computer Science, Vol. 92, 1980.

[P85] A. Pnueli, "In Transition from Global to Modular Temporal Reasoning about Programs". Logics and Models of Concurrent Systems, (K. Apt, ed.), Vol. 13 of NATO ASI Series F: Computer and System Sciences, Springer-Verlag, 123-144, 1985.

[QS81] J.P. Quielle, and J. Sifakis, "Specification and Verification of Concurrent Systems in CESAR", Proc. of the 5th International Symposium in Programming, 1981.

DIMACS Series in Discrete Mathematics
and Theoretical Computer Science
Volume 3, 1991

Tool Support for the Refinement Calculus

D.A. CARRINGTON AND K.A. ROBINSON

ABSTRACT. This paper explores the concept of formal refinement of specifications to executable programs and the role for mechanized support. The paper uses the refinement calculus as the method of refinement. A prototype editor built using the Cornell Synthesizer Generator is described and based on our experiences, we suggest features that are desirable in future tools.

1. Introduction

Software development requires a process to transform an abstract specification into an executable substitute, that is, a computer program that exhibits the required behaviour. The transformation process is often referred to as refinement and normally requires a sequence of steps. These steps represent the design decisions leading to the choice of algorithms and of data representations. Formalizing the refinement process is important to establish the correctness of the product. A program is *correct* if it is a satisfactory implementation of the original specification. Correctness may be established by testing (almost always infeasible), by constructing a program proof (usually difficult if attempted after the program has been constructed), or by using a refinement strategy that ensures that each step in the development process maintains the original requirements.

The starting process for refinement is a specification. A major function of a specification is to assist communication between the developer and the client with the objective that both parties understand the intended behaviour of the product to be created.

There is growing acceptance of the benefits of formal specification methods that avoid the ambiguity of natural language. The objective of the specification is to define *what is required* of the product and not *how it is to be achieved.* A

1980 *Mathematics Subject Classification* (1985 *Revision*). Primary 68Q60; Secondary 68N05.

This research is partially supported by an Australian Research Council grant.

specification language should encourage precision, conciseness, and abstraction. It is also desirable to be able to manipulate the specification and prove properties or consequences of the specification. In this way, confidence in the specification as an accurate reflection of the requirements can be established. Unless the specification is expressed in a formal notation, it is not possible to have a systematic method of verifying that an implementation is correct.

Mathematical specification techniques using predicate logic, such as the Vienna Definition Method (VDM) [**1**, **8**] and Z [**6**, **17**] describe program behaviour as a relation between the input and output states. This allows us to capture the essential behavioural characteristics of a system or algorithm without concern for the implementation. Such specifications have the desirable properties listed above but are not suitable for expressing the concrete algorithms and data structures of the executable program. VDM also incorporates refinement techniques and is one of the few sustained research efforts to tackle this problem.

Having produced a specification that is considered to meet the informal requirements, we want to be able to refine the specification by changing the data structures and introducing programming language constructs until we have an implementation: an executable substitute. The aim of a formal refinement methodology is to ensure that at all times the refinement is in some essential way *consistent* with the initial specification. There are considerable advantages from using uniform notation throughout the refinement process. At the very least, it avoids the difficulties associated with converting between notations. There are two possible approaches: adding programming language elements to a specification language, or adding specification constructs to a programming language. Whichever approach is taken, it is crucial to have a formal semantics that encompasses the extended language. The refinement calculus method adopts the approach of adding specification to a programming language with the extended language defined using Dijkstra's weakest pre-condition semantics [**3**]. Section two provides an overview of the refinement calculus and the specification statement on which it is based.

As well as ensuring that the final program is a correct implementation of the initial specification, the refinement steps also provide an excellent design history by documenting the choices made during the development. In that sense, the refinement calculus provides a more rigorous version of the development style that Knuth [**9**] calls *literate programming*. In section three, we consider suitable roles for tools to assist in the application of the refinement calculus.

Section four reviews a prototype refinement editor, a tool for developing programs from specifications based on the rules of the refinement calculus. Based on our experiences with this tool, we make some suggestions for future tools.

2. The refinement calculus

The refinement calculus developed by Morgan and Robinson [**12, 13, 10**] provides a rigorous technique for deriving an executable program from an abstract specification. The essential ingredients of the calculus are

- a programming language (with a formal semantics) that we use as our target language. We presume that this language can be executed on a computer.
- a specification construct that is abstract *i.e.* it must allow the specification of programs without being restricted by implementation concerns that are present in any *real* programming language. Our formal semantics must include the specification construct.
- a formal definition of the notion of specification transformation, that we call *refinement.*

We now consider each of these in turn.

2.1. A programming language. We follow Dijkstra and use a weakest precondition semantics. We also use a guarded command language similar to that used by Dijkstra. It should be noted that any programming language could be used, provided that it is given a weakest pre-condition semantics.

If we consider procedural programs, then associated with each program is a *program state* that consists of a set of variables which we denote $\vec{v}$. A program is a mechanism for affecting a change of state, from an *initial state* to a *final state.* A program state can be specified by a predicate and an early specification of programs was given by Hoare [7] as

$$\{P\}\; S\; \{R\}$$

where P is a predicate describing the program's initial state, R is a predicate describing the final state and S is a program that, when executed in a state satisfying P will terminate in a state satisfying R. The predicate P is frequently called the *pre-condition* and R the *post-condition.*

For any program (or more correctly in what follows, any *specification*) S and post-condition R, the weakest pre-condition of S with respect to R is P_{weak} such that for any P satisfying

$$\{P\}\; S\; \{R\}$$

then $P \implies P_{weak}$. Thus P_{weak} places the weakest possible constraints on the initial state for S to be able to achieve R. In our work we denote the weakest pre-condition of S with respect to R by the functional application of S to R, *i.e.*

$$S\, R \;\; \triangleq \;\; \mathrm{wp}(S, R)$$

We introduce the constructs of the language as we introduce the refinement rules.

2.2. A specification construct. The specification statement [11] captures the essential behaviour of a program or part of a program, without concern for implementation issues. It is denoted

$$\vec{w}\colon [\,Pre \;/\; Post\,]$$

and denotes a program that, executed in a state satisfying Pre, will terminate in a state satisfying $Post$ while modifying only variables in $\vec{w}$ (where $\vec{w}$ is a subset of the program variables $\vec{v}$). Pre is a predicate in $\vec{v}$ while $Post$ is a predicate in $\vec{v}$ and $\vec{v}_0$ where the decorated variables $\vec{v}_0$ refer to the corresponding values in the initial state.

The weakest pre-condition for $\vec{w}\colon [\,Pre \;/\; Post\,]$ to terminate in a state satisfying R must be at least as strong as Pre, and every program state that satisfies $Post$ must also satisfy R. The formal definition is

$$\vec{w}\colon [\,Pre \;/\; Post\,]\; R \quad \triangleq \quad Pre \wedge \forall\,(\vec{w} \bullet Post \implies R)_{[\vec{v}_0 \setminus \vec{v}]}$$

where $E_{[\vec{x} \setminus \vec{e}]}$ is the expression E with all free occurrences of each of the variables in $\vec{x}$ replaced by the respective expression in $\vec{e}$. We assume that the $\vec{v}_0$ are local to $Post$ and hence are not contained in R; otherwise, a further renaming is required to avoid confusion.

Two syntactic variations are

- $\vec{w}\colon [\,Post\,]$ where the implicit pre-condition is $\exists\,(\vec{w} \bullet Post)_{[\vec{v}_0 \setminus \vec{v}]}$, the weakest condition for the existence of a program state satisfying $Post$.
- $\vec{w}\colon [\,Pre \,/\, I \,/\, Post\,]$ where I is a predicate in $\vec{v}$ that is invariant, that is, it is true in both the initial and the final state. This is a shorthand for $\vec{w}\colon [\,Pre \wedge I \;/\; I \wedge Post\,]$.

Two primitive specifications are **abort** and **magic** defined as follows:

$$\textbf{abort}\,R \quad \triangleq \quad \mathsf{false} \qquad \{\textbf{abort} = \vec{w}\colon [\,\mathsf{false} \;/\; Post\,]\}$$

$$\textbf{magic}\,R \quad \triangleq \quad \mathsf{true} \qquad \{\textbf{magic} = \vec{w}\colon [\,\mathsf{true} \;/\; \mathsf{false}\,]\}$$

abort is the easiest specification to implement: any implementation will do; while **magic** is the most difficult: it is impossible to implement.

We introduce an example that is used subsequently to illustrate some refinement rules. The example is very simple and is chosen to demonstrate the principles involved. We wish to implement $\mathsf{Min}(b, N, m)$ that computes the minimum value in the array $b[0..N-1]$, passing back the result in m. This is represented by the following specification statement: [1]

$$\mathsf{Min}(b, N, m) \quad \triangleq \quad m\colon [\,N > 0 \;/\; Post\,]$$

$$\text{where} \quad Post \quad \triangleq \quad \begin{pmatrix} \exists\,(i\colon 0..N-1 \bullet m = b[i]) \\ \forall\,(i\colon 0..N-1 \bullet m \leq b[i]) \end{pmatrix}$$

[1] We use vertical stacking of conjuncts to denote conjunction.

2.3. Refinement. With the refinement calculus, the development of a program is a process of replacing specification statements by other programming constructs using the rules of the calculus to perform the transformations. To ensure that the resulting program is a valid implementation of the initial specification, we require an ordering between specifications that captures the idea that one specification (S_i) may be replaced by another (S_j) in any context. This ordering is denoted $S_i \sqsubseteq S_j$. If in all initial states in which S_i does not abort, the set of final states of S_j is contained in the set of final states of S_i, then we say S_i is refined by S_j. The formal definition uses weakest pre-conditions:

$$S_i \sqsubseteq S_j \text{ iff } (S_i\ R) \implies (S_j\ R) \text{ for all } R$$

The ordering $\sqsubseteq$ is a partial ordering over specifications and defines a complete lattice. Thus for any specification S

$$\mathbf{abort} \sqsubseteq S \sqsubseteq \mathbf{magic}$$

Based on our concept of refinement, we can depict the design process as the sequence

$$S_0 \sqsubseteq S_1 \sqsubseteq \cdots \sqsubseteq S_n$$

where S_0 is the initial specification and S_n is our executable program.

3. Refinement rules

The rules of the refinement calculus express valid transformations of specifications. Many of the rules have an associated applicability condition that specifies when the rule may be used. Rather than attempting to supply a comprehensive set of rules, we give an example refinement, introducing rules as required. Each refinement rule is denoted by

$$\frac{\textit{conditions}}{\textit{refinement}}$$

which permits the *refinement* to be carried out provided the *conditions* are satisfied.

3.1. New variables. A little contemplation of the specification statement for Min will suggest that we need an additional variable to control the computation. The construct $[\![\ \mathbf{var}\ \vec{t}\ \bullet\ S\ ]\!]$ [2] introduces new variables $\vec{t}$ into the program state inside the block delimited by the symbols $[\![$ and $]\!]$. It is defined by its weakest pre-condition

$$([\![\ \mathbf{var}\ \vec{t}\ \bullet\ S\ ]\!])\ R \qquad \triangleq \qquad \forall\,(\vec{t}\bullet S\ R)$$

[2] In this example, and in the discussion in general, we ignore the type of variables. Introducing typed variables adds a set membership constraint to the pre- and post-conditions.

and is introduced by the rule

$$\frac{\vec{v} \cap \vec{t} = \{\}}{S \sqsubseteq [\![\ \mathbf{var}\ \vec{t}\ \bullet\ \vec{t}{:}S\ ]\!]}$$

where $\vec{t}{:}S$ is S with $\vec{t}$ added to the variables changeable by S.

Applying this rule to our example, we get

$$m\textbf{:}\,[N > 0\ /\ Post] \quad \sqsubseteq \quad [\![\mathbf{var}\ j \bullet m, j\textbf{:}\,[N > 0\ /\ Post]]\!]$$

3.2. Sequential composition. We define the sequential composition $S_1 ; S_2$ of two specifications by

$$(S_1; S_2)R \qquad \triangleq \qquad S_1(S_2\ R)$$

and we introduce a rule that allows the sequential decomposition of a specification

$$\frac{\vec{w}\textbf{:}\,[Pre\ /\ S_2\ Post] \sqsubseteq S_1}{\vec{w}\textbf{:}\,[Pre\ /\ Post] \sqsubseteq S_1; S_2}$$

The above rule allows us to decompose a specification as a sequential composition in which the second component of the composition is known and the first component can be computed.

We will now refine the inner specification to a sequential composition using the intermediate predicate

$$I \quad \triangleq \quad \begin{pmatrix} 0 < j \leq N \\ \exists\,(i : 0..j-1 \bullet m = b[i]) \\ \forall\,(i : 0..j-1 \bullet m \leq b[i]) \end{pmatrix}$$

giving

$$m, j\textbf{:}\,[N > 0\ /\ Post] \quad \sqsubseteq \quad m, j\textbf{:}\,[N > 0\ /\ I]\ ;\ m, j\textbf{:}\,[I\ /\ Post]$$

3.3. Assignment. The assignment $\vec{w} := \vec{e}$ changes the variables $\vec{w}$ to the values $\vec{e}$.

$$(\vec{w} := \vec{e})R \qquad \triangleq \qquad R_{[\vec{w}\ \backslash\ \vec{e}]}$$

and is introduced by the rule

$$\frac{Pre \implies Post_{[\vec{w}, \vec{v}_0\ \backslash\ \vec{e}, \vec{v}]}}{\vec{w}\textbf{:}\,[Pre\ /\ Post] \sqsubseteq \vec{w} := \vec{e}}$$

The component $(m, j\textbf{:}\,[N > 0\ /\ I])$ is then refined to the assignment

$$m, j := b[0], 1$$

We intend to refine the second component to a loop, so we will give some of the refinement rules leading up to refinement to a loop.

3.4. Guarded commands. We define the semantics of an isolated guarded command $B \rightarrow S$, where B is a predicate, called the guard, and S is a specification.

$$(B \rightarrow S)\, R \qquad \triangleq \qquad B \Rightarrow (S\, R)$$

and the rule

$$\frac{B \implies (P \sqsubseteq Q)}{P \sqsubseteq B \rightarrow Q}$$

Notice that the semantics of a guarded command define $B \rightarrow S$ to behave like **magic** when B is false, and for that reason uncomposed guarded commands do not exist in real programming languages.

Having refined a specification to a number of guarded commands we allow a refinement to a non-deterministic union of guarded commands

$$\frac{\forall\,(i : 1..n \bullet S \sqsubseteq B_i \rightarrow S_i)}{S \sqsubseteq \mathop{[\!]}_{i} (B_i \rightarrow S_i)}$$

3.5. The if-statement. We define the if-statement (Dijkstra's alternative command) to be an encapsulation of a union of guarded commands that behaves like **abort** in any state that does not satisfy any of the guards. Thus we can refine a specification to an if-statement provided the initial state satisfies at least one of the guards. An if-statement behaves like **abort** when the embedded union behaves like **magic**.

$$\frac{\begin{array}{c} S = \vec{w} : [\, PreS \;/\; PostS \,] \\ S \sqsubseteq \mathop{[\!]}_{i=1,n} (B_i \rightarrow S_i) \\ PreS \implies \bigvee_{i=1,n} (B_i) \end{array}}{S \sqsubseteq \textbf{if} \mathop{[\!]}_{i=1,n} (B_i \rightarrow S_i)\ \textbf{fi}}$$

3.6. The do-statement. There is not space to elaborate here, but we define refinements that lead to recursion and regard recursion rather than iteration as fundamental. The do-statement (Dijkstra's repetitive command) is defined as a recursive if-statement:

$$\textbf{do} \mathop{[\!]}_{i=1,n} (B_i \rightarrow S_i)\ \textbf{od} \;\; \triangleq \;\; DO : [\![\textbf{if}\ B_0 \rightarrow \textbf{skip} \;[\!]\; \mathop{[\!]}_{i=1,n} (B_i \rightarrow S_i\ ;\ DO)\ \textbf{fi}]\!]$$

$$\text{where } B_0 \;\triangleq\; \neg \bigvee_{i=1,n} (B_i)$$

The recursive block introduced into the refinement carries an obligation to establish termination.

A rule can be derived that allows us to refine any specification to a do-statement, either via recursion

$$\frac{S \sqsubseteq P : \left[\!\left[\textbf{if } B_0 \rightarrow \textbf{skip} \; [\!] \; \underset{i=1,n}{[\!]} \; (B_i \rightarrow S_i \; ; P) \textbf{ fi} \right]\!\right]}{S \sqsubseteq \textbf{do} \underset{i=1,n}{[\!]} (B_i \rightarrow S_i) \textbf{ od}}$$

or directly:

$$\frac{\begin{array}{c} S = \vec{w} : [\, I \;/\; Post \,] \\ (I \wedge \neg \bigvee_{i=1,n} (B_i)) \implies Post_{[\vec{v}_0 \backslash \vec{v}]} \end{array}}{S \sqsubseteq \textbf{do} \underset{i=1,n}{[\!]} \left(B_i \rightarrow \vec{w} : \left[B_i \Big/ I \Big/ \begin{array}{c} Post' \\ \tau \prec \tau_0 \end{array} \right] \right) \textbf{od}}$$

$$\text{where} \quad Post' \;\triangleq\; \forall \left(\vec{w} \,\bullet\, Post_{[\vec{v}_0 \backslash \vec{v}'_0]} \implies Post \right)_{[\vec{v}'_0 \backslash \vec{v}]}$$

$$\text{and} \quad \tau_0 \;\triangleq\; \tau_{[\vec{v} \backslash \vec{v}_0]}$$

τ is a variant function defined on the program variables and whose range, restricted by I, is some set well-ordered on $\prec$. The significance of the well-ordering $\prec$ is that, for any given $\vec{v}_0$, the set $\{\tau \mid I \wedge \tau \prec \tau_0\}$ is finite.

We are now in a position to resume the refinement of the example. If *Post* does not contain $\vec{v}_0$, the above rule gives

$$m, j \colon [\, I \;/\; Post \,] \quad \sqsubseteq \quad \textbf{do} \underset{i=1,n}{[\!]} (B_i \rightarrow m, j \colon [\, B_i \;/\, I \,/\; \tau < \tau_0 \,]) \textbf{ od}$$

We have to choose a set of guards $\{B_i\}$ and a variant function. Given the condition on the above refinement we choose one guard $(j \neq N)$ observing that

$$I \wedge j = N \Rightarrow Post$$

For the variant function and order $\prec$ we choose

$$\tau \;\triangleq\; N - j \qquad \text{and} \qquad \prec \;\triangleq\; <$$

This leaves us to refine

$$m, j \colon [\, j \neq N \;/\, I \,/\; N - j < N - j_0 \,] = m, j \colon [\, j \neq N \;/\, I \,/\; j > j_0 \,]$$

It is obvious that we can satisfy the requirement $j > j_0$ by incrementing j, so we will use the composition rule to refine to a sequential composition in which the second component is a specification that increments j.

$$\begin{array}{l} m, j \colon [\, j \neq N \;/\, I \,/\; j > j_0 \,] \\ \sqsubseteq \quad m, j \colon \left[j \neq N \wedge I \Big/ \begin{pmatrix} I \\ j > j_0 \end{pmatrix}_{[j \backslash j+1]} \right] ; j \colon [\, j = j_0 + 1 \,] \end{array}$$

The second component refines to the assignment $j := j + 1$ and the first component is simplified, then refined by deleting the variable j from the window and then reorganized to reveal an invariant component as follows:

$$
\begin{array}{lll}
 & m, j\!: \left[j \neq N \wedge I \;\middle/\; \begin{pmatrix} I \\ j > j_0 \end{pmatrix}_{[j \backslash j+1]} \right] & \\
= & m, j\!: \left[j \neq N \wedge I \;\middle/\; \begin{array}{c} 0 < j+1 \leq N \\ \exists\,(i : 0..j \bullet m = b[i]) \\ \forall\,(i : 0..j \bullet m \leq b[i]) \\ j+1 > j_0 \end{array} \right] & \\
\sqsubseteq & m\!: \left[j \neq N \wedge I \;\middle/\; \begin{array}{c} 0 < j+1 \leq N \\ \exists\,(i : 0..j \bullet m = b[i]) \\ \forall\,(i : 0..j \bullet m \leq b[i]) \end{array} \right] & \\
\sqsubseteq & m\!: \left[\exists\,(i : 0..j-1 \bullet m = b[i]) \;/\; I' \;\middle/\; \begin{array}{c} \exists\,(i : 0..j \bullet m = b[i]) \\ m \leq b[j] \end{array} \right] & \\
\text{where} \quad I' & \triangleq & \begin{pmatrix} 0 < j < N \\ \forall\,(i : 0..j-1 \bullet m \leq b[i]) \end{pmatrix}
\end{array}
$$

We now refine the final specification obtained above in four ways:

(i) to the guarded command $m \leq b[j] \rightarrow$ **skip**;
(ii) to the guarded command $m \geq b[j] \rightarrow m := b[j]$;
(iii) to the union of the two guarded commands;
(iv) to an if-statement containing that union.

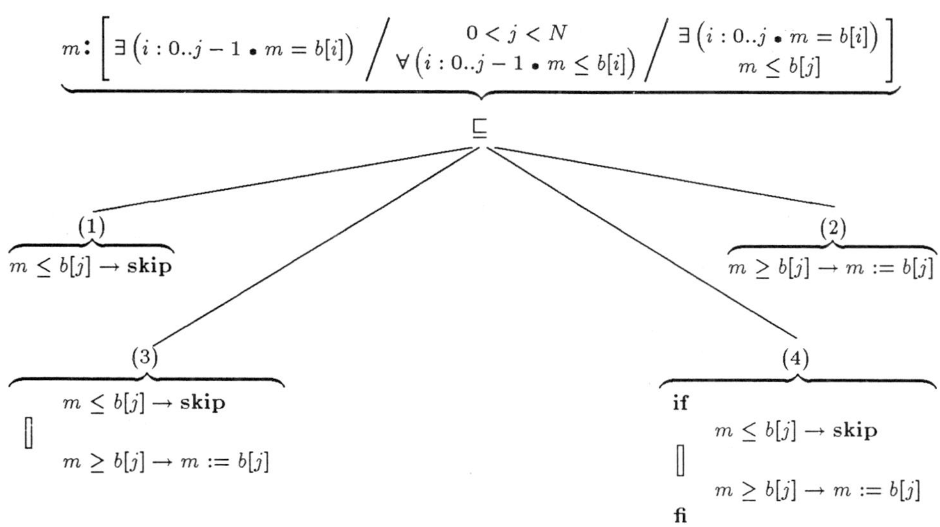

Assembling the refinements we get the following program:

$$
\begin{array}{l}
[\![\ \mathbf{var}\ j \bullet m, j := b[0], 1; \\
\qquad \mathbf{do}\ j \neq N \rightarrow \\
\qquad\qquad \mathbf{if}\ m \leq b[j] \rightarrow \mathbf{skip} \\
\qquad\qquad [\!]\ m \geq b[j] \rightarrow m := b[j] \\
\qquad\qquad \mathbf{fi}; \\
\qquad\qquad j := j + 1 \\
\qquad \mathbf{od} \\
]\!]
\end{array}
$$

3.7. Data refinement. The preceding example dealt exclusively with *procedural* refinement. The refinement calculus also contains rules for *data* refinement where abstract types in a program are replaced by more concrete types. For more details see [**13**].

4. Tool roles

The refinement calculus is intended to improve the quality of the software development process. However applying the calculus involves many small steps where each step may require the computation of a new specification and/or the checking of an applicability condition associated with a rule. While most of these are very straight-forward, there is scope for errors to be introduced by the number of steps.

From this observation we envisage two major roles for refinement calculus tools:

- correctly applying the rules of the refinement calculus as directed by the tool user, and
- recording the sequence of refinement steps in a software development.

The first role requires establishing that the applicability condition associated with each rule is satisfied whenever it is used and performing the computations of the calculus. We do not expect to automate software development; our aim is to provide systematic support. Thus the human designer retains the initiative in the design process with the tool playing the *careful assistant* role with responsibility for confirming the viability of each refinement step. Since the proof obligation of each step is a lemma in the first order predicate calculus, we cannot expect automatic confirmation of all steps. While some are easy to prove, many rely on domain knowledge. Some form of interactive proof editor seems necessary. This approach to tools for software development is not new; Floyd [**4**] predicted it and many people since have adopted this style for co-operative interaction. Perhaps the best known work is the programmer's apprentice project at MIT [**16, 18**].

The recorded design history based on the refinement steps is a valuable artifact of the design process. A tool simplifies the task of capturing this information systematically and accurately. The design history can serve several purposes:

- to explain and document the current design,

- for future modification by incremental changes to part of the refinement sequence,
- for re-use in different contexts.

5. A prototype refinement editor

A prototype refinement editor [2] has been built as an interactive tool for the refinement calculus. It was constructed using the Synthesizer Generator [**14, 15**] which generates language-based editors from descriptions of an abstract syntax and display representation. Attribute grammars are used to incorporate context-sensitive constraints and incremental re-evaluation algorithms to ensure that a consistent document exists after each editing operation.

The development of a prototype was simplified by using parts of Bill Pugh's pv demonstration editor supplied with the Synthesizer Generator distribution. The pv editor is a preliminary program verification editor for Dijkstra's guarded command language. The editor generates verification conditions and incorporates an interactive simplifier for proving them. Using pv, programs are entered statement by statement with the verification conditions confirming the correctness of the program.

Our refinement editor manipulates similar objects but the development style is quite different. From pv we have been able to use the basic structures for predicates, predicate simplification and expression substitution for computing weakest pre-conditions. We also chose the same domain of discourse: integer arithmetic with either simple or array variables. The prototype is restricted to procedural refinement.

The abstract syntax for the refinement editor represents the refinement process as a tree with the initial specification as the root node and each edge corresponding to a refinement step. When the refinement is complete, the leaf nodes correspond to constructs of the executable programming language. Proof subtrees are attached to nodes to verify the applicability condition of each refinement rule. Some of these can be immediately simplified but there is the capability for user interaction for more difficult cases.

The refinement rules of the calculus are provided as transformation commands that act on selected nodes in the tree by extending the tree with an additional refinement step. The Synthesizer Generator displays a sub-window of valid transformation commands whose contents depend on the current edit selection.

The editor uses two display unparsing schemes to present two views of the program under development:

- the primary view is a linear exposition of the refinement steps based on a top-down depth-first traversal of the tree.
- the alternative view is a consolidation of the refinement tree to extract the "final" program. This displays only the leaf nodes. Parts of the program that are not fully refined are displayed as specification statements.

The editor user can dynamically choose either view for independent subtrees so partial compression or expansion is possible.

5.1. Experiences. While our first refinement editor is very rudimentary, it has provided considerable insight into the challenge of building support tools for the refinement calculus. Basing the support tool on the editing paradigm has been a good idea since it encourages an exploratory style where the user investigates refinements, deleting those that are not successful. We have used it to explore the refinement of many small programming examples and we are enthusiastic about the potential of the approach.

As an illustration, after using the editor to develop the Min example, we investigated how easily it could be modified to compute Max instead. The necessary editing changes were to

(i) the initial specification (which then propagated through the complete refinement tree),
(ii) the guards within the if command(or equivalently the guarded statements), and
(iii) the intermediate predicate used for the first sequential composition.

The last step was required because this predicate was entered by hand rather than by computation based on the goal post-condition. It would certainly be feasible to incorporate rules to transform predicates in this manner. This example highlights the benefits of recording the design steps and having the capability to incrementally modify individual steps and observe the consequences. Entering information once and then propagating it wherever it is required is convenient compared to performing refinements by hand where there is potential for transcription errors.

There are, of course, shortcomings with the prototype in the scope of problems that can be tackled and with performance. The Synthesizer Generator provides a realistic prototype very quickly and effectively but we encountered some limitations in the flexibility of the user interface.

6. Future tools

In future research into tools for the refinement calculus we plan to investigate the following issues:

Notation: If the use of the predicate calculus is to be successful, we need to abbreviate our pre- and post-conditions. The normal way to do this is by defining names (possibly parameterized) for subcomponents of the predicates. Any more substantial tool needs this capability but must also be able to manipulate predicates that contain such abbreviations without requiring complete expansion. The work of Griffin [5] is of interest here.

Interactive proof techniques: A realistic refinement editor will rely heavily on some form of interactive theorem prover for first-order predicates. The ability to manipulate predicates is crucial to the viability of a refinement tool. Also necessary are methods to establish and use theories about

- program objects — both abstract data types used in specifications such as sets and sequences and the concrete data types of our programming language.
- problem domain knowledge

User Interface: The user interface for a refinement tool is extremely important. How to present the developing refinement sequence and provide facilities for navigating though and modifying that sequence is a significant challenge. Some of the problems are similar to those addressed by hypertext systems and we are interested to see if we can use results from that research. The user of a refinement tool needs to focus on different concerns at different stages of the refinement task. Once a refinement step has been established by verifying the applicability of the rule, the proof is normally of little concern and need not be displayed. By comparison, during the proof construction, we are unlikely to be interested in the refinement path leading to this proof.

Data refinement: Extending our work to include data refinement will require a slightly different approach. Instead of focusing on the refinement of a single specification, data refinement is distributed over a larger scope and requires the transformation of each component of the scope. The transformations are driven by an abstraction invariant that relates the concrete variables to the abstract ones.

Storage and retrieval: We see a need to go beyond a monolithic design document for each program. If the potential for reusing the design history is to be achieved, more flexible methods of storage and (particularly) retrieval are required. The possible application of object-oriented data base technology will be investigated.

Calculus rules: The refinement calculus rules are not a inviolate set. There is a common core of rules which will be augmented and extended by new rules as experience with the calculus grows. Future tools should make it easy to add new rules and package derived rules (combinations of more basic rules). This provides potential difficulties as a refinement sequence is dependent on the rules used in its development. It may be necessary to store the rules used with each refinement sequence.

7. Conclusions

The development process is at least as important as the product, the executable software, since it serves to document the design and provides a method for checking the correctness of the program (with respect to the initial specification). The refinement calculus is a formal method for refining specifications.

Applying the calculus effectively requires computer-based support to manage the many steps from the initial abstract specification to the final executable program. In the paper we have summarized our experiences with tool support for the refinement calculus and made suggestions about goals for future tools.

8. Acknowledgements

We gratefully acknowledge the influence and inspiration of the research on specification at the Programming Research Group, Oxford University. The refinement calculus was developed during Ken Robinson's study leave at PRG in 1985/86. Special thanks are owed to Carroll Morgan for his collaboration on the development of the refinement calculus.

References

1. D. Bjorner and Jones C.B. *Formal Specification and Software Development.* Prentice-Hall, 1982.
2. D.A. Carrington and K.A. Robinson. A prototype program refinement editor. In *Australian Software Engineering Conference*, pages 45–63. ACS, 1988.
3. Edsgar W. Dijkstra. *A Discipline of Programming.* Prentice-Hall, 1976.
4. R.W. Floyd. Toward interactive design of correct programs. In *IFIP*, pages 7–10, 1971.
5. T.G. Griffin. *Notational Definition and Top-Down Refinement for Interactive Proof Development Systems.* PhD thesis, Dept. of Computer Science, Cornell University, 1988.
6. I.J. Hayes, editor. *Specification Case Studies.* Prentice-Hall, 1987.
7. C.A.R. Hoare. An axiomatic basis for computer programming. *Comm. ACM*, 12(10):576–580, October 1969.
8. C.B. Jones. *Systematic Software Development using VDM.* Prentice-Hall, 1986.
9. D.E. Knuth. Literate programming. *The Computer Journal*, 27(2):97–111, 1984.
10. C.C. Morgan. *Programming from Specifications.* Prentice-Hall, 1990.
11. C.C. Morgan. The specification statement. *ACM Transactions on Programming Languages and Systems*, 10(3):403–419, July 1988.
12. C.C. Morgan and K.A. Robinson. Specification statements and refinement. *IBM Journal of Research and Development*, 31(5):546–555, September 1987.
13. C.C. Morgan, K.A. Robinson, and P. Gardiner. On the refinement calculus. Technical Report PRG-70, Programming Research Group, 8-11 Keble Road, Oxford OX1 3QD, UK, 1988.
14. T.W. Reps and T. Teitelbaum. *The Synthesizer Generator.* Springer-Verlag, 1989.
15. T.W. Reps and T. Teitelbaum. *The Synthesizer Generator Reference Manual.* Springer-Verlag, third edition, 1989.
16. C. Rich and H.E. Shrobe. Initial report on a LISP programmer's apprentice. *IEEE Transaction on Software Engineering*, 4(6):456–467, 1978.
17. J.M. Spivey. *The Z Notation: A Reference Manual.* Prentice-Hall, 1989.
18. R.C. Waters. The programmer's apprentice: Knowledge based program editing. *IEEE Transaction on Software Engineering*, 8(1):1–11, 1982.

Key Centre for Software Technology, Department of Computer Science, University of Queensland, St Lucia 4067, Australia

E-mail: davec@uqcspe.cs.uq.oz.au

Department of Computer Science, University of New South Wales, Kensington 2033, Australia

E-mail: kenr@spectrum.cs.unsw.oz.au

DIMACS Series in Discrete Mathematics
and Theoretical Computer Science
Volume 3, 1991

A Unified Approach to the Deadlock Detection Problem in Networks of Communicating Finite State Machines

WUXU PENG
Dept of Computer Science
Southwest Texas State University
San Marcos, TX 78666

S. PURUSHOTHAMAN
Department of Computer Science
The Pennsylvania State University
University Park, PA 16802

Abstract

We consider the deadlock detection problem (DDP) of networks of communicating finite state machines (NCFSMs). The DDP problem is known to be undecidable for NCFSMs. In this paper, we provide a characterization of those subclasses of networks for which the deadlock problem is decidable. We also provide a proof technique based on our characterization and illustrate our technique on an example.

1 Introduction

Communicating finite state machines are a very useful abstract model for specifying, verifying and synthesizing communication protocols [2, 3]. In this model a system of communicating finite state machines can communicate typed messages asynchronously with each other over uni-directional, unbounded FIFO channels.

A central issue in this model is whether a network of communicating finite state machines (NCFSMs) is free of progress errors. Several widely addressed progress properties are: freedom from deadlocks, freedom from unspecified receptions, and freedom from unbounded communication. The problem of checking for non-progress in NCFSMs is known to be undecidable [1, 2, 3]. Because of this negative conclusion a natural question is: "For what classes of NCFSMs is the progress problem decidable?" A large amount of literatures (e.g. [8, 6, 7]) has been devoted to identifying classes of NCFSMs for which some of the progress problems are decidable. Specific classes of NCFSMs are usually obtained by placing restrictions on the structure of the systems. For instance, the number of machines in the system, the number of message types allowed, and channel capacity (maximum number of pending messages allowed in channels) etc. However the underlying question

For what classes of NCFSMs are the progress problems decidable?

has not been answered before.

In this paper, we investigate the deadlock detection problem from the language-theoretic point of view. Techniques developed here are applicable to other progress problems. Specifically, we give a necessary and sufficient condition on execution sequences of a class of network such that the deadlock detection problem is decidable for that class of NCFSMs. We believe that our work provides insight into the underlying nature of decidability of deadlock detection problem. We provide an example to illustrate our technique.

This paper is organized as follows: In Section 2 we introduce necessary notations and definitions, we present our main result in Section 3, in Section 4 we give an example to illustrate the main result and conclude in Section 5.

2 Preliminaries

A communicating finite state machine (CFSM) is a labeled directed graph with a distinguished initial state, where each edge is labeled by an event. The events of a CFSM **are send** and **receive** commands over a finite set of message types M. The communication between CFSMs is assumed to be asynchronous (i.e., non-blocking sends and blocking receives). Consequently, we assume the availability of an infinite buffer between each pair of machines to store pending messages.

Let $I = \{1, \cdots, n\}$, where $n \geq 2$ is some constant. Formally, we have

Definition 2.1 *(CFSM) A CFSM P_i is a four-tuple $(S_i, \langle M_{i,j}\rangle_{j\in I} \cup \langle M_{j,i}\rangle_{j\in I}, \delta_i, p_{0i})$, where*

(1) *S_i is the set of local states,*

(2) *$M_{i,j}$ is the set of message types that P_i can send to machine P_j, and $M_{j,i}$ is the set of message types that P_i can receive from machine P_j. It is assumed that $M_{i,i} = \emptyset$, since P_i can not directly send messages to or receive messages from itself.*

(3) *Let $-M_{i,j} = \{-m | m \in M_{i,j}\}$ and $+M_{j,i} = \{+m | m \in M_{j,i}\}$.*
δ_i is a partial mapping, δ_i: $S_i \times (\langle -M_{i,j}\rangle_{j\in I} \cup \langle +M_{j,i}\rangle_{j\in I}) \times I \longrightarrow 2^{S_i}$. $\delta_i(p, -m, j)$ is the set of new states that machine P_i can possibly enter after sending message of type m to machine P_j, and $\delta_i(p, +m, j)$ is the set of new states that machine P_i can possibly enter after receiving message of type m from machine P_j.

(4) *p_{0i} is the initial local state.*

A state p in P_i is said to be a send (receive, resp.) state iff all of its outgoing edges are send (receive, resp.) edges. p is said to be a mixed state iff it has both outgoing send and receive edges. Let $RMsg(p)$ be the set of message types that can be received in state p, i.e. $RMsg(p) = \{m | \exists p' \ \exists j \ p' \in \delta_i(p, +m, j)\}$. Define

$$
\begin{aligned}
&M_i = \cup_{j\in I}(M_{i,j} \cup M_{j,i}), \\
&M = \cup_{i\in I} M_i, \text{ and} \\
&-M_i = \cup_{j\in I} - M_{i,j}, \qquad +M_i = \cup_{j\in I} + M_{j,i}, \\
&\pm M_i = -M_i \cup +M_i, \\
&-M = \cup_{i\in I} - M_i, \qquad +M = \cup_{i\in I} + M_i, \\
&\pm M = +M \cup -M.
\end{aligned}
$$

Without loss of generality, we assume that $M_{i,j} \cap M_{k,l} = \emptyset$ if $(i,j) \neq (k,l)$. Due to this assumption, for any $a \in \pm M_i$ we can simplify the notation $\delta_i(p,a,j)$ to $\delta_i(p,a)$.

Definition 2.2 (*Network of communicating finite state machines*) *A network of communicating finite state machines (NCFSM) is a tuple* $N = \langle P_1, \cdots, P_n \rangle$, *where each* P_i $(i \in I)$ *is a CFSM.*

A global state of N *is a tuple* $[\langle p_i \rangle_{i \in I}, \langle c_{i,j} \rangle_{i,j \in I}]$, *where* p_i *is a local state of machine* P_i, $c_{i,j}$ *is the sequence of messages in the channel from machine* P_i *to* P_j. *Let* V *be the cartesian-product of the sets* $S_1, \cdots, S_n$, *i.e.* $V = S_1 \times \cdots \times S_n$, *and let* C *be the cartesian-product of the sets* $M^*_{1,2}, \cdots, M^*_{1,n}, M^*_{2,1}, \cdots, M^*_{n,n-1}$.

Initially, N *is in its initial state* $[\langle p_{0i} \rangle_{i \in I}, \langle c_{i,j} \rangle_{i,j \in I}]$, *where* $c_{i,j} = \varepsilon$ $(i \neq j)$. *Let* $[\langle p_i \rangle_{i \in I}, \langle c_{i,j} \rangle_{i,j \in I}]$ *be a global state. The global state transition function* $\delta_N : (V \times C) \times \pm M \longrightarrow 2^{V \times C}$ *is a partial function defined as:*

(1) if $\exists i,j \in I$ $(i \neq j)$ *such that* $p'_i \in \delta_i(p_i, -m, j)$ *then*
$[\langle p'_i \rangle'_{i \in I}, \langle c'_{i,j} \rangle'_{i \in I}] \in \delta_N([\langle p_i \rangle_{i \in I}, \langle c_{i,j} \rangle_{i,j \in I}], -m)$, *where*
$p_k = p'_k$ $(k \neq i)$, $c_{k,l} = c'_{k,l}$ $(k \neq i$ *or* $l \neq j)$, *and* $c'_{i,j} = c_{i,j}.m$.

(2) if $\exists i,j \in I$ $(i \neq j)$ *such that* $p'_i \in \delta_i(p_i, +m, j)$ *then*
$[\langle p'_i \rangle'_{i \in I}, \langle c'_{i,j} \rangle'_{i \in I}] \in \delta_N([\langle p_i \rangle_{i \in I}, \langle c_{i,j} \rangle_{i,j \in I}], +m)$, *where*
$c_{k,l} = c'_{k,l}$ $(k \neq j$ *or* $l \neq i)$, *and* $m.c'_{j,i} = c_{j,i}$.

We use $P_i \rightarrow P_j$ to denote the channel from P_i to P_j. In essence, the first case in Definition 2.2 denotes the event that P_i sends a message m to P_j, which causes the message m to be appended to the end of channel $P_i \rightarrow P_j$. The second case represents the event that P_i receives a message of type m sent by P_j, which has the effect of removing the first message (which must be of type m, or error (unspecified reception) would occur) in the channel $P_j \rightarrow P_i$. In both cases, after the successful completion of the event, P_i enters local state p'_i while all other machines remains in the same local states and the contents of all other channels are unchanged.

To simplify the expressions, we will use the notation $[v, c]$ to denote a global state whenever necessary, where by convention $v = \langle p_i \rangle_{i \in I} \in V$, and $c = \langle c_{i,j} \rangle_{i,j \in I} \in C$. $[v_0, c_0]$ will be used to denote the initial state.

Definition 2.3 (*Reachability function*) *Let* $N = \langle P_1, \cdots, P_N \rangle$ *be an NCFSM. The global state transition function* δ_N *can be easily extended to the following reachability function* $\delta^*_N : (V \times C) \times \pm M^* \longrightarrow 2^{V \times C}$,

(1) $\delta^*_N([v,c], \varepsilon) = \{[v,c]\}$.

(2) $\delta^*_N([v,c], e.a) = \{[v',c'] \mid \exists [v'',c''] \in \delta^*_N([v,c],e),\ [v',c'] \in \delta_N([v'',c''],a)\}$.

We often write $\delta^*_N([v_0,c_0],e)$ *as* $\delta^*_N(e)$. *Furthermore, we will drop the subscripts in* δ_N *and* δ^*_N *if no confusion arises.*

Given a reachability function δ^*, we define the set of all states reached as a result of some execution to be the *reachability set*. More formally,

Definition 2.4 (*Reachability sets*) *Let $N = \langle P_1, \cdots, P_n \rangle$ be an NCFSM. The reachability set $RS(N)$ is the set of all reachable global states, $RS(N) = \{[v,c] \mid [v,c] \in \delta^*(e), e \in \pm M^*\}$.*

In the rest of this paper sequences of events will form the back bone of our discussions. Hence, we will use the following abbreviations:

Event Sequence A string $e \in \pm M^*$ is an *event sequence.*

Executable An event sequence e is *executable*, notated as $\delta^*(e) \neq \emptyset$, if $\delta^*(e)$ is defined.

Feasible An event sequence $e \in \pm\Sigma^*$ is *feasible*, if [1]

1. $\forall e' \in \mathit{pref}(e)\ \forall g \in \Sigma\ |e'|_{+g} \leq |e'|_{-g}$; and
2. $\forall i,j \in I$, if $+g_{i,j}$ is the kth receive event in $f_{i,j}(e)$, then $-g_{i,j}$ is the kth send event in $f_{i,j}(e)$.

Feasibility is a very strict requirement. We can easily show that feasible sequences are context-sensitive. We will use $F(N)$ to denote the set of all feasible event sequences of a network N.

Stable An event sequence e is *stable*, if it is feasible and in addition it contains the same number of send and receive events of any message type, i.e., $\forall a \in \Sigma \mid e_{+a} \mid = \mid e_{-a} \mid$. We will use $SE(N)$ to denote the set of all stable event sequences of a network N.

It should be clear to the reader that a definition of deadlock (or any progress error) can be stated in terms of execution sequences. Of course, execution sequences capture the semantics (and causality) of processes and buffers present in a network. The causality constraints among possible actions in a network can be split into those imposed by the behavior of FIFO buffers and those imposed by the sequencing constraints of the processes. The notion of feasible and stable (event) sequences defined above capture the constraints of the FIFO buffers in the network. The sequencing constraints of the network can be easily captured as a shuffle of the sequencing constraints of the individual processes. Formally, we define:

Definition 2.5 (*Shuffle-product of NCFSMs*) *Let $N = \langle P_1, \cdots, P_n \rangle$ be an NCFSM. The shuffle-product of N, notated as SP(N), is a four-tuple (V, M, Δ, v_0), where*

(1) $V = S_1 \times S_2 \times \cdots \times S_n$.

(2) $v_0 = [p_{01}, p_{02}, \cdots, p_{0n}] \in V$.

(3) The transition function Δ: $V \times \pm M \longrightarrow 2^V$ *is defined as*

[1] For a string w, $|w|$ is the length of w, $|w|_a$ is the number of occurrence of letter a in w and $pre(w)$ is the set of all prefixes of w.

$$v' \in \Delta(v,a),\ \textit{where}\ a \in \pm M_i \subseteq \pm M,\ \textit{iff}\ v'_j = v_j\ (j \in I\ \&\ j \neq i)\ \textit{and}\ v'_i \in \delta_i(v_i, a).$$

The shuffle-product $SP(N)$ can be viewed as a (nondeterministic, in general) finite automaton by identifying some subset of V as final state set. We use $SP(N)(F)$ to notate the finite automaton obtained from the shuffle-product with $F \subseteq V$ as the final state set.

A tuple $v \in V$ is a *receive* node if for each $i \in I$ v_i is a receive state in P_i. $REV(N)$ denotes the set of all receive nodes in V.

A number of progress properties have received wide attention and one of the well known progress properties is the deadlock detection problem (DDP).

Definition 2.6 (*Deadlock*) *Let* $N = \langle P_1, \cdots, P_n \rangle$ *be an NCFSM and* $[v,c] \in RS(N)$ *be a global state.*

$[v,c]$ *is a* deadlock state *if the predicate*

$$d([v,c]):\ v \in REV(N)\ \&\ c = c_0$$

holds.

The network N *is* free of deadlocks, *if the predicate*

$$\forall [v,c] \in RS(N)\ (\textbf{not}\ d([v,c]))$$

holds.

It is well known that in general it is undecidable whether an NCFSM is free of deadlocks ([2, 3]). We state this fact in following theorem.

Theorem 2.1 *DDP is undecidable.*

3 A Unified Approach to DDP

Theorem 2.1 states that the problem of detecting deadlocks in a network is undecidable in general. To cope with this negative result, many special classes of NCFSMs have been identified for which the DDP is decidable. This usually involves finding sufficient conditions under which the problem becomes decidable.

In this section we take a different approach to this problem. Instead of trying to find special classes of NCFSMs with decidable DDP, we give a necessary and sufficient condition under which the DDP of a given class of NCFSM is decidable.

First let us formalize the concept of *classes* of NCFSMs.

A *class* $\mathcal{N}$ of NCFSMs is a tuple (Q, Σ, T), where

1. Σ is a finite (or countably infinite) set of message types,
2. Q is a collection of NCFSMs each of which draws message types from Σ,

3. T is a predicate which characterizes the properties of NCFSMs in the Q. For instance, T may be the predicate: "each $N \in Q$ has only two CFSMs", which is the class of NCFSMs with two CFSMs.

Let $\mathcal{N} = (Q, \Sigma, T)$ be a class of NCFSMs, and let $N = \langle P_1, \cdots, P_n \rangle \in \mathcal{N}$. We say that the DDP is *decidable* for a network class $\mathcal{N}$, if the predicate

$$D(\mathcal{N}) : \forall N \in \mathcal{N}\ \exists [v, c] \in RS(N)\ (v \in RV(N) \& c = c_0).$$

is decidable. In the following we would like to relate the decidability of deadlock to conditions on execution sequences. It is easy to see that if an execution sequence e leads to a deadlock state, then it should have the property that for every send event in e there should be a corresponding receive event. In fact, e should be a stable event sequence.

In general the set $SE(N)$ is context-sensitive, since we can easily construct a linear bounded automaton (LBA) that accepts $SE(N)$. Even for some trivial classes of NCFSMs $SE(N)$ remains context-sensitive. However in order to check if DDP is decidable for N, it is not necessary to know every member in $SE(N)$. This observation is based on the fact that there a number of event sequences that are really interleavings of the same set of actions of the processes, and hence lead to the same global state. We should therefore consider these interleavings as being equivalent. Formally we have, two event sequences e_1 and e_2 are *equivalent*, notated as $e_1 \simeq e_2$, iff

1. e_1 and e_2 are permutations of each other,
2. $\delta^*(e_1) = \delta^*(e_2)$.

Since we are really interested in stable event sequences, we will say two event sequences e_1 and e_2 are *stable equivalent*, notated as $e_1 \simeq_{st} e_2$, iff $e_1 \simeq e_2$ and both e_1 and e_2 are stable.

Define $class(e) = \{e' \mid e' \simeq_{st} e\}$. It is easy to see that $\simeq_{st}$ is an equivalence relation on $SE(N)$ and $class(e)$ is an equivalence class.

As mentioned earlier, we only need a *representative* from each equivalence class of $SE(N)$ to check for existence of deadlock in N. To that end, we define a language $C_N \subseteq SE(N)$ to be a *stable cover set* for N if

$$\forall e \in SE(N)(equiv(e) \cap C \neq \emptyset). \tag{1}$$

With the concepts of stable cover sets and shuffle-product automata, we have the following theorem regarding the decidability of DDP for classes of NCFSMs.

Theorem 3.1 *Let $\mathcal{N} = (Q, \Sigma, T)$ be a class of NCFSMs. Let $\mathcal{R} = \{L(M_v) \mid M_v = SPA(N, \{v\}) : v \in V_N\}$, i.e. $\mathcal{R}$ is the collection of shuffle-product automata, each of which has some node $v \in V_N$ as the single final state. The DDP is decidable for $\mathcal{N}$ if and only if for every network $N \in \mathcal{N}$, there exists a stable cover set $C_N \subseteq SE(N)$ such that for every $M_v \in \mathcal{R}$, the predicate $C_N \cap L(M_v) = \emptyset$ is decidable.*

Proof: Let $\mathcal{N} = (Q, \Sigma, T)$ be a class of NCFSMs.

If. Let N be a member in $\mathcal{N}$, and assume that we have a stable cover set $C_N \subseteq SE(N)$ which satisfies the condition listed in the theorem.

Claim: $[v, c_0]$ is a deadlock state if and only if $C_N \cap L(M_v) \neq \emptyset$.

Proof of the claim:

Assume that $C_N \cap L(M_v) \neq \emptyset$. Since C_N is a stable cover, each $e \in C_N$ is stable. Moreover C_N contains at least one element from each equivalence class $equiv(e)$. As $L(M_v)$ contains all executable event sequences which can lead to a reachable global state of the form $[v, c]$, there must be at least one executable event sequence $e \in C_N \cap L(M_v)$. such that $[v, c_0] \in \delta(e)$.

Conversely assume that $[v, c_0]$ is a reachable deadlock state. Therefore there exists an executable sequence e where $[v, c_0] \in \delta(e)$. As C_N is a stable cover, there must exist $e' \simeq_{st} e$ and $e' \in C$. Obviously $e' \in L(M_v)$, hence $C_N \cap L(M_v) \neq \emptyset$.

By definition DDP is decidable for N. Since DDP is decidable for any $N \in \mathcal{N}$ we conclude that DDP is decidable for the class $\mathcal{N}$.

Only If. Assume that DDP is decidable for $\mathcal{N}$. We must show that for every network $N \in \mathcal{N}$ there exists a stable cover set C_N for N such that the emptiness problem of $C_N \cap L(M_v)$ is decidable for all $v \in V_N$. In the following argument, the symbol $\Longrightarrow$ is used to express logical implication. "$A \Longrightarrow B$" means that the decidability of A implies the decidability of B. We have following logical reasoning:

$$
\begin{aligned}
&D(\mathcal{N})\\
&\Longrightarrow \forall\, N \in \mathcal{N}\ (\exists\, [v,c] \in RS(N)\ (v \in RV(N)\ \&\ c = c_0))\\
&\Longrightarrow \forall\, N \in \mathcal{N}\ (\exists\, e \in NL(N)\ \exists\, [v,c] \in \delta(e)\ (v \in RV(N)\ \&\ c = c_0))\\
&\Longrightarrow \forall\, N \in \mathcal{N}\ (\exists\, e \in SE(N)\ \exists\, [v,c_0] \in \delta(e))\\
&\Longrightarrow \forall\, N \in \mathcal{N}\ (\exists\, e \in SE(N)\ \exists\, v \in RV(N)\ (e \in L(M_v))\\
&\Longrightarrow \forall\, N \in \mathcal{N}\ (\exists\, v \in RV(N)\ (e \in SE(N) \cap L(M_v)))\\
&\Longrightarrow \forall\, N \in \mathcal{N}\ (\exists\, v \in RV(N)\ (SE(N) \cap L(M_v) \neq \emptyset)).
\end{aligned}
$$

However, $SE(N)$ itself is a stable cover set.

□

It is obvious that the problem of finding a cover set with the stated properties is undecidable. However Theorem 3.1 presents a unified view of DDP for NCFSMs. It sheds new light on the decidability of DDP for NCFSMs in following sense: Given a specific class $\mathcal{N}$ of NCFSMs, to test if DDP is decidable for $\mathcal{N}$ we try to find a cover set satisfying the theorem. If we can find such a cover, we can conclude that DDP is decidable for $\mathcal{N}$. We shall illustrate this idea in next section.

In the methodology engendered by this characterization, we expect to be able to define the cover set C_N for a network N independent of the transitions (or semantics) of a particular network N. Furthermore, we expect to be able to check that such a cover set has the necessary properties. Based on the fact that the language of a shuffle-product automaton is regular, we have the following:

Corollary 3.1 *For a class of networks $\mathcal{N}$, let $\mathcal{L}$ be a family of languages such that the cover set C_N for every network $N \in \mathcal{N}$ belongs to $\mathcal{L}$. If*

1. *$\mathcal{L}$ is closed under intersection with regular languages, and*
2. *The emptiness problem is decidable for $\mathcal{L}$*

then the deadlock detection problem is decidable for the class of networks $\mathcal{N}$.

The Corollary given above provides a very tight sufficient condition to show that a class of networks has decidable DDP.

4 Applications

An NCFSM $N = \langle P_1, \cdots, P_n \rangle$ is *cyclic* if the topology graph of the network N is a simple cycle. More formally, N is a cyclic network if there exists a permutation $\{i_1, i_2, \cdots, i_n\}$ of the set I such that P_{i_j} can only send message to $P_{i_{j+1}}$ and receiving message from $P_{i_{j-1}}$ ($1 \leq j \leq n$, module $n+1$).

All two-machine networks are cyclic. As it is known that DDP is even undecidable for the class of two-machine networks, so DDP is undecidable for general cyclic networks.

However DDP is decidable for cyclic networks where only one channel is unbounded. Let $\mathcal{N}^1_{1-u-cyc}$ be the class of cyclic networks of which only one channel is unbounded (referred to as 1-U cyclic NCFSMs, for short).

Let $N = \langle P_1, \cdots, P_n \rangle$ be a 1-U cyclic NCFSM. To simplify the discussion we assume without loss of generality that P_i can only send to P_{i+1} and receive from P_{i-1} ($1 \leq i \leq n$, module $n+1$) and only the channel $P_1 \rightarrow P_2$ is unbounded. Let $L = (A^*B)^*H$, where

$A = (\cup_{2 \leq i \leq n-1} - \Sigma_{i,i+1}) \cup (-\Sigma_{n,1}) \cup (\cup_{2 \leq i \leq n-1} + \Sigma_{i,i+1}) \cup (+\Sigma_{n,1})$,

$B = (-\Sigma_{1,2})(+\Sigma_{1,2})$,

$H = (A \cup (-\Sigma_{1,2}))^*$.

The following lemma is a generalization of a lemma from [4].

Lemma 4.1 *Let $N = \langle P_1, \cdots, P_n \rangle$ be a 1-U cyclic NCFSM. For each executable event sequence e there exists another event sequence $e' \simeq e$ and $e' \in L$.*

Proof: Let $N = \langle P_1, \cdots, P_n \rangle$ be a 1-U cyclic NCFSM. The proof is by induction on the number k of receive events from $+\Sigma_{1,2}$.

Basis: $k = 0$. Since e does not contain events from $+\Sigma_{1,2}$, $e \in H$. Clearly the conclusion is true.

Induction: Assume that the conclusion holds for some $k > 0$. Let e be an executable event sequence which contains $k+1$ receive events from $+\Sigma_{1,2}$. Let $s_{1,2}$ be the first send event from $-\Sigma_{1,2}$ and $r_{1,2}$ be the first receive event from $+\Sigma_{1,2}$ in e. We can rewrite e as $e = w_1.s_{1,2}.w_2.r_{1,2}.w_3$, where w_1 and w_2 do not contain any events from $(+\Sigma_{1,2})$.

As the network is cyclic (P_2 can only send to P_3 and receive from P_1), we can move all the send events from $(-\Sigma_{2,3})$ in w_2 before $s_{1,2}$. Let $w = w_1'.s_{1,2}.w_2'.r_{1,2}.w_3$ be the event sequence after such a reordering of e, where w_2' does not contain any events from $(-\Sigma_{2,3}) \cup (+\Sigma_{1,2})$. It is easy to see that $w \simeq e$. Since w_2' does not contain events initiated by P_2, we can move the event $r_{1,2}$ as the immediate successor of $s_{1,2}$. Let $w' = w_1'.s_{1,2}.r_{1,2}.w_2'.w_3$ be the event sequence after such reordering of w. Still $w' \simeq e$ holds.

Since w_3 contains k receive events from $+\Sigma_{1,2}$, by induction hypothesis there exists another event sequence $w_3' \simeq w_3$ and $w_3' \in L$. Hence $e \simeq e' \in L$, where $e' = w_1'.w_2.s_{1,2}.r_{1,2}.w_3'$.

□

Lemma 4.1 says that for any executable sequence e there is another sequence $e' \simeq e$ such that when e' is executed, there is at most one pending message in the only unbounded channel $P_1 \to P_2$ as long as P_2 can still receive. Therefore to check for deadlocks, we need only concentrate on the behavior of other bounded channels.

Theorem 4.1 *There exists a regular stable cover set for each network $N \in \mathcal{N}_{1-u-cyc}$.*

Proof: Let $N = \langle P_1, \cdots, P_n \rangle$ be a 1-U cyclic NCFSM.

To find a stable cover set for N, we need to construct a language which contains at least one element from each equivalent class $SE(e)$ where e is a stable event sequence.

Let $L' = (A^*B)^*A^*$, where A, B are defined as above. Notice that each $e \in L'$ contains the same number of events from $-\Sigma_{1,2}$ and $+\Sigma_{1,2}$. We can conclude from Lemma 4.1 that for each stable event sequence e, there exists $e' \simeq e$ and $e' \in L'$.

We can construct a finite state automaton F to accept the set of all stable events in L' as follows. Let each state in F record the number of pending event types, message types and message positions it has seen so far. The only exception is that when F sees an event $-g \in -\Sigma_{1,2}$, it will expect another event $+g \in +\Sigma_{1,2}$ in its next step. Notice that since all the channels except $P_1 \to P_2$ are bounded and the number of message types each machine can send is finite, the number of states in F is finite. The final state set in F includes only those states in which F has seen a stable event sequence (only the start state need be in the final state set). Therefore $L(F)$, the language accepted by F is a cover set for N.

□

By Corollary 3.1, the DDP is decidable for $\mathcal{N}_{1-u-cyc}$.

5 Conclusions

Deadlock detection problem for networks of communicating finite state machines has been known to be undecidable. The undecidability stems from the fact that even a two-machine NCFSM has the same computing power as a Turing machine. With special restrictions, specific classes of NCFSMs have been found for which the deadlock detection problem is decidable. However the underlying question "For what classes of NCFSMs is the DDP decidable," has not been answered before.

In this paper we considered the deadlock detection problem in NCFSMs from the formal language point of view. We have given a necessary and sufficient condition to show decidability of detecting deadlocks in classes of NCFSMs. We believe our work reveals the nature of decidability vs. undecidability of DDP.

The language concepts were first (as far as the authors know) introduced in the analysis of NCFSMs by K. Okumura [5]. The work in [5] concentrated mainly on establishing correspondence between the languages and the networks, in analogy with the traditional formal languages theory. However, the executable event sequences are context-sensitive even for many trivial networks. It appears that a direct use of the executable event sequences would not aid the analysis. This was our main motivation for introducing the concept of cover set for a particular property (DDP in this paper). Although we only discussed the decidability of DDP in this paper, the idea can also be applied to the decidability of detecting other properties such as unspecified receptions and unboundedness.

References

[1] G. Bochmann. Finite State Description of Communication Protocols. *Computer Networks*, Vol.2, 1978, pp.361–371.

[2] D. Brand and P. Zafiropulo. On Communicating Finite-state Machines. *JACM*, 30(2), 1983, pp.323–342.

[3] M. Gouda, E. Manning, and Y. T. Yu. On the Progress of Communication between Two Finite State Machines. *Information and Control*, 63(3), 1984, pp.308–320.

[4] M. Gouda, E. M. Gurari, Ten-Hwang Lai, and L. E. Rosier. On Deadlock Detection in Systems of Communicating Finite State Machines. *Computers and Artificial Intelligence*, Vol.6, 1987, No.3, pp.209–228.

[5] K. Okumura. Protocol Analysis from Language Structure. In *Protocol Specification, testing, and verification VIII*, S.Aggarwal and K.Sabnani (Editors), North-Holland, 1988, pp.113–124.

[6] Jan Pachl. Protocol Description and Analysis Based on a State Transition Model with Channel Expressions, In *Protocol Specification, testing, and verification VII*, H. Rubin and C. H. West (Editors), North-Holland, 1987, pp.207–219.

[7] W. Peng and S. Purushothaman. Analysis of Communicating Processes for Non-Progress. In *Proc of the 9th IEEE International Conference on Distributed Computing Systems*, June 1989, pp.280–287.

[8] Y. T. Yu and M. Gouda. Deadlock Detection for a Class of Communicating Finite State Machines. *IEEE Transactions on Communications*, Dec 1982, pp.2514–2518.

DIMACS Series in Discrete Mathematics
and Theoretical Computer Science
Volume 3, 1991

A Computer-Aided Verification Tool for Finite State Controller Systems *

Mark Bickford
Mandayam Srivas
Odyssey Research Associates, Inc.,
301A Harris B. Dates Drive
Ithaca, NY 14850.

Abstract

This paper describes a computer-aided verification tool for specifying and verifying a class of *finite state controller systems.* The tool customizes a general purpose functional language based verification system for this domain of application. The tool reduces the effort of verifying systems in this class by helping the designer to generate the specification in a structured fashion and reason about the system at a higher level. The tool has been used to verify a large realistic microprocessor design.

1 Introduction

Computer-Aided theorem proving has been and continues to be an active area of research within computer science. Several theorem proving systems have been built in the past couple of decades. Some of these have been used successfully for verifying the correctness of large realistic computer system designs. They range from inference-based proof generating

*This work was funded by the Air Force Systems Command at Rome Air Development Center under Contract No. F30602-86-C-0115.

systems, such as HOL [4] to interactive theorem provers, such as the Boyer-Moore prover [3] and Clio [1].

Most of the effort expended in using a general-purpose theorem prover for a verification task goes towards specifying (modeling) the system, formulating the correctness conditions and guiding the prover to prove the conditions. A substantial part of this effort can be automated or at least interactively computer-aided for a specific domain of applications by developing a tool that "customizes" the theorem prover to the domain. Such a computer-aided verification tool is intended to run in conjunction with the theorem prover.

In this paper we describe a computer-aided verification tool for the specification and verification of *finite state controller* systems. The tool is designed to be used in conjunction with Clio [1], which is a functional language based verification system. However, the ideas behind the design of the tool are useful in specializing any general purpose theorem prover, especially those that are based on functional/equational logic, to finite state controller systems.

A *finite state controller* system is one in which a single component, called the *controller*, plays a central role in determining the overall behavior of the system. Every other component of the system acts as a "slave" taking commands from the controller to perform stated actions. The controller is a finite state machine which schedules actions, possibly concurrently, on the components of the system during a transition. Most hardware systems designed as synchronous sequential circuits, in which the actions on the components are triggered by some synchronizing event of a clock, can be viewed as finite state controller systems at a certain level of abstraction. Register transfer level designs of microprocessors as well as that of control-dominated application specific IC's fall in this class making the domain a useful one for consideration.

The tool provides support in the following areas. *Specification:* It automatically generates all the text (in the logic/language of the theorem prover) that describes the structural aspects of the system design from information supplied interactively by the designer using a graphic interface. It also automatically generates those (auxiliary) parts of the specification which are common to every system in the class. *Verification:* It turns out that most properties that one wants to prove about finite state controller systems can be proved by *analyzing* the controller state graph. The tool helps the user to formulate suitable ***invariant*** and ***advance*** conditions in terms of the controller state graph. Often, the structure of the proofs of properties of such systems is related to the structure of the controller graph. The tool helps the user to guide the theorem prover based on the controller graph structure.

The tool has been used on several examples including a realistic design of a pipelined RISC processor called ***MiniCayuga***. A slightly different version of ***MiniCayuga*** had been verified [2] earlier without the help of the tool. The original ***MiniCayuga*** exercise inspired the work on the tool described here. To keep the exposition simple we will use a small example to present the details of the tool, and only provide a brief summary of the use of the tool in the verification of ***MiniCayuga***.

The next section describes a finite state controller system by introducing a simple system designed to compute the greatest common divisor (gcd) of two numbers. Section 3 gives an overview of the tool. Most of the details of the tool are described in section 4 in the context of the gcd example. Section 5 summarizes the application of the tool on ***MiniCayuga***. Section 6 presents our concluding remarks.

2 Finite State Controller Systems: An Example

Figure 1 shows a finite state controller system intended for computing the greatest common divisor of two natural numbers in hardware. The circuit accepts four input signals denoted `a`,`b`,`ok`, and `reset`, where `a` and `b` code natural numbers and `ok` and `reset` are single bits (`booleans`). There are two output signals, `done` and `out`, where `out` codes a natural number and `done` is a single bit. The circuit operates in one of two modes: `idle` and `busy`. When `idle`, if the signal `ok` is set to true in a cycle, the values on `a` and `b` are read, `done` becomes false and the circuit becomes `busy`; if `ok` is false the circuit continues to `idle`. When `busy`, the circuit computes the gcd of the signals read in the last `idle` cycle. Eventually, the `done` signal becomes true and the circuit moves back to `idle`, at which point the value of `out` will be the gcd of `a` and `b`. This, provided `a` and `b` are positive. Note that the circuit is not required to compute the gcd when either `a` or `b` is zero. At any time when the circuit is `busy`, if the `reset` signal becomes true then the circuit will move to `idle`.

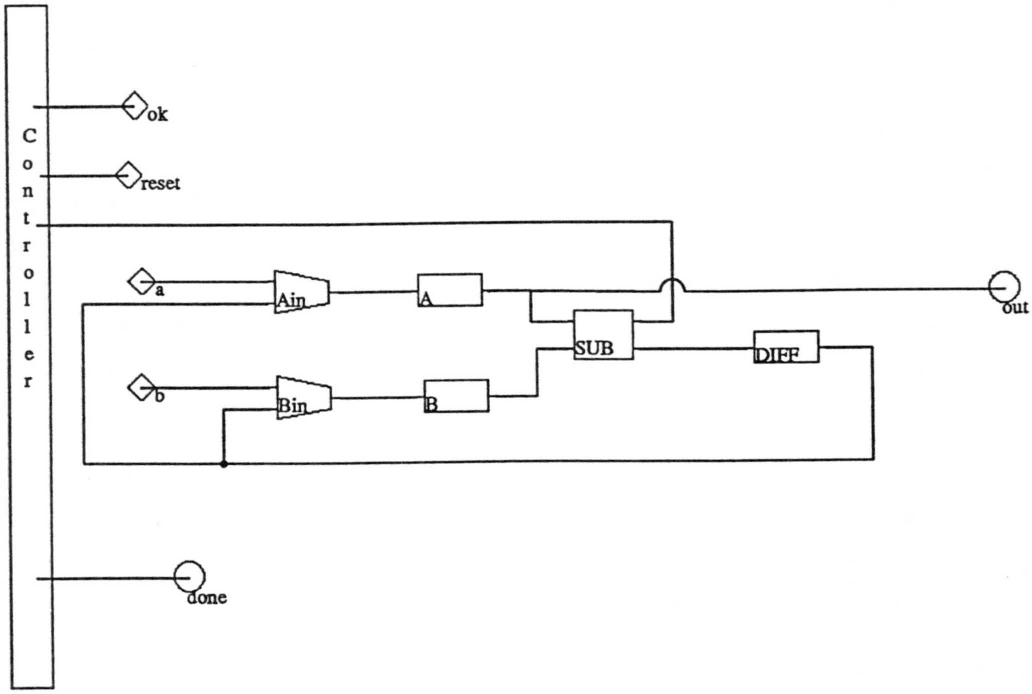

Figure 1: Gcd Data Path

Besides the `Controller`, the design consists of three latches, named `A`, `B` and `DIFF`, two multiplexers, named `Ain`, `Bin`, and a special subtractor component named, `SUB`. `SUB` has two inputs `A` and `B` and two outputs. The first output (`diff`), which goes into `DIFF`, is the absolute value of the difference of `A` and `B`. The second one (`compare`), which goes to the controller, can take one of three values, denoted LESS, EQUAL, or GREATER, corresponding to $\mathtt{A} < \mathtt{B}$, $\mathtt{A} = \mathtt{B}$, and $\mathtt{A} > \mathtt{B}$, respectively. Thus, `SUB` is actually both a subtractor and a comparator.

Figure 1 shows the *data path* architecture, i.e., everything excluding the controller, of the system. A circular icon denotes an output port, and a diamond shaped icon denotes an input port. The controller state machine is shown in figure 2. The diamond shaped icon in the controller denotes a conditional branch. (The circular icon with a cross will be described later.)

The design works as follows. When `ok` is true, the latches `A` and `B` are set from a

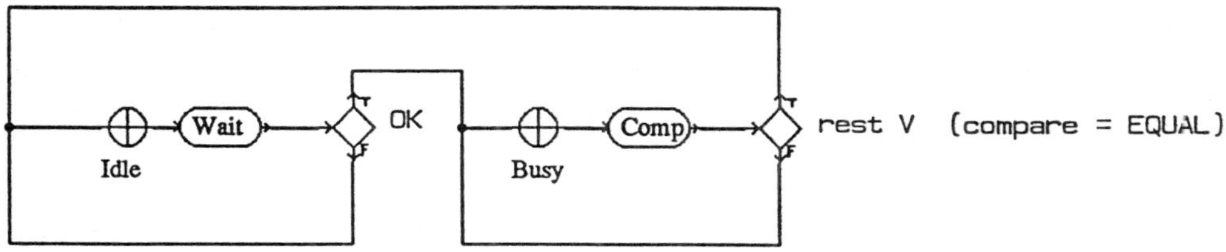

Figure 2: Gcd Controller State Machine

and b and the controller goes into the Comp state. When ok is false and the controller is in the Comp state, it initiates the subtract & compare action on SUB, latches the output diff of SUB into DIFF, and then if A > B (compare is GREATER) the output of DIFF is latched into A, if A < B, DIFF is latched into B and the cycle repeats. If A = B then the controller goes into the Wait state and done is made true.

We assume the set of states shown in the finite state machine includes every possible state the controller of the system may assume. Every transition of the controller (*cycle*) denotes an integral multiple (chosen by the designer) of the unit of discrete time (*phase*). For the gcd circuit we use four phases for every cycle. The controller may trigger actions in each of the phases inside a given state. The delay of an action, which is measured in integral number of phases, is the time required to complete an action.

2.1 Correctness of GCD

A *behavior* of a finite state controller system can be denoted by a trace (possibly *infinite*) of the system state and the outputs over *time* as the controller of the system makes a series of transitions. The state of the system is a *tuple* of the states of all the components including the controller.

In general, one is interested in the set of all traces generated by a system from a class of possible initial states for a class of possible changes of input signals over time. The correctness of a finite state controller system is formulated as a property that a specified set of behaviors of the system must satisfy. Typically, a correctness property takes one of the following forms: a condition that must be true in every state of a trace or a condition that must eventually become true in a state. For example, some of the properties that a gcd circuit behavior is expected to satisfy are stated below informally.

1. In every state the circuit is either `idle` (`done` is `True`) or `busy` (`done` is `False`).
2. If `reset` is `True` in a state then the circuit becomes `idle` in the next cycle unless it is already `idle`.
3. If the circuit is in a "properly loaded" state, then eventually `done` will become `True` at which time the signal at `out` will be equal to the gcd of the contents of the latches `A` and `B`. A circuit is properly loaded if `done` is `False` and `A` and `B` contain positive numbers. One way to ensure the circuit is properly loaded is to `reset` and then input positive signals at `a` and `b`.

A proof of such correctness properties is usually dependent on first identifying (and proving) a suitable set of *invariants* and *advance conditions* associated with one or more states of the controller. An invariant associated with a controller state is a condition which must be true "every time" the controller is in that state. That is, the invariant condition must true when the controller is in the concerned state if the condition was true the previous time the controller was in that state. An advance condition is associated with a pair of controller states `S1` and `S2`. It is a condition which relates the system state (and also possibly the inputs and outputs) when the control is at `S1` to the system state (and, possibly, the outputs) when the control reaches `S2` after making a specified number of transitions.

Let us consider some of the interesting invariant and advance conditions for the gcd circuit. We identify two invariants - one associated with `Wait` (`Idle`) and the other with `Comp` (`Busy`). We identify two advance conditions - one associated with a single step transition from `Idle` to `Busy` and the other with a single step transition from `Busy` to `Busy`.

Idle_Invariant: "done is True"

Busy_Invariant: "A and B are positive and done is False"

Busy_Busy_Advance: "The gcd of A and B equals the gcd of A' and B' and the maximum of A' and B' is less than the maximum of A and B, where A, B denote the old values and A', B' denote the new values of the latches A and B."

Busy_Idle_Advance: "reset at `busy` is True or (out at `idle` = A at `busy` = B at `busy`)."

Note that the above invariant and advance conditions form some of the crucial lemmas needed for proving the correctness properties of gcd. For example, the third correctness property of gcd stated earlier can be proved by induction on the maximum of the contents of `A` and `B`. The above conditions can be used in getting the basis and inductive steps of the proof.

3 Overview of the Tool

The rationale behind the design of the tool and the nature of support it must provide was based on the experience we gained in directly using Clio (without the tool) for the verification of *MiniCayuga*. Our goal was to identify the generic aspects of the specification and verification techniques used in the *MiniCayuga* exercise, and incorporate them into an environment which aided in verifying other similar systems. Another design goal was to make user-interface graphical and interactive. The tool provides a window-based graphical interface, which is implemented using SUN View.

In the specification area, we noticed that about 70 percent of the specification of the design was devoted to formalizing the *structural* and *generic* aspects of the design. The structural aspects include information regarding names of components, their ports and actions, connections between them, and the structural details of the controller state diagram. The generic aspects include the types and functions which help to derive the composite behavior of a design from the structural description and specification of the behavior of the controller and the components. Our goal was to have the tool generate the structural and generic parts automatically. The tool provides a friendly interface for the user to input the structural information of the design graphically. The only parts of the design for which the user must generate textual specification on his own are: (1) the effect of actions on the components, (2) the conditions, if any, upon which the transitions and the schedule of the controller are determined. The tool obtains the textual specification from the user interactively. (In general, the textual specification that the user provides might be dependent on the function symbols that are automatically generated by the tool. If this is the case, the tool displays the necessary symbols for the benefit of the user at appropriate points.)

In the verification area we focused our attention on providing support for proving a restricted class of conditions: *bounded path* invariant and advance conditions. These are similar to the general invariant and advance conditions we discussed in the last section, but are required to be true only for a user-selected set of finite and bounded paths through the control state diagram. They form the building blocks for proving the general invariant and advance conditions using a suitable induction. We chose to restrict our attention to this class because it increases the chances of automating or at least standardizing their proof.

The tool allows the user to mark selected points in the controller state diagram as *cutpoints* (denoted by $\oplus$ in Figure 2), and associate invariant and advance predicates with the cutpoints. The placement of the cutpoints also determines the paths for which the invariant and advance predicates must be true. (The paths determined by the cutpoints will be described precisely later.) The tool automatically generates the verification conditions that must be proved to establish the invariant and advancement properties associated with the cutpoints.

The Spectool does not yet provide any direct support for proving the verification conditions. That is, the user must use the Clio system for this purpose after loading into Clio the text generated by the Spectool. The Spectool, however, increases the likelihood of a *standard proof strategy* succeeding on the verification conditions because of the following reasons. Firstly, the specification generated by the tool is structured in a "standard" fashion. Secondly, the Spectool generates a set of control annotations to the Clio theorem prover and a few generic lemmas which are needed for the standard proof strategy to work.

3.1 A Summary of the Clio Verification System

Clio is a system for verifying properties of programs written in a lazy functional language called Caliban. Caliban is similar to Turner's functional language Miranda[1] [7]. It supports most of the features found in modern functional languages, such as higher-order functions, strong-typing, automatic type inferencing, polymorphism, pattern matching.

The properties to be proved are expressed in the Clio *assertion language* and proved using the Clio theorem prover. In our application the system specifications are expressed in Caliban and the properties to be verified are expressed in the assertion language. An assertion can be an arbitrary formula in first order predicate calculus logic. The atomic assertions include equations of the form `'term1' = 'term2'`, where `term1` and `term2` enclosed inside the back quotes can be arbitrary Caliban expressions. The quantified variables can range over arbitrary Caliban objects including functions. The semantics of Caliban is provided denotationally by a Scott domain [6], which is a complete partial order with a *least* element denoted by the Caliban constant `bottom`. Symbols defined by a Caliban program denote the least fixed point of their definitions. The equality means that `term1` and `term2` have the same semantics.

The Clio theorem prover is "rewrite-based" and interactive. It is closer in spirit to the Boyer-Moore prover [3] than tactic-based proof systems, such as HOL [4]. The basic proof technique employed by the Clio theorem prover is *normalization*, i.e., simplifying Caliban expressions on the two sides of an equation to an identical form. The prover supports a set of *proof tactics* which are useful in conjunction with normalization. The prover can be used in an interactive or automatic mode. In the interactive mode it prompts the user with the various choices in proof tactics available to him until it proves the assertion or discovers a contradiction. The user must make the appropriate selection. In the automatic mode it makes the selection on its own based on a builtin strategy.

Universally quantified formulas are proved by induction. The prover supports two forms of induction: *Structural induction* is used on types (user-defined or built-in) with constructors, such as `NAT` (natural numbers), trees, etc. *Fixed point* induction can be used on types with "limit objects" in them, such as functions and infinite (lazy) lists. A feature that distinguishes Clio from HOL and Boyer-Moore is that it can define partial functions

[1] Miranda is a trademark of Research Software Limited.

and reason about them at the assertion level.

4 Using the Tool on GCD

The main steps involved in using the tool on a finite state controller system are the following:

1. Construct the data path.
2. Construct the controller finite state diagram.
3. List the actions scheduled in each control state.
4. Associate invariants and advance conditions with chosen points in the control diagram.
5. Prove the derived verification conditions.

A Caliban specification of the design may be generated automatically at the end of the third step. The verification conditions that must be proved may be generated automatically at the end of the fourth step. The tool allows the user to iterate through the steps as appropriate. The tool provides a window-based graphical user-interface to help the user perform each of the above steps. Figure 5 given at the end of the paper shows the state of the window as soon as the user enters the tool. As the reader can see in the figure, there is a separate button on the menu to execute every step involved. In the following we illustrate each of the steps on the gcd example.

The figures given at the end of the paper give a series of snapshots of the interaction with Spectool which we went through in constructing a finite state controller circuit diagram for the gcd example. In the following, while displaying Caliban and Clio assertion level texts, we indicate the parts supplied by the user by underscoring them or enclosing them in a box to distinguish them from the parts automatically generated by the tool.

4.1 Constructing the Data Path

The first task in constructing the data path is to define the *component classes*, i.e., the different kinds of components, used in the data path. The second task is to place instances of these on the screen and connect them.

4.1.1 Specifying Component Classes

When defining a new component class, a window comes up with several areas which are filled in to specify the component class. The user gives the class name; for the gcd example we define three component classes, `Latch`, `Mux`, and `Subtractor`. He also draws an icon to represent this component class and indicates which points in the icon represent input ports and which represent output ports, giving each port a name. For example, Figure 6 shows a snapshot of the window which was used for defining a `LATCH` class.

The *local state* of a component consists of the values on its output ports and the value of its (optional) internal state. If the component class has an internal state, the user gives the type of the internal state by writing a Caliban type expression. This can be one of the predefined Caliban types such as `NAT`, `char`, `bool`, `[bool]` (list of bool), `NAT->bool`, etc. or, it can be a constructed type defined by the user in an auxiliary window reserved for auxiliary definitions that the user would want to create. For the gcd example, only the class, `Latch`, has an internal state, and we define it to be the predefined Caliban type `NAT`, the natural numbers. In the same way, as the user names the output ports, he specifies a Caliban type expression for the values on that port. In the gcd example, the output ports of all components were given the type `NAT` except for the `compare` output of the `Subtractor` which has type `Tri ::= GREATER | EQUAL | LESS`.

4.1.2 Specifying Actions

To complete the definition of a component class, we must specify the *actions* that it supports. We give each action a name and then give Caliban expressions, in terms of the input ports and internal state, that define each of the output ports and the new internal state. For example, for the `Latch` class, its only action `set` is specified by filling the template that the tool displays in a window with Caliban texts as shown below. Note that `set` causes the internal state of a `Latch` to be set to the value, in of its input port.

```
name = set
If input is <<in>> give a Caliban expression for the following:
new_state = in
output_out = in
delay = 1
```

The delay associated with each action measures in phases, which is a user defined (integral) subdivision of a cycle, the time required to complete an action. In the gcd circuit, we have four phases per cycle and we assign the `set` action on `Latch` a delay of one phase, the `subtract` action on the `Subtractor` a delay of two phases, and the simple actions `choose1` and `choose2` on the `Mux` have zero delay.

4.1.3 Connections

Having specified the component classes to construct the data path the user must create instances of the components, place them on the screen and then connect their ports. To create an component instance, the user selects (Figure 7) a component class from the menu of available classes, and then places them (Figure 8) at the desired location in the canvas area reserved for data path circuit using the mouse. Input ports may only be connected to output ports and vice versa, (Type-checking is done when the generated Caliban specification is compiled.) There is a box labeled `Controller` on the screen, and the output of a component may be connected to this. When this is done, the user gives a name to this controller input. These names will then be used when specifying the controller. The diamond shaped icon represents an external input and the circle shaped icon represents an external output. The tool automatically generates all the (rectilinear) segments of a connection (Figure 9), but the user must guide the tool by clicking at the source and the corners until he gets to the destination. A data path construction generated using the tool for the gcd circuit is shown in figure 1.

4.2 Specifying the Controller

The controller is specified after the data path has been specified so that the set of controller inputs is known. Then the controller finite state diagram is drawn (Figure 10) in the same fashion as the data path using a predefined set of icons provided for the construction of the controller. There is an icon for a controller state and icons for the two conditional control structures supported: `if-then-else`, `case`. (The "cutpoint" icon ⊕ will be described later.) The state icons are placed in the diagram and named. This will generate the finite enumerated type of controller states. The "output" of each state is connected, directly or through conditional branches, to another state icon. The conditions in the branch icons are specified by clicking on the icon and then entering a Caliban expression in terms of the controller inputs.

After the finite state diagram has been drawn and the number of phases per cycle has been specified, the `scheduler` mode is selected. In this mode, a controller state and a phase are selected, and the list of actions, to be initiated in the chosen phase when the controller is in the chosen state, is specified. This list is conveniently specified by clicking (Figure 11) on a component in the data path diagram. This pops up the list of actions defined on the component. The user then selects the desired action of the component to be added to the list of actions for the phase. In addition, branches can be inserted into a list of actions by entering a boolean expression in terms of the controller inputs, and then specifying action lists for the two branches.

4.3 Specifying Invariants

When drawing the controller finite state diagram, an icon representing a *cutpoint* (⊕) can be inserted in the controller diagram. A cutpoint represents an abstract "point in the flow of control" rather than any control logic and is used as a convenient means to associate invariants and advance conditions with controller states. It may be either inserted before a state icon, in which case it will be associated with that state, or before a conditional icon, in which case it will be associated with every state at the end of the branches of the conditional. In the gcd example, the points labeled `idle` and `busy` are the cutpoints. Usually, it is necessary to have a cutpoint to cut every loop in the controller machine.

In the invariant mode, the tool prompts (Figure 12) the user to enter an invariant predicate on the design state for every cutpoint and an advance predicate, a relation on a pair of design states, for every pair of cutpoints. The user may then enter a Clio assertion level text to define the appropriate predicate. The predicates default to true if the user chooses not to specify them. Attaching an invariant predicate (`I`) to a cutpoint (`p`) implies that the following condition must hold for every controller state (`cs`) associated with the cutpoint: If `I` is true of the design state at a point when the controller was in `cs`, then `I` must be true again the next time controller gets back to a state associated with `p` by means of a minimal length controller path. That is, any path that goes through no cutpoints other than `p` exactly once.

Attaching an advance predicate (`A`) with a pair of cutpoints (`p1`, `p2`) implies that the following condition must be true for every pair of controller states (`cs1`, `cs2`) associated with the cutpoints: Suppose

1. the design was in state `s1` at a point when the controller was in `cs1`,
2. the invariant associated with `cs1` was true of `s1`, and
3. the design state is `s2` when the controller moves to `cs2` by means of a path that goes through no cutpoints other than `p2` exactly once.

Then, `A` must be true of `s1` and `s2`.

A formal definition of the invariant and advance conditions associated with the cutpoints in the gcd controller are given below. (The functions **`get_A_state`**, **`get_B_state`**, etc., which extract the internal state of the corresponding components, are among the symbols automatically generated by the tool.) For example, `Idle_Invariant` must be proved to hold only for the path which goes directly from `Idle` back to `Idle`. `Busy_Invariant` must be proved for the path leading from `Busy` directly back to `Busy`, etc. The verification condition, which is automatically generated by the tool, necessary to establish the invariant and advance properties for the gcd circuit is shown later in section 4.4.

```
Idle_Invariant 's' :==  'done_out s'='True'

Busy_Invariant 's' :== 'MIN s > #0'='True' & 'done_out s'='False'
```

```
Idle_Busy_Advance_Relation 's1' 's2' 'ins' 'outs':==
     'get_A_state s2'='a_input (hd ins)'
     ___________________________________
         & 'get_A_state s2'='a_input (hd ins)'
         _____________________________________

Busy_Busy_Advance_Relation 's1' 's2' 'ins' 'outs':==
 'GCD s1'='GCD s2' & 'MAX s2 < (MAX s1)'='True'
 ______________________________________________

Busy_Idle_Advance_Relation 's1' 's2' 'ins' 'outs':==
 'reset_input (hd ins)'='True' \/ ('get_A_state s1'='get_B_state s1'
 ___________________________________________________________________
                                 & 'gcd_out s2'='get_A_state s1')
                                 ________________________________
```

4.4 Specification Generated

A sampling of the specification generated by the tool for gcd is shown below. It should give the reader an idea of the syntactic details that the user is spared by using the tool. Even the small gcd example generated a 274 line Caliban file, and Caliban is a very terse language. The length is due to all the details of timing and structure that are built-in to the model. The specification consists of the following parts:

Component classes specification: Every component class is defined as a (generic) type. The elements of this type, which will be specified in the structural specification part, are the instances of this class used in a particular design.

Structural specification: The type `COMP` is the labeled union of the component classes, the type `LOCAL_STATE` is the labeled union of the various local states of the component classes, and the type `ACTION` is the labeled union of the actions for each class. The constructors, `C_name`, `S_name`, and `A_name` are the labels for the labeled unions.

Controller specification: The controller is specified by means of two functions `nextstate` and `scheduler`. The `nextstate` function gives the next controller state as a function of the current state and controller inputs. The `nextstate` function is used at the end of each cycle to advance the controller state. The function `scheduler` gives a list of actions as a function of the controller state, controller inputs, and the phase (an implicit clock input to the controller).

Generic part: The function which defines the composite behavior of the design for a single clock cycle is called **execute**. The state of the design in a given phase within a clock cycle is represented as a function type STATE, which maps every component instance in the design to its LOCAL_STATE. The function **execute** is defined hierarchically in terms of **do_phases**, **do_phase**, **do_actions**, and **do_act**. The STATE of the design is made into a PROCESS (a function from TIME, i.e., clock phases, to STATE) before passing it to **do_phases** to keep track of the progression of the state of the design over the four phases.

```
|| Component Classes Specification
||*****************************************************
|| The component class LATCH.
LATCH :: type
LATCH_localstate ::~ <<NAT,NAT>>
LATCH_ACT ::= set !LATCH
LATCHdelay (set c) = 1
LATCHcomp (set c) = C_LATCH c
LATCHout (set c) s in = in
LATCHstate (set c) s in = in

|| Structural Specification part
MUX ::= Ain | Bin
LATCH ::= A | B | DIFF
SUBTRACTOR ::= SUB
COMP ::= CONTROLLER | EXTERNAL |
         C_Latch Latch | C_Mux Mux | C_Subtractor Subtractor
LOCAL_STATE ::= S_Controller Controlstate | S_External extstate
                S_Latch Latch_localstate | S_Mux Mux_localstate
                S_Subtractor Subtractor_localstate
ACTION ::=   A_LATCH LATCH_ACT | A_MUX MUX_ACT
           | A_Subtractor Subtractor_act
NUM_PHASES = 4
```

```
|| Connections
MUXinput s (C_MUX Ain) =
  <<external_input_a (s External), getLATCHout0 (s (C_LATCH DIFF))>>
LATCHinput s (C_LATCH A) = getMUXout0 (s (C_MUX Ain))
....

||Controller Specification
CONTROLSTATE ::= Wait | Comp
nextstate Wait in = ((okof in))->(Comp);(Wait)
nextstate Comp in = ((resetof in)\/((compareof in) = Equal))->(Wait);(Comp)
scheduler Comp <<ok, reset, compare, ok, reset>> 0 =
                                [A_SUBTRACTOR(Subtract SUB)]
scheduler Comp <<ok, reset, compare, ok, reset>> 2 = [A_LATCH(set DIFF)]
scheduler Comp <<ok, reset, compare, ok, reset>> 3 = (SET_GREATER compare)
....

||***************************************************
|| The generic Execute function.
PROCESS ::~ TIME->STATE
STATE ::~ COMP->LOCAL_STATE
Execute s x = advance_controller(do_phases 0
                               (make_process (set_ext x s)))
do_phases t p = p t , t = NUM_PHASES
                do_phases (t+1) (do_phase t p)
do_phase t p = do_actions t (scheduler (controllerstate (p t))
                                        (controllerinput (p t)) t) p
do_actions t [] p = p
do_actions t (a:x) p = do_actions t x (do_act a t p)
```

4.5 Verification Conditions Generated

From the specified invariants and advance conditions for all the cutpoints, the tool can create a set of *verification conditions.* There is one verification condition for each path, and it states that if the conditions for traversing that path and the invariant for the starting point of the path hold, then after executing the path, the invariant for the end point will hold and the specified relation between the starting and ending states will hold. The tool generates a single predicate VC that conjoins the verification of each of the paths.

The tool will determine the set of all minimal paths through the control state machine that start and end at cutpoints. It will generate a predicate which states the conditions under which that path will be executed. In the gcd example, the tool finds and names four such paths: (1) idlePATH0 from idle to idle, (2) idlePATH1 from idle to busy, (3) busyPATH0 from busy to idle, (4) busyPATH1 from busy to busy.

The tool generates the following Caliban functions and Clio assertion level predicates to characterize these paths. these predicates and functions are used in the formulation of the generic VC: (1)
tt path_precond: an assertion level predicate which defines the precondition that can be assumed at the start of each path; (2)
tt path_postcond: an assertion level predicate which defines the postcondition that must be true at the end of each path; (3) path_condition: a predicate which defines the condition under which a give path will be taken in a given state; (4) path_length, path_end, path_inputs: Caliban functions which define the length, the ending cutpoint and the inputs consumed for each of the paths.

A part of the definition of the VC generated by the tool for the gcd example is given below.

```
VC 'path' 's':=
 (in) path_precond 'path' 's' 'in' => path_postcond 'path' 's' 'in'

path_precond 'path' 's' 'in' := '!path'='True'
                              & 'input_ok (path_length path) in'='True'
                              & path_condition 'path' 's' 'in'
```

```
                        & Invariant `path_start path` `s`
path_postcond `path` `s` `in` :=
   Invariant `path_end path` `next s (path_inputs path in)`
   & Advance_Relation `path_start path` `path_end path`
                      `s` `next s (path_inputs path in)`
                      `path_inputs path in`
                      `next_outs s (path_inputs path in)`

VC_ok `s` :==  (path) VC `path` `s`
VC_ok_lemma :== (s) Proper_state `s` => VC_ok `s`
```

4.5.1 The Proper-state Invariant

In certain situations the Caliban type associated to model the internal state of a component or the signal at a port may not be constraining enough. For instance, we might want to model the output of a component as a bit vector, for which we could use the Caliban type for list of boolean, `[bool]`, but we might want to further restrict this type so that is consists only of bit vectors of length 32. This can be specified by means of a Caliban predicate which must be true of every element in the subtype.

The tool allows the user to give a "properness" predicate for each output port. The default is the Caliban predicate `!` which stands for "is not bottom" and which means, essentially, "is defined". In addition to a properness predicate for each output port, an additional assertion can be given which states some invariant satisfied by the local state of the component. In the gcd example, we asserted that the internal state and output of a `Latch` are always equal.

Since the properness predicates are not used by the type-checker, they must be checked during the verification. The tool generates an assertion language predicate called *Proper_state*. This predicate says, among other things, that the values on the output ports of all the components satisfy their properness predicates. The predicate, *Proper_state*, then becomes one of the *invariants* which we establish during the verification.

5 Microprocessor Application

MiniCayuga is a 3-stage instruction pipelined RISC processor. It consists of 32 general purpose registers (each 32-bit wide) organized into a register file (REGFILE), and several special purpose registers including NXPC ("next program counter"). It is a scaled down version of the machine *Cayuga* designed by the *Cayuga* Group [5] at Cornell University for studying RISC architectures.

The design of *MiniCayuga*, which was simplified to help focus our study on the formal analysis of pipelining and interrupts, differs from that of *Cayuga* in the following respects. It supports a subset of the instruction set of *Cayuga*. The instruction set is chosen so that every instruction class of *Cayuga* is represented and is computationally universal. The interface of *MiniCayuga* to the memory is simpler than that of *Cayuga*. It supports a (32-bit wide) word at a time transfer between the memory and the cpu. *MiniCayuga* assumes that the cache is off-chip, and that the memory operates synchronously with the processor. It supports a single kind of interrupt while *Cayuga* has multiple prioritized interrupts.

5.1 MiniCayuga Instructions

The instructions of *MiniCayuga* are grouped into four classes: - *arithmetic*, *load*, *store*, *jump*. Instructions belonging to the load/store class are the only ones that move data between memory and registers; the rest are all register to register. The instruction in the arithmetic and jump classes take three clock cycles to complete, while the ones in the load and store classes take four cycles. However, since the processor uses a 3-stage instruction pipeline up to three instructions are executed simultaneously in a cycle. This allows one instruction to be completed every cycle except when the instruction is in the load/store class, in which case it takes an extra cycle.

The only exception to this normal sequence of execution occurs when there is an interrupt. An interrupt may arrive at any time during the execution of an instruction. *MiniCayuga* supports just one kind of interrupt request: intrpt. In the event of an interrupt, the processor goes through a "complete-clean_up-handler_fetch" sequence of actions. It completes the *current* instruction (i.e., the one most advanced in the pipeline), aborts the other two instructions, and starts fetching instructions from a predetermined location in memory where the program for handling the interrupt is stored. The processor takes 2 additional cycles to complete this sequence. The processor automatically disables, i.e., ignores, interrupts for these 2 cycles.

Note that it is not important what the interrupt handler does from the point of view of the correctness of the hardware circuit that deals with the interrupt. One may program the interrupt handler as appropriate for the application. The requirement imposed, and verified by our proof of correctness, is that the design implement the "complete-clean_up-handler_fetch' sequence as described above. Hence, we leave the interrupt handler program unspecified.

The instructions are of fixed length (32-bit) and use a fixed format: `opcode i? src1 src2 dst`, where `dst` and `src1` always denote (5-bit) indexes into `REGFILE`, and `src2` can be either a register index or a 16-bit immediate constant depending on the bit `i?`. In the following `R[j]` denotes the `j`th register in the register file; `oprnd1` denotes the contents of `R[src1]`; `oprnd2` denotes 'if `i?` then `src2` else the contents of `R[src2]`'; and, `msb[R[i]]` denotes the most significant bit of `R[i]`.

```
    Instruction                      Effect
   -------------                   ----------
ARITHMETIC CLASS (ACLASS)
ADD i? src1 src2 dst            R[dst] <-- oprnd1 + oprnd2
SUB i? src1 src2 dst            R[dst] <-- oprnd1 - oprnd2
CNE i? src1 src2 dst    msb[R[dst]] <-- if oprnd1 = oprnd2 then 0 else 1

LOAD CLASS (LCLASS)
LOAD i? src1 src2 dst         R[dst] <-- mem[oprnd1 + oprnd2]

STORE CLASS (SCLASS)
STR i? src1 src2 dst           mem[oprnd1 + oprnd2] <-- R[dst]

JUMP CLASS (JCLASS)
JMP i? src1 src2 dst               NXPC <-- oprnd1+oprnd2
JIT i? src1 src2 dst    NXPC <-- if msb[R[dst]]=1 then oprnd1 + oprnd2
JIF i? src1 src2 dst    NXPC <-- if msb[R[dst]]=0 then oprnd1 + oprnd2
```

Table 4

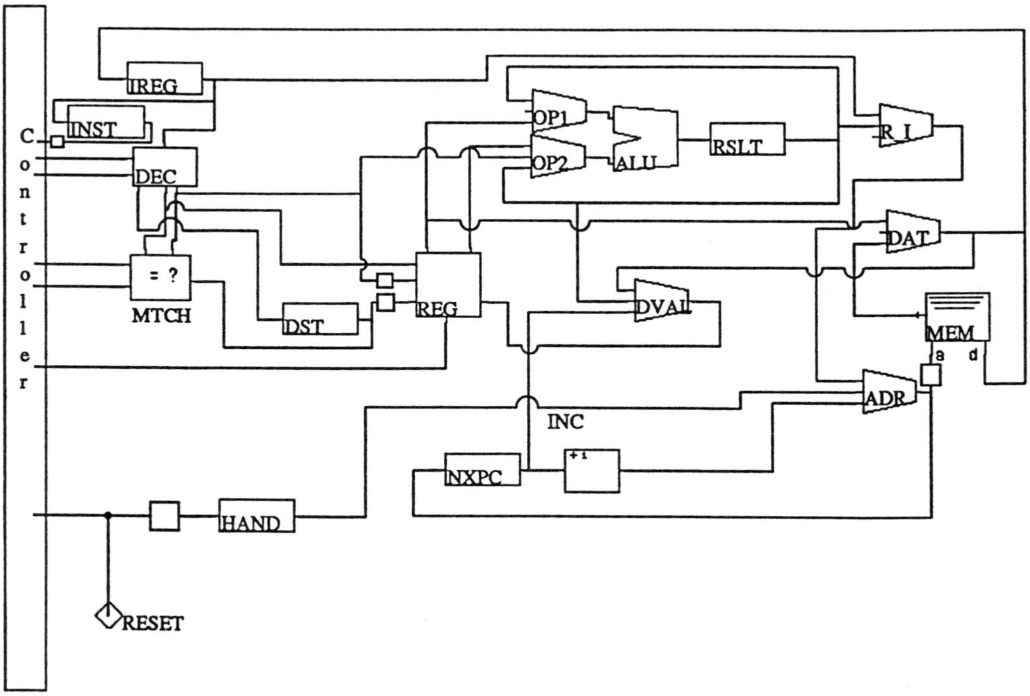

Figure 3: MiniCayuga Data Path

5.2 MiniCayuga Design

A block diagram of the *MiniCayuga* design is shown in Figure 3. The memory, which is shown as part of the block diagram for the sake of clarity, is external to the *MiniCayuga* design. The major blocks of the design are CONTROLLER and the components of the *data path*: REGFILE, which contains all the general purpose registers, ALU, DECODER, and INC, which is an incrementer, MTCH?, which is a special-purpose comparator, IREG, INSTRN, NXPC, RESULT, and DST, which are latches, MUX1, ... , MUX6, which are 3-way multiplexors. The CONTROLLER uses a nonoverlapping 4-phase clock to synchronize the actions on the data path elements.

The actions involved in the execution of every instruction of *MiniCayuga* can be partitioned into three stages: *instruction-fetch*, *compute*, and *write-back*. The instruction-fetch stage consists of the actions involved in fetching the instruction: a memory read (delay: 2 clock phases) and latching the instruction into IREG (1 phase). The compute

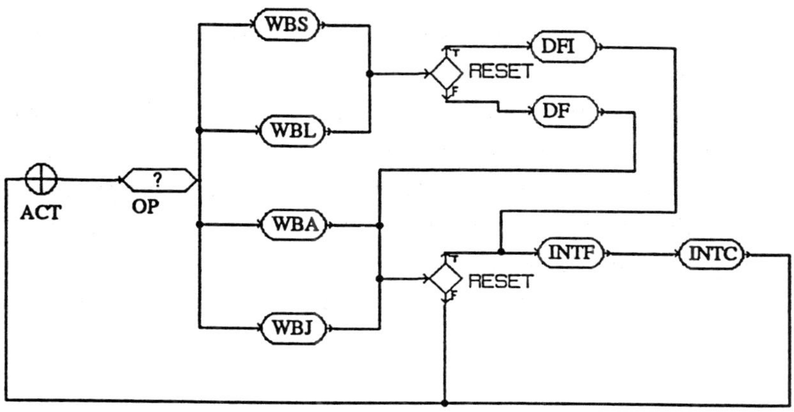

Figure 4: MiniCayuga State Machine

stage includes all the actions necessary to compute the "result" of the instruction: reading (1 phase) the REGFILE to set up the operands for ALU, completing an ALU operation (2 phases) and storing the result in RESULT (1 phase). For load/store instructions the result denotes the memory address of the destination/source operand. The write-back stage consists of the actions necessary to "move the result of the instruction to its destination." For load/store class instructions this involves a memory access, for others it only involves writing into REGFILE. Hence, the duration of write-back is longer for load/store instructions (3 phases) than for the others (1 phase).

In any given clock cycle the processor can potentially be executing in parallel the actions involved in three different stages of three distinct instructions in a program. For example, in an execution of the sample program given in Table 5, in the clock cycle in which the processor will be writing-back the results of the *current* instruction (I1) it will also be performing the compute stage of the (*next* instruction) I2 and fetching the (the *next to next* instruction) I3. During such a cycle INSTRN contains a copy of I1; RESULT contains the result of I1; DST contains the destination index (R3) for the write-back operation; IREG contains the instruction I2; NXPC+1, which is the value on the address bus, is the address of I3;

```
I1: ADD 0 R1 R2 R3      write-back I1
I2: CNE R2 R1 R4        compute I2
I3: JIT 0 R2 R1 R4      fetch I3
```

Table 5

The diagram by the side of the program depicts the overlapping of the actions of the three stages for that cycle. The write-back is scheduled during the latter part of the compute stage to avoid clashing with the reading of the register file. The controller will be in the state WBA ("write-back arithmetic") in this cycle. As shown in the controller diagram in figure 4, the processor will move back to WBA in the next cycle, provided there is no interrupt, since the next instruction to be written back also belongs to arithmetic class. If there is an interrupt, it takes two extra cycles before moving to one of the WBS/L/A/J states. The precise state it moves to depends on the class to which the first instruction in the interrupt handler belongs to.

The write-back stage cannot always be scheduled in parallel with the other two stages. For example, consider the sample program given in Table 6, where the instruction whose write-back is being completed is a LOAD instruction. The write-back, which now involves a memory operation, clashes with instruction fetch and also with compute stage since a `REGFILE` operation is needed. The solution adopted in *MiniCayuga* is to perform the the write-back alone in the current cycle. The compute and instruction-fetch stages are postponed for the following cycle. The pipeline is said to be ***suspended*** in a cycle in which a write-back alone is scheduled. The pipeline is said to be ***active*** otherwise. The controller must be in one of the `WB/S/L/A/J` states when it is active.

```
I1: LOAD 0 R1 R2 R3   write-back I1
I2: CNE  0 R2 R1 R4   compute I2
I3: JIT  0 R2 R1 R4   fetch I3
```

Table 6

The controller will be in state `WBL` ("write-back load") in this cycle and moves to `DF` ("delayed fetch"), to perform the instruction fetch exclusively, before moving to `WBA`, provided there is not interrupt. As before, an interrupt will cause the controller to take additional two cycles to move back to one of the actives states.

Note that the sources of the operands for the `ALU` operation are normally obtained from the `REGFILE` indexed by the appropriate field of the `DECODER` output. In certain situations, however, the `REGFILE` must be "by-passed" and the operand must be ***internally forwarded*** to the `ALU` from `RESULT`. This happens, for example, if in the sample program given in Table 5, `R2` in instruction `I2` were `R3`, the same as the destination of `I1`. `MTCH?` compares the destination of the current instruction (`I1`) with the sources of the next instruction (`I2`), and passes the result to the `CONTROLLER` to help determine if internal forwarding is necessary.

5.3 Proving a MiniCayuga Invariant

The active states of the controller, i.e., the ones enclosed in the dotted box, are the states in which an instruction execution gets completed by the processor. In a sense they denote the most externally significant points in the execution of a program by the processor. So, we create a cutpoint, called `active`, common to the active states.

There is an invariant condition which must be satisfied every time the control reaches any one of the active states for the processor to be executing a program correctly. The condition states that the hidden registers of the processor, which hold the intermediate state of the three partially executed instructions in the pipeline, have the correct information in them. The invariant predicate is defined with respect to three "reference points" in the program being executed introduced earlier in section 5.2: the current, next and next-to-next instructions. It formalizes the observation made informally in section 5.2 while explaining how the pipeline works for the program fragment shown in Table 5. A formal definition of the invariant in the Clio assertion language is given below.

The advance condition associated with the cutpoint, which relates the the externally visible part of the current state (s1) at the cutpoint with that of the state (s2) when the control is at the cutpoint the next time around, ensures that the effect of the completion of the current instruction is as "intended."

```
ACT_Invariant 's' :==
    'next_instrn s' = 'get_IREG_state s'
        & 'next-next-instrn s' = '(get_MEM_state s) get_MUX1_out s'
        & 'current_instrn s' = 'get_INSTRN_state s'
        & 'current_dst s' = 'get_DST_state s'
        & 'current_result s' = 'get_RESULT_state s'

ACT_ACT_Advance_Relation 's1' 's2' 'ins' :==
    'External_part s2'='Intended_effect (External_part s1) ins'
```

6 Concluding Remarks

We have described the design of a computer-aided verification tool for the specification and verification of finite state controller systems. The tool specializes a general purpose functional language based verification system, Clio, to the task of verifying this class of systems. We were motivated to undertake this work after we had used Clio for verifying a large realistic design of the microprocessor *MiniCayuga*. Our experience with the *MiniCayuga* effort suggested that a significant part of the technology that was used in the effort was applicable to other systems in the same class. The tool is designed by embedding this technological knowledge in it.

We have used the tool on several small examples as well as on a slight variant of *MiniCayuga*. The reverification of *MiniCayuga* using the tool was completed in a couple of weeks as compared to the original verification effort which took about 8 months. This, of course, is not a fair comparison because the tool was designed based on *MiniCayuga* design structure. A better evaluation can be made only after the tool further on systems which are quite different from *MiniCayuga* in their design structure. There is no doubt that the tool reduces the time spent on specifying the design. The question is how much of a saving would be obtained in the verification area. In other words, the question is how reusable and general are the proof tactics that the tool provides. They have been successful in all the small examples that we have tried. We plan to use the tool on other significant examples to get a better idea of the scope of the tool.

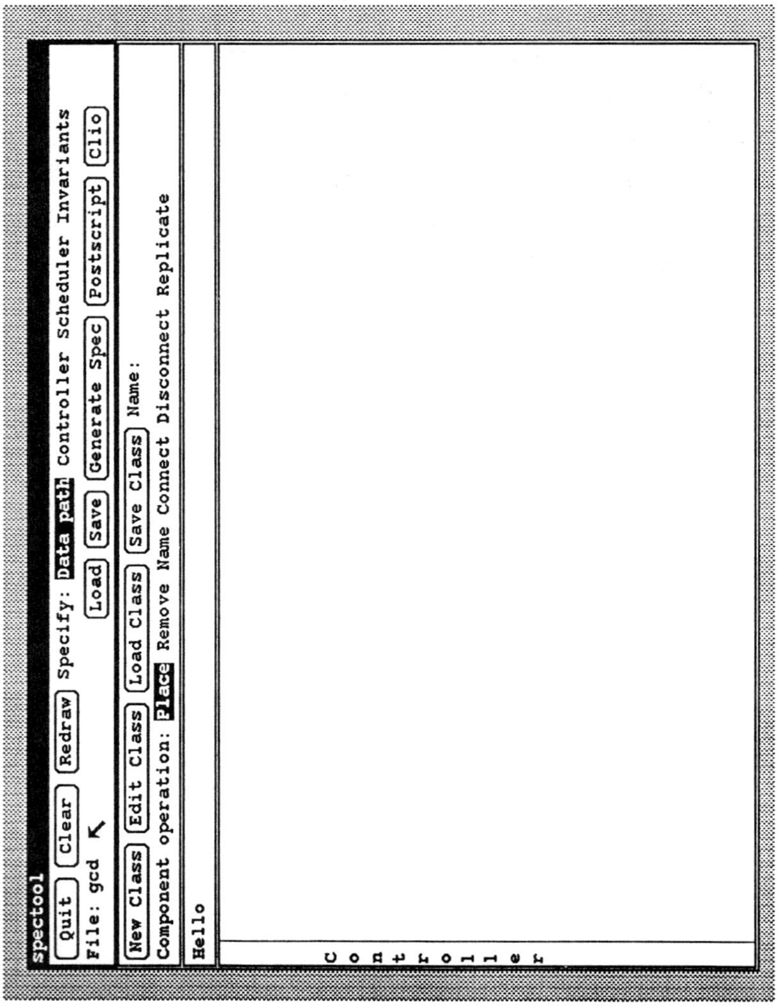

Figure 5: Initial Canvas

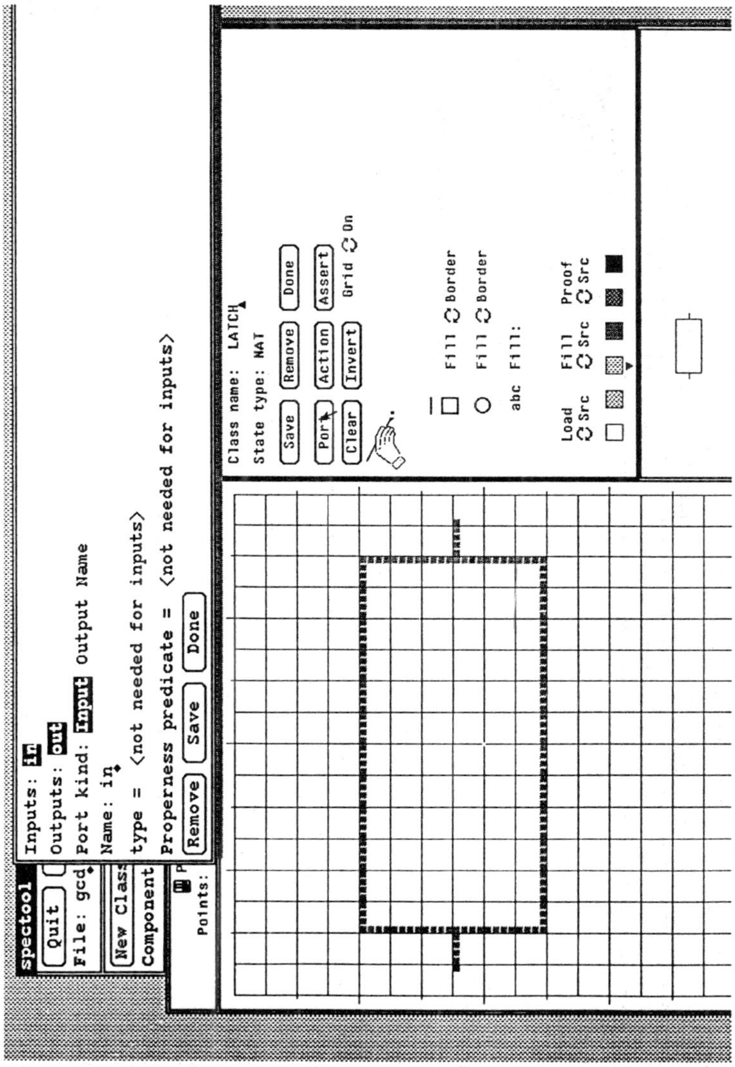

Figure 6: Defining a new class

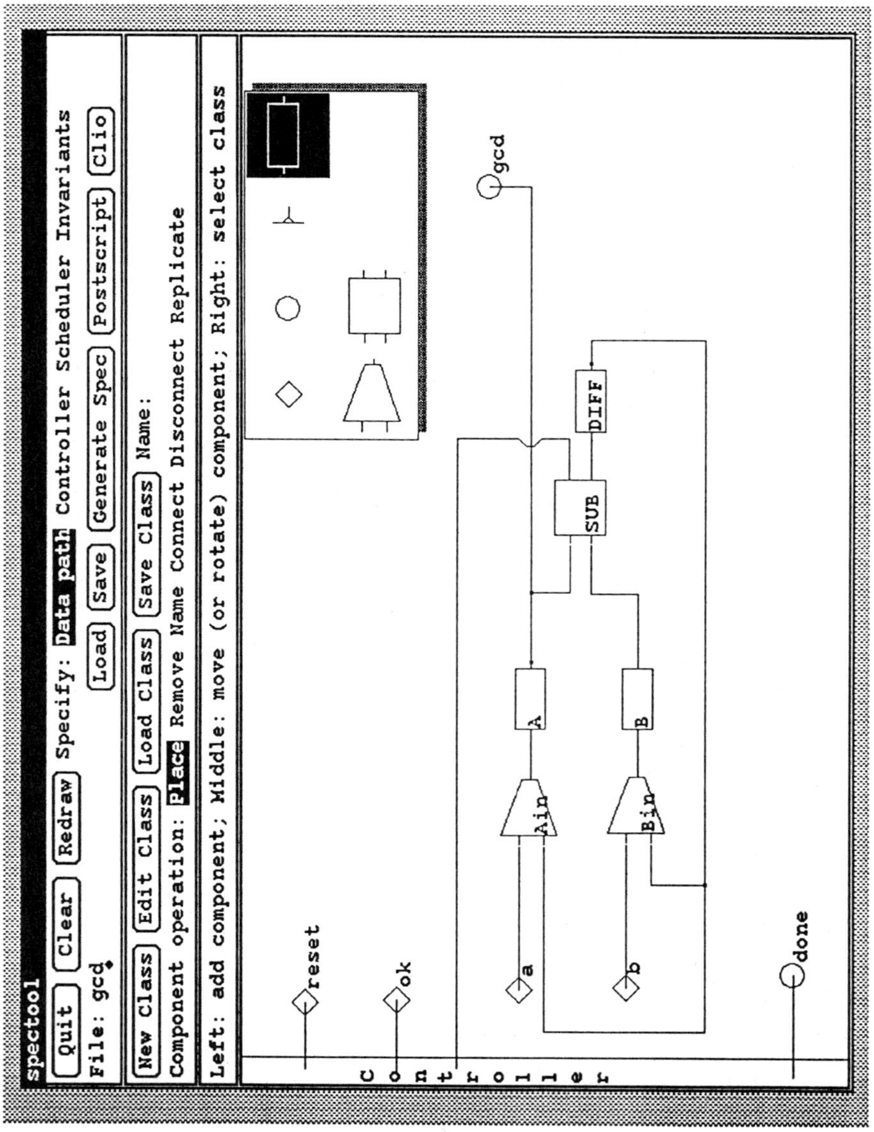

Figure 7: Selecting a class

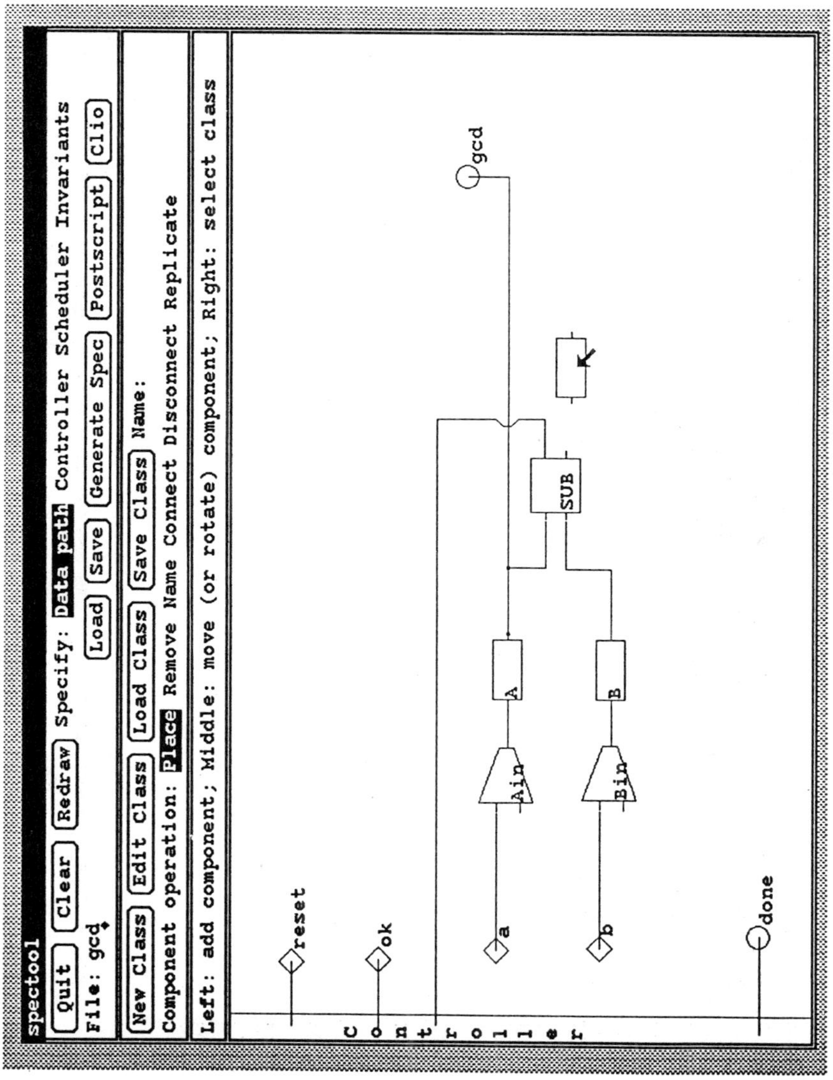

Figure 8: Placing the components

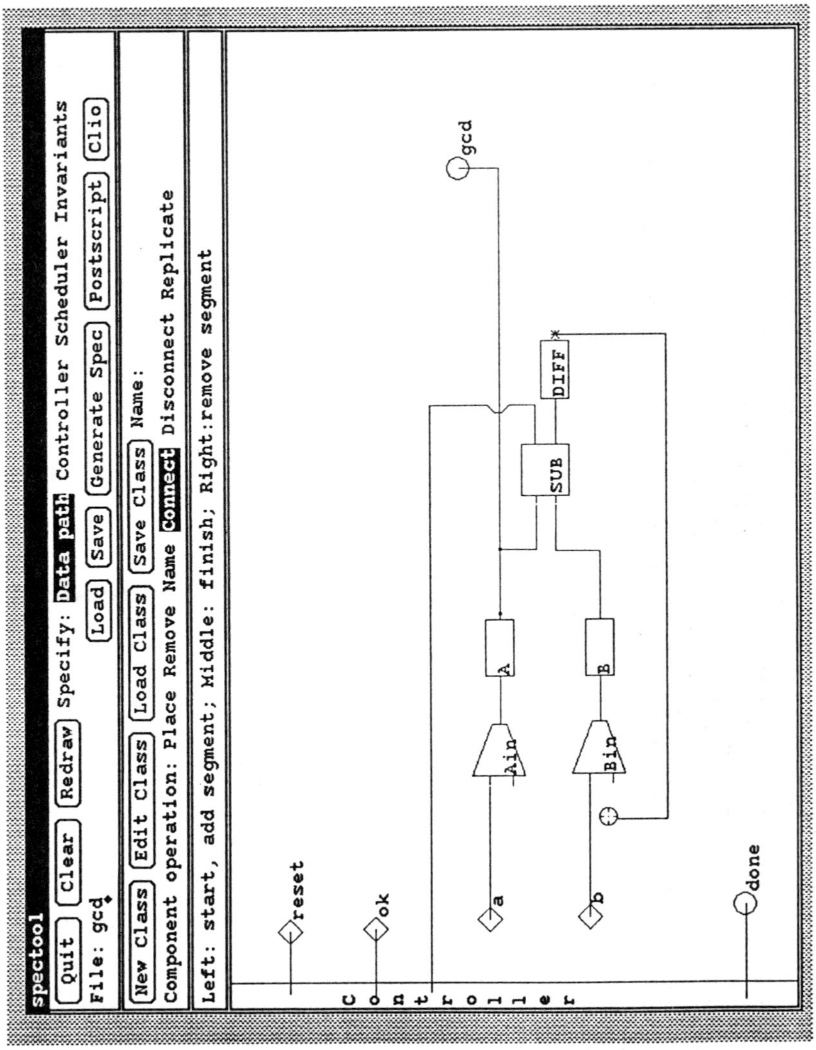

Figure 9: Connecting the components

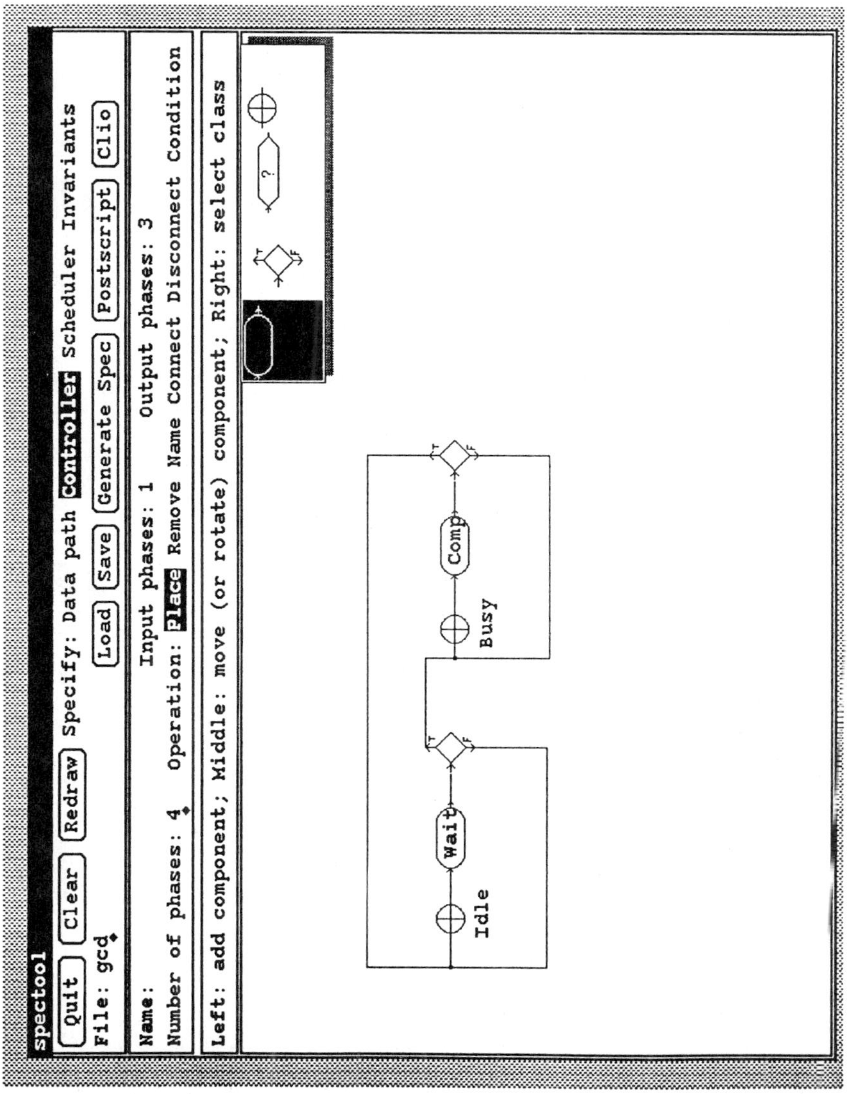

Figure 10: Constructing the controller diagram

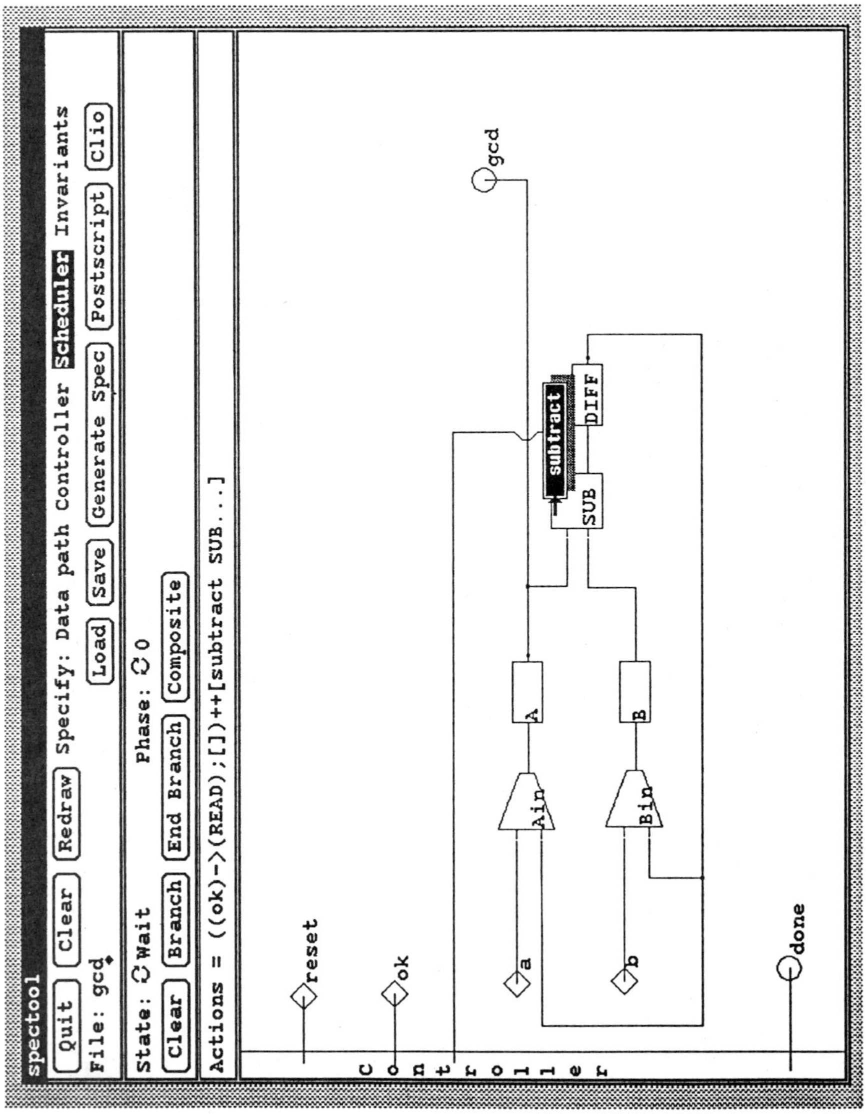

Figure 11: Specifying the Scheduler

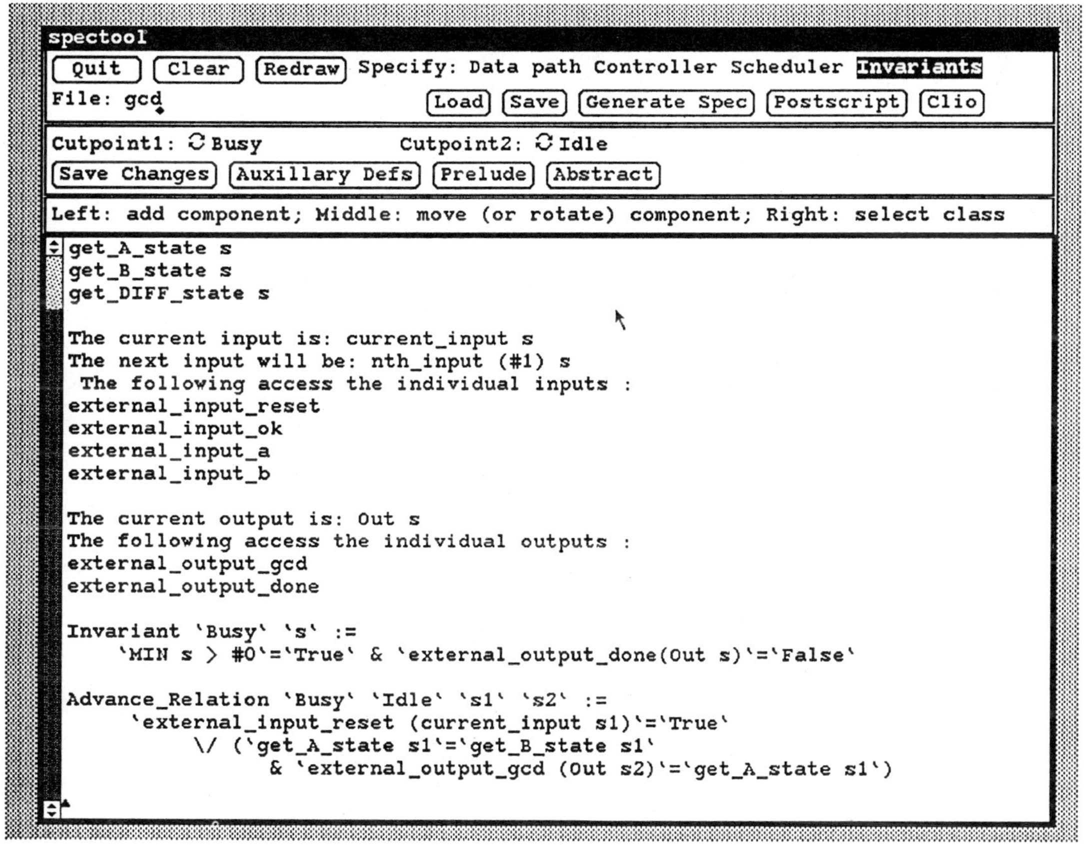

Figure 12: Annotating the controller

References

[1] M. Bickford, C. Mills, and E.A. Schneider. "Clio: An Applicative Language-Based Verification System". Technical Report TR 15-7, Odyssey Research Associates, Inc., 301A Harris B. Dates Drive, Ithaca, NY 14850, March 1989.

[2] Mark Bickford and Mandayam Srivas. "Verification of a Pipelined Microprocessor Using Clio". In *Workshop on Hardware Specification, Verification, and Synthesis: Mathematical Aspects*, Cornell University, Ithaca, NY 14853, July 1989.

[3] Robert S. Boyer and J Strother Moore. *ACM Monograph Series: A Computational Logic.* Academic Press, Inc., 1979.

[4] Mike Gordon. "A Machine Oriented Formulation of Higher Order Logic". Technical Report 68, Computer Laboratory, University of Cambridge, Cambridge, U.K., July 1985.

[5] The Cayuga Group. "Cayuga Chip Specification: Release 2.2". Technical report, Departments of Computer Science and Electrical Engineering, Cornell University, Ithaca, NY 14853, December 1988.

[6] Joseph E. Stoy. *Denotational Semantics: A Scott-Strachey Approach.* MIT Press, Cambridge, MA 02139, 1977.

[7] D.A. Turner. "An overview of Miranda". *ACM SIGPLAN Notices*, 21(12):158–166, December 1986.

DIMACS Series in Discrete Mathematics
and Theoretical Computer Science
Volume 3, 1991

Program Verification by Symbolic Execution of Hyperfinite Ideal Machines

James M. Morris and Mark Howard*
Odyssey Research Associates
301A Harris B. Dates Drive
Ithaca, New York 14850-1313
jmorris%oravax.uucp@cu-arpa.cs.cornell.edu

1 Abstract

Ariel, a program verification system based on symbolic execution and non-standard analysis, is described. Programs are verified by showing that the traces resulting from their execution on a hyperfinite abstract interpreter satisfy their specifications. An Ariel abstract interpreter is written in a functional programming language, Caliban, for which we have a theorem prover. Caliban is similar to Miranda. Ariel is especially useful for verifying programs that compute with real numbers. The real data type of an Ariel abstract interpreter is taken to be a hyperfinite subset of the non-standard reals. The imprecision of real arithmetic is modelled by assuming that real arithmetic operations compute results that differ from the ideal ones by an infinitesimal. Programs are required to compute results that differ by an infinitesimal from the ideal. Programs that have this property are said to be *asymptotically* correct. If a program is asymptotically correct, then it can be shown that there *exists* a sufficiently precise machine on which the program can be run to obtain results of a desired accuracy. The theory does not predict how accurate that machine has to be to achieve the desired results, but does provide a means of verification that catches a large number of programming errors. A prototype Ariel system has been built and is in use.

*Funded by the U.S. Air Force, RADC contract F30602-86-C-0116 and the STARS program contract BOA #3695.STARS-043.

2 Introduction

This paper describes Ariel, a program verification system based on the symbolic execution of hyperfinite ideal machines. By *ideal machine* we mean a transition system that gives an operational semantics for the language in which the programs we wish to verify are written. The terms of a transition system can be thought of as the states of an ideal (mathematical as opposed to electronic) machine, tailored to the language in question, which executes the program to be verified. The meaning of the program is identified with the execution traces of the transition system or ideal machine. A program is verified by showing that its traces satisfy its specification.

Our transition systems are defined in an applicative programming language for which we have a theorem prover. This applicative language, called Caliban, is lazy, higher-order, and polymorphic. The theorem prover, called Clio, provides a meta-language for making assertions about the meanings of Caliban terms. One can write a Caliban expression whose meaning is the trace of a (sub-)program beginning in some initial state. The pre- and postconditions that the initial and final states of such a trace respectively are to satisfy are written in Clio's meta-language. A (sub-)program is verified by using Clio to show that its precondition implies its postcondition. Clio's proof rule for implication attempts to prove the postcondition under assumption of the precondition. As Clio's prover is basically a term rewriting engine, proving the postcondition is basically a matter of expanding the symbols in the postcondition until no more expansion is possible. Here two properties of Caliban are important. First, Caliban is lazy. It is possible to write Caliban terms that represent non-terminating computations. Only those components of a term (state) of the transition system needed to determine that term's (state's) successor are actually evaluated. This is important for verifying sub-programs independently of their invocations. Secondly, Caliban expressions are executable. This means that much of the calculation involved in determining the next state of an Ariel transition system can actually be carried out automatically by Clio, which has the effect of reducing the amount of "bookkeeping" that must be dealt with in an Ariel verification. Ariel can execute programs symbolically. This provides us with the means to define a trace that represents all those traces in which the program's variables have values satisfying its precondition.

A hyperfinite ideal machine is a transition system in which some of the components of terms are defined using constructs from non-standard analysis. The original version of Ariel was designed primarily for the verification of numerical programs written in a subset of C. We formulated a notion of correctness of numerical programs, called *asymptotic correctness*, using concepts from non-standard analysis (see [5] for a report on the theory of asymptotic correctness). This led to the formulation of non-standard models of computation which have proved to be of general utility in program verification.

The theory underlying Ariel uses non-standard analysis to formalize intuitive notions of what it means for a numerical program to compute a correct result. Numerical programs are not usually handled by verification systems because a sufficiently abstract yet tractable model of real arithmetic is difficult to design. When one attempts to model the amount of error introduced into a numerical calculation by finite arithmetic, one quickly gets bogged down in the intricacies of numerical analysis. The sort of verification we're interested in starts out with a program that purportedly implements some numerical method. We provide the means for showing that the program computes the desired result when run on a sufficiently accurate machine, i.e. that it's *logically* correct. Logical correctness turns out to be equivalent to asymptotic correctness. Asymptotic correctness captures programming errors that don't involve the *magnitude* of the error in computed results. Simple examples of such errors are : the sign of some variable is wrong, an expression uses the wrong operator, etc. A more subtle logical error is the following: comparing the signs of two quantities for equality by multiplying them and checking whether the product is positive. If the quantities are of different sign but small enough that the multiplication underflows to zero, the sign comparison will be incorrect.

Ariel has proven to be useful for verifying non-numerical programs. Indeed, we now take as a principal thesis of our research that hyperfinite models provide the basis for a general method of proving limit properties of computer programs. An example of the sort of limit property we have in mind is the following. Consider any resource which limits the computational power of a finite machine, e.g. storage. Given a program, P, we would like to formalize a notion of correctness that has that P is correct with respect to a specification, if P's behavior satisfies the specification provided that the machine on which P is run has sufficient storage. To verify that a program is correct in this way, one has to show that its behavior "in the limit" is correct.

The symbolic execution technique we have devised for Ariel is also generally useful. Defining a semantic model in an executable specification language and then "running" the model symbolically under control of a theorem prover for the specification language has proven to be very effective. This is largely due to the theorem prover's facilities for dealing automatically with much of the detail involved in reasoning about a symbolic execution. The result is a system in which one may use one's operational understanding of how a program works as a guide to that program's verification without being overwhelmed by the details of a mathematical formalization of the program's execution.

3 The Use of Non-standard Analysis in Ariel

A program is asymptotically correct if, given *any* measure of the accuracy of the result to be computed, there *exists* a machine which will compute the result to the degree of accuracy given. Note first that this is weaker than the sort of properties the numerical analysts deal with. Numerical analysis tries to answer questions about how much accuracy in the machine operations and (most importantly) in the representation of the input data is required in order that the program compute a result that satisfies a given accuracy requirement. The problem with numerical analysis is its difficulty: questions of the sort just mentioned are hard to answer in general.

A program that is asymptotically correct will not contain errors of a logical nature. So, if a program has been specified to implement some numerical method, the theory of asymptotic correctness can be used to show that the program implements the method, as long as the specification doesn't require of the results some particular accuracy.

Why not just posit an ideal computing device that has the mathematical reals as a data type? The operations of such a machine would then deliver exact results when operating on exact data. The trouble is no worldly machine has such a data type. Moreover, most numerical methods only compute exact results in the limit. Therefore, their (straightforward) implementations would be non-terminating on such a machine.

What we would like to model is an ideal machine whose representational power and accuracy are limited, but without having to say to what degree. Intuitively, we would like to say things about the model like "its operations compute results which are close to ideal when their operands are not too big", which is the translation , in our general setting, of the idea that the operations of a worldly machine compute results of a certain accuracy when their operands are not so large as to cause overflow.

Analysis is the mathematics of approximation. It formalizes, in a precise way, our intuitive notions of closeness or approximation. Consider *continuity*. Intuitively, a real function f is continuous if, whenever two values x and y in its domain are close (approximately equal), then $f(x)$ and $f(y)$ are close. Conventional analysis formalizes this intuitive notion by paraphrasing it in terms of the epsilon-delta argument we all learned in elementary calculus: f is continuous at x if, for all $\epsilon > 0$, there is a $\delta > 0$ such that $\mid f(y) - f(x) \mid < \epsilon$ for all y for which $\mid y - x \mid < \delta$. Non-standard analysis gives a more direct formalization by idealizing "close" by the non-standard notion of *infinitely close*([3] is a general introduction to non-standard analysis). The non-standard formalization of continuity is: f is continuous at a if, whenever x is infinitely close to a, $f(x)$ is infinitely close to $f(a)$. The direct way in which non-standard analysis expresses intuitive notions makes it easier to understand than conventional analysis and more easily implemented in theorem provers. Non-standard analysis can be expressed naturally in terms of operations rather than predicates and with fewer quantifiers, which makes it especially compatible with term-rewrite systems like Ariel's theorem prover.

Non-standard analysis is formed by adding the new undefined predicate *standard* to the usual language of mathematics. Other formal counterparts of informal concepts of non-standard analysis are defined from the predicate *standard*. A real number is *infinite*, corresponding to the informal notion "large", if its absolute value is larger than any standard real number; a real number is *infinitesimal*, i.e. "small", if its absolute value is smaller than any standard positive real number. Two real numbers are *infinitely close*, i.e. "close", if their difference is infinitesimal.

Concepts that can be defined without use of the predicate *standard* are said to be *internal*. These are the concepts of conventional mathematics. Those which cannot be defined without use of *standard* are called *external*, because they formalize notions which are outside conventional mathematics. The point of non-standard analysis is that it provides precise rules for manipulating the external notions. The rules of most interest to us here are:

1. there exists a real number which is non-standard,
2. the collection of standard real numbers satisfies the same internal properties as the collection of all real numbers.

Rule 2 is called the *transfer principal*; it is central to the theory underlying Ariel.

Numerical programs are useful in spite of the fact that they only compute approximations to ideal results, because the results they compute (one hopes) are close enough to ideal, i.e. they compute "good" approximations. For a given program, how good the approximation is depends on the accuracy of the machine on which its executed. If the program is correct, we should get increasingly accurate results, as we run it on increasingly accurate machines and, "in the limit", we should get exact results. This is what it means for a program to be asymptotically correct. Using the standard formalization of continuity as a guide, we might paraphrase asymptotic correctness in terms of an epsilon-delta argument: for any desired degree of accuracy ϵ of the result of running a program P, there exists a degree of accuracy δ of representational power and arithmetic operations such that, whenever P is run on a machine M of accuracy δ, the results will be within ϵ of the ideal ones. Such a formalization would be as unwieldly to reason with as it is to state. Instead, we take the non-standard formalization of continuity as our guide. We let the machine M of the preceeding paraphrase be *hyperfinite* — the non-standard version of finite. In particular, its real data type is a hyperfinite subset of the non-standard reals. M's real data type will contain all the standard real numbers plus nonstandard reals infinitely close to each standard real. The non-standard values close to a standard real number x correspond to sequences of reals that converge to x. In our theory such sequences are the approximations computed by a program when run on machines of increasing accuracy. We then restate asymptotic correctness: a program P is asymptotically correct if it delivers results that are infinitely close to ideal when run a hyperfinite machine with finite data. This restatement is justified by the transfer principle. Let $\mathcal{M} =< M_0, M_1, M_2, \ldots >$ be a sequence of ideal machines of increasing accuracy and $\mathcal{R} =< r_0, r_1, r_2, \ldots >$ be the results of running P on the members of $\mathcal{M}$. If executing P on a hyperfinite member M_ω of $\mathcal{M}$ produces a result r_ω which is infinitely close to ideal (let r denote the ideal result), then for any standard real value ϵ, the following property is true of $\mathcal{R}$ ($^*\mathbf{N}$ is the non-standard natural numbers)

$$\Phi(\epsilon, {}^*\mathbf{N}) \leftrightarrow \exists n \in {}^*\mathbf{N}\, \forall m \in {}^*\mathbf{N}(m \geq n \rightarrow \mid r - r_m \mid < \epsilon)$$

By transfer, $\Phi(\epsilon, \mathbf{N})$ is true. Since ϵ is arbitrary, $\Phi(\epsilon, \mathbf{N})$ is asymptotic correctness.

4 Clio

Ariel is built around a theorem proving system called Clio. Clio is designed for proving properties of expressions written in a lazy higher-order polymorphic applicative language called Caliban. Caliban is similar to Haskell and Miranda. Theorems about Caliban expressions are stated in Clio's metalanguage, which is essentially first-order logic with Caliban expressions as terms. Clio's prover is based on term rewriting augmented with other proof techniques such as structural induction, case splitting over variables of constructed type, fixed point induction, etc.

4.1 Caliban

Caliban is so similar to the functional languages mentioned above that we needn't describe much of it here. We mention in passing that a lazy applicative language like Caliban is an apt notation for specifying operational semantics. Caliban's constructed types and definitions by pattern matching provide the means for creating very concise descriptions of the components of ideal machines. In addition, the fact that Caliban type constructor functions are taken to be non-strict means that infinite objects, such as the trace of a non-terminating computation, can be defined.

Since it was designed to be a specification language, Caliban also provides a means for working with arbitrary sets. A set together with a bottom element becomes a flat domain called a *sort*. A sort is introduced by declaring its name along with the names and signatures of the functions on it. Properties of the sort and functions are specified by means of axioms. The functions are assumed to be strict and total. The non-standard real numbers we work with in Ariel are represented as a sort.

4.2 Clio's Metalanguage

Clio's metalanguage is essentially the language of first order predicate calculus with Caliban expressions as terms and bounded quantifiers that range over Caliban types. For Ariel's purposes, atomic formulas are of the form $A = B$, where A and B are Caliban expressions enclosed in back quotes. The usual logical connectives are available. Both universal and existential quantifiers are available. All quantifiers are bounded, i.e. they are of the form $(x :: \tau)$ or $(Ex :: \tau)$ where τ is a Caliban type. Metalanguage predicates may be defined.

4.3 Clio's Prover

Clio's prover is basically a term rewriting engine. Whenever an assertion is proved or assumed it is made into a collection of rewrite rules which are added to Clio's rule base. Rules are used in two ways, as conditional rewrite rules and in a limited form of resolution (unit clause resolution when the result is a ground clause). Clio's rule manager also allows the user, interactively, to instantiate existing rules and to expand them (replace symbols by their definitions).

The basic strategy used by the prover is to skolemize an assertion to be proved and then reduce it to conjunctive normal form, i.e. a conjunction of clauses each of which is a disjunction. Each disjunct is a (possibly negated) atomic formula. The prover then attempts to prove each atomic formula by rewriting its left and right sides to the same normal form.

The prover's basic strategy is augmented by a number of additional proof techniques. If the assertion is a conjunction, Clio will break it up into its constituent conjuncts, each of which will have to be proved. If the assertion is an implication, Clio will assume the antecedent and add the corresponding rules to the rule base and then present the consequent to the user to prove (or try to prove it automatically). Existential assertions may be proved by supplying a witness to replace the quantified variable and then proving the resulting assertion. Proof by contradiction may also be used. Universal sentences may be proved either by generalization or structural induction (if the type of the variable of quantification is a constructed type). Disjunctions are proved by proving one of the equivalent assertions: $\neg B \Rightarrow A$ or $\neg A \Rightarrow B$, if the original assertion was $A \vee B$. Negations are proved by first pushing the negation inside the assertion. Then the usual suite of techniques is made available. A case split is available; one may case split both on Caliban and metalanguage expressions.

5 The Ariel Formalization of Numerical Computation

We formalize numerical computations as execution traces of an abstract interpreter. We do so for a number of reasons. First, we intend to expand our semantic model to include concurrent execution and the verification techniques for concurrent programs we find most attractive are based on operational models. Second, we wanted to model, to a certain extent, the possibility that a program was executed on a distributed system comprised of machines with different arithmetic, e.g. different rounding rules. On such a system the time at which a value was computed becomes significant — the approximate value resulting from evaluation of a given expression in the same context would, in principle, be different for different times. Two such values would differ only by an infinitesimal, but they wouldn't be the same. Third, an operational formalization seems most compatible with the concepts of non-standard analysis we use to model real arithmetic. The problem has to do with ensuring that one does not attempt to prove external properties of an execution trace by induction or by means of a proof rule whose soundness depends on an induction on the length of traces. An example of the latter is the theorem about **while**-loops that says that if I is an invariant of the loop body when the loop guard is true, then I and the negation of the loop guard is true when the loop terminates (if it does).

The semantic model used by Ariel is a transition system along the lines of G. Plotkin's work on structured operational semantics [4]. Ariel is based on an abstract interpreter, i.e. a transition system whose configurations consist of a syntactic representation of a program and some data that represent the current state of the program's execution. A program is represented by its abstract syntax tree coded as a Caliban constructed type. A program's static semantics are checked by the parsing program that builds its abstract syntax tree. The static semantic property most pertinent to execution is the resolution of ambiguous identifier references. In an abstract syntax tree an applied occurrence of a program variable is replaced by the abstract syntax tree for the declaration at the binding occurrence of the identifier determined by the scope rules of the language in which the program is expressed. Clio actually implements this replacement as a reference, so the size of the abstract syntax tree does not grow.

A state of the interpreter consists of a stack of procedure activation records called a *context*. Each activation record consists of a representation of which procedure corresponds to the activation record, where control is in that procedure, a partially evaluated expression, and a local store. As an expression is evaluated, its constituent parts are replaced by the results to which they evaluate. A store is a function from abstract locations to machine values (represented by its graph — a list of location-value pairs). A context is an element of a Caliban constructed type.

A next-state function is defined in Caliban. Traces of the abstract interpreter are defined by iterating the next-state function starting in some initial state. Thus, the meaning of a program or a piece of a program can be thought of as a function from initial states to traces. The next-state function may be iterated a fixed number of times or until some point in the program is reached depending on what one wants to verify. Since Caliban is a lazy language, just those parts of a trace required to prove a program's specification will actually be computed.

To illustrate how proofs of asymptotic correctness are mechanized, we must first describe the formalizationm of real arithmetic. The natural numbers are a built-in Caliban type, `NAT`. They are modelled as a constructed type generated by the zero-ary constructor `ZERO` and the strict unary constructor `SUCC`:

`NAT ::= Zero | Succ !NAT`

(The `!` indicates strictness in an argument.) Operations on elements of `NAT` are defined by recursion in the usual way. Caliban also contains built-in types of truth values, `BOOL`, and characters, `CHAR`. Ariel uses the latter two Caliban types for its universes of truth values and characters, respectively. Ariel's integers are constructed from `NAT`. The Caliban definition is

`INT ::= Intzero | Pos !NAT | Neg !NAT`

Note that `Pos Zero` is the first positive integer, namely 1, and analogously for `Neg Zero`. The successor and predecessor functions on `INT` are defined in terms of the successor function on `NAT` in the obvious way.

The non-standard reals are formalized as a Caliban sort. Recall that a Caliban sort is essentially a set (made into a flat domain by appending $\perp$ below all the set's elements) plus functions on the set which are constrained to be strict and total. Sorts are incorporated into the Caliban domain by making their elements new atomic elements of the latter. The definition of the non-standard reals as a Caliban sort is as follows.

```
REAL :: sort including Int_to_real !INT
        with
rstp :: REAL -> REAL
            rfinite :: REAL -> BOOL
            rst :: REAL -> BOOL
            rzero :: REAL
            rplus :: REAL->REAL->REAL
            rminus:: REAL -> REAL -> REAL
            rneg :: REAL -> REAL
            rmult :: REAL -> REAL -> REAL
```

```
                rrecip :: REAL -> REAL
                rless :: REAL -> REAL -> BOOL
rclose :: REAL -> REAL -> BOOL
```

The `including Int_to_real !INT` clause indicates here that the sort contains all the integers plus additional elements; `Int_to_real` may be thought of as an injection function which also may be used as a constructor. It may be used in definitions by pattern matching, for example.

The functions on `REAL` are declared to satisfy a set of axioms — the usual axioms for the reals, minus the least-upper-bound axiom (which property the non-standard reals lack).

The axioms for non-standard analysis are based on the standard-part map, `rstp`, and the predicates `rfinite`, `rst`, and `rclose` (finite, standard and close, respectively). Ideally, there should be a single, polymorphic, undefined predicate *standard*, which should commute with applications of standard functions. This would yield a completely general transfer principle. However, such a predicate could not, in general, be continuous on non-flat domains such as function spaces, since its standardness would have to depend on only finitely many of its values, which is not the case. Therefore, it could not be part of the Caliban domain. Instead, our axioms model consequences of the transfer principle. One of these consequences is that the usual operations on the reals (plus, times, reciprocal, max, min, etc.) commute with the standard-part map. For example,

```
AXIOM
  (x)(y)'rstp (rplus x y)'='rplus (rstp x) (rstp y)'
```

(The back quotes set off Caliban expressions in metalanguage formulae; `(x)` and `(y)` indicate that `x` and `y` are universally quantified.) There are similar axioms for the other arithmetic operations. Other axioms give the relationships between the standard-part map and the predicates mentioned above. For example,

```
AXIOM
  (x)(y)'rclose x y' = '(rstp x) = (rstp y)'
```

As our main purpose here is to describe our program verification system, we will omit further details of our formalization of non-standard analysis.

We have specified that our hyperfinite ideal interpreter computes results of real arithmetic operations that differ from the ideal ones by an infinitesimal. This state of affairs is modelled by a function, `machine`, of type `[NAT]->REAL->REAL`. The first argument is to be interpreted as a time stamp, i.e. when the value was computed. Time stamps are represented as lists of natural numbers for the following reason. Each activation record in a context contains a count of the number of steps performed so far in the interpretation of the procedure corresponding to the activation record. The global time of the interpreter can conveniently be taken to be the ensemble of these local times collected together into a list in the order in which they appear in the context. Such a representation of global time turns out to be useful for differentiating the multiple incarnations of local variables that arise when a procedure is activated recursively. The second argument to `machine` is the ideal value approximated by the machine value. The meaning of this function is captured by the following axiom:

```
AXIOM 'rstp (machine n x)'='(rstp x)'.
```

6 Symbolic Execution Using Clio

The basic difference between the Ariel approach to symbolic execution and others(cf. [2]) is that Ariel executes symbolically a complete semantic model of the program to be verified. Each state of an Ariel ideal machine represents all pertinent semantic information and each state of a symbolic execution trace is available directly to the theorem prover. The utility of having this much information available is that it allows us to verify *any* program written in the language supported, e.g programs that use aliasing or constructs for which proof rules or predicate transformers are not available. It allows us to verify security or real time properties which require reasoning about of sequences of states. It allows us to prove *total* correctness.

The Ariel system is interactive and integrated with its theorem prover, Clio. Given a specification for a procedure, one verifies it by proving that its symbolic execution trace satisfies its specification. If the program is specified by a precondition and a postcondition, one proves that the final state of such a trace satisfies the postcondition under the assumption that the initial state satisfied the precondition. This is presented to Clio as the assertion that the precondition implies the postcondition. Clio's built-in proof rule for implication causes the precondition to be assumed and a proof of the postcondition to be started; assumption of the precondition causes it to be transformed into a collection of rewrite rules.

The first step Clio takes to prove the postcondition is to send it to the reducer to be put into normal form. This invokes symbolic execution.

A typical postcondition might say that the value returned by a function-procedure call (perhaps via a parameter) has some property. Assume, for the sake of discussion, that the postcondition of the procedure we wish to verify is

```
Post 'x' 's' 'v' := 'Return_state x s' = 'v'
```

where **Return_state** is a Caliban function that iterates (in the sense of repeated function application) the next-state function beginning in state **s** until the current command is a return. Assume **x** is one of the procedure's parameters and **v** denotes some machine value. Clio puts **Post** into normal form by applying to it all the applicable rewrites in its rule base. When the definitions specifying the Ariel abstract machine are compiled by Clio, they are transformed into rewrite rules. So, the process of normalizing **Post** is essentially a matter of carrying out the repeated application of the next-state function beginning with a rule base that reflects the assumptions of the precondition. If the procedure to be verified consists of no more than a sequence of assignment statements, then when **Post** is normalized, the store of the resulting state will reflect the effect of the sequence of assignments and the postcondition is proved by "looking up" **x** in the final store and seeing whether its value is **v**. Clio should be able to do this automatically.

The usual treatment of conditional statements in symbolic execution is to bifurcate the current path in the symbolic execution tree and attach the conditional statement's guard and its negation to the pathconditions of the respective newly created paths. In the Ariel system, this is handled by a case split. Clio detects the need for a case split automatically, when it encounters a Boolean subexpression whose value it is unable to determine in the process of normalizing an assertion. What happens is that the truth of the subexpression is assumed (and made into rules) and the proof of the assertion continued and similarly for the assumption that the subexpression is *false*. Ariel represents the path condition of other symbolic execution schemes by Clio's rule base, which makes all Clio's rule manipulation facilities available for arguments involving the path condition.

Loops are treated as follows. An assertion is attached to the beginning of the loop statement. This assertion should state

1. that each time control is at the beginning of the loop, the loop guard is *true*,

2. some invariant having to do with what the loop is to compute, and

3. that some function of the variables manipulated by the loop is strictly decreasing.

We then prove that this assertion cannot hold of any unbounded trace of a hyperfinite interpreter. The proof uses a theorem to the effect that in any unbounded trace of a hyperfinite machine, there are two states in which all identifiers are assigned the same values. This contradicts 3. Thus we first prove that the loop terminates provided that the invariance of the loop assertion can be established. Clio makes this into a rule (or rules). The postcondition for the loop should consist of 2 and the negation of 1, i.e. when the loop ends (because its guard evaluates to *false*), it establishes what we wanted to compute. This is familiar. We prove the postcondition by contradiction. Having assumed its negation, we then prove the loop assertion by induction to get the required contradiction. If parts 2 and 3 of the loop assertion are actually invariant, then the only problem is with a state in which the loop guard is *false*. In this case, symbolic execution is invoked from the state in which control is at the beginning of the loop for the ith time, which allows Clio to prove the postcondition — a contradiction. If, in the state at the beginning of the ith iteration, the loop guard evaluates to *true*, the induction step can be proved directly.

Subroutines may be verified independently. A subroutine is specified by giving a pre- and postcondition for it. Using Clio, one proves that the subroutine terminates in a state satisfying its postcondition when started up in a state satisfying its precondition. Having successfully carried out the verification, one has Clio make the specification into rewrite rules. The latter will essentially be conditional rewrites which add the content of the postcondition to the rule base, whenever it's possible to prove the precondition.

7 A Worked Example

We shall use as an example a Newton's-method iteration for computing approximations to square roots. This choice is appropriate for a number of reasons. First of all, it is an eminently practical way to compute square roots: it converges very quickly (assuming one starts reasonably close to the desired square root—which isn't hard to arrange). Second, it illustrates the points we want to make in favor of our methods. If the ideal machine on which one computes a Newton iteration has the standard real numbers as its real data type, then the iteration never terminates; it produces closer and closer approximations to the desired square root, forever. On the other hand, a precise numerical analysis of Newton's method, given some machine with standard finite arithmetic, is difficult.

Our method for computing the square root of a positive, non-zero real number a is to find a zero of the function $f(x) = x^2 - a$ using Newton's method. In general, one has the following recurrence relation for computing successive iterates using Newton's method

$$x_{n+1} = x_n - \frac{f(x_n)}{f'(x_n)}.$$

In our case, x_{n+1} is computed according to the following relation

$$\begin{aligned} x_{n+1} &= x_n - \frac{x_n^2 - a}{2x_n} \\ &= \frac{x_n^2 + a}{2x_n} \\ &= \frac{1}{2}(x_n + \frac{a}{x_n}). \end{aligned} \tag{1}$$

We begin with an initial approximation $a + 1$; it need only be greater than $\sqrt{a}$, since both $f(x) > 0$ and $f'(x) > 0$ for $x > \sqrt{a}$. A C program that implements the Newton's Method approximation to the square root function is given in Figure 1.

To verify `square_root`, we first need a specification for it. Informally, the specification is straightforward: when passed a positive real number as argument, `square_root` should compute an approximation that is "close to" the square root of its argument. However, we want to verify the asymptotic correctness of `square_root`. Thus, we need to specify that when it's passed an argument that approximates a value in the domain of the standard square-root function which is representable on the machine it is to be run on, `square_root` computes a result that approximates the ideal square root of its argument. This translates to a requirement that the argument be positive by more than an infinitesimal and that it be finite. If the argument were a positive infinitesimal, then it is "too small" — it underflows to zero. Similarly, if the argument were infinite, we say that it is "too large" — it overflows. In either case the argument is beyond the representational power of the machine on which we want to run `square_root`. We want to verify that when passed an argument that is neither too small nor too large, `square_root` computes a result that is close to ideal.

```
double square_root(a)
double a;
{
  double x;
  double y;

  x = a + 1.0;
  y = (x + (a/x))/2.0;
  while ((y < x) && (a<y*y)) {
    x = y;
    y = (x + (a/x))/2.0;
  };
  return (x);
}
```

Figure 1: A C program for computing square roots using Newton's Method

Our specification for `square_root` will be formalized in terms of a pre- and post-condition expressed in Clio's metalanguage. The pre-condition will assert that in the context of a call on `square_root` its actual parameter has the properties described above; its post-condition will assert that, likewise, the value it returns has the property described above. The definition of `square_root`'s precondition in Clio's assertion language is as follows.

```
precond 'start' '[a]' :==
  '(IS_Real a)'='True' &
  '(finite a)'='True' &
  '(rless rzero (_real a))'='True' &
  '(rless rzero (rstp (_real a)))'='True'
```

The first conjunct of the definition expresses the (static semantic) requirement that `square_root`'s actual parameter be of type *real*. This is checked by the routine that builds the abstract syntax tree. It is reasserted in the precondition to inform Clio of the fact. The remaining conjuncts state the requirements paraphrased above, namely that the actual parameter be positive by more than an infinitesimal and finite.

The intuitive content of the postcondition for `square_root` is that the square of the value it returns is "close-to" the value of the actual parameter of its call. In Clio's assertion language postcond has the following definition.

```
postcond 'start' '[a]' :=
  'ustp (utimes
         (return_value (call start square_root [a]))
         (return_value (call start square_root [a])))'
   =  'Real (rstp (_real a))'
```

The Caliban function return_value symbolically executes a function call by iterating the next-state function beginning with the context supplied it as argument until a return command is executed. It then extracts the value to be returned from the call from the final state of the trace of the function execution. The function call constructs the initial context of the function from the context passed it as argument, the abstract syntax tree for the function to be called, and the actual parameters of the call. In words, postcond says that the square of the value returned by square_root is infinitely close to the value passed as actual parameter in the just-terminating call.

Our specification of square_root is that its pre-condition implies its post-condition. This is formalized by means of a metalanguage predicate correctness:

```
correctness 'function' precond postcond 'l' :==
             (initial) ('is_well_defined initial' = 'True' &
                        precond 'initial' 'l' ) ->
                        postcond 'initial' 'l'
```

This predicate simply asserts that beginning in any well-defined initial context satisfying a precondition (which is given as an argument) a procedure (again given as an argument) will terminate in a context satisfying a post-condition. Note that we are specifying *total* correctness. The specification for square_root is then

```
spec :== correctness 'square_root' precond postcond '[input]'
```

where input is the symbolic value whose square root we will "compute".

Our strategy for proving the total correctness of square_root is to assume the negation of spec and then prove by induction on the length of traces an invariant which contradicts a basic theorem about hyperfinite machines. In particular, this invariant will assert that the values assigned to x, the successive approximations to $\sqrt{a}$, decrease unboundedly. That this can't be the case for a hyperfinite machine follows by transfer from the fact that it's not true for a standard finite machine. The heart of the proof establishes the invariance of the invariant. The intuitive content of the invariant is that the following are true at the outset of each iteration of the **while**-loop:

- the values assigned to `x` so far are strictly decreasing and
- the value assigned to `a` is unchanged.

The invariant is formalized as follows.

```
invariant 'start' 'i':=
    reached 'start' 'looptop' 'i' &
    '(val a start looptop i = (val a start looptop #0)) &
    (IS_Real (val y start looptop i)) &
    (IS_Real (val x start looptop i))'='True' &
    'decreasing (val x start looptop) i'='True' &
    'loop_guard start looptop i'='boolean True'
```

Here `looptop` denotes the root of the abstract syntax tree for the **while**-statement in `square_root` and `i` is a natural number that represents the number of times control has reached `looptop` in a trace whose initial context is `start` ; `reached 'start' 'looptop' 'i'` asserts this. `val a start looptop i` gives the value stored in the abstract address to which `a` is bound when `reached 'start' 'looptop' 'i'` holds and analogously for `val x start looptop i` and `val y start looptop i`. The predicate `decreasing` takes as arguments a function f from natural numbers to machine values and a natural number n. The latter is taken to be the endpoint of an initial segment of natural numbers. `decreasing` is true of f and n, just in case $f(i) > f(i+1)$ for $0 < i < n-1$. In `invariant`, `decreasing` is passed `val x start looptop` which, since Caliban functions are Curried, is a function from natural numbers to machine values, as required. The last clause of `invariant` asserts that the **while**-statements's guard always evaluates to *true*; our assumption that `spec` is false will allow us to prove this and obtain the desired contradiction.

There are four lemmas that we will use in the verification of our example. They summarize the mathematics underlying the method on which the subroutine is based. We will not give their proof here.

The first says that provided an approximation is greater than but not too "close to" the square root of a, the next approximation computed by `square_root` will be strictly less than the previous one and greater than than $\sqrt{a}$. The ideal sequence is strictly decreasing and bounded below by

$\sqrt{a}$; this lemma states that the computed values have this property until the round-off error in the computation of the next approximation becomes significant relative to the difference between $\sqrt{a}$ and the current approximation. The problem is to define "close to", i.e. to compute the amount of error introduced in the computation of the next approximation. Here we see the utility of the simplifications of the asymptotic paradigm. The error introduced in the calculation of the next approximation in `square_root` is an infinitesimal, since the error introduced in each operation in its computation is an infinitesimal and there are only finitely many such operations. In contrast, the term for the error in the next approximation for a standard finite machine is complicated and the complexity of such terms grows rapidly with the complexity of their associated expressions.

Now the question is does this formalization model the asymptotic behavior of `square_root`, when the approximations it computes are close to $\sqrt{a}$? Yes. Let $F(a,x) = \frac{1}{2}(x + \frac{a}{x})$ be the ideal value of the approximation computed from x and let $machine(x)$ be the machine representation of the ideal value x for a given machine. On a standard finite machine, $machine(F(a,x))$ is contained in an interval $(F(a,x) - \epsilon, F(a,x) + \epsilon)$ where ϵ is the error introduced in the computation of $machine(F(a,x))$. This is also the case on a hyperfinite machine except that ϵ is an infinitesimal. All we know is that the machine value is in the monad of the ideal one; we don't know whether it is larger or smaller than the ideal value. To show that a program is asymptotically correct, we must show that it computes the desired result in spite of this uncertainty.

In the following, *rstp* denotes the standard-part map and F and *machine* are as above.

Lemma 1 *If a and x are finite, $rstp(a) > 0$, and $rstp(a) < rstp(x^2)$, then $a < machine(F(a,x)^2)$ and $machine(F(a,x)) < x$.* □

The second lemma asserts that overflow doesn't occur during the computation of the next approximation in `square_root`. In the asymptotic paradigm, overflow is represented by a an infinite computed value. Lemma 2 asserts that a real number is finite, if it is bounded above and below by finite reals. It is expressed in terms of a predicate *bounded_above* which is true of a real number x if there exists a finite y such that $x \leq y$.

Lemma 2 *If $a < machine(y^2)$, $y < x$, and $bounded\_above(x)$, then y is finite.* □

The third lemma essentially says that the negation of the loop guard in `square_root` implies that the value stored in **x** is infinitely close to the square root of the value stored in **a**.

Lemma 3 *If either of the following are true, then* $rstp(x^2) = rstp(a)$*:*

(i) $a < machine(x^2)$, $rstp(y) = rstp(F(a,x))$, *and it is not the case that* $a < machine(y^2)$, *or*

(ii) $a < machine(x^2)$, $rstp(y) = rstp(F(a,x))$, *and it is not the case that* $y < x$.

□

The fourth lemma states that for any node in a program's abstract syntax tree and any of the program's variables, when the program is interpreted by a hyperfinite machine, the value of the variable when control is at the node cannot decrease unboundedly. In the statement of the lemma we will use a Caliban function *nth_value* which takes a context c, an abstract-syntax-tree node n, an abstract location l, and a natural number i and yields the value stored in l when control is at n for the ith time beginning in c. The Caliban function *decreasing* was introduced above.

Lemma 4 *For all contexts* c*, nodes* n *and abstract addresses* x*, there exists a natural number* i *such that*

$$decreasing\ (nth\_value\ c\ n\ x)\ i$$

is false. □

The invariant is proved by induction on the number of times control is at the beginning of the loop.The base case is straightforward; it follows from Lemma 1 with $a+1$ put for x. The induction step follows essentially from Lemmas 2 and 3. Lemma 2 is necessary to show that the newly computed approximation is finite. The interpreter requires that this be established to guarantee that evaluation of the expression that computes the new approximation does not overflow. We cannot include in **invariant** a conjunct asserting that **y** is finite, because such an assertion is external and, hence, can't be proved by induction. Instead, we require Lemma 2 which states a fact about about the nonstandard reals. The main thing to be shown in the proof of the induction step is that the loop guard evalutes to *true*. If it doesn't, symbolic execution takes control to the return statement where

Lemma 3 implies that `postcond` is satisfied contradicting our assumption to the contrary.

Having established the invariance of `invariant`, Clio then gets a contradiction from Lemma 4 and the conjunct `'decreasing (val x start looptop) i'='True'` of the invariant that proves the total correctness of the program.

8 Conclusions

We have described Ariel and the approach to program verification upon which it is based. We have sketched a sample verification of a numerical program. We would hope to have convinced the reader that symbolic execution and nonstandard analysis provide the means for constructing a viable verification system for numerical programs. We have built a prototype verification system and are in the process of trying it out on a suite of production numerical routines. These routines implement a continuous simulation of a rocket trajectory. They comprise about 1400 lines of C code (translated from the original FORTRAN) and calculate such things as a solution to the two-body problem, a Runge-Kutta integration, a Newton-Raphson iteration to solve a version of Kepler's equation, etc. Our prototype verifys programs written in a subset of C; we are extending the C system to handle Unix system calls and are in the process of designing a new system for (a subset of) Ada.

9 Acknowledgments

The notion of asymptotic correctness and the use of nonstandard analysis to formalize it is due to Ian Sutherland based on a suggestion of Richard Mansfield. We would like to acknowledge the other members of the Ariel group, past and present, who contributed to the work reported here: Ian Sutherland, Doug Hoover, and Steve Brackin. We thank Garrel Pottinger and an anonymous referee for their comments on an earlier draft of this paper.

References

[1] Mark Bickford, Charlie Mills, and Edward A. Schneider. *Clio: An Applicative Language-Based Verification System.* Odyssey Research Asso-

ciates Technical Report TR 15-7. April, 1989.

[2] S.L.Hantler and J.C.King. An introduction to proving the correctness of programs. *ACM Computing Surveys*, 8(3):331–353, September, 1976.

[3] Albert E. Hurd and Peter A. Loeb, *An Introduction to Nonstandard Real Analysis.* Academic Press, 1985.

[4] Gordon D. Plotkin. *A Structural Approach to Operational Semantics.* DAIMI FN-19. Computer Science Department. Aarhus University, Denmark. September, 1981.

[5] Ian Sutherland. *A Mathematical Theory of Asymptotic Computation.* Rome Air Development Center Technical Report RADC-TR-87-261, Air Force Systems Command, Griffiss Air Force Base, NY. 1987.

DIMACS Series in Discrete Mathematics
and Theoretical Computer Science
Volume 3, 1991

On Automatically Distinguishing Inequivalent Processes*

Rance Cleaveland
Department of Computer Science
North Carolina State University
Raleigh, North Carolina 27695-8206
USA

Abstract

This paper describes a technique for generating a logical formula that differentiates between two bisimulation-inequivalent finite-state systems. The method works in conjunction with a partition-refinement algorithm for computing bisimulation equivalence and yields formulas that are often minimal in a precisely defined sense. As a number of other behavioral equivalences may be characterized in terms of bisimulation equivalence, our technique may be used to generate appropriate distinguishing information for these relations also.

1 Introduction

A popular technique for verifying finite-state systems involves the use of a *behavioral equivalence*. In this approach, specifications and implementations are formalized as finite-state machines, and verification consists of establishing that an implementation is equivalent to, in the sense of *behaving the same as*, its specification. A number of equivalences have been proposed in the literature [1, 3, 7, 8, 12, 14], and several automated tools include facilities for computing them [2, 5, 6, 13].

One particularly interesting equivalence is *bisimulation equivalence* [14]. In addition to the fact that a number of other equivalences may be described in terms of it [4], the relation has a *logical* characterization: two systems are equivalent exactly when they satisfy the same formulas in a simple modal logic due to Hennessy and Milner [9]. This fact suggests a useful diagnostic methodology for tools that compute bisimulation equivalence: when two systems are found not to be equivalent, one may explain why by giving a formula satisfied by one and not the other.

The purpose of this paper is to develop a technique for determining a Hennessy-Milner formula that distinguishes two bisimulation-inequivalent finite-state systems. To this end, we show how to use information generated by the partition-refinement algorithm of Kanellakis and Smolka [11] to compute such a formula efficiently. On the basis of our results, the

*Research supported by NSF/DARPA research grant CCR-9014775.

tools mentioned above may be modified to give users diagnostic information in the form of a formula when a system is found not to be equivalent to its specification.

Hillerström [10] also gives a technique for computing a Hennessy-Milner formula. However, his method relies on the use of a backtracking-based algorithm that is less efficient that the more popular partition-based algorithms we are interested in.

The remainder of the paper is organized as follows. The next section defines bisimulation equivalence and examines the connection between it and the Hennessy-Milner Logic. Section 3 then describes a modification of Kanellakis-Smolka bisimulation algorithm that computes distinguishing formulas; a small example is also presented to illustrate the workings of the new algorithm. The section following discusses how our technique may be adapted to various other versions of the Kanellakis-Smolka algorithm, as well as to the partition-refinement algorithm of Paige and Tarjan [15], and it considers how the formulas that are generated may be used to give relevant information when other equivalences are being computed. The final section contains our conclusions and directions for further research.

2 Transition Graphs, Bisimulations and Hennessy-Milner Logic

Finite-state systems may be represented as *transition graphs.* Vertices in these graphs correspond to the states a system may enter as it executes, with one vertex being distinguished as the start state. The edges, which are directed, are labeled with the actions and represent the state transitions a system may undergo. The formal definition is the following.

Definition 2.1 *A* transition graph *is a quadruple* $\langle Q, q, Act, \rightarrow \rangle$, *where:*

- Q *is a set of states (vertices);*
- $q \in Q$ *is the start state;*
- Act *is a set of actions; and*
- $\rightarrow \subseteq Q \times Act \times Q$ *is the derivation relation (set of labeled edges).*

We shall often write $q_1 \xrightarrow{a} q_2$ to indicate that there is an edge labeled a from state q_1 to state q_2; in this case, we shall sometimes say that q_2 is an *a-derivative* of q_1. When a graph does not have a start state indicated, we shall refer to the corresponding triple as a *transition system.* A state in a transition system gives rise to a transition graph in the obvious way: let the given state be the start state, with the other three components of the transition graph coming from the transition system.

Reactive systems [16] compute by interacting with their environment. For such systems, the traditional language equivalence of automata theory is insufficiently discriminating, since the resolution of nondeterministic choices may leave a system in states that react differently to stimuli offered by the environment. Bisimulation equivalence remedies this shortcoming by requiring that equivalent systems have state sets that "match up" appropriately: the start states must be matched, and if two states are matched then they must have matching a-derivatives for any action a either is capable of. These intuitions may be formalized in terms of ***bisimulations*** on a ***single*** transition system.

$$
\begin{aligned}
[\![tt]\!]_T &= Q \\
[\![\neg\Phi]\!]_T &= Q - [\![\Phi]\!]_T \\
[\![\Phi_1 \wedge \Phi_2]\!]_T &= [\![\Phi_1]\!]_T \cap [\![\Phi_2]\!]_T \\
[\![\langle a\rangle\Phi]\!]_T &= \{\, q \in Q \mid \exists q'.\, q \xrightarrow{a} q' \wedge q' \in [\![\Phi]\!]_T \,\}
\end{aligned}
$$

Figure 1: The semantics of formulas in Hennessy-Milner Logic.

Definition 2.2 *Let $\langle Q, Act, \rightarrow\rangle$ be a transition system. Then a relation $R \subseteq Q \times Q$ is a* bisimulation *if, whenever $q_1 \; R \; q_2$, the following hold.*

1. *If $q_1 \xrightarrow{a} q_1'$ then there is a q_2' such that $q_2 \xrightarrow{a} q_2'$ and $q_1' \; R \; q_2'$.*

2. *If $q_2 \xrightarrow{a} q_2'$ then there is a q_1' such that $q_1 \xrightarrow{a} q_1'$ and $q_1' \; R \; q_2'$.*

Two states and in a transition system are *bisimulation equivalent* if there is a bisimulation relating them. When q_1 and q_2 are bisimulation equivalent we shall write $q_1 \sim q_2$.

Let $G_1 = \langle Q_1, q_1, Act, \rightarrow_1\rangle$ and $G_2 = \langle Q_2, q_2, Act, \rightarrow_2\rangle$ be two transition graphs satisfying $Q_1 \cap Q_2 = \emptyset$. Then G_1 and G_2 are bisimulation equivalent exactly when the two start states, q_1 and q_2, are equivalent in the transition system $\langle Q_1 \cup Q_2, Act, \rightarrow_1 \cup \rightarrow_2\rangle$. This definition may be generalized to arbitrary transition systems (i.e., ones whoses state sets are not disjoint), at the cost of a slightly more complicated definition for the transition system in which bisimulation equivalence is to be computed. In the remainder of the paper we only consider the problem of determining a formula that distinguishes two inequivalent states in a transition system.

A number of behavioral equivalences may be characterized in terms of bisimulation equivalence on suitably transformed transition systems [4, 5, 6]. For example, if the transition system is *deterministic*, meaning that every state has at most one a-derivative for a given a, then bisimulation equivalence coincides with *language equivalence* from formal language theory. To determine if two states in an arbitrary transition system are language equivalent, it suffices to apply a "determinizing" transformation to the transition system and then determine whether their corresponding states in this new system are bisimulation equivalent. Other equivalences, including testing equivalence [7, 8], failures equivalence [3], and observational equivalence [14], may be computed in an analogous fashion.

Bisimulation equivalence also has a *logical* characterization in terms of Hennessy-Milner Logic (HML) [9]: two states are equivalent exactly when they satisfy the same HML formulas. The syntax of HML is defined as follows, where $a \in Act$.

$$\Phi ::= tt \mid \neg\Phi \mid \Phi \wedge \Phi \mid \langle a\rangle\Phi$$

Given a transition system $T = \langle Q, Act, \rightarrow\rangle$, the interpretation of the logic maps each formula to the set of states for which the formula is "true"; Figure 1 gives the formal definition. In the remainder of the paper we shall omit explicit reference to the transition system used to interpret formulas when it is clear from the context. Intuitively, the formula tt holds of any state, and $\neg\Phi$ holds of a state if Φ does not. $\Phi_1 \wedge \Phi_2$ holds of a state if both Φ_1 and Φ_2 do, while the modal proposition $\langle a \rangle \Phi$ holds if the state has an a-derivative for which Φ holds. We shall say that a state q in transition system T *satisfies* formula Φ if $q \in [\![\Phi]\!]_T$.

Let $H(q)$ be the set of HML formulas that a state q satisfies:

$$H(q) = \{ \Phi \mid q \in [\![\Phi]\!] \}.$$

The next theorem is a corollary of a result proved in [9].

Theorem 2.3 *Let $\langle Q, Act, \rightarrow \rangle$ be a finite-state transition system, with $q, q' \in Q$. Then $H(q) = H(q')$ if and only if $q \sim q'$.*

It follows that if two states in a (finite-state) transition system are inequivalent, then there must be a HML formula satisfied by one and not the other. This is the basis of the following definition of distinguishing formula.

Definition 2.4 *Let $\langle Q, Act, \rightarrow \rangle$ be a transition system, and let $S_1 \subseteq Q$ and $S_2 \subseteq Q$. Then HML formula Φ* distinguishes *S_1* from *S_2 if the following hold.*

1. $S_1 \subseteq [\![\Phi]\!]$.

2. $S_2 \cap [\![\Phi]\!] = \emptyset$.

So Φ distinguishes S_1 from S_2 if every state in S_1, and no state in S_2, satisfies Φ. Theorem 2.3 thus guarantees a formula that distinguishes $\{q_1\}$ from $\{q_2\}$ if $q_1 \not\sim q_2$.

Finally, we shall adopt the following criterion in assessing whether a distinguishing formula contains extraneous information.

Definition 2.5 *Let Φ be a HML formula distinguishing S_1 from S_2. Then Φ is* minimal *if no Φ' obtained by replacing a non-trivial subformula of Φ with the formula tt distinguishes S_1 from S_2.*

Intuitively, Φ is a minimal distinguishing formula for S_1 with respect to S_2 if each of its subterms plays a role in distinguishing the two.

3 Computing Distinguishing Formulas

In this section, we describe an algorithm for computing bisimulation equivalence [11] and show how to alter it to compute distinguishing formulas. We then consider a small example that illustrates the behavior of the modified algorithm.

3.1 The Kanellakis-Smolka Algorithm

The Kanellakis-Smolka algorithm exploits the fact that an equivalence relation on a set of states may be viewed as a *partition*, or set of pairwise-disjoint subsets (called *blocks*) of the state set whose union is the state set. In this representation blocks correspond to the equivalence classes—so two states are equivalent exactly when they belong to the same block. Beginning with the partition containing one block (representing the trivial equivalence

```
function split(B, a, B') =
  {{s ∈ B | ∃s' ∈ B'. s -a-> s'}, {s ∈ B | ¬∃s' ∈ B'. s -a-> s'}} − {∅};

algorithm bisim(Q, Act, →);
  begin
    P1 := {Q};
    P2 := ∅;
    while P1 ≠ P2 do begin
      P2 := P1;
      P1 := ∅;
      foreach B ∈ P2 do P1 := P1 ∪ split(B, a, B');
    end
  end
```

Figure 2: The partition refinement algorithm for bisimulation equivalence.

relation consisting of one equivalence class), the algorithm repeatedly *refines* this partition (by splitting blocks) until the associated equivalence relation becomes a bisimulation. In order to determine whether the partition needs further refining, the algorithm looks at each block in turn. If a state in a block B has an a-derivative in a block B' and another state in B does not, then the algorithm splits B into two blocks, one containing the states having an a-derivative in B' and the other containing the states that do not. When no more splitting is possible, the resulting equivalence relation corresponds exactly to bisimulation equivalence on the given transition system. The algorithm is given in Figure 2. Function `split` is used to split one block with respect to another; notice that $\mathtt{split}(B,a,B') = \{B\}$ (i.e. B is not split with respect to a and B') if either all the states in B, or none of them, have an a-derivative in B'. It should also be pointed out that $P_1 = P_2$ exactly when no more splits in P_1 are possible. The worst-case complexity of *bisim* is $O(|\rightarrow| * |Q|)$.

3.2 Generating Distinguishing Formulas

One straightforward way to compute distinguishing formulas is to associate a formula, $\Phi(B)$, with each block B in the partition in such a way that that the following hold.

- $B \subseteq [\![\Phi(B)]\!]$.
- $B' \cap [\![\Phi(B)]\!] = \emptyset$ if $B' \neq B$.

In the initial partition, $\{Q\}$, $\Phi(Q)$ is set to *tt*. Now suppose a block B is split, i.e. suppose there is an action a and another block B' such that $split(B,a,B') = \{B_1, B_2\}$, with every state in B_1 having a transition into B' and no state in B_2 having one. Then $\Phi(B_1)$ may be set to $\Phi(B) \wedge \langle a \rangle \Phi(B')$, while $\Phi(B_2)$ becomes $\Phi(B) \wedge \neg\langle a \rangle \Phi(B')$. Arguing inductively, it is easy to establish that for any block B, a state satisfies $\Phi(B)$ exactly when it is contained in B. Since two states that are not bisimulation equivalent will eventually wind up in different blocks, it is a simple matter to compute a formula that distinguishes such states: just return the formula associated with one of the containing blocks.

Although intuitively appealing, this approach has a drawback; it generates very large formulas. In general, the size of a formula associated with a block will grow in size as 2^r, where r is defined as the number of iterations of the `while` loop in *bisim*. In certain cases $r = |Q|$, so the formulas obtained using this method may be exponential in the size of the state space. There is, however, a polynomial-size representation using a set of propositional equations, so this complexity is not as severe as it seems; moreover, it is the case that an exponential-size formula may also be minimal. More importantly, the formulas include a large amount of extraneous information: not only does such a formula distinguish one state from another inequivalent state, it also distinguishes it from every state to which it is inequivalent. In fact, the formulas generated this way are rarely minimal, and because of this, they are not useful from a diagnostic standpoint.

Another Approach

We now describe a better technique for generating distinguishing formulas. The method uses information computed by a slightly altered version of *bisim* that, in addition to computing the partition as described above, retains information about how blocks are split. Then, a postprocessing step constructs a formula that generates a formula distinguishing the states in one block from the states in another.

Bisim is modified as follows. Rather than discarding an old partition after it is refined, the new procedure constructs a "tree" of blocks as follows. The children of a block are the new blocks that result when the block is split; accordingly, the root is labeled with the block Q, and after each iteration of the `foreach` loop the leaves of this tree represents the current partition. When a block B is split (by $split(B, a, B')$), we place the new block $B_1 = \{ s \in B \mid \exists s' \in B'.\ s \xrightarrow{a} s' \}$ as the left child and the new block $B_2 = \{ s \in B \mid \neg\exists s' \in B'.\ s \xrightarrow{a} s' \}$ as the right child, and we label the arc connecting B to B_1 with a and B'. Recall that every state in B_1 has an a-transition into B' and that no state in B_2 does. If a block is not split during an iteration of the `foreach` loop, it is assigned a copy of itself as its only child.[1] Figure 5 contains an example of such a tree.

Given a block tree computed by the new version of *bisim*, and two states s_1 and s_2 that are inequivalent and hence in different blocks, the postprocessing step builds a formula $\delta(s_1, s_2)$ that distinguishes $\{s_1\}$ from $\{s_2\}$. We describe how $\delta(s_1, s_2)$ is computed in two steps. We first give a simpler method that conveys the intuition behind our approach, and then we show how to modify it to generate smaller formulas. The simpler computes a formula $\Delta(B_1, B_2)$ distinguishing block B_1 from block B_2 using the following general idea. Suppose that B_1 and B_2 are siblings, with B_1 being to the left of B_2. Let a and B' be the labels on the arc from the parent to B_1. From the way that the block tree is generated, we know that every state in B_1 has an a-derivative in B' and that no state in B_2 does. Accordingly, if one recursively builds formulas that distinguish B' from each block containing an a-derivative of B_2 and takes their conjunction (call it Φ), then every state in B_1 satisfies $\langle a \rangle \Phi$ and no state in B_2 does. The details of $\Delta(B_1, B_2)$ for general B_1 and B_2 are the following.

1. Let P be the deepest common ancestor of B_1 and B_2; let L be the left child of P and

[1] **Strictly speaking, this is not necessary; these blocks may be left childless. We include these spurious children to simplify our inductive argument of correctness.**

R the right child. (Notice that either $B_1 \subseteq L$ and $B_2 \subseteq R$, or $B_2 \subseteq L$ and $B_1 \subseteq R$.) Let a and B' be the labels on the arc from P to L.

2. Compute $R_a = \{ C \mid \exists s \in R, s' \in C.\ s \xrightarrow{a} s' \wedge C \cap B' = \emptyset \wedge \forall C' \supset C.\ C' \supset B' \}$; these are the blocks that contain an a-derivative of some state in R (first conjunct) and have a sibling that is a (not necessarily proper) ancestor of B' (second and third conjuncts). This last requirement also ensures that each $C \in R_a$ is at least one level higher in the tree than R.

3. Set $\Phi = \bigwedge_{C' \in R_a} \Delta(B', C')$, where $\bigwedge_{C' \in \emptyset} \Delta(B', C') = tt$.

4. If $L = B_1$, then return the formula $\langle a \rangle \Phi$; otherwise, return the formula $\neg \langle a \rangle \Phi$.

We now have the following.

Theorem 3.1 *$\Delta(B_1, B_2)$ distinguishes B_1 from B_2.*

Proof. By induction on the depth of B_1 and B_2 in block tree. □

As a rule, $\Delta(B_1, B_2)$ is much shorter than the formula generated using the naive method mentioned previously; it will always be no larger than this formula. It may also be used to construct a formula distinguishing two inequivalent states s_1 and s_2; first locate disjoint blocks B_1 and B_2 such that $s_i \in B_i (i = 1, 2)$, then build $\Delta(B_1, B_2)$. In general, however, $\Delta(B_1, B_2)$ is *not* minimal. Roughly speaking, one reason for this is that in step (2), $\Delta(B', C_i)$ may also distinguish B' from C_j for some other C_j in R_a, meaning that $\Delta(B', C_j)$ need not be included as a conjunct. Moreover, even if $\Delta(B_1, B_2)$ were a minimal formula distinguishing B_1 from B_2, it would not necessarily be a minimal formula distinguishing $\{s_1\}$ from $\{s_2\}$, where $s_i \in B_i$.

We now describe our method for computing $\delta(s_1, s_2)$. Although this formula will not necessarily be minimal either, it will in general be much smaller than the formula computed using Δ as described above; it is guaranteed to be no larger. The details are as follows.

1. Determine the deepest block P in the block tree such that $s_i \in P$ for $i = 1, 2$. Let L and R be the left and right children of P, with a, B' labeling the arc from P to L. Note that either $s_1 \in L$ and $s_2 \in R$, or vice versa. Let s_L be the state in L, and s_R the state in R.

2. Execute the code in Figure 3. The idea is the following. For each state in B' that is an a-derivative of s_L we will generate a minimal set of formulas satisfied by s_L whose conjunction is satisfied by no derivative of of s_R. We will then take the set yielding the smallest conjunction. *Size* is the variable used to record the size of the current shortest conjunction, while Γ contains the current collection of formulas being built.

3. If $s_L = s_1$ then return $\langle a \rangle \Phi$; otherwise, return $\neg \langle a \rangle \Phi$.

Theorem 3.2 *The formula $\delta(s_1, s_2)$ distinguishes $\{s_1\}$ from $\{s_2\}$.*

```
Size := ∞;
S_L := { s' | s_L -a-> s' } ∩ B';
S_R := { s' | s_R -a-> s' };
foreach s'_L ∈ S_L do begin
  Γ := ∅;
  foreach s'_R ∈ S_R do begin
    Φ' := δ(s'_L, s'_R);
    Γ := Γ ∪ {Φ'};
  end;
  foreach Φ_i ∈ Γ do begin
    S_i := { s' ∈ S_R | s' ∉ [[Φ_i]] ∧ ∀Φ_j ∈ Γ. i ≠ j ⇒ s' ∈ [[Φ_j]] };
    if S_i = ∅ then Γ := Γ − {Φ_i};
  end;
  if |⋀Γ| < Size then begin
    Size := |⋀Γ|;
    Φ := ⋀Γ;
  end;
end;
```

Figure 3: Code for generating conjunctions.

Proof. By induction on the depth of the deepest block in the block tree containing both s_1 and s_2. □

In general, $\delta(s_1, s_2)$ will not be minimal. However, it is possible to characterize situations when $\delta(s_1, s_2)$ will be minimal, as the following result indicates.

Theorem 3.3 *Suppose that in each recursive call to δ generated by $\delta(s_1, s_2)$ ($s_1 \not\sim s_2$) the following holds.*

1. *Let Γ be the set of conjuncts used to build Φ and s'_L the state in s_L that was used to create Γ. Then for each $\Phi' \in \Gamma$ there is an $s_j \in S_R$ such that Φ minimally distinguishes $\{s'_L\}$ from $\{s_j\}$, and $s_j \in [\![\Phi'']\!]$ for all other $\Phi'' \in \Gamma$.*
2. *$\{ s' \mid s_L \xrightarrow{a} s' \} - B' \subseteq S_R$.*

Then $\delta(s_1, s_2)$ is minimal

Proof. By contradiction. Of importance is the fact that Φ is the shortest length formula that δ can build to distinguish an a-derivative of s_L from all a-derivatives of s_R. □

It is also the case that a minimizing procedure can be applied to $\delta(s_1, s_2)$ once it has been computed; the result of this would be a minimal formula. The minimizing procedure is straightforward: repeatedly replace subformulas in the formula by *tt* and see if the resulting formula still distinguishes s_1 from s_2. If so, the subformula may either be omitted (if it is one of several conjuncts in a larger conjunction) or left at *tt*. The computational tractability of this procedure remains to be examined, however.

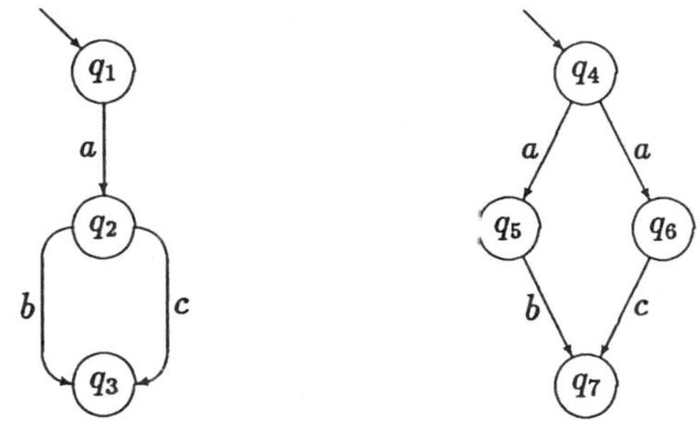

Figure 4: Two inequivalent transition systems.

It should be noted that δ may still generate exponential length formulas. However, one may represent such a formula (as a set of propositional equations) in space proportional to $|Q|^3$. This results from the fact that there can be at most $|Q|^2$ total recursive calls generated by the above procedure and the fact that each distinguishing formula is of the form $(\neg)\langle a \rangle \Phi$, where Φ contains at most $|Q|$ conjuncts, each of the form $\delta(s_i, s_j)$ for some s_i and s_j. By saving information appropriately and modifying the procedure for δ so that the semantic information of the formula computed is also returned, we may establish the following bound on the amount of computation needed to compute such a series of equations.

Theorem 3.4 *An equational representation of $\delta(s_1, s_2)$ may be calculated in $O(|Q|^5)$ time, once the tree of blocks has been computed.*

Proof. Follows from the fact that determining the equation for each recursive call of δ requires $O(|Q|^3)$ work. □

We close this subsection with some general remarks about our method. One feature of our approach is that the overhead involved in maintaining the block tree is minimal; nodes need not be labeled with the corresponding sets of states, except at the leaves. Also, the postprocessing step is only invoked after equivalence is computed—so if the states of interest are found to be equivalent, then the postprocessing may be avoided altogether.

3.3 An Example

To illustrate our algorithm we consider a well-known example of two transition graphs that are not bisimulation equivalent. Figure 4 shows the transition system that includes the two transition graphs. State q_1 is the start state of one graph, while state q_4 is the start state of the other. Figure 5 contains the tree of blocks generated by the altered *bisim* algorithm. Notice that $q_1 \not\sim q_4$, as they are in different blocks.

In order to build $\delta(q_1, q_4)$, the algorithm first locates the lowest common ancestor of the two blocks (B_2, in this case). The left child is B_{10}, the right child is B_{11}, the action causing the split is a, and the block causing the split is B_6. The formula that will be returned, then, will be

$$\langle a \rangle (\delta(q_2, q_6) \wedge \delta(q_2, q_5));$$

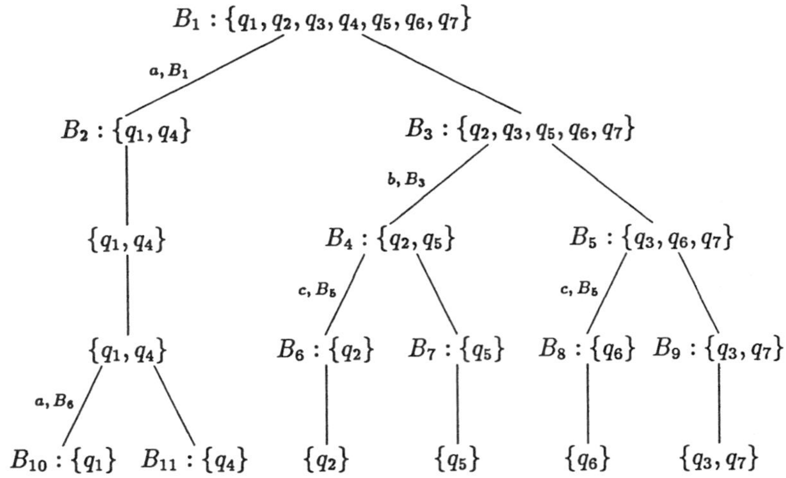

Figure 5: The tree of blocks generated by *bisim*.

this formula holds of q_1 and not q_4. By repeating this process, it turns out that

$$\begin{aligned}\delta(q_2, q_6) &= \langle c \rangle tt \text{ and}\\ \delta(q_2, q_5) &= \langle b \rangle tt.\end{aligned}$$

So the formula distinguishing q_1 from q_4 is

$$\langle a \rangle(\langle c \rangle tt \wedge \langle b \rangle tt).$$

This formula states that q_1 and q_4 are inequivalent because q_1 may engage in an a-transition and evolve into a state from which both b- and c-transitions are available. Note that this formula is minimal. By way of contrast, the formula generated by the first naive method would be the following.

$$\langle a \rangle tt \wedge \langle a \rangle(\neg\langle a \rangle tt \wedge \langle b \rangle\neg\langle a \rangle tt \wedge \langle c \rangle(\neg\langle a \rangle tt \wedge \neg\langle b \rangle\neg\langle a \rangle tt))$$

This formula is clearly not minimal, since, for example, the formula obtained by substituting tt for subformula $\langle a \rangle tt$ is still distinguishes q_1 from q_4.

4 Implementation Issues

The version of the Kanellakis-Smolka algorithm used in the previous section is naive in several respects. In this section we briefly consider the application of our technique to more sophisticated partition-refinement strategies. We also show how the formulas generated when using ***bisim*** to compute ***language equivalence*** may be converted into words accepted by one state and not another.

4.1 Other Refinement Strategies

There are several ways in which the bisimulation equivalence algorithm in the previous section may be improved. We consider some of them here and briefly indicate how the computation of δ may be adapted to suit them.

1. *Early termination.* Early termination may be used to improve the speed (although not the complexity) of the Kanellakis-Smolka algorithm when one is interested in the equivalence of two particular states. If the two states ever wind up in different blocks, then they will never be deemed equivalent, and one may terminate the equivalence-checking procedure. Our method is directly applicable in this situation, since it does not require that bisimulation equivalence on the transition system be completely determined in order to generate a formula.

2. *Early partition updating.* Another method for speeding up the the equivalence checking algorithm is to *replace* a block that is split by the resulting subblocks in the partition under consideration. Our formula-generation technique may be used as is; only the inductive proofs of correctness and minimality would change.

3. *Paige-Tarjan splitting.* Paige and Tarjan [15] develop a partition-refinement algorithm whose complexity is $O(|\rightarrow| * \log|Q|)$ for transition system $\langle Q, Act, \rightarrow\rangle$. The essence of their approach is a more elaborate splitting procedure that, in certain instances, enables blocks to be split with respect to two blocks by only examining one of the two blocks. Their algorithm also uses the inverse of the $\rightarrow$ relation. Although it is beyond the scope of this paper to consider this approach fully, we remark that our technique may be adapted to computing distinguishing formulas in this setting.

4.2 Language Equivalence

A number of other behavioral equivalences may be seen as instances of bisimulation equivalence applied to appropriately transformed transition systems. For these equivalences, the formulas that are generated during the bisimulation-checking step may be interpreted in ways that convey more appropriate diagnostic information for the equivalence under consideration. To illustrate this, we briefly indicate how the formulas generated using our approach may be used to distinguish states that are not language equivalent. Other equivalences may be treated analogously.

Language (also called *trace*, or *may*) equivalence on states in a transition system corresponds to bisimulation equivalence on a "determinized" version of the transition system [4]. As two states are equivalent in this sense exactly when they are capable of engaging in the same sequences of actions, the most appropriate way to distinguish two inequivalent states is to produce a sequence of actions that one state is capable of and the other is not. Based on the formula constructed during the bisimulation checking phase of the algorithm, it is possible to generate such a sequence on the basis of the following properties of minimal distinguishing formulas for deterministic transition systems.

- A minimal distinguishing formula for two states will contain no subformulas involving conjunctions. This results from the fact that for such systems, $[\![\langle a\rangle(\Phi_1 \wedge \Phi_2)]\!] =$

$[\![\langle a\rangle\Phi_1 \wedge \langle a\rangle\Phi_2]\!]$; thus, if $\langle a\rangle(\Phi_1 \wedge \Phi_2)$ is a distinguishing formula then either Φ_1 or Φ_2 is redundant.

- If $\langle a\rangle\neg\Phi$ is a minimal distinguishing formula for state q_1 with respect to state q_2, then so is $\neg\langle a\rangle\Phi$. This results from the fact that q_1 must satisfy $\neg\langle a\rangle\Phi$ and the fact that the minimality of $\langle a\rangle\neg\Phi$ ensures that q_2 has an a-derivative.

The latter property is especially important, since it allows negations to be "driven out" of the bodies of modal formulas. Accordingly, it is possible to derive a formula of the form $(\neg)\langle a_1\rangle\cdots\langle a_n\rangle tt$ from the one generated; this indicates that one of the states is capable of performing the sequence $a_1\cdots a_n$ and the other is not. Note that it is important that the distinguishing formula generator generate minimal formulas for states (not just blocks).

5 Conclusions and Future Work

This paper has shown how it is possible to alter partition-refinement based algorithms for computing bisimulation equivalence to compute a formula in the Hennessy-Milner Logic that distinguishes two inequivalent states. The generation of the formula relies on a postprocessing step that is invoked on a tree-based representation of the information computed by the equivalence algorithm. The formulas are often minimal in a certain sense, and the postprocessing step has a $O(|Q|^2)$ effect on the worst-case complexity of the equivalence-checking algorithm.

There are several avenues for future work to be pursued. Clearly, the complexity of the minimization procedure mentioned in passing at the end of Section 3 needs to be analyzed fully; if this procedure is efficient enough, then it may be incorporated into the distinguishing formula generation procedure. Another area of investigation would involve an implementation of our technique; we plan to incorporate this distinguishing formula capability into the Concurrency Workbench [5, 6], a tool for the analysis of finite-state systems. Yet another involves determining appropriate ways of using formulas computed in the course of checking equivalences other than bisimulation equivalence. Of particular interest is *testing (or failures) equivalence* [3, 7, 8]. These equivalences may be characterized in terms of the tests a process may pass and must pass. One method for distinguishing states that are not testing equivalent would be to build a test based on the formula computed by the bisimulation equivalence checker that one state may (or must) pass and that the other must (or may) not. Finally, it may be possible to extend our techniques to the computation of distinguishing formulas in the context of *preorder* checking. Another method of verifying processes involves the use of a behavioral preorder; in this setting, an implementation satisfies a specification if the implementation is "greater than" (intuitively: "behaves at least as well as") the specification. One interesting preorder is the *prebisimulation preorder*, which has a logical characterization in terms of an intuitionistic variant of the Hennessy-Milner logic: one state is "greater than" another if it satisfies all the formulas satisfied by the latter. This property could serve as the theoretical basis for computing diagnostic information in the same way that the logical characterization of bisimulation equivalence served as the theoretical basis for the techniques described in this paper.

Acknowledgement

I would like to thank Henri Korver for spotting errors in, and for his helpful comments on, previous drafts of this paper.

References

[1] Bloom, B., S. Istrail and A. Meyer. "Bisimulation Can't Be Traced." In *Proceedings of the ACM Symposium on Principles of Programming Languages*, 1988.

[2] Boudol, G., V. Roy, R. de Simone and D. Vergamini. "Process Algebras and Systems of Communicating Processes." In *Proceedings of the Workshop on Automatic Verification Methods for Finite-State Systems*, Lecture Notes in Computer Science 407, 1–10. Springer-Verlag, Berlin, 1990.

[3] Brookes, S.D., C.A.R Hoare and A.W. Roscoe. "A Theory of Communicating Sequential Processes." *Journal of the ACM*, v. 31, n. 3, July 1984, pp. 560-599.

[4] Cleaveland, R. and M. Hennessy. "Testing Equivalence as a Bisimulation Equivalence." In *Proceedings of the Workshop on Automatic Verification Methods for Finite-State Systems*, Lecture Notes in Computer Science 407, 11–23. Springer-Verlag, Berlin, 1990.

[5] Cleaveland, R., J. Parrow and B. Steffen. "A Semantics-Based Tool for the Verification of Finite-State Systems." In *Proceedings of the Ninth IFIP Symposium on Protocol Specification, Testing and Verification*, 287–302. North-Holland, Amsterdam, 1990.

[6] Cleaveland, R., J. Parrow and B. Steffen. "The Concurrency Workbench." In *Proceedings of the Workshop on Automatic Verification Methods for Finite-State Systems*, Lecture Notes in Computer Science 407, 24–37. Springer-Verlag, Berlin, 1989.

[7] DeNicola, R. and Hennessy, M. "Testing Equivalences for Processes." *Theoretical Computer Science*, v. 34, 1983, 83-133.

[8] Hennessy, M. *Algebraic Theory of Processes*. MIT Press, Boston, 1988.

[9] Hennessy, M. and R. Milner. "Algebraic Laws for Nondeterminism and Concurrency." *Journal of the Association for Computing Machinery*, v. 32, n. 1, January 1985, 137-161.

[10] Hillerström, M. *Verification of CCS-processes*. M.Sc. Thesis, Computer Science Department, Aalborg University, 1987.

[11] Kanellakis, P. and Smolka, S.A. "CCS Expressions, Finite State Processes, and Three Problems of Equivalence." In *Proceedings of the Second ACM Symposium on the Principles of Distributed Computing*, 1983.

[12] Larsen, K. and A. Skou. "Bisimulation through Probabilistic Testing." Proceedings of the ACM Symposium on Principles of Programming Languages, 1989.

[13] Malhotra, J., Smolka, S.A., Giacalone, A. and Shapiro, R. "Winston: A Tool for Hierarchical Design and Simulation of Concurrent Systems." In *Proceedings of the Workshop on Specification and Verification of Concurrent Systems*, University of Stirling, Scotland, 1988.

[14] Milner, R. *Communication and Concurrency.* Prentice Hall, 1989.

[15] Paige, R. and Tarjan, R.E. "Three Partition Refinement Algorithms." *SIAM Journal of Computing*, v. 16, n. 6, December 1987, 973-989.

[16] Pnueli, A. "Linear and Branching Structures in the Semantics and Logics of Reactive Systems." Lecture Notes in Computer Science 194, 14-32. Springer-Verlag, Berlin, 1985.

DIMACS Series in Discrete Mathematics
and Theoretical Computer Science
Volume 3, 1991

Auto/Autograph

Valérie Roy *
Robert de Simone †

January 15, 1991

Abstract

We describe the Auto and Autograph tools for verification and analysis of concurrent systems in their more recent developments. Auto is dedicated to a philosophy of "verification by reduction", based on automata morphisms and quotients. Autograph provides a graphical lay-out on which to display both terms and informations on terms, back and forth to Auto. We stress the openness aspects of both systems and their interface formats to the outside world. We see this as a contribution to the evergrowing collaborative trends in between similar tools, mostly under the pressure of national or european Esprit projects.

1 Introduction

Collaborative work in the field of verification tools for process calculi is steadily increasing, mostly under the pressure of european Esprit Projects. Interconnections in between such tools, which appear to be highly complementory, is now becoming a reality which should altogether avoid anarchical outburst and strict bureaucratic overplanning. While describing advances in the design of Auto and Autograph, we insist on their particular forms of interface formats, some of them already common with other tools[GLZ 89,Fer 87,BG 87,CPS 89,Arn 89,BFHP 89].

But the major part of this paper is aimed at describing new features of the Auto and Autograph systems, mostly concerned with the way they deal with recollection of results from analysis after reductions. These informations should be reported on the algebraic process terms themselves (or even better: on their graphical display), while experiments are conducted on their underlying transition systems.

*ENSMP-CMA Sophia-Antipolis

†INRIA Sophia-Antipolis

2 Auto

2.1 Presentation

Auto [SV 89] is a support system for verification, and more generally analysis, of process calculi terms. It deals only with terms with finite model representations in the semantics of transition systems, and so bases itself on automata transformation algorithms, hereafter called **reductions**. The main activities in Auto may be split in four:

1. Structural Automata Construction

2. Automata Reduction (with Abstraction)

3. Formal Analysis: Equivalence and other Comparisons.

4. Retrieval of Results and Diagnostics

Of course those activities are to a large extend entwined, and separated only for the sake of description. Reductions may take place at intermediate levels of construction, due to congruence properties of the equivalence directing the reduction. Partial results lifted back from certain directions of analysis may be reinjected along other lines. By *retrieval of results* we mean the uplifting of informations from the reduced structures back to the early processes. We shall stress this aspect further in a following section.

The general syntaxic means of reduction in Auto is embodied in *abstraction*, using so-called *abstract actions*. Abstract actions represent an higher level class of behaviours. The idea is to consider certain (terminated) sequences of concrete behaviours (signal emissions/receptions) as altogether equivalent and atomic, and to call such a set simply an abstract action. A process which may succeed in completing any of these "runs" at the more concrete level will perform the abstract action then. This framework generalises *observational behaviours* defining weak bisimulation, where any sequence from $\tau^{\star}.a.\tau^{\star}$, with τ a private actions, are to be seen as equal to a simple a.

Abstract actions gather in a new alphabet, called an *Abtraction Criterion*. Abstracting a process results in a system which is conceptually far simpler, and where meaningful activities have been isolated. Pragmatical use of abstraction in "everyday" verification often follows these lines: abstract actions are split in between "positive" ones, leading to configurations of the system to be inquired; then some "negative" ones do represent faulty behaviours which should certainly not take place there. The abstraction is then a refutation attempt. But if the dreaded behaviours do nevertheless occur, they identify a number of (abstract) questionable states, and also questionable behaviours leading to them.

2.2 Data Objects and Types

Auto uses but a restricted number of objects classes, specially meaningful in the approach of *verification by reduction*. They are:

1. Actions and Processes
2. Abstract Actions and Criteria
3. Automata
4. Partitions
5. Pathes

Processes and *Abstract Actions* are user-provided and enjoy an external syntax for input to the system. Current algebraic *processes* syntax is adapted from Meije[Bou 85], a language that extends CCS[Mil 80] to cope with the possibility of unrelated signals taking place simultaneously. This leads to a full commutative group of actions in a single step duration. On the other hand efforts are made to allow *any* process calculus as legible syntax to Auto [MV 90]. Current *abstract actions* syntax is based on regular expressions notations, with as generators predicates on signal names. Words from the entailed rational languages should be seen as sequences of behaviours. Full details may be found in [BRSV 89]. These syntaxes may be output by the Autograph interface (described later), from simple graphical representations.

Pathes are just sequences of transitions, alternating states with performed actions, while starting and ending with a state.

Partitions are (most often disjoint) sets of sets of states, representing for instance equivalence classes amongst all states of a given automaton. A partition is thus almost always linked to a related automaton. In case the partition does not recover all states, missing ones are supposed to share in the same additional class. A one-class partition thus represents a predicate (or property) selecting a subset of an automaton state.

The relation from data structures to conceptual Auto activities is as follows:

Processes are expanded into *automata* by **structural automata construction**, as defined in the SOS rules formalism.

Abstract actions help define the **reductions** to be applied on these automata. Fortunately many short-hand notations for specific useful abstractions are also implemented in Auto, so one rarely needs defining them from scratch. Still, dedicated abstractions add power to the expressiveness of reductions.

One subsequently extract either *partitions* or *pathes* through **Analysis** and **Equivalence** checking, as informative structures. *Partitions* collect classes of equivalent states, or lists of states enjoying a given behavioural property. *Pathes* are generated in particular as indications of faulty behaviours by application of "abstraction as refutation".

Partitions and *Pathes* are reinterpreted and carried back to ther original process algebraic description by **Retrieval of Results**. In this sense both should have in the future a syntactic description to be handed back to graphical display.

Another structure which is important to picture is the reduced automaton itself, so that one checks whether the abstract system conforms the expectation, or not.

2.3 Auto as a system

Auto is a small command language consisting of a main toplevel loop, storing results of parsings and computations in global identifiers. *Parsing* commands, are used for entering processes and abstract actions into Auto, possibly from Autograph; *set* commands are used for computing (composition) of functions on objects already stored in the environment, binding the result to new identifier names. All this was described more thoroughly in [BRSV 89].

2.4 Input/output

We stress here to a certain level of detail the formats Auto may use to exchange files with various companion systems. Some were devised in the framework of the french C^3 project, while others bridge and adapt to various european verification tools in the framework of Esprit projects.

Auto supports textual "easily" readable syntaxic formats for reading and writing terms and automata in Meije (or CCS) syntax. While pleasant to deal with, this format suffers two drawbacks: at reading, the parsing is time-consuming –specially on large automata–; at writing, pretty-printing procedures make it space-consuming instead. This format is the one currently used from Autograph to Auto.

Concerning automata, there exists a specific "non-human-readable" format containing exactly the lisp object, which is used for saving and restoring automata, or from the reactive real-time language Esterel to Auto [Res 90]. But future directions lay in the definition of a common format (called `fc` in french) with the tools Mec[Arn 89] and Aldebaran[Fer 87]. The format needs hardly parsing and proves fairly complete to carry additional optional informations (and in the case of Auto to embed pathes and partitions in their related automaton structure). This format is fully described in annex.

Now the situation is far less clear in the case of networks of processes, specially since there is a broad variety of operators used there. First considerations on basic needs shall be expressed while describing Autograph.

2.5 Reductions

All reductions are performed by merging states (of the same automaton), constructing quotients from behavioural equivalence relations. These relations are possibly computed on transformed automata, after abstraction of behaviours has taken place.

2.5.1 Abstractions

Behaviours abstraction functions in Auto are:

1. **abstract**, which uses an abstract criterion as parameter.

2. **tau-sature**, which completes transitions according to prefix or postfix *tau* transitions (building $\tau^\star$ and $\tau^\star.a.\tau^\star$).

`Abstract` results in an automaton on a new abstract alphabet, constructed while searching iteratively the concrete one for sequences matching the abstract actions. `tau-sature` is only a useful short-hand for computing weak bisimulation.

2.5.2 Quotients

Semantical equivalences are defined on these refined automata and their behaviours. They use fixpoints of relations, like bisimulations and other branching time semantics, or treat full sequences of behaviours as experiments, like trace languages and other linear time semantics [Gla 88]. All such semantics shall be congruences for all operators of the form of [GV 89]. Such functions in Auto are:

1. **mini** for strong bisimulation congruence.

2. **refined-mini**, starting from an initial partition.

3. **obs** for weak observational bisimulation congruence.

4. **tau-simpl**, which collapses τ-cycles behaviours.

5. **trace** for trace language congruence.

6. **dterm** for trace determinisation (without minimisation). congruence.

All these reductions take optional signal lists parameters, indicating which subset stays visible. Other signals are renamed down to τ. We could add a `refined-trace` functions, using as parameter a one-class partition indicating terminal states. `Obs` is not primitive and could be obtained as `mini(tau-sature(tau-simpl(....`

2.6 Comparisons and Equivalences

We already described some equivalences used for minimizing automata. The same functions may theoretically be used to compare two transition systems, depending whether both initial states lie equivalent.

But here we want to collect informations in case of non equivalence. Supposing both processes were first reduced to normal form, then the final partition retains only singleton or couples as equivalent classes: couples represent equivalent states (one from each automaton), and singletons represent states without match (from either automaton). This structure should be further queried, possibly interactively, or the subset of equivalent states could be displayed graphically, or longest strings of unmatched behaviours running only through singletons could be searched. More actual experiments are still due here.

The functions providing equivalence informations are few in Auto. One should remember that they may be combined with previous reduction functions to compose richer equivalence checking.

1. `eq` provides a "yes/no" answer on bisimulation equivalence.

2. `strong-partition` provides the list of equivalence classes.

Equivalence is only an instance of comparaison, observation and testing on an automata under analysis. Promising other directions are *graphical specifications* [BL 89].

2.7 Retrieval of Results and Diagnostics

Results put to light by reductions and further analysis should now be lifted back to the initial process terms. Pathes and state partitions will thus be provided "inverse images", so to speak. As a way to attain this, we build a side structure from the shape of the results and the actual combinations of reductions performed. We shall call *guideline* such a structure.

A guideline is a generalized automaton-like observer, safe that it includes additional info in its states, possibly drawn from the process itself. In contrast, an observer is usually required to be defined independantly from its observed system. Typical such info consists in states that are allowed as configurations of the observed process while the guideline reaches this state. Guidelines may also contain another sort of state information, namely places it allows as terminated endpoints of abstract behaviours.

Term or Automaton	*reduction function* $\longrightarrow$	Reduced Automaton
$\Downarrow$		$\Downarrow$ *analysis*
Source info		Pathes or Partition
$\Uparrow$		$\Downarrow$ *synthesis*
Uplifted Guideline	*retrieval* $\longleftarrow$	Guideline

As pictured above, the guideline shall be built going backward stucturally along the chain of reductions performed. Some reductions will change its state space, others shall add or transform transitions. In the end there shall always be the possibility of **applying** it on the corresponding observed system of the same stage. Applying a guideline on a process (or an automata) will consist in "simulating" the observed process in a way that will satisfy the guideline, or equivalently in a way constrained by the guideline. This will identify which states, or behaviours, of the term under analysis, shall indeed project themselves by reduction down to the chosen pathes or partition classes.

With this reconstruction philosophy the only information which needs be kept through reductions is the relation from a state to the class of states that were mapped to it taking the quotient. These states in Auto are formally encoded in a vector description of states in the sequential automata components occurring in the early algebraic process description.

The operation induced by reduction functions on guidelines are actually quite simple:

quotients replace in each state the set of allowed configurations of the analysed system. It is replaced by its inverse image for the morphism associated with the quotient.

Abstraction replaces actual actions by their concrete actions regular sets counterparts. This induces new states in the guideline, and a need to indicate which sets may be terminal as representative behaviours of the analysed process.

3 Autograph

3.1 Presentation

Autograph [RS 89] is a plain, non syntaxic graphical editor for pictures that are later interpreted into process calculi terms. By this we mean that the drawings produced are not structural -safe in particular spots-, and that coherency and structure are checked only on demand, for translation into textual algebraic form. This allows the graphical description to depart from strict ties to the textual one, including elliptic representation conventions. This allow also semantics of drawings generated from few simple graphical object types to be endowed with scarcely variant semantics, depending on the target formalism and language. Autograph was started with the Meije algebra in mind, but could give graphical versions to (process algebraic simple kernels of) Esterel[BG 87] and Lotos[BB 89].

3.2 Graphical Data Objects and Meanings

3.2.1 Graphics

Autograph works from menus and mouse selection of functions to be applied on graphical objects. As in Auto, the object classes were kept to a bare minimum, highly common, and endowed with graphical links properties which are fairly *ad-hoc*. They are:

1. Boxes
2. Ports
3. Vertices
4. Edges
5. Labels (not a graphical object)

Roughly, these few graphical elements represent all that was informally used in graphical description of parallel systems (communicating automata) even before efforts were made to embed them into an algebraic syntax formally, which was to lead to process calculi theory.

Boxes are rectangles and represent subprocesses. Boxes next to one another are supposedly set in parallel. Ports are circles, which may only be located on the boxes frames. They figure the connexion points for communication. Possible communications are represented by edges drawned in between ports. So boxes, ports and edges build up *networks* of processes, and recollect the *structural operators* features from process calculi. Figure 1 shows a network example.

Vertices are round-shaped and figure automata states. Transitions are figured as edges (just like connections in between ports). So edges are broken lines, whose ends imperatively fall both into states or ports. Vertices and edges thus create *regular*

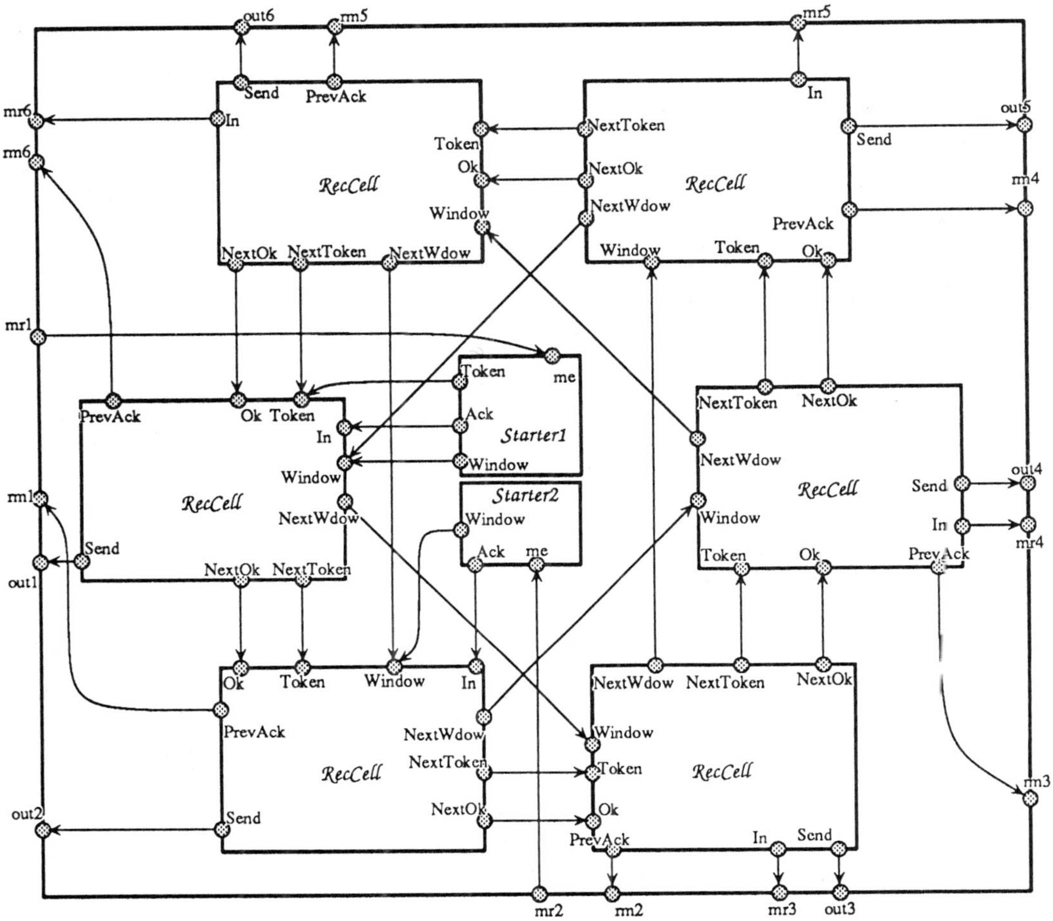

Figure 1: a sliding window protocol receiver

individual processes, and recollect the *dynamic behavioural operators* features from process calculi.

3.2.2 Namings

All these elements may be named (or tagged) using labels. Labels appear on the screen but are not conceived as graphical objects, that is they do not share the same uniform internal treatment as other objects (in particular they may not be themselves named!). Naming is not always imperative. Names serve different purposes depending on the type of object they are applied on:

• on boxes, they provide a global or local naming facility so that a description may be shared in between several windows, and subterms instanciated in several locations. Recursive terms (using themselves as subterms) may also be generated this way;

• on ports, they indicate the local name of the signal to be communicated upon. In particular, all ports of named boxes referring to subsystems drawned elsewhere should be named, and no other need be. This implies that Autograph takes care of all renamings to avoid names clash and generate proper communications at translation into textual form. Naming other ports still may help control this translation;

• on vertices, they help control the translation by providing user-defined identifiers for the state names, which then become the variables occurring in the (guarded right linear) recursive definition of automata term; they may also occur several time in the same automaton (if only once with outgoing edges) and save drawing long backheaded arrows through a multirepresentation of the same state.

• on edges, they provide on one hand labels for transitions, that are imperative (an edge may actually be labelled several times, thus representing more than one transition); on the other hand they allow naming of connections inter ports, to save drawings of line connections and secure user-provided names at translation altogether. All ports locally sharing connections with the same names are transitivally supposed to be connected. See [Roy 89] for details. All connections pervading to the outside need to be named at this level, to figure the way signals show externally and provide the sort of the description.

To recall, names are either: **imperative**, in the case of transition labels, innermost boxes ports and empty subterms boxes; **short-hand conventions** allowing recourse to textual pointers links rather than common physical line joining, in the case of automata states and communication connections; **naming controls** on the identifiers used at translation into process calculus syntax in the case of states, intermediate levels ports and connections.

3.2.3 Additional semantics

There is a default semantics (expressed as Meije or CCS terms) to the drawings, at least the correct ones. Notions of correct drawings, together with all abuse of representation allowed by Autograph, are out the scope of this paper and described in [Roy 89]. Briefly: Boxes next to one another represent parallel substerms, containment of boxes represent the subterm relation, edges without outside ports generate signal restrictions and edges in between ports with different names generate relabellings. Automata are even more simply interpreted.

But this semantics may be altered, using a system of semantical annotations by functions from a special menu, each time dedicated to a specific new calculus to represent. This holds mostly of networks.

Justification for these annotations is drawned from the Ecrins [MdS 87] system, where a syntaxic formalism is defined for creating new calculi through new operators. A new operator is introduced by syntaxic considerations (name, binding power, type) and SOS rules for semantics. The type is always of the form $(Signals^m \otimes Process^n) \longrightarrow Process$. So we may superpose a semantics (and even several if they may be disambiguated by binding powers) with semantical markings: on a box (the target process). its ports (the *Signals* parameters), its son boxes (the process arguments), their ports (these arguments behaviours).

This part of Autograph is in development and shall be developped hopefully in the final paper.

3.3 Autograph's Interface

3.3.1 Menus and Functions

Autograph contains a number of menus (in its current version). Half are classical screen and window management menus, including cut-and-paste of subdrawings and Postscript dump of improved representations of both networks and automata (with edges splined up for instance). The other half is more specific to the edition of objects. There are four such menus: `Nets` for boxes and ports, `Automata` for vertices, `Edges` for transitions and connections, and `Labels` for namings. They all contain functions like `select`, `move`, `create`, `kill` and so on...

There is a `completion` function in the `Windows` menu of Autograph, which checks for coherency of drawings and construct the following records at nodes, whose fields may be variant according to the **Type** field:

1. **Type**, the name of the corresponding operator (e.g. PAR, SEQ, AUTOMATON, VAR, LOCALDECL)
2. **Name** (mostly in case of variables and local process declarations).
3. **Fathernode** for navigation.
4. **SigParameters** for the relevant arguments to the operator.
5. **SigRename** because signals renamings, which introduce process instantiation, have a specific representation in Autograph (by edges drawing).
6. **Internals** for signals local scope binding, as in CCS restriction.
7. **Sort** giving optionally the set of all visible signals at this subterm level

8. **SubProcessesGraph** identifying the subprocesses and optionally a relation in between them

9. **LocalDeclaredProcesses** for local process definitions that are used in the subprocesses graph (mostly in case of LOCALDECL operators).

10. **Automata** for sequential automata components. These automata will correspond after actual translation to entries in a "automata fc" table.

11. **Pragmas** for uninterpreted info (like graphical positions remainders).

3.3.2 Input/output

There is an additional variable menu in Autograph, or more exactly there are distinct versions of Autograph, depending on a menu providing the semantic annotating and translation functions to a given calculus format, or a companion verification systems (*e.g.* Auto). In the future translation functions should all produce instances of a "common format", prealgebraic. This is largely started for automata with the format described in annex, far less for networks. But globally the format should closely follow the records structure produced by the previous `completion` function.

There is also a specific `explore` function, whose mode is in particular started while inputting a description file which lacks geometrical features, for instance as provided by Auto. Exploration of automata is an established functionality of Autograph. Exploration of networks is being completed, and should be described in the full paper, as well as exploration of *pathes* and *partitions* on already displayed process descriptions. These functions are immediate, provided an "fc" input format for these objects is agreed on.

A Common Automata Format

This format was devised in connection with designers of the Aldebaran and Mec systems.

A.1 Basics

An *Atom* is any ASCII string, as in `"foo"`.

A *Monomial* is a dot separated list of atoms.

A *Label* is a + separared list of monomials.

An integer list is always preceded by its length, if not less or equal to one.

A.2 States

A state is a record with the following fields: *number, action, property* and *name*. All but the first are optional. The number is an integer, while others are of *Label* type. They contain information under the form required by any given system. Actions are primary informations, while names are only decorations meant for source recollections, and properties are additional *pragmas*.

A.3 Macrotransitions

They lead from a state to a series of states. They store their information in records similar to these of states (but without the *number*). Here labels have the intuitive meaning that they represent a union of actions behaviours (as monomials), these behaviours being themselves simultaneous co-occurrences of atomic signals.

A.4 Automata

They are represented as a series of tables, preceded by the header

```
automaton label
states integer
[initial integer list]
[terminal integer list]
```

and followed by the `end` keyword.

Atoms, monomials and labels tables are each introduced by their keyword and size, with the definition list following, an element in a row, prefixed by a line numbering. For instance:

```
monomials integer
0: a0.a1
1: a1.a1.a3
...
```

Notice that atoms are refered as the letter a glued to a line number. Similarly one refer to monomial `m18` or label `132`.

States are kept in a table where entries are:

`state` *integer* [`action` *label*] [`prop` *label*] [`name` *label*] `trans` *nbtrans trans*

where *trans* is a liste of macrotransitions, whose number is recollected as *nbtrans*. Macrotransitions entries are:

integer: [*label*] `- >` `state list` [`name` *label*] [`prop` *label*]

where the first label recalls *Actions*.

References

[Arn 89] A. Arnold *"Mec: a System for Constructing and Analysing Transition Systems", in Acts of the Workshop on Automatic Verification Methods for Finite State Systems, LNCS (1989)*

[BFHP 89] B. Backlund, P. Forslund, O. Hagsand, B. Pehrson, "Generation of Graphic Language-oriented Design Environments", 10th IFIP Protocol Specification, Testing and Verification", Twente, North-Hooland (1989)

[BG 87] G. Berry, G. Gonthier, *"The Esterel Synchronous Programming Language: Design, Semantics, Implementation", to appear in Comp. Sci. Prog. (1989)*

[Bou 85] G. Boudol *"Notes on algebraic calculi of processes", Logics and Models of Concurrent Systems, NATO ASI Series F13,K. Apt, Ed.(1985)*

[BRSV 89] G. Boudol, V. Roy, R. de Simone, D. Vergamini, *"Process Calculi, from Theory to Practice: Verification Tools", in Acts of the Workshop on Automatic Verification Methods for Finite State Systems, LNCS (1989)*

[BL 89] G. Boudol, K. Larsen, *"Graphical Versus Logical Specifications"*, INRIA Research Report 1104 (1989).

[BB 89] E. Brinskma, T. Bolognesi, *"Introduction to the ISO Specification Language Lotos", The Formal Description Technique Lotos, North-Holland (1989)*

[CES 83] E.M. Clarke, E. Emerson, A. Sistla, *"Automatic Verification of Finite-State Concurrent Systems using Temporal Logic Specifications: a practical approach"*, Proc. 10th ACM POPL (1983).

[CPS 89] R. Cleaveland, J. Parrow, B. Streffen, *"A semantics Based Verification Tool for Finite State Systems"*, in Proceedings of the Ninth International Symposium on Protocol Specification, Testing, and Verification, North-Holland (1989).

[Fer 87] J.C. Fernandez, *"Aldebaran: un système de vérification par réduction de processus communicants*, French Thesis, IMAG Grenoble (1987).

[Gla 88] R. van Glabbeek *"The Linear Time - Branching Time Spectrum", CWI Report RvG 8801*

[GW 89] R. van Glabbeek, W. Weijland, *"Branching Time and Abstraction in Bisimulation Semantics"*, Proceedings 11th IFIP World Comp. Congress, San Francisco (1989).

[GV 89] J.F. Groote, F. Vaandrager, *"Structured Operational Semantics and Bisimulation as a Congruence"*, Proceedings 16th ICALP, Stresa, LNCS (1989).

[GLZ 89] J. Godskesen, K. Larsen, M. Zeeberg, *"TAV (Tools for Automatic Verification) Users Manual"*, University of Aalborg R 89-19 (1989).

[MV 90] E. Madelaine, D. Vergamini, *"Finiteness Conditions and Structural Construction of Automata for all Process Algebras", this volume (1990)*

[Mil 80] R. Milner *"A Calculus of Communicating Systems", LNCS 92, Springer-Verlag (1980)*

[Mil 83] R. Milner *"Calculi for Synchrony and Asynchrony"*, TCS 25, (1983)

[MdS 87] E. Madelaine, R. de Simone, *"Ecrins, un Laboratoire de Preuve pour les Calculs de Processus"*, (in french), Inria Research Report 672 (1987).

[Res 90] A. Ressouche, *"Ocauto"*, Esterel Technical Documents, CMA-Ecoles des Mines (Sophia-Antipolis) (1990).

[Roy 89] V. Roy, *"Autograph: un Outil d'Analyse Graphique de Processus Parallèles Communicants", French Thesis, Université de Nice (1990)*

[RS 89] V. Roy, R. de Simone, *"An Autograph Primer", I.N.R.I.A. Technical Report 112 , (1989)*

[SV 89] R. de Simone, D. Vergamini *"Aboard Auto", I.N.R.I.A. Technical Report 111 (1989)*

DIMACS Series in Discrete Mathematics
and Theoretical Computer Science
Volume 3, 1991

A Data Path Verifier for Register Transfer Level Using Temporal Logic Language Tokio

HIROSHI NAKAMURA, YUJI KUKIMOTO, MASAHIRO FUJITA
AND HIDEHIKO TANAKA

ABSTRACT. A data path verifier for register transfer level is presented in this paper. The verifier checks if all the operations and the data transfers in a behavioral description can be realized on a given data path without any scheduling conflicts. Temporal logic based language *Tokio* is adopted as a behavioral description language in this verifier. In *Tokio*, designers can directly describe concurrent behaviors controlled by more than one finite state machine without unfolding parallelism. The verifier checks for the consistency between a behavior and a structure automatically and lightens the load of designers. The actual LSI chip which consists of 18,000 gates on CMOS gate array has been successfully verified. This verifier is concluded to have the ability to verify practical hardware design.

1. Introduction

Recently, extensive studies have been carried out on the derivation of efficient and error-free data paths in register transfer level assistance. High-level synthesis [5] is one of the solutions to this problem. The approach adopted in high-level synthesis is to synthesize a data path automatically from a given behavioral description. The derived data paths with this approach, however, are not as satisfactory as those which are designed manually yet .

On the other hand, designers initially have the image of a data path to be designed in an actual design process, because they have designed many similar circuits. In addition, they seldom develop hardware which is completely different from the one ever designed. In such situations, it is better to utilize the designers' experience positively instead of synthesizing a data path automatically. It is very important to construct an effective design assistance system based on this approach. However this has not been discussed well so far. Thus we propose a practical assistance system of register transfer level design based on this approach and present a data path verifier, which is the core part of the system.

Figure 1 shows the design flow of the proposed assistance system. The basic idea in this flow is to construct final data paths by modifying initial data paths given by designers.

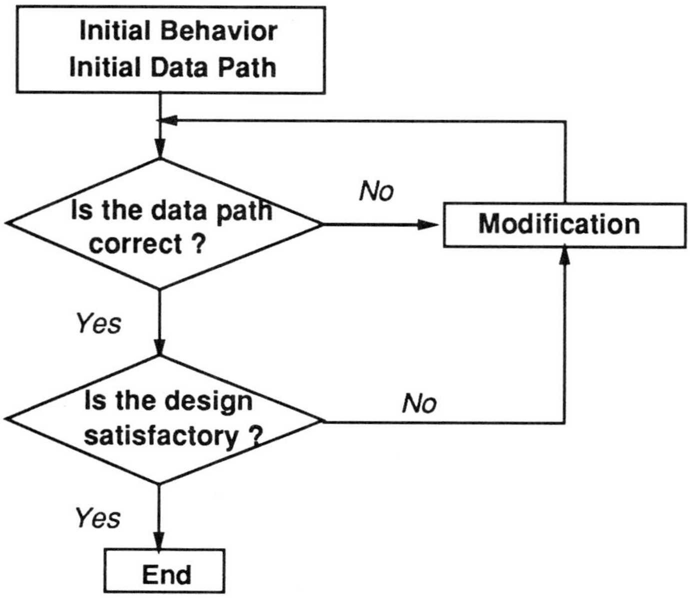

FIGURE 1. Flow Chart of Practical Design Assistance

At first, a designer gives an initial behavioral description and an initial structure of a data path to be designed. The initial structure is formed through the designer's experience. Now we must verify whether the given behavior can be realized on the given structure. The proposed data path verifier points out when and where scheduling conflicts happen, i.e., which paths and functional units are doubly allocated at which time slot. The structure and the behavior are modified to compensate for that error, and again they are verified. If there is no design error, the performance of the behavior is simulated and the cost of the structure is estimated. If the behavior and the structure satisfy required performance and permitted cost, the design at register transfer level finishes and it proceeds to the lower level design such as logic synthesis. If the cost/performance constraints are not satisfied, the design flow is followed by the stage of modification and the process of verification. This design flow of Figure 1 is justified because the initial data path given by a designer is fairly good in many cases. If the initial data path is fairly good, the revised data path will be very efficient. The data path verifier proposed in this paper plays an important role in this assistance system.

2. An Example

In this section, we explain the semantics of Tokio by showing some simple Tokio programs. After that, taking the example of the circuit which computes square roots, we show how this verifier is effectively used in the process of deriving the proper data path for a pipelined behavior from the data path for a sequential behavior with modification.

2.1. Behavioral Description Language Tokio. Tokio [1][2][3] is a logic programming language based on first-order interval temporal logic [4]. Intuitively, Tokio is regarded as an extension of Prolog with temporal operators. Since Tokio has the notion of time in its own semantics, the various algorithms of hardware can be described flexibly. The essential notion of hardware, such as concurrency and sequentiality, can be specified accurately and simply. Using Tokio, designers can directly describe concurrent behaviors controlled by more than one finite state machine without unfolding parallelism. Parallel behaviors such as pipelined execution can be easily described.

Now we explain the semantics of temporal operators. The expression

```
head :- p,q.
```

denotes that the predicates p and q are executed in the same time interval where the predicate head is defined. Concurrency is represented with a *comma* operator.

Sequentiality is expressed as follows.

```
head :- p && q.
```

The *chop* operator (&&) specifies the sequential execution of the two predicates p and q. This operator divides the interval where the predicate head is defined into two sub-intervals as shown in Figure 2. The predicate p is executed in the former interval and q is executed in the latter interval.

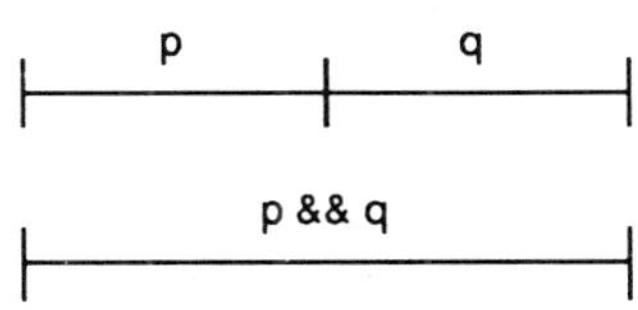

FIGURE 2. Chop Operator

Conditional branches are described by using *cut* operators(!).

```
head :- condition1 ,!, p.
head :- condition2 ,!, q.
head :- !,r.
```

The semantics of these predicates is as follows.

if condition1 = true then execute p
else if condition2 = true then execute q
else execute r

Let us take a simple example shown in Figure 3-(a). The statement "`*tmp <= *c + *tmp`" represents that the data of register `*c` is added to the data of register `*tmp` and that the result of the computation is stored in register `*tmp` at the next clock. The predicate `sub` denotes the parallel execution of the computation described above and the decrement of the data of register `*c`. The predicate `main` denotes that the computation of the predicate `sub` is repeated until the data of register `*c` is equal to zero. Figure 3-(b) shows a timing chart of this sample program.

```
main :- *c=0 ,!,*output <= *tmp.
main :- !,sub && main.
sub :- *tmp <= *c + *tmp, *c <= *c - 1.
```

(a) Sample Tokio Program

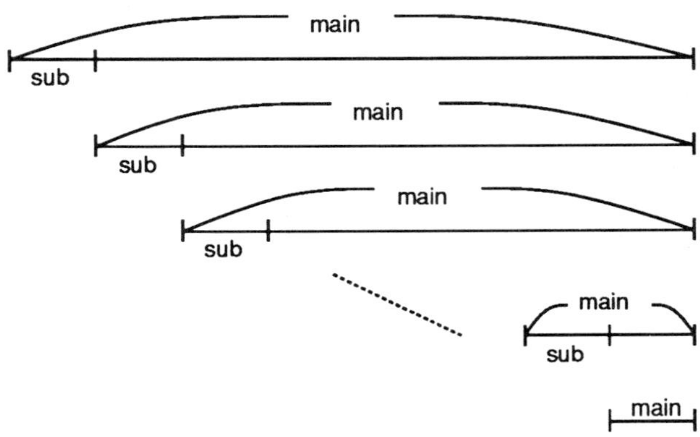

(b) Timing Chart of Sample Program

FIGURE 3. Sample Tokio Program

As seen from the example, the description in Tokio is based on a top-down approach. This approach is very useful for describing hardware because designers can give behavioral descriptions with high modularity. Using this programming style, we need not unfold concurrency of hardware for every clock cycle. On the other hand, it becomes rather difficult for designers to check the consistency between a behavior and a data path. Our strategy is to adopt a flexible behavioral description language and to verify the consistency automatically.

2.2. Computing Square Roots by Sequential Execution. The example to be taken is the circuit which calculates square roots by using Newton's method. The algorithm of Newton's method is shown in Figure 4-(a). The behavioral description in Tokio and the structure of a data path are shown in Figure 4-(b) and Figure 4-(d) respectively. In this behavioral description, the computation of Newton's method is realized by sequential execution for each input data. In this case, the verification of the data path is not so difficult because there is

```
Y := 0.222222 + 0.888889 * X;
C := 0;
DO UNTIL C = 2 LOOP
  Y := (Y + X / Y) / 2;
  C := C + 1;
ENDDO;
```

(a) Algorithm of Newton's Method

```
main :- *adr = 8,!,true.
main :- !,input && stage1 && stage2 && main.
input :- !, *input1 <= *memory(*adr), *adr <= *adr + 1
     && *reg1 <= 0.222222 + 0.888889 * *input1.
stage1 :-!, *reg2 <= *input1 / *reg1
     &&  *reg2 <= *reg2 + *reg1
     &&  *reg3 <= *reg2 / 2,  *input2 <= *input1.
stage2 :- !, *reg4 <= *input2 / *reg3
     &&  *reg4 <= *reg4 + *reg3
     &&  *output <= *reg4 / 2.
```

(b) Sequential Behavioral Description in Tokio

```
main :- *adr = 8,!,true.
main :- !,input && stage1 && main,(stage2 && true).
```

(c) Pipelined Behavioral Description in Tokio

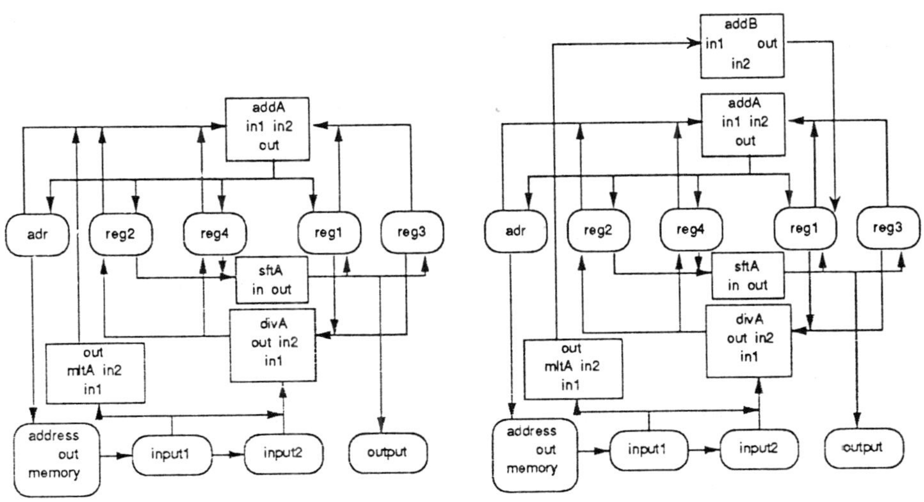

(d) Data Path for Sequential Execution (e) Data Path for Pipelined Execution

FIGURE 4. Computation of Square Roots

little concurrency in the description. We have only to verify the data path for each local interval and need not check whether some of these intervals occur concurrently.

2.3. Deriving a Data Path for Pipelined Execution. Now we modify the sequential behavioral description of the circuit computing square roots and derive a pipelined behavioral description from it. It is quite easy to get a pipelined description from the original one by using temporal operators. The pipelined description is shown in Figure 4-(c). We have only to change the top-level predicate **main**. The second **main** predicate denotes that **stage1** is followed by both **main** and **(stage2 && true)**. Since **main** starts the computation for the next input data, this behavior represents a pipelined execution. Figure 5 shows a timing chart for the pipelined execution.

At this point, we must construct a data path for the pipelined behavior, because the data path for sequential computation may not be sufficient. To construct a completely new data path, however, is not efficient. It is better to modify the original data path and to get the proper one. In such a case, we can use this verifier effectively. At first, the verifier checks whether the pipelined behavior can be realized on the original data path. If there are some scheduling conflicts, the behavioral description or the data path is modified considering the output of the verifier in order to avoid the conflicts. This process is repeated until there are no conflicts and the proper data path is constructed.

From the result of the verification, we can say that adder addA is doubly used in the second interval of **input** and the second interval of **stage2** and that these intervals occur concurrently. Thus we add one more adder to the original structure and the final data path is obtained. The data path for the pipelined behavior is shown in Figure 4-(e).

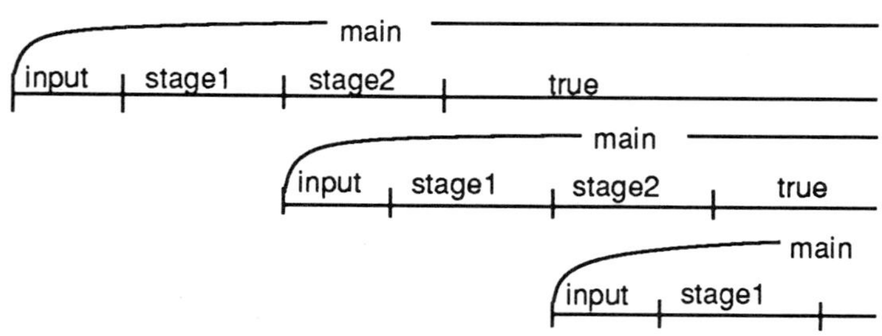

FIGURE 5. Timing Chart for Pipelined Execution

3. Verification Method

3.1. Overview. The data path verifier proposed in this paper checks if all the operations and the data transfers in a behavioral description can be realized on a given data path without any scheduling conflicts. The structure of this

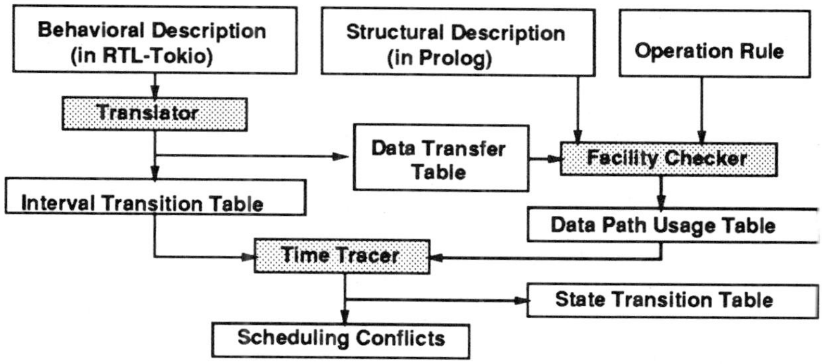

FIGURE 6. Structure of Data Path Verifier

verifier is shown in Figure 6. The inputs of this system are a behavioral description in RTL-Tokio (Register Transfer Level Tokio, which is a subset of Tokio), a structural description in Prolog, and operation rules in Prolog. The operation rule is a rule for linking an operation in the behavior with a functional unit in the data path.

The process of the verification is divided into two stages.

Linking a behavior with a data path: To find a set of paths and functional units for each data transfer and operation from the structural description. In this stage, all the data transfers and operations in the behavior are linked with appropriate data path elements in the structure using operation rules, and the link information between them is recorded in a data path usage table. This stage is executed by a *translator* and a *facility checker*.

Detection of scheduling conflicts: To verify whether any paths or functional units are doubly allocated in concurrent time intervals. The concurrency of the behavioral description in RTL-Tokio is unfolded in this process. The state transition table of a control part is also extracted. This stage is executed by a *time tracer*.

3.2. Linking a Behavior with a Data Path. The link information between the behavior and the structure is derived in accordance with the following steps.

(1) To find a set of functional units which realize an operation of each data transfer from the structure.
(2) To search for a data path from a source register to the input of the functional unit and that from the output of the unit to a destination register.

The function of each unit is defined by operation rules. At the end of this stage, the link information is recorded in a data path usage table. From this table, we can know which data path element is used by each data transfer and operation. The names of the registers and those of the memories in the behavior are assumed to be the same as those in the structure.

3.3. Detection of Scheduling Conflicts. In this stage, scheduling conflicts in data paths are detected by unfolding the concurrency of a behavior to actual data transfers with a time tracer. The time tracer makes clear what operations and data transfers are done concurrently by traversing all the transitions. After that we check if any paths or functional units are doubly allocated in concurrent time intervals by using a data path usage table. If no path or functional unit is doubly allocated, the given data path turns out to be correct. Otherwise designers modify the behavior and the data path to avoid the detected scheduling conflicts. We have implemented this time tracer in two ways: a forward tracer and a backward tracer.

- *Forward Trace*: At first all the concurrent time intervals are searched for by tracing transitions forward. In the second stage, all these intervals are checked whether they use the same data path element. The scheduling conflicts of data paths are detected in this stage.
- *Backward Trace*: In a backward trace, all the pairs of time intervals which use the same data path element are searched for at first, and they are verified whether they occur concurrently by tracing transitions backward in accordance with an interval transition table.

In this paper, only the algorithm of the forward trace is explained. The algorithm of the backward trace is similar to that of the forward trace except for a direction of tracing.

3.3.1. Unfolding Concurrency of a Behavior. In the first stage, all the concurrent time intervals are searched for with a forward tracer. It traverses all the transitions from an initial state in depth-first. A state in RTL-Tokio description is defined by [*interval-name*,*clock-number*].

step1: An initial interval I_{init} and an initial clock I_{clock} (usually 0) are selected.
$S_0 = \{[I_{init}, I_{clock}]\}$.

step2: (Unfolding the concurrency of a behavior)
If S_i has predicate calls, all these concurrent intervals are added to S_i and S_i' is obtained. S_i' is a list of the intervals which occur concurrently at the i-th clock.

step3: (Proceeding to the next clock)
S_{i+1} is obtained from S_i' by proceeding to the next clock. The computation of this process is as follows.

(1) Increment the clock number I_{clock} for each element of S_i' if the obtained clock number is not larger than the length of that interval.
(2) Transfer to the next interval after *chop* operator &&.

Step3 is followed by step2.

Detection of the iteration of a search: Let S_n' be the newly obtained state in step2. If $0 \leq \exists\, i < n$, $S_n' \subseteq S_i'$ holds or S_{n+1} next to S_n' cannot be obtained, the iteration of a search is detected.

3.3.2. Detection of Scheduling Conflicts. All the intervals in each S_i' occur concurrently unless they have exclusive transitive conditions. Using a data path usage table, all the concurrent intervals are checked whether they use the same data path resource or not. If they use the same path or the same functional unit, it results in scheduling conflicts of data paths. In this stage, we check if the data path conflicts can be avoided by using some alternative data paths.

4. Experimental Results and Evaluation

The data path verifier is implemented on SUN4/260 using SICStus-Prolog (about 80KLIPS).

4.1. General-purpose Processor. The processor cited from [6] is adopted as an example. This computer has 16 instructions. The computation of each instruction consists of an instruction-fetch part and an execution part. For all the instructions except add and adc, the next instruction-fetch stage begins after the execution stage has finished. In the case of add and adc, however, the next instruction-fetch begins one clock earlier. Thus the instruction-fetch stage and the last clock cycle of the execution stage are executed concurrently.

The data path structure is shown in Figure 7. The bit-width of the structure is 8 except for the lines explicitly indicated. The results are shown in Table 1.

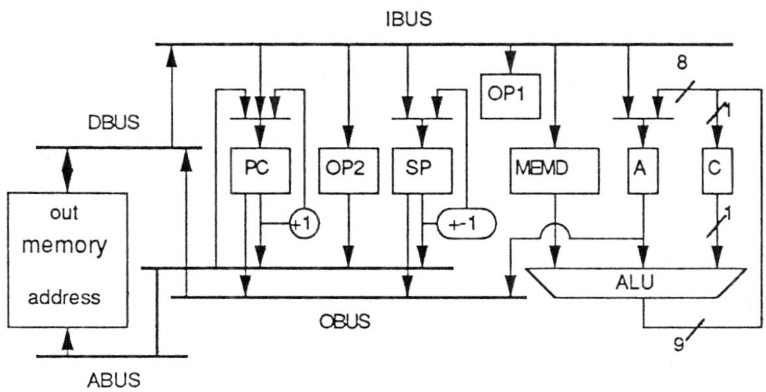

FIGURE 7. Data Path Structure of Processor

The backward trace can test whether the given two intervals occur concurrently or not. It is also possible to select one interval and to search for the intervals which occur concurrently with it. In this example, we have only to pay attention to concurrent intervals with the last clock cycle of the execution part in tracing backward. The required CPU time for this backward trace was 1.38 sec.

CPU time (sec)				Number of Derived States	Number of State Transitions
Translator	Facility Checker	Forward Trace	Backward Trace		
2.22	6.51	17.3	775	26	198

TABLE 1. Verification Results of General-purpose Processor

4.2. Network Interface Processor. The next example is a network interface processor (NIP) in PIE64 [7]. PIE64 is a parallel inference machine which executes knowledge information processing in parallel, and is under development in our laboratory. This machine consists of 64 inference units and two high speed interconnection networks. The processor manages data transfers and process synchronization between inference units.

The processor has already been designed and is going to be implemented on CMOS gate array. The total number of the gates is about 18,000 including both a data path and a control part. The structure of NIP is divided into four parts. The data path verifier has been applied to the main parts which consist of about 11,000 gates. Since the required performance of this processor is very severe, the behaviors of NIP are 6-stage pipelined. Figure 8 shows the data path structure of the processor. The verification results of the two parts of the processor are shown in Table 2.

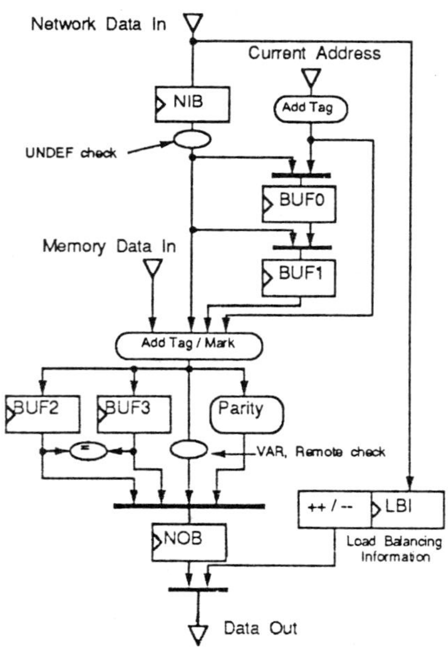

(a) Data Transfer Part

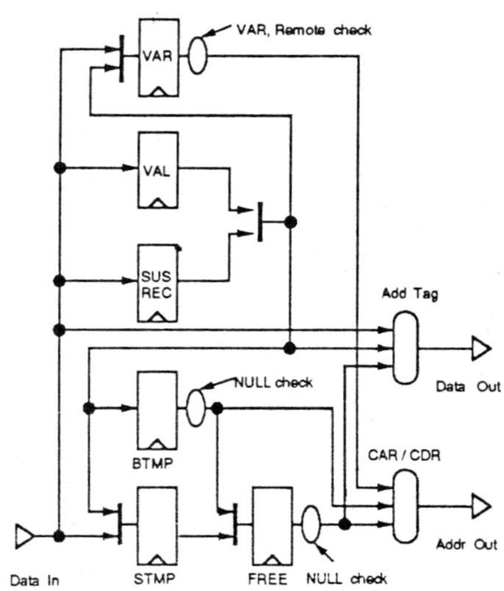

(b) Process Synchronization Part

FIGURE 8. Data Path Structure of Network Interface Processor

	CPU time (sec)				Number of Derived States	Number of State Transitions
	Translator	Facility Checker	Forward Trace	Backward Trace		
Data Transfer Part (Upper-Part)	0.80	1.07	2.36	40.1	14	53
Data Transfer Part (Lower-Part)	3.23	4.66	183.3	>3600	81	279
Process Synchronization Part	5.97	7.85	1488	>3600	236	1168

TABLE 2. Verification Results of Network Interface Processor

4.3. Evaluation. Most of the CPU-time is spent in a *time trace* part. Since a forward tracer traverses all the state transitions, its cost is proportional to the number of transitions. The cost of detecting the iteration of the search is proportional to the number of states, because traced states are presently recorded with enumeration. Thus the computational complexity for the forward trace is $O(n \times m)$, where n is the number of state transitions and m is the number of states.

In a backward trace, ${}_NC_2$ pairs of intervals are listed in the worst case, where N is the number of intervals. The worst case occurs when a certain data path element (such as a bus) is used in all the intervals. The backward trace, however, is suitable for partial verification as explained in section 4.1. Therefore, it is recommended that the important part of design is tested partially using the backward trace at first, and then the whole design are verified using the forward trace.

5. Conclusion

We have presented a data path verifier at register transfer level and proposed a design assistance system based on this verifier. The verifier has been successfully applied to a real LSI chip. Practical hardware design can be assisted with this verifier if an interactive tool for improvement is provided. Our current research aims at the assistance of this improvement stage.

REFERENCES

1. M.Fujita, S.Kono, H.Tanaka and T.Moto-oka, *Aid to Hierarchical and Structured Logic Design Using Temporal Logic and Prolog*, IEE Proceedings, Vol.133, Pt.E, IEE, (1986), 283–294.
2. T.Aoyagi, M.Fujita and T.Moto-oka *Temporal Logic Programming Language Tokio*, Logic Programming '85, LNCS-221, Springer-Verlag, (1985), 128–137.
3. S.Kono, T.Aoyagi, M.Fujita and H.Tanaka, *Implementation of Temporal Logic Programming Language Tokio*, Logic Programming '85, LNCS-221, Springer-Verlag, (1985), 138–147.

4. B.Moszkowski, *A Temporal Logic for Multi-Level Reasoning about Hardware*, Proceedings of CHDL '83, IFIP, (1983).
5. M.C.McFarland, A.C.Parker and R.Camposano, *Tutorial on High-Level Synthesis*, Proceedings of 25th Design Automation Conference, ACM/IEEE, (1988), 330–336.
6. O.Karatsu, *VLSI Design Language Standardization Effort in Japan*, Proceedings of 26th Design Automation Conference, ACM/IEEE, (1989), 50–55.
7. H.Koike and H.Tanaka, *Multi-Context Processing and Data Balancing Mechanism of the Parallel Inference Machine PIE64*, Proceedings of Fifth Generation Computer Systems Conference, ICOT, (1988), 970–977.

Institute of Information Sciences and Electronics, The University of Tsukuba, 1-1-1 Ten'nou-dai, Tsukuba, Ibaraki 305, Japan

E-mail: nakamura@arch3.is.tsukuba.ac.jp

Department of Electrical Engineering, The University of Tokyo, 7-3-1 Hongo, Bunkyo-ku, Tokyo 113, Japan

E-mail: kuki@mtl.t.u-tokyo.ac.jp

E-mail: tanaka@mtl.t.u-tokyo.ac.jp

Fujitsu Laboratories Ltd., 1015 Kamikodanaka, Nakahara-ku, Kawasaki 211, Japan

E-mail: fujita@flab.fujitsu.co.jp

DIMACS Series in Discrete Mathematics
and Theoretical Computer Science
Volume 3, 1991

Model Checking and Graph Theory in sequential ATPG *

P. Camurati M. Gilli P. Prinetto M. Sonza Reorda

Dipartimento di Automatica e Informatica
Politecnico di Torino
Corso Duca degli Abruzzi 24
I-10129 Turin Italy

Abstract

Particular design environments may prevent Design for Testability techniques from reducing testing to a combinational problem: ATPG for sequential devices remains a challenging field. Random and deterministic structure-oriented techniques are the state-of-the-art, but there is a growing interest in methods where the function implemented by the circuit is known. This paper shows how a test pattern may be generated while trying to disprove the equivalence of a good and a faulty machine. The algorithms are derived from Graph Theory and Model Checking. An example is analyzed to discuss the applicability and the cost of such an approach.

1 Introduction

Despite the introduction of numerous Design for Testability techniques, test pattern generation for synchronous sequential circuits is still needed for some design environments [Moto90]. Sequential ATPG is therefore receiving an increasing attention as a still very open field for academic research and industrial application. Sequential testing's inherent complexity is due to the need for a "homing sequence" to initialize the circuit to a known state [Micz83] and to the fact that a fault is not in general simultaneously controllable and observable, but only through a sequence of time frames in the past and in the future, respectively.

The extension of fault simulation techniques to test generation for sequential devices has been the object of research [ACAg89], but reduced CPU time and high fault coverage for very complex circuits still cannot be guaranteed.

Other works focused on deterministic algorithms: starting from Huffmann's model, which transforms sequential behaviour over time frames into combinational behaviour

*Partially supported by the EEC under contract ESPRIT BRA 3216 "CHARME"

on iterated structures, the basic method extends the D-algorithm [PuRo71]. Many enhancements to it have been proposed, limiting the number of copies [Fuji85], introducing a 9-valued algebra [Muth76], searching for a path to be sensitized [Marl86], [Chen88], exploiting the benefits of techniques used in combinational ATPG [ScAu89].

A common feature of both random and deterministic ATPGs is that they are all *structure-oriented*, i.e., the only knowledge of the device they have is its topology. Some efforts have shown a growing interest on algorithms where a certain amount of reasoning is performed or the *function* implemented by the circuit is known. A recent paper [HuSe89] describes an application of Temporal Logic: its operators are used both to describe future behaviour and to justify the past one and a reasoning process yields a test pattern. The state-transition table is used in [MDNS88] to justify backward the values needed to control an observable fault until the initial state and the primary inputs are reached. The need to know the function restricts the application domain to medium-sized circuits, since it must be extracted from the structure. Whenever the same approach is embedded within a synthesis system, this limitation is easily overcome [ChJo90].

This paper presents a method to generate test pattern sequences for synchronous circuits modeled as Mealy or Moore Finite State Machines (FSMs). This method is based on a strong interaction between Graph Theory and Model Checking. The basic hypothesis is that the function realized by the device is known in terms of a graph specifying state transitions and outputs. The faulty circuit is also modeled as a graph, where the effect of the single stuck-at fault has been taken into account. If we can prove that the two machines are equivalent then the fault is undetectable, otherwise the counterexample is the test pattern. Equivalence is proved resorting to the product machine of the good and the faulty ones. The resulting graph is normally too big for other than trivial circuits, so we use Graph Theory [HoUl79] and Model Checking [BCDM86], keeping the product machine as a mere conceptual frame of reference. We build, without any any loss of information, a Non Deterministic Finite Automaton (NFA) bound to the product machine, using ad hoc techniques to keep its size manageable. We then transform it into a DFA that has the same number of states as the NFA. This DFA recognizes a language which is only a part of the language associated to the NFA, but it is sufficient to determine the test pattern. The DFA is a Moore Machine and it is a suitable input for the Model Checker of the MCB system [CESi86]. The use of NFAs and DFAs reduces the size of the graphs, so that this approach is applicable to medium-sized circuits, but, above all, it overcomes the Model Checker's limitation to Moore FSMs, at least for testing applications. The test detection condition is easily stated as a CTL formula and the counterexample facility of MCB is used to compute the test pattern.

2 Modeling and testing FSMs

Finite State Machines are extensively used as building blocks in a variety of applications, ranging from VLSI devices to network controllers.

Definition 1

A Finite State Machine $\mathcal{M}$ is defined by a 6-tuple [Koha70]:

$$\mathcal{M} = (I,\ O,\ S,\ \delta,\ \lambda,\ s^0)$$

where I, O, and S are finite, nonempty sets of inputs, outputs, and states, respectively, $\delta : I \times S \rightarrow S$ is the state transition function, $\lambda : I \times S \rightarrow O$ is the output function, and s^0 is the initial state.

Definition 2

Given two completely specified FSMs $\mathcal{M}_1$ and $\mathcal{M}_2$ operating on the same input, output, and state sets, they are *equivalent* if and only if for all input sequences their output sequences are equal.

Definition 3

Given two FSMs $\mathcal{M}_1 = (I,\ O,\ S_1,\ \delta_1,\ \lambda_1,\ s_1^0)$ and $\mathcal{M}_2 = (I,\ O,\ S_2,\ \delta_2,\ \lambda_2,\ s_2^0)$, the product machine $\mathcal{M}_{12}$ is defined as a 6-tuple:

$$\mathcal{M}_{12} = (I,\ O_{12},\ S_{12},\ \delta_{12},\ \lambda_{12},\ s^0)$$

where

$O_{12} = O \times O$

$S_{12} = S_1 \times S_2$

$\delta_{12} : I \times S_1 \times S_2 \rightarrow S_1 \times S_2$ is defined by: $\delta_{12}(s_1, s_2, i) = (\delta_1(s_1, i),\ \delta_2(s_2, i))$,

$\lambda_{12} : I \times S_1 \times S_2 \rightarrow O \times O$ is defined by: $\lambda_{12}(s_1, s_2, i) = (\lambda_1(s_1, i),\ \lambda_2(s_2, i))$,

where

$s_1 \in S_1;\ s_2 \in S_2;\ i \in I$

$s^0 = (s_1^0, s_2^0)$.

The test generation problem for a FSM may be stated as follows: given a good machine $\mathcal{M}_1$ and a single stuck-at-x ($x \in \{0, 1\}$) fault $\mathcal{F}$, generate a faulty machine $\mathcal{M}_2$, operating on the same input, output, and state sets, the fault $\mathcal{F}$ changing the δ and/or λ functions. If we can demonstrate that the two FSMs are **not** equivalent, i.e., that there exists an input sequence $i_1, i_2, \ldots, i_k$ such that the k-th elements of the two output sequences differ, i.e., $o_{k,\mathcal{M}_1} \neq o_{k,\mathcal{M}_2}$, then a test pattern has been found. If equivalence can be proven, then fault $\mathcal{F}$ is undetectable.

A method of demonstrating the equivalence of two FSMs $\mathcal{M}_1$, $\mathcal{M}_2$ is to construct the product machine $\mathcal{M}_{12}$ [PoMC64] and to prove that:

$$\forall \text{ input sequences,}$$
$$\forall \text{ elements of the output sequences } o_{\mathcal{M}_1} = o_{\mathcal{M}_2}$$

In order to make this approach applicable to real examples, it is necessary to simplify the product machine $\mathcal{M}_{12}$ reducing the number of states. As a result we will obtain a NFA, which recognizes a language corresponding to all the test patterns for the fault $\mathcal{F}$. If only one pattern is needed, we will give an algorithm to construct a further reduced NFA.

Given two machines $\mathcal{M}_1$ and $\mathcal{M}_2$ and their product machine $\mathcal{M}_{12}$, we consider the subset Q_{12} of S_{12}, defined as:

$$Q_{12} = \{\, q_{12} = (q_1, q_2) \in S_{12} : \forall\, i,\ \lambda_1(q_1, i) = \lambda_2(q_2, i) \,\}$$

while we call Z_{12} the subset of S_{12}, whose states are not in Q_{12} and namely:

$$Z_{12} = S_{12} - Q_{12}$$

It follows:

Theorem 1

Two FSMs $\mathcal{M}_1$ and $\mathcal{M}_2$ are equivalent iff in the product machine $\mathcal{M}_{12}$ the subset Z_{12} is either unreachable from the initial state $s^0 = (s_{01}, s_{02})$ or empty.

We omit the proof of Theorem 1 because it is quite obvious. By using Theorem 1 we observe that only the states of Q_{12} are needed to prove the equivalence of two FSMs and all the states of Z_{12} can be replaced by a single state; this state is a failure state and it indicates that the FMSs are not equivalent: in other words it indicates that a test pattern for the fault has been found. We will clarify this concept by means of the definition of the NFA associated to a product machine $\mathcal{M}_{12}$:

Definition 4

Given two Mealy FSMs $\mathcal{M}_1$ and $\mathcal{M}_2$ and their product machine $\mathcal{M}_{12}$, if $s^0 \in Q_{12}$, the NFA associated with $\mathcal{M}_{12}$ is the 5-tuple:

$$NFA(\mathcal{M}_{12}) = (I,\, Q_{12} \cup F' \cup f,\, \delta_{q12},\, s^0,\, f)$$

where f is the final state of the NFA, F' is a set of states differing from Q_{12} because they take into account the last term of the test sequence; δ_{q12} has a domain which is a subset of $(Q_{12} \cup F') \times I$, while it ranges over subsets of $((Q_{12} \cup F') \cup f)$ and namely:

if $t \in Q_{12}$ and $\forall\, i\; \delta_{12}(t,i) \in Q_{12}\; : \delta_{q12}(t,i) = \delta_{12}(t,i)$

if $t \in Q_{12}$ but $\exists$ i: $\delta_{12}(t,i) \notin Q_{12}$

for this i : $\delta_{q12}(t,i) = \{f'_j : f'_j \in F'$ and

$\lambda_{12}(\delta_{12}(t,i),j) = (o_1, o_2)$ with $o_1 \neq o_2\}$

for the other i: $\delta_{q12}(t,i) = \delta_{12}(t,i)$

if $t \in F'$: $\delta_{q12}(t = f'_j, i) = f$, if $i = j$, otherwise it is not defined.

By using Definition 4 and Theorem 1 we obtain Theorem 2.

Theorem 2

Given two FSMs $\mathcal{M}_1$ and $\mathcal{M}_2$, representing the good and the faulty machines, respectively, every string of the language recognized by DFA($\mathcal{M}_{12}$) is a test pattern. □

We note that for Moore machines we can replace in Definition 4 the last input with the empty string; in other words we can substitute all the states in $F' \cup f$ with the state f; it is easy to show that we will obtain a DFA instead of NFA.

We can further reduce the NFA of Definition 4 by eliminating the unreachable states. Pruning the NFA reduces the memory required to store it and makes the approach applicable to bigger circuits.

This pruning operation can be done by using only the connections of the directed graph obtained by the NFA without the input labels. Obviously, if the final state is not reachable, the two machines $\mathcal{M}_1$ and $\mathcal{M}_2$ are equivalent and there is no test pattern.

The previous results are valid for two FSMs operating on the same input and output sets and with different state sets; we can obtain some simplifications by noting that, in the testing problem, the state set of the good machine equals the state set of a faulty one; this entails that the two machines are nearly isomorphic. We have:

$$\mathcal{M}_1 = (I,\, O,\, S,\, \delta_1,\, \lambda_1,\, s^0)$$
$$\mathcal{M}_2 = (I,\, O,\, S,\, \delta_2,\, \lambda_2,\, s^0)$$

If δ_2 differs from δ_1 and λ_1 differs from λ_2 in n points of their domains, which are the same being the state set the same, we restrict investigation to the states in which δ_1 or λ_1 are modified. They are at most n. We indicate with M the set of these states.

We denote with R_1 the set of states reachable from M in one or more steps by the function δ_1 and with R_2 the set reachable in one or more steps by δ_2. We can change the definition of Q_{12} in this way:

$$Q_{12} = \{ q_{12} = (q_1, q_2) \in S_{12} : \\ ((q_1 \in R_1 \ \wedge \ q_2 \in R_2 \ \wedge \ q_1 \neq q_2) \vee (q_1 = q_2)) \wedge (\forall\, i,\, \lambda_1(q_1, i) = \lambda_2(q_2, i))\}$$

Using this new set Q_{12} we can obtain a simpler NFA by means of Definition 4 and it is easy to show that for this new NFA Proposition 2 is still valid. It is possible to eliminate also in this case the unreachable states in the NFA.

The construction of the NFA($\mathcal{M}_{12}$) associated to the product machine of two FSMs leads to the determination of all the test patterns for a fault; sometimes it is sufficient to find only one pattern, possibly the shortest one. In order to do this we use MCB [CESi86]. MCB has as an input the graph of any Moore FSM and a formula of Branching Time Temporal Logic called CTL, describing the property to be verified. We must reduce the NFA of Definition 4 to a structure that can be accepted by MCB; it is well known that every NFA is equivalent to a DFA [HoUl79] but in this case it is sufficient to replace every subset $\{f_j\}$ of Definition 4 by only one of its elements; in fact our goal is to find only one test pattern. The DFA is a Moore FSM in which the output associated with all the states except the final state is 0, while the output associated with the final state is 1. It is important to note that we reduce the problem of the equivalence of two Mealy Machines to that of a simple DFA, which is a trivial Moore Machine; thus we are able to use MCB in finding a test pattern for a generic FSM. The formula to be verified by the MCB asserts that the output of the DFA has to be different from 1. Since the logic is branching in time, the **AG** operator is used to assert that the formula is true for all possible evolutions in the future, i.e., for every path and at every node on the path.

The equivalence formula, if O is the output of the DFA, is:

$$\mathbf{AG}(\tilde{}\ \ O).$$

3 An example

In this section we analyze the example of Fig. 1, originally presented in [MDNS88].

The functional description of the FSM expresses the inputs to the D-FFs in terms of the state variables $x_0\, x_1$ and of the input i:

$$\delta_1((x_0, x_1), i) = ((\bar{i} \cdot x_0 + i \cdot x_1), \overline{(i \oplus x_1)})$$

The primary output depends on the state variables $x_0\, x_1$ and on the primary input i: $\lambda_1((x_0, x_1), i) = i \cdot \overline{x_0}$.

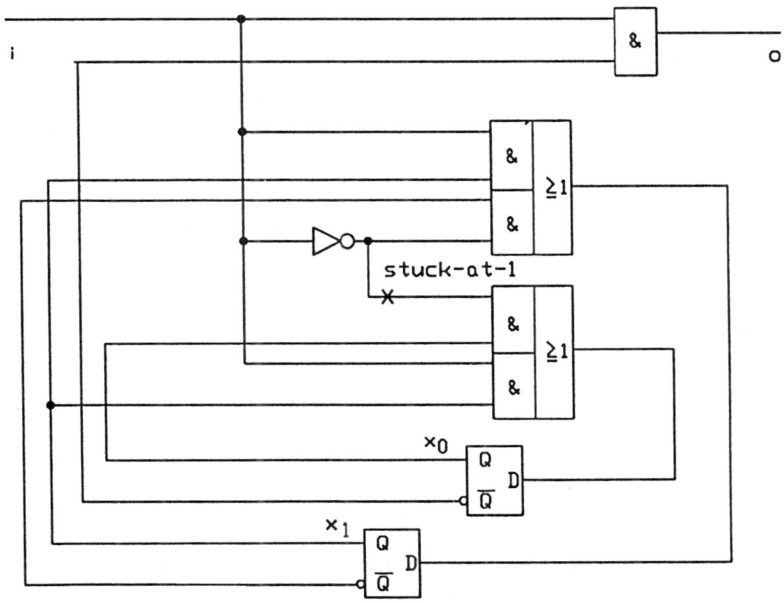

Figure 1 – An example

Fig. 2(a) shows the state transition graph of the good machine. Let us consider a stuck-at-1 fault, as indicated in Fig. 1. The good machine's graph is modified to yield the faulty machine's one (Fig. 2(b)). The output function is not modified:

$$\lambda_2(x_0, x_1), i) = i \cdot \overline{x_0}$$

The corresponding automaton is in this case the DFA of Fig. 3.

Only 6 states of this automaton are reachable from the initial state; the remaining 4 are unreachable and are not useful to determine test patterns. This DFA recognizes the language consisting of all the possible test patterns. Two of the 6 states (f_1' and f) are only introduced to formally define the DFA, thus only 4 states are really needed, compared with 16 states for the product machine. The testing sequences correspond to the following language (which is a regular expression on the alphabet {0,1}):

$$T = (00+1)^{\star}\ 01\ (00+1)^{\star}\ 011$$

Obviously, the shortest testing sequence is 01011.

By using the fact that the two machines operate on the same state set, we can eliminate from the DFA the states 0001 and 0100, because the δ_1 function is only modified in state 10 and the states 00 and 01 are not reachable from 10 in the faulty machine. After reduction we obtain the same DFA without the unreachable states of Fig. 3.

If we use MCB with Moore FSM corresponding to the DFA of Fig. 3 as an input, the CTL formula to be disproved is:

$$\mathbf{AG}\ (\tilde{}\ O).$$

The MCB finds that the formula is false and the counterexample sequence coincides with the test pattern.

4 Conclusions

This paper presented an approach to ATPG for synchronous circuits modeled as Moore or Mealy FSMs. The graph representing state and output transitions is first extracted from the structural description of the circuit, then the effect of the single stuck-at fault is injected, resulting in a faulty graph. The test pattern generation algorithm is based on techniques used to prove the equivalence of FSMs by considering the product machine as a mere conceptual framework. Transforming the product machine's graph into a NFA without unreachable states reduce memory requirements, at the expense of CPU time. In general, we accept to pay this price to make the approach applicable to bigger circuits. The advantage of the NFA with respect to the product machine resides in the drastic reduction of the number of states, which in the best case are almost linear

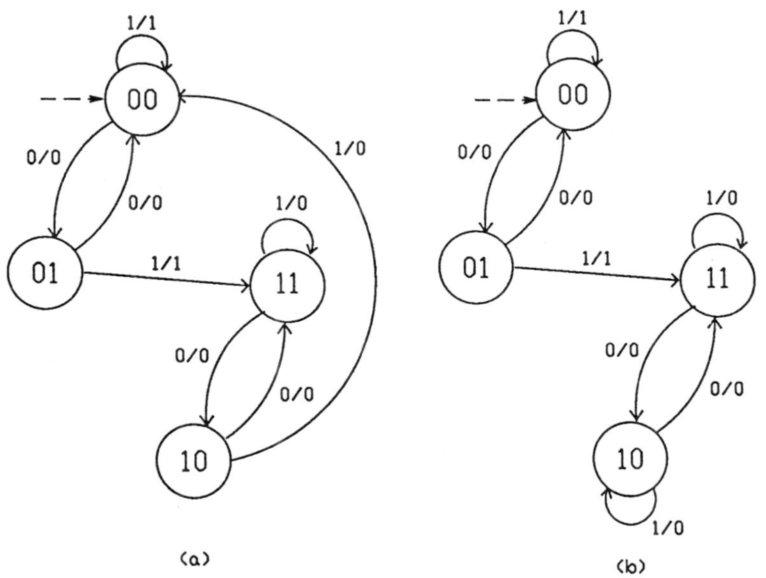

Figure 2 – The good and faulty automata for the example of Fig. 1

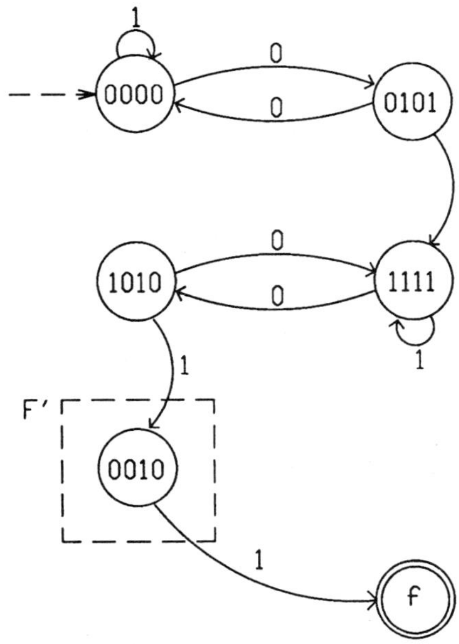

Figure 3 – The DFA corresponding to the product machine of Fig. 3

in the number of states of the good machine. The Model Checking algorithm is used to disprove, when possible, a CTL formula and MCB's counterexample facility returns the test pattern. A fault simulator [CGSR89] serves the purpose of fault dropping. Preliminary experimental data, on an international benchmark set [BBKo89] (Tab. 1) indicate that the good and faulty machine's graphs are considerably smaller than their upper bound and that the CPU time to create them is reasonable. The main limit is now imposed by MCB, which is able to deal only with a reduced number of states. Results on fault coverage are encouraging: on the s298 circuit 100% coverage for detectable faults is reached.

Circuit	Good Machine			Faulty Machines (reachable states)			
	total states	reachable states	CPU time (s)	avg	max	min	CPU time (s)
s27	8	6	0.05	4	7	3	0.05
s208	256	17	7.81	17	256	1	12.81
s298	16384	218	0.82	189	347	2	0.96
s344	32768	2625	400.17	3216	19693	11	775
s349	32768	2625	405.81	3171	19693	11	768
s382	2M	8865	64.61	5999	13858	2	52.01
s386	64	13	1.37	11	22	2	1.51
s400	2M	8865	66.56	6027	15790	2	53.54
s444	2M	8865	42.70	5929	15790	2	38.71
s526	2M	8868	38.07	6884	14580	2	41.45
s526n	2M	8868	37.47	6879	14580	2	41.22

Table 1 – Graphs statistics

The novelty of our research resides in modeling both the good and the faulty circuit as graphs, so that the fault detection condition can easily be stated. The use of the NFA and efficient graph simplification algorithms make this approach more efficient. Although the current version refers to the single stuck-at, the algorithm is independent of the fault model, since the latter intervenes only during the generation of the faulty graph. It is thus possible to investigate the use of other models, namely functional ones, whose goodness may be assessed giving a figure of merit in terms of single stuck-at fault coverage.

The programs to extract graphs, to build and reduce the NFA, the combinational ATPG, and the fault simulator have been implemented in separate packages, amounting to a total of 4000 C-code lines and together with MCB, they run on a SUN Sparcstation-1.

References

[ACAg89] V.D. Agrawal, K.T. Cheng, P. Agrawal: "A directed search method for test generation using a concurrent fault simulator," IEEE Transactions on Computer-Aided Design, Vol. 8, n. 2, February 1989, pp. 131-138

[BBKo89] F. Brglez, D. Bryan, K. Koźmiński: "Combinational profiles of sequential benchmark circuits," ISCAS'89: IEEE International Symposium on Circuits And Systems, Portland, OR (USA), May 1989, pp. 1929-1934

[BCDM86] M. Browne, E.M. Clarke, D. Dill, B. Mishra: "Automatic verification of sequential circuits using temporal logic," IEEE Transactions on Computers, Vol. C-35, n. 12, December 1986, pp. 1035-1044

[CESi86] E.M. Clarke, E.A. Emerson, A.P. Sistla: "Automatic verification of finite-state concurrent systems using temporal logic specifications," ACM Transactions on Programming Languages and Systems, Vol. 8, n. 2, April 1986, pp. 244-263

[CGSR89] G. Cabodi, S. Gai, M. Sonza Reorda: "Partitioning Techniques in Multiprocessor Simulators," ESM-89: European Simulation Multiconference, Rome (Italy), June 1989, pp. 311-317

[Chen88] W.T. Cheng: "The Back Algorithm for sequential test generation," ICCD'88: IEEE International Conference on Computer Design, Rye Brook, NY (USA), October 1988, pp. 66-69

[ChJo90] K-T. Cheng, J-Y. Jou: "Functional test generation for Finite State Machines," ITC'90: International Test Conference 1990, Washington, DC (USA), September 1990, pp. 162-168

[Fuji85] H. Fujiwara: "Logic testing and design for testability," The MIT Press, Cambridge, MA (USA), 1975

[HoUl79] J.E. Hopcroft, J.D. Ullman: "Introduction to automata theory, languages, and computation," Addison-Wesley, Reading, MA (USA), 1979

[HuSe89] R.V. Hudli, S.C. Seth: "Temporal Logic based test generation for sequential circuits," IFIP TC 10/WG 10.2 Working Conference on CAD systems using AI techniques, Tokyo (Japan), June 1989, pp. 91-98

[Koha70] Z. Kohavi: "Switching and finite automata theory," Computer Science Series, Mc Graw Hill, New York, NY (USA), 1970

[Marl86] R. Marlett: "An effective test generation system for sequential circuits," DAC-23: 23th IEEE/ACM Design Automation Conference, Las Vegas, NV (USA), June 1986, pp. 250-256

[MDNS88] H.K.T. Ma, S. Devadas, A.R. Newton, A. Sangiovanni-Vincentelli: "Test generation for sequential circuits," IEEE Transactions on Computer-Aided Design, Vol. 7, n. 10, October 1988, pp. 1081-1093

[Micz83] A. Miczo: "The sequential ATPG: a theoretical limit," ITC'83: IEEE International Test Conference, Philadelphia, PA (USA), October 1983, pp. 143-147

[Moto90] A. Motohara: "Design for Testability of ASICs in Japan," IEEE 13th Annual Workshop on Design for Testability, Vail, CO (USA), April 1990, *(Oral presentation; no proceedings available)*

[Muth76] P. Muth: "A nine-valued circuits model for test generation," IEEE Transactions on Computers, Vol. C-25, n. 6, June 1976, pp. 630-636

[PoMC64] J.F. Poage, E.J. McCluskey: "Derivation of optimum test sequences for sequential machines," 5th Annual Symposium on Switching Theory and Logical Design, 1964

[PuRo71] G.R. Putzolu, J.P. Roth: "A heuristic algorithm for the testing of asynchronous circuits," IEEE Transactions on Computers, Vol. C-20, n. 6, June 1971, pp. 639-647

[ScAu89] M.H. Schulz, E. Auth: "ESSENTIAL: an efficient self-learning test pattern generation algorithm for sequential circuits," ITC'89: International Test Conference 1989, Washington, DC (USA), September 1989, pp. 28-37

DIMACS Series in Discrete Mathematics
and Theoretical Computer Science
Volume 3, 1991

Compositional Design and Verification of Communication Protocols, using Labelled Petri Nets

JEAN CHRISTOPHE LLORET
VERILOG, 150 rue Nicolas Vauquelin, 31081 Toulouse cedex,
PIERRE AZÉMA, FRANÇOIS VERNADAT
LAAS-CNRS, 7 Avenue Colonel Roche, F-31077 Toulouse cedex

Abstract

This paper deals with the design and verification of telecommunication protocols. A methodology based upon structured Labelled Petri Nets is proposed. A multi-gate rendez-vous operator is introduced for net composition such that global properties can be derived from properties of component nets. Protocol conformance with respect to service specification is a main property. The specification and verification of a protocol is developed.

1 Introduction

The aim of ISO formal description techniques is to increase the designer confidence in system critical parts. Labelled Predicate Transition nets (LPrT) are particularly relevant for the design of telecommunication protocols. This paper proposes a methodology to specify and verify telecommunication protocols by means of LPrT nets. In this paper, the following two principles are used as guidelines.

Structured Specification The structured system decomposition is based upon communication primitives. Then, design modularity results from LPrT module specification in a top down manner via sub-net composition. The rendez-vous communication paradigm of ISO language LOTOS is extended to multi-gate rendez-vous. Several input/output events may appear on a single transition: Petri Net transitions are labelled by sets of communicating events.

Incremental Description A single system is described with respect to several levels of abstraction. Each new abstraction level supplies a more detailed model and new properties are to be verified. Communication by multi-rendez-vous is a first means of abstraction. A second means concerns the data part. To facilitate data abstraction in a LPrT net, logic programming (Prolog) is used as a declarative and prototyping language. A set of Horn clauses specifying data operations is associated with a labeled PrT net. This set, or Prolog database, is easily updated to take into account different levels of assumption.

The main contributions of this paper concern the support of former principles, and the definition of a LPrT multi-rendez-vous operator with value-passing by logic unification.

Multi-rendez-vous is introduced in a stepwise approach, from basic semantical models, that is Labelled transition systems (LTS) and Labelled Petri Nets (LPN), to labelled Predicate Transition nets. The labelled Predicate Transition nets provide the highest abstraction level and are the user interface model. This stepwise definition presents the following characteristics:

didactical: Synchronization aspects are studied in the context of labelled Petri Nets. Communication with value passing is only introduced for Labelled Predicate transition nets.

analytical: Verification techniques are based on the analysis of the LPrT model behaviour. A specific technique, so-called projection or service computation, is derived from observational equivalence [Mil80]). The global service results from the composition of sub-net services. A modular design of labelled PrT nets entails a modular verification.

The composition of Labelled Petri Nets is described in Section 2. A value passing mechanism and the parameterization of Petri Nets are developed in Section 3.

Section 4 is devoted to a specific application: specification and verification of application layer of protocol telereport, designed by the french electricity board EDF[1].

2 Composition of multi-event Labelled Petri Nets

This section introduces Labelled Petri Nets and multi-rendez-vous operator $|_{LPN}$ as the core language (no value passing) and the basic composition operator, respectively.

The expressive power of this composition operator results from the ability to specify rendez-vous among several (more than two) transitions, on the same or distinct interaction points. This multiple-rendez-vous is an abstraction with respect to implemented communication mechanisms. In the context of the advocated progressive modelling, multi-rendez-vous enables the design of abstract and very compact models.

The design of OSI protocols by means of multi-event-rendez-vous is presented in subsection 1. The rendez-vous of Labelled Transition Systems is introduced in subsection 2. Labelled Petri nets and composition operator $|_{LPN}$ are presented in subsections 3 and 4, respectively.

2.1 Multi-events in Protocol Design

OSI reference model: The Open System Interconnection (OSI) architecture, as defined at the International Organization for Standardization (ISO), is adopted as a guideline for protocol design. A layered model may be structured as a system of three subnets (see Fig 1), composed of two application entities A_{us} and A_{pr} which communicate through a session service S.

According to the previous architecture, the design of a protocol layer involves both the service and the protocol. The (A) protocol and the (S) service it uses will be described by labelled transition systems.

Service primitive SP and service access point SAP are respectively mapped on interactions and gates (i.e. interaction points) of Labelled Transition Systems.

[1]Electricite' de France,DER Clamart, 1 Av. G. De Gaulle, 92141 Cedex, France

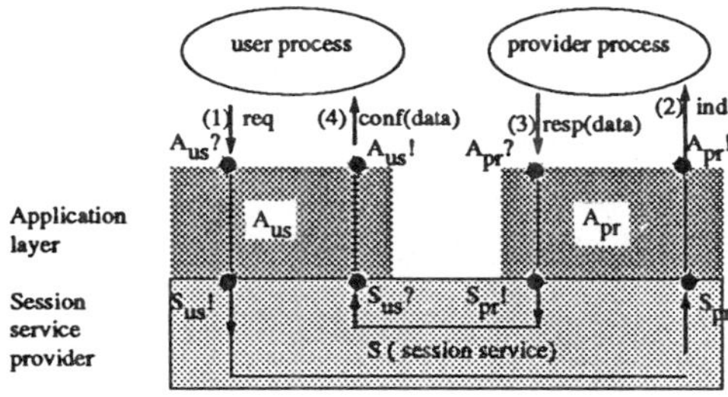

Figure 1: Telereport application layer in OSI architecture

gates : **A process modelled by a LTS is regarded as a black box communicating with its environment through gates. Like in ISO language LOTOS [vEVD89], gates are** not oriented — symbols ! and ? in gates $A_{us}!$ and $A_{us}?$ are used for comments and denote what the designer wants them to do: respectively output and input value from the point of view of entity A_{us} —.

The set of gates of process *proc* is a priori defined, it is denoted α_{proc}. The set of gates of entity A_{us} is $\alpha_{A_{us}} = \{A_{us}?, A_{us}!, S_{us}?, S_{us}!\}$.

interactions : interactions are represented by ground terms (terms without variables) constructed from constants (i.e. identifiers of data values) and function symbols. The most external function symbol of an interaction (such as "*resp*" in "*resp*(*data*)") is the name of the abstract service primitive, and the arguments, e.g. *data*, are the service primitive parameters.

events : An event is the most elementary communication unit. Let α be the set of gates, let V be the set of interactions. An event is a couple (g, v) which consists of gate g and interaction v. Event (g, v) is denoted $g(v)$. Let *gate* be the function which returns the gate of an event $(gate(g(v)) = g)$ and let $\mathcal{E}$ be the set of events.

Transition labelling is now defined according to the following principles:

1. Multi-event action: *an action describes several communication events that are performed synchronously on different gates.* A transition is labelled by an action. Because an action refers to a set of events, the expressiveness is increased with respect to the reference languages CCS [Mil80] and LOTOS [vEVD89] in which an action is either a single (observable) event or internal action i.

2. A single event per gate that is gate is an unshared resource. A service access point is dedicated to a single entity. Consequently, *two distinct events on the same gate do not belong to the same action.*

Formally, an action A is a subset of events $(A \subseteq \mathcal{E})$ such that,

$\forall e_1, e_2 \in A, e_1 \neq e_2 \Rightarrow gate(e_1) \neq gate(e_2)$.

Action $\emptyset$ is denoted i (no event = internal action).

Definition 2.1 Labelled Transition system labelling Let *proc* be a labelled transition system; let $\mathcal{T}$ be *proc* transition set.

Let $\alpha_{proc} \subseteq \alpha$ be a subset of gates, let V be the set of interactions.

An event e, connected to a gate in α_{proc}, is a couple $(gate, interaction) \subset \alpha_{proc} \times V$. e

is denoted *gate*(*interaction*).
proc labelling is couple $(\alpha_{proc}, l_{proc})$ where l_{proc} is the label function. The domain of label function l_{proc} is transition set $\mathcal{T}$. The range of function l is the subset of actions constituted by events connected in α_{proc}.
In the sequel, $\mathcal{L}_{proc}$ denotes couple $(\alpha_{proc}, l_{proc})$.

2.2 Multi-rendez-vous Composition of Labelled Transition Systems

Labelled Transition Systems supply an operational semantics to Labelled Petri Nets. The multi-rendez-vous of labelled transition systems is the interpretation of multi-rendez-vous of Petri nets.

Definition 2.2 *Labelled Transition System*
A labelled transition system *proc* is a 5-tuple $(S, \mathcal{T}, -t \rightarrow_{t \in \mathcal{T}}, s_0, \mathcal{L})$ where:

- S set of states.
- T set of transitions.
- $-t \rightarrow \subset S \times S$ state change performed by transition t.
- $s_0 \in S$, initial state.
- $\mathcal{L}$ is a labelling as defined above.

Two labelled Transition systems $proc_1$ and $proc_2$ are composed relative to their common gates.

LTS common gates : the set of common gates $\alpha_{proc_1} \cap \alpha_{proc_2}$ is denoted $\alpha_\cap$.

events to be synchronized : events to be synchronized of label $l(t)$ are events whose gate is a shared gate between $proc_1$ and $proc_2$. Notation is $sync_{\alpha_\cap}(l(t))$:
$sync_{\alpha_\cap}(l(t)) = \{e \in l(t) \mid gate(e) \in \alpha_\cap\}$.

Definition 2.3 *LTS composition operator* $|_{LTS}$
Let $proc_i = (S_i, \mathcal{T}_i, -t_i \rightarrow_{t_i \in \mathcal{T}_i}, s_{i,0}, \mathcal{L}_i)_{i=1,2}$ be two LTS.
Composed LTS $proc_1 \; |_{LTS} \; proc_2$ is $(S_1 \mid S_2, \mathcal{T}_1 \mid \mathcal{T}_2, -t \rightarrow_{t \in \mathcal{T}_1 | \mathcal{T}_2}, s_{1,0} \mid s_{2,0}, \mathcal{L}_1 \mid \mathcal{L}_2)$ where:
$\mathcal{L}_1 \mid \mathcal{L}_2$ is $proc_1 \mid proc_2$ labelling defined by set of gates $(\alpha_{proc_1} \cup \alpha_{proc_2})$ and labelling function l defined on domain $\mathcal{T}_1 \mid \mathcal{T}_2$.
Function l is defined together with composed states and composed transitions by the following derivation rules (s_i, s'_i and t_i are states and a transition of $proc_i$ respectively):

Independant execution

$$(1)\ \frac{s_1 - t_1 : l_1 \rightarrow s'_1}{s_1 \mid s_2 - t_1 : l_1 \rightarrow s'_1 \mid s_2} \quad (sync_{\alpha_\cap}(l_1) = \emptyset)$$

$$(2)\ \frac{s_2 - t_2 : l_2 \rightarrow s'_2}{s_1 \mid s_2 - t_2 : l_2 \rightarrow s_1 \mid s'_2} \quad (sync_{\alpha_\cap}(l_2) = \emptyset)$$

Synchronization

$$(3)\ \frac{s_1 - t_1 : l_1 \rightarrow s_1',\ s_2 - t_2 : l_2 \rightarrow s_2'}{s_1 \mid s_2 - t_1 \mid t_2 : (l_1 \cup l_2) \rightarrow s_1' \mid s_2'} \quad (sync_{\alpha_\cap}(l_1) = sync_{\alpha_\cap}(l_2))$$

2.3 Labelled Petri Nets

Firstly, Labelled Petri Nets are defined as Place/Transition nets of capacity one, associated with a labelling as defined above.
The behaviour of a labelled Petri Nets is then introduced as a labelled transition system. The marking graph and the step graph are two possible candidates for describing a net behaviour. In the step graph, parallel executions are explicitly taken into account and thus a more precise representation of the net behaviour is given. Labelled Petri Net operator $|_{LPN}$ is interpreted on step graphs by Labelled Transition System operator $|_{LTS}$ in such a way that diagram 2 commutes.

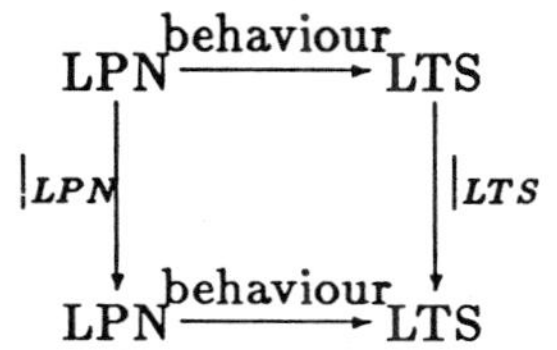

LPN: Labelled Petri Nets
LTS: Labelled Transition Systems
$\mathcal{B}(N)$: net N behaviour (step graph)

$\forall N_1, N_2 \in LPN, \mathcal{B}(N_1 \mid_{LPN} N_2) = \mathcal{B}(N_1) \mid_{LTS} \mathcal{B}(N_2)$

Figure 2: Semantics of Labelled Petri Nets composition "$|_{LPN}$"

Definition 2.4 *Labelled Petri Net*
A labelled Petri Net is a 6-tuple $N = (P, \mathcal{T}, pre, post, M_0, \mathcal{L}_N)$ where:

- $\mathcal{P}$ is a set of places;
- $\mathcal{T}$ is a set of transitions, $P \cap \mathcal{T} = \emptyset$
- $pre, post : \mathcal{T} \rightarrow \mathcal{P}(P)$ are two mappings which connect a transition to two sets of places called preconditions and postconditions respectively.
- M_0 is a set of places called initial marking.
- $\mathcal{L}_N$ is the net labelling defined by couple (α_N, l_N) as introduced previously.

Labelled Petri Nets enable a concise description of behaviours; furthermore states are decomposed as sets of marked places: such a refinement leads to a precise and graphic representation of conflicts and parallelism.
Example: net N_1, Fig 3, depicts concurrent transitions t_1 and t_2 (places are circles, transition boxes inscribed by a transition name; if a transition box is connected to a gate, the edge is annoted by the value of the corresponding transition event; the preconditions and postconditions of a transition are the input and output places respectively; places with a token belong to the initial marking).

Definition 2.5 *Firing rule* The transition firing rule proposed for "augmented condition/event nets" in [PD87] is adopted. But the definition of parallel transitions differs because an interaction point is a non shared resource: with respect to labelled Petri nets, two transitions connected to a common interaction point cannot be fired in parallel.

Let M be a marking (or set of places) and T a set of transitions (set T is called *step*). T is M enabled (notation $M[T>$) iff

- T transitions are *pairwise independent* that is two distinct transitions $t_1, t_2 \in T$ share neither a place (*place independance*: $(pre(t_1) \cup post(t_1)) \cap (pre(t_2) \cup post(t_2)) = \emptyset$) nor a gate (*gate independance*: $gate(l_N(t_1)) \cap gate(l_N(t_2)) = \emptyset$, where $gate(l_N(t)) = \{gate(e) \mid e \in l_N(t)\}$).
- preconditions of T transitions are fulfilled ($pre(T) \subset M$, with $pre(T) = \bigcup_{t \in T} pre(t)$) and postconditions which are not preconditions are not fulfilled ($M \cap (post(T) \setminus pre(T)) = \emptyset$).

Marking M' results from firing step T in marking M (notation $M[T > M'$) iff step T is enabled in M and $M' = (M \setminus pre(T)) \cup post(T)$.

The behaviour of a LPN is a LTS such that a state is a reachable marking and a LTS transition is a firable step.

Definition 2.6 *LPN Behaviour*
Let $N = (P, \mathcal{T}, pre, post, M_0, \mathcal{L}_N)$ be a LPN. Reachable markings and firable steps of net N are respectively the smaller sets $M_0[>$ and $M_0[$ defined by:

- $M_0 \in [M_0 >$
- If $M \in [M_0 >$ and T enabled in M with $M[T > M'$ then $M' \in M_0[>$ and $T \in M_0[$.

Net behaviour is LTS, $\mathcal{B}(N) = (M_0[>, M_0[, -T \rightarrow_{T \in M_0[}, M_0, \mathcal{L}_B)$ where:

- change of state in $\mathcal{B}(N)$ corresponds to step execution:$M, M' \in M_0[>, M - T \rightarrow M' \Leftrightarrow M[T > M'$.
- Behaviour labelling $\mathcal{L}_B = (\alpha_B, l_B)$ directly follows net labelling: the gate set of $\mathcal{L}_B$ is $\mathcal{L}_N$ one, i.e. $\alpha_B = \alpha_N$. Labelling function l_B is the canonical extension to sets of the net transition labelling function l_N: $\forall T \in M_0[, l_B(T) = \bigcup_{t \in T} l_N(t)$.

2.4 Multi-rendez-vous Composition of Labelled Petri Nets

Operator $|_{LPN}$ is defined in order to simulate composition of behaviours. The three following characteristics of net multi-rendez-vous are worth noticing:
1 *Parallelism between transitions is taken into account* When composing two labelled Petri nets, two transitions of a net, which fire in parallel, may be merged with a transition of the other net (on Fig 3 transition $\{t_1, t_2\} |_{LPN} \{t_3\}$ results from the merging of transition sets $\{t_1, t_2\}$ and $\{t_3\}$).
2 *Minimal mergings* The possibility of running synchronously (transitions are in parallel) is distinguished from a necessary synchronization (transitions are merged). In the

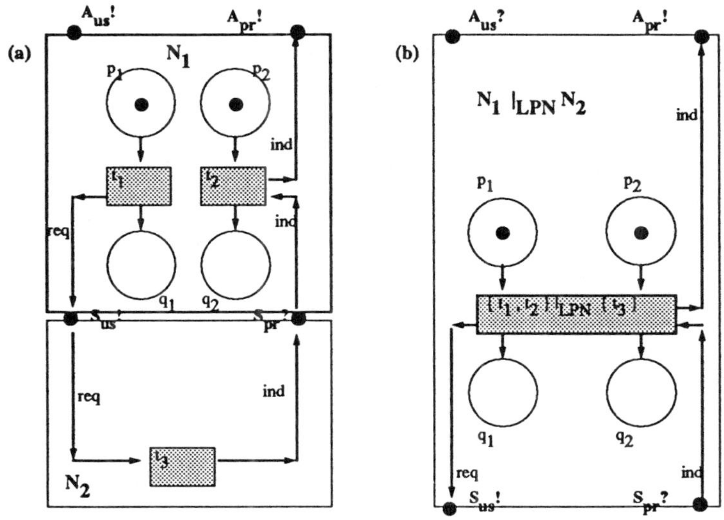

Figure 3: Parallel composition of Labelled Petri Nets

composition of two nets, all synchronization possibilities are preserved but the number of mergings is minimal.
3 The derivation of a composed net $N_1 |_{LPN} N_2$ does not require computation of behaviours $\mathcal{B}(N_1)$ and $\mathcal{B}(N_2)$.

Let $N_i = (P_i, \mathcal{T}_i, pre_i, post_i, M_{0,i}, \mathcal{L}_i)_{i=1,2}$ be two LPN with labelling $\mathcal{L}_i = (\alpha_i, l_i)_{i=1,2}$ and let $\alpha_\cap$ be the common gate alphabet.

Preliminary definitions:

Transition to be merged: Transition t of net N_1 or N_2 is to be merged iff there is at least one event to be synchronized in t label (recall that an event to be synchronized is an event whose gate belongs to common gate set $\alpha_\cap$).

Mergeable Transition sets: Two transition sets $T_1 \subseteq \mathcal{T}_1$ and $T_2 \subseteq \mathcal{T}_2$ are mergeable iff T_1 (resp. T_2) is a set of transitions *to be merged*, pairwise *independent* and the set of events to be synchronized of T_1 transitions equals the one of T_2 transitions (i.e. $sync_{\alpha_\cap}(l_1(T_1)) = sync_{\alpha_\cap}(l_2(T_2))$).

Minimal mergeable Transition sets: T_1 and T_2 are minimal mergeable transition sets iff T_1 are mergeable transition sets and if it is not possible to partition T_1, T_2 merging in smaller ones that is: for all $T_1' \subset T_1$ and $T_2' \subset T_2$, T_1' and T_2' are not mergeable.

Definition 2.7 *LPN composition operator* $|_{LPN}$

Composed net $N_1 |_{LPN} N_2$ is $(P_1 \cup P_2, \mathcal{T}_1 |_{LPN} \mathcal{T}_2, pre, post, M_{1,0} \cup M_{2,0}, \mathcal{L}_1 |_{LPN} \mathcal{L}_2)$ where:

- $\mathcal{T}_1 |_{LPN} \mathcal{T}_2$ transitions are, on the one hand, $\mathcal{T}_1$ and $\mathcal{T}_2$ transitions which are not to be merged and, on the other, merged transitions $T_1 |_{LPN} T_2$ where T_1, T_2 are minimal mergeable transition sets.

- Preconditions and postconditions of a nonmerged transition are unchanged in composed net. Pre and postconditions of merged transition $T_1 \mid_{LPN} T_2$ are defined as the set union of the respective pre and postconditions: $pre(T_1 \mid_{LPN} T_2) = pre_1(T_1) \cup pre_2(T_2)$ and similarly for postconditions.

- New labelling $\mathcal{L}_1 \mid_{LPN} \mathcal{L}_2$: gate sets of the composed nets are added; the labelling of a nonmerged transition is unchanged; labelling of merged transition $T_1 \mid_{LPN} T_2$ is $l_1(T_1) \cup l_2(T_2)$.

Example: figure 3b depicts composed net $N_1 \mid_{LPN} N_2$.

Proposition 2.1 *Properties of multi-rendez-vous composition*
Operators $\mid_{LTS}$ and $\mid_{LPN}$ are commutative and associative (up to state and transition bijective renaming).
The behaviour of a composed net is the composition of behaviours: $\mathcal{B}(N_1 \mid_{LPN} N_2) = \mathcal{B}(N_1) \mid_{LTS} \mathcal{B}(N_2)$ (up to state and transition bijective renaming).

3 Net parameterization and value passing

This section introduces Labelled Predicate Nets as a parameterized version of LPN. LPrT nets are PrT nets [Gen88] extended with labels, and featuring, in particular, direct execution in Prolog. PrT Net parameterization is introduced in subsection 1. Operator $\mid_{LPN}$ is extended to LPrT nets in subsection 2..

3.1 Labelled Predicate Transition Nets

Definition 3.1 A LPrT net consists of an LPN annotated by first-order language terms. Places are "dynamic" predicates: their truth value varies according to the current marking; constraints on logical variables are described by "static" predicates; A LPrT $N_{\mathcal{I}}$ is a structure $N = (P, T, pre, post, cond, M_0, \mathcal{L}_N)$ instanciated by interpretation $\mathcal{I}$, which specifies truth value of static predicates and domains for logical variables, where:

Places are dynamic predicates.

Transitions $\mathcal{T}$ such that $\mathcal{T} \cap P = \emptyset$.

Pre and Post : transition pre and postconditions are sets of place instances.

cond : is a binding of static predicates to each transition. This set of static predicates is a constraint on the variables which occur in pre, postconditions and events.

Initial Marking M_0 is a set of place ground instances, i.e., initially true propositions.

Labelling $\mathcal{L}_N$ is an LPN labelling with the extension that event values are no longer necessarily constants but may be ***arbitrary terms.***

Interpretation I is a set of Horn clauses, that is logical implications with at the most one conclusion. These clauses describe variable types, static logical conditions and procedures.

Definition 3.2 LPrT Net unfolding Unfolding consists of instanciating logical variables such that data constraints are true with respect to axioms and configuration. Unfolded net $\mathcal{U}(N_I)$ is computed as the smallest net (with respect to places, transitions sets and characteristic functions) such that the following rule is fulfilled:

Let $t : l(t);\ pre(t) \rightarrow post(t)\ IF\ cond(t)$ be a N transition. Let σ be a substitution that replaces $l(t), pre(t)$ and $post(t)$ variables by constants and such that:

no multiple arcs : σ creates no multiple pre or postconditions: two distinct places in $pre(t)$ (resp. $post(t)$) remain distinct after instanciation by σ.

constraint satisfaction : Predicates in $cond(t)$ after instanciation by σ are logic consequences of axioms and configuration: $\mathcal{I} \models \sigma(cond(t))$.

Unfolded net $\mathcal{U}(N_I)$ consists of LPN transition $t_\sigma : \sigma(l(t)); \sigma(pre(t)) \rightarrow \sigma(post(t))$.

3.2 Composition of Labelled Predicate Transition Nets

Operator $|_{LPrT}$ extends in a straightforward manner $|_{LPN}$. The Labelled PrT net semantics supplied by the unfolding procedure is preserved, that is diagram 4 commutes. Moreover, communication with value passing may be described.

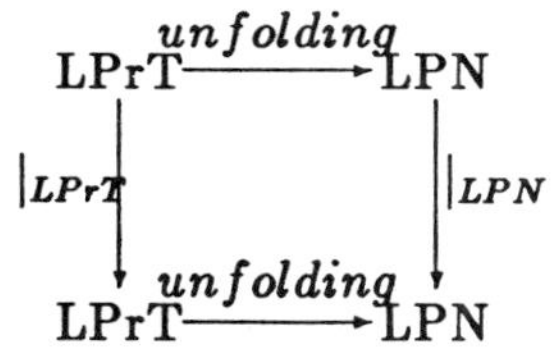

LPrT: Labelled Predicate Transition Nets
$\mathcal{U}(N)$: LPrT net N unfolding
$\mathcal{I}$: any LPrT net interpretation
$\forall N, N' \in LPrT, \mathcal{U}((N\ |_{LPrT}\ N')_I) = \mathcal{U}(N_I)\ |_{LPN}\ \mathcal{U}(N_I)$

Figure 4: Semantics of Labelled PrT composition "$|_{LPrT}$"

The computation of global LPrT net $N\ |_{LPrT}\ N'$ is independent of any interpretation. This global net is highly compact because the merging of two sets of parameterized transitions is a single parameterized transition. All modes of communication, namely value passing, will be modelled by logic unification.

Definition 3.3 Communication via Unification:

Merging of sets of parameterized transitions T_1 and T_2 is based on the computation of most general unifier σ_{T_1,T_2} of events to be synchronized.

Two (parameterized) actions A_1 and A_2 may communicate through gates $\alpha_\cap$ iff the respective (parameterized) events to synchronize $sync_{\alpha_\cap}(A_1)$ and $sync_{\alpha_\cap}(A_2)$ can be unified. Then, the most general unifying substitution σ_{A_1,A_2} of $sync_{\alpha_\cap}(A_1)$ and $sync_{\alpha_\cap}(A_2)$ is used to compute resulting communication action $A_1\ |_{LPrT}\ A_2$, that is, $\sigma_{A_1,A_2}(A_1 \cup A_2)$.

Remark: The communication substitution is unique, for actions A_1 and A_2, — up to variable renaming — because two distinct parameterized events in an action have distinct gates, and consequently an event to be synchronized in A_1 can be unified with a single event in A_2.

Definition 3.4 LPrT Composition

Let $N_i = (P_i, \mathcal{T}_i, pre_i, post_i, cond_i, M_{0,i}, \mathcal{L}_i)_{i=1,2}$ be two LPrT with labelling $\mathcal{L}_i$ defined by (α_i, l_i).
Mergeable Transition sets: $T_1 \subset \mathcal{T}_1$ and $T_2 \subset \mathcal{T}_2$ are mergeable iff labels $l_1(T_1)$ and $l_2(T_2)$ are actions which communicate via substitution σ and T_1 (respectively T_2) is a set of transitions *to merge*, pairwise *gate independent* and pairwise *place independent with respect to* σ. Transitions t_1, t'_1 are place independent with respect to σ iff $\sigma(pre_1(t_1) \cup post_1(t_1)) \cap \sigma(pre_1(t'_1) \cup post_1(t'_1)) = \emptyset$).

Composed net $N_1 |_{LPrT} N_2$ is $(P_1 \cup P_2, \mathcal{T}_1 |_{LPrT} \mathcal{T}_2, pre, post, cond, M_{0,1} \cup M_{0,2}, \mathcal{L}_1 |_{LPrT} \mathcal{L}_2)$ where sets and functions are defined as in the case of basic operator $|_{LPN}$ (see definition 2.7).
Let $T_i : l_i(T_i); pre_i(T_i) \rightarrow post(T_i)\ IF\ cond(T_i)$, $(i = 1, 2)$ be two minimal mergeable transition sets;
let σ be a communication substitution for $l_1(T_1)$ and $l_2(T_2)$.
Then merging $(T_1 |_{LPrT} T_2)$ is defined by the following rule:

$$T_1 |_{LPrT} T_2 : \sigma(l_1(T_1) \cup l_2(T_2)); \quad \sigma(pre_1(T_1) \cup pre_2(T_2)) \rightarrow \quad \sigma(post_1(T_1) \cup post_2(T_2))$$
$$IF\ \sigma(cond(T_1) \cup cond(T_2) \cup ind_{T_1,T_2})$$

where ind_{T_1,T_2} is a set of inequalities among logical variables whose purpose is to preserve the independence of transitions T_1 and T_2 after unfolding.

4 Application

A remote reading mechanism, the so-called Telereport [HRJ89] application layer, is first modelled then analysed. A specific property concerns the retransmission aspect; it is important to verify that the provided service is independent of transmission errors provided the latter are not too numerous.

The LPrT net model is highly concise, since the global net (after transition merging) exhibits 8 transitions only. Such a compactness results from the use of the multi-gate rendez-vous.

Case analysis leads to consider three configurations. A hybrid approach, i.e., compositional and global, is proposed to compute the corresponding services.

4.1 Models

Application entities A_{us} and A_{pr} cooperate through a nonperfect session service S in order to provide a remote reading service to the user process (see Fig 1).

Application user service: The interface between the user process and entity A_{us} is of particular interest since it determines the external service provided to the user process: user request $req(C)$ is parameterized by requested data code C and is issued on SAP, $A_{us}?$; confirmation $conf(Mess)$ is issued on SAP $A_{us}!$ and can be of 3 types:
$Mess$ value is either "*error*": transmission by session service has failed,
or "*nak*": no data of code C is available to the provider process,
or C read value Val_C.

Entity A_{us} is in charge of recovery if a transmission error occurs in the session service. Let NR_{Max} be the maximum number of consecutive recoveries with respect to the same read request. To study the correctness of the recovery mechanism a specific model of session service has been designed: session service may lose consecutively at most NE_{Max} messages. The provided service depends on relative values of parameters NR_{Max} and NE_{Max}.

Configuration Three configuration classes are distinguished:

- Perfect session service (no error $NE_{max} = 0$)
- Faulty session service but less errors than recoveries ($NE_{max} \leq NR_{max}$)
- Unreliable service: more errors than recoveries ($NE_{max} > NR_{max}$)

The requested code may either be available to the provider (expected confirmation with value) or not (expected confirmation: "*nak*").

The set of potential configurations, and the database facts which belong to net Horn clause interpretation, are the following:

	$conf_0$	$conf_1$	$conf_2$	*database*
NE_{Max}	0	succ(0)	succ(succ(0))	$MaxErr(NE_{max})$.
NR_{max}		succ(0)		$MaxRecover(succ(0))$.
code c_1		c_1 available $\rightarrow$ value v_1		$codeType(c_1)$. $codeVal(c_1, v_1)$.
code c_2		c_2 not available $\rightarrow$ *nak*		$codeType(c_2)$.

Session Service Net (see Fig 5, net S)

A request $req(C)$ or a response $resp(Mess)$ may be conveyed without error from user entity A_{us} to provider entity A_{pr} respectively and vice-versa. Primitive *req* becomes *ind* and *resp* becomes *conf*.

The corresponding normal transitions are *transmitReq* and *transmitResp*.

Request $req(C)$ or response $resp(Mess)$ may either be lost or incorrectly transmitted. This corresponds to transitions *loseReq* and *loseResp*. The number of consecutive errors before a reset signal (transition *reset*) is saved by variable K and bounded by logical condition $infMaxErr(K)$. Predicate $infMaxErr(K)$ is defined by the following clause:

/* $infMaxErr(K)$ is true if $K < NE_{max}$ */
$infMaxErr(K)$:- $maxErr(NE_{max})$,   /* database fact */
                    $inf(K, NE_{Max})$.   /* true if $K < NE_{max}$ */

The occurrence of an error results in sending message $conf(error)$ to user entity A_{us} (transition *error*).

User entity of Application protocol (see Fig 5, net A_{us})

The normal transition sequence is (1) *init* receipt of user process request $req(C)$, (2) *send* request transmission to session service, (3) *end* confirmation $conf(Mess)$ is received and returned to the user process, if $Mess \neq error$, that is $Mess$ is a code value or "*nak*").

When a session service error is detected, via primitive $conf(error)$), recovery may take place, i.e. user process request is repeated. Transition *recover* is enabled if less than NR_{max} errors have occured; this enabling condition, $infMaxRecover(K)$, is defined by

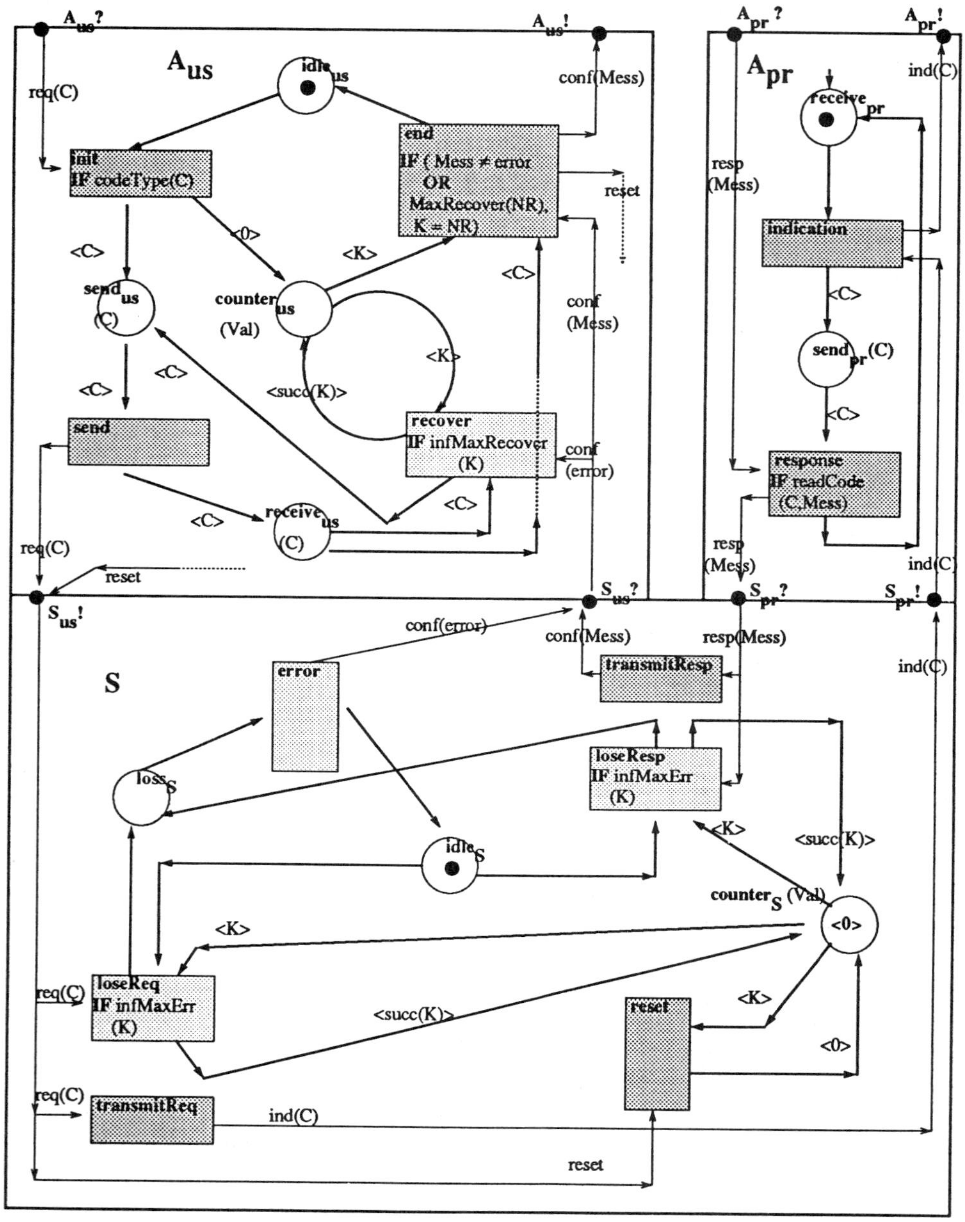

Figure 5: LPrT nets of Application entities A_{us} and A_{pr} and session service S

the following clause:

$$infMaxRecover(K) :- \quad MaxRecover(NR_{max})$$
$$inf(K, NR_{max}).$$

In case of confirmation *conf(error)* and if the maximum number of recoveries is exceeded, transition *end* fires. Signal *reset* is sent to session service.

Provider entity of Application protocol (see Fig 5, net A_{pr})

Provider entity is initially ready to receive read indication *ind(C)*: transition *indication* is enable; procedure *readCode(C, ValOrNak)* of transition *response* performs a read action; when C is available (*codeVal(C, Val)* in database) the value of *Val* is substituted for variable *ValOrNak*; if C is not available variable *ValOrNak* takes value *nak*, negative acknowledge. Procedure *readCode* is defined by the following net clause:

$readCode(C, Val) :-$	$codeVal(C, Val).$	code available
$readCode(C, nak) :-$	$\backslash + codeVal(C, Val).$	code not available

4.2 Verification

An "easy to use" verification technique is to compute the service provided by the protocol and to compare it with the expected one.

The global service provided by a behaviour (LTS) is a reduced LTS, minimal with respect to the state number, which preserves properties of observed communication actions; with respect to gate set G, a local service is derived from the behaviour after the hiding of events which occur on gates outside G.

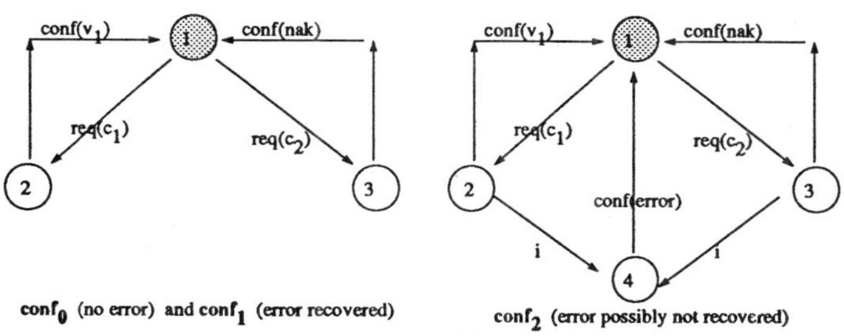

Figure 6: Application service with respect to user process

The standard observational equivalence [Mil80] is used to compute the service; observable actions sequences and deadlocks are preserved.

Service Derivation

Let N be a LPrT system composed of three sub-nets: $N = N_1 \,|_{LPrT}\, N_2 \,|_{LPrT}\, N_3$.

Net service $S(N)$ is defined as the service provided by the behaviour of N, i.e. $S(\mathcal{B}_{LPrT}(N))$.

There are two basic ways to compute service $S(N)$; a third one may also be derived in a combined manner.

Service composition : individual subnet services are computed first, then they are composed. We have, $S(N) \approx S(N_1) \mid_{LTS} S(N_2) \mid_{LTS} S(N_3)$. Partial verifications may be conducted on individual services, and then reused for the verification of the global service. Furthermore, as services are reduced behaviours, composition of individual services can be more efficient than computation of the global netbehaviour.

LPrT composition : sub-nets are composed to obtain a global net whose behaviour and service are derived in a second step. This approach is mandatory when individual behaviours are unbounded, even if the global behaviour is bounded. For example, unbounded LPrT net may be a fifo queue model.

Combined approach : The former computations produce an equivalent result because diagrams 2 and 4 commute and because Milner observational equivalence $\approx$ is a congruence with respect to $\mid_{LTS}$. A combined approach may still be useful to gain a better insight into protocols. For example, $S(N) \approx S(N_1 \mid_{LPrT} N_2) \mid_{LTS} S(N_3)$: on the one hand N_3 individual service is computed, and on the other net composition $N_1 \mid_{LPrT} N_2$ may be performed.

Telereport Service Local service provided to user process is depicted in Fig 6 (transitions represent application service primitives).

As long as errors are relatively few, i.e., in configurations $conf_0$, $conf_1$, the provided service is error-free: the recovery mechanism plays the intended role.

Configuration $conf_3$ is a degraded configuration: errors may be too numerous, and the application service may not deliver the expected answer (v_1 or nak). Correct status *error* is however delivered to user process. Confirmation $conf(error)$ follows an internal transition after the last (unrecovered) error.

5 Conclusion

The introduced composition and verification techniques are implemented in the available software PIPN. The tool enforces structured and incremental approaches. It includes a graphical SADT-like editor that defines system architecture. The compiler computes global nets, as a result of subnet composition. State space can be interactively explored by means of the simulator. Finally, the complete behaviour can be computed and then reduced to protocol service (see Figure 7). A model checker [Pap86] for CTL logic is also available. A current issue is to combine logic and service verification in a combined approach [PA89].

References

[Gen88] Hartmann J. Genrich. Equivalence transformation of PrT-nets. In *Workshop on Application and Theory of Petri Nets*, pages 229–248, 1988.

[HRJ89] P. Haudebourg, G. Revelaud, and T. Journey. Spécification du protocole de téléreport. Technical Report HR/23-2077 and HR23-1943, Electricité De France, DER, Clamart, 1989.

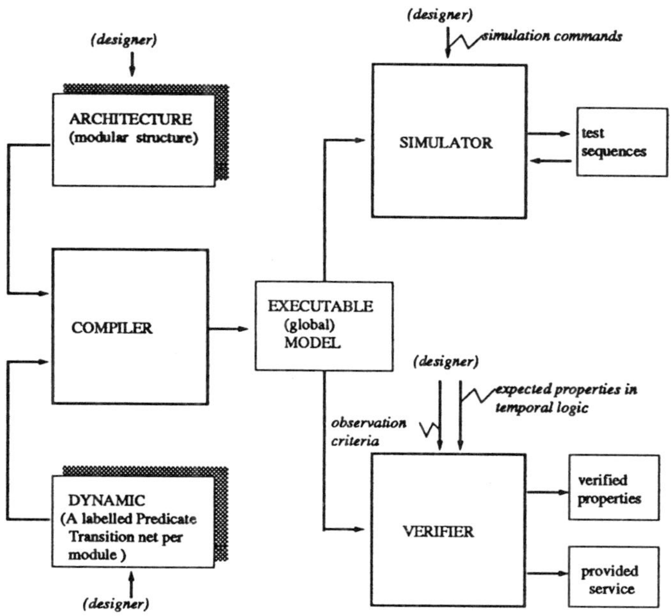

Figure 7: PIPN tool architecture

[Mil80] Robin Milner. *A Calculus of Communicating Systems*, volume 92 of *Lecture Notes in Computer Science.* Springer-Verlag, 1980.

[PA89] F. Vernadat P. Azema, J.C. Lloret. Requirement analysis for communication protocols. In *Workshop On Automatic Verification Methods for Finite State Systems, Grenoble-France*, 1989.

[Pap86] Giorgos Papapanagiotakis. *Réseaux de Prédicats et Logique Temporelle pour la vérification des applications réparties.* PhD thesis, Laboratoire d'Automatique et d'Analyse des Systèmes du C.N.R.S, 7,Avenue du Colonel Roche 31077 Toulouse Cedex, 1986.

[PD87] U. Montanari P. Degano, R. De Nicola. A distributed operational semantics fot ccs based on condition/event systems. Technical Report 111, Instituto di Elaboratione dell 'Informazione, C.C.R, Pisa, 1987.

[vEVD89] P.H.J. van Eijk, C.A. Vissers, and M. Diaz, editors. *The Formal Description Technique LOTOS.* Elsevier Science Publishers B.V. (North Holland), 1989.

DIMACS Series in Discrete Mathematics
and Theoretical Computer Science
Volume 3, 1991

Liveness Analysis and the Automatic Generation of Concurrent Programs

UGO BUY AND ROBERT MOLL

1. Introduction and Background

The automatic synthesis of programs is one of the most challenging activities in computational logic. Deductive approaches to the synthesis of sequential programs have met with only limited success, in part because of the relative inadequacy of proposed specification languages for such programs. That is, specifications often turn out to be less enlightening and more difficult to construct than the code they specify. This is not the state of affairs with synchronization code. Desired program properties, such as liveness or freedom from deadlock, are relatively easy to include in a specification but are quite difficult to discern in code. Indeed, it is precisely this feature of concurrent programming that makes automated program synthesis an attractive goal. In this paper we discuss an aspect of the automatic synthesis of synchronization code for asynchronous processes.

Our synthesis approach conforms to the following paradigm:

(1) a specification is written in a nonconstructive specification language;
(2) that specification is analyzed in an attempt to establish that crucial concurrency properties are respected; and
(3) if the concurrency properties of the specification are established, then Ada code is generated.

Here we report on the most difficult part of this process: establishing that a program specification satisfies a crucial concurrency requirement, namely liveness. (Other related aspects of our proposed approach, such as safety properties and deadlock avoidance, are discussed elsewhere [**4**]). By a liveness property we mean a condition that must be satisfied at some point during the execution of a program. A liveness property contrasts with a safety property, a condition that must hold continuously during program execution.

Underlying our approach to synthesis is a model of a concurrent program in which processes communicate by accessing and modifying shared resources. This *resource-oriented model* uses temporal logic for specification and analysis. By a *process* we mean an abstract state machine defining an active computation. A process computation generally requires the allocation of shared resources. By a *synchronizer* we mean an abstract state machine that defines a set of encapsulated resources [**12**]. A synchronizer specification states the safety and liveness properties that the synchronizer must satisfy. Our approach to program specifications is similar to that presented in [**12**], which, however, does not address the issue of liveness of synchronizer operations. Synchronizer liveness is the main focus of this paper.

We believe our approach to synthesis and, in particular, our approach to liveness analysis can be automated for two reasons. First, by identifying indiscernible synchronizer computation states, we drastically reduce the size of the resulting synchronizer state space graph. Second, by using this graph to drive the formal analysis of program specifications, we limit the need for general theorem proving capabilities.

An experimental prototype has been implemented, which automates a significant portion of our analysis. At present we are working on extending the prototype to automate the remaining aspects of the analysis.

This paper is organized as follows. Synchronizer specifications are described in section 2. A graph model of synchronizer behavior is discussed in section 3. A proof lattice-based method for liveness analysis is given in section 4 along with a portion of the analysis of the familiar readers/writers example. We describe a significant portion of the target code for the readers/writers example in section 5. Some conclusions are presented in section 6.

2. Synchronizer Specifications

Our specification language is built around the concept of the state of the resources encapsulated in a synchronizer. Specifications are formulated in terms of a set of state variables and a set of operations that can access and modify those variables. Linear-time temporal logic is used to define the semantics of the constructs appearing in a synchronizer specification. Thus, a synchronizer specification consists of a set of clauses defining the state transitions performed by each synchronizer operation. A specification also gives the safety and liveness conditions governing the execution of synchronizer operations. An example, the specification of synchronizer *buffer* from the readers/writers problem, is given in Figure 1. In this problem a set of reader and a set of writer processes must access a shared memory area simultaneously. Synchronizer *buffer* controls access and modifications to the shared memory on behalf of the reader and writer processes.

The **control_resources** clause of the specification defines two resources for synchronizer *buffer*, namely *read_permission* and *write_permission*. The clause

```
Synchronizer Buffer
Control_resources write_permission allocated_by start_write;
                                   deallocated_by end_write;
                  read_permission  allocated_by start_read;
                                   deallocated_by end_read;
State_variables
  reader_#: 0..10 initially 0 ;
  writer_#: 0..1 initially 0 ;
Operations
  start_read, start_write, end_read, end_write ;
Operation_preconditions
  ∀s∈start_write: (reader_# = 0) ∧ (writer_# = 0)
  ∀s∈start_read: (writer_# = 0)
Operation_state_changes
  ∀s∈start_read: reader_# ← reader_# + 1
  ∀s∈start_write: writer_# ← 1
  ∀e∈end_read: reader_# ← reader_# − 1
  ∀e∈end_write: writer_# ← 0
Operation_priorities
  start_write > start_read
Liveness
  ∀s∈(start_write, start_read) s@request → ◇s@terminate
```

FIGURE 1. Specification of synchronizer *buffer* in the readers-writers problem

asserts that, in order for a reader (or a writer) process to access the shared memory, the resource *read_permission* (or *write_permission*) must be allocated first. Allocation is performed by operations *start_read* and *start_write*. Operations *end_read* and *end_write* perform the corresponding deallocations.

The **state_variables** clause defines the type and initial value of the variables *reader_#* and *writer_#*. The synchronizer uses these variables to represent the state of the encapsulated resources. Together they capture the number of reader and writer processes that are in the synchronizer at any point in time. The **operations** clause declares the operations that units outside synchronizer *buffer* can invoke to perform allocation and deallocation of the resources encapsulated in this synchronizer. Whenever a synchronizer operation is invoked, a model of operation execution is used in which the operation invocation goes through a sequence of three phases. In the *request* phase the operation has been invoked and is waiting to be selected for execution; in the *service* phase the operation has been selected for execution and will be executed next; in the *terminate* phase the operation has completed execution. For invocation o of operation O the propositions *o@request*, *o@service* and *o@terminate* are true when o is in the corresponding phase and false otherwise.

The **operation_preconditions** clause specifies the conditions to be satisfied in order for each operation request to be executed. Temporal operators are not allowed in this clause. A request for operation *start_write* can be serviced only if the synchronizer state satisfies:

$$(reader\_\# = 0) \wedge (writer\_\# = 0)$$

The **operation_state_changes** clause specifies the changes in the value of the state variables resulting from the execution of a given operation. Whenever a request for operation *start_read* is executed, the value of the state variable *reader_#* is incremented. The **operation priorities** clause defines the order in which instances of invocations of different operations are to be considered for service. In synchronizer *buffer*, operation *start_write* has higher priority than operation *start_read*.

Finally, the **liveness** clause specifies that invocations of operations *start_read* and *start_write* must eventually be serviced. As with most temporal logic systems, the formula $\Box p$ asserts that predicate p is true in the current state and any subsequent state. The formula $\Diamond p$ asserts that p is true either in the current state or in at least one subsequent state.

Linear-time temporal logic underlies the computational model of a synchronizer specification. Thus, time is viewed as a totally ordered sequence of *time instants*. Each time instant corresponds to a well-defined synchronizer state (i.e. the assignment of a value to every synchronizer state variable). In general a synchronizer operation O can be invoked an arbitrary number of times by the units contained in the concurrent program. Each operation invocation o can be in the *request* phase for an arbitrary (and possibly indefinite) number of consecutive time instants, while waiting for the appropriate preconditions and priority conditions to become true. When these conditions become true, invocation o is *enabled*, in which case the predicate *enabled*(o) is true.

When an invoked operation o is selected for execution, o goes into the *service* phase for exactly one time instant. At the following instant operation o is in the *terminate* phase. At this instant the synchronizer state variables have been modified to reflect the changes caused by the execution of o. Operation execution in a synchronizer is strictly sequential: only one operation invocation can be in the service phase at any time instant. Once an operation request reaches the *terminate* phase it remains in that phase forever.

3. Graph Model of Synchronizer Behavior

We model synchronizer behavior using an augmented finite state machine called a Reduced State Transition Graph or RSTG. This model underlies the proof lattice mechanism we introduce to establish liveness. Proof lattice inferences are aimed at identifying valid transition sequences between states of an RSTG.

The RSTG of a synchronizer is a graph $G = (N, E)$. Nodes N in this graph represent sets of synchronizer states; arcs E represent transitions resulting from the execution of enabled operation requests. Each node N_i is associated with a set of synchronizer states X_i, and each edge E_j is associated with a synchronizer operation O_j and a predicate P_j on the state of the synchronizer, subject to the following conditions. First, the states associated with N_i are indiscernible in that they enable the same operations. Second, if edge E_j is labeled (P_j, O_j) leading from node N_i to N_k, if operation O_j is executed when the synchronizer is in one of the states associated with N_i, and if predicate P_j is true, then the synchronizer is in a state associated with N_k after O_j has been executed. Third, state sets associated with distinct nodes are disjoint.

Construction of an RSTG $G = (N, E)$ is carried out in two phases. First, the nodes N of the graph are defined. Each node corresponds to a conjunctive normal form combination of the predicates enabling synchronizer operations. Then the edges E are defined. This is accomplished by identifying the operations that are enabled at each node and by tracking the state transitions caused by the execution of each enabled operation.

For instance, to define the nodes in the RSTG of synchronizer *buffer* we proceed as follows. First, the predicates enabling the synchronizer's operations are identified. Here the predicates are $(writer\_\# = 0)$ and $(reader\_\# = 0)$. They are trivially in conjunctive normal form. To these predicates we add additional conditions which take into account the invariant assertions implied by the specification. Invariant assertions are state predicates that must hold of a synchronizer in every time instant. Here the declaration of variable *reader_#* implicitly defines the following two invariant assertions:

$$(1) \qquad (reader\_\# \geq 0) \quad \text{and} \quad (reader\_\# \leq 10)$$

Likewise, the declaration of variable *writer_#* implicitly defines the following two invariant assertions:

$$(2) \qquad (writer\_\# \geq 0) \quad \text{and} \quad (writer\_\# \leq 1)$$

Consequently, a *buffer* operation can be enabled only if its execution upholds these four assertions, assuming the assertions are true before the operation is executed.

In general, to guarantee that a synchronizer invariant assertion is satisfied, the following two conditions must be true. First, the assertion must hold in the initial synchronizer state. Second, the preconditions for each synchronizer operation must be augmented in such a way as to preserve the truth of the invariant assertion as operations are executed and the synchronizer moves from state to state. This activity can be performed by use of the notion of *weakest precondition* as defined by Dijkstra in [7]. Given a synchronizer operation O and an invariant assertion I, $WP(O, I)$ the weakest precondition of O and I, is the most general predicate P such that, if P holds before O is executed, then I will hold after O is executed. In the resource-oriented model, the derivation of weakest preconditions is simplified because loop constructs cannot appear as

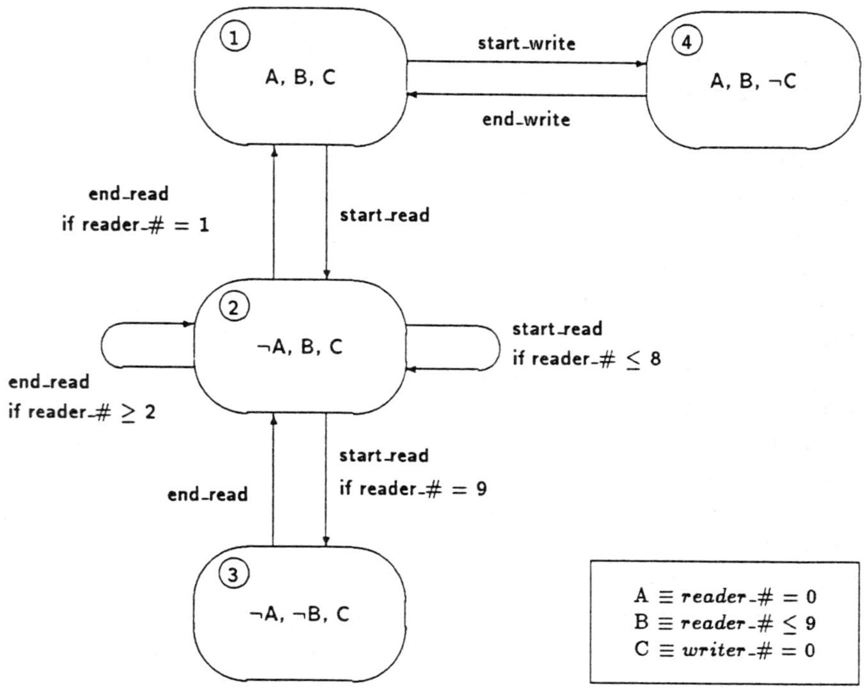

FIGURE 2. RSTG of synchronizer *buffer*

operation state changes in a synchronizer specification and temporal operators are not allowed in invariant assertions.

In synchronizer *buffer*, variables *reader_#* and *writer_#* are both initially zero, and so invariant assertions (1) and (2) are satisfied in the initial synchronizer state. Now the weakest preconditions for these invariant assertions and each synchronizer operation must be computed. As an example, let us consider the derivation of the weakest precondition for operation *start_read* and assertion ($reader\_\# \leq 10$). Given that the execution of *start_read* increments the variable *reader_#*, the weakest precondition is given by the predicate ($reader\_\# \leq 9$). This predicate is then added as a conjunct to the preconditions for operation *start_read*.

After the activity of computing the weakest preconditions for *buffer* has been completed, a simplification procedure can be applied to remove redundant predicates. For instance, the weakest precondition for operation *end_read* and assertion ($reader\_\# \leq 10$) is the predicate ($reader\_\# \leq 11$). The latter predicate is implied by the former invariant assertion and therefore can be assumed to be true in every time instant associated with *buffer*. As a result, the following three predicates are identified as the conjuncts appearing in the augmented preconditions for this synchronizer: ($reader\_\# = 0$), ($reader\_\# \leq 9$), and ($writer\_\# = 0$). Consequently, each node in the RSTG of synchronizer *buffer* corresponds to a particular combination of values of these three predicates. For the purpose of exposition we call these predicates A, B, and C.

After the node set for synchronizer *buffer* has been defined, the RSTG is built incrementally, beginning from the node corresponding to the initial synchronizer state. In the initial state predicates A, B, and C are true. Consequently, the node labeled by these three predicates is added to the RSTG for *buffer*. Then the following expansion procedure is applied. First, the operations enabled in the initial node are identified, based on the truth values of the predicates labeling the node. When A, B, and C are true operations *start_read* and *start_write* are enabled. For each enabled operation the state changes resulting from the execution of the operation are considered. For instance, operation *start_write* changes the value of variable *writer_#* to 1. As a result, after this operation is executed, predicates A and B are still true but C is false. Therefore the node labeled A, B and $\neg C$ is added to the RSTG. In addition, an edge labeled *start_write* is created, leading from the initial node to the new node. In a similar fashion the execution of operation *start_read* when *buffer* is in the initial state is represented by an arc labeled *start_read* leading from the initial node to the node labeled $\neg A$, B and C.

The expansion procedure is applied iteratively to every node added to the RSTG, until all reachable nodes have been expanded. Note that from a given node the execution of an operation may lead to different nodes, depending on the actual synchronizer state. For instance, the execution of operation *start_read* when *buffer* is in a state associated with the node labeled $\neg A$, B and C may lead to a state associated with the same node or to a state associated with the node labeled $\neg A$, $\neg B$ and C, depending on the actual value of variable *reader_#*.

The RSTG for synchronizer *buffer* is shown in Figure 2, along with the predicate truth values corresponding to each node. Note that while eight nodes are possible, only four actually appear in the RSTG. Node 1 corresponds to the initial synchronizer state. Operations *start_read* and *start_write* are enabled in this state, as shown by the edges leaving node 1. Node 2 represents the synchronizer states in which resource *read_permission* has been allocated to a number of reader processes that is less than the maximum number of readers allowed. Consequently both *start_read* and *end_read* are enabled in the states corresponding to this node. The execution of the former operation leads to node 3 when the variable *reader_#* is one short of the maximum reader number; otherwise it leaves the synchronizer in a state still contained in node 2. Weakest precondition analysis is used here to determine the conditions required for transitions to different states. Likewise, the execution of operation *end_read* can either lead to node 1 or back to node 2, depending on whether variable *reader_#* is one or greater than one. Node 3 corresponds to the synchronizer state in which *reader_#* equals the maximum number of readers. Only operation *end_read* is enabled in this state, leading to state 2. Finally, node 4 represents the state in which a writer process is in the synchronizer. Only operation *end_write* is enabled in this state.

An implementation of the RSTG construction described here is currently under way. In particular, we are focusing on two crucial aspects in this construction, namely the derivation of weakest preconditions and the computation of the nodes reached by executing a given operation when in the synchronizer is in a given RSTG node. Although in general these aspects can be quite difficult to compute mechanically, we believe that in practice RSTG construction can be fully automated for a broad class of problems. In particular, we are convinced that automation is possible for such traditional examples in the area of concurrent programming as the readers/writers, the producers/consumers and the sleepy barber [**8**]. In addition, we are investigating the use of existing theorem provers to carry out significant portions of this analysis [**3, 6**]. For instance, we have found that the simplification procedure applied to the predicates associated with RSTG nodes becomes relatively straightforward if the Boyer-Moore theorem prover is used [**3**].

4. Proof Lattice-based Analysis

We now describe our deductive method for establishing the following liveness condition:

$$\forall o \in O\ \Box(o@request \rightarrow \Diamond\ o@service) \tag{3}$$

We prove liveness using the notion of a *proof lattice*. Proof lattices were introduced in [**9**] to prove first-order formulas. They were used in [**11**] to analyze a subset of first-order temporal logic. A proof lattice is a finite directed acyclic graph in which each node is labeled with an assertion and such that:

(1) There is a single entry node with no incoming edges
(2) There is a single exit node with no outgoing edges
(3) If a proof lattice node labeled P has outgoing arcs to nodes labeled R_1, R_2, ... R_n, the following formula holds for the synchronizer:

$$\Box(P \rightarrow \Diamond(R_1 \vee R_2 \vee \ldots \vee R_n))$$

Proof lattices can be combined with an axiomatic system for a given computational model to control the set of deductions required for the proof of liveness properties in that model. In particular, every arc in a proof lattice represents a logical deduction in the chosen axiomatic system. As an example, an axiomatization for a concurrent dialect of the language Algol is given in [**11**]. The computational model underlying our proof system is the RSTG of a given synchronizer. In particular, the structure of an RSTG is used to drive the deductions performed during proof lattice construction. Although a complete description of our axiomatization for the RSTG model is beyond the scope of this paper, we outline the major aspects of this approach and illustrate a case of proof lattice construction for the readers/writers example.

Our approach to liveness is based on several assumptions. First we assume that the implementation of a synchronizer uses a *fair scheduler*, that is, a scheduler that does not ignore an operation invocation that is enabled infinitely often:

$$(4) \qquad \Box\Diamond enabled(o) \rightarrow \Diamond o@service$$

Second, we assume that a resource allocation is always followed by the corresponding deallocation. This assumption is realistic because deadlock analysis is independent of the liveness analysis described here [**4**]. Thus, if operations A and D allocate and deallocate a given resource, any invocation a of operation A is eventually followed by an invocation d of operation D:

$$\Box(a@request \rightarrow \Diamond d@request)$$

Third, we assume that, whenever a synchronizer state enables an operation O, an invocation o of O has been requested and is waiting for service. This implies that a synchronizer cannot be blocked indefinitely by the lack of invocations of one or more of its operations. In particular, the readers/writers example satisfies this assumption. Fourth, we make use of the notion of *conformity* between a state variable x and a resource R. We say x is *conformal* with R if the value of x reflects the number of open allocations of R (i.e. the number of executions of the operation that allocates R that have not been followed by matching deallocations). In the *buffer* example variable *reader_#* is conformal with resource *read_permission*.

The following theorem establishes the validity of the proof lattice-based approach. A proof is given in [**4**]. The predicate $In(N_i)$ indicates whether the current synchronizer state belongs to an RSTG node N_i.

Fundamental Theorem

Given an operation O defined in synchronizer S and an RSTG for S, if for every node N in the RSTG a proof lattice can be built whose entry node is labeled $In(N) \wedge o@request$ and whose exit node is labeled $enabled(o)$, then the following assertion is true for every synchronizer state

$$\Box(o@request \rightarrow \Box\Diamond enabled(o))$$

The conclusion of the theorem is that, under the stated hypotheses, every invocation o of operation O is enabled infinitely often. In the presence of the fair scheduler assumption, this guarantees that the liveness condition (3) above is satisfied. Thus, every invocation of operation O is eventually serviced. Consequently the liveness analysis of synchronizer operations can be performed by constructing a suitably labeled proof lattice for each node N in the RSTG of the synchronizer.

Proof lattice construction rules fall into three categories. Rules in the first group tie proof lattice deductions to RSTG transitions. For example, if an RSTG node N has arcs leading to nodes $N_1,\ldots, N_k$, a proof lattice node L labeled $In(N)$ has descendants labeled $In(N_1), \ldots In(N_k)$.

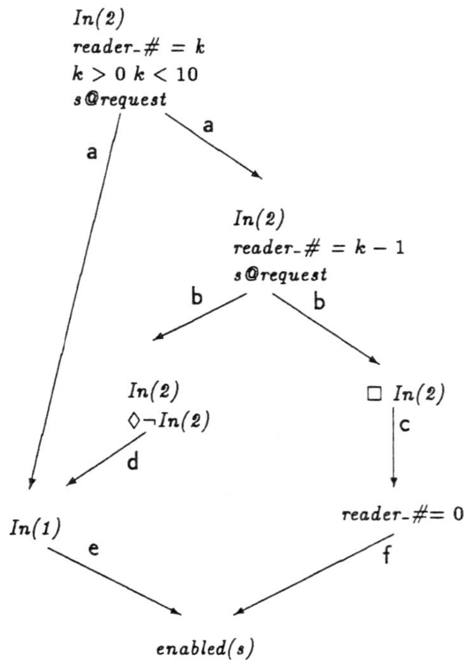

FIGURE 3. Proof lattice for operation *start_write* for RSTG node 2

Rules in the second group are based on axioms of linear-time temporal logic. For example, suppose there is a proof lattice node L labeled $\Box p$. In this case, predicate p can be added to the label of any descendant of node L, based on these two axioms of temporal logic: $(\Box p \rightarrow p)$ and $(p \rightarrow \Diamond p)$.

Rules in the third group reflect the behavior of resource allocation and deallocation induced by our synchronizer model. As an example, consider the *allocation completion* rule. Suppose that a state variable x of synchronizer S is conformal with resource R, which is allocated and deallocated by operations A and D. Then a proof lattice node L labeled $\Box\neg enabled(A)$ has a descendant labeled $(x = \bar{x})$, where $\bar{x}$ is the initial value of x. The validity of this rule is proved by use of the assumption of complete resource deallocation. Since operation A is never enabled, eventually all executions of this operation will be followed by executions of operation D. Roughly speaking, the effects of each execution of operation A on variable x will eventually be "undone" by the corresponding execution of operation D. Consequently, variable x will eventually return to its initial value.

A schematic description of a proof lattice for operation *start_write* is given in Figure 3. This proof lattice is aimed at proving that the operation is eventually serviced, assuming that it is invoked when synchronizer *buffer* is in a state corresponding to RSTG node 2. Thus, the entry node is labeled by the predicates $In(2)$ and $s@request$, assuming s is an invocation of operation *start_write*.

Step (a) is performed by applying the first set of construction rules to the entry node in the proof lattice. Note that four edges leave node 2 in the RSTG; however, the edges labeled *start_read* are temporarily disabled, because this operation cannot be serviced due to the pending invocation of higher-priority operation *start_write*. Consequently, only two edges are created in the proof lattice, corresponding to the RSTG edges labeled *end_read*.

Step (b) is performed by applying the second set of rules and temporal logic axiom $(\Box p \vee \Diamond\neg p)$, where p is instantiated to the predicate $In(2)$. So the formula $(In(2) \rightarrow (\Box In(2) \vee \Diamond\neg In(2)))$ is also valid. Consequently, the node under consideration has two descendants, one for each of the disjuncts in the temporal logic axiom.

Step (c) is performed by noting that *end_read* is enabled when $In(2)$ is true, whereas *start_read* is not enabled. Moreover, these two operations deallocate and allocate resource *read_permission*, respectively. Since variable *reader_#* is conformal with this resource, the allocation completion rule can be applied to produce a descendant labeled (*reader_#* $= 0$). This predicate contradicts the earlier assumption $\Box In(2)$. As a result, a rule leading directly to the exit node, whose label is *enabled*(s), can be applied in the following proof step (f).

Step (d) is performed by applying a construction rule from the first group. This rule is similar to the rule applied in step (a); however, here the label of a proof lattice node has the additional predicate $\Diamond\neg In(2)$. The execution of an enabled operation can lead from node 2 to node 1 or node 3; however, the edge leading to node 3 is disabled due to the priority specification. Consequently, the formula $(\neg In(2) \rightarrow \Diamond In(1))$ is valid and a descendant labeled $In(1)$ is created for the proof lattice node under consideration. Finally, a rule leading to the exit node can be applied in the step (e), concluding the proof lattice construction.

The above example has shown the liveness of operation *start_write*, assuming this operation is invoked when synchronizer *buffer* is in a given state subset. Similar proofs have been made for the other synchronizer operations and states in an effort to complete the liveness proof for synchronizer *buffer*. These proofs have been generally successful. However, difficulties have arisen in the proof of liveness for operation *start_read*, which has a lower priority than operation *start_write* in the specification of synchronizer *buffer*. Because of this priority condition, invocations of *start_read* may "starve" due to the continuous presence of *start_write* invocations.

To avoid the possibility of starvation of operation *start_read*, the specification of synchronizer *buffer* can be modified by removing the *operation priorities* clause appearing in Figure 1. In this case, all proof lattice constructions for *start_read* are successful; however, the constructions for *start_write* and RSTG nodes 2 and 3 would fail. In particular, *start_write* invocations could be prevented from ever becoming enabled by a continuous presence of *start_read* requests. Nevertheless, an analysis of the failed proof lattice constructions leads to the conclusion that invocations of *start_read* and *start_write* should be placed in the same

FIFO queue. In this case, all the proof lattices for synchronizer *buffer* can be constructed successfully, thus establishing the liveness of this synchronizer.

Our approach to liveness analysis has been applied successfully to a variety of traditional examples in the domain of concurrent programs, such as the producers/consumers, the sleepy barber [**8**] and a simplified version of a memory controller. While the full expressive power of this approach is still under active investigation, a prototype that partially automates the proof lattice construction activity shown here has been implemented. The prototype takes as input an RSTG for a given synchronizer, a synchronizer operation O and an RSTG node N_i. The goal of the prototype is to construct a proof lattice for operation O and the input RSTG such that the entry node is labeled $In(N_i)$. An implementation of the Boyer-Moore theorem prover [**3**] is used to prove theorems that do not contain temporal operators. So far, the prototype has been applied successully to a variety of cases from the above example set. In particular, all the proof lattices for the readers/writers and the sleepy barber example have been constructed automatically.

Although in general proof lattice construction requires the full power of a theorem prover for first-order temporal logic, which is undecidable [**1**], in practice the prototype has performed satisfactorily in the case of the above example set. In our opinion there are several reasons for this. First, the state structure of an RSTG is used to drive the activity of proof lattice construction. Second, the liveness analysis is rather stylized, thus sidestepping the need to use the full deductive power of a temporal logic system during proof lattice construction. Third, our assumptions about the behavior of resource allocation and deallocation simplify proof lattice construction.

The results obtained automatically for the readers/writers and the sleepy barber examples are of particular significance because these examples contain only one synchronizer. Proof lattice construction is also applicable to examples including multiple synchronizers; however, in that case, the interactions among these synchronizers cannot be analyzed completely using the approach described here. Additional techniques for extending the proof lattice based analysis to program specifications containing multiple synchronizers are under active investigation [**4**].

5. Target Code Generation

In our approach, target code generation is a relatively straightforward translation activity that maps a program specification into the corresponding target program. In addition to a program specification, the code generator is supplied with the results of the analyses that were done in earlier phases. Augmented operation preconditions found during RSTG construction are particularly important. For instance, the predicate ($reader\_\# \leq 9$) is added to the preconditions of operation *start_read* in the target code of synchronizer *buffer*.

The proof lattice constructions also play a role in code generation. They can identify queue structures that make it possible for synchronizer target code to satisfy the liveness requirements appearing in the corresponding specification. For instance, in the modified version of the *buffer* specification shown in Figure 1 the failed proof lattice constructions for operation *start_write* suggest that invocations of this operation and operation *start_read* should be placed in the same request queue.

A subset of Ada is used as our target language [**2**]. Target program construction is performed by instantiating suitable templates of target code for the processes and the synchronizers contained in the corresponding specification. Each process and synchronizer appearing in a program specification is implemented as an Ada task. Roughly speaking, an Ada task is a program unit that can be executed in parallel with other units (see [**2**] for a more detailed description of the Ada language).

Here we limit our discussion to the generation of synchronizer code. For a discussion of process code generation, see [**4**].

The generation of the Ada task implementing a given synchronizer specification is accomplished by template instantiation. The template of target code for a synchronizer S must perform the following activities:

(1) wait for operation invocations by other units;
(2) store information about such operation invocations in suitable queues;
(3) issue calls to allocate and deallocate resources required by operations defined in S;
(4) check queues defined in S for enabled operations; and
(5) select enabled operations for service and perform the corresponding state changes.

To perform the above activities the Ada task implementing synchronizer S defines variables that correspond to the state variables of S. In addition, the Ada task defines queue structures for storing information about invoked operations that are waiting for service. An example of target code, namely the Ada task *buffer*, which implements the modified version of synchronizer *buffer*, is shown in Figure 4.

The target code for synchronizer *buffer* consists of a *task specification* and a *task body*. The task specification defines an Ada entry for each *buffer* operation. The task body defines local variables and executable statements of task *buffer*. In general, the executable code for a synchronizer consists of a main loop. The first statement in the loop is an Ada **select** statement which waits for a task entry to be called (i.e. for a synchronizer operation to be requested). When an entry is actually called, a record describing the request is created and placed on a suitable queue. The queue structures in the synchronizer are examined next to determine whether there are any enabled operation requests. If one is found, then the following actions are executed. First, the request is removed from its queue. Next, the changes to the value of the state variables that appear in the

```
task buffer is
   entry start_read(reader_id: in reader_range) ;
   entry end_read(reader_id: in reader_range) ;
   entry start_write(writer_id: in writer_range) ;
   entry end_write(writer_id: in writer_range) ;
end buffer;
task body buffer is
   reader_#: integer range 0..10 := 0;
   writer_#: integer range 0..1 := 0;
   use queue_manager;
   start_queue, end_read_queue, end_write_queue: operation_queue ;
   reader_id: reader_range; writer_id: writer_range;
begin
   loop
      select accept start_read(reader_id: in reader_range) do
            enqueue(start_read, reader_id, start_queue);
         end
      or accept start_write(writer_id: in writer_range) do
            enqueue(start_write, writer_id, start_queue);
         end
      or accept end_read(reader_id: in reader_range) do
            enqueue(end_read, reader_id, end_read_queue);
         end
      or accept end_write(writer_id: in writer_range) do
            enqueue(end_write, writer_id, end_write_queue);
         end
      or terminate
      end select ;
      loop
         if not empty(end_read_queue)
         then
            dequeue(reader_id, end_read_queue);
            reader_# := reader# - 1;
            <send_acknowledgement_to_requestor>;
         elseif not empty(end_write_queue)
         ...
      end loop
   end loop
end buffer
```

FIGURE 4. Target code for synchronizer *buffer*

synchronizer specification are executed. Finally, the queues are checked again to determine whether other operation requests can be selected for service. Note that the *buffer* code makes use of a standard queue package to handle records describing operation invocations.

6. Conclusions

In this paper we have briefly sketched the workings of a proposed automatic synthesis system for synchronization code and we have described one particular crucial feature, a deductive system for establishing the liveness of synchronizer operations.

We believe our method is automatable, and is even potentially practical, for several reasons. First, the absence of loop constructs in operation state changes means that loop invariants need not be generated during RSTG construction, thus making this activity relatively straightforward. This is crucial because the RSTG reduces the excessive state information present in system behavior to a small, tractable body. Secondly, we have wedded our proof lattice machinery to the RSTG construction. This means that proof flow is directed by RSTG state transitions, thus controlling the size of the proof search space. Finally, proof lattice construction is facilitated by the absence of loop constructs in operation state transitions, by the absence of temporal operators in operation preconditions, and by the stated assumptions about the behavior of resource allocation and deallocation. As a result, only a limited subset of temporal logic has been required to build the proof lattices in the example set considered so far.

We believe that the practicality of the proposed approach is enhanced by the structure of program specifications. First, our approach to synchronizer specification is close to the way people think about resources (i.e. in terms of operations performing resource allocations and deallocations). Second, high-level support is provided, so that specifications can include such features as operation preconditions and priorities. Third, these properties are clearly separated in a program specification. We believe that these aspects of our approach compare favorably with previous work that does not support the specification of these properties explicitly [**5**, **10**].

References

1. M. Abadi, *Temporal Theorem Proving*, PhD Dissertation, Dept. of Computer Science, Stanford University, Stanford, California, 1987.
2. Ada Language Reference Manual, Military Standard 1815A, US Department of Defense, 1983.
3. R. S. Boyer, J. S. Moore, *A Computational Logic Handbook*, Academic Press, Boston, Massachusetts, 1988.
4. U. Buy, *Automatic Synthesis of Resource Sharing Concurrent Programs*, PhD Dissertation, Dept. of Computer and Information Science, University of Massachusetts, Amherst, Massachusetts, September 1990.

5. E. M. Clarke, E. A. Emerson, Design and Synthesis of Synchronization Skeletons Using Branching-Time Temporal Logic, *Lecture Notes in Computer Science 131*, Springer-Verlag, New York, New York, 1981.
6. R. L. Constable *et al*, *Implementing Mathematics with the Nuprl Proof Development System*, Prentice-Hall, Englewood Cliffs, New Jersey, 1986.
7. E. W. Dijkstra, Guarded Commands, Nondeterminacy and Formal Derivation of Programs, *Communications of the Association for Computing Machinery, 18*, 8, August 1975.
8. E. W. Dijkstra, *Cooperating Sequential Processes*, Technical Report EWD-123, Technological University, Eindhoven, The Netherlands, 1965.
9. L. Lamport, Proving the Correctness of Multiprocess Programs, *IEEE Trans. on Software Engineering, SE-3*, 2, March 1977.
10. Z. Manna, P. Wolper, Synthesis of Communicating Processes from Temporal Logic Specifications, *ACM Transactions on Programming Languages and Systems, 6*, 1, January 1984.
11. S. S. Owicki, L. Lamport, Proving Liveness Properties of Concurrent Programs, *ACM Transactions on Programming Languages and Systems, 4*, 3, July 1982.
12. K. Ramamritham, Synthesizing Code for Resource Controllers, *IEEE Transactions on Software Engineering, SE-11*, 8, August 1985.

Department of Electrical Engineering and Computer Science, University of Illinois, Chicago, Illinois 60610

E-mail: buy@uicbert.eecs.uic.edu

Department of Computer and Information Science, University of Massachusetts, Amherst, Massachusetts 01003

DIMACS Series in Discrete Mathematics
and Theoretical Computer Science
Volume 3, 1991

Branching Time Regular Temporal Logic for Model Checking with Linear Time Complexity

KIYOHARU HAMAGUCHI, HIROMI HIRAISHI AND SHUZO YAJIMA

ABSTRACT. Firstly in this paper, we propose a branching time logic BRTL (Branching time Regular Temporal Logic) which has automata connectives as temporal operators. BRTL is more expressive than CTL proposed by Clarke et.al. and it is modest in terms of model checking, i.e. it has a model checking algorithm which runs in time proportional both to the size of a given Kripke structure and to the length of a given formula, as shown in the paper.

Secondly, in order to improve the succinctness of the temporal logic formulas, we introduce the mechanism of *substitutions to Boolean variables* and *references to the Boolean variables*. We propose EBRTL(Extended BRTL), i.e. BRTL with the substitution mechanism, and show examples of descriptions of some temporal properties. We develop its model checking algorithm whose time complexity is linear both in the size of a given Kripke structure and in the length of a given formula and exponential in the number of the Boolean variables used in the formula.

1. Introduction

In order to verify correctness of finite state systems, researchers developed various kinds of propositional temporal logics, applied model checking methods on Kripke structures and succeeded in verifying finite state systems of medium size[**1**, **2**, **3**].

CTL proposed in [**1**] has a model checking algorithm which runs in time $O(\mathrm{Size}(S) \times \mathrm{Len}(f))$ ($\mathrm{Size}(S)$ and $\mathrm{Len}(f)$ are the size of the given Kripke structure and the length of the given CTL formula respectively.) The expressive power of CTL is, however, restricted, because only $\mathcal{U}$(until) and X(next) with path quantifiers are allowed. ECTL[**4**] is an extension of CTL, which have *automata connectives*[**5**] as temporal operators and can express an arbitrary ω regular set. The model checking algorithm shown in [**4**] is exponential in the size of the number of the states of automata connectives, because Muller automata are used as their bases.

1980 *Mathematics Subject Classification* (1985 *Revision*). Primary 03B45, 68M15.

One of the two concerns of this paper is to develop a temporal logic system which is more tractable, in terms of model checking, than ECTL and more expressive than CTL. We propose a branching time logic BRTL (Branching Time Regular Temporal Logic) and show a model checking algorithm which runs in time $O(\text{Size}(S)\times\text{Len}(f))$, i.e. linear in the number of the states of automata connectives used in f.

The automata connectives of BRTL are based on restricted deterministic ω finite automata of Büchi type. Although this means that BRTL cannot express all ω regular sets, it can express the properties of repetition of some events as shown by examples.

The other concern is to improve the succinctness of the temporal logic formulas by introducing the mechanism of *substitutions to Boolean variables* and *references to the Boolean variables.* We propose EBRTL(Extended BRTL), i.e. BRTL with the above mechanism, and show examples of descriptions of some temporal properties. We develop its model checking algorithm whose time complexity is $O(\text{Size}(S)\times\text{Len}(f)\times 2^{|V|})$ ($|V|$ is the number of the different Boolean variables used in f).

In Section 2, BRTL is defined and its model checking algorithm is shown. In Section 3, BRTL is extended by introducing the mechanism stated above.

2. Branching Time Regular Temporal Logic

The truth values of propositional logic are represented by T(true) and F(false). X^ω and $X^\dagger$ represent sets of infinite and finite sequences of elements of a set X respectively. $|X|$ is the number of the elements of X. V_T and V_F are *tautology* and *invalid* formulas respectively.

2.1. Syntax and Semantics.

DEFINITION. 2.1. A logic-type deterministic ω finite automaton (ldo-fa) $A = (Q, \Sigma, P, Br, \delta, q_0, F)$ *is defined as follows.*

Q and $\Sigma = \{T, F\}^n$ are sets of finite number of states and input symbols respectively. $P = \{p_1, \cdots, p_n\}$ is a set of atomic propositions. F is a set of accepting states and the elements of $Q - F$ are called rejecting states. q_0 is the initial state.

Let BF be a set of all propositional formulas constructed from the elements of P. $Br : Q \times Q \to BF$ is a partial function which satisfies the following two conditions.

We define $BF_q \stackrel{def}{=} \{f \mid \exists q'.Br(q, q') = f\}$, for each $q \in Q$.

For any $q \in Q$, (1) $f_1 \wedge f_2 \equiv V_F$ for any $f_1, f_2 \in BF_q$ and (2) $\bigvee_{f\in BF_q} f \equiv V_T$

We define Edges(A) $\stackrel{def}{=} \{(q, q') \mid \exists f.Br(q, q') = f\}$.

$\delta : Q \times \Sigma \to Q$ is a state transition function such that, for $q, q' \in Q$ and $v \in \Sigma$, $\delta(q, v) = q' \Leftrightarrow f = Br(q, q')$ is defined and $f(v) = T$.

For $x \in \Sigma^\omega$, we define $Inf(x)$ as the set of the states through which A goes infinitely often. Then the set of words accepted by A is $\{x | Inf(x) \cap F$ is not empty $\}$ and is described by $|A|$. □

Although complementation of an ω finite automaton is not easy in general, even if it is deterministic, an automaton that satisfies the following Cond. 1 can be complemented easily, (i.e. by exchanging the accepting states and the rejecting states.)

Cond. 1 Let A be a ldo-fa $(Q, \Sigma, P, Br, \delta, q_0, F)$. There is no input sequence which transits A from a rejecting state $q_r \in (Q - F)$ to the state q_r via some accepting state $q_a \in F$. □

DEFINITION. 2.2. *A logic-type deterministic ω finite automaton A which satisfies Cond. 1 is called* ldo-fa-1. □

It is easy to prove the following two lemmas.

LEMMMA. 2.3. *For an ldo-fa-1 $A = (Q, \Sigma, P, Br, \delta, q_0, F)$, an ldo-fa-1 $\overline{A}$ which accepts $\Sigma^\omega - |A|$ is obtained by exchanging accepting states and rejecting states of A.*

LEMMMA. 2.4. *For ldo-fa-1's $A_i = (Q_i, \Sigma, P, Br_i, \delta_i, q_{i0}, F_i)$ $(i = 1, 2)$, an ldo-fa-1 $A_1 | A_2$ which accepts $|A_1| \cup |A_2|$ is obtained as follows:*

- $Q = Q_1 \times Q_2 = \{(q_1, q_2) | q_1 \in Q_1 \ , q_2 \in Q_2\}$
- *$Br : Q \times Q \to BF$ is defined by $Br((q_1, q_2), (q_1', q_2')) = Br_1(q_1, q_1') \wedge Br_1(q_2, q_2')$, where $q_i, q_i' \in Q_i$ $(i = 1, 2)$ and both of $Br_1(q_1, q_1')$ and $Br_1(q_2, q_2')$ are defined.*
- $q_0 = (q_{10}, q_{20})$.
- $F = \{(q_1, q_2) | q_1 \in F_1$ *or* $q_2 \in F_2\}$.

□

The lemmas mean that the languages defined by ldo-fa-1's are closed under Boolean operations over sets.

Let AP be a set of atomic propositions.

DEFINITION. 2.5. *$S = \langle \Sigma, I, R, \Sigma_0 \rangle$ is called a Kripke structure, where Σ is a set of states, $I : \Sigma \to 2^{AP}$ is an assignment function, $R \subseteq \Sigma \times \Sigma$ is a* total *relation, $\Sigma_0 \subseteq \Sigma$ is a set of initial states. The size of a given Kripke structure, denoted by Size(S), is defined as $|\Sigma| + |R|$.* □

DEFINITION. 2.6. *Syntax and Semantics*

The syntax of BRTL(Branching Time Regular Temporal Logic) is as follows:

<u>BRTL formulas:</u> The set of all BRTL formulas are described by BF.

- *$p \in AP$ is a BRTL formula.*
- *If f and g are BRTL formulas, then $\neg f$ and $f \vee g$ are BRTL formulas.*
- *If A is an automaton connective , then $\exists A$ is a BRTL formula.*

Automata connectives

If A' is an ldo-fa-1, then $A'[AP/B]$ and $\neg A'[AP/B]$ are automata connectives, where $A'[AP/B]$ is obtained by replacing each atomic proposition $p_i \in AP$ ($1 \leq i \leq n$) with $f_i \in B \subseteq BF$ corresponding to p_i simultaneously. The numbers of the elements in AP and B have to be equal.

The semantics of BRTL is defined on a Kripke structure $S = \langle \Sigma, I, R, \Sigma_0 \rangle$. $S, s \models f$ means that the BRTL formula f holds at the state s on S. In the following, $p \in AP$, f and g are BRTL formulas and A is an automaton connective.

- $S, s \models p$ *iff* $I(s) \ni p$
- $S, s \models f \vee g$ *iff* $S, s \models f$ *or* $S, s \models g$
- $S, s \models \neg f$ *iff* $S, s \not\models f$
- $S, s \models \exists A$ *iff there exists an infinite sequence* $\sigma = s_0 s_1 s_2 \cdots$ *starting from* s *on* S *and a run (a sequence of states)* $q_0 q_1 q_2 \cdots$ *in* Q *such that,* $S, s_i \models BR(q_i, q_{i+1})$ *holds and all the states which appear infinitely in the run are in* F.
- $S, s \models \exists \neg A$ *iff* $S, s \models \exists \overline{A}$

If $\forall s \in \Sigma_0 . S, s \models f$, *then we describe* $S \models f$. □

The Boolean operators $\wedge$, $\equiv$ and $\Rightarrow$ are also used. Besides we define $\forall A \stackrel{def}{=} \neg\exists\neg A$ and $\forall\neg A \stackrel{def}{=} \neg\exists A$.

For BRTL formulas $f = \exists A$ or $\exists\neg A$, where $A = A'[AP/B]$ for some ldo-fa-1 A' and BRTL formulas B, the length of f, denoted as $\text{Len}(f)$ is defined as follows:

$\text{Len}(A) \stackrel{def}{=} |Q| + |\text{Edges}(A')| + \sum_{f \in Br'(Q \times Q)} \text{Len}(f) + \sum_{g \in B} \text{Len}(g)$

$Br'(Q \times Q)$ represents the set of propositional formulas labeled to transitions of A'. Intuitively the length means the length of the description of the ldo-fa-1 A' and the set of B.

2.2. Comparison with CTL.

DEFINITION. 2.7. *The truth value of a given CTL formula is defined on each state on a given Kripke structure.*

While BRTL uses path quantifiers $\exists$ and automata connectives, $\forall Xf$, $\exists Xf$, $\forall[f_1 \mathcal{U} f_2]$ and $\exists[f_1 \mathcal{U} f_2]$ are allowed to be used[1].

The semantics of these formulas are as follows:

- $S, s \models \forall Xf$ *($\exists Xf$) iff* $S, s_1 \models f$ *for all (some)* s_1 *such that* $(s, s_1) \in R$
- $S, s \models \forall[f_1 \mathcal{U} f_2]$ *($\exists[f_1 \mathcal{U} f_2]$) iff* $S, s_i \models f_1$ ($i = 0, 1, 2, \cdots, n$) *and* $S, s_{n+1} \models f_2$ *for all (some) infinite sequences* $s = s_0 s_1 s_2 \cdots$ *starting from* s *and an integer* $n \geq 0$.

□

Since the automata connectives are not easy to write, we introduce informally a description language like programming languages. For example, the automata

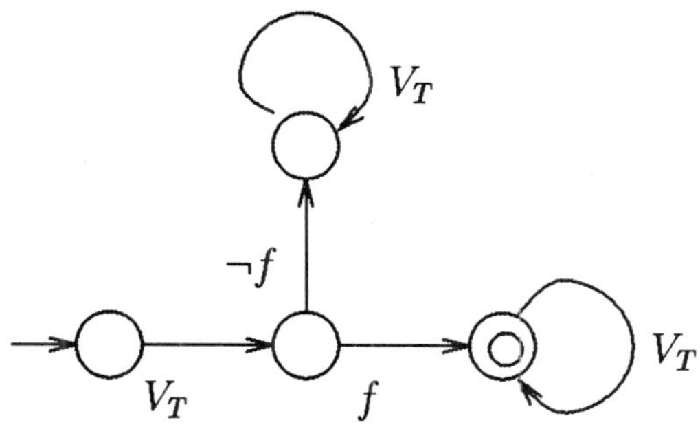

FIGURE 1. $Next(f)$

connectives of Fig. 1 and Fig. 2 are described by the descriptions (1) and (2) in the following.

- *state:* [a] and [r] represent an accepting state and a rejecting state of an automaton connective respectively. [ac] and [re] are special states. [ac] ([re]) represents an accepting (rejecting) state from which the automaton never transits to the other states for any input.
- *branch:* **if** S_1 **else if** $S_2 \cdots$ **else** S_n **endif** represents a branch from a state. S_i has a higher *precedence* than S_{i+1}.
- *loop:* **loop** S **endloop** represents a loop from a state to the loop.
- *label and goto:* 'L:' is a label to be put on a state and **goto** L means the transition to the state.

If no transition is possible from a state, it is assumed to transit to a rejecting state. [x][x] is interpreted as [x] V_T [x] (x = a, r, ac or re). The first state is assumed to be the initial state of the automaton.

With the above notations, we can define the temporal operators used in CTL.

Let $Next(f)$ and $Until(f_1, f_2)$ be automata connectives such that

$$Next(f) \stackrel{def}{=} ([r][r]f[ac]) \tag{1}$$

$$Until(f,g) \stackrel{def}{=} (\texttt{[r]}\,\textbf{if loop}\ f_1 \wedge \neg f_2\ \textbf{endloop}\ \textbf{else if}\ \neg f \wedge \neg f_2\ \texttt{[re]}\ \textbf{else if}\ f_2\ \texttt{[ac]}\ \textbf{endif}\,) \tag{2}$$

Then,

$$\forall[f_1 \mathcal{U} f_2] \stackrel{def}{=} \forall Until(f_1, f_2) \quad \exists[f_1 \mathcal{U} f_2] \stackrel{def}{=} \exists Until(f_1, f_2)$$

$$\forall X f \stackrel{def}{=} \forall Next(f) \qquad \exists X f \stackrel{def}{=} \exists Next(f)$$

Temporal properties containing the concept of repetition can also be expressed. For example, a BRTL formula $f = \forall A_R$, where A_R is shown in Fig. 3, means "for a given $m \geq 2$, p is true on every state s_i ($i = km, k = 0, 1, \cdots$)".

It can be shown that this property cannot be expressed by any CTL formula

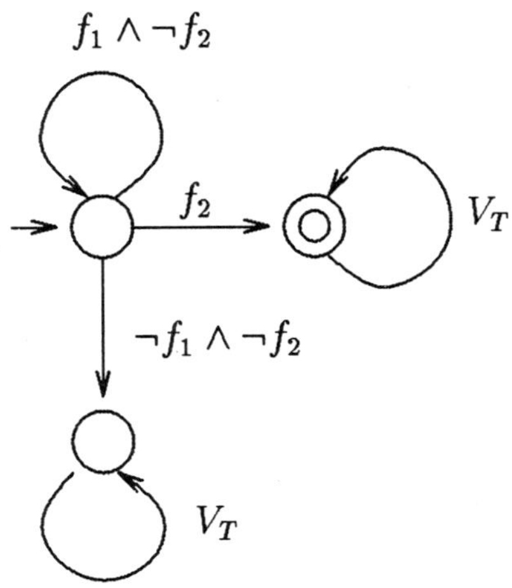

FIGURE 2. $Until(f_1, f_2)$

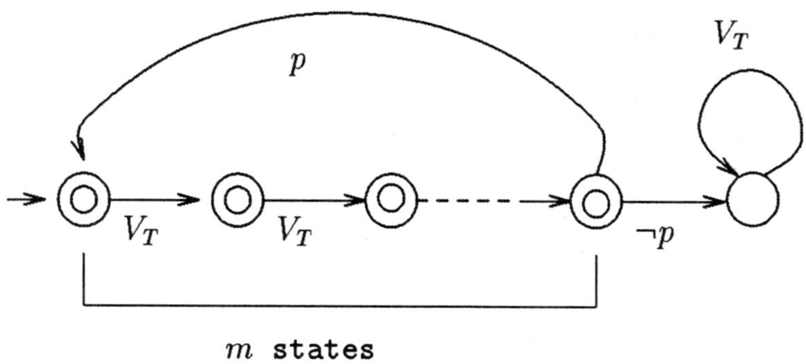

FIGURE 3. An automata connective representing repetition

by extending the proof by Wolper in [5] to Kripke structures.

For a Kripke structure $S = \langle \Sigma, I, R, \Sigma_0 \rangle$ and a state $s_0 \in \Sigma$,

$$Re_i(S, s_0) \stackrel{def}{=} \{s_i | (s_j, s_{j+1}) \in R, j = 0, 1, 2, \cdots i - 1\}$$

LEMMMA. 2.8. *Let p be an atomic proposition. C_i is defined as a set of pairs of a Kripke structure $S = \langle \Sigma, I, R, \Sigma_0 \rangle$ and a state s, i.e. (S, s), satisfying the following conditions.*

(1) $\forall s' \in Re_j(s).S, s' \models p \quad (j = 0, 1, \cdots, i)$,
(2) $\forall s'' \in Re_{i+1}(s).S, s'' \models \neg p$,
(3) $\forall s''' \in Re_j(s).S, s''' \models p \quad (j = i + 1, i + 2, \cdots)\}$

If CTL formula f has at most n 'X's , then, for any $(S_i, s_i) \in C_i$, $(S_{i'}, s_{i'}) \in C_{i'}$ $(i, i' > n)$,

$$S_i, s_i \models f \Leftrightarrow S_{i'}, s_{i'} \models f \qquad (*)$$

In other words, for any $i > n$, the truth value of f is same for $(S_i, s_i) \in C_i$.

(Sketch of Proof) This is proved by induction on subformulas. We investigate the following four cases. The others are straightforward.

- $f = \forall X g$: For a pair $(S, s) \in C_i$,
  $S, s \models \forall X g \Leftrightarrow S, s' \models \forall g$ for all pairs (S, s') such that $(s, s') \in R$.
  Since g has at most $n - 1$ 'X's, $(S, s') \in C_{i-1}$ and $i - 1 > n - 1$, whether $S, s' \models \forall g$ or not is independent of i.
  For any s' such that $(s, s') \in R$, $S, s \models \forall X g \Leftrightarrow S, s' \models \forall g$.
  Thus whether $S, s \models \forall X g$ or not is determined independent of i.
- $f = \forall [g_1 \mathcal{U} g_2]$: Let $(S, s_i) \in C_i$. $S, s_i \models \forall [g_1 \mathcal{U} g_2] \Leftrightarrow$
  For any infinite sequence $s_i s_{i-1} s_{i-2} \cdots s_{n+1} \sigma_n$ starting from s_i, where $s_j \in \Sigma$ and σ_n is an infinite sequence starting from s_n such that $(s_{n+1}, s_n) \in R$,

$$\begin{gathered}(S, s_i \models g_2) \text{ or } ((S, s_i \models g_1) \text{ and} \\ ((S, s_{i-1} \models g_2) \text{ or } ((S, s_{i-1} \models g_1) \text{ and} \cdots \\ ((S, s_{n+1} \models g_2) \text{ or } ((S, s_{n+1} \models g_1) \text{ and} \\ (S, s_n \models \forall [g_1 \mathcal{U} g_2]))) \cdots))\end{gathered}$$

  By the induction hypothesis, for $k = 1, 2$,
  $S, s_j \models g_k \ (j = n + 1, n + 2, \cdots, i - 1, i) \Leftrightarrow S, s_{n+1} \models g_k$
  Thus whether $S, s_i \models \forall [g_1 \mathcal{U} g_2]$ or not is determined independent of i.
- $f = \exists g$: This is reduced to Case 1 , 2 or 4, because $S, s \models \exists g \Leftrightarrow S, s \models \forall g$ for any $(S, s) \in C_i$. □

THEOREM. 2.9. *Given an integer $m \geq 2$, $D \stackrel{def}{=} \{(S, s) | \forall s' \in Re_{km}. S, s' \models p, (k = 0, 1, \cdots)\}$, where S is a Kripke structure and s is a state on S. There exists no CTL formula f which satisfies*

$$(S, s) \in D \Leftrightarrow S, s \models f$$

(Proof) Let us assume that a CTL formula f satisfies the above condition and f has l 'X's.
Let $km - 1 \geq l$. From Lemma 2.8, $S, s \models f \Leftrightarrow S', s' \models f$ holds, for any $(S, s) \in C_{km}$ and any $(S', s') \in C_{km-1}$. Besides $(S, s) \in D$ and $(S', s') \notin D$. Thus there exists no CTL formula that can express the property. □
ECTL[4] has as automata connectives deterministic ω finite automata of Muller type, the class of which is equivalent to ω regular sets, as automata connectives. This means that there are some properties which cannot be expressed in BRTL, but can be expressed in ECTL.
From the above arguments, the relation among CTL, BRTL and ECTL in terms of expressive power is as follows:

$$CTL < BRTL < ECTL$$

2.3. A Model Checking Algorithm. A model checking algorithm is shown in this section. The algorithm is constructed, based on the next lemma.

LEMMMA. 2.10. *The necessary and sufficient condition for* $S, s \models \exists A$ *is that there exists a finite sequence* $(s_0, q_0) \cdots (s_k, q_k)(s_{k+1}, q_{k+1}) \cdots (s_n, q_n)$ *satisfying the following conditions:*

(1) $(s_i, s_{i+1}) \in R$ $(i = 0, 1, \cdots n - 1)$
(2) $S, s_i \models Br(q_i, q_{i+1})$ $(i = 0, 1, \cdots n - 1)$
(3) $s_k = s_n$ *and* $q_k = q_n$
(4) $q_j \in F$ $(j = k, k + 1, \cdots n)$

ALGORITHM 2.11. – *Input: A Kripke structure S and a BRTL formula f*
– *Output:* $S \models f$ *?*
– *Method:*

(1) *The automata connectives of the form* $\neg A$ *occurred in* f *are transformed to* $\overline{A}$ *by exchanging the accepting states and rejecting states.*
(2) *By the following (a)-(c), every subformula* g *of* f *is labeled to every state* $s \in \Sigma$ *if and only if* $S, s \models g_i$, *in a bottom up manner.*
The output is 'yes' if and only if f *is labeled to every* $s_0 \in \Sigma_0$.

(a) *When* g *is an atomic proposition, its truth value on each* $s \in \Sigma$ *is given.*
(b) *When* g *is* $h_1 \vee h_2$ *or* $\neg h$ *for some BRTL formulas* h_1, h_2 *or* h, *the truth value of* g *on each state can be obtained by the truth values of* h_1, h_2 *or* h *which are already labeled.*
(c) *When* g *is* $\exists A$ *for some automata connective* A, *the following (i)-(iv) are performed.*

(i) *A graph* $G = (V, E)$ *is constructed, where*
$V = \{(s, q) | s \in \Sigma, q \in Q\}$ *and*

$E = \{((s,q),(s',q'))|(s,s') \in R$ *and* $S, s \models Br(q,q')\}$.

(ii) *The set of vertices of all the strongly connected components on the subgraph* G' *of* G *whose vertices are in* $V' = \{(s, q_F)|q_F \in F\}$ *is obtained. It is described as* V_C

(iii) *The set of the vertices in* G *which can reach some vertex in* V_C *is obtained. It is described as* V_R

(iv) $\exists A$ *is labeled to every* $s \in \Sigma$ *such that* $(s, q_0) \in V_R$ *(*q_0 *is the initial state of* A*).*

□

PROPOSITION. 2.12. *Algorithm* 2.11 *runs in time* $O(Size(S) \times Len(f))$.

(Sketch of Proof) This is proved by the following facts:

- In 1, the transformation from $\neg A$ contained in f to $\overline{A}$ can be performed in time $O(\text{Len}(f))$.
- In 2.(b), if the truth values of h_1, h_2 or h are determined on each state $s \in \Sigma$, then the truth values of $h_1 \vee h_2$ and $\neg h$ can be determined in time $O(|\Sigma|)$.
- In 2.(c).i, the determination of $S, s \models Br(q,q')$ for all $s \in \Sigma$ and all $q, q' \in Q$ can be performed in time $\text{O}(\text{Size}(S) \times (\sum_{f \in Br'(Q \times Q)} \text{Len}(f) + \sum_{g \in B} \text{Len}(g)))$. Then the construction of the graph G costs $\text{O}(\text{Size}(S) \times (|Q| + |\text{Edges}(a)|))$ in time, thus in total, $\text{O}(\text{Size}(S) \times (\text{Len}(A)))$.
- In 2.(c).ii, iii and iv, the construction of strongly connected components and the calculation of the vertices reachable to the components can be performed in linear to the size of the graph. (Here the size is the sum of the numbers of edges and vertices.)

□

3. Temporal Logic with Substitutions and References

3.1. Succinctness of Descriptions. Some temporal properties expressed in the linear time temporal logic with □ (always) and ◯ (next) as temporal operators cannot be described easily in CTL or BRTL. For example, an assertion for a sequence detector such that the value of a signal line z is 1 iff it recognizes the sequence 110 on a signal line x, is described in the linear time temporal logic as follows.

$$\Box((x \wedge \bigcirc(x \wedge \bigcirc \neg x)) \equiv \bigcirc \bigcirc z)$$

Using BRTL, this property is expressed as

$$\forall \Box \forall A,$$

where $\Box f$ is equivalent to $\neg Until(V_T, \neg f)$ and $A \stackrel{def}{=}$

```
[r]  if x1 [r] {
        if x1 [r] {
           if ¬x1 ≡ z [ac]
           else if x1 ∧ ¬z [ac]
           else [re]
        }
        else L: [r]  {
           if ¬z [ac]
           else [re]
        }
     else [r] V_T goto L
```

By allowing substitution to Boolean variables and references to them, the property can be expressed succinctly as

$$\forall\Box\forall(\ [\mathtt{r}]\ V_T, v_1 := x[\mathtt{r}]V_T, v_2 := x\ [\mathtt{r}]\ \ (v_1 \wedge v_2 \wedge \neg x) \equiv z\ [\mathtt{ac}]\ )$$

Here v_1 and v_2 are Boolean variables and $v_1 := x$ means that the value of x is substituted to v_1.

DEFINITION. 3.1. *Syntax of EBRTL(Extended BRTL)*
Let $V = \{v_1, v_2, \cdots, v_m\}$ *be a set of Boolean variables. AP of BRTL contains V. EBRTL formulas are defined similarly to BRTL formulas. Automata connectives are similar to those of BRTL except that*

- *Formulas labeled to the transitions are composed of EBRTL formulas.*
- *Substitutions* $v_i := f$ *(f is an EBRTL formula) can be labeled to the transitions. For a transition, substituting to a variable* v_i *is allowed only once.*

More precisely, the function Br is redefined as follows:
$Br : Q \times Q \rightarrow BF \times SUB$ *is a partial function which satisfies the following three conditions, where SUB is a class of sets of substitutions.*
We define $BF_q \stackrel{def}{=} \{f | \exists q'.Br(q, q') = (f, sub), sub \in SUB\}$ *for each* $q \in Q$. *For any* $q \in Q$, *(1)* $f_1 \wedge f_2 \equiv V_F$ *for any* $f_1, f_2 \in BF_q$, *(2)* $\bigvee_{f \in BF_q} f \equiv V_T$ *and (3) each element* $sub \in SUB$ *has at most one substitution of the form* $v_i := f$ *for each* $v_i \in V$. *(sub can be empty.)* □

Before defining the semantics, we show a sequence along which $\exists(\ [\mathtt{r}]\ v_1 := x\ [\mathtt{r}]\ v_2 := x\ [\mathtt{r}]\ (v_1 \wedge v_2 \wedge \neg x) \equiv z\ [\mathtt{ac}]\ )$ holds in Fig. 4. The tuple labeled to each state stores the values of v_1 and v_2. We have to give initial values of the Boolean variables. In the example, $v_1 = F$ and $v_2 = F$ are the initial values.
Note that the change of the values of the Boolean variables caused by substitutions at a state occurs at its next state. This prevents oscillations of the values at each state.

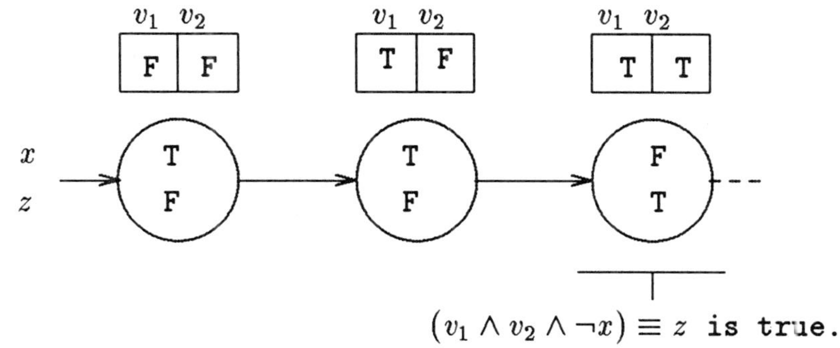

FIGURE 4. Intuitive Semantics of an EBRTL formula

DEFINITION. 3.2. *Semantics*
The semantics of EBRTL is defined for a Kripke structure S. When an EBRTL formula f holds at a state s on S with initial values of Boolean variables represented by a vector notation $\vec{v} = (v_1, v_2, \cdots, v_m)$ ($v_j = F$ *or* T), *we denote as* $S, s, \vec{v} \models f$. $\vec{v}[j]$ *represents the j-th component of the vector* $\vec{v}$.
While the truth value of each atomic propositions except Boolean variables are determined for $(S, s, \vec{v})$ *by* $I(s)$, *that of the Boolean variable* v_j $j = 1, 2, \cdots, m$ *is the value of the j-th component of* $\vec{v_j}$.
Here we define the semantics for $S, s, \vec{v} \models \exists A$ *(A is an automaton connective).*
$S, s, \vec{v} \models \exists A$ *iff there exists an infinite sequence* $\tau = t_0 t_1 t_2 \cdots$ *that satisfies the following conditions.*
Here $t_i = (s_i, \vec{v_i}, q_i) \in \Sigma \times \{T, F\}^m \times Q$.

- $(s_i, s_{i+1}) \in R$
- *Let* $(f, sub) = Br(q_i, q_{i+1})$. *If* $(\vec{v_i}[j] := g) \in sub$, *then* $\vec{v_{i+1}}[j] = T$ *iff* $S, s_i, \vec{v_i} \models g$. *Otherwise* $\vec{v_{i+1}}[j] = \vec{v_i}[j]$.
- *Let* $(f, sub) = Br(q_i, q_{i+1})$. $S, s_i, \vec{v_i} \models f$.

□

3.2. A Model Checking Algorithm for EBRTL. *The outline of a model checking algorithm of EBRTL is shown.*

ALGORITHM 3.3.

- *Input: A Kripke structure S, a BRTL formula f and initial values of the Boolean variables* $\vec{v}$
- *Output:* $S, s, \vec{v} \models f$ *for all the initial states of S ?*

- *Method: We modify the model checking algorithm shown in Algorithm* 2.11 *as follows:*
  - *In constructing the graph G, tuples of the form $(s, q, \vec{v})$, where $s \in \Sigma$, $q \in Q$ and $\vec{v}$ is the vectors of the truth values for $v_j \in V$, are used instead of (s, q).*

□

PROPOSITION. 3.4. *The model checking algorithm for EBRTL runs in time $O(Size(S) \times Len(f) \times 2^{|V|})$.*

(Sketch of Proof) This is proved by the fact that the number of the tuples of the form $(s, q, \vec{v})$ is $Size(S) \times Len(f) \times 2^{|V|}$. □

4. Conclusion

We proposed a temporal logic BRTL which is more expressive than CTL and is more tractable than ECTL in terms of model checking. Furthermore, we provided an extension of BRTL which has the mechanism of substitutions to Boolean variables and references to them.
The linear time complexity of the model checking algorithms of CTL or BRTL is obtained by excluding nondeterminism from their descriptions. CTL or ETL can express, in a sense, the concept of nondeterminism by allowing formulas of the form $\forall(f \vee g)$ (f and g are the formulas without the path quantifiers $\forall$ and $\exists$) and $A \vee B$ (A and B are automata connectives based on nondeterministic finite automata) respectively.*
The mechanism of substitutions and references to Boolean variables also introduces nondeterminism to BRTL, in a sense. In the example of the sequence detector, the branches of the automaton connective of BRTL formula are "folded" by using substitutions. This means that the substitution mechanism is another way to introduce nondeterminism.
Future problems are as follows:

- *Determination of the expressive power of EBRTL.*
- *Determination of the complexity of model checking problem for EBRTL.*

REFERENCES

1. E. M. Clarke, E. A. Emerson, and A. P. Sistla. Automatic Verification of Finite State Concurrent Systems Using Temporal Logic Specifications: A Practical Approach. In *10th ACM Symposium on Principles of Programming Languages*, pages 117–126, January 1983.
2. M. C. Browne, E. M. Clarke, D. L. Dill, and B. Mishra. Automatic Verification of Sequential Circuits Using Temporal Logic. *IEEE Transactions on Computers*, C-35(12):1035–1044, December 1986.
3. H. Hiraishi. Design Verification of Sequential Machines Based on a Model Checking Algorithm of ϵ-free Regular Temporal Logic. In *Computer Hardware Description Languages and their applications*, pages 249–263, June 1989.

4. E. M. Clarke and O. Grümberg and R. P. Kurshan. A Synthesis of Two Approached for Verifying Finite State Concurrent Systems. Technical report, Carnegie Mellon University, 1987. manuscript.
5. P. Wolper. Temporal Logic Can Be More Expressive. In *Proceedings of 22nd Annual Symposium on Foundations of Computer Science*, pages 340–348, 1981.

DEPARTMENT OF INFORMATION SCIENCE, FACULTY OF ENGINEERING, KYOTO UNIVERSITY, KYOTO, 606, JAPAN.

E-mail: hama%kuis.kyoto-u.ac.jpvma.cc.cmu.edu

DIMACS Series in Discrete Mathematics
and Theoretical Computer Science
Volume 3, 1991

Issues Arising in the Analysis of L.0

Linda Ness
Bellcore, Morristown, NJ
linda@bellcore.com

1. Introduction

L.0 is a programming language, which is currently being experimentally used at Bellcore as an executable specification language for communications software. It's design was driven by the special nature of communications systems. The most fundamental aspect of communications systems is that many fairly simple things are happening at the same time, for all time. It is precisely this aspect that is so difficult to specify in languages based on a sequential or asynchronous model. It is easy to specify this in L.0 because L.0 has an abstract concurrent-read-common-concurrent-write shared memory model, a notion of *next*, and because the basic composition operator is *conjunction*. Another fundamental aspect of communications systems is that many of the things happening simultaneously are very similar. Thus, another basic feature of L.0 is quantification, which permits parametrized specifications.

The semantics of the language given originally[1] was a combination of declarative and operational. Now that there is evidence that it is fairly easy to specify communications software in L.0, it is important to try to develop analysis tools. For this it is preferable to have declarative versions of the semantics given in several different standard paradigms so that analysis tools or algorithms previously developed might be applied to L.0.

One goal of this paper is to present the fundamental semantic constructs of L.0 in the framework of *linear predicate temporal logic*. Temporal logic was chosen because there is a fairly natural fit, and because there has been a great deal of work on its application to the specification and

1052-1798/91 $1.00 + $.25 per page

verification of reactive systems[2] [3]. Most of communications software can be viewed as a reactive system. The algorithm for translating L.0 syntax into the constructs given here is not given in this brief paper. It is however given in[4]. Neither is the focus of this paper is not on examples. They may be found in a number of papers[1] [5] [6] [7].

The main reason, that this presentation of the semantics does not immediately lead to the applicability of the various decision procedures and model checkers developed for temporal logic, is that L.0 is based on a subset of predicate (not propositional) temporal logic, and hence permits restricted forms of quantification. Thus, the second goal of this paper is to indicate the uses of quantification in L.0. The simple approach to developing some analysis tools for L.0, is to simply develop them for programs, in which the quantification is restricted to be finite, and the state spaces constructed during execution, are restricted to be contained in a finite universe. This approach is very unsatisfying because quantification is one of the main reasons that L.0 is so expressive. Furthermore, even if this restriction is made, the size of the finite universe, arising in applications of interest is likely to be huge, relative to the size that can be handled by existing tools. Kurshan[8] has developed a formal notion of homomorphism as one approach to handling this problem. An approach would be to develop tools to perform more restricted analysis which would work in presence of less stringents restrictions on quantification.

There has been quite a bit of work on the identification of executable subsets of temporal logic, and the development of executable temporal logic languages, i.e. languages in which a program is a temporal logic formula, and whose execution constructs a model for that formula. The subset of temporal logic exploited by L.0 is very similar to that identified by Dov Gabbay[9] and developed in MetateM. [10] The languages Tempura[11], Lustre[12] and Artic[13] can each be viewed as executable temporal logic languages. Tempura is based on interval temporal logic, rather than propositional, and uses quantification in a more limited way. Unlike L.0 it does not assume any solution to the frame problem. It has been used primarily to specify hardware. The semantics of Lustre can be given in terms of linear temporal logic. However, it appears that quantification is more restricted, since boolean Lustre may be compiled into a finite automaton. Arctic was designed for the specification and implementation of real-time computer systems, and has been applied to music. Its primary data-type is a real-valued function of time.

2. The fundamental constructs of L.0 in terms of temporal logic

Recall that a predicate temporal logic formula is constructed from predicates using the standard boolean operators, the temporal future operators *next* and *until* and universal and existential *quantification*. In addition, temporal past operators may be used, although it has been shown that the past operators do not add expressiveness[14]. The only temporal past operator that will be used here is *previous*.

Here the *until* used will be the weak *until*. In other words it is not required that the second argument of *until* is eventually true on a state. Precisely,

$S_i \models f \ until \ g$ if

$S_j \models f$ for all $j \geq i$
or ($\exists\, k \geq i$ such that $S_k \models g$ and $S_j \models f$ for all j such that $i \leq j < k$)

The temporal future operators *always* and *eventually* may be defined in terms of *until*, using the predicate *false* and negation.

2.1 Predicates on history and on extended history

These are the simplest components of L.0 programs. A *predicate on extended history* is inductively defined to be either:

1. *a predicate or*

2. *a boolean combination of predicates on extended history or*

3. *the past temporal operator previous applied to a predicate on extended history or*

4. *a universally quantified formula of the form* $r => f$, *where r is a restricting predicate and f is a predicate on extended history or*

5. *an existentially quantified formula of the form* $r \wedge f$, *where r is a restricting predicate and f is a predicate on extended history.*

A *predicate on history* has the form *previous p*, where *p* is a predicate on extended history. The *restricting predicates* are a subset of the predicates on history. They may be used to ensure the the quantification is finite but unbounded.

The "truth" of a predicate on history on a state in a model, may be deduced by examining only the interpretations for predicates in the previous states, i.e. the *history*. For predicates on extended history, the interpretation for predicates in the current state, may need to be examined as well. In other words, no future interpretations need to be examined. This is the key to executability, since the future states of the model need not have been constructed yet.

2.2 Cause-effect formulas

The second fundamental building block of L.0 programs is the cause-effect formula. Causality is specified using formulas of the form:

cause => effect

where the *cause* is restricted to be a *predicate on history*, and the effect is any L.0 program. Such a formula is true on a state in a sequence, if either the cause was not true on the history, or the cause was true on the history and the effect is true on the current state. If general formulas were allowed for the cause, the intuition would not correspond to causality, for the truth of the cause might depend on the current and future states.

2.3 A restricted class of until formulas

A restricted class of until formulas is used to specify that several cause-effect formulas apply until the first occurrence of at least one of a set of events, and to also specify the effects of some of those events. Thus, the only until formulas allowed have the form:

- *the binary until operator, applied to a conjunction of cause-effect formulas and a deactivator formula*

where a *deactivator formula d* either has form:

cause(d) and effects(d)

where

$cause(d) = (c_0 \text{ or } c_1 \text{ or } \ldots \text{ or } c_n)$

and

$effects(d) = \bigwedge_{i:\, i \in S} (c_i => e_i)$

Here $S \subset \{1, \ldots, n\}$, each of the conjuncts of *effects(d)* is a cause-effect formula, and *cause(d)* is a predicate on history. If none of the events in *cause* (*d*) have specified effects, the deactivator specification has the simpler form.

$d = cause(d)$

2.4 Universally quantified cause-effect formulas

Universal first order quantification of cause-effect formulas is allowed, providing that the cause contains a conjunct which is a restricting predicate.

2.5 Names of formulas

Formulas may be referred to by name, so that formulas may be structured modularly. Definitions of formula names may be recursive, providing the recursive reference is "over time". This can be expressed in temporal logic using second order quantification.

3. An inductive definition of temporal formulas permitted in L.0

The temporal formulas permitted in L.0 programs are either:

- a predicate
- a conjunction of L.0 programs

- a cause-effect formula where the cause is a predicate on history and the effect is an L.0 program
- a binary until operator, applied to a conjunction of cause-effect formulas and a deactivator formula[1]
- the next operator applied to an L.0 program
- a universally quantified cause-effect formula, where the cause contains a conjunct which is a restricting predicate
- the name of an L.0 program

The original until operator in L.0 had a slightly different semantics in the case of nested untils. However, as in proven in[4], there is an algorithm for mapping these non-standard until formulas to the until formulas of temporal logic, which applies in all reasonable cases.

4. Only safety properties are expressible in L.0

It is not possible to express *eventually* in this subset of temporal logic, because negation of programs is not permitted. The omission of eventually, means that it is not possible to express liveness properties in L.0. However, this omission is not as alarming as it first may seem, for it is possible in L.0 to express pseudo-random sequences, using predicates. This is probably more practical than specifying sequences via the *eventually* operator. Also, the notion of fairness, for which *eventually* is crucial, is not essential to L.0, because L.0 does not base its semantics on the interleaving of atomic events. Instead, the semantics of L.0 is fundamentally the semantics of temporal logic, which is synchronous, and which permits simultaneous occurrence of "events".

4.1 An execution strategy for basic L.0 programs

The subset of temporal logic, consisting of basic L.0 programs is executable, in the sense that there is an algorithm for inductively determining from a basic L.0 program, the obligations it imposes on the current state, and the obligation it imposes on the subsequent sequence of states. This strategy extends to an execution algorithm for general L.0 programs. The decomposition of a basic L.0 program into current obligations and future obligations uses a recursive semantic tautology for basic L.0 until formulas. In fact, an operational semantics for L.0 can be given in terms of a dynamically changing rule set[15].

A fundamental characteristic of the execution strategy is that history, once constructed will never be altered. The key to this it that cause formulas must be predicates on history.

An L.0 program may impose several obligations on the current state, in the sense that a conjunction of predicates must be true on the current state. It is in this sense that L.0 exploits the synchronous semantics of temporal logic. If these obligations conflict, execution of the L.0 program halts. Since there is more than one possible state which satisfies the current obligations, the execution strategy is intrinsically non-deterministic.

5. The current restrictions on the data domain, expressions, and states

Currently the data domain is restricted to be the set of all trees with labeled edges, with the property that all of the child edges of a node have unique labels. The labels are restricted to be from the alphabet of strings of aschii characters. Such a tree is called a namespace. States in L.0 are restricted to be namespaces. Each non-root node of a namespace has a *name*, which is the sequence of labels along the path from the root to the node. Because of the restriction on namespaces, the name of a non-root node in a namespace uniquely identifies it. In general, a *names* is a sequence of labels. Note that namespaces may be equivalently characterized as prefix-closed sets of names. The *suffix-value* of a name on a namespace is the set of strings prefixed by that name in the namespace.

An expression in L.0 programs is either a *name*, a function symbol applied to expressions, a concatenation or union of expressions, or the *suffix value of a name*. The *suffix value* of a name on a namespace is just the set of strings prefixed by the name in that namespace. Geometrically, if the namespace is finite, the value is the tree with labeled edges rooted at the name. The suffix value expression permits simulation of standard variable-value programming, within the more general declarative paradigm. Note that indirection is possible because the suffix value of a namespace is a namespace(which might be a name).

The value of an expression in an L.0 program is a namespace. The algorithm used to interpret expressions is the expected one. The expressions in an L.0 program are only used as arguments of predicates or functions.

6. A restriction on the interpretations of predicate symbols

In an L.0 model, each predicate symbol must be assigned a *predicate definition*. The simplest type of definition is for a predicate symbol of arity 0. Then the definition is either *undefined* or consists of a single *set-theoretic predicate*, which consists of two sets: a *domain*, which is a set of names, and a set of *solutions*, each of which is a subset of the domain[15]. Such a predicate is interpreted as being true on a state, if the intersection of the domain, with the state, is one of the solutions. This is equivalent to a very strong disjunctive normal form, where each element of the domain or its negation occurs in each disjunct. Negation is interpreted as non-existence. There is no restriction that the domain must be finite, here. (With this set-theoretic interpretation, finding a solution of a conjunction of predicates, can be interpreted as finding a global section of a sheaf.)

The definition of an n-ary predicate symbol *p*, is a mapping from the n-fold cartesian product of *Namespaces(Labels)* to the set of such set-theoretic predicates ∪ *undefined.*

6.1 Restrictions on the set of predicate symbols permitted

Currently, the predicate definitions allowed are equality of expressions, existence and non-existence of a name, *true* and *false*, and membership in a set of consecutive integers, or in an explicitly given set of names. Furthermore, only limited kinds of equality may be used in effects, and even that in a restricted way: equality of the suffix values of two names, and equality of the suffix value of a name and an expression.

Restricting predicates must contain a conjunction which is either a set membership predicate, or an exists predicate.

Users may provide interpretations for function symbols in "C".

7. A frame assumption

A model for a temporal formula consists of a data domain, an interpretation for expressions, an interpretation for predicate symbols, and a sequence of states. Each of these have been restricted. However, there is one final restriction imposed on a sequence of states, that is to be a model for an L.0 program. This restriction is that successive states are related by one of the replacement rules determined by the predicate, which is the conjunction of predicates, which are to be true on the next state, i.e. the current program obligations. The point is that each predicate (when specific values are supplied for its arguments), has an associated domain and set of solutions. Each solution determines a replacement rule for namespaces, namely replace the intersection of the domain with the namespace by one of its solutions. For a restricted set of predicates (such as are currently in L.0), the result is again prefix-closed, and hence a namespace.

This is nothing but a generalization of the usual frame assumption in sequential programming: namely replace the value of the variable by the new value being assigned. However, this generalization permits the size of the state space to be dynamic. For if the intersection of the domain of a predicate with the previous state space is empty, the set of names making up a solution are added. Conversely, if the intersection of the domain of a predicate with the previous state space is "too large" to be a solution, some of the names are removed to obtain a solution. One of the conveniences in using L.0 as a specification language lies is this concise dynamic allocation and deallocation of "space". To apply analysis algorithms developed for logical formulas, it would be necessary to be able to explicitly specify the frame within an L.0 program, and then consider only the class of programs which explicitly specified it. Fortunately, it appears that this can be done easily, and does not restrict the class of programs, because of the restriction that each predicate must have a specified domain.

8. Uses of quantification in L.0

Universal quantification is extremely powerful. For, as is shown in[4] it permits specification of parameter passing by value, indirection, and the analogue of a set of simultaneous "calls" to the same "procedure". When all three of these uses are combined, one can program in a table driven manner. Thus it is easy to specify a generic non-deterministic finite state machine[1], request handlers that are able to handle a finite but unbounded set of requests each time[16] and specify reconstruction of predicates of particular restricted types[17] In fact using universal quantification, one can easily specify SIMD parallelism.

As seems to be well-known[11], encapsulation can be added by adding existential quantification. Finally, since L.0 permits equality of names, pass by reference can be specified using equality of names and existential quantification. Thus, L.0 augmented by existential quantification, provides permits programmers to program, in a well-structured manner, in temporal logic.

9. Remarks about conjunction and equality

Conjunction permits functional decomposition of specifications. The communication is via shared variables. It seems that well structured L.0 programs are based on a clearly articulated dynamic read-write protocol, among functional components. This permits writing of observer programs which may, for example, filter data, watch for bugs, or write to the screen, to animate the program. Conjunction also makes programs exponentially shorter.

It is interesting to note that when programs are written using encapsulation, it seems possible to restrict the use of equality of names to equality between names local to the parent and child modules. Thus, apart from parameter passing, the uses of equality that seem necessary are standard assignment, and one-way derivations, which deduce that the suffix-value of a name can be the value of a function applied to arguments, which may refer to other current suffix-values.

10. Acknowledgments

L.0 was developed primarily through the joint efforts of Jane Cameron, David Cohen, B. Gopinath, and the author. Prem Uppaluru and Diane Sonnenwald also made contributions to the language, as did a number of the users. B. Gopinath was also the head of the IC* project, during the period when the first version of L.0 was developed. The language implementation was done by David Cohen and Bill Keese (on a sequential machine). The most recent debugger was done by Tim Guinther.

A connection between L.0 and temporal logic has also been recognized by Bob Kurshan, Fred Schneider, Ambuj Singh, and Prem Uppaluru.

References

1. E.J. Cameron, D.M. Cohen, L.A. Ness, H.N. Srinidhi, "L.0: A Language for Modeling and Prototyping Communications Software",(to appear in *Proceedings of the Third International Conference on Formal Description Techniques*, Madrid, November 5-8, 1990.)

2. A. Pnueli, "The Temporal Logic of Programs", *Proceedings of the 18th Annual Symposium on Foundations of Computer Science*(1977) pp. 46-57.

3. A. Pnueli, "The Temporal Logic of Programs", *Proceedings of the 18th Annual Symposium on Foundations of Computer Science*(1977) pp. 46-57.

4. L. Ness, "L.O: A Parallel Executable Temporal Logic Language", Bellcore Public Released TM-ARH-014974 September, 1989.

5. E. J. Cameron, N. H. Petschenik, L. Ruston, S. Shah, H. Srinidhi, "From Description to Simulation to Architecture: An Approach to Service-Driven System development", *Proceedings of the First International Conference on Systems Integration*, Morristown, N.J. April 23-26, 1990.

6. D. M. Cohen, T. M. Guinther, L. Ness, "Rapid Prototyping of a Communication Protocol Using a New Parallel Language", *Proceedings of the First International Conference on Systems Integration*, Morristown, N.J. April 23-26, 1990.

7. S. Aggarwal, F.S. Dworak, and P.Obenour, "An Environment for Studying Switching System Software Architecture", *Proceedings of IEEE Global Telecommunications Conference*, 1988.

8. Kurshan, R.P., "Reducibility in Analysis of Coordination", *Discrete Event Systems: Models and Applications*, LNCIS 103(1987), pp. 19-39.

9. D. Gabbay, "Declarative Past and Imperative Future: Executable Temporal Logic for Interactive Systems", in A. Galton, editor, In B. Banieqbal, H. Barringer, and A. Pnueli, editors, *Proceedings of Colloquium on Temporal Logic in Specification*, Altrincham, 1987, pages 402-450. Springer-Verlag, LNCS Volume 398, 1989.

10. H. Barringer, M. Fisher, D. Gabbay, G. Gough, R. Owens, "MetateM: A Framework for Programming in Temporal Logic".

11. B. Moszkowski. *Executing Temporal Logic Programs*. Cambridge University Press, Cambridge. 1987.

12. D. Pilaud, N. Halbwachs, "From a synchronous declarative language to a temporal logic dealing with multiform time", *Proc. Symposium on Formal Techniques in Real Time and Fault Tolerant Systems*, Warwick, Sept 88.

13. R. Dannenberg, "Arctic: Functional Programming for Real-Time Systems", *Proceedings of the Nineteenth Annual Hawaii International Conference on System Sciences*, 1986.

14. O. Lichtenstein, A. Pnueli, L. Zuck, "The Glory of the Past", *Proc. Conf. on Logics of Programs*, Springer-Verlag LNCS #193, 1985, pp. 196-218.

15. E.J.Cameron, D.M.Cohen, B.Gopinath, L.Ness, W.M.Keese, P.Uppaluru, J.R.Vollaro, "The IC* Model of Parallel Computation and Programming Environment," *IEEE Transactions on Software Engineering*, Vol. 14, No 3, March 1988, pp. 317-327.

16. E.J. Cameron, D.M. Cohen, B. Gopinath, L. Ness, "IC*: An Environment for Designing Communications Software", *Proceedings of SETSS '90 7th Int'l Conference on Software Engineering for Telecommunication Switch Systems*, Bournemouth, England, July 3-6, 1989.

17. E.J. Cameron, L. Ness, A. Sheth, "A Universal Executor for Flexible Transactions Which Permits Maximal Parallelism".

DIMACS Series in Discrete Mathematics
and Theoretical Computer Science
Volume 3, 1991

AUTOMATED RTL VERIFICATION BASED ON PREDICATE CALCULUS

M. Langevin

Département d'Informatique et Recherche Opérationnelle
Université de Montréal, C.P. 6128, Succ. A
Montréal, Québec, CANADA, H3C 3J7
e-mail: langevin@iro.umontreal.ca
Tel: (514) 343-6111 (#3545)

Abstract

This paper presents a technique for formally verifying synchronous circuits modelled at the Register Transfer Level of abstraction. The circuit's behavior, specification, and hypotheses on its clock sequencing are modelled using a special kind of predicate calculus formulas, named transfer formulas. The proof process consists in applying two general rules of the predicate calculus in a specific order. The automated verification process uses an acyclic graph for representing each transfer formula; in this way, the application of the rules is similar to the manipulation of boolean functions represented by acyclic graphs.

1 Introduction

An error-free design can be assured using good test vectors for simulation, formal verification tools, or silicon compilation. Automated synthesis is the easiest way to design a circuit; unfortunately, no existing tool can handle a wide range of circuit designs. For example, the VTI CAD system can be used to generate the layout of specific components like ROMs, RAMs, PLAs, State Machine, and DataPath, from a high level description [16]; but the compilation of a complete synchronous circuit (datapath and control parts) is not yet available. The validation of these circuits has to be performed with simulation or formal verification tools.

Since the complete simulation of a large circuit may be impossible, the designer needs other tools to ascertain the validity of his design; formal verification has been introduced to fill this gap. Much progress has been made in formal verification until now [7]; for example, BULL's PRIAM system has proven the functional correctness of circuits of up to 20000 transistors [11], while the formal verification of a high level description of the RSRE's VIPER processor has been carried out with the HOL system [8].

This paper presents a formal model to prove that the specification of a synchronous circuit is realized by the interconnexion of its components; the circuit specification and structural description are model at the RT Level, but memory component can be latches (asynchronous parts), i.e. the circuit description is more detailed than the circuits with registers only for memory components. Some formalisms have been proposed to perform automated RTL verification [4,15], but these tools do not handle the presence of latches in the design. The systems HOL [6] and VERIFY [3], are accurate formalisms to model the specification and component behavior of the proposed kind of circuit; however, the proofs of correctness in these systems have to be completely or partially human directed, because of the high flexibility of their underlying formalisms. An important feature of a formal verification system is its automation because the circuit designer is not interest to manage the proof; this goal is also pursued in [17,18], where the Boyer-Moore theorem prover is used to verify synchronous design.

The proposed method is based on a subset of predicate calculus for modelling circuit behavior. The proof of correctness of a circuit consists in always applying certain logical rules in a specific order, which means that the process can be automated. Direct acyclic graphs (DAGs) are used to represent logical formulas since the manipulation of these formulas is reduces to the graph-based manipulation of boolean functions [5].

2 Level of modelling

The proposed formalism is intended for modelling the behavior of synchronous circuit driven with a multiphase clock, where memory elements could be latches or registers [1]. Carriers are used to represent the nodes of the circuit [2]: there are clock carriers, input carriers, output carriers, and internal carriers for a particular circuit. A carrier represents a bit vector with a certain length.

The circuit behavior is modelled at Register Transfer Level [2]; at this level, the combinational parts are instantaneous. The time granularity used to model the behavior of the circuit components corresponds to the clock phases; a time unit thus corresponds to the interval between the middle point of two adjacent active clock phases. For example, Figure 1 presents the meaning of the time for a circuit which operates in a 3-phase clock mode. This time representation is called the phase-time representation. The clock frequency and sequencing is assumed to assure correct timing.

The specification of each circuit's component is a set of transfer functions; a transfer function specifies the possible value that can take a component's output port (an internal or output carrier of the circuit). The components are supposed to be already verified (bottom-up verification), or they will be verified later (top-down verification). Their verification could be performed with other tools of the CAD system: simulators or formal verification tools; in the ideal case, a silicon compiler could be used to generate the components' layout from their specification. A transfer function is defined for any time t, i.e. for any clock phase. A carrier can have memory or not [5]. If a carrier is memoryless, its value at time t+1 depends only on the value of other carriers at time t+1; whereas the value of a memory carrier at time t+1 also depends on the value of carriers at time t (at the previous clock phase).

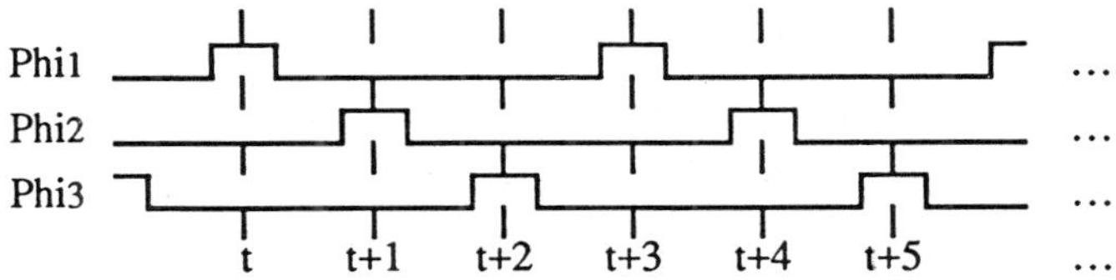

Figure 1

The specification of a circuit consists in providing transfer functions that specified the possible values for each output carrier and for certain internal carriers; these carriers are the observable carriers of the circuit. The time representation used for these transfer functions corresponds to the clock cycle; this is the cycle-time representation. The circuit's specification is a structural and a temporal abstraction of its behavior [13]. In the specification, the value of an observable carrier V at the end of cycle t+1 (at the last phase of this cycle) must depend only on the values of the input carriers during the cycle and the values of the observable carriers at the end of the previous cycle (the cycle t). These transfer functions must be defined for any clock cycle t. To be able to perform the verification of the circuit, the correspondence between the cycle-time representation and the phase-time representation has to be established. Consider the case of a d-phase clock: the nth phase of cycle $t + 1$ $(1 \leq n \leq d)$ occurs at phase-time $d.t + n$, i.e. the end of each cycle occurs at a phase-time which is a multiple of d. Hypotheses on the value of clock carriers have to be made when the circuit behavior is specified. A circuit description is the set of its transfer functions which represent the circuit's behavior, the specification, and the hypotheses on the clock sequencing.

3 An example

Let the behavior of a k-bit latch with input D, output Q, and control line 'load' be modelled. The carriers D and Q are described by time function returning bit vectors of length k, while the 'load' carrier is described by a function returning a one bit value. The latch behavior can be modelled with the following first-order formula:

$$\text{load}(t+1) * \text{EQU}(Q(t+1),D(t+1)) + \overline{\text{load}(t+1)} * \text{EQU}(Q(t+1),Q(t))$$

where '*' stands for logical 'and', '+' for logical 'or', 'EQU' is the equality predicate, and the variable t represents a clock phase (t is considered to be universally quantified). This formula, called transfer formula, specifies the possible value that can take Q at any phase t+1; if load is true at this phase then the latch is transparent, otherwise the latch keeps its last value.

Two of this latches and one multiplexor are interconnected to form the circuit of Figure 2; after a clock cycle, the value of B is a new value iff 'we' (Write Enable) is high during the second phase. The formulas H1, H2, H3 and H4 constrain the possible values for the clock carriers while the circuit specification is the formula S. To prove that the structure of the circuit meets its specification, it must be shown that the formula S is a logical consequence of the formulas H1, H2, H3, H4, and the formulas F1, F2 and F3, extracted from the circuit structure; F1 and F2 are latch specification while F3 is the multiplexor specification.

$$\mathrm{we}(2t+2) * \mathrm{EQU}(B(2t+2),A(2t+2)) + \overline{\mathrm{we}(2t+2)} * \mathrm{EQU}(B(2t+2),B(2t)) \qquad (S)$$
$$\mathrm{EQU}(\mathrm{phi}_1(2t+1),T) \qquad (H_1)$$
$$\mathrm{EQU}(\mathrm{phi}_2(2t+1),F) \qquad (H_2)$$
$$\mathrm{EQU}(\mathrm{phi}_1(2t+2),F) \qquad (H_3)$$
$$\mathrm{EQU}(\mathrm{phi}_2(2t+2),T) \qquad (H_4)$$
$$\mathrm{phi}_1(t+1) * \mathrm{EQU}(X(t+1),B(t+1)) + \overline{\mathrm{phi}_1(t+1)} * \mathrm{EQU}(X(t+1),X(t)) \qquad (F1)$$
$$\mathrm{phi}_2(t+1) * \mathrm{EQU}(B(t+1),Y(t+1)) + \overline{\mathrm{phi}_2(t+1)} * \mathrm{EQU}(B(t+1),B(t)) \qquad (F2)$$
$$\mathrm{we}(t) * \mathrm{EQU}(Y(t),A(t)) + \overline{\mathrm{we}(t)} * \mathrm{EQU}(Y(t),X(t)) \qquad (F3)$$

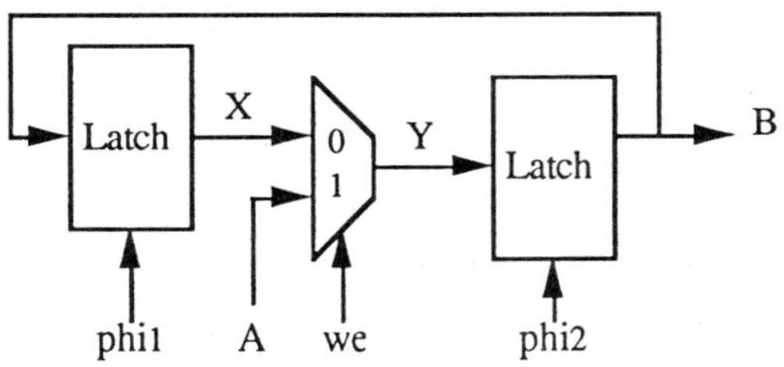

Figure 2

The following proof demonstrates that the circuit's behavior meets its specification. It is straithforward for this small example, but could be tedious for larger circuits (note that this proof holds for any width of the datapath). An algorithm to perform automatically this kind of proof is proposed in this paper.

1. $\mathrm{phi}_2(2t+2) * \mathrm{EQU}(B(2t+2),Y(2t+2)) + \overline{\mathrm{phi}_2(2t+2)} * \mathrm{EQU}(B(2t+2),B(2t+1))$ (F2,[t | 2t+1])
2. $\mathrm{EQU}(B(2t+2),Y(2t+2))$ (1,H4)
3. $\mathrm{we}(2t+2) * \mathrm{EQU}(Y(2t+2),A(2t+2)) + \overline{\mathrm{we}(2t+2)} * \mathrm{EQU}(Y(2t+2),X(2t+2))$ (F3,[t | 2t+2])
4. $\mathrm{we}(2t+2) * \mathrm{EQU}(B(2t+2),A(2t+2)) + \overline{\mathrm{we}(2t+2)} * \mathrm{EQU}(B(2t+2),X(2t+2))$ (2,3)
5. $\mathrm{phi}_1(2t+2) * \mathrm{EQU}(X(2t+2),B(2t+2)) + \overline{\mathrm{phi}_1(2t+2)} * \mathrm{EQU}(X(2t+2),X(2t+1))$ (F1,[t | 2t+1])
6. $\mathrm{EQU}(X(2t+2),X(2t+1))$ (5,H3)
7. $\mathrm{we}(2t+2) * \mathrm{EQU}(B(2t+2),A(2t+2)) + \overline{\mathrm{we}(2t+2)} * \mathrm{EQU}(B(2t+2),X(2t+1))$ (4,6)
8. $\mathrm{phi}_1(2t+1) * \mathrm{EQU}(X(2t+1),B(2t+1)) + \overline{\mathrm{phi}_1(2t+1)} * \mathrm{EQU}(X(2t+1),X(2t))$ (F1,[t | 2t)
9. $\mathrm{EQU}(X(2t+1),B(2t+1))$ (8,H1)
10. $\mathrm{we}(2t+2) * \mathrm{EQU}(B(2t+2),A(2t+2)) + \overline{\mathrm{we}(2t+2)} * \mathrm{EQU}(B(2t+2),B(2t+1))$ (7,9)
11. $\mathrm{phi}_2(2t+1) * \mathrm{EQU}(B(2t+1),Y(2t+1)) + \overline{\mathrm{phi}_2(2t+1)} * \mathrm{EQU}(B(2t+1),B(2t))$ (F2,[t | 2t])
12. $\mathrm{EQU}(B(2t+1),B(2t))$ (11,H2)
13. $\mathrm{we}(2t+2) * \mathrm{EQU}(B(2t+2),A(2t+2)) + \overline{\mathrm{we}(2t+2)} * \mathrm{EQU}(B(2t+2),B(2t))$ (10,12)

4 Transfer formulas

In predicate calculus it is possible to use functions that return bit vectors of fixed length. The value taken by a carrier of the circuit at a given time can be modelled as a time function that returns bit vectors. This kind of function is called a signal, as in [10].

Functions can be used also to model high level operators similar to the ones used in RT level HDL's [2]. An operator is modelled as a function that returns a bit vector, and takes as parameters a (possibly empty) list of bit vectors. The operators have only a syntactic meaning, like in [15], i.e. the function of an operator used in the circuit specification has to be realized by a component of the circuit. Hierarchical verification is favored because the functionality of the operator is verified when the corresponding component is verified.

A signal or an operator that returns a bit vector of length 1 can be considered a predicate; it is true if the returned bit is high, and false if the returned bit is low. Special kind of formulas, called transfer formulas, are used to model the transfer functions. The syntax used for the description of a given circuit is similar to the predicate calculus syntax, and corresponds to a subset of first-order predicate calculus (with equality) [12]. The syntax uses temporal expression with the form (c.t+n) to represent a phase-time; the constant c and n are non-negative integers while the variable t can represent a clock phase or a clock cycle. Also, the terms have the type Quad_n defined as $\{0,1,\Phi,\Delta\}^n$; this type represents a vector of n values defined as 0, 1, undefined or high impedance. Here is the syntax:

A) Signals:

A function V: (time) $\rightarrow$ $Quad_n$, is a signal. Each circuit carrier is represented by a signal.

B) Operators:

1) The function M^k: ($Quad_{n_1} \times \ldots \times Quad_{n_k}$) $\rightarrow$ $Quad_n$, is associated with each operator M of the circuit,
2) Special operators:
   i) The constant T: () $\rightarrow$ $Quad_1$ models the truth value "true", and the bit high;
   ii) The constant F: () $\rightarrow$ $Quad_1$ models the truth value "false", and the bit low;
   iii) The constant UD_n: () $\rightarrow$ $Quad_n$ models a vector of n undefined values;
   iv) The constant HZ_n: () $\rightarrow$ $Quad_n$ models a vector of n high impedance values;
   v) The operator BUS_n: ($Quad_n \times Quad_n$) $\rightarrow$ $Quad_n$ models the connection of two sources (defined like the primitive JOIN in [3]).

C) Terms are defined recursively as:

1) If I is a temporal expression and S is a signal then S(I) is a term,
2) If $V_1,\ldots,V_n$ are terms and M^n an operator, then $M^n(V_1,\ldots,V_n)$ is a term (the type of V_i has to be the same as the specified type of the ith parameter).

D) Predicates:

A term P of type $Quad_1$ is considered a predicate, except for the terms HZ_1 and UD_1.

E) Conditions (propositional formulas):

1) A predicate is a condition,
2) If G and H are conditions then so are
   i) $\overline{G}$ (logical not)
   ii) G * H (logical and)
   iii) G + H (logical or).

F) Transfers formulas (TFs) for the term U are defined recursively as:

1) If V is a term then EQU(U,V) is a TF for the term U (this special predicate is also called an atomic TF). It means that the term U takes on the value of the term V; U is the destination while V is the source (the types of U and V have to be the same).
2) If P is a condition, and F_1 and F_2 are TF for the term U, then $P * F_1 + \overline{P} * F_2$ (IF P THEN F_1 ELSE F_2) is a TF for the term U.

In the following, we will assume that the type restrictions are respected in all TFs. The only variable that could appear in a TF is the time variable t; it is considered to be universally quantified. Therefore, a TF for the term U defines the values that U can take for all t. This is the general form of the TF:

$$cond_1 * EQU(U,U_1) + cond_2 * EQU(U,U_2) + \ldots + cond_n * EQU(U,U_n)$$

where the $cond_i$ are composed from a set $P_1,\ldots,P_m$ of predicates. This formula specifies that term U takes the value of the term U_i if $cond_i$ is satisfied; U can take one and only one value because all the $cond_i$ are mutually disjoint and their union is a tautology. These formulas can be represented with direct acyclic graphs (DAGs) if the predicates are matched to the nodes of the graph and the atomic TFs EQ(X,Y) are matched to the leaves (Figure 3). The difference of TFs with boolean functions is that more than two values can be transferred in the TFs. A boolean TF is a TF where each Ui is T or F; its representation is the same as for a boolean function. The reduction algorithm for the DAGs that represent boolean functions proposed in [5] can be applied in the same way as for the DAGs that represent TFs. Moreover, as in the case of boolean functions, there exists a canonical form for these DAGs.

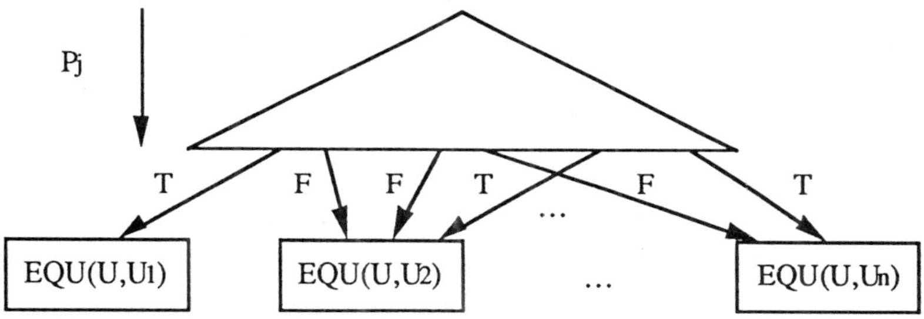

Figure 3

For a given circuit operating in d-phase mode, the description is modelled with specific TFs. For the specification, a TF is defined for each observable carrier at time dt+d (the end of the cycle); its value depends on the value of input carriers at a time between dt+d and dt, and on the value of the observable carriers at the time dt. For the hypotheses on the clock sequencing, a TF is defined for each clock carrier at time dt+n, $1 \leq n \leq d$; its value is either true or false depending on whether or not the clock carrier is active at the nth phase. For the behavior, a TF is defined for each internal and output carrier; if the carrier is a memory carrier, its value at time t+1 depends on the value of the other carriers at time t+1 and t, otherwise, its value at time t depends only on the value of the other carriers at time t.

Example of a circuit description: The circuit NextPC (Figure 4), operating with a 2-phase clock, computes the future value of the PC. At the end of a cycle if the flags' value satisfies the jump condition (JC) then the PC takes on the value of the jump address (JA); otherwise the last value of the PC (the one at the end of the previous cycle) is incremented by 1. The increment operation is modelled with the operator INC while the test condition is modelled with the operator SATISFY; it is assumed that these operators have a verified implementation.

The specification of NextPC is modelled with the transfer formula S while the formulas F1, F2, F3, F4 and F5, extracted from the circuit component, modelled the circuit behavior. The hypothesis on the clock sequencing are the same as for the circuit of Figure 2.

$$\begin{aligned} &\mathrm{SATISFY}(\mathrm{JC}(2t+2),\mathrm{Flag}(2t+2)) * \mathrm{EQU}(\mathrm{PC}(2t+2),\mathrm{JA}(2t+2)) + \\ &\overline{\mathrm{SATISFY}(\mathrm{JC}(2t+2),\mathrm{Flag}(2t+2))} * \mathrm{EQU}(\mathrm{PC}(2t+2),\mathrm{INC}(\mathrm{PC}(2t))) \end{aligned} \quad (S)$$

$$phi_2(t+1) * \mathrm{EQU}(\mathrm{PC}(t+1),\mathrm{A}(t+1)) + \overline{phi_2(t+1)} * \mathrm{EQU}(\mathrm{PC}(t+1),\mathrm{PC}(t)) \quad (F_1)$$
$$phi_1(t+1) * \mathrm{EQU}(\mathrm{X}(t+1),\mathrm{PC}(t+1)) + \overline{phi_1(t+1)} * \mathrm{EQU}(\mathrm{X}(t+1),\mathrm{X}(t)) \quad (F_2)$$
$$jp(t) * \mathrm{EQU}(\mathrm{A}(t),\mathrm{JA}(t)) + \overline{jp(t)} * \mathrm{EQU}(\mathrm{A}(t),\mathrm{Y}(t)) \quad (F_3)$$
$$\mathrm{EQU}(jp(t),\mathrm{SATISFY}(\mathrm{JC}(t),\mathrm{Flag}(t))) \quad (F_4)$$
$$\mathrm{EQU}(\mathrm{Y}(t),\mathrm{INC}(\mathrm{X}(t))) \quad (F_5)$$

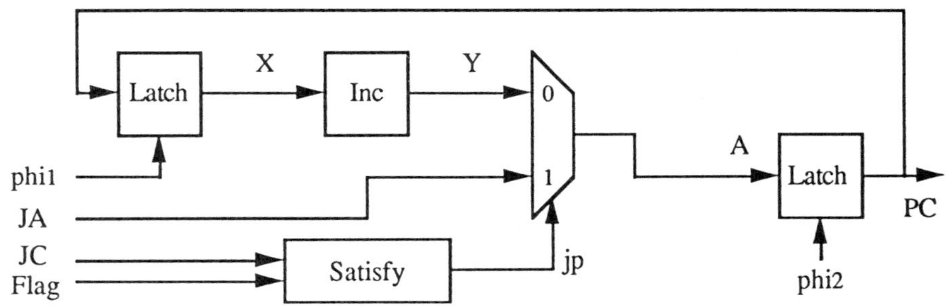

Figure 4

5 Behavioral verification

With the proposed formalism, the proof of correctness of a circuit consists in applying some rules of predicate calculus with the goal of deriving the transfer formulas of the specification from the transfer formulas of the circuit behavior. Let $F_1,\ldots,F_n$ be the TFs of the circuit behavior,

$H_1,\ldots,H_m$ be the hypotheses on clock sequencing, and $S_1,\ldots,S_p$ be the TFs of the circuit specification. The goal is to prove that each of the formulas S_i is a theorem when the formulas $F_1,\ldots,F_n,H_1,\ldots,H_m$ are taken as assumptions. A transfer formula S_i specifies the values that an observable carrier V at time dt+d can take as a function of the values of the input carrier between the times dt and dt+d, and the values of the observable carriers at the time dt (the constant d is the number of phases). This kind of TF is called an observable TF, and is defined as:

Def 1: A term U is observable iff one of the following conditions is true:

i) U has the form S(I) where S is an input signal and I a temporal expression,
ii) U has the form S(dt) where S is an observable signal,
iii) U has the form $M(A_1,\ldots,A_n)$ where M is an n-ary operator and each A_i is observable.

Def 2: A condition C is observable iff each predicate that occurs in C is observable.

Def 3: An atomic transfer formula EQU(U,V) is observable iff the term V is observable.

Def 4: A TF is observable iff each of its conditions and atomic TFs is observable.

An observable TF for a term U specifies the values that U can take as a function of only the values of the observable carrier at time dt and the values of the input carriers at some arbitrary time; the TF S_i is an observable TF for the term V(dt+d). Another TF for V(dt+d) can be extracted from the circuit's behavior since the variable t of the TF for the carrier V can be replaced by any temporal expression. Now, suppose that this TF is transformed in an observable form using logical rules; if this observable formula is identical with S_i then S_i is deduced from the circuit behavior. This is the verification technique proposed in this paper. Two specific rules are used to transform a TF in an observable form:

1) The substitution rule:

Suppose that we have the two following TFs:

$$cond_1 * EQU(U,U_1) + cond_2 * EQU(U,U_2) + \ldots + cond_n * EQU(U,U_n) \quad (1.1)$$

$$G * EQU(P,T) + \overline{G} * EQU(P,F) \quad (1.2)$$

The substitution of 1.2 in 1.1 consists in replacing all the occurrences of P in the conditions of the TF 1.1 by the condition G of the TF 1.2.

2) The transition rule:

Suppose that we have the two following TFs:

$$C_{1,1} * EQU(U,U_1) + \dots + C_{1,n} * EQU(U,U_n) \quad (2.1)$$

$$C_{2,1} * EQU(V,V_1) + \dots + C_{2,m} * EQU(V,V_m) \quad (2.2)$$

The transition of 2.1 with 2.2 consists in replacing all the occurrences of V in the terms U_i of 2.1 by its corresponding value specified in 2.2. First, the logical 'and' is applied between the two TFs and the following formula is obtained:

$$C_{2,1} * C_{1,1} * EQU(U,U_1) * EQU(V,V_1) + \dots + C_{2,1} * C_{1,n} * EQU(U,U_n) * EQU(V,V_1)$$

$$\dots$$

$$C_{2,m} * C_{1,1} * EQU(U,U_1) * EQU(V,V_m) + \dots + C_{2,m} * C_{1,n} * EQU(U,U_n) * EQU(V,V_m)$$

The transitive rule of equivalence is applied to each product of atomic TFs. The rule used is:

$EQU(G,H) * EQU(X,Y) \Rightarrow EQU(G,H')$ where H' is the term H in which the occurrences of X are replaced by Y.

The algorithm used to deduce the observable TF for V(ct+n) consists in applying these two rules in a specific order; this algorithm is presented in Figure 5. The initial TF is taken from the list of TFs of the circuit behavior (if the signal represents an output or an internal carrier), or from the list of TFs of the hypotheses on the clock sequencing (if the signal represents a clock carrier). Each non-observable predicate P in the conditions of this TF are replaced by its value specified by the observable TF for P; this transformation is performed with the application of the substitution rule. After, each non-observable source term X of the atomic TF is replaced by its value specified by the observable TF for X; this transformation is performed with the application of the transition rule. The resulting formula is the observable TF for V(ct+n).

The application of the two deduction rules is similar to boolean function graph manipulation when the TFs are represented with a DAG. The substitution rule consists in replacing a predicate P of a TF E by the value specified in the TF D for P. Since the TF D is in the boolean form, this rule can be applied in the same way as the composition of boolean function. This algorithm generates a graph by applying the IF...THEN...ELSE (ITE) function between three graphs: the graph of D, the graph of E where P is restricted to the value true, and the graph of E where the value of P is restricted to false. When the algorithm reaches the leaves of these graphs, the function ITE is applied on three leaves: a leaf that represents a truth value for P, a leaf of E that represents an atomic TF when P is true, and a leaf of E that represents an atomic TF when P is false. Like for boolean functions, the result leaf depends only on the value of the leaf of P.

The transition of a TF D for U with a TF E for V can be performed in the similar way as the application of the logical 'and'. When this function is applied between two boolean functions, the logical 'and' is applied only to the pair of leafs' value. To perform the transition rule, instead of applying the logical 'and', the equality rule is applied to the two leaves EQU(U,U_i) and EQU(V,V_j); the occurrences of the term V in the term U_i are replaced by the term V_j. Unlike the application of the logical 'and', the equality rule is applied in a strict order.

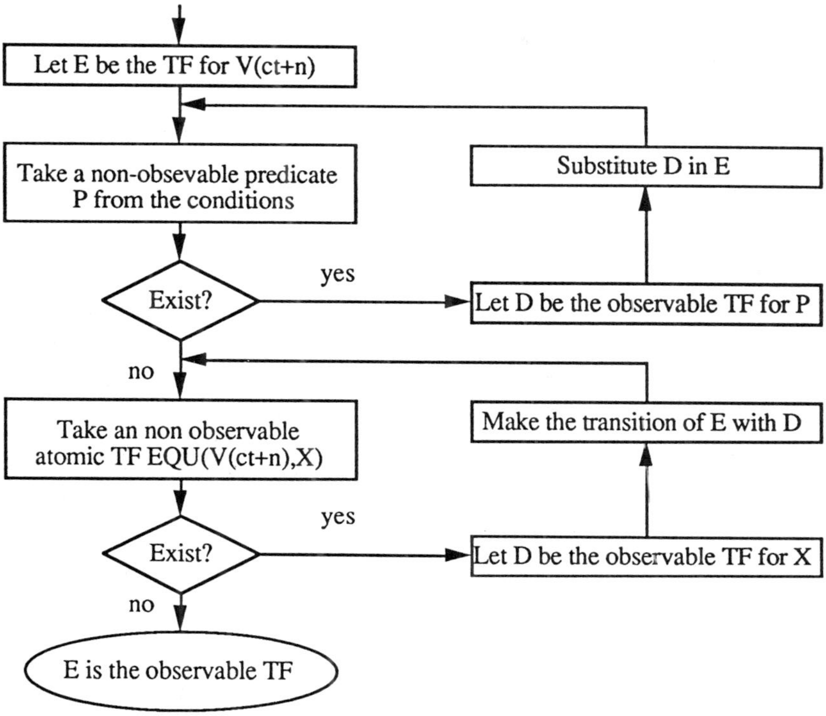

Figure 5

Closed loops in the circuit can be detected when the circuit is verified. During the recursive application of the algorithm to deduce the initial TF for U, if the term U is met, a closed loop in the circuit is detected since the value of the term U depends on the value of the term U [1].

With this algorithm, the way to prove that the specification TF for the term V(dt+d) is a theorem consists in deducing the observable TF for the term V(dt+d) from the implementation and in showing that it is equivalent to the specified TF. This comparison is carried out by comparing the two DAGs as in [5]. Also, the inclusion of formulas can be used to show the correctness of an incomplete specification.

6 Experimental Results

A prototype of the proposed verification technique is under development in Common Lisp. The example of section 4 was successfully verified in about 30 seconds on a Mac Plus. This tools was used in the WORP project [9] at the EPFL, Switzerland; WORP (Watch Oriented RISC Processor) is a RISC processor because it has only one level of programming and each instruction is executed in 4 phases [14]. Its general structure is of the type Harvard, i.e. data and code reside in different memories. It has a 2K of 48-bit instructions saved in a ROM, and 256 of 8-bit words of RAM that can be accessed in a relative or absolute mode. It supports 7 interrupt levels, and can address up to 3 external devices.

The circuit is built from the interconnexion of a ROM, a latches bank, an ALU, some working registers, adders, multiplexors, buses, etc. Only a partial verification of this processor has been performed up to now. We specify the possible values that can be stored in the Index Latch of the processor after the execution of any instruction; these values depend on certain fields of the instruction (the ALU code operation, the two source registers for the ALU operation, etc.). A closed loop in the circuit's behavior was detected during this verification. Once this error corrected, the specification has been successfully verified; a trace of roughly 10 pages is produced during the verification and it takes about 7 minutes of real time on a Mac II.

7 Future work

Some improvements to our proposed verification system are planned. First of all, an RT level HDL language could be used to describe the circuit specification, component and hypotheses; the transfer formulas should be deduced directly from the syntax of the HDL (the partial specification of the WORP processor needs 200 lines of edit text and no syntactical analysis is performed). The type validation will be performed directly from the HDL circuit description. Secondly, like the operator BUS, it seems possible to use other predefined operators that have a specific associated meaning. This will be essential in the verification of microcode such as [10]; this requires to associate rewriting rules with the operators, as in [4]. Also to extend the range of verifiable circuits, the pipeline behavior and precharged component will be studied. The study of several circuit examples would help to establish the limitations of our approach.

We try to improve the graph algorithm processing performance by using techniques such as proposed in [11]. Last but not least, the system has to be integrated in a complete CAD system in order to assure the consistency of component models with their real behavior.

Acknowledgements

We want to thank Professor Jacques Zahnd from the EPFL for his help in the formalization of the model, and for his remarks on this research in general. Part of this work was done by the author at the EPFL, Switzerland.

References

[1] F. Anceau, "The Architecture of Microprocessors", Addison-Wesley, 1986.
[2] M. R. Barbacci, "A Comparison of Register Transfer Languages for Describing Computers and Digital Systems", IEEE Trans. on Comp., Vol. C-24, No. 2, February 1975.
[3] H. G. Barrow, "Verify: A Program for Proving Correctness of Digital Hardware Designs", Artificial Intelligence, Vol. 24, 1984.
[4] A. Bartsch, H. Eveking, H.-J. Faerber, M. Kelelatchew, J. Pinder, U. Schellin, "LOVERT - A Logic Verifier of Register-Transfer Level Description", Proceedings of the IMEC IFIP International Workshop on "Applied Formal Methods for Correct VLSI Design", November 1989.
[5] R. E. Bryant, "Graph-Based Algorithms for Boolean Function Manipulation", IEEE Trans. on Comp., Vol. C-35, No. 8, August 1986.
[6] A. Camilleri, M. Gordon, T. Melham, "Hardware Verification Using Higher Order Logic", in From HDL Descriptions to Guaranteed Correct Circuit Designs, ed. D. Borrione, North-Holland, 1987.
[7] P. Camurati, P. Prinetto, "Formal Verification of Hardware Correctness: Introduction and Survey of Current Research", Computer, July 1988.
[8] A. Cohn, "A Proof of Correctness of the VIPER Microprocessor: the First Level", in VLSI Specification, Verification, and Synthesis, eds. G. Birtwistle and P. A. Subrahmanyam, Kluwer Academic Publishers, 1988.
[9] C. Iseli, "WORP: A Watch Oriented RISC Processor", B. Sc. Thesis, Laboratoire de Systèmes Logiques, EPFL, December 1987.
[10] J. Joyce, G. Birtwistle, M. Gordon, "Proving a Computer Correct in Higher Order Logic", Technical Report No. 100, Computer Laboratory, University of Cambridge, December 1986.
[11] J. C. Madre, F. Anceau, M. Currat, J. P. Billon, "Formal Verification in an Industrial Design Environment", to appear in 'The international Journal on Computer Aided VLSI Design'.
[12] Z. Manna, "Mathematical Theory of Computation", McGraw-Hill, 1974.
[13] T. F. Melham, "Abstraction Mechanisms for Hardware Verification", in VLSI Specification, Verification, and Synthesis, eds. G. Birtwistle and P. A. Subrahmanyam, Kluwer Academic Publishers, 1988.
[14] W. Stallings, "Tutorial: Reduced Instruction Set Computers", IEEE Computer Society Press, Order Number 713, 1986.
[15] R. Vemuri, "A Formal Model for Register Transfer Level Structures and Its Applications in Verification and Synthesis", Proceedings of the IMEC IFIP International Workshop on "Applied Formal Methods for Correct VLSI Design", November 1989.
[16] "VTI User's Guide", VLSI Technology, Inc, 1988.
[17] A. Bronstein, C. L. Talcott, "Formal Verification of Pipelines Based on String-Functional Semantics", Proceedings of the IMEC IFIP International Workshop on "Applied Formal Methods for Correct VLSI Design", November 1989.
[18] L. Pierre, "The Formal Proof of Sequential Circuits Described in CASCADE Using the Boyer-Moore Theorem Prover", Proceedings of the IMEC IFIP International Workshop on "Applied Formal Methods for Correct VLSI Design", November 1989.

DIMACS Series in Discrete Mathematics
and Theoretical Computer Science
Volume 3, 1991

The algebraic feedback product of automata. A state machine based model of concurrent systems.

VICTOR YODAIKEN

ABSTRACT. Finite automata provide faithful mathematical models of simple digital systems, but seem inadequate for such multi-level complex systems as complex circuits, networks, and operating systems. We define an algebraic *feedback product* over automata that allows for modular composition of automata without requiring conventions about scheduling or inter-component communication. By keeping automata in product form, we also damp the "state explosion" problem: we never need to construct the product of the state sets of the components. In order to allow for symbolic definition and analysis of product form automata, we develop a *modal primitive recursive arithmetic*. The arithmetic facilitates compact, symbolic definition of automata and automata products, and allows for high-level reasoning about these objects in a style similar to that of the temporal logics [**EH83, MP79**]. We claim that the multi-level, finite state semantics provided by product form automata has several mathematical and intuitive advantages over standard computational models.

1. Introduction

Finite state automata provide a clear and faithful mathematical representation of digital devices and programs. Since large-scale digital systems are generally constructed by interconnecting less complex components, the problem of representing such systems in terms of interconnected automata has received a great deal of attention in the computer science literature, e.g., [**LM86, GC86, COGB86, Har84, Kur85, Mil79**]. In this paper we describe an algebraic automata product, called the *feedback product* which can be employed to emulate the intricate control and component interconnection of a wide variety of digital systems[1]. A classical Moore machine or finite state transducer [**Moo64, HU79**]

This research was funded in part by the Office of Naval Research under contract N00014-85-K-0398.

[1] The feedback product used here is a variation of the general product described in [**Gec86**].

provides a mathematical model of a discrete finite state mechanism with output. A feedback product form transducer provides a mathematical model of an interconnected system of finite state mechanisms. A product form transducer contains some number of *component* transducers representing sub-systems. Each component transducer is associated with a *feedback function*. When input is provided to the product form machine, each feedback function generates an input sequence for its associated transducer. The generated sequence depends on the input provided to the product machine, the global system state, and the outputs of some or all of the other components. Thus, each system transition corresponds to parallel computations in each of the components.

- Generated sequences can be of differing lengths (or empty), so components may change state at varying rates.
- A feedback product form transducer can be "multiplied-out" to obtain a classical transducer, we remain within the domain of finite state machine theory.
- Since components may, themselves, be in product form, we can model multi-layer systems.
- There is no need to make the alphabets of components and the product transducer overlapping or disjoint — the feedback functions, rather than the transition labels, define interactions.

The feedback product provides several advantages over inter-connection techniques previously described in the computer science literature. Among these advantages are: a direct (non-interleaved) model of concurrency, a model for systems containing components that change state at differing rates, and a natural model of encapsulation and information hiding. The feedback product also provides a formal framework for describing concurrent systems *without* making any assumptions about underlying computation environment: e.g., scheduling, storage, or communication mechanisms. We are interested in specification of operating systems, circuits, and real-time controllers. These systems are not implemented within environments that provide uniform, fixed communication primitives, and these systems may contain components which have radically different granularities of state change. We are hesitant to accept, for example, the paradigm of a concurrent algol-like programming language as fundamental. And we are dubious about the prospects of verifying the behavior of systems which implement scheduling, storage and communication mechanisms in terms of a formal framework which assumes the previous existence and character of these mechanisms.

Direct analysis and definition of product form automata would be rather awkward. In preference, we define a modal formal language based on the primitive recursive functions [**Pet67, Goo57**]. The modal primitive recursive (m.p.r.) functions are evaluated in the implicit context of a product form transducer and *trace* (sequence of transition symbols). The trace is intended to represent the sequence of state transitions which have driven the transducer from its initial state.

We never explicitly construct either traces or transducers. Instead, we describe these objects in terms of the values that they confer on m.p.r. functions. For example, the m.p.r. boolean function **enable** is defined so that **enable**$(a) = 1$ in the context of transducer P and trace w if and only if the trace $w :: a$ obtained by appending a to w does not drive P to an undefined state.

We say a transducer P *satisfies* a m.p.r. expression E iff E is non-zero in the context of (P, w) for every w that does not drive P to an undefined state. In other words, P satisfies E iff E is true in every reachable state of P. Every m.p.r. boolean expression is, thus, a specification of the class of transducers which satisfy it. We can show that some fairly simple extensions to the primitive recursive functions are sufficient to obtain a powerful and expressive language in which to describe product form transducers. We will show that:

- There is an expression E for every transducer P, so that P' satisfies E iff P' is strongly equivalent to P (same alphabet, same behaviors, same output). Furthermore, given E, there is a deterministic algorithm for constructing a transducer which satisfies E. Thus, m.p.r. expressions provide an alternative to explicit or regular language based definitions of transducers.
- There are m.p.r. functionals which are analogous in behavior to the operators of branching time [**EH83, BAPM83**] and interval [**MM83**] temporal logic. The m.p.r. functionals, however, have s semantics based on the pumping-lemma of regular languages [**HU79**] in place of a fixed-point semantics. Furthermore, the m.p.r. analogs of the temporal operators have a well defined multi-level semantics. For example, the assertion that *within board b within sub-circuit c eventually gate g is high, implies that eventually within board b′ Y will be true*, can be easily formalized in the m.p.r. arithmetic.
- The m.p.r. functions obey the proof rules, and are amenable to the proof techniques of integer mathematics — they are not segregated within a formal logic separate from the main body of mathematics.
- Detailed operation of algorithms can be defined with m.p.r. functions, without use of programming languages or other distinct *action languages*.

Related formalisms. In some respects, our work is closest to that of Clarke *et al* [**CAS83, EH83, Sis83**] and Ostroff [**OW87**], who have also used modal formalisms to describe finite automata. We are interested in verifying a similar class of systems, including circuits, architectures, and communication protocols. In some sense, our work can be seen as providing an alternative semantics for temporal logic based formalism. There certainly appears to be no barrier to the extension of model checking methods to product form automata. Our methodology, however, is based on intuitive and mathematical models that are quite different from those of temporal logic. Statecharts [**Har84**] are an extension of state diagrams to allow for description of concurrent and composed systems.

While statecharts are quite expressive, they suffer from some of the same limits as traditional state diagrams, e.g., we cannot extend the statechart of a 2 bit shift register to obtain the statechart of a 3 bit shift register. Furthermore the formal semantics of statecharts is either given in terms of a cartesian product of automata [**Har85**] or outside of automata theory altogether [**Har87**]. Alternatively, Lynch, Gouda and others [**LM86, GC86**] have proposed extending automata with some communication abstraction and embedding them in a network. Such an approach, however, requires us to abandon the mathematical simplicity of finite automata, and to accept rather complex conventions about the network operation.

Outline. The remainder of this paper is in 4 sections. Section 2 defines product form automata, the m.p.r. functions, and an interpretation. In section 3 we show the expressive range of the m.p.r. functions and develop some proof methods. Section 4 provides an application of the m.p.r. functions and semantics to the definition of temporal logic style functionals. We conclude with a brief summary.

Notational Preliminaries. Much of the body of this paper is concerned with sequences and paths. Let $\langle\rangle$ denote the empty sequence and let juxtaposition denote sequence concatenation: $\langle a_1,\dots,a_n\rangle\langle b_1,\dots,b_n\rangle = \langle a_1,..,a_n,b_1,\dots,b_n\rangle$. We define single element appending so that $a : u \stackrel{def}{=} \langle a\rangle u$ and $u :: a \stackrel{def}{=} u\langle a\rangle$. The length function can then be defined recursively as follows: $length(\langle\rangle) \stackrel{def}{=} 0, length(a : u) \stackrel{def}{=} 1 + length(u)$. We also take this opportunity to introduce our use of dot-index notation as a supplement to standard subscript notation. In some cases it will be clearer to describe an indexed set $x.1,\dots,x.n$ in preference to $x_1,\dots,x_n$.

2. Feedback products and modal functions.

2.1. Transducers. Product form transducers are classical transducers presented as a product of transducers. Each of the factors represent a component sub-system and there is additional structure representing the coordination of these sub-systems. Since the factors are also product form transducers, we can consider product form transducers as trees with leaves being transducers representing systems that are atomic (systems with no components).

We define product form transducers inductively. First we define the class $\mathbf{P}_0$ of flat product form transducers — those with no factors. The elements of this class are essentially standard Moore machines [**Moo64, HU79**]. The class $\mathbf{P}_{i+1}$ of transducers consists of those transducers containing at least one factor belonging to $\mathbf{P}_l$, and no factors of belonging to $\mathbf{P}_j$ for $j > l$. The class $\mathbf{P}$ is the union over all $\mathbf{P}_i$.

2.1.1. *Flat transducers.* It is convenient to consider all alphabets to consist of the integers $\{1,\dots,k\}$ for some k, and to consistently reserve 0 for the null transition that leaves state unchanged.

DEFINITION 2.1. *The flat product form transducers*
The class $\mathbf{P}_0$ *of flat product form transducers is the smallest set containing all tuples of the form:*

$$P = (A, S, start, O, \lambda, \delta)$$

- $A = \{1, \ldots n\}$ *is a finite transition alphabet, with 0 reserved for use as the null transition that leaves state unchanged,*
- $O = \{1, \ldots k\}$ *is a finite output alphabet,*
- $S = \{1, \ldots h\}$ *is a finite state set,*
- $start \in S$ *is a distinguished start state,*
- $\lambda : S \to O$ *is an output function,*
- $\delta : S \times A \to S$ *is a transition function, with* $\delta(s, 0) = s$.

We can define a function $\Delta : \mathbf{P}_0 \times A^* \to S$ so that $\Delta(P, w)$ is the state to which w drives P from the initial state.

DEFINITION 2.2. Δ *for flat transducers*

$$\Delta(P, \langle\rangle) \stackrel{def}{=} initial$$
$$\Delta(P, w :: a) \stackrel{def}{=} \delta(\Delta(P, w), a)$$

2.1.2. *Product form transducers.* A general product form transducer P is also defined as a tuple including a state set, input and output alphabets, and transition and output functions. But, each of these product form transducers also contains a tuple of *factor* product form transducer and *feedback functions*: $(\phi_1, P_1, \ldots, \phi_r, P_r)$. Factors are connected by making the input to each factor depend on the input to the product form automaton, the state of the product form automaton, and the outputs of all of the factors. When a product form automaton accepts a single input symbol, each factor is provided with a sequence of 0 or more input symbols, representing the parallel activity of all the components. For example, suppose we have a product form automaton representing a network of computers, and containing factor transducers representing each individual site. A single state transition representing a message transfer might induce *receive* state transitions for some of the factors, a *send* transition for one factor, and no transitions for factors representing sites not involved in the transfer.

FIGURE 2.1. *A product form transducer.*

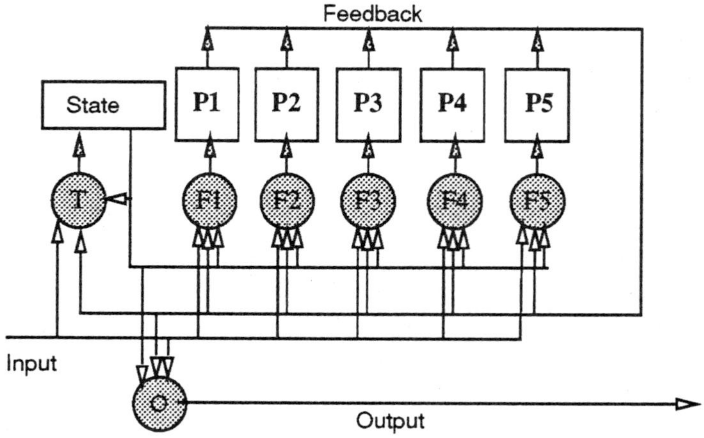

Γ=14

$P1 - P5$ are factors, T is a transition function, and $F1 - F5$ are feedback functions.

Notation. When we define a product form automaton, we need a convenient notation to distinguish elements of the product, such as alphabet and state set, from the elements of its factors. Thus, we write δ to refer to the transition function of P, and write $\delta.i$ to refer to the transition function of the ith factor of P. Similarly, if O is the output alphabet of a transducer, then $O.i$ is the output alphabet of factor P_i.

DEFINITION 2.3. *The class of product form transducers*

The class $\mathbf{P}$ of product form transducers is the infinite union of the classes $\mathbf{P}_i$ for $i \geq 0$. Each class $\mathbf{P}_{l+1}$ of product form transducers is the smallest set containing all tuples of the form:

$$P = (A, O, S, start, \lambda, \delta, (\phi_1, P_1, \dots, \phi_r, P_r))$$

- *$A = \{1, \dots n\}$ is a finite transition alphabet, with 0 reserved for use as the null transition that leaves state unchanged,*
- *$O = \{1, \dots k\}$ is a finite output alphabet,*
- *$S = \{1, \dots h\}$ is a finite state set,*
- *$start \in S$ is a distinguished start state,*
- *Each P_i $(0 < i \leq r)$ is a product form transducer, with $l = max\{l' : (\exists 0 < i \leq r) P_i \in \mathbf{P}_{l'}\}$.*
- *$\lambda : S \times O.1 \dots \times O.r \to O$,*
- *$\delta : S \times O.1 \dots \times O.r \times A \to S$ is a transition function, with $\delta(s, o_1, \dots, o_r, 0) = s$.*
- *Each $\phi_i : S \times O.1 \dots \times O.r \times A \to (A.i)^*$ is a feedback function, with $\phi_i(s, o_1, \dots, o_r, 0) = \langle\rangle$.*

Note that the transition and output functions can only use the output of the factors, and do not see the internal state of factors.

The product state set of a flat transducer P, is the state set of P. If P is not flat, then the product state set of P, denoted $PS(P)$ is given as follows:

DEFINITION 2.4. *The product state set $PS(P)$*

$$PS(P) \stackrel{def}{=} S \times \prod_{i=1}^{r} PS(P_i).$$

A *configuration* is a tuple $c = (s, o_1, \dots, o_r) \in S \times O.1 \dots \times O.r$. The start configuration is the tuple consisting of *start* and the the outputs of each factor in its own start configuration. Since the start configuration of a flat transducer is *start*, the recursion must terminate. Let $st = (start, \lambda.1(st.1), \dots, \lambda.r(st.r))$ denote the start configuration. After an a transition, we will get a new configuration $c' = (\delta(c, a), o'.1, \dots o'.r)$ where each $o'.i$ is of the configuration which P_i reaches by following $\phi_i(c, a)$ from its own current configuration.

Intuitively, we intend that when a product form automaton accepts a transition a, each factor will traverse transition sequence $\phi_i(c, a)$. We formalize the concept of current configuration by defining two functions. The first is a map Δ so that $\Delta(P, w)$ is the configuration of P after it has accepted the path w as input. The second is ϕ_i^*, the reflexive transitive closure of ϕ_i. We define Δ inductively on the classes comprising $\mathbf{P}$. For P in $\mathbf{P}_0$, we have already defined $\Delta(P, w)$. For P in $\mathbf{P}_{l+1}$ we need to also use the configurations of the factors. We assume that Δ is defined for each of the factors. Let $\mathbf{c}$ be a configuration. Then we can define ϕ_i^* as follows.

DEFINITION 2.5. *Reflexive, transitive closure of ϕ*

$$\phi_i^*(c, \langle\rangle) \stackrel{def}{=} (c)_i$$
$$\phi_i^* \ (c, a : w) \stackrel{def}{=} \phi_i(c, a)\phi_i^*(\delta(c, a), \lambda.1(\Delta(P_1, a)), \dots \lambda.r(\Delta(P_r, a)), w))$$

The empty sequence $\langle\rangle$ induces an empty sequence for the factors. Each a transition in configuration c will take the transducer to a new configuration $(\delta(c, a), \lambda.1(\Delta(P_1, a)), \dots \lambda.r(\Delta(P_r, a)), w))$. Thus, the sequence induced by $a : w$ in configuration $\mathbf{c}$ is obtained by concatenating the sequence induced by a to the sequence induced by w in the configuration resulting from a. To avoid notational clutter we can define a function $\psi(w, i)$ which gives the sequence induced for i by w from the start configuration: $\psi(w,i) \stackrel{def}{=} \phi_i^*(\mathrm{st}\ ,w)$.

DEFINITION 2.6. *Δ for product form transducers*

$$\Delta(P,\langle\rangle) \stackrel{def}{=} (start, start.1, \ldots, start.r)$$
$$\Delta\ (P, w :: a) \stackrel{def}{=}$$
$$(\delta(\Delta(P,w),a), \lambda.1(\Delta(P_1,\psi(w :: a, 1))), \ldots, \lambda.r(\Delta(P_1,\psi(w :: a, r)))$$

Note that $\Delta(P,w)$ is defined only if $\Delta(P_i, \phi_i^*(\mathrm{st},w))$ is also defined. This means that any trace of a product form transducer will not drive any of the factors into undefined states. We can now, formally, define the language of traces $\mathcal{L}(P)$

DEFINITION 2.7. *The trace language $\mathcal{L}(P)$*

$$\mathcal{L}(P) \stackrel{def}{=} \{w \in A^* : \Delta(P,w) \textit{ is defined }\}$$

2.2. Transition precedence. In standard computational formalisms, state is generally considered to be represented by an assignment function. We find assignment functions to be difficult to work with, and hard to compose. The prospect of finding the assignment function of a feedback product is unappealing. We propose *relative transition precedence* as an alternative representation of state.

Given a trace $w = \langle a_1, \ldots, a_n\rangle$ we say that a_i represents an event that preceded a_j iff $i < j$. The "most recent" event represented in w is given by a_n, the "2^{cd} most recent" is given by a_{n-1}, and the " i^{th} most recent event" is given by $a_{n-(i-1)}$. It is convenient to have an imaginary event a_0 to represent events which have not yet occurred, i.e., the kth most recent event for $k > n$. We can also consider the "i^{th} most recent b event" for some b in the alphabet to be that a_j so that $a_j = b$ and there are exactly i events $a_k = b$ with $k \geq j$. Again, we will find it convenient to let a_0 stand for the kth most recent b if there are less than k b's in the sequence. The boolean function $RelOrder(w,a,i,b,j)$, is true iff the ith most recent a transition recorded in w *preceded* the jth most recent b transition recorded in w. For example, in a site on a computer network, we might test to see if a message has been sent more recently than its acknowledgment has been received, by evaluating $RelOrder(w, receive.ack_m, 1, send.m, 1)$. In a real-time circuit, we might test to see if at least $(k+1)$ time units have passed since the logic level was most recently raised by evaluating $RelOrder(w, raise, 1, tick, k)$. In order to evaluate $RelOrder$, we can define $Place(w,a,i)$ to be 0 if there are fewer than $(i+1)$ a's in w, or if $a = 0$, and let $Place(a,i) = j$ such that $(w)_j = a$ and $\sum_{k=j}^{n}((w)_k = a) = i$, otherwise. By convention 0 denotes the null transition which leaves state unchanged, and we find it convenient to just ignore the 0's in the trace. Thus, $RelOrder(w,a,i,b,j)$ will be true of w iff $Place(w,a,i) < Place(w,b,j)$. We find the following conventions useful:

- Recall that the alphabets of state machines are sets $\{1, \ldots k\}$, and 0 is reserved for the null transition which leaves state unchanged. It is convenient, to consider all null transitions as less recent than any non-null transitions, so we let $place(w, 0, i) = 0$.
- The rightmost a transition in w is considered to be the 1st most recent a transition, so $place(w, a, 1)$ is the greatest j such that $(w)_j = a$.
- If there are fewer than i a transitions in w, then $place(w, a, i) = 0$, the null transition.

DEFINITION 2.8. *Place*
The "place" of a pair (a, i) in a trace.

$$place(\langle\rangle, a, i) \stackrel{def}{=} 0$$

$$place(u :: b, a, i) = \begin{cases} 0 & if\ a = 0 \\ & or\ if\ i = 0 \\ length(u :: b) & else\ if\ a = b \bigwedge i = 1 \\ place(u, i-1) & else\ if\ a = b \\ place(u, a, i) & otherwise. \end{cases}$$

We say that (a, i) precedes (b, j) in w iff $place(w, a, i) < place(w, b, j)$.

DEFINITION 2.9. *RelOrder*

$$RelOrder(w, a, i, b, j) \stackrel{def}{=} \begin{cases} 1 & iff\ place(w, a, i) < place(w, b, j); \\ 0 & otherwise. \end{cases}$$

Each expression $RelOrder(w, a, i, b, j)$ defines a finite indexed congruence on A^*. Define the congruence as follows.

DEFINITION 2.10. *A congruence on RelOrder*

$$u \equiv v \bmod (a, i, b, j) \Leftrightarrow (\forall x, y) RelOrder(xuz, a, i, b, j) = RelOrder(xvz, a, i, b, j)$$

DEFINITION 2.11. *The congruence classes induced by RelOrder*

$$[w]_{(a,i,b,j)} = \{u : u \equiv w \bmod (a, i, b, j)\}$$

THEOREM 2.1. *For any fixed (a, i, b, j) the set $\{[w]_{(a,i,b,j)}\}$ contains less than $\sum_{k=0}^{i+j} n^k$ distinct elements, where n is the cardinality of A.*

Proof. *Let $length(w) > i + j$. If w contains any elements that are equal to neither a or b, those elements can be discarded to obtain a shorter, congruent sequence. If w contains only a's and b's, then there are either more than i a's or more than j b's. If there are more than i a's we can discard the leftmost a to obtain a shorter, congruent sequence, and similarly for the b's. Thus any sequence of length greater than $i + j$ must be congruent to a shorter sequence.*

The Myhill-Nerode theorem [**Myh57, Ner58, HU79**] states that any language which is defined as the union of finite indexed congruence relations is regular. Thus theorem 2.1 implies that the language $L \subset A^*$ where $L = \{w : RelOrder(w, a, i, b, j) = 1\}$ is a regular language. In fact, L is a particularly simple type of regular language, an aperiodic language. A language is aperiodic iff there is some t so that $wa^{t+j}v$ is in the language iff $wa^t v$ is in the language [**Pin86**]. This condition, obviously, holds in L.

2.3. The Modal Primitive Recursive Functions. The class MPR of modal primitive recursive functions includes a small set of initial functions, and all those functions definable from the initial functions with a finite number of applications of an even smaller set of function construction rules. The m.p.r. functions form an extension to the class PR of primitive recursive functions [**Pet67, Smo85**]. Because the initial primitive recursive functions are also initial m.p.r. functions, and because both of the primitive recursive rules for defining new functions are also m.p.r. function definition rules, everything expressible in primitive recursive arithmetic is also expressible in m.p.r. arithmetic.

DEFINITION 2.12. *The class of m.p.r. functions*
The class MPR of the modal primitive recursive functions consists of all the functions which can be generated after a finite number of steps using the following rules[a].

(1) ***Zero:*** $f(\vec{x}) \stackrel{def}{=} 0$,
(2) ***Successor:*** $f(x) \stackrel{def}{=} x + 1$,
(3) ***Projection:*** $f(y, \vec{x}) \stackrel{def}{=} \vec{x}_y$
(4) ***Substitution:*** $f(\vec{x}) \stackrel{def}{=} h(g_1(\vec{x}), \ldots g_n(\vec{x}))$,
(5) ***Primitive Recursion:*** $f(0, \vec{x}) \stackrel{def}{=} g(\vec{x})$, $f(r + 1, \vec{x}) \stackrel{def}{=} h(r, \vec{x}, f(r, \vec{x}))$,
(6) ***Alphabet:*** $f \stackrel{def}{=}$ ***Alphabet***,
(7) ***Output function:*** $f(x) \stackrel{def}{=}$ ***Out***(x),
(8) ***Component names:*** $f \stackrel{def}{=}$ ***Components***,
(9) ***Pumping number:*** $f \stackrel{def}{=}$ ***pump_number***,
(10) ***Enabling:*** $f(x) \stackrel{def}{=}$ ***enable***(x),
(11) ***Precedence:*** $f(x, y, x', y') \stackrel{def}{=}$ ***precedes***(x, y, x', y'),
(12) ***Feedback:*** $f(x, y) \stackrel{def}{=}$ ***effect***(x, y),
(13) ***Path offset:*** $f(y, \vec{x}) \stackrel{def}{=}$ $($***after*** $y)g(\vec{x})$,
(14) ***Component selection*** $f(y, \vec{x}) \stackrel{def}{=}$ $($***inn*** $y)g(\vec{x})$.

[a] *The form of this presentation is adapted from* [**Smo85**].

Note that every p.r. function is also a m.p.r. functions. The defining equations for p.r. functions are quite informative: if f is p.r. then we can reduce an expression $f(n)$ to a unique integer k by a finite number of rewritings using the defining equations. The defining equations for functions which are m.p.r., but not p.r. are not so informative. We remedy this situation by developing a map $\rho : MPR \rightarrow PR$, which reduces each m.p.r. function to a primitive recursive function that exposes the implicit dependencies on state machines and traces. We will *define* the value of a m.p.r. expression $f(n)$ in the context of P and w, to be $(\rho f)(P, w, n)$. This interpretation is the subject of the next section.

2.4. Interpretation. One of the advantages of basing our language of functions on the p.r. recursive functions is that we can describe mappings on MPR iteratively. Every m.p.r. function is either an initial function, or defined from simpler m.p.r. functions by substitution, primitive recursion, transition offset, or component selection. Thus, we can define ρ by first defining $\rho : InitialFunctions \rightarrow PR$, and then, supposing that ρ is defined on all MPR functions constructed using n or fewer applications of the defining rules, define (ρf) for a function constructed using one of the composing rules. Definition 2.14 provides a recursive definition of ρ.

DEFINITION 2.13. *The value of a m.p.r. function*
The value of a m.p.r. function f in the context (P, w) of an automaton P and trace w is defined to be $(\rho f)(P, w, \vec{x})$.

THEOREM 2.2. *The functional ρ is an effectively 1-1 map from the m.p.r. functions to the p.r. functions.*
Proof. *Note that (ρf) is obviously a p.r. function when f is an initial m.p.r. function. Proceeding by induction we can see that (ρf) must be p.r. for all functions $f \in MPR$. The function is "effectively" 1-1 because for every p.r. function f, there is a m.p.r. function, f itself, so that $f'(\vec{x}) = (\rho f)(P, w, \vec{x}) = f(\vec{x})$. That it, the map is 1-1 modulo hiding the extra, meaningless arguments.*

COROLLARY 2.1. *If f is p.r. then $(\rho f)(P, w, x) = f(x)$.*

COROLLARY 2.2. *For arbitrary m.p.r. function f, $(\exists P, w, \vec{x})(\rho f)(P, w, \vec{x}) > 0$ is semi-decidable, but not decidable.*
Proof: *Undecidability follows from the well known undecidability of the p.r. functions* [**Goo57**] *and the inclusion of the p.r. functions in the m.p.r. functions. Semi-decidability follows from a simple dovetailing argument; the set of possible argument vectors is enumerable, so we simply test each vector until/if we find a satisfying case.*

COROLLARY 2.3. *$(\rho f)(P, w, \vec{m}) = g(n)$ is decidable in time bounded by Ackermann's function for arbitrary m.p.r. functions f and g and constants m and n.*
Proof: *Follows from the well known decidability of closed p.r. terms* [**Goo57**] *and reduction using ρ.*

DEFINITION 2.14. *The evaluation functional, ρ*
Let $P = ((A, O, S, start, \lambda, \delta, (\phi_1, P_1, \ldots \phi_r, P_r))$

$$\begin{array}{ll}
\textit{If} & f(x) \stackrel{def}{=} 0 \\
& (\rho\ f)(P, w, x) \stackrel{def}{=} 0 \\
\textit{If} & f(x) \stackrel{def}{=} x + 1 \\
& (\rho\ f)(P, w, x) \stackrel{def}{=} x + 1 \\
\textit{If} & f(y, \vec{x}) \stackrel{def}{=} \vec{x}_y \\
& (\rho\ f)(P, w, y, \vec{x}) \stackrel{def}{=} \vec{x}_y \\
\textit{If} & f \stackrel{def}{=} \textbf{\textit{Alphabet}} \\
& (\rho\ f)(P, w) \stackrel{def}{=} A \\
\textit{If} & f \stackrel{def}{=} \textbf{\textit{Components}} \\
& (\rho\ f)(P, w) \stackrel{def}{=} \{1 \ldots, r\} \\
\textit{If} & f \stackrel{def}{=} \textbf{\textit{pump_number}} \\
& (\rho\ f)(P, w, x) \stackrel{def}{=} size(S) * \prod_i size(PS(P.i)) \\
\textit{If} & f(x) \stackrel{def}{=} \textbf{\textit{Out}}(x) \\
& (\rho\ f)(P, w, x) \stackrel{def}{=} (\lambda(\Delta(P, w))_x \\
\textit{If} & f(x) \stackrel{def}{=} \textbf{\textit{enable}}(x) \\
& (\rho\ f)(P, w, x) \stackrel{def}{=} \begin{cases} 1 & if\ w :: x \in \mathcal{L}(P) \\ 0 & otherwise. \end{cases} \\
\textit{If} & f(x) \stackrel{def}{=} \textbf{\textit{precedes}}(x, y, x', y') \\
& (\rho\ f)(P, w, x, y, x', y') \stackrel{def}{=} RelOrder(w, x, y, x', x') \\
\textit{If} & f(x) \stackrel{def}{=} \textbf{\textit{effect}}(x, y) \\
& (\rho\ f)(P, w, x, y,) \stackrel{def}{=} \begin{cases} \phi_y(\Delta(P, w), x) & if\ 0 < x \leq r; \\ \langle\rangle & otherwise. \end{cases} \\
\textit{If} & f(\vec{x}) \stackrel{def}{=} h(g_1(\vec{x}), \ldots, g_n(\vec{x})) \\
& (\rho f)(P, w, \vec{x}) = (\rho h)(P, w, \ldots (\rho g_i)(P, w, \vec{x}), \ldots) \\
\textit{If} & f(0, \vec{x}) \stackrel{def}{=} g(\vec{x}), f(r+1, \vec{x}) \stackrel{def}{=} h(r, \vec{x}), f(r, \vec{x})) \\
& (\rho f)(P, w, 0, \vec{x}) \stackrel{def}{=} (\rho g)(P, w, \vec{x}) \\
& (\rho f)(P, w, r+1, \vec{x})) \stackrel{def}{=} (\rho h)(P, w, r, \vec{x}, (\rho f)(P, w, r, \vec{x})) \\
\textit{If} & f(y, \vec{x}) \stackrel{def}{=} (\textbf{\textit{after}}\, y) g(\vec{x}) \\
& (\rho f)(P, w, y, \vec{x}) \stackrel{def}{=} (\rho g)(P, wy, \vec{x}) \\
\textit{If} & f(y, \vec{x}) \stackrel{def}{=} (\textbf{\textit{inn}}\, y) g(\vec{x}) \\
& (\rho f)(P, w, y, \vec{x}) \stackrel{def}{=} \begin{cases} (\rho g)(P_y, \psi(P, w, y), \vec{x})) & if\ 0 < y \leq r; \\ 0 & otherwise. \end{cases}
\end{array}$$

3. Proof system

Given a m.p.r. expression E, we want to be able to deduce properties of the transducers that satisfy E from the syntactic structure of E and from general rules about transducers. Goodstein [**Goo57**] has described a purely syntactic proof system for p.r. functions, and this system is sound for m.p.r functions by the construction of ρ. We need , however, to provide for modal deductions as well as purely arithmetic deductions. We use the convention that any non-zero valued term is *true*, and any 0 valued term is *false*. Terms are implicitly evaluated in the current state, but we have methods for forcing a term to be evaluated in the initial state. A *theorem* is a term (possible open) that is satisfied by all product form automata. Since proof rules and theorems are all derived from semantic considerations, issues of soundness are trivial. Completeness is not of particular interest either: Clearly, no proof system for a class of functions that includes the primitive recursive functions can be complete. Incompleteness, however, does not appear to be a major obstacle in practice. As for "expressive completeness", we show, in section3.3 that for every minimal product form automaton, there is a specification which specifies only that automaton.

Notational preliminaries. We will write $(\mathbf{after}\, a)f$ in place of $(\mathbf{after}\langle a\rangle)f$ whenever it is clear that we intend a to be a single transition. We will also find it convenient to be able to apply **enable** and **effect** to sequences (paths).

DEFINITION 3.1. *Extensions to paths.*

$$\boldsymbol{enable}(\langle\rangle) \stackrel{def}{=} 1$$
$$\boldsymbol{enable}(a:u) \stackrel{def}{=} \boldsymbol{enable}(a) * (\boldsymbol{after}\langle a\rangle)\, \boldsymbol{enable}(u)$$

$$\boldsymbol{effect}(\langle\rangle, c) \stackrel{def}{=} \langle\rangle$$
$$\boldsymbol{effect}(a:u, c) \stackrel{def}{=} \boldsymbol{effect}(a,c)(\boldsymbol{after}\, a)\, \boldsymbol{effect}(u,c)$$

3.1. Elementary proof methods.

We find it convenient to be able to treat expressions as un-named functions. That is, we might write $(\mathbf{inn}\, c)(E \bigwedge E')$ to force evaluation of the entire expression $t \bigwedge t'$ within the context of component c. Similarly, we write $(\mathbf{after}\, a)(E \bigwedge E')$ to force the evaluation of the expression $E \bigwedge E'$ in the state reached via an a transition. We can now list some simple results. The first result is that state independent functions can be "factored" out of the scope of **inn** and **after**.

THEOREM 3.1. *Distributive law for modal functionals.*

$$(\boldsymbol{after}\, u)(f(E_1, \ldots, E_n\}) = (\boldsymbol{after}\, u)f((\boldsymbol{after}\, u)E_1, \ldots, (\boldsymbol{after}\, u)E_n)$$

and

$$(\boldsymbol{inn}\, c)(f(E_1, \ldots, E_n)) = (\boldsymbol{inn}\, c)f((\boldsymbol{inn}\, c)E_1, \ldots, (\boldsymbol{inn}\, c)E_n)$$

The second, trivial, result is that primitive recursive functions are state independent.

THEOREM 3.2. *Distributive law for non-modal functions. If f is a primitive recursive function then,*

$$(\textbf{\textit{after}}\, u)f(\vec{x}) = f(\vec{x}) = (\textbf{\textit{inn}}\, c)f(\vec{x}).$$

A consequence of these two theorems is that many primitive recursive functionals can be factored out of expressions. For example:

$$(\textbf{after}\, u)\sum f(x) = \sum(\textbf{after}\, u)f(x).$$

A more interesting theorem allows *inversion* of expressions containing nested modal modifiers.

THEOREM 3.3. *Inversion.*

$$(\textbf{\textit{after}}\, u)(\textbf{\textit{inn}}\, c)f(\vec{x}) = (\textbf{\textit{inn}}\, c)(\textbf{\textit{after effect}}(u, c))f(\vec{x})$$

This theorem states that the result of advancing the entire product state machine to the state reached by following u, and then evaluating $f(\vec{x})$ in the context of component c, is equal to the result of simply advancing component c by the sequence induced by u, and then evaluating $f(x)$.

The "compositional proof rules" that are the staple of process based formal methods are not necessary in m.p.r. arithmetic. We note that if $E \rightarrow E'$ is a theorem, then $(\textbf{inn}\, c)(E \rightarrow E')$ must also be a theorem. This follows from elementary semantical considerations. Using the distributive laws, we can also deduce $(\textbf{inn}\, c)E \rightarrow (\textbf{inn}\, c)E'$.

3.2. Example. We illustrate m.p.r. specifications with the specification of a single bit latch and a shift register composed from latches. We illustrate m.p.r proof methods by verifying the correct operation of the shift register.

The alphabet of the latch should be *load*.0 and *load*.1, its output function should be $Data \in \{0, 1\}$, and it should obey $(\textbf{after}\, load.x)Data = x$. A modal grammar for the latch can't include the specification $(\textbf{after}\, load.x)Data = x$ (since modal grammars cannot make use of **after**), so we use **precedes** instead. We write $Data = \textbf{precedes}(load.0, 1, load.1, 1)$ to make $Data = 1$ if the most recent transition was a *load*.1 transition: if the most recent *load*.0 preceded the most recent *load*.1. Note, that in the initial state, $Data = 0$. The complete grammar is in figure 3.1:

FIGURE 3.1. *A single bit latch.*

$$
\begin{aligned}
&cell \stackrel{def}{=} \{ \\
&\qquad \boldsymbol{Alphabet} = \{load.x \;:\; x \in \{0,1\}\} \\
&\qquad \bigwedge \boldsymbol{Outputs} = \{Data \in \{0,1\}\} \\
&\qquad \bigwedge Data = \boldsymbol{precedes}(load.0, 1, load.1, 1) \\
&\qquad \bigwedge \boldsymbol{enable}(load.x) = 1 \\
&\}
\end{aligned}
$$

Thus, *cell*() is a boolean function which is satisfied onl;y by transducers that behave like storage cells. We write $\Box f(x)$ to assert that $f(x) > 0$ in all states. A precise definition of $\Box$ can be found in section refsec:temporal. We use $\Box$ in constructing a shift register that contains cells. When we write $(\mathbf{inn}\, c)\Box cell$, we are asserting that the value of the function $\Box cell$ is positive in the context of the component state machine named c. Thus, $(\mathbf{inn}\, c)\Box cell$ defines component c to be a state machine which obeys the specification *cell*.

The output of a n bit shift register will be an integer value between 0 and $2^n - 1$. It is sometimes advantageous to be a little more careful in distinguishing between bits and integers, we could define the output to be a vector of binary values, but there is no need for such care here. We will use the same name for the output function of the shift register as we used for the latch — *Data*. The alphabet of such a device will consist of symbols $Left.0$, and $Left.1$, which represent a single bit shift left, and symbols $Right.0$, and $Right.1$, representing single shifts to the right. The high order bits are the rightmost bits, so $left.x$ should divide the current output by 2, and then add $2^{n-1} * x$. Similarly, $right.x$ should multiply the current output by 2, add x and then mod the result by 2^{n+1}. We develop a straightforward implementation based on n components, $latch.0, \ldots, latch.(n-1)$, where each component satisfies *cell*. The value of *Data* will depend on the values of the outputs of the components:

$$Data = \sum_{i=0}^{n-1} 2^i * (\mathbf{inn}\, latch.i)Data.$$

Note that $(\mathbf{inn}\, latch.i)Data$ is the name of a function defined within sub-system $latch.i$, while *Data* is the name of a function in the current context.

We consider $latch.0$ to be the leftmost latch. When the shift register computes $Right.x$ we want $latch.0$ to load x, and for each $0 < i < n$, we want $latch.i$ to load the current output of its neighbor to the right. Let $R\_neighbor(i+1) \stackrel{def}{=} (\mathbf{inn}\, latch.i)Data$. Now we define the effect of $Right.x$ as follows:

$$\mathbf{effect}(Right.x, latch.i) = \begin{cases} \langle load.x\rangle & if\ i = 0 \\ \langle load.R\_Neighbor(i) & otherwise. \end{cases}$$

A similar mapping defines the effects of a transition labeled $Left.x$ on the component cells. The complete specification appears in figure 3.2

FIGURE 3.2. *A n bit shift register.*

$$
\begin{array}{l}
ShiftRegister(n) \stackrel{def}{=} \{ \\
\bigwedge \textbf{Alphabet} = \{Left.x, Right.x : x \in \{0,1\}\} \\
\bigwedge \textbf{Components} = \{latch.i : i \in \{0,\ldots,n-1\}\} \\
\bigwedge (\forall i)(\textbf{inn}\, latch.i)(\Box cell) \\
\bigwedge \textbf{Outputs} = \{Data \in \{0,\ldots 2^n - 1\}\} \\
\bigwedge Data = \sum_{i=0}^{n-1} 2^i * (\textbf{inn}\, latch.i)Data \\
\bigwedge InData(i) = (\textbf{inn}\, latch.i)Data \\
\bigwedge \textbf{effect}(a.x, latch.i) = \begin{cases} \langle load.x \rangle & if \;\; i = 0 \\ & \quad \bigwedge a = Right.x; \\ & or \;\; i = n-1 \\ & \quad \bigwedge a = Left.x; \\ \langle load.InData(i-1) \rangle & else\ if\ a = Right; \\ \langle load.InData(i+1) \rangle & otherwise. \end{cases} \\
\bigwedge \textbf{enable}(Left.x) = 1 \\
\bigwedge \textbf{enable}(Right.x) = 1\}
\end{array}
$$

We can easily build a longer shift register from n bit shift registers by connecting them serially. That is we define a grammar with k components $Reg.0, \ldots, Reg.(k-1)$ each satisfying $ShiftRegister(n)$. We then let $\textbf{effect}(Right.x, Reg.0) = Right.x$, and let $\textbf{effect}(Right.x, Reg.(i+1)) = Right.b_i$ where b is the rightmost bit of $Reg.i$. We could, likewise, build a fifo queue for k bit words by connecting $k+1$ shift registers in parallel. We would use k of the shift registers to hold the individual bits of each queued word, and use the other register to keep track of the depth of the queue (e.g. right shifting 1 into the final register every time a new element was enqueued).

We can use the proof rules introduced here to determine that the shift register described in figure 3.2 satisfies the property: $(\textbf{after}\, Left.x)Data = Data\ div\ 2 + x * 2^{n-1}$. Note that most of the proof uses simple combinatoric properties of integers.

1. $(\textbf{after}\, Left.x)Data$
2. $=$ *From the specification*
   $(\textbf{after}\, Left.x)2^i * (\sum_{i=0}^{n-1}(\textbf{inn}\, latch.i)Data)$
3. $=$ *By state independence of* $\sum$
   $(\sum_{i=0}^{n-1} 2^i * (\textbf{after}\, Left.x)(\textbf{inn}\, latch.i)Data)$
4. $=$ *By inversion*
   $(\sum_{i=0}^{n-1} 2^i * (\textbf{inn}\, latch.i)(\textbf{after}\, \textbf{effect}(Left.x, latch.i))Data)$
5. $=$ *By arithmetic*
   $2^{n-1} * (\textbf{inn}\, latch.(n-1))(\textbf{after}\, \textbf{effect}(Left.x, latch.(n-1)))Data)$
   $+(\sum_{i=0}^{n-2} 2^i * (\textbf{inn}\, latch.i)(\textbf{after}\, \textbf{effect}(Left.x, latch.i))Data)$
6. $=$ *From the specification of the shift register*
   $2^{n-1} * (\textbf{inn}\, latch.(n-1))(\textbf{after}\, load.x))Data)$
   $+(\sum_{i=0}^{n-2} 2^i * (\textbf{inn}\, latch.i)(load.(\textbf{inn}\, latch.(i+1))Data))Data)$
7. $=$ *From the specification of cell*
   $2^{n-1} * x + (\sum_{i=0}^{n-2} 2^i * (\textbf{inn}\, latch.(i+1))Data$
8. $=$ *From arithmetic*
   $2^{n-1} * x + (\sum_{i=1}^{n-1} 2^{i-1} * (\textbf{inn}\, latch.i)Data)$
9. $=$ *From arithmetic*
   $2^{n-1} * x + (\sum_{i=1}^{n-1} 2^i * (\textbf{inn}\, latch.i)Data)\ \textbf{\textit{div}}\ 2$
10. $=$ *Since* $\textbf{\textit{div}}2$ *ignores the low order bit*
    $2^{n-1} * x$
    $+((\sum_{i=1}^{n-1} 2^i * (\textbf{inn}\, latch.i)Data) + (\textbf{inn}\, latch.0)Data)\ \textbf{\textit{div}}\ 2$
11. $=$ *From arithmetic*
    $2^{n-1} * x + (\sum_{i=0}^{n-1} 2^i * (\textbf{inn}\, latch.i)Data)\ \textbf{\textit{div}}\ 2$
12. $=$ *From the definition of Data*
    $2^{n-1} * x + Data\ \textbf{\textit{div}}\ 2$

3.3. Modal Grammars . The most unconventional aspect of our proof theory, is the concept of syntactic proofs of finite state realizability. We will show that there are certain syntactic restrictions, so that if f meets the restrictions, then $f(\vec{m})$ defines exactly one (minimal) product form automaton. Thus, the construction of f is a proof that the system specified by $f(\vec{m})$ is realizable as a finite state machine. Functions of this restricted form are called *modal grammars*. A modal grammer is a symbolic, and highly compact definition of a family of deterministic finite state machine indexed by the parameters. Both the specifications in the example are modal grammars.

Modal grammars provide both a formal basis for exact specifications, and a style of specification that we find intuitive. The grammars rely on the use of restricted sub-sets of the m.p.r. functions. We write $mpr(x_1, \ldots x_n)$ to denote the extension of the primitive recursive functions with the functions and functionals names in the argument list. Thus. $mpr(\textbf{precedes})$ consists of all the functions which can be constructed from the initial primitive recursive functions, **precedes** and the primitive recursive functional.

Unitary grammars. The unitary modal grammars are defined below.

DEFINITION 3.2. *Any function $f(\vec{x})$ is a unitary modal grammar iff:*

$$\begin{aligned} f(\vec{x}) \stackrel{def}{=} \{ & \\ & \boldsymbol{Alphabet} = f_{alphabet}(\vec{x}) \\ & \bigwedge \boldsymbol{Out}_0 = out_0(\vec{x}) \\ & \ldots \\ & \bigwedge \boldsymbol{Out}_k = out_k(\vec{x}) \\ & \bigwedge (\forall a \in \boldsymbol{Alphabet})\ \boldsymbol{enable}(a) = f_{enable}(a, \vec{x}) \\ & \} \end{aligned}$$

where $f_{alphabet}$, $output_i$, and f_{enable} are all functions in $mpr(\mathbf{precedes})$.

In the sequel we will omit the $\bigwedge$ symbols connecting clauses of a specification, leaving the conjunction to be implicit. We will also allow arbitrary names for output functions, i.e. we will not limit ourselves to **Outputs**. We will define a set of output function names, $\mathbf{Outputs} = \{f, g, h, \ldots\}$, which act as indexes so that f abbreviates $\mathbf{Outputs}_f$.

THEOREM 3.4. *If $f(\vec{x})$ is a unitary modal grammar, then there is exactly one minimal automaton which satisfies $f(\vec{m})$ for each appropriate vector of arguments $\vec{m}$. Furthermore, the satisfying automaton is counter free (aperiodic)*[2]

The proof of theorem 3.4 follows from the observation that only the i rightmost a's and j rightmost b's in a trace w are of interest in the evaluation of $\mathbf{precedes}(a, i, b, j)$. We say i is the bound on a induced by $\mathbf{precedes}(a, i, bj)$, and similarly say that j is the bound on b. Given a modal grammar and appropriate arguments, we collect all the bounds for each symbol in the alphabet. Take the sum of all the largest of such bounds for each symbol. Now any string of length greater than that bound will contain elements that are not of interest in the evaluation of any sub-term $\mathbf{precedes}(a, i, b, j)$ in the grammar. Take the set of all strings of length bounded by this sum as a state set, and let $\delta(w, a) = w :: a$ iff $f_{enable}(a)$ is true for trace w. The resulting state machine will satisfy the grammar, and any other satisfying state machine will be equivalent under minimization.

Composite grammars. Let $(\mathbf{inn}\, x)\, \mathbf{Outputs}_i$ denote the restriction of the functional **inn** so that it can be applied only to the output functions of the components. Thus $(\mathbf{inn}\, c)\, \mathbf{precedes}(a, i, b, j)$ is forbidden under this restriction. Modal grammars for composite systems may only use the outputs of components, they are forbidden from extracting other state information.

[2] We conjecture that a model checking system might be able to exploit the aperiodic structure of the automata specified by unitary modal grammars [**Pin86**]. Loops in such automata have a very simple form.

DEFINITION 3.3. *Any function $f(\vec{x})$ is a composite modal grammar iff,*

$$
\begin{aligned}
f(\vec{x}) \stackrel{def}{=} \{ & \\
& \textbf{\textit{Alphabet}} = f_{alphabet}(\vec{x}) \\
& \bigwedge \textbf{\textit{Components}} = f_{component}(\vec{x}) \\
& \bigwedge (\forall c \in \textbf{\textit{Components}})(\textbf{\textit{inn}}\, c)(\Box G_c)(\vec{x}) > 0 \\
& \bigwedge \textbf{\textit{Outputs}}_0 = output_0(\vec{x}) \\
& \ldots \\
& \bigwedge \textbf{\textit{Outputs}}_k = output_k(\vec{x}) \\
& \bigwedge (\forall a \in \textbf{\textit{Alphabet}})(\forall c \in \textbf{\textit{Components}})\ \textbf{\textit{effect}}(a, c) = f_{f\_effect}(a, c, \vec{x}) \\
& \bigwedge (\forall a \in \textbf{\textit{Alphabet}})\ \textbf{\textit{enable}}(a) = f_{enable}(a, \vec{x}) \\
& \}
\end{aligned}
$$

where $f_{alphabet}$, $output_i$, f_{enable}, $f_{components}$, and f_{f_effect} are functions in $mpr((\textbf{inn}\, x)\ \textbf{Outputs}_i, \textbf{precedes})$, and each G_i is either a unitary or a composite modal grammar.

THEOREM 3.5. *If $f(\vec{x})$ is a composite modal grammar, then there is exactly one minimal automaton which satisfies $f(\vec{m})$ for each appropriate vector of arguments $\vec{m}$. Furthermore, there is a composite modal grammar for every product form finite automaton.*

It follows from theorem 3.5 and theorem 2.2 that there is a recursive procedure for deciding $\Box\mathcal{G} \rightarrow t$ for closed modal grammar $\mathcal{G}$ and arbitrary closed term t.

4. Temporal logic style functionals

The appeal of the temporal logic in computer science derives from the natural temporal logic characterizations of such important concepts as liveness and safety [**MP79, Kro87**]. There are several drawbacks to temporal logic, however. Among these are a semantics that is resistant to composition (c.f. [**Pnu85, COGB86, Lam83**]), fixed time granularity (c.f. [**BCDM86**]), and awkward treatment of algorithms (temporal logic is often used in conjunction with an algorithmic *action language*). The temporal logics also suffer from the notational and deductive opacity that seems inherent in formal logics [**Bou88**]. We believe that the m.p.r. arithmetic provides an alternative basis for temporal logic style reasoning that overcomes these problems.

4.0.1. *Safety and intervals.* In temporal logic a proposition $\bigcirc p$ is true iff p is true in the "next state". In m.p.r. arithmetic we have an analogous operator of the same symbol, so that $\bigcirc f(\vec{x})$ is true iff $f(\vec{x}) > 0$ in every state reachable via one transition.

$$\bigcirc f(\vec{x}) \stackrel{def}{=} (\forall a \in \textbf{Alphabet})(\textbf{enable}(a) \rightarrow (\textbf{after}\, a) f(\vec{x}) > 0).$$

We will need some *interval operators* for assertions about paths. We let $(\textbf{sometimes}\, u)f(n)$ be true iff $f(n) > 0$ *sometime* during the computation of u. Similarly, we let $(\textbf{always}\, u)f(n)$ be true iff $f(n) > 0$ every state visited by u. Finally, we might want to get the cumulative value of a function during the computation of a path, and we write $(\textbf{cumu}\ u)f(n)$ for that value. If a state is visited during the computation of a path u, there must be some prefix of u which terminates at that state. We write $v \prec u$ to denote that v is a prefix of u, i.e., $v \prec u \leftrightarrow (\exists z)u = vz$. Note that $u \prec u$ and $\langle\rangle \prec u$.

DEFINITION 4.1. *The interval operators.*

$$(\textbf{\textit{sometimes}}\, u)f(\vec{x}) \stackrel{def}{=} (\exists v \prec u)(\textbf{\textit{after}}\, v)f(\vec{x}) > 0$$
$$(\textbf{\textit{always}}\, u)f(\vec{x}) \stackrel{def}{=} (\forall v \prec u)(\textbf{\textit{after}}\, v)f(\vec{x}) > 0$$
$$(\textbf{\textit{cumu}}\ u)f(\vec{x}) \stackrel{def}{=} \sum_{v \prec u}(\textbf{\textit{after}}\, v)f(\vec{x})$$

Let $Paths(i)$ be the set of enabled paths of length i: $Paths(i) = \{u : \textbf{enable}(u) \bigwedge length(u) = i\}$. We can now state the theorem that makes m.p.r. analogs of temporal logic operators possible.

THEOREM 4.1. *There is a total map plen* : $MPR \to MPR$ *so that:*

$$(\exists u)\ (\textbf{\textit{after}}\, u)f(\vec{x}) > 0$$
$$\to (\exists u \in Paths((plen\ f)(\vec{x})))(\textbf{\textit{sometimes}}\, u)f(\vec{x})$$
$$\{\text{and }\}$$
$$(\exists u \in Paths((plen\ f)(\vec{x})))(\textbf{\textit{always}}\, u)f(x) > 0$$
$$\leftrightarrow (\forall i)(\exists u in Paths(i))(\textbf{\textit{always}}\, u)f(x)$$

Proof sketch. If $(plen\ \textbf{enable})(x) = (plen\ \textbf{effect})(x, y) = \textbf{pump\_number}$, the theorem will hold. Inspection of ρ shows that both **enable** and **effect** depend only on the configuration, and there are at most **pump_number** configurations. If $(plen\ \textbf{precedes})(a, i, b, j)$ is $i + j$, the theorem must also hold by theorem 2.1. If $f(x, y) = (\textbf{after}\, x)g(y)$, then we can let $(plen\ f)(x, y) = (\textbf{after}\ x)(plen\ g)(y)$, and a similar construction suffices for **inn**.

The implications of theorem 4.1 can be illustrated by considering the concepts of eventuality and henceforth. Suppose that we wish to show that $f(x) > 0$ is inevitable (eventual). Thus, we need to show that there is some i so that no enabled path of length i or more keeps $f(x) = 0$. The second part of the theorem tells us that $f(x)$ is inevitable iff there is no enabled path $u \in Paths((plen\ f)(x))$ so that $(\textbf{always}\, u)(f(x) = 0)$. Similarly, $f(x) > 0$ after every enabled path iff there is no enabled path $u \in Paths((plen\ f)(x))$ so that $(\textbf{sometimes}\, u)(f(x) = 0)$.

4.0.2. *Branching style functionals.* In this section we define operators analogous to those of the branching temporal logic [**BAPM83, EH83**]. We will mainly follow Ben-Ari's UB logic [**BAPM83**] so that we can appropriate the proof system he describes. We have the following theorem:

THEOREM 4.2. *All the proof rules (except necessitation, $\vdash p \Rightarrow \vdash \Box p$) and all of the axioms of UB are sound and valid in m.p.r. arithmetic.*

The proof is an almost immediate consequence of the definitions and theorem 4.1.

DEFINITION 4.2.

$$\Box f(\vec{x}) \stackrel{def}{=} (\forall u)(\textit{enable}(u) \rightarrow (\textit{after}\, u) f(\vec{x}) > 0)$$
$$\Diamond f(\vec{x}) \stackrel{def}{=} (\forall u \in \textit{Paths}((\textit{plen}\ f)(\vec{x})))(\textit{sometimes}\, u) f(\vec{x}))$$
$$\boxdot\ f(\vec{x}) \stackrel{def}{=} (\exists u \in \textit{Paths}((\textit{plen}\ f)(\vec{x})))(\textit{always}\, u) f(\vec{x}))$$
$$\diamond\ f(\vec{x}) \stackrel{def}{=} (\exists u \in \textit{Paths}((\textit{plen}\ f)(\vec{x})))(\textit{sometimes}\, u) f(\vec{x}))$$

We can, now , define a m.p.r. analog of the temporal logic operator ***until***. We want E ***until*** E' to be true iff every enabled path, either keeps E non-zero, or makes E' non-zero.

$$E\ \textit{until}\ E' \stackrel{def}{=} (\forall u)(\mathbf{enable}(u) \rightarrow (\mathbf{always}\, u)E \bigvee (\mathbf{sometimes}\, u)E').$$

We can show that this definition accords with the usual, recursive, definition of ***until*** without much difficulty. Suppose that E ***until*** E' is true. Then, by definition, we have $(\forall u)(\mathbf{enable}(u) \rightarrow (\mathbf{always}\, u)E \bigvee (\mathbf{sometimes}\, u)E')$. We can rewrite this to distinguish the empty path from the other paths as follows:

$$\begin{aligned}&\mathbf{enable}(\langle\rangle) \rightarrow (\mathbf{always}\langle\rangle)E \bigvee (\mathbf{sometimes}\langle\rangle)E')\\ &\bigwedge(\forall a:u)(\mathbf{enable}(a:u) \rightarrow (\mathbf{always}\, a:u)E \bigvee (\mathbf{sometimes}\, a:u)E').\end{aligned}$$

The first clause can be greatly simplified, because $\mathbf{enable}(\langle\rangle) = 1$, $(\mathbf{always}\langle\rangle)E = E$, and $(\mathbf{sometimes}\langle\rangle)E' = E'$, all by definition. We have shown, therefore:

$$\begin{aligned}&(E \bigvee E')\\ &\bigwedge(\forall a:u)(\mathbf{enable}(a:u) \rightarrow (\mathbf{always}\, a:u)E \bigvee (\mathbf{sometimes}\, a:u)E').\end{aligned}$$

Now we simplify the second clause, by splitting the empty prefix case from both the **always** and **sometimes**.

$$\begin{aligned}(E \bigvee E') \bigwedge(\forall a:u)(\mathbf{enable}(a:u) \rightarrow\ & (E \bigwedge(\mathbf{after}\, a)(\mathbf{always}\, u)E)\\ & \bigvee(E' \bigvee(\mathbf{after}\, a)(\mathbf{sometimes}\, u)E').\end{aligned}$$

If E' is false, then E must be true, so we conclude:

$$\begin{aligned}E' \bigvee(E \bigwedge(\forall a:u)(\mathbf{enable}(a:u) \rightarrow\ & (\mathbf{after}\, a)(\mathbf{always}\, u)E)\\ & \bigvee(\mathbf{after}\, a)(\mathbf{sometimes}\, u)E').\end{aligned}$$

Since $\mathbf{enable}(a:u) = (\mathbf{enable}(a) \bigwedge(\mathbf{after}\, a)\, \mathbf{enable}(u)$ we get:

$$\begin{aligned}E' \bigvee(E \bigwedge(\forall a)(\mathbf{enable}(a) \rightarrow (\mathbf{after}\, a)(\forall u)\, \mathbf{enable}(u) \rightarrow (\mathbf{after}\, a)(\mathbf{always}\, u)E)\\ \bigvee(\mathbf{after}\, a)(\mathbf{sometimes}\, u)E').\end{aligned}$$

This reduces, by definition, to $E' \bigvee(E \bigwedge \bigcirc E\ \textit{until}\ E')$ which is the standard definition of the temporal logic weak ***until***.

5. Conclusion

The modal primitive recursive arithmetic described in this paper is an attempt to extend the range of systems that we can comfortably describe in terms of deterministic finite automata. We claim:

- A mathematical model of composite digital systems which reflects the natural structure of those systems;
- A simple, yet expressive, notion of state that enables us to avoid the tedious task of state set enumeration;
- An expressive functional notation for specifying and verifying the behavior of finite state systems.

Attempts to address the problems of "state-space explosion" and modular composition are related in our work through the feedback product. State machines representing composite systems are kept in product form so that the mathematical structure of the automata reflects the design structure of the system, and so that we can avoid explicit consideration of the composite state set. The arithmetic in general, and modal grammars in particular, provide us with compact symbolic descriptions of product form automata. We have carried out "by hand" specifications and proofs of quite complex examples, and are reasonably satisfied with the results. It remains to be seen whether or not the modularity and compactness of m.p.r. will prove to be an advantage when we seek to automate the verification process.

References

[BAPM83] M. Ben-Ari, A. Pnueli, and Z. Manna. The temporal logic of branching time. *Acta Informatica*, 20, 1983.

[BCDM86] M.C. Browne, E. M. Clarke, D. L. Dill, and B. Mishra. Automatic verification of sequential circuits using temporal logic. *IEEE Transactions on Computers*, C-35(12), 1986.

[Bou88] R. T. Boute. On the shortcomings of the axiomatic approach as presently used in computer science. In *Compeuro 88 Systems Design: Concepts Methods, and Tools*, 1988.

[CAS83] E. M. Clarke, Emerson A., and A.P. Sistla. Automatic verification of finite-state concurrent systems using temporal logic specifications: A practical approach. In *Proceedings of the 10th Annual Symposium on Principles of Programming Languages*, pages 117–119, 1983.

[COGB86] E.M. Clarke, M.C. O. Grumberb, and Browne. Reasoning about networks with many identical finite state processes. Technical Report cmu-cs-86-155, Carnegie-Mellon University, October 1986.

[EH83] E. A. Emerson and J. Y. Halpern. Sometimes and 'not never' revisited: on branching versus linear time temporal logic. *Journal of the ACM*, 33(1), January 1983.

[GC86] G. Gouda, M. and C. Chang. Proving liveness for networks of communicating finite state machines. *ACM Transactions on Programming Languages and Systems*, 8(1), January 1986.

[Gec86] Ferenc Gecseg. *Products of Automata*. Monographs in Theoretical Computer Science. Springer Verlag, 1986.

[Goo57] R. L. Goodstein. *Recursive Number Theory*. North Holland, Amsterdam, 1957.

[Har84] D. Harel. Statecharts: A visual formalism for complex systems. Technical report, Weizmann Institute, 1984.

[Har85] D. Harel. On the development of reactive systems. In K. Apt, editor, *Logics and Models of Concurrent Systems*. Springer-Verlag, 1985.

[Har87] D. Harel. On the formal semantics of statecharts. In *Proceedings of the Symposium on Logic in Computer Science*, Ithaca, June 1987.

[HU79] John E. Hopcroft and Jeffry D. Ullman. *Introduction to Automata Theory, Languages, and Computation*. Addison-Welsey, Reading MA, 1979.

[Kro87] F. Kroger. *Temporal Logic of Programs*. EATCS Monographs on Theoretical Computer Science. Springer-Verlag, 1987.

[Kur85] Kurshan. Modelling concurrent processes. In *Proc of Symposia in Applied Mathematics.*, 1985.

[Lam83] L. Lamport. What good is temporal logic? In *Proceedings of the IFIP 9th World Computer Congress*, pages 657–667, Paris, September 1983.

[LM86] A. Lynch, N. and M. Merrrit. Introduction to the theory of nested transactions. Technical Report TR-367, Laboratory for Computer Science, MIT, 1986.

[Mil79] R. Milner. *A Calculus of Communicating Systems*, volume 92 of *Lecture Notes in Computer Science*. Springer Verlag, 1979.

[MM83] B. Moszkowski and Z. Manna. Reasoning in interval temporal logic. Technical Report STAN-CS-83-969, Stanford University, July 1983.

[Moo64] E.F. Moore, editor. *Sequential Machines: Selected Papers*. Addison-Welsey, Reading MA, 1964.

[MP79] Z. Manna and A. Pnueli. The modal logic of programs. In *Proceedings of the 6th International Colloquium on Automata, Languages, and Programming*, volume 71 of *Lecture Notes in Computer Science*, New York, 1979. Springer-Verlag.

[Myh57] J. Myhill. Finite automata and the represntation of events. Technical Report WADD TR-57-624, Wright Patterson Air Force Base, 1957.

[Ner58] A. Nerode. Linear automaton transformations. In *Proc. AMS*, 1958.

[OW87] J.S. Ostroff and W.M. Wonham. Modelling, specifying, and verifying real-time embedded computer systems. In *Symposium on Real-Time Systems*, Dec 1987.

[Pet67] Rozsa Peter. *Recursive functions*. Academic Press, 1967.

[Pin86] J.E. Pin. *Varieties of Formal Languages*. Plenum Press, New York, 1986.

[Pnu85] A. Pnueli. Applications of temporal logic to the specification and verification of reactive systems: a survey of curent trends. In J.W. de Bakker, editor, *Current Trends in Concurrency*, volume 224 of *Lecture Notes in Computer Science*. Springer-Verlag, 1985.

[Sis83] A.P. Sistla. *Theoretical Issues in the Design and Verification of Distributed Systems*. PhD thesis, Harvard University, 1983.

[Smo85] C. Smorynski. *Self-Reference and Modal Logic*. Springer-Verlag, 1985.

Department of Computer and Information Science, University of Massachusetts, Amherst, Massachusetts 01003

E-mail: yodaiken@cs.umass.edu

DIMACS Series in Discrete Mathematics
and Theoretical Computer Science
Volume 3, 1991

Results on the Interface Between Formal Verification and ATPG *

Hyunwoo Cho Gary Hachtel Seh-Woong Jeong
Bernard Plessier Eric Schwarz Fabio Somenzi
University of Colorado at Boulder
Department of Electrical and Computer Engineering
Boulder, Colorado 80309

February 28, 1991

Abstract

We present algorithms for FSM verification and image computation which improve on [6], giving 1-4 orders of magnitude speedup. New features include: **Primary input splitting**—this PODEM feature enlarges the search space but shortens search due to implications. **Identical subtree recombination**—effective for **iterative networks** (e.g., serial multiplers). **Free-variable recognition**—prevents unbalanced bipartitioning tree in tautological subspaces. **Reached set pruning**—significant when the image contains large numbers of previously reached states.

1 Introduction

This paper describes a new tool for formal verification of the combinational and sequential assertions, for example, that the result of applying some or all of the tools available for FSM synthesis is equivalent to the original specification. [5], [11], [3, 10], [15, 16]. In a recent paper, Coudert *et al.* discussed an algorithm [6, 7], called the "SIAM Algorithm" in the sequel. This algorithm introduced the ideas of a) BDD-based symbolic breadth first traversal of an implicitly defined STG, and b) a means of function simplification which has been later called the "generalized cofactor." In this paper we extend the SIAM approach to a more general type of recursion, analogous in a broad sense to the operation of PODEM, in which Shannon cofactors are taken with respect to input, as well as output variables, and show how this leads to the significant extensions listed in the abstract. We also introduce methods for subproblem recognition, which

*This work was supported in part by NSF/DARPA grant MIP-8719546.

enables us to identify large subproblems of the recursion as either trivial or identical to previously solved subproblems. We show that the resulting pruning of the recursion tree can be quite significant in reducing cpu time and storage requirements, in comparison to [6, 8, 18]. We further discuss the application of this new technology to sequential ATPG, and show how minimum length test patterns may be obtained, and how to use the new traversal technology to significantly reduce justification and fault simulation expenses in sequential ATPG. Our results indicate that traversal (and thus verification) is normally "easy" for any machine for which both the next state function and the set of reachable states can be represented as "compact" BDD's. This is the case for a sizeable class of "iterative" networks (e.g., adders, some multipliers, shifters, etc.).

2 Preliminaries

We focus on the equivalence or compatibility of incompletely specified deterministic finite state machines (FSM's for short). We assume that all don't cares are in the outputs of the FSM (any next state don't cares have been mapped into a dummy state). A Mealy FSM $\mathcal{M}$ is defined by a 7-tuple $(\Sigma, O, S, S^0, \delta, \lambda, \xi)$, where $\Sigma(O)$ is the input (output) alphabet, S is the finite set of states of the machine, S^0 is **a set** of initial states, δ is the next state function ($\delta : S \times \Sigma \rightarrow S$), λ is the output function ($\lambda : S \times \Sigma \rightarrow O$) and ξ is the output don't care function ($\xi : S \times \Sigma \rightarrow O$). Note δ, λ, and ξ are multiple-output boolean functions, or, as referred to in [6], **boolean functional vectors**. We shall use these two denotations interchangeably, and describe such objects as BFV's. Also, the components of λ correspond one to one to the components of ξ. Thus $\mathcal{M}$ is incompletely specified if, for some j, $\xi_j \neq \emptyset$. These BFV's implicitly define the **State Transition Graph** (STG) of the given FSM.

Two machines are equivalent if the output of their product machine is identically 1, for every applicable input sequence, starting from S^0. That is, every edge of the STG of the product machine has a boolean "1" as its output label. Past methods, [13, 14, 8, 6], have focussed on implicitly exploring the STG, while storing at most fragments of the computed STG. We have seen essentially no published data on product machine traversal.

In this paper the **image** of a function $F : B^m \rightarrow B^n$, **over a specified subdomain** $C \subseteq B^m$ is defined as $Img(F, C) = \{v \mid \exists x \in C \wedge F(x) = v\}$. Similarly, we define the **pre-image** of a set $V \subseteq B^n$ as $Pre(F, V) = \{x \mid F(x) \in V\}$. Suppose the **image** of $F : B^m \rightarrow B^n$ is V. Then the **pre-image** of a cube $c \in V$ is $\bigcap_{i=1}^{n} f_i^{c_i}$, where if $c_i = 1, f_i^{c_i} = f_i$ (i.e., the ith component of F), if $c_i = 0, f_i^{c_i} = \bar{f}_i$ (i.e., the complement of f_i), and if $c_i = -, f_i^{c_i} = 1$.

A **Generalized Cofactor** of F with respect to C is defined as any function, $F \uparrow C$, which satisfies

$$F = CF \uparrow C + \bar{C} F \uparrow \bar{C} \tag{1}$$

Note that if $F \uparrow C$ satisfies this equation it also satisfies the ordinary cofactor equality $CF \uparrow C = CF$. An **Image Restrictor** of F with respect to C is defined as any function, $F \downarrow C$, which satisfies

$$Img(F \downarrow C, B^m) = Img(F, C). \tag{2}$$

The SIAM paper described a simple BDD operation, denoted there as $F \downarrow C$, which satisfies both (1) and (2). To emphasize the taxonomy, we denote this operator instead by $F \updownarrow C$. Thus, $F \uparrow C$ is a (generalized) cofactor and $F \updownarrow C$ is an **image restricting (generalized) cofactor**.

3 FSM Equivalence Checking

As reported, SIAM verifies the equivalence of two completely specified FSM's, by constructing the product of the two machines and checking that the output is 1 for all input sequences. Procedure FSM_TRAVERSAL of Figure 1 is based on symbolic breadth-first search of the STG of a specified FSM. The algorithm consists of a **for** loop, indexed by a level counter i, with most statements involving either high-level ROBDD operations or generalized cofactor computations [2]. Starting from a set S^0 of initial states, the algorithm partitions the reachable states of $\mathcal{M}$ into disjoint subsets $new^i, \ldots, new^{L-1}$, where L is the value of i on termination. Let l_u be the length of the shortest path in the STG from any node in the set S^0 to node u. Then $u \in new^{l_u}$. The computation of new^{i+1} from new^i and π are discussed below in relation to subprocedure IMAGE. The procedure returns the pair (L, χ^L) when the traversal is completed. If the comparison machines are equivalent, the algorithm completes when $new^L = \emptyset$. In this case, $\chi^L = \emptyset$, $reached^L = \bigcup_{i<L} new^i$ is the set of all states of the FSM reachable from S^0, and $L = \max_{\{v \in reached^L\}}(l_v) + 1$. The procedure makes use of the projection of g along variable x_k, which is $g \perp x_k = g_{x_k} + g_{\bar{x}_k}$, where g_{x_k} is the ordinary cofactor of g with respect to x_k. Similarly, the projection of g with respect to $x_1, \ldots, x_K$, is recursively defined by $g \perp (x_1, \ldots, x_K) = (g \perp x_1) \perp (x_2, \ldots, x_K)$.

3.1 Distinguishing Sequences

If the machines are not equivalent, the procedure returns from the **if** block of Line 2. Note that the test $F \not\equiv 1$ can be performed in constant time on the BDD representation of F. The ITE_TEST operation at Line 3 is a specialization of the ITE_CONSTANT operation [2], and returns the first exception it finds to the proposition that the input BDD is the tautology (identically zero) if the second argument is 0 (1). In this case, the algorithm computes a **shortest distinguishing sequence** of input vectors. That is, a sequence $\chi^L = (x^1, \cdots, x^L$ has been found such that when applied to a reset state, the corresponding output sequence has the property that $\Lambda(x^1) = 1, \Lambda(x^2) = 1, \ldots, \Lambda(x^{L-1}) = 1$ but $\Lambda(x^L) = 0$. This situation is analogous to the search phase of Dijkstra's algorithm for the shortest path in a graph. We shall, following [8], refer to s^L

Procedure FSM_TRAVERSAL$(\lambda, \delta, \xi, S^0, \pi)$ {
1 $\lambda = \lambda \cup \xi$; $from = reached = new^0 = S^0$
  for $(i = 1; ; i{+}{+})$ {
2   **if**($\lambda \Downarrow from \not\equiv 1$) {
3     $(x^i, s^i) =$ ITE_TEST$(\lambda \Downarrow from, 0)$
      $\chi^i =$DIST_SEQ$(\delta, i, x^i, s^i, (new^0, \ldots, new^{i-1}))$
      return(i, χ^i) }
4   $r =$IMAGE$(\delta \Downarrow from, 0, \pi, reached)$
5   $new^i = r - reached$
6   **if**$(new^i \equiv \emptyset)$ **return**$(i, \emptyset)$
    $reached = reached + new^i$
7   $from =$BEST_BDD$(new^i, r, reached)$ } }

Figure 1: Breadth-First Search for FSM Traversal

as the **excitation state**, and x^L as **excitation input** x^L.

Figure 2 gives Procedure DISTINGUISHING_SEQUENCE for finding the sequence σ.

On the i-th iteration of FSM_TRAVERSAL, a BDD representing new^i is saved for use by DIST_SEQ. DIST_SEQ first calculates H, the pre-image of s^L. Then it finds $F = H \bigcap new^{L-1}$. F is guaranteed to be non-empty. A cube of the ON-set of F is then selected arbitrarily by ITE_TEST. This cube is (x^{L-1}, s^{L-1}). The process is then iterated until S^0 is reached. which computes $F = \delta \bigcap new^i$ Then it finds an arbitrary vertex $(x^-, s^-) \in Pre(F, s^+)$, which is guaranteed to exist. This process iterates until S^0 is reached. In effect, the distinguishing sequence finds a path in the implicitly defined STG for the machine, such that each x^i is contained by the edge input label of the ith edge of this path. We note that $s^1 \in s0$, that is the first excitation state in the distinguishing sequence is in the initial state set $s0$.

Theorem 3.1 (Shortest Paths)

a) The length of the distinguishing sequence computed by DISTINGUISHING_SEQUENCE *is the shortest possible, and the*

b) The time complexity of this postprocessing procedure is $O(N_M^2 \times n \times L)$, where N_M is the size of the largest BDD encountered and n is the dimension of the range space.

c) The storage complexity of DISTINGUISHING_SEQUEN-CE *is $O(N_M \times L)$. Note N_M bounds the size of the BDD representations of the L sets new^i, $i = 1, 2, \cdots, L$, the functional vector Δ, and the generalized cofactors of Δ with respect to the new^i.* ∎

As the experimental results in [6] first showed, the required storage is affordable even for FSM's with large ($> 10^6$) numbers of states, **assuming a**

```
Line Procedure DISTINGUISHING_SEQUENCE(δ, S^0, L, x^L, s^L, new^1, ···, new^L) {
    new^0 = S^0
    χ = (x^L) ;  s^+ = s^L
    for(i = (L − 1, L − 2, ···, 1)) {
      F = δ ⋃ new^i
      (x^−, s^−) = PRE_IMAGE(F, s^+)
      χ = (x^−), χ ;  s^+ = s^−
    }
    return(χ)
  }
```

```
Line Procedure 1_PRE_IMAGE_PT(F, s) {
    H = 1
    for(i = 1, ···, |s|) {
      if(s_j ≡ 1)H = H ⋃ F_j
      if(s_j ≡ 0)H = H ⋃ F̄_j
      if(H = ∅) return(error_out)
    }
    return(ITE_TEST(H, 1))
  }
```

INPUTS: S^0 — Set of initial states.

δ — Next state transition function of given FSM.

L— Level of breadth first.

$new^i, i = 1, 2, \cdots, L$ — Sets of states first reached at each level of breadth first search.

OUTPUT: A shortest distinguishing sequence of FSM input vectors,guaranteed to exist.

Figure 2: Backtrace Procedure for Finding a Shortest Distinguishing Sequence

reasonable BDD variable ordering. As discussed in the results section, the IMEC MIN_MAX benchmarks, [6], have about $(2^{3q})/6$ reachable states, but with the proper BDD ordering, *reached* requires, on termination of FSM_TRAVERSAL, only $13(2^q - 1)$ nodes in the associated BDD, exclusive of the initial state.

3.2 Comparison to SIAM

FSM_TRAVERSAL differs from the analogous SIAM algorithm as follows. First, SIAM used an operation called "CHAR_TO_ VECT" followed by BDD "COMPOSE" operations which have been replaced here by the indicated generalized cofactor operation. Second, our algorithm has an improved stopping condition (Line 5) which can save a level of breadth first search by using projection operations. Briefly, when *new* differs from *reached* only in variables outside the support of the transition function δ, it can be predicted that $new^i = \emptyset$ without computing it. Second, our version computes the shortest distinguishing sequence as discussed above. Third, we have an improved algorithm for image computation, as discussed below.

As in SIAM [7], Procedure FSM_TRAVERSAL of Figure 1, is an algorithm for traversing the product of two or more machines. The specification of multiple machines enables the specification of detailed sequential predicates which can be verified instead of simple machine equivalence. That is, a machine can be constructed which represents a predicate, rather than an implementation machine. The product of the predicate machine and the behavioral machine produces a true output sequence for all applicable input sequences if and only if the the behavioral machine satisfies the sequential predicate.

Sequential output don't cares have been handled in the same way as they have been in combinational logic synthesis, namely by specifying a don't care function for each primary output, [1]. Our formulation also permits us to handle **Temporal Don't Cares**, as discussed in [7]. Also, there is an extra step required when dealing with incompletely specified machines which was not covered in [6]. If the specification machine is specified by a cube table representation of its STG, and if an outgoing edge is not specified for every primary input minterm for any given node, then for any given STG node, outgoing edges are added until the union of the input labels of these edges becomes the tautology. We assume that an incompletely specified cube table is given for the FSM.

4 Image Computation

The SIAM paper [6] discussed an algorithm, called "Algorithm 2" (we refer to this as VECT_TO_CHAR). Procedure IMAGE of Figure 3 extends this algorithm, with ideas from PODEM [9]. The inputs include a splitting variable index k, a splitting order (permutation) π, and a set R, denoting states which have been reached in previous levels of traversal, and hence are no longer of interest in the current image computation. Note π contains indices with respect to a combined variable $z = (x, y)$, of length $m + n$.

IMAGE is based on recursive Shannon expansion, and is distinguished from other such algorithms by SPECIAL_CASES and IMPLICATIONS. The latter (not shown) finds implications of assigning the positive or negative phase of the selected splitting variable, which could be an input or an output variable of F. The value of $k^+(k^-)$ returned by IMPLICATIONS is one less than the index in π of the next undetermined variable in the splitting order. The $R^+(R^-)$ returned is the input R cofactored with respect to the literals corresponding to output variables whose values have just been implied. Thus all implications are accounted for in the functions passed in the recursive calls to IMAGE of Line 4. Currently IMPLICATIONS is based on the $\Downarrow$ operator. We are considering using the ITE_CONSTANT operator as in [2].

SPECIAL_CASES is distinguished from its counterpart in VECT_TO_CHAR as follows. First, the hash table is searched for a pointer vector matching the BFV F (represented by a vector of ROBDD pointers [2]). If so, TBL_MATCH returns the image IMG which was computed when this subproblem was first solved (Line 1). This innovation is called **subtree recombination**, because, like the BDD operation REDUCE, it converts a binary recursion tree into a graph. We are currently studying an extension of the recombination mechanism whereby a match is declared by TBL_MATCH if two pointer vectors are the same modulo complementation and/or permutation. That is, the subtree is pruned if a) BFV ROBDD pointer vector is identical to one already in storage (in which case it returns an unaltered stored image); b) the pointer vector matches a stored one modulo complementation (in which case it returns an image with the appropriate literals complemented); or c) it matches a stored vector modulo complementation and/or permutation (in which case it returns an image with the appropriate literals re-indexed and/or complemented).

If R is the tautology in the current subspace of the recursion, we know that all states in this subspace have already been reached on previous levels of search (reached set pruning—Line 2). If for a given subspace of the binary recursion a BFV F has been reduced by implications until it becomes identical to a BFV of free variables (e.g., $F_1 = x_1,\ F_2 = x_2, \cdots$), the image of F is the tautology in the given subspace (free-variable recognition). Other terminal cases and topological partitioning are as in [6].

The key to efficient computation for iterative networks is subtree recombination. We emphasize that our approach is not passive in this respect. First we determine a proper ordering according to the following rules (the MIS2.2 heuristic is not effective in such cases). (1) Separate the control variables from the data variables and place them first in the splitting order; (2) Order the variables of the bit slices in either ascending or descending order of bit significance; (3) BDD variable order should be compatible to image recursion splitting order. Input variable splitting is required to create identical subproblems by exploiting latent symmetry, and implications are required to prevent meaningless nodes

```
Procedure IMAGE(F, k, π, R) {
/*y ≡ F(x) = (f_1(x), ···, f_n(x)) : B^m → B^n*/
  k = k + 1
1 (T, IMG) =SPECIAL_CASES(F, k, π, R)
  if(T ≡ false) {
2    if(LAST_OUTPUT(k, π))return(1)
     else {
3       j = π[k]
        (F^+, k^+, R^+) =IMPLICATIONS(j, 1, F, k, π, R)
        (F^-, k^-, R^-) =IMPLICATIONS(j, 0, F, k, π, R)
4       IMG = y_j IMAGE(F^+, k^+, π, R^+) ∪
                  ȳ_j IMAGE(F^-, k^-, π, R^-) } }
5 HASH_TBL_STORE(F, IMG, unique_subproblems)
  return(IMG) }
```

```
Procedure SPECIAL_CASES(F, k, π, R) {
  (T, IMG) =TBL_MATCH(F, unique_subproblems)
1 if(T ≡ true) return(true,IMG)
2 if(R ≡ 1) return(true,∅)
3 if(NO_UNDETERMINED_INPUTS(k, π))
     return(true,F(x))
4 if(NO_UNDETERMINED_OUTPUTS(k, π))
     return(true,y)
  (𝓕, 𝓚, 𝓡) = FULL_PARTITIONING(F, π, R)
5 if(FREE_VARIABLE_SUBPROBLEM(𝓕)) return(true,1)
6 if(|𝓕| > 1) return(true, ∩_i IMAGE (𝓕_i, 𝓚_i, π, 𝓡_i))
7 return(false,∅) }
```

Figure 3: Procedure for Finding the Image of a BFV.

name	states	time	depth	size	calls
mclc	35	3.76	14	102	120
planet	48	0.71	23	13	73
sand	32	0.57	5	1	75
scf	115	1.64	16	26	212
styr	30	0.53	9	6	95
tbk	32	0.83	3	1	85
tlc	400	1.12	21	8	375
others	441	6.59	198	167	1110
SUM	1133	15.75	289	324	2145

Table 1: Traversal of MCNC Benchmarks.

from appearing in the recursion tree.

5 Results

All experiments were run on a DEC3100, and are summarized in four tables. The term 'depth' denotes the number of symbolic cycles (L of FSM_TRAVERSAL); 'size' represents the number of nodes in the BDD representation of $reached^L$; and 'calls' the number of recursive calls to IMAGE. The tables represent the following experimental results. **Table 1**: Traverse the 40 MCNC FSM benchmarks and two other controllers. **Table 2**: Traverse 22 of the 31 ISCAS sequential ATPG benchmarks [4]. The initial state was the all zero state in all cases. **Table 3**: Traverse adders, serial multipliers [12], MIN_MAX circuits, counters [6, 7], and benchmarks used in [8]. **Table 4**: Verify two implementations of the same circuit against each other (for a selection of the circuits in Tables 1–3). The entries for ISCAS benchmarks refer to pairs of equivalent circuits (e.g., s349 refers to (s349, s344)).

Since the MCNC benchmarks have no specified reset state, the traversal and verification was done by selecting the first encountered symbolic state as the reset state. Of course this was a weak assumption, and for SCF we performed self-verification for every pair of possible starting states, as discussed in [17]. All the MCNC examples can be verified equivalent to the result of logic synthesis in 24 seconds. We have so far traversed 22 of the 31 ISCAS benchmarks. For most of the remaining circuits, we could not build the BDD's for λ and δ, possibly because of a poor variable ordering.

Finally we show the results of head-to-head comparisons, where comparative data is available, between our results, Berkeley's, and those of Bull Research, Paris. Table 5 shows a traversal comparison between Boulder and Bull, with our numbers adjusted for Sun Sparc1 performance, without inclusion of setup times for building BDD's. Table 6 shows a self-verification comparison between

name	states	time	depth	size	call
s1238	2616	30.61	3	1057	6515
s1494	48	1.63	22	12	96
s349	2625	7.90	7	1061	3213
s382	8865	13.94	151	148	6525
s400	8865	13.96	151	148	6525
s444	8865	12.27	151	198	4775
s526	8868	18.98	151	198	5927
s641	1544	14.86	7	125	3815
s713	1544	15.66	7	125	3801
others	14542	234.43	333	2471	16309
SUM	58382	364.24	983	5543	57501

Table 2: Traversal of ISCAS Benchmarks.

name	states	time	depth	size	call
add16	65536	0.95	2	1	31
add32	$4.29\cdot10^9$	4.55	2	1	63
mpy8	255	1.47	9	10	163
mpy16	65535	14.34	17	18	663
mpy24	$1.68\cdot10^7$	63.32	25	26	1483
mpy32	$4.29\cdot10^9$	189.96	33	34	2623
mpy8b	16384	4.28	9	1	278
mpy16b	$1.07\cdot10^9$	336.41	17	1	10660
mod10	1024	2.75	1024	1	1024
mod12	4096	12.90	4096	1	4096
mM9	$2.25\cdot10^7$	12.21	4	107	506
mM15	$5.86\cdot10^{12}$	40.40	4	185	630
mM20	$1.92\cdot10^{17}$	83.43	4	250	732
mM25	$6.30\cdot10^{21}$	148.13	4	315	838
mM30	$2.06\cdot10^{26}$	237.78	4	380	940
Fig_6	$2.52\cdot10^7$	25.42	257	43	14003
key	$7.21\cdot10^{16}$	6.19	2	1	2
stage	$1.84\cdot10^{19}$	48.65	2	1	34
pewd	$5.19\cdot10^{33}$	18.47	9	85	374
dsip	$2.70\cdot10^{67}$	141.31	16	225	1509
mult4	2625	11.55	7	534	900
SUM	$2.7\cdot10^{67}$	1404.47	5547	2220	41552

Table 3: Traversal of Iterative Circuits.

name	states	time	depth	size	call
s1238	2616	71.92	3	2498	6515
s1494	48	4.46	22	42	88
s349	2625	16.26	7	2300	3213
s400	8865	37.17	151	434	6525
s526n	8868	48.36	151	586	5927
s713	1544	29.13	7	302	3815
s832	25	1.85	11	31	85
MpyA.B	16384	76.82	9	63	1217
scf	209	26.06	17	348	507
mM9.9'	$2.25 \cdot 10^7$	69.88	4	369	1633
SUM	$2.25 \cdot 10^7$	381.91	382	6973	29525

Table 4: Verification of Pairs of Implementations.

Boulder and Berkeley, with our numbers Given for a DEC5000, and Berkeley's for an IBM6000 model530. Note these computers are not directly comparable, but so far as we know, the IBM 6000 is a 25-30 mip machine, although IBM claims it to be a 40 mip machine. We rate the DEC5000 as a 22mip machine (twice as fast as a Sun SPARC1). Thus, adjustments to the numbers in Table 6 should probably be in our favor.

6 Conclusions

Our results indicate that the state of the art in formal FSM verification is now on a par with most of the current sequential synthesis tools. That is, we can normally verify anything that we can (near-optimally) synthesize. We have defined the taxonomy and applicability of generalized cofactor operations. Based on the presented theory and experiments, we may draw the following conclusions. We have demonstrated that our modifications of the SIAM Algorithm have been very successful—our implementation is 1-4 orders of magnitude faster than that published in [6]. As predicted by the theory, these speedups were gained mainly by the power of the subtree recombination mechanism, which can account for a power of two speed up at each level of recursion, in concert with the PODEM technique of enlarging the search space for image computation by including inputs in the search space. We use a hash table lookup procedure for identifying previously solved subproblems. If the number of distinct subproblems grows large, a hash-based cache [2] helps containing memory requirements. In most cases an FSM whose valid state set can be represented by a small BDD, can be traversed efficiently.

In direct comparisons with the referenced methods of Bull and Berkeley our approach has proven superior in most cases. Some of the results reported (for

Bull-Boulder Traversal Comparsion (Sun SPARC1)		
name	time	time
	(Bull)	(Boulder)
mm10	4.4	8.12
mm20	12	38
mm30	36	94
s444	14.6	14
add16	0.6	0.4
add32	2.2	1.6
key	2	0.4
stage	5mn47s	1mn
sbc	9mn37	325

Table 5: Traversal comparison: Boulder vs. Bull.

example, the SBC numbers of Table 5 and 6) use a new technique for choosing the splitting variable in the image recursion, based on the PODEM single path backtrace heuristic. This was not discussed in the text and will be the subject of future investigation. Also, in a private communication, the Bull authors suggested that we use isomorphic subtree recombination, rather than just identical subtree recombination. We have implemented this but have yet to show any gain from it, despite the fact that Bull has found a significant improvement.

The results point to future research on the interaction between the HDL specification and the verification. To reap the benefits of the pruning mechanisms most efficiently, the HDL specification should identify the **data/ control dichotomy**, as well as associated, **bit-sliced registers**. This is necessary not only for pruning but as an advisory to the BDD ordering heuristics, that otherwise could fail catastrophically on arithmetic problems like MIN_MAX benchmarks. Further, on examples like the counters, it is clearly advantageous to use $from = new$ in the traversal (because for pulse counters it is not even necessary to call VECT_TO_CHAR). On the other hand, for problems with STate explosion like MIN_MAX, it is beneficial to use $from = reached$. Of course preprocessing of desired temporal verification predicates at the HDL level is another aspect of such interaction. We have found it advantageous to separate data from control. While this may eliminate only a few variables from the search space, it may have the additional effect of regularizing the image being constructed, and thus make symbolic computations simpler. Again, it would be helpful to provide directives in this regard from the HDL level.

We have also contributed efficient, algorithms, of expected **polynomial** complexity, for finding shortest distinguishing sequences when verification fails—as in the case of sequential test generation. When applied to the state justification part of sequential ATPG, this should lead to test sequences of significantly

Berkeley-Boulder Self-Verification Comparsion		
name	time	time
	(Berkeley)	(Boulder)
	(IBM6000 M530)	(DEC5000)
sand	11.0	1.10
scf	13.8	2.92
s344	17.9	11.98
s444	58.8	20.39
s526	30.1	33.25
s713	31.7	22.18
add32	8.5	32.19
mm30	190.1	179.0
sbc	1084.3	1468.0

Table 6: Self-Verification comparison: Boulder vs. Berkeley.

reduced length. For example traversal of SBC required only 10 cycles so the length of the justification sequence is bounded from above by 10. In contrast, STEED found test sequences as long as 50. It wasn't stated what part of the long sequences were due to justification, but obtaining shorter sequences looks like a real possibility.

Acknowledgements

Our code was built on Berkeley's MIS2.2 implementation of the BDD package of Bryant and co-workers [2], without which our implementation task would have been prohibitive. The generalized cofactor concept was revealed to us by Bob Brayton and his students at the Boulder-Berkeley-Stanford Synthesis Research Seminar, January, 1990, where we were introduced to the SIAM paper in a talk by Hervé Touati. We acknowledge helpful discussion with R. Brayton, O. Coudert, R. Jacoby, J. C. Madre, C. Pixley, and H. Touati.

References

[1] K.A. Bartlett, R. Brayton, G. Hachtel, R. Jacoby, C. Morrison, R. Rudell, A. Sangiovanni-Vincentelli, and A. Wang. Multi-level logic minimization using implicit don't cares. *IEEE Transactions on Computer-Aided Design of Integrated Circuits and Systems*, CAD-7(6):723–740, June 1988.

[2] K. S. Brace, R. L. Rudell, and R. E. Bryant. Efficient implementation of a BDD package. In *Proceedings of the 27th Design Automation Conference*, pages 40–45, June 1990.

[3] R. Brayton, R. Rudell, A. Sangiovanni-Vincentelli, and A. Wang. Mis: Multiple-level interactive logic optimization system. *IEEE Transactions on Computer-Aided Design of Integrated Circuits and Systems*, CAD-6(6):1062–1081, November 1987.

[4] F. Brglez, D. Bryan, and K. Koźmiński. Combinational profiles of sequential benchmark circuits. In *Proceedings of the 1989 International Symposium on Circuits and Systems*, pages 1929–1934, Portland, OR, May 1989.

[5] G. Colon-Bonet, D. Bostick, G. Hachtel, E. Schwarz, and M. Lightner. Fast algorithms for extraction of topologically irredundant boolean equations and don't care sets from hardware description languages. In *IEEE International Conference on Computer-Aided Design*, November 1989.

[6] O. Coudert, C. Berthet, and J. C. Madre. Verification of sequential machines using boolean functional vectors. In L. Claesen, editor, *Proc. IFIP International Workshop on Applied Formal Methods for Correct VLSI Design*, pages 111–128, Leuven, Belgium, November 1989.

[7] O. Coudert, C. Berthet, and J. C. Madre. Formal boolean manipulations for the verification of sequential machines. In *Proceedings of the 1st European Design Automation Conference*, pages 57–61, March 1990.

[8] A. Ghosh, S. Devadas, and A. R. Newton. Test generation for highly sequential circuits. In *Proceedings of the IEEE International Conference on Computer Aided Design*, pages 362–365, November 1989.

[9] P. Goel. An implicit enumeration algorithm to generate tests for combinational logic circuits. *IEEE Transactions on Computers*, C-30(3):215–222, 1981.

[10] G. Hachtel, M. Lightner, R. Jacoby, C. Morrison, P. Moceyunas, and D. Bostick. BOLD: The Boulder Optimal Logic Design system. In *Hawaii International Conference on System Sciences*, 1988.

[11] G. D. Hachtel, R. M. Jacoby, J.-K. Rho, and F. Somenzi. Reducing finite state machines for optimal encodability. In *submitted to ICCAD-90*, November 1990.

[12] P. N. Lowenstein. Formal verification of state-machines using higher-order logic. In *IEEE Internatinal Conference on Computer Design*, pages 204–207, Cambridge, MA, October 1989.

[13] H.-K. T. Ma, S. Devadas, and A. R. Newton. On the verification of finite state machines at differing levels of abstraction. *IEEE Transactions on CAD*, Vol-7:713–722, June 1988.

[14] H.-K. T. Ma, S. Devadas, A. R. Newton, and A. Sangiovanni-Vincentelli. Test generation for sequential circuits. *IEEE Transactions on CAD*, Vol-7:1081–1093, October 1988.

[15] S. Malik, E. M. Sentovich, and R. K. Brayton. Retiming and resynthesis: Optimizing sequential networks with combinational techniques. In *Proceedings of the Hawaii International Conference on Systems Science*, pages 397–406, January 1990.

[16] G. De Micheli and R. Yip. Logic transformations for synchronous logic design. In *Proceedings of the Hawaii International Conference on Systems Science*, pages 407–416, January 1990.

[17] C. Pixley. A computational theory and implementation of sequential hardware equivalence. In R. Kurshan and E. Clarke, editors, *Proc. CAV Workshop (also DIMACS Tech. Report 90-31)*, Rutgers University, NJ, June 1990.

[18] H. Touati, H. Savoj, and R. Brayton. Implicit enumeration of finite state machines using bdd's. In *Proceedings of the IEEE International Conference on Computer Aided Design*, November 1990.